T0231973

Introduction to Nonlinear Control

Introduction to Nonlinear Control

Stability, Control Design, and Estimation

Christopher M. Kellett and Philipp Braun

PRINCETON UNIVERSITY PRESS

PRINCETON AND OXFORD

Copyright© 2023 by Princeton University Press

Princeton University Press is committed to the protection of copyright and the intellectual property our authors entrust to us. Copyright promotes the progress and integrity of knowledge. Thank you for supporting free speech and the global exchange of ideas by purchasing an authorized edition of this book. If you wish to reproduce or distribute any part of it in any form, please obtain permission.

Requests for permission to reproduce material from this work
should be sent to permissions@press.princeton.edu

Published by Princeton University Press
41 William Street, Princeton, New Jersey 08540
99 Banbury Road, Oxford OX2 6JX

press.princeton.edu

All Rights Reserved

ISBN 9780691240480
ISBN (e-book) 9780691240497

British Library Cataloging-in-Publication Data is available

Editorial: Hallie Stebbins and Kiran Pandey
Production Editorial: Theresa Liu
Jacket/Cover Design: Wanda España
Production: Erin Suydam
Publicity: William Pagdatoon
Copyeditor: Bhisham Bherwani

Jacket Image: Panther Media GmbH / Alamy Stock Photo

This book is composed in LATEX. The publisher would like to acknowledge the authors of this volume for providing the print-ready files from which this book is printed.

Printed on acid-free paper. ∞

Printed in the United States of America

10 9 8 7 6 5 4 3 2 1

MATLAB© and Simulink© are registered trademarks of TheMathWorks Inc. and are used with permission. The MathWorks does not warrant the accuracy of the text or exercises in this book. This book's use or discussion of MATLAB©, Simulink©, or related products does not constitute an endorsement or sponsorship by The MathWorks of a particular pedagogical approach or particular use of the MATLAB© and Simulink© software.

The authors dedicate this book:
to Katrina (C.M.K.)
to Ana (P.B.)

Contents

Preface

The idea of writing this book originated from a core third year course in the Mechatronics Engineering programme at the University of Newcastle, which was also available as an elective course for Electrical Engineering students and Mathematics students (who were recommended to be in their fourth year).

Prior to this course, the engineering students had nominally had a second-year course that would be similar in content to that of the book by Åström and Murray [12] and a third-year course in linear control systems, possibly similar in content to that of the book by Hespanha [63]. However, one of our aims in presenting the material in this book is to not rely too heavily on assumed knowledge from either [12] or [63] so that the book is sufficiently self-contained and accessible to any student who has a bit of background in ordinary differential equations (ODEs) and some basic linear algebra. In particular, our hope is that the material is accessible to students in the latter years of Science degree programs in general, as well as to industry practitioners. Courses teaching material covered in this book that build upon the content in Åström and Murray [12] and Hespanha [63] can skip Chapters 3 and 4, for example. In this case, Chapters 3 and 4 can be given to students for self-study to refresh their knowledge required for the subsequent topics.

Our initial intention in writing this book was to present only material which could be covered in a single semester, as we did when we delivered the course at the University of Newcastle. However, we quickly realized that the variety of topics in nonlinear control made this goal impossible, and even the current selection of topics is incomplete. Indeed, even the two times we co-taught the course our journey through the material differed and we covered slightly different topics each time, once we had covered the fundamental material. While some sections are critical for the understanding of following chapters, most topics only rely on background knowledge of only some preceding chapters and thus can be studied and taught independently. This gives educators the flexibility to cover topics based on their own preference and to assign research projects to keen students on chapters not covered in a particular course.

The topics covered here were selected based on a few overarching criteria.

1. When considering nonlinear systems, Lyapunov's second method remains the workhorse of the field. It is also the tool required to put the technique of linearization on a firm footing and to talk about and estimate the region of attraction for an equilibrium point. This then forms our starting point and becomes a thread that runs through the entire book. Consequently, Chapter 2 is a prerequisite for all other chapters.

2. Since this book should be accessible for undergraduate Mechatronics Engineering students, it should provide useful tools for those students who do not go on to pursue a Master's or a PhD. In particular, this motivated the inclusion of the absolute stability material, the antiwindup techniques, and nonlinear model

predictive control. The individual chapters, as they are discussed in this book, should give the students an intuitive idea of the topic, which can be applied in industry or further studied in detail in graduate courses or during a PhD.

3. For similar reasons to the above, we neither present results in their greatest generality nor worry about completely rigorous proofs. Rather, we have tried to present the most relevant and/or intuitive results, with references provided for more general results and complete proofs. Students who want to learn more about antiwindup designs, sliding mode control, or model predictive control can proceed with the monographs [168], [137] and [128], respectively, for example.

4. While we have tried not to assume the reader has undertaken prior study in control topics, for students familiar with linear systems as presented in [12] and/or [63], there are numerous points of contact with that material such as, again, the absolute stability material, and also material such as the connection between ISS, $\mathcal{L}_2$-gain, and $\mathcal{H}_\infty$ norms, or the linear quadratic regulator. In the same context, Chapters 3, 4, and 5 are included for completeness, but most of the content is assumed to be known by students starting to study the book. These chapters can thus be skipped by readers with a basic knowledge of control theory and they can be studied selectively if necessary for results discussed later in the book.

We are grateful to several colleagues who provided useful comments and insights at different stages of the writing process and in our development of the aforementioned course, including Joel Ferguson, Miroslav Krstić, Lorenzo Marconi, Andrea Serrani, Adrian Wills, and Luca Zaccarian.

Canberra, Australia   Christopher M. Kellett
Canberra, Australia   Philipp Braun
December 20, 2022

OVERVIEW OF THE CONTENT BOOK

Part I: Dynamical Systems

CHAPTER 1: NONLINEAR SYSTEMS—FUNDAMENTALS AND EXAMPLES

- Chapter 1 introduces necessary definitions and notation used throughout the book. Additionally, simple examples motivating the concepts of the following chapters are discussed.

CHAPTER 2: NONLINEAR SYSTEMS—STABILITY NOTIONS

- In Chapter 2 stability concepts for dynamical systems are introduced and corresponding necessary and sufficient conditions are derived. Here, we rely heavily on Lyapunov functions and Lyapunov results. The chapter is fundamental for all other chapters in the book.

CHAPTER 3: LINEAR SYSTEMS AND LINEARIZATION

- Chapter 3 discusses linear systems stability and results for nonlinear systems deduced from linearization. A reader familiar with linear systems can skip this chapter and return to it if necessary for results discussed in the remainder of the book. Section 3.5.3 covers controller design in terms of pole placement. With the knowledge of Chapter 3 the reader can study Luenberger observers (see Section (16.1)), which provides the reader with the basic concepts necessary to perform a controller design for a given plant.

CHAPTER 4: FREQUENCY DOMAIN ANALYSIS

- While dynamical systems in the frequency domain are not the focus of this book, basic results are covered in Chapter 4. The results are needed for the derivations in Chapter 6. The reader familiar with linear system theory in the frequency domain can skip this chapter.

CHAPTER 5: DISCRETE TIME SYSTEMS

- Chapter 5 translates the results in Chapters 2 and 3 from the continuous time setting to the discrete time setting. In addition, relations between continuous time systems and their discrete time counterparts are discussed. A basic knowledge on discrete systems is needed for Chapter 15 and for discrete time filter designs in Chapters 16 and 17.

CHAPTER 6: ABSOLUTE STABILITY

- In Chapter 6 absolute stability is discussed, i.e., a robust stability concept for linear systems. Absolute stability is motivated through the Lur'e problem. To study absolute stability, the reader should be familiar with linear systems in both the time and frequency domains (see Chapters 3 and 4).

CHAPTER 7: INPUT-TO-STATE STABILITY

- Chapter 7 discusses a robust stability concept, Input-to-State Stability (ISS), for nonlinear systems. Since input-to-state stability extends nominal stability concepts, the reader should have a good understanding of the content in Chapter 2. This chapter also makes the relation between ISS and dissipativity and passivity explicit.

Part II: Controller Design

CHAPTER 8: LMI-BASED CONTROLLER AND ANTIWINDUP DESIGNS

- Chapter 8 illustrates how linear matrix inequalities can be used in the controller design for linear systems. By phrasing specific control problems in terms of semidefinite programming problems, additional criteria such as robustness properties or maximization of the region of attraction can be embedded in the controller design. Before studying the chapter, the reader should digest the content in Chapter 3 and understand $\mathcal{L}_2$-stability (see Chapter 7). For a deeper understanding of the topic we refer the reader to [168].

CHAPTER 9: CONTROL LYAPUNOV FUNCTIONS

- Chapter 9 extends Lyapunov functions and stability (see Chapter 2) to control Lyapunov functions and stabilizability. In addition to necessary and sufficient conditions for the existence of control Lyapunov functions, two methods (namely backstepping and forwarding) to construct control Lyapunov functions are discussed. We refer the reader to [31] and references therein for more general results on the existence of control Lyapunov functions.

CHAPTER 10: SLIDING MODE CONTROL

- Chapter 10 introduces the main concepts of sliding mode controllers and introduces the concept of finite time stability. Additionally, it provides the tools for sliding mode observers (see Section 18.2). For a reference with more general results, we refer the reader to [137].

CHAPTER 11: ADAPTIVE CONTROL

- Chapter 11 describes some of the limits of static state feedback laws and introduces adaptive control as a form of dynamic feedback. We include an introduction to model reference adaptive control and present adaptive backstepping (requiring Section 9.4 as a prerequisite) and follow with the tuning function approach to address the issue of overparameterization. The introduction of dynamic feedback also provides a bridge to the observer designs in Part III.

CHAPTER 12: INTRODUCTION TO DIFFERENTIAL GEOMETRIC METHODS

- Chapter 12 introduces the basic concepts used in differential geometric approaches to nonlinear control. This is arguably the only chapter in the book that does not require Chapter 2 as a prerequisite. However, we believe that it is strongly beneficial to place this material later in a course rather than sooner to allow students to develop a familiarity with nonlinear systems. The material here makes frequent reference to the linear case, and hence Chapters 3 and 4 are useful prerequisites.

CHAPTER 13: OUTPUT REGULATION

- Chapter 13 presents the important internal model principle in order to solve the robust output regulation problem. This relies on some of the concepts in Chapter 12. While perhaps obvious to experienced practitioners, the derivation of the classical PI controller is included as an illustrative example of the internal model principle. The use of dynamic, rather than static, feedback controllers is similar in some ways to adaptive control (Chapter 11) and, like adaptive control, presages the design of observers in Part III.

CHAPTER 14: OPTIMAL CONTROL

- Chapter 14 discusses optimal control for nonlinear systems and optimal control for linear systems in terms of the linear quadratic regulator. Additionally, results are derived in both the continuous time and discrete time contexts. In addition to Chapter 2, the content derived here relies on knowledge on linear systems (see Chapter 3). For the derivations in the discrete time setting Chapter 5 is a prerequisite. Chapter 14 additionally motivates model predictive control discussed in Chapter 15. In terms of observer designs, optimal control motivates the minimum energy estimator and the Kalman filter (see Chapters 16 and 17). For a monograph on optimal control, we refer the reader to [96].

CHAPTER 15: MODEL PREDICTIVE CONTROL

- Chapter 15 discusses a variety of different model predictive control schemes and corresponding notations. Model predictive control extends ideas from optimal control (see Chapter 14) and is derived for discrete time systems (see Chapter 5) here. For monographs on model predictive control, we refer the reader to [128] and [58].

Part III: Observer Design and Estimation

CHAPTER 16: OBSERVER DESIGN FOR LINEAR SYSTEMS

- Chapter 16 discusses basic observer design for linear systems, including the Luenberger observer, the minimum energy estimator for continuous time systems, and the Kalman filter for discrete time systems. To understand the concepts in this chapter, Chapter 3 and Chapter 14 should be studied. For the discrete time Kalman filter, knowledge on discrete time systems (see Chapter 5) is essential.

CHAPTER 17: EXTENDED AND UNSCENTED KALMAN FILTER AND MOVING HORIZON ESTIMATION

- Chapter 17 discusses extensions of the observers in Chapter 16 for nonlinear systems. In addition, moving horizon estimation is briefly discussed, which is the dual to model predictive control (see Chapter 15).

CHAPTER 18: OBSERVER DESIGN FOR NONLINEAR SYSTEMS

- Chapter 18 discusses two observer design concepts for nonlinear systems: high-gain observers and sliding mode observers. The sliding mode observer designs rely on results discussed in in Chapter 10.

Glossary

ACRONYMS

- 0-GAS   0-input global asymptotic stability
- BIBO   bounded-input, bounded-output
- CLF   control Lyapunov function
- CVX   Matlab Software for Disciplined Convex Programming
- EKF   extended Kalman filter
- iISS   integral input-to-state stability
- ISS   input-to-state stability
- LMI   linear matrix inequality
- LP   linear program
- LQR   linear quadratic regulator
- MHE   moving horizon estimation
- MPC   model predictive control
- NLP   nonlinear program
- OCP   optimal control problem
- ODE   ordinary differential equation
- PBH   Popov-Belevitch-Hautus (test)
- PD   proportional differential (controller)
- PDE   partial differential equation
- PID   proportional integral differential (controller)
- SPR   strictly positive real
- UKF   unscented Kalman filter
- QP   quadratic program

LIST OF SYMBOLS

- $\mathbb{N}$ natural numbers
- $\mathbb{N}_0$ natural numbers including zero
- $\mathbb{N}_{[p_1,p_2]}$ $\{p_1, \ldots, p_2\}$ for $p_1, p_2 \in \mathbb{N}_0 \cup \{\infty\}$
- $\mathbb{Z}$ integers
- $\mathbb{R}$ real numbers
- $\mathbb{R}_{\geq c}$ $\{x \in \mathbb{R} : x \geq c\}$ for $c \in \mathbb{R}$ (same definition for $>, \leq, <$)
- $\mathbb{C}$ complex numbers
- $\mathbb{C}_+$ $\{x \in \mathbb{C} : \operatorname{Re}(x) > 0\}$ (same definition for $-, <$)
- $\overline{\mathbb{C}}_+$ $\{x \in \mathbb{C} : \operatorname{Re}(x) \geq 0\}$ (same definition for $-, \leq$)
- $\mathcal{S}^n$ $\{P \in \mathbb{R}^{n \times n} : P^T = P\}$ (vector space of symmetric matrices)
- $\mathcal{S}^n_{>0}$ $\{P \in \mathcal{S}^n : x^T P x > 0 \; \forall x \neq 0\}$ (positive definite matrices)
       (same definition for $\geq, \leq, <$)

- $\mathbb{R}[s]$ polynomial ring in s over the field $\mathbb{R}$
- $\mathcal{R}_f(x^e)$ region of attraction   (see Definition 2.28)
- $\mathcal{P}$ class-$\mathcal{P}$ functions   (see Definition 1.5)
- $\mathcal{K}$ class-$\mathcal{K}$ functions   (see Definition 1.6)
- $\mathcal{K}_\infty$ class-$\mathcal{K}_\infty$ functions   (see Definition 1.7)
- $\mathcal{L}$ class-$\mathcal{L}$ functions   (see Definition 1.8)
- $\mathcal{KL}$ class-$\mathcal{KL}$ functions   (see Definition 1.9)
- $\mathrm{Re}(\cdot)$ real part of a complex number
- $\mathrm{Im}(\cdot)$ imaginary part of a complex number
- j $j = \sqrt{-1}$
- $\mathrm{int}(\cdot)$ $\mathrm{int}(\mathcal{A}) = \{x \in \mathbb{R}^n : \exists \varepsilon > 0 : \mathcal{B}_\varepsilon(x) \subset \mathcal{A}\}$ (interior of a set)
- $\overline{\cdot}$ closure of a set
- $\partial\cdot$ boundary of a set
- $\overline{x}$ $\overline{a+jb} = a - jb$, $a, b \in \mathbb{R}$ (complex conjugate)
- $\mathrm{He}(\cdot)$ He $A = A + A^T$, $A \in \mathbb{R}^{n \times n}$ (see Equation (8.20))
- $|\cdot|$ $|x| = \sqrt{\overline{x}^T x}$ (Euclidean norm)
- $\|\cdot\|$ matrix norm (induced by Euclidean norm)
- $\|\cdot\|_{\mathcal{L}_2}$ $\mathcal{L}_2$-norm   (see Equation (4.8))
- $\|\cdot\|_{\mathcal{L}_\infty}$ $\mathcal{L}_\infty$-norm   (see Equation (4.9))
- $\|\cdot\|_\infty$ $\mathcal{H}_\infty$-norm   (see Equation (4.10))
- $\mathcal{B}_\varepsilon(x)$ $\{y \in \mathbb{R}^n : |x - y| < \varepsilon\}$ (open ball of radius $\varepsilon > 0$ around $x \in \mathbb{R}^n$)
- $\overline{\mathcal{B}}_\varepsilon(x)$ $\{y \in \mathbb{R}^n : |x - y| \le \varepsilon\}$ (closed ball of radius $\varepsilon > 0$ around $x \in \mathbb{R}^n$)
- $\overline{\mathcal{B}}_\varepsilon, \mathcal{B}_\varepsilon$ closed, open ball of radius $\varepsilon > 0$ around the origin
- I identity matrix of appropriate dimension
- $\mathrm{sat}(\cdot)$ saturation function   (see Equation (6.4))
- $\mathrm{dz}(\cdot)$ deadzone function   (see Equation (6.5))
- $\mathrm{sign}(\cdot)$ sign function   (see Equation (6.7))
- $\mathrm{rank}(\cdot)$ rank of a matrix (i.e., number of linear independent rows)
- $\mathrm{trace}(\cdot)$ trace of a square matrix (i.e., the sum of the complex eigenvalues)
- $\mathrm{diag}(\cdot)$ block diagonal matrix
- $\mathrm{dim}(\cdot)$ dimension of a matrix, vector or vector space
- $\mathrm{span}\{\cdot\}$ $\mathrm{span}\{v_1, \ldots, v_m\} = \{\sum_{i=1}^m \lambda_i v_i | \lambda_i \in \mathbb{R}, i \in \{1, \ldots, m\}\}$, $v_i \in \mathbb{R}^n$
- $\mathcal{O}[\cdot]$ Big $\mathcal{O}$ notation   (see Definition 7.7)
- $L_f V(x)$ Lie derivative   (see Equation (9.5))
- $[f, g](x)$ Lie bracket   (see Definition 12.19)
- $\mathrm{ad}_f^k g(x)$ repeated Lie bracket   (see Equation (12.19))

Part I

Dynamical Systems

Chapter One

Nonlinear Systems—Fundamentals and Examples

1.1 STATE SPACE MODELS

State space models refer to a system of first-order ordinary differential equations

$$\dot{x}(t) = f(x(t)), \tag{1.1}$$

where $f : \mathbb{R}^n \to \mathbb{R}^n$ and the dot refers to differentiation with respect to time; i.e., $\dot{x}(t) = \frac{d}{dt} x(t)$. In other words, x is an n-dimensional vector describing the system "state" and $f(\cdot)$ defines how each state evolves in time. A more cumbersome notation can be used to write out individual components as

$$\dot{x}_1 = f_1(x_1, x_2, \ldots, x_n)$$
$$\dot{x}_2 = f_2(x_1, x_2, \ldots, x_n)$$
$$\vdots$$
$$\dot{x}_n = f_n(x_1, x_2, \ldots, x_n).$$

A *solution* of (1.1) is an absolutely continuous function that satisfies Equation (1.1) almost everywhere.

The expression in (1.1) defines a *time-invariant* or *autonomous* (in time) system. This is because the vector field $f(x)$ defining how the system evolves is independent of time. By contrast, a system of the form

$$\dot{x}(t) = f(t, x(t)) \tag{1.2}$$

is referred to as *time-varying* or *non-autonomous* due to the time dependence in the vector field $f(t, x)$.

It is important to be aware of the conditions that guarantee that differential equations such as (1.1) or (1.2) have a reasonable mathematical meaning, particularly with regard to the existence and uniqueness of solutions.

Theorem 1.1 ([86, Theorem 3.1]). *Given $x_0 \in \mathbb{R}^n$, $r > 0$, and $0 \le t_0 < t_1$, let $f(t, x)$ be piecewise continuous in t and satisfy the (local) Lipschitz condition*

$$|f(t, x) - f(t, y)| \le L|x - y|$$

for an $L > 0$, for all $x, y \in \{\xi \in \mathbb{R}^n : |\xi - x_0| \le r\}$ and $t \in [t_0, t_1]$. Then there exists $\delta > 0$ so that

$$\dot{x}(t) = f(t, x(t)), \quad x(t_0) = x_0$$

has a unique solution over $[t_0, t_0 + \delta]$.

The above could be considered the first standard result in the theory of ordinary differential equations. Subsequent results frequently include that if the solutions remain in a compact set and the Lipschitz condition holds on that compact set, then for each initial condition in the compact set, the resulting solution is unique and exists for all subsequent time. Another common result is that if the Lipschitz condition holds globally (so that a single L can be chosen for all $x, y \in \mathbb{R}^n$) and if the function f is uniformly bounded (with respect to t), then a unique solution exists from all initial conditions.

While important, a detailed investigation of these issues would take us too far from our main goals. The important observation here is that our subsequent tools and techniques rely on having models that satisfy certain assumptions and, when applying those tools and techniques to a specific model, these assumptions should be checked. For example, our results are not applicable to models where the right-hand side f is discontinuous. Pointers to additional material on ordinary differential equations are provided at the end of the chapter in Section 1.4.

A state space model, or a system of first-order ordinary differential equations, is equivalent to an nth-order ordinary differential equation. For example, consider the differential equation

$$\frac{d^n}{dt^n} y(t) = \phi\left(y(t), \dot{y}(t), \dots, \frac{d^{n-1}}{dt^{n-1}} y(t)\right), \tag{1.3}$$

where $\phi : \mathbb{R}^n \to \mathbb{R}$. With the definition

$$x_1 = y, \quad x_2 = \dot{y}, \quad x_3 = \ddot{y}, \ \dots \ , \ x_n = \frac{d^{n-1}}{dt^{n-1}} y, \tag{1.4}$$

Equation (1.3) can be written as n first order ordinary differential equations as

$$\dot{x}_1(t) = x_2(t),$$
$$\dot{x}_2(t) = x_3(t),$$
$$\vdots$$
$$\dot{x}_{n-1}(t) = x_n(t),$$
$$\dot{x}_n(t) = \phi(x_1(t), x_2(t), \dots, x_n(t)).$$

Our primary focus will be systems with external inputs

$$\dot{x}(t) = f(x(t), u(t)), \tag{1.5}$$

where $f : \mathbb{R}^n \times \mathbb{R}^m \to \mathbb{R}^n$. The external input will be used to model different conceptual setups. For example, with $u : \mathbb{R}_{\geq 0} \to \mathbb{R}^m$ as a function of time, the input can represent an exogenous time-dependent signal which leads to a time-varying system $\dot{x}(t) = f(x(t), u(t)) \doteq \tilde{f}(t, x(t))$. Such a time-varying exogenous input may be an unknown disturbance acting on the system or may be a predefined reference signal. In the case of a pre-defined reference signal, $u(t)$ often corresponds to *open-loop control* or *steering*. Open-loop control is susceptible to disturbances and measurement noise and will not be the focus of this book.

Alternatively, $u : \mathbb{R}^n \to \mathbb{R}^m$ can be a degree of freedom manipulating the dynamics

$$\dot{x}(t) = f(x(t), u(x(t))) \doteq \bar{f}(x(t)). \tag{1.6}$$

The definition of *state feedback laws* $u(x)$, such that the closed-loop dynamics (1.6) satisfy specific properties, will be the focus of multiple chapters. This is referred to as *closed-loop controller design*. The simplest control problem in this context is *regulation*, where, for example, the mismatch between the current state and a constant reference value is used to define the control signal $u(x(t))$ in such a way that the closed-loop solution converges to a neighborhood around the reference value and stays there indefinitely despite disturbances caused through actuators and measurement noise caused through sensors. The regulation problem can be extended to *reference tracking, path-following,* ᴠor *optimal steady-state operation,* for example.

Example 1.2. Consider the mass-spring system shown in Figure 1.1, where a block of mass m slides horizontally on a surface and is connected to a wall via a spring. The block is subject to the restoring force of the spring (F_{sp}), friction forces (F_f), and an external driving force or input (F). Denote the displacement from the natural resting position of the block by y.

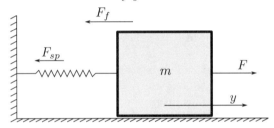

Figure 1.1: Mass-spring system.

Using Newton's second law of motion, we obtain the second-order differential equation

$$m\ddot{y} = F - F_f - F_{sp} = F - c\dot{y} - ky, \tag{1.7}$$

where $F_f = c\dot{y}$ is viscous friction in air and $F_{sp} = ky$ is a linear spring. Let $x_1 = y$ and $x_2 = \dot{y}$ and the driving force $u = F$. Then

$$\dot{x}_1(t) = x_2(t) \tag{1.8a}$$
$$\dot{x}_2(t) = -\tfrac{k}{m}x_1(t) - \tfrac{c}{m}x_2(t) + \tfrac{1}{m}u(t), \tag{1.8b}$$

which gives us the system of first-order differential equations in (1.5). In other words, this provides the state space model for the mass-spring system. We refer to x_1 and x_2 as the system *states* and u as the system *input*. If the driving force F or the input u is assumed to be zero, the mass-spring system (1.8) is of the form of the first-order differential Equation (1.1).

Consider now the expressions for potential and kinetic energy for the mass-spring system with input $u = 0$. The potential energy is given by

$$\tfrac{1}{2}ky^2 = \tfrac{1}{2}kx_1^2$$

and the kinetic energy is given by

$$\tfrac{1}{2}mv^2 = \tfrac{1}{2}m(\dot{y})^2 = \tfrac{1}{2}mx_2^2,$$

where v denotes the velocity of the block.

We now pose a question: how does the total energy of the mass-spring system evolve with time? The total energy is given by the sum of the potential and kinetic energy:

$$E(x_1, x_2) = E(x) = \tfrac{1}{2}kx_1^2 + \tfrac{1}{2}mx_2^2.$$

To see how the total energy evolves over time, we can take the derivative with respect to time of $E(x)$. However, x evolves with time and should more precisely be written as $x(t)$ so that, to be entirely precise, the total energy is $E(x(t))$. Therefore, taking the time derivative of E involves an application of the chain rule

$$\begin{aligned}
\tfrac{d}{dt}E(x(t)) &= \tfrac{d}{dt}\left(\tfrac{1}{2}kx_1(t)^2 + \tfrac{1}{2}mx_2(t)^2\right) \\
&= kx_1\dot{x}_1(t) + mx_2\dot{x}_2(t) \\
&= kx_1(t)x_2(t) - kx_1(t)x_2(t) - cx_2(t)^2 \\
&= -cx_2(t)^2,
\end{aligned}$$

where the third equality is obtained by substituting the system equations (1.8) for $\dot{x}_1$ and $\dot{x}_2$.

We see that the total energy is always positive since it is the sum of two squared quantities. We also see that, as long as the block is moving, so that the velocity $x_2 = \dot{y} \neq 0$, the total energy E is decreasing since its time derivative is negative. From these two facts we can infer that, eventually, the block must stop moving. While our intuition suggests that the block stops moving at the natural rest position $x_1 = y = 0$, the above arguments do not guarantee that this is the case.

In the above example, we referenced the *natural resting position of the block*, by which we meant a configuration of states that remain constant for all time. In other words, at some point in the state space, call it $x^e \in \mathbb{R}^n$,

$$\tfrac{d}{dt}x(t) = f(x(t)) = f(x^e) = 0,$$

which implies that, if the system is at the point x^e, it will remain there indefinitely. We call a point $x^e \in \mathbb{R}^n$ such that $f(x^e) = 0$ an *equilibrium*.

Definition 1.3 (Equilibrium). *Consider the systems (1.1), (1.2), and (1.5). The point $x^e \in \mathbb{R}^n$ is called an equilibrium of the system (1.1) if $0 = f(x^e)$ is satisfied. If $0 = f(t, x^e)$ for all $t \in \mathbb{R}_{\geq 0}$ then x^e is an equilibrium of the non-autonomous system (1.2). Additionally, the pair $(x^e, u^e) \in \mathbb{R}^n \times \mathbb{R}^m$ is called an equilibrium pair of the system (1.5) if $0 = f(x^e, u^e)$ holds.*

We will generally assume that the equilibrium of interest is at the origin; i.e., $x^e = 0$. This poses no loss of generality as we can always perform a simple translation of coordinates so that the translated system has its equilibrium at the origin. To see this, assume $x^e \in \mathbb{R}^n$ with $x^e \neq 0$. Define $z = x - x^e$. Since x^e is a constant, when we write the system equations for the state z we obtain

$$\tfrac{d}{dt}z(t) = \tfrac{d}{dt}x(t) - \tfrac{d}{dt}x^e = f(x(t)) = f(z(t) + x^e).$$

Defining $\hat{f}(z) \doteq f(z + x^e)$ yields the state space system

$$\dot{z} = \hat{f}(z), \tag{1.9}$$

where, with $z^e = 0$, $\hat{f}(z^e) = f(z^e + x^e) = f(x^e) = 0$. In other words, the origin $z^e = 0$ is an equilibrium for the system (1.9). A similar translation $z = x - x^e$ and $v = u - u^e$ can be performed for (1.5) to shift an equilibrium pair to the origin.

Returning to the mass-spring system above, we see that the equilibrium without input (i.e., with $u = 0$) is at $x_1 = y = 0$ and $x_2 = \dot{y} = 0$. Indeed, for the right-hand side of (1.8) to be zero, (1.8a) immediately implies that $x_2 = 0$. With $x_2 = 0$, (1.8b) implies that $x_1 = 0$. For the system with nonzero input, we obtain the equilibrium pairs $(x^e, u^e) = ([ka, 0]^T, a)$ for any $a \in \mathbb{R}$.

Example 1.4. Consider the ideal pendulum shown in Figure 1.2, where ℓ is the length of the pendulum, m is a point mass concentrated at the end of the pendulum, and θ is the angle of the pendulum with respect to the downward vertical axis.

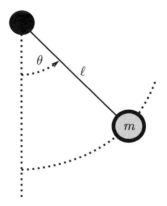

Figure 1.2: Pendulum system.

A model of the pendulum can be derived by, again, balancing forces:

$$m\ell\ddot{\theta} = -mg\sin\theta - k\ell\dot{\theta}, \tag{1.10}$$

where $k > 0$ is a friction coefficient. Let $x_1 = \theta$ and $x_2 = \dot{\theta}$. Then

$$\dot{x}_1(t) = x_2(t), \tag{1.11a}$$
$$\dot{x}_2(t) = -\tfrac{g}{\ell}\sin x_1(t) - \tfrac{k}{m}x_2(t). \tag{1.11b}$$

To find the equilibria for the pendulum, we set the time derivatives to zero (so the state cannot change) and solve

$$0 = x_2$$
$$0 = -\tfrac{g}{\ell}\sin x_1 - \tfrac{k}{m}x_2,$$

which implies $x^e = [n\pi, 0]^T$ for $n = 0, \pm1, \pm2$, etc.. This is consistent with our intuition in that the pendulum will remain stationary if it starts perfectly still in either the downward or upward position, i.e., at $[0, 0]^T$ or $[\pi, 0]^T$, respectively. Furthermore, rotations of 2π simply place us again at an equilibrium point.

Repeating the energy considerations of Example 1.2, we can write the potential energy of the pendulum as

$$mg\ell(1 - \cos\theta) = mg\ell(1 - \cos x_1)$$

and the kinetic energy of the pendulum as

$$\tfrac{1}{2}mv^2 = \tfrac{1}{2}m\ell^2\dot{\theta}^2 = \tfrac{1}{2}m\ell^2 x_2^2.$$

As before, we examine the time evolution of the total energy by taking the derivative with respect to time (remembering to use the chain rule) and obtain

$$
\begin{aligned}
\tfrac{d}{dt}E(x(t)) &= (mg\ell \sin x_1(t))\, \dot{x}_1(t) + m\ell^2 x_2(t)\dot{x}_2(t) \\
&= (mg\ell \sin x_1(t))\, x_2(t) - m\ell^2 x_2(t)\left(\tfrac{g}{\ell}\sin x_1(t)\right) - m\ell^2 x_2(t)\left(\tfrac{k}{m}x_2(t)\right) \\
&= -k\ell^2 x_2(t)^2. \quad (1.12)
\end{aligned}
$$

As with the mass-spring system, as long as the pendulum is moving, $\dot{\theta} = x_2 \neq 0$ and the total energy E is decreasing since its derivative is strictly negative. However, here we need to be a bit more careful as the total energy is *not* strictly positive everywhere except $x_1 = 0$, though it is the case that in a neighborhood around $x_1 = 0$ the total energy is strictly positive. As with the mass-spring system, these energy considerations guarantee that the pendulum stops moving (i.e., $x_2 = \dot{\theta}$ must eventually become zero), but are insufficient to guarantee that the angle where the pendulum comes to rest is $\theta = 0$. We will return to this example in Section 2.3.

While real-world applications can be described through models such as (1.1) or (1.5), full knowledge of the state $x \in \mathbb{R}^n$ is usually not available. For example, for the pendulum in Figure 1.2 a common implementation includes a sensor for measuring the angle $\theta = x_1$ but does not also include a velocity sensor. In other words, we can measure $\theta = x_1$ but not $\dot{\theta} = x_2$. To capture this information, the model (1.1) can be augmented by a second *output* equation,

$$y = h(x), \quad (1.13)$$

where $h : \mathbb{R}^n \to \mathbb{R}^p$ and $y \in \mathbb{R}^p$ represents the measured output. For system (1.5), h can additionally depend on u; i.e., $h : \mathbb{R}^n \times \mathbb{R}^m \to \mathbb{R}^p$,

$$y = h(x, u). \quad (1.14)$$

In the example of the pendulum as discussed above, we would have $h(x) = x_1$ and $y = x_1$.

1.1.1 Notational Conventions

We will make use of a common abuse of notation where we drop the t-argument in (1.1) and write

$$\dot{x} = f(x), \tag{1.15}$$

with an initial state $x_0 \in \mathbb{R}^n$. We will then also denote solutions of (1.15) by $x : \mathbb{R}_{\geq 0} \to \mathbb{R}^n$ such that $x(0) = x_0 \in \mathbb{R}^n$ and the absolutely continuous function $x(\cdot)$ satisfies

$$\tfrac{d}{dt} x(t) = f(x(t))$$

for almost all $t \in \mathbb{R}_{\geq 0}$. It is important to keep in mind that we thus use x as both a point in $\mathbb{R}^n$ and as a solution of (1.15); i.e., x may denote either a vector or a function of time, where the distinction should be clear from the context.

Additionally, when considering the time derivative of energy-like functions, such as E above, several notational conventions are used in the literature. One that we will use most often makes the application of the chain rule explicit by writing the time derivative as an inner product of the gradient of E with $\dot{x}$; that is

$$\tfrac{d}{dt} E(x(t)) = \langle \nabla E(x), \dot{x} \rangle = \langle \nabla E(x), f(x) \rangle,$$

where

$$\nabla E(x) = \begin{bmatrix} \frac{\partial}{\partial x_1} E(x) \\ \frac{\partial}{\partial x_2} E(x) \\ \vdots \\ \frac{\partial}{\partial x_n} E(x) \end{bmatrix}.$$

In Example 1.2, writing out this step yields

$$\begin{aligned} \tfrac{d}{dt} E(x(t)) = \langle \nabla E(x), f(x) \rangle &= \begin{bmatrix} kx_1 & mx_2 \end{bmatrix} \begin{bmatrix} x_2 \\ \frac{1}{m}(-kx_1 - cx_2) \end{bmatrix} \\ &= kx_1 x_2 + x_2(-kx_1 - cx_2). \end{aligned}$$

1.1.2 Rescaling

Similarly to how we shifted an equilibrium to the origin in (1.9), we can sometimes simplify the system of interest via a process of rescaling or normalization. In many cases, this can make the qualitative behavior or the structure of the system more obvious. In some cases, it can improve the numerical conditioning of the problem for the purposes of simulation or further calculations.

First, consider the mass-spring system (1.8) and assume there is no friction; i.e., $c = 0$. Define a rescaling of time by a constant $\omega > 0$, to be chosen later, so that $\tau = \omega t$ which implies the differential relation $d\tau = \omega dt$. Similarly to how we derived the state space model (1.8), let $z_1 = x_1$ and

$$z_2 = \tfrac{d}{d\tau} z_1 = \tfrac{dt}{d\tau} \dot{z}_1 = \frac{\dot{x}_1}{\omega} = \frac{x_2}{\omega}. \tag{1.16}$$

Therefore, we can calculate

$$\tfrac{d}{d\tau} z_2 = \tfrac{dt}{d\tau} \dot{z}_2 = \frac{1}{\omega} \frac{\dot{x}_2}{\omega} = -\frac{k}{m\omega^2} x_1 = -\frac{k}{m\omega^2} z_1 . \tag{1.17}$$

Recall that we have not yet fixed ω, and so we are free to choose $\omega = \sqrt{k/m}$. This choice leads to the simple oscillator system

$$\tfrac{d}{d\tau} z_1 = z_2, \qquad \tfrac{d}{d\tau} z_2 = -z_1, \tag{1.18}$$

which, qualitatively, captures the behavior of all mass-spring systems in the absence of friction. Of course, for a *particular* mass-spring system, we can reverse the above transformations to get back to the specific physical system described by a given mass and a given spring.

Consider now the mass-spring system with friction; that is, (1.8) with $c > 0$. Define z_1 and z_2 in the same way as above. The only change in the above analysis comes with the calculation

$$\tfrac{d}{d\tau} z_2 = \tfrac{dt}{d\tau} \dot{z}_2 = \frac{1}{\omega} \frac{\dot{x}_2}{\omega} = -\frac{k}{m\omega^2} x_1 - \frac{c}{m\omega^2} x_2. \tag{1.19}$$

Making the same choice of $\omega = \sqrt{k/m}$ and defining a constant $\alpha = c\sqrt{m/k}$ we obtain the system equations

$$\tfrac{d}{d\tau} z_1 = z_2, \qquad \tfrac{d}{d\tau} z_2 = -z_1 - \alpha z_2. \tag{1.20}$$

Here, rather than the three parameters $m, k, c > 0$ of (1.8), we have a system with only one parameter $\alpha > 0$.

Finally, consider the pendulum (1.11). In addition to the time rescaling, $\tau = \omega t$, $\omega > 0$, we can also scale the state $z_1 = x_1/\beta$. As before, we postpone fixing values for ω and β. By taking a similar definition as before for z_2 we get

$$z_2 = \tfrac{d}{d\tau} z_1 = \tfrac{dt}{d\tau} \dot{z}_1 = \frac{\dot{x}_1}{\omega \beta} = \frac{x_2}{\omega \beta}. \tag{1.21}$$

We then calculate

$$\begin{aligned}
\tfrac{d}{d\tau} z_2 = \tfrac{dt}{d\tau} \dot{z}_2 &= \frac{1}{\omega} \frac{\dot{x}_2}{\beta \omega} = \frac{1}{\beta \omega^2} \left(-\frac{g}{\ell} \sin x_1 - \frac{k}{m} x_2 \right) \\
&= \frac{1}{\beta \omega^2} \left(-\frac{g}{\ell} \sin(\beta z_1) - \frac{k\omega\beta}{m} z_2 \right) \\
&= -\frac{g}{\ell\beta\omega^2} \sin(\beta z_1) - \frac{k}{m\omega} z_2.
\end{aligned} \tag{1.22}$$

Of the two free parameters, taking $\omega = k/m$ is an obvious choice. If we take $\beta = 1$ and define $\alpha = g/(\ell\omega^2)$ then we obtain the system

$$\tfrac{d}{d\tau} z_1 = z_2, \qquad \tfrac{d}{d\tau} z_2 = -\alpha \sin z_1 - z_2. \tag{1.23}$$

Alternatively, again with $\beta = 1$, we could take $\omega = \sqrt{g/\ell}$ and define $\alpha = k/(m\omega)$

to obtain

$$\tfrac{d}{d\tau} z_1 = z_2, \qquad \tfrac{d}{d\tau} z_2 = -\sin z_1 - \alpha z_2. \tag{1.24}$$

Another alternative would be to keep $\omega = k/m$ and take $\beta = \ell \omega^2 / g$.

As before, the above rescaled systems make certain qualitative elements clearer than they might otherwise be. For example, observe that near $x = 0$, $\sin x \approx x$. Therefore, in a manner to be made precise in Chapter 3, near the point $z_1 = z_2 = 0$, (1.24) behaves approximately like

$$\tfrac{d}{d\tau} z_1 = z_2, \qquad \tfrac{d}{d\tau} z_2 = -z_1 - \alpha z_2, \tag{1.25}$$

which is precisely the rescaled mass-spring system (1.20). In other words, in terms of qualitative behavior, the mass-spring system behaves like the pendulum (at least close to the downward equilibrium).

Reminder: While this rescaling can make a system easier to deal with numerically or analytically (if lengthy calculations by hand are required), it is necessary to reverse the transformations to get back to the specific system of interest. This will be particularly important later when we begin to design controllers that will need to be applied to a specific system.

1.1.3 Comparison Functions

Before moving on, we introduce several classes of functions that will be used repeatedly throughout the subsequent chapters. The example functions below are plotted in Figure 1.3.

Definition 1.5 (Class-$\mathcal{P}$ functions). *A function $\rho : \mathbb{R}^n \to \mathbb{R}_{\geq 0}$ is said to be positive definite ($\rho \in \mathcal{P}^n$) if it is continuous, zero at the origin, and strictly positive away from the origin; i.e., $\rho(0) = 0$ and $\rho(x) > 0$ for all $x \in \mathbb{R}^n \backslash \{0\}$.*

We will frequently use positive definite functions defined on the positive real half line, $\rho : \mathbb{R}_{\geq 0} \to \mathbb{R}_{\geq 0}$, and we will denote this class of functions by $\mathcal{P}$ (rather than $\mathcal{P}^1$). An example of a positive definite function is

$$\rho(s) = \frac{s}{1 + s^2}, \qquad s \geq 0.$$

Note that positive definite functions can be arbitrarily close to zero away from the origin. Furthermore, as the above example demonstrates, it is even possible that a positive definite function approaches zero since

$$\lim_{s \to \infty} \rho(s) = \lim_{s \to \infty} \frac{s}{1 + s^2} = 0.$$

As an example of a positive definite function $\mathcal{P}^n$, recall that a positive definite matrix is defined as a matrix $P \in \mathbb{R}^{n \times n}$ such that

$$x^T P x > 0, \quad \forall x \in \mathbb{R}^n \backslash \{0\}.$$

The function $\rho(x) = x^T P x$ is then class-$\mathcal{P}^n$ for a positive definite matrix $P \in \mathbb{R}^{n \times n}$.

Definition 1.6 (Class-$\mathcal{K}$ functions). *A function $\alpha : \mathbb{R}_{\geq 0} \to \mathbb{R}_{\geq 0}$ is said to be of* *class-$\mathcal{K}$ ($\alpha \in \mathcal{K}$) if it is continuous, zero at zero, and strictly increasing.*

Observe that class-$\mathcal{K}$ functions are those functions of class-$\mathcal{P}$ that are strictly increasing. Examples of such functions are

$$\alpha_1(s) = \frac{s}{s+1} \quad \text{or} \quad \alpha_2(s) = \tanh(s) \quad \text{for} \quad s \geq 0.$$

A nice property of class-$\mathcal{K}$ functions is that they are invertible on their range, continuous, and strictly increasing. For the above example, the hyperbolic tangent maps $s \in [0, \infty)$ to values $\tanh(s) \in [0, 1)$ and the inverse function, $\text{atanh}(r)$ or $\tanh^{-1}(r)$, is defined for arguments $r \in [0, 1)$.

Another nice property of class-$\mathcal{K}$ functions is that the composition of two class-$\mathcal{K}$ functions is again a class-$\mathcal{K}$ function. In other words, for two functions $\alpha_1, \alpha_2 \in \mathcal{K}$, the function $\alpha : \mathbb{R}_{\geq 0} \to \mathbb{R}_{\geq 0}$ given by

$$\alpha(s) \doteq \alpha_1(\alpha_2(s)) = \alpha_1 \circ \alpha_2(s), \qquad s \geq 0,$$

is also a class-$\mathcal{K}$ function.

Definition 1.7 (Class-$\mathcal{K}_\infty$ functions). *A function $\alpha : \mathbb{R}_{\geq 0} \to \mathbb{R}_{\geq 0}$ is said to be of class-$\mathcal{K}_\infty$ ($\alpha \in \mathcal{K}_\infty$) if $\alpha \in \mathcal{K}$ and, in addition, $\lim_{s \to \infty} \alpha(s) = \infty$.*

Examples of class-$\mathcal{K}_\infty$ functions include

$$\alpha_1(s) = s, \qquad \alpha_2(s) = s^2, \qquad \text{and} \qquad \alpha_3(s) = s + \sin(s).$$

The last function highlights a commonly made mistake when using functions that are strictly increasing; namely that a strictly increasing function must have a *strictly positive* derivative. Indeed, while α_3 is strictly increasing, its derivative is given by

$$\alpha_3'(s) = 1 + \cos(s), \qquad s > 0,$$

which is zero at an infinite number of points.

As with class-$\mathcal{K}$ functions, the composition of class-$\mathcal{K}_\infty$ functions yields another class-$\mathcal{K}_\infty$ function. A particularly nice property of class-$\mathcal{K}_\infty$ functions is that they are invertible and the inverse function is also of class-$\mathcal{K}_\infty$.

Note that we have the following relationship between the above defined function classes: $\mathcal{K}_\infty \subset \mathcal{K} \subset \mathcal{P}$.

Definition 1.8 (Class-$\mathcal{L}$ functions). *A function $\sigma : \mathbb{R}_{\geq 0} \to \mathbb{R}_{\geq 0}$ is said to be of class-$\mathcal{L}$ ($\sigma \in \mathcal{L}$) if it is continuous, strictly decreasing, and $\lim_{s \to \infty} \sigma(s) = 0$.*

Examples of class-$\mathcal{L}$ functions include

$$\sigma_1(t) = e^{-t}, \quad \text{and} \quad \sigma_2(t) = \frac{1}{1+t} \quad \text{for} \quad t \geq 0.$$

As with $\mathcal{K}$ and $\mathcal{K}_\infty$ functions, the composition of a class-$\mathcal{K}$ function α with a class-$\mathcal{L}$ function σ satisfies $\alpha \circ \sigma \in \mathcal{L}$. Inverting such functions requires a certain amount of care but, as we will not require the inverse of class-$\mathcal{L}$ functions in this book, we do not pursue this.

Definition 1.9 (Class-$\mathcal{K}\mathcal{L}$ functions). *A function $\beta : \mathbb{R}_{\geq 0} \times \mathbb{R}_{\geq 0} \to \mathbb{R}_{\geq 0}$ is said to*

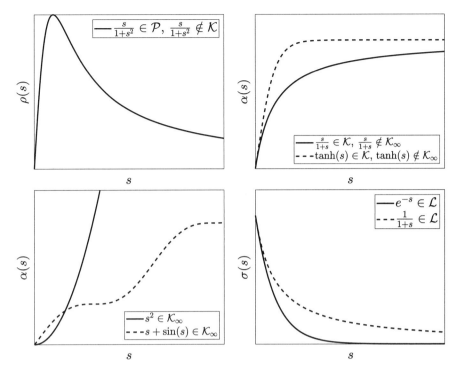

Figure 1.3: Example comparison functions.

be of class-$\mathcal{KL}$ ($\beta \in \mathcal{KL}$) if it is class-$\mathcal{K}_\infty$ in its first argument and class-$\mathcal{L}$ in its second argument. In other words, $\beta \in \mathcal{KL}$ if for each fixed $t \in \mathbb{R}_{\geq 0}$, $\beta(\cdot, t) \in \mathcal{K}_\infty$ and for each fixed $s \in \mathbb{R}_{> 0}$, $\beta(s, \cdot) \in \mathcal{L}$.

Examples of class-$\mathcal{KL}$ functions can easily be constructed from the previous examples of $\mathcal{K}$ and $\mathcal{L}$ functions; for example,

$$\beta(s, t) = se^{-t}, \qquad \beta(s, t) = \frac{s}{1 + t}.$$

The use of these various function classes has become standard in modern stability analysis due to the fact that it is possible to manipulate these functions in a wide variety of ways, particularly for the purposes of proving theorems. Since we will largely avoid technical proofs, the above summary of comparison functions is sufficient for our purposes.

1.2   CONTROL LOOPS, CONTROLLER DESIGN, AND EXAMPLES

When it comes to controller design in subsequent chapters, we focus on *closed-loop dynamics* where the state $x(t)$ or the measured output $y(t)$ is used to define a feedback law $u = \mu(x(t))$ or $u = \mu(y(t))$, respectively. In contrast, *open-loop* (or *non-feedback*) controller designs, also referred to as *steering*, are not the focus of this book. As an example, one should have applications such as reverse cycle air conditioning in mind, where information on the mismatch between the target temperature and the current temperature is used to define a control law u. As a related example of open-loop control consider a toaster, where bread is heated for a prespecified amount of time and using a particular amount of energy. The open-loop control is noticeably independent of the temperature of the bread. Based on experience, the human operator adjusts the heating time at the beginning, but in general does not intervene with the heating process once the toaster is switched on.

The closed-loop controller design methods discussed here can be motivated through the block diagrams in Figure 1.4 with and without an observer. State

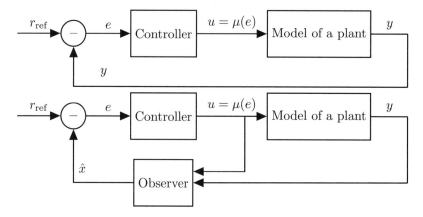

Figure 1.4: Feedback interconnection with and without observers controlling a model of a plant.

observers generating an estimate $\hat{x}$ of the state x based on the measured output y are discussed in Part III. In Figure 1.4 (top) the mismatch between the measured output y and a given reference signal r_{ref} is used to define a feedback law:

$$u = \mu(e) = \mu(y - r_{\text{ref}}).$$

The model of the plant is described through the dynamics (1.2) with output (1.13). Note that this includes $y = x$ as a special case. In general we aim to asymptotically drive the output $y(t)$ or the state $x(t)$ to a constant or time-varying reference signal r_{ref}, which constitutes an induced equilibrium pair (x^e, u^e). Since we are able to shift the equilibrium to the origin through an appropriate coordinate transformation according to the discussion after Definition 1.3, we will ignore the reference value in most of our derivations.

While the derivation or design of the controller is generally based on a *model of a plant* as shown in Figure 1.4, the performance of the controller needs to be evaluated for the closed-loop block diagrams in Figure 1.5, where the actual plant to be controlled replaces the mathematical model used for the design. Here, additional *disturbances v* and *noise w* are taken into account. The disturbances impact the input $u = \mu(e) + v$ and can be interpreted as the mismatch between the model of the plant and the plant in terms of the *actuators*. The actuators translate the virtual input signal generated by the controller into a physical signal applied to the

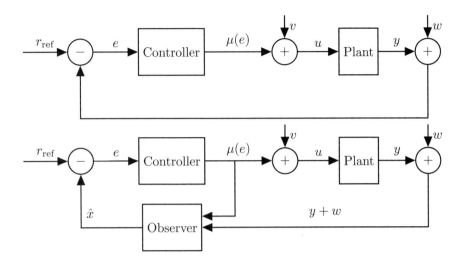

Figure 1.5: Realistic feedback interconnection with respect to the plant dynamics including disturbances v and measurement noise w.

plant.

The noise w captures the mismatch between the internal state x of the model of the plant and the actual plant, and it captures inaccuracies due to *sensors* measuring the output. In particular, while the output of the plant is given by the unknown signal $y(t)$, only the quantity $y(t) + w(t)$ returned by the sensors is accessible for the observer and controller design.

Example 1.10. As a simple example we consider the model of the mass-spring system in Example 1.2 described through the dynamics

$$m\ddot{y} + c\dot{y} + ky = u \tag{1.26}$$

derived in (1.7), where the position y can be measured through a sensor and the actuator acts as a force $u = F$ on the mass.

To stabilize the mass at a specific reference position y^e, we can consider a proportional plus derivative controller (PD-controller) of the form

$$u = k_P(y - r_{\mathrm{ref}}) + k_D(\dot{y} - \dot{r}_{\mathrm{ref}}),$$

for example. The constants k_P and k_D define proportional and derivative gains of the controller. Using the definition in (1.26) together with $\dot{r}_{\mathrm{ref}} = 0$ (since we want to stabilize y at a fixed position) we obtain the closed-loop dynamics

$$m\ddot{y} + (c - k_D)\dot{y} + (k - k_P)y = -k_P r_{\mathrm{ref}}. \tag{1.27}$$

Under the assumption that all solutions of (1.27), i.e., independently of the initial condition, converge to an equilibrium, the derivatives vanish — $\ddot{y}(t) \to 0$, $\dot{y}(t) \to 0$ for $t \to \infty$ — and we are left with the condition

$$(k - k_P)y = -k_P r_{\mathrm{ref}}. \tag{1.28}$$

Thus, (1.28) tells us how to select the reference value $r_{\text{ref}} = -\frac{k-k_P}{k_P}y^e$ for an equilibrium $x^e = [y^e, \ 0]^T$ of interest.

To ensure convergence to x^e, i.e., asymptotic stability discussed in detail in the next chapter, the degree of freedom in the controller in terms of k_P and k_D needs to be used. If $k_D \neq 0$, then information on the derivative $\dot{y}$ is taken into account in the controller and an observer needs to be included to obtain an estimate of $\dot{y}$.

In the presence of disturbances and noise as in Figure 1.5, the dynamics become

$$m\ddot{y} + c\dot{y} + ky = k_P(y + w - r_{\text{ref}}) + k_D(\dot{y} + \dot{w} - \dot{r}_{\text{ref}}) + v.$$

If we again calculate the steady state, where we additionally need to assume that $\dot{w}(t) \to 0$ for $t \to \infty$, it holds that

$$(k - k_P)y = k_P w - k_P r_{\text{ref}} + v,$$

and the steady state satisfies

$$y^e = -\frac{k_P}{k - k_P}r_{\text{ref}} + \frac{k_P w + v}{k - k_P}. \tag{1.29}$$

Since v and w are unknown, r_{ref} can only be selected based on the model of the plant, leading to a steady-state error given by the second term on the right-hand side in (1.29).

If we add an integral term to the controller, i.e., if we consider the proportional-integral-derivative controller (PID-controller)

$$u = k_I \int_0^t (y - r_{\text{ref}}) \ dt + k_P(y - r_{\text{ref}}) + k_D(\dot{y} - \dot{r}_{\text{ref}}),$$

the closed loop corresponding to the model of the plant is given by

$$m\ddot{y} + (c - k_D)\dot{y} + (k - k_P)y - k_I \int_0^t y \ dt = -k_I \int_0^t r_{\text{ref}} \ dt - k_P r_{\text{ref}} - k_D \dot{r}_{\text{ref}}.$$

To get rid of the integral, we take the time derivative and again assume that r_{ref} is constant

$$m\dddot{y} + (c - k_D)\ddot{y} + (k - k_P)\dot{y} - k_I y = -k_I r_{\text{ref}}.$$

Here, we observe that the steady state satisfies $y^e = r_{\text{ref}}$.

If the controller is applied to the setting in Figure 1.5, the overall closed loop satisfies

$$m\dddot{y} + (c - k_D)\ddot{y} + (k - k_P)\dot{y} - k_I y$$
$$= k_I(w - r_{\text{ref}}) + k_P(\dot{w} - \dot{r}_{\text{ref}}) + k_D(\ddot{w} - \ddot{r}_{\text{ref}}) + \dot{v},$$

with steady state

$$y^e = r_{\text{ref}} - w.$$

For the derivation, we have additionally assumed that the disturbance v and the noise w are constant. Interestingly, under these assumptions, only the noise impacts the steady state while the disturbance does not.

Using the coordinate transformations

$$x = \begin{bmatrix} x_1 \\ x_2 \end{bmatrix} = \begin{bmatrix} y \\ \dot{y} \end{bmatrix} \quad \text{and} \quad x = \begin{bmatrix} x_1 \\ x_2 \\ x_3 \end{bmatrix} = \begin{bmatrix} y \\ \dot{y} \\ \ddot{y} \end{bmatrix},$$

respectively, the PD- and the PID-controller can be summarized through the dynamics

$$\begin{bmatrix} \dot{x}_1 \\ \dot{x}_2 \end{bmatrix} = \begin{bmatrix} x_2 \\ -\frac{k}{m}x_1 - \frac{c}{m}x_2 \end{bmatrix} + \begin{bmatrix} 0 \\ \frac{1}{m} \end{bmatrix} \begin{bmatrix} k_P & k_D \end{bmatrix} \begin{bmatrix} x_1 - r_{\text{ref}} \\ x_2 \end{bmatrix} \tag{1.30}$$

$$\begin{bmatrix} \dot{x}_1 \\ \dot{x}_2 \\ \dot{x}_3 \end{bmatrix} = \begin{bmatrix} x_2 \\ x_3 \\ -\frac{k}{m}x_2 - \frac{c}{m}x_3 \end{bmatrix} + \begin{bmatrix} 0 \\ 0 \\ \frac{1}{m} \end{bmatrix} \begin{bmatrix} k_I & k_P & k_D \end{bmatrix} \begin{bmatrix} x_1 - r_{\text{ref}} \\ x_2 \\ x_3 \end{bmatrix}, \tag{1.31}$$

which are both of the form (1.6). In this representation the PD and PID controller are structurally the same. In particular, if we ignore that x_i is a placeholder of the ith derivative of y, then (1.30) and (1.31) are simply proportional controllers and (1.6) is able to represent a variety of different controller designs.

1.2.1   The Pendulum on a Cart

A widely used academic example for control systems is the *pendulum on a cart*, visualized in Figure 1.6. Following [12, Example 2.1], the dynamics of motion are

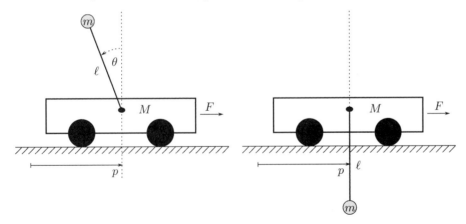

Figure 1.6: Pendulum on a cart.

given by the second-order differential equation

$$M(q)\ddot{q} + C(q,\dot{q}) + K(q) = B(q)u,$$

with inertia matrix $M(q)$, Coriolis forces and damping $C(q,\dot{q})$, potential energy $K(q)$, and external forces $B(q)$.

More precisely, when $q = [p, \theta]^T$, where $p \in \mathbb{R}$ denotes the position of the cart and $\theta \in [-\pi, \pi)$ denotes the angle of the pendulum (with respect to the upright position), the second-order differential equation modeling the pendulum on a cart can be described as

$$
\begin{bmatrix} (M+m) & -m\ell\cos(\theta) \\ -m\ell\cos(\theta) & (J+m\ell^2) \end{bmatrix} \begin{bmatrix} \ddot{p} \\ \ddot{\theta} \end{bmatrix} + \begin{bmatrix} c\dot{p} + m\ell\sin(\theta)\dot{\theta}^2 \\ \gamma\dot{\theta} - mg\ell\sin(\theta) \end{bmatrix} = \begin{bmatrix} F \\ 0 \end{bmatrix}.
$$

The parameters of the second-order differential equation are defined in Table 1.1 (see also [12]). With the assumption $m > 0$, $\ell > 0$, we rescale the dynamics using

M	mass of the cart
m	mass of the pendulum
ℓ	distance from cart to center of mass of the pendulum
J	moment of inertia of the pendulum
c, γ	friction constants
g	gravitational acceleration constant
F	external force

Table 1.1: Parameters of the inverted pendulum on a cart.

the definitions

$$
\bar{M} = \frac{M+m}{m\ell}, \qquad \bar{J} = \frac{J+m\ell^2}{m\ell}, \qquad \bar{c} = \frac{c}{m\ell}, \qquad \bar{\gamma} = \frac{\gamma}{m\ell}, \qquad \bar{F} = \frac{F}{m\ell},
$$

to obtain

$$
\begin{bmatrix} \bar{M} & -\cos(\theta) \\ -\cos(\theta) & \bar{J} \end{bmatrix} \begin{bmatrix} \ddot{p} \\ \ddot{\theta} \end{bmatrix} + \begin{bmatrix} \bar{c}\dot{p} + \sin(\theta)\dot{\theta}^2 \\ \bar{\gamma}\dot{\theta} - g\sin(\theta) \end{bmatrix} = \begin{bmatrix} \bar{F} \\ 0 \end{bmatrix}. \tag{1.32}
$$

As discussed in Section 1.1 higher order differential equations can be rewritten as first-order differential equations. In particular, with $x = [p, \theta, \dot{p}, \dot{\theta}]^T$ and $u = \bar{F}$ the system of first-order differential equations

$$
\begin{bmatrix} \dot{x}_1 \\ \dot{x}_2 \end{bmatrix} = \begin{bmatrix} x_3 \\ x_4 \end{bmatrix}
$$

$$
\begin{bmatrix} \bar{M} & -\cos(x_2) \\ -\cos(x_2) & \bar{J} \end{bmatrix} \begin{bmatrix} \dot{x}_3 \\ \dot{x}_4 \end{bmatrix} = \begin{bmatrix} -\bar{c}x_3 - \sin(x_2)x_4^2 + u \\ -\bar{\gamma}x_4 + g\sin(x_2) \end{bmatrix}
$$

is equivalent to (1.32). Due to the matrix on the left-hand side, the system is not of the form (1.5) yet. The inverse of a matrix $A \in \mathbb{R}^{2\times2}$ can be calculated analytically through the formula

$$
A = \begin{bmatrix} a & b \\ c & d \end{bmatrix} \implies A^{-1} = \frac{1}{ad - cb} \begin{bmatrix} d & -b \\ -c & a \end{bmatrix} \tag{1.33}
$$

in the case $\det(A) = ad - cb \neq 0$.

This implies that in the case of the inverted pendulum, the inverse matrix is given by

$$
\begin{bmatrix} \bar{M} & -\cos(x_2) \\ -\cos(x_2) & \bar{J} \end{bmatrix}^{-1} = \frac{1}{\bar{M}\bar{J} - \cos^2(x_2)} \begin{bmatrix} \bar{J} & \cos(x_2) \\ \cos(x_2) & \bar{M} \end{bmatrix} \tag{1.34}
$$

and thus the dynamical system

$$
\begin{bmatrix} \dot{x}_1 \\ \dot{x}_2 \\ \dot{x}_3 \\ \dot{x}_4 \end{bmatrix} = \begin{bmatrix} x_3 \\ x_4 \\ \dfrac{-\bar{J}\bar{c}x_3 - \bar{J}\sin(x_2)x_4^2 - \bar{\gamma}\cos(x_2)x_4 + g\cos(x_2)\sin(x_2) + \bar{J}u}{M\bar{J} - \cos^2(x_2)} \\ \dfrac{-\bar{M}\bar{\gamma}x_4 + \bar{M}g\sin(x_2) - \bar{c}\cos(x_2)x_3 - \cos(x_2)\sin(x_2)x_4^2 + \cos(x_2)u}{M\bar{J} - \cos^2(x_2)} \end{bmatrix} \tag{1.35}
$$

of the form (1.5) is achieved. Checking the above determinant condition associated with (1.33), the determinant satisfies

$$
\bar{M}\bar{J} - \cos^2(x_2) \geq \left(\tfrac{M+m}{m\ell}\right)\left(\tfrac{J+m\ell^2}{m\ell}\right) - 1 > \tfrac{m}{m\ell}\tfrac{m\ell^2}{m\ell} - 1 = 0 \tag{1.36}
$$

if at least one of the parameters $M, J \in \mathbb{R}_{\geq 0}$ is strictly positive and thus the inverse (1.34) is indeed well defined for all $x \in \mathbb{R}^4$.

To calculate equilibria of the dynamical system we set the right-hand side of (1.35) to zero. Since this implies $x_3 = x_4 = 0$, we are left with the conditions

$$
g\cos(x_2)\sin(x_2) + \bar{J}u = 0, \tag{1.37}
$$
$$
\bar{M}g\sin(x_2) + \cos(x_2)u = 0. \tag{1.38}
$$

Rearranging terms provides $\sin(x_2) = -\tfrac{1}{\bar{M}g}\cos(x_2)u$ and thus

$$
\left(-\frac{1}{\bar{M}}\cos^2(x_2) + \bar{J}\right)u = 0.
$$

Using the estimate (1.36) it follows that

$$
-\frac{1}{\bar{M}}\cos^2(x_2) + \bar{J} \neq 0
$$

for all $x_2 \in \mathbb{R}$ if at least one of the parameters $M, J \in \mathbb{R}_{\geq 0}$ is strictly positive. Thus, for an induced equilibrium (and hence in particular for an equilibrium of the pendulum without input) the external force $u = \bar{F}$ needs to be equal to zero. With (1.38) it follows that $x_2 \in \{k\pi \in \mathbb{R} : k \in \mathbb{Z}\}$, and hence the set of (induced) equilibria is given by

$$
x^e \in \mathbb{R} \times \{k\pi \in \mathbb{R} : k \in \mathbb{Z}\} \times \{0\} \times \{0\} \subset \mathbb{R}^4, \qquad u^e \in \{0\}. \tag{1.39}
$$

This shows that the pendulum can be stabilized at any position ($x_1 \in \mathbb{R}$) where the angle is a multiple of π. For $k \in \mathbb{Z}$ even, $x_2 = k\pi$ captures the pendulum in the upright position, while $k \in \mathbb{Z}$ odd captures the hanging pendulum in Figure 1.6, left. Additionally, at an equilibrium the velocity of the cart, the angular velocity, and the external force need to be equal to zero.

While the pendulum on the cart is an academic example, real-world applications can be interpreted using these dynamics. For example, keeping a rocket, visualized in Figure 1.7 (right), in the upright position is similar to keeping the pendulum in the upright position. While the cart can only move from left to right (i.e., $x_1 \in \mathbb{R}$), the rocket has an additional degree of freedom. The equilibrium of the pendulum on a cart where the pendulum is pointing downward is similar to that of a tower crane, shown in Figure 1.7 (left).

Figure 1.7: Tower crane and rocket as real-world applications of the pendulum on a cart.

If we disregard the position of the cart, we are left with the dynamics

$$\begin{bmatrix} \dot{x}_2 \\ \dot{x}_3 \\ \dot{x}_4 \end{bmatrix} = \begin{bmatrix} x_4 \\ \dfrac{-\bar{J}\bar{c}x_3 - \bar{J}\sin(x_2)x_4^2 - \bar{\gamma}\cos(x_2)x_4 + g\cos(x_2)\sin(x_2) + \bar{J}u}{M\bar{J} - \cos^2(x_2)} \\ \dfrac{-\bar{M}\bar{\gamma}x_4 + \bar{M}g\sin(x_2) - \bar{c}\cos(x_2)x_3 - \cos(x_2)\sin(x_2)x_4^2 + \cos(x_2)u}{M\bar{J} - \cos^2(x_2)} \end{bmatrix}, \quad (1.40)$$

which can be used to describe a segway shown in Figure 1.8. A segway can be

Figure 1.8: Segway as a real-world example of the inverted pendulum on a cart and reference tracking.

controlled by leaning forward or backward, leading to an angle $x_2 \neq 0$. Based on our calculations, a state $x \in \mathbb{R}^4$ of the inverted pendulum with $x_2 \neq 0$ cannot be an equilibrium. However by ignoring the position x_1, equilibrium pairs $([x_2^e, \ x_3^e, \ x_4^e]^T, u^e)$ with $x_2 \neq 0$ can be derived.

In particular, for a given angle x_2^e, a constant input $u^e(x_2^e)$ can be derived such that the segway (1.40) is moving forward or backward with a constant velocity $x_3^e(x_2^e)$. To compute $u^e(x_2^e)$ and $x_3^e(x_2^e)$ we compute the equilibria of (1.40). Here

$\dot{x}_2 = 0$ again implies $x_4^e = 0$ independently of x_2. Thus, we obtain the conditions

$$-\bar{J}\bar{c}x_3^e + g\cos(x_2^e)\sin(x_2^e) + \bar{J}u^e = 0 \tag{1.41}$$

$$\bar{M}g\sin(x_2^e) - \bar{c}\cos(x_2^e)x_3^e + \cos(x_2^e)u^e = 0 \tag{1.42}$$

and x_3^e can be written as

$$x_3^e = \frac{g}{\bar{J}\bar{c}}\cos(x_2^e)\sin(x_2^e) + \frac{1}{\bar{c}}u^e. \tag{1.43}$$

Using this expression obtained from (1.41) in (1.42) leads to

$$u^e(x_2^e) = \frac{\bar{M}g\sin(x_2^e) - \frac{g}{\bar{J}}\cos^2(x_2^e)\sin(x_2^e)}{\cos^2(x_2^e) - \cos(x_2^e)} \tag{1.44}$$

for $x_2 \neq (2k+1)\pi$, $k \in \mathbb{Z}$. If the handle of the segway is at a particular angle x_2^e and the external force is selected as in (1.44), then the segway is driving with constant velocity $x_3^e(x_2^e)$ defined in (1.43). The condition $x_2^e \neq (2k+1)\pi$, $k \in \mathbb{Z}$ in (1.44) shows that with a finite input u an angle of 90° cannot be realized.

Another variation of the pendulum is obtained by removing the cart and by fixing the position of the inverted pendulum. In this case the angle and the angular velocity of the inverted pendulum can be described through the nonlinear dynamics [12, Example 2.2]

$$\frac{d}{dt}\begin{bmatrix} \theta \\ \dot{\theta} \end{bmatrix} = \begin{bmatrix} \dot{\theta} \\ \frac{1}{J+m\ell^2}\left((mg\ell)\sin(\theta) - \gamma\dot{\theta} + \ell\cos(\theta)u\right) \end{bmatrix}, \tag{1.45}$$

and the parameters are again defined in Table 3.1.

The model of the inverted pendulum on a cart and its variations will be used throughout the book to illustrate various concepts.

1.2.2 Mobile Robots—The Nonholonomic Integrator

A simple model of a mobile robot is described through the dynamics

$$\begin{aligned} \dot{x}_1 &= u_1\cos(\phi), \\ \dot{x}_2 &= u_1\sin(\phi), \\ \dot{\phi} &= u_2. \end{aligned} \tag{1.46}$$

Here, $[x_1,\ x_2]^T \in \mathbb{R}^2$ describes the position of the robot in the plane and $\phi \in [-\pi, \pi)$ denotes its orientation. The setting is depicted in Figure 1.9. The position and the orientation of the mobile robot can be manipulated through the input $u \in \mathbb{R}^2$. Even though the dynamics are quite simple, feedback design is not as trivial as it might seem and the model (1.46) is still used in ongoing research.

Depending on the application, different representations of the dynamics are used in the literature. Instead of using the Cartesian coordinates as in (1.46), the

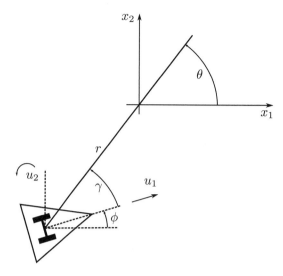

Figure 1.9: A mobile robot in the plane described through its position and orientation.

position components of the model can be represented in polar coordinates,

$$
\begin{aligned}
r &= \sqrt{x_1^2 + x_2^2} = |x| \\
\theta &= \arctan_2(-x_2, -x_1),
\end{aligned}
\tag{1.47}
$$

for example.

Here the function $\arctan_2 : \mathbb{R}^2 \backslash \{0\} \to (-\pi, \pi]$ denotes the inverse tangent where in contrast to $\arctan(y_2/y_1)$, $y_1, y_2 \in \mathbb{R}$, the signs of y_1 and y_2 are taken into account. We follow the implementation in MATLAB for the definition of $\arctan_2$, i.e.,

$$
\arctan_2(x_2, x_1) \doteq \begin{cases}
\arctan\left(\frac{x_2}{x_1}\right), & x_1 > 0, \ x_2 \neq 0, \\
\arctan\left(\frac{x_2}{x_1}\right) + \operatorname{sign}(x_2)\pi, & x_1 < 0, \ x_2 \neq 0, \\
\pi, & x_1 < 0, \ x_2 = 0 \\
\operatorname{sign}(x_2)\frac{\pi}{2}, & x_1 = 0, \ x_2 \neq 0.
\end{cases}
\tag{1.48}
$$

While (1.47) allows us to go from Cartesian coordinates to polar coordinates, the Cartesian coordinates can be computed through

$$
\begin{aligned}
x_1 &= -r \cos(\theta) \\
x_2 &= -r \sin(\theta)
\end{aligned}
\tag{1.49}
$$

for given $r \in \mathbb{R}_{\geq 0}$ and $\theta \in (-\pi, \pi]$ in polar coordinates.

Using (1.47) and (1.49) for $r \neq 0$, the time derivative of $r(t)$ can be expressed as

$$\dot{r} = \frac{d}{dt}\sqrt{x_1^2 + x_2^2} = 2(\dot{x}_1 x_1 + \dot{x}_2 x_2) \cdot \left(\tfrac{1}{2}(x_1^2 + x_2^2)^{-\frac{1}{2}}\right)$$

$$= \frac{x_1 + x_2}{r} = \frac{-u_1 \cos(\phi) r \cos(\theta) - u_1 \sin(\phi) r \sin(\theta)}{r}$$

$$= -u_1 \left(\cos(\phi) \cos(\theta) + \sin(\phi) \sin(\theta)\right)$$

$$= -u_1 \cos(\theta - \phi).$$

The last equation follows from the identity

$$\cos(\alpha \pm \beta) = \cos(\alpha) \cos(\beta) \mp \sin(\alpha) \sin(\beta), \qquad \alpha, \beta \in \mathbb{R}.$$

To derive an expression for $\dot{\theta}$ we take the time derivative of

$$\sin(\theta) = -\frac{x_2}{r},$$

which satisfies

$$\dot{\theta} \cos(\theta) = -\frac{\dot{x}_2 r - x_2 \dot{r}}{r^2}$$

$$= -\frac{u_1 \sin(\phi) r - (-u_1 \cos(\theta - \phi))(-r \sin(\theta))}{r^2}.$$

Rearranging the terms and simplifying the expressions finally provides the dynamics

$$\dot{\theta} = -\frac{u_1}{r} \frac{\sin(\phi) - \cos(\theta - \phi) \sin(\theta)}{\cos(\phi)}$$

$$= -\frac{u_1}{r}(-\sin(\theta - \phi)).$$

The last equation follows again from a trigonometric identity.

Thus, for $r \neq 0$, i.e., away from the origin, the dynamics of the mobile robot in Cartesian coordinates (1.46) can alternatively be written as

$$\dot{r} = -u_1 \cos(\theta - \phi),$$

$$\dot{\theta} = u_1 \frac{\sin(\theta - \phi)}{r}, \tag{1.50}$$

$$\dot{\phi} = u_2,$$

in polar coordinates.

With the definition $\gamma = \theta - \phi$, i.e., $\dot{\gamma} = \dot{\theta} - \dot{\phi}$, the variable ϕ in (1.50) can be eliminated and a third representation of the mobile robot is given by

$$\dot{r} = -u_1 \cos(\gamma),$$

$$\dot{\gamma} = u_1 \frac{\sin(\gamma)}{r} - u_2, \tag{1.51}$$

$$\dot{\theta} = u_1 \frac{\sin(\gamma)}{r}.$$

These coordinate transformations can for example be found in [4].

A different representation of the mobile robot (1.46) is known as the *nonholonomic integrator* or *Brockett integrator*. As in (1.51), we derive the dynamics of the nonholonomic integrator in two steps. As a first step we introduce the coordinate transformation

$$y_1 = \phi,$$
$$y_2 = x_1 \cos(\phi) + x_2 \sin(\phi),$$
$$y_3 = x_1 \sin(\phi) - x_2 \cos(\phi).$$

For y_1 it holds that

$$\dot{y}_1 = \dot{\phi} = u_2. \tag{1.52}$$

With $\cos^2(\phi) + \sin^2(\phi) = 1$, the second component in the new variables can be written as

$$\begin{aligned}
\dot{y}_2 &= \tfrac{d}{dt}\left(x_1 \cos(\phi) + x_2 \sin(\phi)\right) \\
&= \dot{x}_1 \cos(\phi) + x_1(-\sin(\phi))\dot{\phi} + \dot{x}_2 \sin(\phi) + x_2 \cos(\phi)\dot{\phi} \\
&= u_1 \cos^2(\phi) + u_1 \sin^2(\phi) + \dot{\phi}\left(x_1(-\sin(\phi)) + x_2 \cos(\phi)\right) \\
&= u_1 - u_2 y_3.
\end{aligned} \tag{1.53}$$

Finally, the third component satisfies

$$\begin{aligned}
\dot{y}_3 &= \tfrac{d}{dt}\left(x_1 \sin(\phi) - x_2 \cos(\phi)\right) \\
&= \dot{x}_1 \sin(\phi) + x_1 \cos(\phi)\dot{\phi} - \dot{x}_2 \cos(\phi) - x_2(-\sin(\phi))\dot{\phi} \\
&= u_1 \cos(\phi)\sin(\phi) - u_1 \sin(\phi)\cos(\phi) + \dot{\phi}\left(x_1 \cos(\phi) + x_2 \sin(\phi)\right) \\
&= u_2 y_2.
\end{aligned} \tag{1.54}$$

With the additional transformation of the input

$$v_1 = u_2$$
$$v_2 = u_1 - y_3 u_2$$

the dynamics

$$\begin{aligned}
\dot{y}_1 &= u_2 & &= v_1 \\
\dot{y}_2 &= u_1 - u_2 y_3 & &= v_2 \\
\dot{y}_3 &= u_2 y_2 & &= v_1 y_2
\end{aligned} \tag{1.55}$$

are obtained. As a second step we consider the coordinate transformation

$$z_1 = y_1, \qquad z_2 = y_2, \qquad \text{and} \qquad z_3 = -2y_3 + y_1 y_2$$

which leads to the dynamics of the *Brockett integrator*

$$\begin{aligned}
\dot{z}_1 &= v_1, \\
\dot{z}_2 &= v_2, \\
\dot{z}_3 &= z_1 v_2 - z_2 v_1.
\end{aligned} \tag{1.56}$$

While the representations are equivalent, for the analysis of the dynamical system and for the controller design one or the other representation may be beneficial. The model of the nonholonomic integrator is used in several subsequent chapters to illustrate various nonlinear control concepts and limitations.

1.3   EXERCISES

Exercise 1.1. Consider the nth-order differential equations

$$\frac{d^n}{dt^n} y(t) = 1, \qquad \sum_{k=0}^{n} \frac{1}{k+1} \frac{d^k}{dt^k} y(t) = 1 \qquad \text{and} \qquad \prod_{k=0}^{n} \frac{d^k}{dt^k} y(t) = 1.$$

Rewrite the nth-order models in state space form $\dot{x} = f(x)$.

Exercise 1.2. Consider the differential equation

$$\dot{x} = f(x) = \begin{bmatrix} \cos(x_2)\sin(x_1) + x_1^3 - 2x_1 \\ x_1 x_2^2 + x_1^2 + (1+x_2)(2+x_2) - \sin(x_2+2) \end{bmatrix}. \tag{1.57}$$

Show that $x^e = [0,\ -2]^T$ is an equilibrium of the differential equation. Shift the equilibrium x^e to the origin, i.e., define a coordinate transformation $y = x - x^e$ such that $\dot{y} = \tilde{f}(y)$ with $\tilde{f}(0) = 0$. (Note that the ordinary differential equation might have multiple equilibria.)

Exercise 1.3. Show that the composition of two class-$\mathcal{K}$ functions is again a class-$\mathcal{K}$ function. Show that the inverse of a class-$\mathcal{K}_\infty$ function is also a class-$\mathcal{K}_\infty$ function. What can you say about the inverse of a class-$\mathcal{K}$ function?

Exercise 1.4. Show (by constructing a counterexample) that the composition of two class-$\mathcal{L}$ functions is not a class-$\mathcal{L}$ function. What can you say about the inverse of a class-$\mathcal{L}$ function?

Exercise 1.5. Show that the composition of a class-$\mathcal{L}$ function with a class-$\mathcal{K}_\infty$ function is a class-$\mathcal{L}$ function.

Exercise 1.6. Let $\alpha_1, \alpha_2, \alpha_3 \in \mathcal{K}_\infty$ and $\beta \in \mathcal{KL}$. What can you say about the function $\alpha_1(\beta(\alpha_2(\cdot), \alpha_3(\cdot)))$?

1.4   BIBLIOGRAPHICAL NOTES AND FURTHER READING

Standard control-focussed references for ordinary differential equations include [86] and [158]. These texts, as well as [59], are also standard texts for the stability concepts and Lyapunov theory presented in the next chapter.

Most texts on ordinary differential equations will contain some version of Theorem 1.1.

The comparison functions introduced in Section 1.1.3 have become standard in the analysis and design of nonlinear systems, particularly when discussing qualitative properties. It is possible to derive many simple results to combine or separate or bound these functions in various ways. Some of the results specifically used in this text are in Appendix A.3. A comprehensive survey of available results can be found in [81].

The mobile robot model of Section 1.2.2 is quite simple. More complicated dynamics for mobile robots can be found in [154]. This reference also contains the coordinate transformation to move from (1.46) to the Brockett integrator (1.56).

Chapter Two

Nonlinear Systems—Stability Notions

2.1 STABILITY NOTIONS

As we intuitively and naturally understand the term, "stability" is clearly a desirable quality for nearly all engineered systems and is frequently present in natural systems. However, to make use of this observation for

$$\dot{x} = f(x), \tag{2.1}$$

we need to be a bit more precise.

Definition 2.1 (Stability). *The origin is* Lyapunov stable *(or simply* stable*) for system (2.1) if, for any $\varepsilon > 0$, there exists $\delta > 0$ (possibly dependent on ε) such that if $|x(0)| \leq \delta$ then, for all $t \geq 0$,*

$$|x(t)| \leq \varepsilon. \tag{2.2}$$

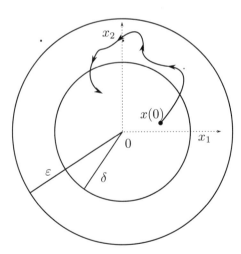

Figure 2.1: The ε-δ stability game. For a given ε-ball, find a δ-ball so that solutions starting inside the δ-ball stay inside the ε-ball.

This definition of stability can be viewed as a game. Our opponent chooses a (possibly small) number $\varepsilon > 0$ and demands that the trajectories of (2.1) stay ε-close to the origin for all future times. We win this game, and declare the origin stable, if we can find another number $\delta > 0$ so that, by starting δ-close to the origin, we satisfy the demand of our opponent. Note that the definition of stability implicitly assumes that solutions of (2.1) exist for all $t \geq 0$ and thus (2.1) is *forward complete*.

As an extremely simple example, consider the one-dimensional system given by

$$\dot{x} = 0, \qquad x(0) = x_0 \in \mathbb{R},$$

for which the solution is $x(t) = x_0$ for all $t \geq 0$. The origin is clearly stable since, for any $\varepsilon > 0$, we can choose $\delta = \varepsilon$ since

$$|x_0| \leq \delta \quad \text{implies} \quad |x(t)| = |x_0| \leq \delta = \varepsilon.$$

As a slightly more complex example, consider the oscillator

$$\dot{x}_1 = x_2, \qquad \dot{x}_2 = -x_1,$$

which has the solution

$$x_1(t) = x_2(0)\sin(t) + x_1(0)\cos(t)$$
$$x_2(t) = -x_1(0)\sin(t) + x_2(0)\cos(t).$$

Using polar coordinates (r, θ), we see that $r(t) = r(0) = \sqrt{x_1(0)^2 + x_2(0)^2} = |x(0)|$ and $\theta(t) = t$. As in the previous game analogy, when presented with the requirement that the trajectory stay ε-close to the origin, we can choose $\delta = \varepsilon$. Then for any initial conditions $|x(0)| = r(0) \leq \delta$ we have that $|x(t)| = r(t) = r(0) \leq \delta = \varepsilon$ and so the origin is stable.

The comparison functions introduced in Chapter 1 can be used to provide an alternative definition of stability. In particular, we could have equivalently defined stability of the origin as the existence of a function $\alpha \in \mathcal{K}$ and an open neighborhood around the origin $\mathcal{D} \subset \mathbb{R}^n$ such that

$$|x(t)| \leq \alpha(|x(0)|), \qquad \forall\, t \geq 0, \ \ \forall\, x_0 \in \mathcal{D}. \tag{2.3}$$

This equivalence is not difficult to show and is left to Exercise 2.1.

Definition 2.2 (Instability). *The origin is* unstable *for system* (2.1) *if it is not stable.*

A simple example of a system with an unstable origin is given by the one-dimensional linear system

$$\dot{x} = x, \qquad x(0) = x_0 \in \mathbb{R},$$

which has the solution $x(t) = x_0 e^t$ for all $t \geq 0$. Clearly we cannot win the ε-δ stability game. For any fixed $\varepsilon > 0$, no matter how small we choose $\delta > 0$, for any initial condition satisfying $|x_0| \neq 0$, $|x_0| < \delta$, eventually $|x(t)| > \varepsilon$. In fact, for this simple example, it is straightforward to compute the time when the norm of the solution crosses the ε threshold.

A slightly more complicated, but also more interesting, example is given by the second-order system

$$\dot{x}_1 = x_1, \qquad \dot{x}_2 = -x_2. \tag{2.4}$$

Note that this is simply two uncoupled first-order systems and the solution of this system is given by

$$x_1(t) = x_1(0)e^t, \quad x_2(t) = x_2(0)e^{-t}, \qquad t \geq 0.$$

Here, we observe that for some initial conditions we get nice behavior in that solutions converge to the origin. In particular, for initial conditions satisfying $[x_1(0), x_2(0)]^T = [0, x_{2,0}]^T$ where $x_{2,0} \in \mathbb{R}$, solutions converge exponentially quickly to the origin.

However, again we see that it is impossible to win the ε-δ stability game no matter how small we choose $\delta > 0$ since even very small, but nonzero, initial values of $x_1(0)$ will lead to $|x(t)|$ eventually exceeding any given value of ε.

Definition 2.3 (Attractivity). *The origin is* attractive *for* (2.1) *if there exists* $\delta > 0$ *such that if* $|x(0)| < \delta$ *then*

$$\lim_{t \to \infty} x(t) = 0. \tag{2.5}$$

It is important to note that attractivity and stability are different notions. We have already seen an example of a system where the origin is stable but not attractive: the origin is stable for $\dot{x} = 0$ since $x(t) = x(0)$, but for $x(0) \neq 0$ it is clear that $x(t)$ does not go to zero.

A classical example of a system for which the origin is attractive but unstable is given by

$$
\begin{aligned}
\dot{x}_1 &= \frac{x_1^2(x_2 - x_1) + x_2^5}{(x_1^2 + x_2^2)\left(1 + (x_1^2 + x_2^2)^2\right)} \\
\dot{x}_2 &= \frac{x_2^2(x_2 - 2x_1)}{(x_1^2 + x_2^2)\left(1 + (x_1^2 + x_2^2)^2\right)} .
\end{aligned}
\tag{2.6}
$$

While a detailed analysis of this system showing that the origin is attractive but unstable is possible, it is not particularly useful in what follows. Rather, in Exercise 2.2 we suggest examining the system behavior via simulation.

While stability and attractivity are distinct concepts, the combination of the two is among the most important concepts in nonlinear systems analysis.

Definition 2.4 (Asymptotic stability). *The origin is* asymptotically stable *for* (2.1) *if it is both stable (as per Definition 2.1) and attractive (as per Definition 2.3).*

An equivalent definition can be given in terms of the comparison functions introduced in Section 1.1.3.

Definition 2.5 ($\mathcal{KL}$-stability). *System* (2.1) *is said to be* $\mathcal{KL}$-stable *if there exist* $\delta > 0$ *and* $\beta \in \mathcal{KL}$ *such that if* $|x(0)| \leq \delta$ *then for all* $t \geq 0$,

$$|x(t)| \leq \beta(|x(0)|, t). \tag{2.7}$$

For ease of reference, we formally state:

Proposition 2.6 ([97, Proposition 2.5]). *The origin is asymptotically stable for* (2.1) *if and only if* (2.1) *is* $\mathcal{KL}$-stable.

One direction of this proof is relatively straightforward. Indeed, the stability formulation in terms of a class-$\mathcal{K}$ function given by (2.3) is satisfied from (2.7) by taking $\alpha(s) = \beta(s, 0)$. Furthermore, that (2.7) implies attractivity follows from the fact that $\mathcal{KL}$ functions are strictly decreasing to zero in their second argument.

As is frequently the case for if-and-only-if style results, one direction is easy (as above), while the other direction is significantly more involved. For Proposition 2.6, it is necessary to explicitly construct a $\mathcal{KL}$ function satisfying (2.7) from the properties of stability and attractivity. The interested reader should consult Section 2.8 for further reading.

Finally, as a commonly desired special case, we define:

Definition 2.7 (Exponential stability). *The origin is* exponentially stable *for* (2.1) *if there exist* $\delta, \lambda, M > 0$ *such that if* $|x(0)| \leq \delta$ *then for all* $t \geq 0$,

$$|x(t)| \leq M|x(0)|e^{-\lambda t}. \tag{2.8}$$

It is immediate that exponential stability is a special case of $\mathcal{KL}$-stability since the bound in (2.8) is an explicit $\mathcal{KL}$ function; i.e., exponential stability is $\mathcal{KL}$-stability where the required function in (2.7) is

$$\beta(s, t) = Mse^{-\lambda t}, \qquad s, t \geq 0.$$

2.1.1   Local versus Global Properties

The stability definitions provided above are all *local* conditions. In other words, the definitions refer to properties that hold in some neighborhood of the equilibrium point, but not necessarily throughout the state space. For nonlinear systems, this is particularly important since there may be more than one equilibrium point. For example, recall that the pendulum (Example 1.4) has an infinite number of equilibrium points.

In the definitions above, the local part of the definition refers to the existence of a $\delta > 0$ so that such-and-such a property holds for initial conditions in some neighborhood of the origin. It is worth noting that while for stability we can imagine that it is necessary to choose small values of δ to stay within ε of the origin, the definitions of attractivity and $\mathcal{KL}$-stability (and hence asymptotic and exponential stability) do not necessarily require that δ be small.

This is particularly important for engineered systems where we generally want "good" behavior from as wide a range of initial conditions as possible. In some cases, we may even be able to obtain "good" behavior (e.g., asymptotic or exponential stability) from all possible initial conditions. In this case, we add the additional descriptor "global" to the relevant definition as follows:

Definition 2.8 (Global attractivity). *The origin is* globally attractive *for* (2.1) *if for all* $x(0) \in \mathbb{R}^n$,

$$\lim_{t \to \infty} x(t) = 0.$$

Definition 2.9 (Global $\mathcal{KL}$-stability). *System* (2.1) *is* globally $\mathcal{KL}$-stable *if* (2.7) *holds for all* $x(0) \in \mathbb{R}^n$ *and all* $t \geq 0$.

Definition 2.10 (Global exponential stability). *The origin is* globally exponentially stable *for* (2.1) *if there exist* $M, \lambda > 0$ *such that* (2.8) *holds for all* $x(0) \in \mathbb{R}^n$ *and* $t \geq 0$.

2.1.2 Time-Varying Systems

To this point we have only considered time-invariant or autonomous systems given by (2.1). However, when considering time-varying or non-autonomous systems

$$\dot{x} = f(t, x), \qquad x(t_0) \in \mathbb{R}^n, \ t \geq t_0 \geq 0 \tag{2.9}$$

(i.e., (1.2)), there is an additional complication that needs to be taken into account. In our discussion of time-invariant systems, we have implicitly assumed that the initial time t_0 is at zero and all of the previous stability definitions are with respect to *elapsed time*. For example, we could have written the definition of exponential stability as a bound

$$|x(t)| \leq M|x(t_0)|e^{-\lambda(t-t_0)}, \quad t \geq t_0.$$

However, a critical property of time-invariant systems is that the initial time does not matter. As such, we can always take $t_0 = 0$.

By contrast, for time-varying systems, the initial time cannot be so easily discarded. Take for example the time-varying system

$$\dot{x} = -\frac{x}{t+1}, \qquad x(t_0) \in \mathbb{R}, \ t \geq t_0 \geq 0. \tag{2.10}$$

It is easy to check that the solution is given by

$$x(t) = x(t_0)\frac{t_0 + 1}{t + 1}. \tag{2.11}$$

As can be observed in Figure 2.2, *when* we start the system (t_0) is as important as *where* we start the system $(x(t_0))$. Put another way, the solution depends explicitly on t_0 rather than on the elapsed time $t - t_0$.

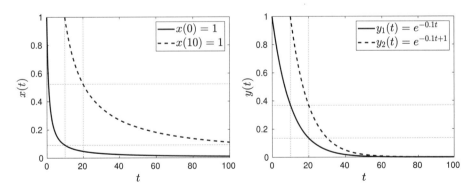

Figure 2.2: On the left, solutions of (2.10) from $x(t_0) = 1$ with $t_0 = 0$ and $t_0 = 10$ are shown. Note the different convergence rates highlighted through the dotted lines showing an elapsed time of 10 seconds and the values of the solutions at that time. As a comparison, solutions of the differential equation $\dot{y} = -0.1y$ with uniform convergence rate are shown on the right.

The system (2.10) still has a certain nice behavior with respect to the origin. For example, (2.11) makes it clear that solutions converge to the origin as $t \to \infty$. In fact, we can extend all of our previous stability and attractivity definitions in a reasonable way to account for the fact that the bounds may depend on the initial time t_0.

Definition 2.11 (Stability). *The origin is* stable *for system* (2.9) *if, for any $\varepsilon > 0$ there exists $\delta(t_0) > 0$ such that if $|x(t_0)| \leq \delta(t_0)$ then, for all $t \geq t_0$,*

$$|x(t)| \leq \varepsilon. \tag{2.12}$$

If $\delta(t_0)$ can be chosen independently of t_0, then the origin is uniformly stable *for system* (2.9).

Returning to (2.11), suppose we are given $\varepsilon > 0$. Then if

$$|x(t_0)| \leq \frac{\varepsilon}{t_0 + 1} \doteq \delta(t_0)$$

then

$$|x(t)| = |x(t_0)| \frac{t_0 + 1}{t + 1} \leq \frac{\varepsilon}{t_0 + 1} \frac{t_0 + 1}{t + 1} \leq \varepsilon$$

for all $t \geq t_0$.

Similar modifications can be made to the other stability and attractivity definitions. We only additionally provide the modification for $\mathcal{KL}$-stability.

Definition 2.12 ($\mathcal{KL}$-stability). *System* (2.9) *is said to be* (globally) $\mathcal{KL}$-stable *if for each $t_0 \geq 0$ there exists $\beta_{t_0} \in \mathcal{KL}$ such that for all $x(t_0) \in \mathbb{R}^n$ and $t \geq t_0$,*

$$|x(t)| \leq \beta_{t_0}(|x(t_0)|, t - t_0). \tag{2.13}$$

If $\beta_{t_0} \in \mathcal{KL}$ can be chosen independently of t_0, then (2.9) *is said to be* uniformly globally $\mathcal{KL}$-stable.

The required $\beta_{t_0} \in \mathcal{KL}$ for (2.11) is given by

$$\beta_{t_0}(s, \tau) = s \frac{t_0 + 1}{\tau + t_0 + 1}.$$

In the above definitions, the uniformity referred to is with regards to the initial time. Of course, as we previously mentioned, for time-invariant systems, the initial time is essentially irrelevant and can always be taken to be zero. Consequently, all of the definitions related to (1.1) are, by default, uniform in the initial time. In the literature, it is not uncommon for authors to consider time-invariant systems (1.1) and to also use the modifier *uniform* in their stability definitions. While technically correct, we will only use *uniform* when explicitly considering time-varying systems (2.9).

2.2 COMPARISON PRINCIPLE

Given that models are only ever an approximation of "real-world" systems, in some sense the exact solution of ordinary differential equations is of limited interest in terms of studying system behavior or guaranteeing good behavior (e.g., asymptotic stability) in the presence of uncertainty. However, differential inequalities allow some wiggle room. Fortunately, in certain cases, differential inequalities can be used to provide bounds on the size of the system trajectories of interest. The following two results are cornerstones of Lyapunov's second method, presented in the next section.

Lemma 2.13 ([81, Lemma 20]). *For any continuous positive definite function $\rho \in \mathcal{P}$ there exists $\beta \in \mathcal{KL}$ such that if $y(\cdot)$ is any locally absolutely continuous function defined on some interval $[0, T]$ with $y(t) \geq 0$ for all $t \in [0, T]$, and if $y(\cdot)$ satisfies the differential inequality*

$$\dot{y}(t) \leq -\rho(y(t))$$

for almost all $t \in [0, T]$ with $y(0) = y_0 \geq 0$, then

$$y(t) \leq \beta(y_0, t), \quad \forall t \in [0, T].$$

While the previous lemma explicitly guarantees the existence of a nice comparison function bound (via $\beta \in \mathcal{KL}$), the following lemma relates the solutions of the differential inequality to the solutions of the related differential equation.

Lemma 2.14 ([86, Lemma 3.4]). *Consider the scalar differential equation $\dot{\psi} = g(\psi)$, $\psi(0) = \psi_0 \in \mathbb{R}$. Let $[0, T)$ be the maximal interval of existence of the solution $\psi(t)$. Let $\phi(t)$ be a continuously differentiable function that satisfies*

$$\dot{\phi}(t) \leq g(\phi(t)), \quad \phi(0) \leq \psi(0).$$

Then $\phi(t) \leq \psi(t)$ for all $t \in [0, T)$.

As an example, consider the differential equation

$$\dot{x} = -(1 + x^2)x, \qquad x(0) = a \in \mathbb{R} \tag{2.14}$$

and let $v(t) = x(t)^2$. Then, by the chain rule,

$$\dot{v}(t) = 2x(t)\dot{x}(t) = -2x(t)^2 - 2x(t)^4 \leq -2x(t)^2 = -2v(t).$$

Consider now the linear differential equation

$$\dot{\psi} = -2\psi, \qquad \psi(0) = a^2,$$

whose solution is given by $\psi(t) = a^2 e^{-2t}$. The comparison principle of Lemma 2.14 yields

$$|x(t)| = \sqrt{v(t)} \leq \sqrt{\psi(t)} = |a|e^{-t}. \tag{2.15}$$

Note that while it may be nontrivial to derive a closed-form solution for (2.14), the bound in (2.15) allows us to conclude that the origin is (globally) exponentially stable for the nonlinear differential equation (2.14).

2.3 STABILITY BY LYAPUNOV'S SECOND METHOD

In Example 1.2 and Example 1.4 we considered the time evolution of the total energy (the sum of the potential and kinetic energy) to attempt a rudimentary stability analysis. In both cases, we saw that the energy was guaranteed to decrease until the mass or the pendulum stopped moving. Despite our intuition about the position of each system when it came to rest, examining the time evolution of the energy left us a bit short of a complete understanding of the stability of the equilibrium point.

In 1892, the Russian mathematician Aleksandr Mikhailovich Lyapunov abstracted away the need to directly consider the energy of a system. Rather, Lyapunov suggested using an energy-like function that allows conclusions to be drawn about (asymptotic, exponential) stability. This idea is a cornerstone of nonlinear systems analysis and the theorems in this chapter will be used repeatedly in our development of nonlinear control tools and techniques.

In the following results, we will refer to a domain $\mathcal{D} \subset \mathbb{R}^n$ where we assume $\mathcal{D}$ is open and contains the origin.

Recall that a positive definite function $\rho \in \mathcal{P}^n$ is a continuous function that satisfies $\rho(0) = 0$ and $\rho(x) > 0$ for all $x \in \mathcal{D} \backslash \{0\}$ (Definition 1.5). It is not difficult to see that, if $\mathcal{D}$ is bounded, then for any function ρ that is positive definite on $\overline{\mathcal{D}}$ (that is, the closure of $\mathcal{D}$) there exists a $\mathcal{K}$ function α so that

$$\alpha(|x|) \leq \rho(x), \qquad \forall x \in \mathcal{D}. \tag{2.16}$$

Theorem 2.15 (Lyapunov stability theorem). *Given* (2.1) *with* $f(0) = 0$, *and a domain* $\mathcal{D} \subset \mathbb{R}^n$, *suppose there exists a continuously differentiable function* $V : \mathcal{D} \to \mathbb{R}_{\geq 0}$ *satisfying* $V(0) = 0$ *and a function* $\alpha_1 \in \mathcal{K}$ *such that, for all* $x \in \mathcal{D}$,

$$\alpha_1(|x|) \leq V(x) \tag{2.17}$$

and

$$\langle \nabla V(x), f(x) \rangle \leq 0. \tag{2.18}$$

Then the origin is stable. If, additionally, $\mathcal{D} = \mathbb{R}^n$ *and* $\alpha_1 \in \mathcal{K}_\infty$, *then the origin is globally stable.*

One proof of Theorem 2.15 follows a straightforward modification of one of the proofs presented below for Theorem 2.16.

Based on the discussion preceding Theorem 2.15, condition (2.17) is a way of stating that the function V is positive definite. When considering $\mathcal{D} = \mathbb{R}^n$, the additional condition that $\alpha_1 \in \mathcal{K}_\infty$ implies that V is *radially unbounded*. In other words, as $x \in \mathbb{R}^n$ grows unbounded in any direction, $|x|$ becomes large and hence $\alpha_1(|x|)$ becomes large and, in turn, $V(x)$ also grows unbounded.

Recalling the notation described in Section 1.1.1, we see that (2.18) is a way of writing the derivative of the time evolution of the function V along the solutions of (2.1). So V, which we call a *Lyapunov function*, is a generalized energy function. That is, it is a function that is positive everywhere except at the origin, where it has a minimum, and, to conclude that the origin is a stable equilibrium, we require that the "energy" not increase over time.

Theorem 2.15 is extremely powerful in that it allows us to ascertain stability of the origin *without the need to solve* (2.1). Indeed, to directly appeal to Definition 2.1, we would need to solve (2.1) and show that all solutions from some set of initial conditions satisfy a bound as in (2.2). By contrast, Theorem 2.15, and in particular (2.18), only depends on the vector field f defining (2.1).

Theorem 2.16 (Lyapunov asymptotic stability theorem). *Given* (2.1) *with* $f(0) = 0$, *and a domain* $\mathcal{D} \subset \mathbb{R}^n$, *suppose there exists a continuously differentiable function* $V : \mathcal{D} \to \mathbb{R}_{\geq 0}$ *satisfying* $V(0) = 0$, *a function* $\alpha_1 \in \mathcal{K}$, *and a positive definite function* $\rho \in \mathcal{P}$ *such that, for all* $x \in \mathcal{D}$, (2.17) *holds and*

$$\langle \nabla V(x), f(x) \rangle \leq -\rho(|x|). \tag{2.19}$$

Then the origin is asymptotically stable. If, additionally, $\mathcal{D} = \mathbb{R}^n$ *and* $\alpha_1 \in \mathcal{K}_\infty$, *then the origin is globally asymptotically stable.*

Note that, from Proposition 2.6, Theorem 2.16 equivalently can be used to conclude (global) $\mathcal{KL}$-stability.

It is critically important to keep in mind that Theorem 2.16 is a *sufficient* condition. In other words, if we can find an appropriate Lyapunov function then we can conclude that the origin is asymptotically stable. However, the failure of a *particular* function to be a Lyapunov function does not allow us to conclude anything.

The only difference between Theorem 2.15 and Theorem 2.16 lies in the difference between the decrease conditions (2.18) and (2.19). The intuitive difference is that (2.18) implies that the (generalized) energy in the system cannot increase, while (2.19) implies that the energy is decreasing everywhere except at the origin. This result is sufficiently important that we will shortly provide two proofs for Theorem 2.16.

Theorem 2.17 (Lyapunov exponential stability theorem). *Given* (2.1) *with* $f(0) = 0$, *and a domain* $\mathcal{D} \subset \mathbb{R}^n$, *suppose there exists a continuously differentiable function* $V : \mathcal{D} \to \mathbb{R}_{\geq 0}$ *and constants* $\lambda_1, \lambda_2, c > 0$ *and* $p \geq 1$ *such that, for all* $x \in \mathcal{D}$,

$$\lambda_1 |x|^p \leq V(x) \leq \lambda_2 |x|^p \tag{2.20}$$

and

$$\langle \nabla V(x), f(x) \rangle \leq -cV(x). \tag{2.21}$$

Then the origin is exponentially stable. If, additionally, $\mathcal{D} = \mathbb{R}^n$, *then the origin is globally exponentially stable.*

Example 2.18. Recall the pendulum system from Example 1.4:

$$\dot{x}_1 = x_2$$
$$\dot{x}_2 = -\frac{g}{\ell} \sin x_1 - \frac{k}{m} x_2,$$

where $k \geq 0$ is a friction coefficient.

With $k = 0$ our intuition is that the downward equilibrium is stable, but not asymptotically stable. Observe that the total energy

$$V(x) = mg\ell(1 - \cos x_1) + \tfrac{1}{2}m\ell^2 x_2^2$$

is positive definite in a domain around the origin given by $\mathcal{D} = (-\frac{\pi}{2}, \frac{\pi}{2}) \times \mathbb{R}$. Taking the total energy as a candidate Lyapunov function, we compute

$$\langle \nabla V(x), f(x) \rangle = \begin{bmatrix} mg\ell \sin x_1 & m\ell^2 x_2 \end{bmatrix} \begin{bmatrix} x_2 \\ -\frac{g}{\ell} \sin x_1 \end{bmatrix}$$
$$= mg\ell x_2 \sin x_1 - mg\ell x_2 \sin x_1 = 0.$$

Therefore, Theorem 2.15 confirms that the origin is stable.

When we additionally consider the friction coefficient $k > 0$, the analysis in Example 1.4 was incomplete since the right-hand side of (1.12) is not negative definite (since it is zero for $x_2 = 0$ and any value of x_1). Hence, the total energy cannot be used in Theorem 2.16 to conclude asymptotic stability of the origin.

We can reverse-engineer a candidate Lyapunov function as follows. First, we recognize the value of the potential energy term in V above since it leads to the cancellation of an otherwise troublesome term. The quadratic term in x_2 is also beneficial since, as we saw in Example 1.4, it leads to a negative definite term in x_2. Therefore, we conjecture that something similar to the total energy

$$V(x) = \tfrac{1}{2}\left(ax_1^2 + bx_1x_2 + x_2^2\right) + \tfrac{g}{\ell}(1 - \cos x_1) \tag{2.22}$$

might be a Lyapunov function, where the constants $a, b > 0$ are to be determined. We first try to compute a and b to guarantee (2.19) and then check that these values also guarantee that V is positive definite.

We compute the inner product

$$\langle \nabla V(x), f(x) \rangle$$
$$= \begin{bmatrix} ax_1 + \frac{b}{2}x_2 + \frac{g}{\ell}\sin x_1 & \frac{b}{2}x_1 + x_2 \end{bmatrix} \begin{bmatrix} x_2 \\ -\frac{g}{\ell}\sin x_1 - \frac{k}{m}x_2 \end{bmatrix}$$
$$= ax_1x_2 + \frac{b}{2}x_2^2 + \frac{g}{\ell}x_2 \sin x_1$$
$$\quad - \frac{b}{2}\frac{g}{\ell}x_1 \sin x_1 - \frac{b}{2}\frac{k}{m}x_1x_2 - \frac{g}{\ell}x_2 \sin x_1 - \frac{k}{m}x_2^2$$
$$= -\frac{b}{2}\frac{g}{\ell}x_1 \sin x_1 - \left(\frac{k}{m} - \frac{b}{2}\right)x_2^2 + \left(a - \frac{b}{2}\frac{k}{m}\right)x_1x_2.$$

The cross-term x_1x_2 can never be sign-definite, and so we choose

$$a = \frac{b}{2}\frac{k}{m}$$

so that the coefficient of the cross-term is zero. We also need to ensure that the coefficient of x_2^2 is negative. A simple choice is

$$b = \frac{k}{m}.$$

Therefore, with these choices for a and b, we see that the inner product satisfies

$$\langle \nabla V(x), f(x) \rangle = -\frac{gk}{2\ell m}x_1 \sin x_1 - \frac{k}{2m}x_2^2 < 0$$

for all $x \in \mathcal{D}\backslash\{0\}$.

Finally, we check that V is positive definite for these choices of a and b. Note that we can write the quadratic portion of V as $\frac{1}{2}x^T Px$ where

$$P = \begin{bmatrix} a & \frac{1}{2}b \\ \frac{1}{2}b & 1 \end{bmatrix} = \begin{bmatrix} \frac{1}{2}\left(\frac{k}{m}\right)^2 & \frac{1}{2}\frac{k}{m} \\ \frac{1}{2}\frac{k}{m} & 1 \end{bmatrix}.$$

We observe that P is a positive definite matrix (since, for example, the leading principal minors are all positive) and therefore

$$V(x) = \tfrac{1}{2}x^T Px + \tfrac{g}{\ell}(1 - \cos(x_1))$$

is a Lyapunov function. By Theorem 2.16, the origin is asymptotically stable.

The example above highlights both the strength and weakness of Lyapunov's second method. As already mentioned, it was not necessary to solve the differential equations in order to determine stability or asymptotic stability of the origin. On the other hand, finding an appropriate Lyapunov function was not entirely straightforward. In fact, for a particular system, finding a Lyapunov function can frequently involve a process of trial and error starting from a few commonly successful forms, such as the quadratic form used above.

Proof of Theorem 2.16: We give two proofs of Theorem 2.16 here; an analytical proof and a geometrical, intuitive proof.

For simplicity, we provide the analytical proof in the case of global asymptotic stability. In other words, assume $\mathcal{D} = \mathbb{R}^n$ and $\alpha_1 \in \mathcal{K}_\infty$, remembering that α_1 in (2.17) indicates that V is positive definite. Since $V : \mathbb{R}^n \to \mathbb{R}_{\geq 0}$ is continuous and satisfies $V(0) = 0$, there exists a function $\alpha_2 \in \mathcal{K}_\infty$ so that, for all $x \in \mathcal{D}$,

$$\alpha_1(|x|) \leq V(x) \leq \alpha_2(|x|). \tag{2.23}$$

Let $\hat{\alpha} \in \mathcal{K}_\infty$, $\sigma \in \mathcal{L}$ come from Lemma A.7 applied to $\rho : \mathbb{R}^n \to \mathbb{R}_{\geq 0}$. We observe that the Lyapunov function decrease condition (2.19) can be rewritten using the bounds (2.23) as

$$\langle \nabla V(x), f(x) \rangle \leq -\rho(|x|) \leq -\hat{\alpha}(|x|)\sigma(|x|)$$
$$\leq -\hat{\alpha}(\alpha_2^{-1}(V(x)))\sigma(\alpha_1^{-1}(V(x))) \leq -\hat{\rho}(V(x)), \tag{2.24}$$

where we have defined the continuous positive definite function $\hat{\rho} \in \mathcal{P}$ by

$$\hat{\rho}(s) \doteq \hat{\alpha}(\alpha_2^{-1}(s))\sigma(\alpha_1^{-1}(s)), \quad \forall s \in \mathbb{R}_{\geq 0}.$$

That $\hat{\rho}$ is positive definite follows from the facts that the composition of $\mathcal{K}_\infty$ functions are $\mathcal{K}_\infty$, the composition of an $\mathcal{L}$ function and a $\mathcal{K}_\infty$ function is an $\mathcal{L}$ function, and the product of a $\mathcal{K}_\infty$ function and an $\mathcal{L}$ function is positive definite. We rewrite (2.24) as

$$\tfrac{d}{dt}V(x(t)) = \langle \nabla V(x(t)), f(x(t)) \rangle \leq -\hat{\rho}(V(x(t)))$$

and let $\hat{\beta} \in \mathcal{KL}$ come from Lemma 2.13 with $\hat{\rho}$. Therefore

$$V(x(t)) \leq \hat{\beta}(V(x(0)), t), \quad \forall t \geq 0.$$

Using the bounds in (2.23) we obtain the $\mathcal{KL}$-stability estimate

$$\alpha_1(|x(t)|) \leq V(x(t)) \leq \hat{\beta}(V(x(0)), t) \leq \hat{\beta}(\alpha_2(|x(0)|), t).$$

Defining the $\mathcal{KL}$ function $\beta(s, t) \doteq \alpha_1^{-1}(\hat{\beta}(\alpha_2(s), t))$ for all $s, t \in \mathbb{R}_{\geq 0}$ we see that system (1.1) is globally $\mathcal{KL}$-stable. By Proposition 2.6 we obtain that the origin is globally asymptotically stable, which completes the analytical proof.

For the geometrical proof, we construct several sets shown in Figure 2.3. Given $\varepsilon > 0$, define

$$\overline{\mathcal{B}}_\varepsilon = \{x \in \mathbb{R}^n : |x| \leq \varepsilon\}.$$

In the event that $\overline{\mathcal{B}}_\varepsilon$ is not a subset of $\mathcal{D}$, we can simply shrink $\overline{\mathcal{B}}_\varepsilon$ by choosing some $\overline{\mathcal{B}}_{\hat{\varepsilon}} \subset \mathcal{D}$ with $0 < \hat{\varepsilon} < \varepsilon$. Hence, without loss of generality, we assume $\overline{\mathcal{B}}_\varepsilon \subset \mathcal{D}$.

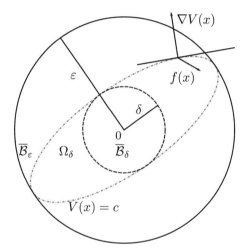

Figure 2.3: Geometry of Lyapunov's second method.

Let $a = \min_{|x|=\varepsilon} V(x)$ and take $c \in (0, a)$. Define

$$\Omega_c = \{x \in \mathcal{D} : V(x) \leq c\}$$

and observe that $\Omega_c \subset \mathcal{B}_\varepsilon$. Note that for x on the boundary of Ω_c, $\nabla V(x)$ is the outward-facing normal vector. Furthermore, the condition (2.18) can be seen to be

$$\langle \nabla V(x), f(x) \rangle = |\nabla V(x)||f(x)| \cos(\theta) \leq 0, \tag{2.25}$$

where θ is the angle between the outward-facing normal $\nabla V(x)$ and the vector field $f(x)$ defining the system (2.1). The inequality implies that $|\theta| \geq \frac{\pi}{2}$, which means that the vector field $f(x)$ either points to the interior of Ω_c or, at most, tangentially along its boundary. Consequently, Ω_c is *invariant*; in other words, solutions starting in Ω_c will always remain in Ω_c.

In order to complete our ε-δ game, we choose $\delta > 0$ so that

$$\overline{\mathcal{B}}_\delta = \{x \in \mathcal{D} : |x| \leq \delta\}$$

satisfies $\overline{\mathcal{B}}_\delta \subset \Omega_c$. Therefore, if $|x(0)| \leq \delta$ we have $x(0) \in \mathcal{B}_\delta \subset \Omega_c$ and, since Ω_c is invariant, $x(t) \in \Omega_c \subset \overline{\mathcal{B}}_\varepsilon$; in other words, $|x(t)| \leq \varepsilon$ for all $t \geq 0$. This proves Theorem 2.15.

The asymptotic stability in Theorem 2.16 relies on the strict negativity in the bound (2.19) (other than at the origin). In particular, the inequality in (2.25) becomes *strictly* less than zero, which implies that $|\theta| > \frac{\pi}{2}$, and hence the vector field $f(x)$ points to the interior of Ω_c. In other words, the strict negativity means the vector field $f(x)$ cannot be tangent to Ω_c at x.

Finally, to obtain the global asymptotic stability result, since $V(x)$ is radially unbounded, by taking larger and larger values for ε we can take larger and larger values for c. This follows from the fact that, since $V(x)$ is unbounded, $a = \min_{|x|=\varepsilon} V(x)$ is also unbounded, allowing us to choose c unbounded. In this way, we are eventually able to obtain convergence to the origin from any point in $\mathcal{D} = \mathbb{R}^n$. $\qquad\square$

Remark 2.19. As mentioned in the analytical proof above, $V(0) = 0$ and $V : \mathbb{R}^n \to \mathbb{R}_{\geq 0}$ continuous imply the existence of an $\alpha_2 \in \mathcal{K}_\infty$ such that $V(x) \leq \alpha_2(|x|)$ for all $x \in \mathcal{D}$. Given that the upper bound is frequently useful, it is not uncommon to define a Lyapunov function as a function satisfying a particular decrease condition such as (2.18) or (2.19), and both upper and lower bounds; i.e., by replacing (2.17) with the existence of $\alpha_1, \alpha_2 \in \mathcal{K}_\infty$ such that, for all $x \in \mathcal{D}$,

$$\alpha_1(|x|) \leq V(x) \leq \alpha_2(|x|).$$

In this case, it is not necessary to additionally assume that $V(0) = 0$ since this follows immediately from $\alpha_1(0) = \alpha_2(0) = 0$. Note that the matched polynomial upper and lower bounds in (2.20) are critical to the conclusion of exponential stability in Theorem 2.17.

We conclude our introduction to Lyapunov's second method with a couple of observations on Lyapunov functions.

Theorem 2.20 (Rescaling of Lyapunov functions). *Let $\alpha \in \mathcal{K}_\infty$ be continuously differentiable on $\mathbb{R}_{>0}$ and $\alpha'(s) > 0$ for all $s > 0$. If $V : \mathbb{R}^n \to \mathbb{R}_{\geq 0}$ is a Lyapunov function for (2.1), then $W : \mathbb{R}^n \to \mathbb{R}_{\geq 0}$, defined by*

$$W(x) \doteq \alpha(V(x)), \quad \forall x \in \mathbb{R}^n,$$

is also a Lyapunov function for (2.1).

This theorem tells us that if we have one Lyapunov function for a system (2.1), we can construct an infinite number of Lyapunov functions for (2.1). This is reassuring as it indicates that the search for a Lyapunov function is not quite the same as searching for a needle in a haystack. However, there is still the issue of finding one Lyapunov function. We return to this question in Section 2.5 and again in Section 3.4.

Proof of Theorem 2.20: To show that W is a Lyapunov function, we must demonstrate that W is positive definite, that $W(0) = 0$, and that W satisfies the decrease condition (2.19). Define $\hat{\alpha}_1 \doteq \alpha \circ \alpha_1 \in \mathcal{K}_\infty$. Then we see that, for all $x \in \mathbb{R}^n$,

$$\hat{\alpha}_1(|x|) = \alpha \circ \alpha_1(|x|) \leq \alpha(V(x)) \tag{2.26}$$

and hence W is positive definite. That $W(0) = 0$ follows immediately from $V(0) = 0$ and $\alpha \in \mathcal{K}_\infty$. Note that if we have an upper bound as discussed in Remark 2.19, the upper bound on W is trivial to derive.

An application of the chain rule allows us to see that, for all $x \in \mathbb{R}^n \backslash \{0\}$,

$$\langle \nabla W(x), f(x) \rangle = \alpha'(V(x)) \langle \nabla V(x), f(x) \rangle \leq -\alpha'(V(x)) \rho(x).$$

Since $\alpha'(s) > 0$ for all $s \in \mathbb{R}_{>0}$, as a consequence of the bound $\alpha_1(|x|) \leq V(x)$, for all $x \in \mathbb{R}^n \backslash \{0\}$ we see that $\alpha'(V(x)) \rho(x) > 0$. Furthermore, since α' and ρ are both continuous, and since $\rho(0) = 0$, we see that $\alpha'(V(x)) \rho(x)$ is positive definite. Therefore, W satisfies the decrease condition

$$\langle \nabla W(x), f(x) \rangle \leq -\hat{\rho}(x) \doteq -\alpha'(V(x)) \rho(x), \quad \forall x \in \mathbb{R}^n$$

and, with the upper and lower bounds derived in (2.26), is hence a Lyapunov function for (2.1). $\qquad \square$

Recall that a continuously differentiable $\alpha \in \mathcal{K}_\infty$ does not necessarily satisfy $\alpha'(s) > 0$ for all $s > 0$. For example, $\alpha(s) = \sin(s) + s$ is of class-$\mathcal{K}_\infty$ but satisfies $\alpha'(s) = 0$ for infinitely many $s > 0$.

Theorem 2.21 (Exponentially decreasing Lyapunov functions). *If there exists a Lyapunov function for system* (2.1) *satisfying* (2.17) *and* (2.19), *then there exists a continuously differentiable function* $W : \mathbb{R}^n \to \mathbb{R}_{\geq 0}$ *with* $W(0) = 0$ *and* $\hat{\alpha}_1 \in \mathcal{K}_\infty$ *so that, for all* $x \in \mathbb{R}^n$,

$$\hat{\alpha}_1(|x|) \leq W(x) \tag{2.27}$$

and

$$\langle \nabla W(x), f(x) \rangle \leq -W(x). \tag{2.28}$$

This theorem indicates that if we have one Lyapunov function, not only can we find an infinite number of Lyapunov functions (via Theorem 2.20), but we can find a Lyapunov function that decreases exponentially fast. To see this, applying the comparison principle (Lemma 2.14) to (2.28), we consider

$$\dot{w} \leq -w,$$

which yields $w(t) \leq w(0)e^{-t}$, and hence the Lyapunov function decreases exponentially.

This exponential decrease can be a very useful property as it is relatively easy to manipulate. However, exponential decrease of the Lyapunov function is not the same as exponential decrease of the solution of (2.1). Indeed, (2.27) yields

$$\hat{\alpha}_1(|x(t)|) \leq W(x(t)) \leq W(x(0))e^{-t},$$

which implies that $|x(t)| \leq \hat{\alpha}_1^{-1}(W(x(0))e^{-t})$. Since $\hat{\alpha}_1 \in \mathcal{K}_\infty$ is, in general, non-linear, the exponential decrease of W does not translate to an exponential decrease of $|x|$.

Proof of Theorem 2.21: As in the analytical proof of Theorem 2.16, there exists

$\hat{\rho} \in \mathcal{P}$ so that

$$\langle \nabla V(x), f(x) \rangle \leq -\hat{\rho}(V(x)), \quad \forall x \in \mathbb{R}^n.$$

Lemma A.8 yields a continuously differentiable $\alpha \in \mathcal{K}_\infty$ satisfying $\alpha'(s) > 0$ for all $s > 0$ and

$$\alpha(s) \leq \alpha'(s)\hat{\rho}(s), \quad \forall s \in \mathbb{R}_{\geq 0}.$$

Defining $W(x) = \alpha(V(x))$ for all $x \in \mathbb{R}^n$, Theorem 2.20 yields that W is a Lyapunov function. We then calculate the decrease condition as

$$\begin{aligned}
\langle \nabla W(x), f(x) \rangle &= \alpha'(V(x))\langle \nabla V(x), f(x) \rangle \\
&\leq -\alpha'(V(x))\hat{\rho}(V(x)) \; \leq \; -\alpha(V(x)) = -W(x)
\end{aligned}$$

for all $x \in \mathbb{R}^n$.   □

2.3.1   Time-Varying Systems

Lyapunov theory for time-varying systems is quite a bit more subtle than for time-invariant systems. We limit ourselves to two of the most important sufficient conditions, with an additional condition left to the exercises. Note that, in general, when dealing with time-varying systems (2.9) we also need to allow the associated Lyapunov functions to be time-varying as well.

Theorem 2.22 (Lyapunov uniform asymptotic stability [86, Theorem 4.8]). *Given the time-varying system* (2.9) *with* $f(t, 0) = 0$ *for all* $t \geq t_0 \geq 0$, *if there exist a continuously differentiable function* $V : \mathbb{R}_{\geq 0} \times \mathcal{D} \to \mathbb{R}_{\geq 0}$ *and functions* $\alpha_1, \alpha_2 \in \mathcal{K}$ *and* $\rho \in \mathcal{P}$ *such that, for all* $x \in \mathcal{D}$ *and* $t \geq t_0 \geq 0$,

$$\alpha_1(|x|) \leq V(t, x) \leq \alpha_2(|x|) \qquad and \qquad (2.29)$$
$$\tfrac{d}{dt}V(t, x) = \nabla_t V(t, x) + \langle \nabla_x V(t, x), f(t, x) \rangle \leq -\rho(|x|), \qquad (2.30)$$

then the origin is uniformly asymptotically stable. If additionally $\mathcal{D} = \mathbb{R}^n$ *and* $\alpha_1, \alpha_2 \in \mathcal{K}_\infty$, *then the origin is uniformly globally asymptotically stable.*

   Comparing the conditions in Theorem 2.22 and Theorem 2.16, other than the function V being time-varying, the major difference is the upper bound in (2.29). As we pointed out above (2.23), for time-invariant Lyapunov functions, this upper bound always exists. However, for time-varying Lyapunov functions, it is necessary to explicitly assume this upper bound to provide a bound that is independent of t. The property captured by this upper bound is sometimes called *decrescent*.

   Warning: A common error when considering time-varying Lyapunov functions is to forget to take the partial derivative with respect to time; i.e., to leave out the term $\nabla_t V(t, x)$.

   Having highlighted the importance of the upper bound in Theorem 2.22, we might ask what happens when this bound is removed. A partial answer is given in the following theorem and an interesting related case is discussed in [59, Chapter VII, §53].

Theorem 2.23 (Lyapunov equiasymptotic stability theorem). *Given the time-*

varying system (2.9) *with* $f(t, 0) = 0$ *for all* $t \geq t_0 \geq 0$, *if there exist a continuously differentiable function* $V : \mathbb{R}_{\geq 0} \times \mathcal{D} \to \mathbb{R}_{\geq 0}$, $V(t, 0) = 0$ *for all* $t \geq 0$, *a function* $\alpha \in \mathcal{K}$ *and* $\lambda > 0$ *such that, for all* $x \in \mathcal{D}$ *and* $t \geq t_0 \geq 0$,

$$\alpha(|x|) \leq V(t, x), \tag{2.31}$$

and

$$\tfrac{d}{dt} V(t, x) = \nabla_t V(t, x) + \langle \nabla_x V(t, x), f(t, x) \rangle \leq -\lambda V(t, x), \tag{2.32}$$

then the origin is asymptotically stable. If additionally $\mathcal{D} = \mathbb{R}^n$ *and* $\alpha \in \mathcal{K}_\infty$, *then the origin is globally asymptotically stable.*

Proof. The decrease condition (2.32) implies

$$V(t, x(t)) \leq V(t_0, x(t_0)) e^{-\lambda(t - t_0)}.$$

The function $\alpha \in \mathcal{K}$ (or $\mathcal{K}_\infty$) is invertible on its range, which with the above expression yields

$$|x(t)| \leq \alpha^{-1}\left(V(t, x(t))\right) \leq \alpha^{-1}\left(V(t_0, x(t_0)) e^{-\lambda(t - t_0)}\right).$$

$\square$

It is important to note that the bound achieved in the proof is dependent not just on the elapsed time, $t - t_0$, but also explicitly on the initial time as seen in the first argument of the function V. The reader can verify that replacing (2.31) with (2.29) allows a continuation of the final calculation in the above proof that yields an upper bound that is independent of the initial time.

2.3.2 Instability

Based on our development of stability concepts and their relation to decreasing energy, or a generalized energy in the form of a Lyapunov function, it is reasonable to extend this same thinking to the definition of instability. One immediate form of this is to simply change the sign of the decrease condition (2.18) in Theorem 2.15.

Theorem 2.24 (Lyapunov theorem for instability [59, Theorem 25.4])**.** *Given* (2.1) *with* $f(0) = 0$, *suppose there exist a continuously differentiable positive definite function* $V : \mathbb{R}^n \to \mathbb{R}_{\geq 0}$ *and an* $\varepsilon > 0$ *such that*

$$\langle \nabla V(x), f(x) \rangle > 0 \tag{2.33}$$

for all $x \in \mathcal{B}_\varepsilon \backslash \{0\}$. *Then the origin is unstable.*

This is not the most general instability theorem and, in fact, cannot be applied in many cases of practical interest. Recall system (2.4),

$$\dot{x}_1 = x_1, \qquad \dot{x}_2 = -x_2. \tag{2.34}$$

Changing the direction of the inequality in (2.25), we see that (2.33) implies that the angle between the outward-facing normal $\nabla V(x)$ and $f(x)$ must be less than

$\frac{\pi}{2}$. It is clear that this is not possible for all points on the axis $x_2 \neq 0$ and $x_1 = 0$. The phase portrait of (2.34) is shown in Figure 2.4.

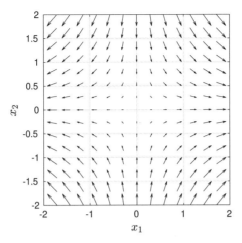

Figure 2.4: Phase portrait of the linear system (2.34) around the origin.

Unfortunately, it is not an unusual situation where a system may exhibit (asymptotically) stable behavior in some directions and unstable behavior in others. Equilibria satisfying Theorem 2.24 are usually called *completely unstable* to distinguish them from unstable equilibria with stable behavior in some directions. Fortunately, there is a more refined energy-like test for instability.

Theorem 2.25 (Chetaev's theorem [86, Thm. 4.3]). *Given* (2.1) *with* $f(0) = 0$, *let* $V : \mathbb{R}^n \to \mathbb{R}$ *be a continuously differentiable function with* $V(0) = 0$ *and* $\mathcal{O}_r = \{x \in \mathcal{B}_r(0) | V(x) > 0\} \neq \emptyset$ *for all* $r > 0$. *If for certain* $r > 0$,

$$\langle \nabla V(x), f(x) \rangle > 0 \qquad \forall \, x \in \mathcal{O}_r, \tag{2.35}$$

then the origin is unstable.

The sets in Theorem 2.25 are indicated in Figure 2.5.

Example 2.26. Consider again system (2.4),

$$\dot{x}_1 = x_1, \qquad \dot{x}_2 = -x_2,$$

and take

$$V(x) = \tfrac{1}{2}x_1^2 - \tfrac{1}{2}x_2^2. \tag{2.36}$$

We see that $V(x) > 0$ for all $|x_1| > |x_2|$. In particular, taking $x_0 = [x_1, x_2]^T$, we see that as long as $|x_1| > |x_2|$, we have $V(x_0) > 0$ even for $|x_0|$ arbitrarily small. Then the expression (2.35) is

$$\langle \nabla V(x), f(x) \rangle = [x_1 \quad -x_2] \begin{bmatrix} x_1 \\ -x_2 \end{bmatrix} = x_1^2 + x_2^2 > 0 \tag{2.37}$$

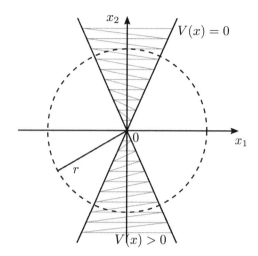

Figure 2.5: The sets involved in Theorem 2.25.

for all $x \neq 0$ so that, in particular, this expression is strictly positive where $V(x) > 0$.

2.3.3   Partial Convergence and the LaSalle-Yoshizawa Theorem

Based on the results so far, there is a gap between Theorem 2.15 guaranteeing stability, i.e., boundedness of solutions, and Theorem 2.16 guaranteeing asymptotic stability, i.e., convergence of all state variables x_i, $i \in \{1, \dots, n\}$, to an equilibrium. In particular, if a subset of the states x_i, $i \in \{1, \dots, n\}$ is converging to a stable equilibrium, neither Theorem 2.15 nor Theorem 2.16 can be used to capture this asymptotic behavior. The gap between these two results is occupied by the *LaSalle-Yoshizawa theorem* which we present in its general form for time-varying systems (2.9) here.

Theorem 2.27 (LaSalle-Yoshizawa). *Consider the time-varying system* (2.9) *with* $f(t, 0) = 0$ *for all* $t \geq t_0 \geq 0$. *Additionally, assume that* f *is locally Lipschitz in* x *uniformly in* t, *i.e., for all* $\mathcal{D} \subset \mathbb{R}^n$ *compact, there exists* $L > 0$ *such that*

$$|f(t, x_1) - f(t, x_2)| \leq L|x_1 - x_2| \qquad \forall\, x \in \mathcal{D},\ \forall t \geq t_0.$$

If there exist a continuously differentiable function $V : \mathbb{R}_{\geq 0} \times \mathbb{R}^n \to \mathbb{R}_{\geq 0}$, *functions* $\alpha_1, \alpha_2 \in \mathcal{K}_\infty$, *and* $W : \mathbb{R}^n \to \mathbb{R}_{\geq 0}$ *continuous such that, for all* $x \in \mathbb{R}^n$ *and* $t \geq t_0 \geq 0$,

$$\alpha_1(|x|) \leq V(t, x) \leq \alpha_2(|x|) \qquad and$$
$$\frac{d}{dt}V(t, x) = \nabla_t V(t, x) + \langle \nabla_x V(t, x), f(t, x) \rangle \leq -W(x), \qquad (2.38)$$

then all solutions of (2.9) *are globally uniformly bounded and satisfy*

$$\lim_{t \to \infty} W(x(t)) = 0. \qquad (2.39)$$

In the case that $W : \mathbb{R}^n \to \mathbb{R}_{\geq 0}$ is positive definite, there exists $\rho \in \mathcal{P}$ so that $\rho(|x|) \leq W(x)$ for all $x \in \mathbb{R}^n$, i.e., Theorem 2.27 reduces to Theorem 2.22. However, if $W(x)$ is positive semidefinite but not positive definite, then convergence to the set $\{x \in \mathbb{R}^n | W(x) = 0\}$ is guaranteed by (2.39), while Theorem 2.22 is not applicable. Thus, for example, if the assumptions of Theorem 2.27 are satisfied for a time-invariant system (2.1) and with

$$W(x) = x^T \begin{bmatrix} 1 & 0 \\ 0 & 0 \end{bmatrix} x \geq 0 \qquad \forall\, x \in \mathbb{R}^2,$$

then Theorem 2.15 only guarantees stability of the origin, while Theorem 2.27 additionally implies that $x_1(t) \to 0$. In particular, it is guaranteed that a subset of the states (i.e., x_1) is converging to the origin while the remaining states (i.e., x_2) stay bounded.

For a proof of Theorem 2.27 we follow the exposition in [93, Theorem A.8].

Proof of Theorem 2.27: Since $\dot{V}(t,x) \leq -W(x) \leq 0$, the function $V(\cdot, x(\cdot))$ is nonincreasing along solutions. Thus, it follows from $|x(t)| \leq \alpha^{-1}(V(t, x(t))) \leq \alpha^{-1}(V(t_0, x(t_0)))$ for all $t \geq t_0$ and for all $x(t_0) \in \mathbb{R}^n$ that $x(\cdot)$ is uniformly globally bounded, i.e., $|x(t)| \leq r \in \mathbb{R}$ for all $t \geq t_0$.

Since $|x(t)| \leq r$ and since $V(t, x(t))) \geq 0$ is nonincreasing, we can conclude that the limit $V_\infty = \lim_{t \to \infty} V(t, x(t)) \in \mathbb{R}$ exists. Integrating the decrease condition (2.38), it holds that

$$\lim_{t \to \infty} \int_{t_0}^t W(x(\tau))\, d\tau \leq -\lim_{t \to \infty} \int_{t_0}^t \dot{V}(\tau, x(\tau)))d\tau \leq -[V_\infty - V(t_0, x(t_0))]$$

and thus the limit $\lim_{t \to \infty} \int_{t_0}^t W(x(\tau))\, d\tau$ exists and is finite. (The existence follows from the monotonicity $0 \leq \int_{t_0}^{t_1} W(x(\tau))\, d\tau \leq \int_{t_0}^{t_2} W(x(\tau))\, d\tau$ for all $t_0 \leq t_1 \leq t_2$.)

Since $|x(t)| \leq r$ and f is locally Lipschitz in x uniformly continuous in t, for all $t \geq t_0 \geq 0$, it holds that

$$|x(t) - x(t_0)| = \left| \int_{t_0}^t f(x(\tau), \tau)\, d\tau \right| \leq \int_{t_0}^t |x(\tau)|\, d\tau \leq Lr|t - t_0|,$$

where L denotes the Lipschitz constant of f on the set $x \in \overline{\mathcal{B}}_r$. Choosing $\delta(\varepsilon) = \frac{\varepsilon}{Lr}$, it holds that

$$|x(t) - x(t_0)| \leq \varepsilon, \qquad \forall |t - t_0| \leq \delta(\varepsilon),$$

which implies that $x(\cdot)$ is uniformly continuous. Moreover, since W is continuous, it is uniformly continuous on compact sets. From the uniform continuity of $x(\cdot)$ and $W(\cdot)$ we can thus conclude the uniform continuity of $W(x(\cdot))$. We can thus apply Barbalat's lemma (see Lemma A.4) showing that $W(x(t)) \to 0$ for $t \to \infty$. $\qquad\square$

2.4   REGION OF ATTRACTION

We have seen in the previous sections that stability and attractivity are in general *local* properties of equilibria x^e of differential equation (2.1). Asymptotic stability (defined through stability and attractivity) of an equilibrium requires the existence of a domain around the equilibrium such that all solutions of (2.1) starting in this domain converge to the equilibrium. However, we have not addressed how large the *region of attraction* (or *domain* or *basin of attraction*) of an equilibrium is.

Definition 2.28 (Region of attraction). *Consider* (2.1) *with an asymptotically stable equilibrium* $f(x^e) = 0$, $x^e \in \mathbb{R}^n$. *The region of attraction of* x^e *is defined as*

$$\mathcal{R}_f(x^e) = \{x \in \mathbb{R}^n : x(t) \to x^e \text{ as } t \to \infty, x(0) = x\}. \tag{2.40}$$

For $x^e = 0$, we will use the shorthand notation $\mathcal{R}_f = \mathcal{R}_f(0)$. The region of attraction is an open, connected, invariant set.

The computation of the region of attraction is far from trivial. We illustrate two methods to estimate the region of attraction based on an example. Consider the system

$$\begin{aligned}
\dot{x}_1 &= -x_2 \\
\dot{x}_2 &= x_1 + (x_1^2 - 1)x_2
\end{aligned} \tag{2.41}$$

with a unique equilibrium at the origin. Note that the origin is locally asymptotically stable. Before proceeding, the reader should attempt to verify this using the common quadratic Lyapunov function candidate $V(x) = x_1^2 + x_2^2$. Why does this fail?

Example 2.29. We start by illustrating how Lyapunov's second method (or direct method) can be used to obtain an approximation of the region of attraction around the origin.

Let $\lambda_{\min}$ and $\lambda_{\max}$ denote the minimum and maximum eigenvalues, respectively, of the positive definite symmetric matrix P. We leave it to the reader to verify that the function $V(x) = x^T P x$ defined through the matrix

$$P = \begin{bmatrix} \frac{3}{2} & -\frac{1}{2} \\ -\frac{1}{2} & 1 \end{bmatrix}$$

satisfies the inequality

$$\lambda_{\min}|x|^2 \le V(x) \le \lambda_{\max}|x|^2 \tag{2.42}$$

for $\lambda_{\min} = 0.69$ and $\lambda_{\max} = 1.81$ and is a Lyapunov function for the system (2.41) with respect to the origin.

The time derivative of $V(x(t))$ satisfies the equation

$$\tfrac{d}{dt}V(x) = -x_1^2 - x_2^2 - x_1^3 x_2 + 2x_2^2 x_1^2.$$

Young's inequality (Lemma A.4) provides the estimate

$$\tfrac{d}{dt}V(x) \le -x_1^2 - x_2^2 + x_1^6 + \tfrac{1}{4}x_2^2 + x_1^4 + x_2^4 = -x_1^2 \left(1 - x_1^2 - x_1^4\right) - x_2^2 \left(\tfrac{3}{4} - x_2^2\right),$$

which implies that $\dot{V}(x) < 0$ whenever

$$1 - x_1^2 - x_1^4 > 0 \qquad \text{and} \qquad \tfrac{3}{4} - x_2^2 > 0.$$

These inequality constraints can be translated into the box constraints

$$\mathcal{C} = \{x \in \mathbb{R}^2 : -0.79 < x_1 < 0.79, \ -0.89 < x_2 < 0.89\}, \tag{2.43}$$

which are shown in Figure 2.6 as the black rectangle. Even though V is a Lyapunov

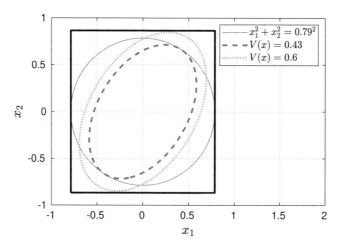

Figure 2.6: Estimate of the region of attraction through level sets of Lyapunov functions.

function with $\dot{V}(x) < 0$ in the set $\mathcal{C}$ defined in (2.43), $\mathcal{C}$ cannot be used as an estimate of the region of attraction since it is not a forward-invariant set. To this end, we need to define a sublevel set of the function V contained in $\mathcal{C}$. Observe that the inclusions

$$\{x \in \mathbb{R}^2 : x^T P x \leq \lambda_{\min}\} \subset \{x \in \mathbb{R}^2 : x^T x \leq 1\},$$
$$\{x \in \mathbb{R}^2 : x^T x \leq 0.79^2\} \subset \mathcal{C}$$

are satisfied, which can be combined to obtain

$$\{x \in \mathbb{R}^2 : x^T P x \leq 0.79^2 \lambda_{\min}\} \subset \{x \in \mathbb{R}^2 : x^T x \leq 0.79^2\} \subset \mathcal{C}. \tag{2.44}$$

Thus, the forward-invariant sublevel set $\{x \in \mathbb{R}^2 : V(x) \leq 0.43\}$ is contained in $\mathcal{C}$ and hence can be used as an estimate for the region of attraction, i.e.,

$$\{x \in \mathbb{R}^2 : V(x) \leq 0.43\} \subset \mathcal{R}_f.$$

The corresponding sets are shown in Figure 2.6. As visualized through the dotted line, in the two-dimensional setting, a better estimate of the region of attraction can be obtained by increasing the level set $V(x) = c$, $c > 0$ until $\{x \in \mathbb{R}^2 : V(x) \leq c\}$ is no longer contained in $\mathcal{C}$. This is, however, in general only possible for systems of dimension $n \leq 2$, while the estimate (2.44) may be applicable regardless of the dimension of the system.

Since the estimate of the region of attraction is based on a level set of a Lyapunov function, the estimate automatically depends on the particular choice of V.

This example shows how an estimate of the region of attraction can be obtained from a quadratic Lyapunov function. If it is possible to visualize the Lyapunov function and the decrease, the estimate can be improved in general. However, the estimate derived in Example 2.29 is very conservative, as we will show in the next example. Here we show how simulating the system in backward time can be used to estimate the region of attraction. However, this approach will in general be limited to systems in $\mathbb{R}^2$.

Example 2.30. We consider again the asymptotically stable origin of the system (2.41) and look at the solution $x(\cdot)$ directly. However, rather than considering $t \to \infty$, consider simulating backwards in time; i.e., take $t \to -\infty$. To see the effect of this, let $\tau = -t$, which implies $d\tau = -dt$ and

$$\frac{d}{d\tau}x(\tau) = -\frac{d}{dt}x(-t) = -f(x(-t)) = -f(x(\tau)). \tag{2.45}$$

In other words, simulating the system backwards in time merely requires changing the sign of the vector field. Choosing an initial condition close to the origin and simulating backwards in time provides a continuum of initial conditions that converge to the origin in forward time. In $\mathbb{R}^2$ when the region of attraction is bounded, the backwards-in-time simulated trajectory converges to the boundary of the region of attraction.

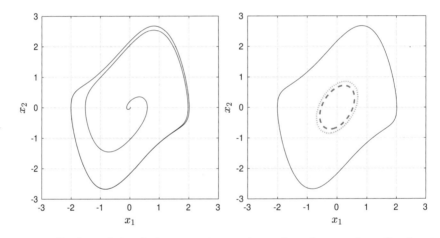

Figure 2.7: Backward simulation to estimate the region of attraction. On the right, additionally the level sets from Figure 2.6 are visualized for comparison.

In Figure 2.7 a solution of the system (2.41) in backward time, starting close to the equilibrium, is visualized on the left. On the right only the tail of the solution is shown, together with the level sets from Figure 2.6. In this example the tail of the solution provides an arbitrarily good approximation of the region of attraction of the origin $\mathcal{R}_f$.

2.5   CONVERSE THEOREMS

We have indicated that once we have one Lyapunov function we can construct infinitely many more Lyapunov functions (Theorem 2.20) and even Lyapunov functions which decrease exponentially (Theorem 2.21). However, the question remains: how do we find a first Lyapunov function that then leads us to all these others?

Unfortunately, this is a difficult problem. While there are some frequently used functions, such as quadratic forms, looking for a Lyapunov function remains something of a mysterious art. We can, though, assert that if a stability property holds, then we are guaranteed the existence of a Lyapunov function. Such theorems are referred to as *Converse Lyapunov Theorems*.

Theorem 2.31 (Converse Lyapunov theorem [59, Theorem 49.4]). *If the origin is uniformly globally asymptotically stable for* (2.9), *then there exist a (smooth) function* $V : \mathbb{R}_{\geq 0} \times \mathbb{R}^n \to \mathbb{R}_{\geq 0}$, *functions* $\alpha_1, \alpha_2 \in \mathcal{K}_\infty$, *and a function* $\rho \in \mathcal{P}$ *such that, for all* $x \in \mathbb{R}^n$ *and all* $t \geq t_0 \geq 0$,

$$\alpha_1(|x|) \leq V(t,x) \leq \alpha_2(|x|) \tag{2.46}$$

and

$$\nabla_t V(t,x) + \langle \nabla_x V(t,x), f(t,x) \rangle \leq -\rho(|x|). \tag{2.47}$$

If $f(t,x)$ *is periodic in* t, *then there exists* $V(t,x)$ *periodic in* t. *If* $f(t,x)$ *is time-invariant, then there exists* $V(t,x)$ *independent of* t.

A similar result can be given with respect to local uniform asymptotic stability for a function V defined on a neighborhood around the origin.

While the above theorem does not tell us how to construct a Lyapunov function, it is nonetheless reassuring that the search for a Lyapunov function is not futile. Additionally, the above result allows us to pursue a certain form of modular feedback design whereby we assume a stabilizing feedback is available for some portion of the system of interest. Theorem 2.31 then guarantees that a Lyapunov function is available that can be used for subsequent design.

It is beyond the scope of our discussions here to prove Theorem 2.31, but it is reasonable to wonder how one proves Theorem 2.31 and yet does not end up with a usable Lyapunov function. The reason for this is that the constructed Lyapunov function relies on solutions of the system (2.1). For example, a building block in a standard converse Lyapunov theorem for exponential stability is the function

$$V(x) = \int_0^\infty |x(\tau)| e^\tau d\tau, \qquad x = x(0) \in \mathbb{R}^n,$$

which requires knowledge of the solutions of (2.1) from every initial condition $x \in \mathbb{R}^n$. However, solving (2.1) is precisely what we are trying to avoid by using a Lyapunov function. Hence, we see that the proof of Theorem 2.31 does not really provide a starting point for construction of a Lyapunov function.

The assumption of exponential stability, along with an assumption on the vector field, allows us to derive a Lyapunov function with a few extra properties.

Theorem 2.32 (Converse Lyapunov theorem [86, Theorem 4.14]). *Suppose the origin is globally exponentially stable for* (2.1). *Furthermore, assume* $f(\cdot)$ *is continuously differentiable and the Jacobian matrix* $[\partial f / \partial x]$ *is bounded. Then there exist constants* $a_1, a_2, a_3, a_4 > 0$ *and a continuously differentiable function* $V : \mathbb{R}^n \to \mathbb{R}_{\geq 0}$

such that, for all $x \in \mathbb{R}^n$,

$$a_1|x|^2 \leq V(x) \leq a_2|x|^2, \tag{2.48}$$

$$\langle \nabla V(x), f(x) \rangle \leq -a_3|x|^2, \tag{2.49}$$

and

$$|\nabla V(x)| \leq a_4|x|. \tag{2.50}$$

2.5.1 Stability

Converse theorems are also available for Theorem 2.15 (stability) and Theorem 2.25 (instability). Interestingly, in contrast to Theorem 2.31, where time-invariant systems with an asymptotically stable origin always admit time-invariant Lyapunov functions, this is not the case for time-invariant systems with merely a stable origin.

Consider the second-order system

$$\dot{x}_1 = x_2, \qquad \dot{x}_2 = \sin^2\left(\frac{\pi}{x_1^2 + x_2^2}\right)x_2 - x_1 \tag{2.51}$$

which has *periodic orbits* given by

$$\Gamma_n = \left\{ x \in \mathbb{R}^2 \mid x_1^2 + x_2^2 = \tfrac{1}{n} \right\}, \qquad n = 1, 2, \ldots$$

To see this, note that (2.51) reduces to the oscillator

$$\dot{x}_1 = x_2, \qquad \dot{x}_2 = -x_1$$

on Γ_n. Furthermore, trajectories spiral outward between periodic orbits; i.e., for initial conditions between periodic orbits Γ_n and Γ_{n+1}, solutions converge to the outer periodic orbit Γ_n.

Suppose there exists a continuous function $V(x)$ that decreases on any periodic orbit. For any initial condition on a periodic orbit Γ_n, call it $x_n(0)$, there exists a time $T > 0$ so that $x_n(0) = x_n(T)$. The fact that $V(x)$ is decreasing then implies

$$V(x_n(T)) < V(x_n(0)) = V(x_n(T))$$

yielding a contradiction. Hence, $V(x)$ must be constant on any periodic orbit.

Since $V(x)$ is nonincreasing along trajectories and satisfies $V(0) = 0$, we have

$$V|_{\Gamma_1} \leq V|_{\Gamma_2} \leq \cdots \leq V|_{\Gamma_n} \leq \cdots \leq V(0) = 0,$$

which contradicts the requirement that $V(x)$ be positive definite. Consequently, despite the fact that the origin is a stable equilibrium point and (2.51) is time-invariant, the system does not admit a time-invariant Lyapunov function.

2.6 INVARIANCE THEOREMS

In our original, energy-based analysis of the mass-spring system (Example 1.2) and the pendulum (Example 1.4), we were unable to definitively prove that the system came to rest at the origin. Recast in the theory of Lyapunov functions from Section 2.3, the problem was that the time derivative of the Lyapunov function was only negative *semidefinite* rather than negative *definite*. Our intuition, though, is that these systems do come to rest at the origin. Invariance theorems provide a tool that allows us to draw conclusions about asymptotic stability from Lyapunov functions that have a negative semidefinite time derivative.

2.6.1   Krasovskii-LaSalle Invariance Theorem

The following theorem was developed independently in the Soviet Union by Krasovskii and in the West by LaSalle. Hence, in the English language literature it is sometimes referred to as LaSalle's Invariance Theorem.

Theorem 2.33 (Krasovskii-LaSalle invariance theorem [158, Thm. 5.3.77]). *Suppose there exists a positive definite and continuously differentiable function $V : \mathbb{R}^n \to \mathbb{R}_{\geq 0}$ such that, in an open domain containing the origin $0 \in \mathcal{D} \subset \mathbb{R}^n$, it holds that*

$$\langle \nabla V(x), f(x) \rangle \leq 0. \tag{2.52}$$

Choose a constant $c > 0$ such that the level set $\Omega_c = \{x \in \mathbb{R}^n : V(x) \leq c\}$ is bounded and contained in $\mathcal{D}$. Let $S = \{x \in \Omega_c : \langle \nabla V(x), f(x) \rangle = 0\}$ and suppose no solution other than the origin can stay identically in S. . Then the origin is asymptotically stable.

We illustrate the use of Theorem 2.33 on the pendulum and a mass-spring-damper example.

Example 2.34. Recall the pendulum system given by

$$\dot{x}_1 = x_2$$
$$\dot{x}_2 = -\tfrac{g}{\ell} \sin x_1 - \tfrac{k}{m} x_2$$

and defined on the domain $\mathcal{D} = (-\pi, \pi) \times \mathbb{R}$. The total energy of the pendulum is given by

$$V(x) = mg\ell(1 - \cos x_1) + \tfrac{1}{2} m\ell^2 x_2^2,$$

which satisfies

$$\langle \nabla V(x), f(x) \rangle = -k\ell^2 x_2^2.$$

Observe that this expression is only negative semidefinite since the right-hand side is equal to zero for $x_2 = 0$ regardless of the value of x_1.

We see that $\langle \nabla V(x), f(x) \rangle = 0$ implies $x_2 = 0$, so in Theorem 2.33,

$$S = \{x \in \mathcal{D} : x_2 = 0\}.$$

In order for x_2 to remain at 0, $\dot{x}_2$ also needs to be zero, which implies $x_1 = 0$. Also, $x_2 = 0$ implies $\dot{x}_1 = 0$. Therefore, the only solution that can remain in S is $x_1(t) = 0$, $x_2(t) = 0$. Hence, consistent with our intuition, in the presence of friction ($k > 0$), the downward equilibrium is asymptotically stable.

Example 2.35. Consider the mass-spring-damper system, shown in Figure 2.8, with a nonlinear spring

$$m\ddot{y} + b\dot{y}|\dot{y}| + k_0 y + k_1 y^3 = 0.$$

With $x_1 = y$ and $x_2 = \dot{y}$ we obtain the state space model

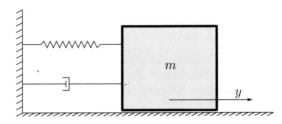

Figure 2.8: Mass-spring-damper system.

$$\dot{x}_1 = x_2$$
$$\dot{x}_2 = \frac{1}{m}\left(-k_0 x_1 - k_1 x_1^3 - b x_2 |x_2|\right).$$

Consider the candidate Lyapunov function

$$V(x) = \frac{k_0}{2m}x_1^2 + \frac{k_1}{4m}x_1^4 + \frac{1}{2}x_2^2. \qquad (2.53)$$

Then

$$\langle \nabla V(x), f(x) \rangle = \frac{k_0}{m}x_1 x_2 + \frac{k_1}{m}x_1^3 x_2 - \frac{k_0}{m}x_1 x_2 - \frac{k_1}{m}x_1^3 x_2 - \frac{b}{m}x_2^2 |x_2|$$
$$= -\frac{b}{m}x_2^2 |x_2| \leq 0.$$

As in the pendulum example, $\langle \nabla V(x), f(x) \rangle = 0$ implies $x_2 = 0$ and hence

$$S = \{x \in \mathbb{R}^2 : x_2 = 0\}.$$

In S, $\dot{x}_1 = 0$ and in order to stay at $x_2 = 0$, we require $\dot{x}_2 = 0$. This implies

$$0 = -\frac{1}{m}(k_0 x_1 + k_1 x_1^3) \quad \Rightarrow \quad \left[x_1 = 0 \ \text{ or } \ x_1 = \pm j\sqrt{\frac{k_0}{k_1}}\right].$$

Therefore, $x = 0$ is asymptotically stable.

2.6.2   Matrosov's Theorem

Theorem 2.33 only applies to time-invariant systems and does not appear to be directly generalizable to time-varying systems. However, with the use of a second function, the following result was provided by Matrosov. Similarly to what is done in Theorem 2.33, denote the set where the time derivative of the time-varying Lyapunov function V is zero by

$$S = \{x \in \mathbb{R}^n : \nabla_t V(t, x) + \langle \nabla_x V(t, x), f(t, x) \rangle = 0\}. \qquad (2.54)$$

Theorem 2.36 (Matrosov invariance theorem [59, Theorem 55.3]). *Given continuously differentiable functions $V, W : \mathbb{R}_{\geq 0} \times \mathbb{R}^n \to \mathbb{R}$, suppose that*

1. *V is positive definite and decrescent; that is, there exist $\alpha_1, \alpha_2 \in \mathcal{K}_\infty$ so that for all $x \in \mathbb{R}^n$ and all $t \geq 0$,*

$$\alpha_1(|x|) \leq V(t, x) \leq \alpha_2(|x|);$$

2. *the time derivative of V along solutions of (2.9) is negative semidefinite; that is,*

$$\nabla_t V(t, x) + \langle \nabla_x V(t, x), f(t, x) \rangle \leq 0;$$

3. *W is bounded; that is, there exists $h \geq 0$ so that for all $x \in \mathbb{R}^n$ and all $t \geq 0$*

$$|W(t, x)| \leq h;$$

4. *the time derivative of W along solutions of (2.9) is bounded away from zero on S in the following sense: for every $a > 0$ there exist $r, b > 0$ so that*

$$|\nabla_t W(t, x) + \langle \nabla_x W(t, x), f(t, x) \rangle| > b$$

for all $t \geq 0$ and all x in the set

$$\{x \in \mathbb{R}^n : |x| > a \text{ and } |x|_S < r\}.$$

Then the origin is uniformly globally asymptotically stable.

Here, $|x|_S = \inf_{y \in S} |x - y|$ denotes the distance to the set S. The intuition behind Theorem 2.36 is as follows. The negative semidefinite time derivative of V indicates that solutions of (2.9) converge toward the set S where the time derivative is zero. In a neighborhood of the set $S \backslash \{0\}$ as well as in S, however, the function W necessarily grows (either positive or negative) because its time derivative is bounded away from zero. Furthermore, W is bounded, which implies that W cannot grow indefinitely. The conclusion, then, is that eventually every solution needs to approach the origin, where the time derivative of both V and W are zero.

2.7 EXERCISES

Exercise 2.1. Prove that stability as phrased in Definition 2.1 is equivalent to the existence of $\alpha \in \mathcal{K}$ satisfying (2.3).

Hint: The fact that for any continuous and positive function $\rho \in \mathcal{P}$ there exists a function $\alpha \in \mathcal{K}$ such that $\rho(s) \leq \alpha(s)$ for all $s \in \mathbb{R}_{\geq 0}$ might be useful, [81, Lemma 1]. Moreover, you can assume that δ in Definition 2.1 depends continuously on ε.

Exercise 2.2. Consider the differential equation (2.6) with unique equilibrium at the origin. In this exercise we numerically investigate attractivity and stability of the origin.

1. Write a MATLAB function

```
dx = odeVinograd(t,x)
```

capturing the dynamics of the ordinary differential equation (2.6).

2. Solve the ordinary differential equation (2.6) for different initial values x_0 using `ode45.m` and visualize the solutions $(x_1(t), x_2(t))$ in the x_1-x_2-plane. For the numerical solutions, select the time span long enough so that the behavior $x(t) \to [0, 0]^T$ can be observed.

3. Visualize the phase portrait of the ordinary differential equation (2.6) in a neighborhood around the origin by using the function `quiver.m`.

4. Explain based on your visualizations and based on the ε-δ-stability criterion (Definition 2.1) why the origin of the ordinary differential equation is unstable.

Exercise 2.3. Modify the analytical proof of Theorem 2.16 to obtain the proof of Theorem 2.17.

Exercise 2.4. Consider the dynamics of the pendulum (1.11) together with the Lyapunov function

$$V(x) = \tfrac{1}{2} x^T \begin{bmatrix} \tfrac{1}{2} \left(\tfrac{k}{m} \right)^2 & \tfrac{1}{2} \tfrac{k}{m} \\ \tfrac{1}{2} \tfrac{k}{m} & 1 \end{bmatrix} x + \tfrac{g}{\ell} (1 - \cos(x_1)) \qquad (2.55)$$

derived in Example 2.18.

Construct the sets in the geometric proof of asymptotic stability visualized in Figure 2.3 for the Lyapunov function (2.55). In particular, using the parameters $g = 9.81$, $m = 1$, $\ell = 4$, and $k = 0.1$, for $\varepsilon = 3$ numerically find $c > 0$ and $\delta > 0$ such that

$$\overline{\mathcal{B}}_\delta(0) \subset \{x \in [-\pi, \pi] \times \mathbb{R} : V(x) \leq c\} \subset \overline{\mathcal{B}}_\varepsilon(0)$$

is satisfied. Visualize the sets (as in Figure 2.3) together with the phase portrait of the ordinary differential equation (1.11).

Exercise 2.5. Consider the one-dimensional differential equations

$$\dot{v} = v, \qquad \dot{w} = 0, \qquad \dot{x} = -x^3, \qquad \dot{y} = -y,$$

which all have the origin as unique equilibrium.

Investigate the stability properties of the origin of the differential equations through Lyapunov or Lyapunov-like functions. In particular, what is the difference between instability, stability, asymptotic stability and exponential stability?

Exercise 2.6. Consider the ordinary differential equation

$$\dot{x}_1 = x_1 - x_1 x_2, \qquad \dot{x}_2 = -x_2 + x_1 x_2. \qquad (2.56)$$

Use Theorem 2.25 and the function $V(x) = \tfrac{1}{2} x_1^2 - \tfrac{1}{2} x_2^2$ to show that the origin of the differential equation is unstable (2.56).

Exercise 2.7. Consider the functions

$$V_1(t, x) = x_1^2 (1 + \sin^2(t)) + x_2^2 (1 + \cos^2(t))$$
$$V_2(t, x) = x_1^2 + x_2^2 (1 + t).$$

Show that V_1 is decrescent while V_2 is not decrescent.

Exercise 2.8. Consider the differential equation

$$\dot{x}_1 = x_2^3, \qquad \dot{x}_2 = -x_1^3 - x_2. \tag{2.57}$$

Use Theorem 2.33 together with the function $V(x) = \frac{1}{4}x_1^4 + \frac{1}{4}x_2^4$ to show that the origin of (2.57) is asymptotically stable.

Exercise 2.9. In Example 2.18 we have derived the Lyapunov function

$$V(x) = \frac{1}{2}x^T P x + \frac{g}{\ell}(1 - \cos(x_1)), \qquad P = \begin{bmatrix} \frac{1}{2}\left(\frac{k}{m}\right)^2 & \frac{1}{2}\frac{k}{m} \\ \frac{1}{2}\frac{k}{m} & 1 \end{bmatrix},$$

with parameters $g, \ell, k, m \in \mathbb{R}_{>0}$, with respect to the origin $x^e = 0$ of the differential equation

$$\dot{x}_1 = x_2, \qquad \dot{x}_2 = -\frac{g}{\ell}\sin(x_1) - \frac{k}{m}x_2. \tag{2.58}$$

1. Write MATLAB functions

$$dx = odePendulum(t,x,parameters)$$

   and

$$Vx = LyapunovPendulum(x,parameters)$$

   capturing the dynamics. The Lyapunov function and the parameters g, ℓ, k, m are stored in `parameters`.

2. Visualize the Lyapunov function on the domain $[-\pi, \pi]^2$ by using the command `surf.m`. Solve the differential equation with respect to the initial condition $x(0) = [1, 1]^T$ and the parameters $g = 9.81$, $k = 0.1$, and $\ell = m = 1$. Visualize the solution $(x_1(t), x_2(t), V(x(t)))$ using `plot3.m` in the same figure as the Lyapunov function.
   Hint: The additional option `'linestyle','none'` in `surf.m` might improve the plot. To ensure that the solution is visible, use the option `'linewidth',2` and `'color','red'` in `plot3.m`.

3. Use `plot.m` to visualize $(t, (x_1(t))$ and $(t, x_2(t))$ (for $x(0) = [1, 1]^T$, $t \in [0, 50]$, and the parameters $g = 9.81$, $k = 0.1$, and $\ell = m = 1$).

4. Use `plot.m` to visualize $(t, V(x(t)))$ and $(t, |x(t)|^2)$ (for $x(0) = [1, 1]^T$, $t \in [0, 50]$, and the parameters $g = 9.81$, $k = 0.1$, and $\ell = m = 1$). Is $\tilde{V}(x) = |x|^2$ a Lyapunov function for the pendulum?

2.8   BIBLIOGRAPHICAL NOTES AND FURTHER READING

Aleksandr Mikhailovich Lyapunov published both his first and second methods for stability analysis in his doctoral dissertation in 1892 [102]. We have focused on his second method (also sometimes referred to as the direct method) in this chapter.

   This chapter has largely focused on the *analysis* of nonlinear systems as opposed to the *synthesis* or *design* of feedback systems. Standard texts with a more comprehensive coverage of analysis topics include [86] and [158]. Despite its age, [59] remains an excellent text for topics in stability theory.

The short monograph [31] uses the same notation as this book and covers Lyapunov and control Lyapunov results for differential inclusions, a more general class of systems than the ones discussed here. Control Lyapunov functions are introduced later in this book in Chapter 9.

The time-invariant system in Section 2.5.1 with a stable origin that requires a time-varying Lyapunov function is from [22, Example 4.11]. The system of Exercise 2.2 possessing an attractive but unstable origin is from [59, Sec. 40], where attribution is given to a 1957 paper (in Russian) by R. E. Vinograd.

In the context of time-varying systems, we have used the term *asymptotic stability* to implicitly include uniformity in the initial state and *uniform asymptotic stability* to cover uniformity in both the initial state and initial time. We have done so because non-uniformity in the initial state appears to be extremely rare. However, what we have termed asymptotic stability is sometimes referred to as equiasymptotic or non-uniform in time stability in order to reserve the term asymptotic stability for a stability and convergence property that is uniform neither in time nor in the initial state (see [59]). Additional texts on nonlinear systems analysis with a significant coverage of time-varying systems include [131] and [167].

The history of converse theorems (Section 2.5) captures much of the history of state space methods, particularly in relation to initial developments in the Soviet Union and the West trying to rapidly catch up following the launch of Sputnik. See [82].

More general versions of Theorem 2.33 are possible whereby V need not be positive definite. Furthermore, rather than requiring the negative semidefinite decrease on all of $\mathbb{R}^n$, attention can be restricted to an invariant set and convergence is guaranteed to a (smaller) invariant set. See [86, Theorem 4.4].

Chapter Three

Linear Systems and Linearization

All real-world systems are nonlinear. The clearest example of this is that constraints are inherently nonlinear. However, a linear model in many cases provides a very good approximation, particularly when restricted to some region of the state space. This is advantageous because linear systems provide a significant amount of structure that can be exploited in both analysis and design. For example, we can derive closed-form solutions for linear ordinary differential equations and there are several constructive and algebraic methods for analysis and design.

Indeed, many books have been written on linear systems theory, and some references are provided in the bibliographical notes in Section 3.7. Here, we present only those results necessary for our subsequent development of nonlinear topics.

In addition to linear systems, Lyapunov functions for polynomial systems obtained through sum of squares programming are discussed in Section 3.4.

3.1 LINEAR SYSTEMS REVIEW

As the simplest possible example, consider a one-dimensional system

$$\dot{x} = ax,$$

with initial state $x(0) \in \mathbb{R}$ and constant $a \in \mathbb{R}$. It is easy to verify that the solution is given by

$$x(t) = x(0)e^{at}, \qquad t \geq 0,$$

since $\frac{d}{dt}x(t) = ax(0)e^{at} = ax(t)$. Furthermore, the origin is:

- (uniformly) globally exponentially stable if and only if $a < 0$;
- globally stable but not exponentially stable if and only if $a = 0$; and
- unstable if and only if $a > 0$.

Finally, it is trivial to see that, when $a < 0$, $V(x) = x^2$ is a Lyapunov function that guarantees global exponential stability (recall Theorem 2.17). While quite simple, this example contains many of the core elements of linear systems theory, particularly in terms of stability theory and Lyapunov functions.

Consider now the linear system

$$\dot{x} = Ax \tag{3.1}$$

with initial condition $x(0) \in \mathbb{R}^n$ and $A \in \mathbb{R}^{n \times n}$ (that is, A is an $n \times n$ matrix consisting of real elements). The solution of (3.1) depends on the matrix exponential

$$x(t) = e^{At}x(0) = \left(\sum_{k=0}^{\infty} \frac{1}{k!}(At)^k\right)x(0).$$

3.1.1 Stability Properties for Linear Systems

Suppose that the matrix $A \in \mathbb{R}^{n \times n}$ is diagonalizable. Then there exists an invertible matrix $T \in \mathbb{C}^{n \times n}$ so that $\Lambda = T^{-1}AT$, where $\Lambda \in \mathbb{C}^{n \times n}$ is a diagonal matrix with the eigenvalues of A, denoted by λ_i, on the diagonal and the columns of T contain the corresponding eigenvectors of the matrix A.

We first observe that

$$A^k = (T\Lambda T^{-1})(T\Lambda T^{-1})\cdots(T\Lambda T^{-1}) = T\Lambda^k T^{-1}.$$

Furthermore, since Λ is diagonal, raising it to a power is the same as raising each diagonal element of Λ to the same power. Therefore,

$$e^{At} = \sum_{k=0}^{\infty} \frac{t^k}{k!}A^k = T\left(\sum_{k=0}^{\infty} \frac{t^k}{k!}\Lambda^k\right)T^{-1}$$

$$= T\begin{bmatrix} e^{\lambda_1 t} & 0 & \cdots & 0 \\ 0 & e^{\lambda_2 t} & \cdots & 0 \\ \vdots & \vdots & \ddots & \vdots \\ 0 & 0 & \cdots & e^{\lambda_n t} \end{bmatrix}T^{-1}. \tag{3.2}$$

It is immediate that in this case (i.e., A diagonalizable) stability properties of the origin can be characterized based on the location of the eigenvalues in the complex plane. For example, if all the eigenvalues have strictly negative real parts, then the matrix-vector product $e^{At}x(0)$ converges to the zero vector exponentially quickly.

Before stating a general result, we must consider what happens when matrices are not completely diagonalizable. In this case, we rely on the Jordan normal form and, for discussion purposes, we restrict attention to the 2×2 Jordan block

$$J = \begin{bmatrix} \lambda & 1 \\ 0 & \lambda \end{bmatrix}$$

and examine the matrix exponential e^{Jt}. It is not difficult to see that

$$J^k = \begin{bmatrix} \lambda^k & k\lambda^{k-1} \\ 0 & \lambda^k \end{bmatrix}$$

and therefore, in the infinite sum defining the matrix exponential, the diagonal elements sum to $e^{\lambda t}$ as in (3.2). A little manipulation of the upper diagonal element yields

$$\sum_{k=0}^{\infty} \frac{kt^k}{k!} \lambda^{k-1} = t \sum_{k=1}^{\infty} \frac{t^{k-1}}{(k-1)!} \lambda^{k-1} = t \sum_{\ell=0}^{\infty} \frac{t^\ell}{\ell!} \lambda^\ell = te^{\lambda t}.$$

Therefore, the matrix exponential of the Jordan block J yields

$$e^{Jt} = e^{\lambda t} \begin{bmatrix} 1 & t \\ 0 & 1 \end{bmatrix}.$$

Here, we see that if the real part of λ is strictly negative, we would still obtain convergence to the origin since $e^{\lambda t}$ will converge faster than t will diverge. However, if λ has zero real part (i.e., is purely imaginary), then $e^{\lambda t}$ is oscillatory, rather than converging, and hence $te^{\lambda t}$ will grow to infinity.

Higher order Jordan blocks yield a similar structure (see Exercise 3.2) and lead us to the following result.

Theorem 3.1 (Stability of linear systems [63, Theorem 8.1]). *For the linear system (3.1), the origin is*

1. *stable if and only if the eigenvalues of A have negative or zero real parts and all the Jordan blocks corresponding to eigenvalues with zero real parts are 1×1;*
2. *unstable if and only if at least one eigenvalue of A has a positive real part or zero real part with the corresponding Jordan block larger than 1×1;*
3. *exponentially stable if and only if all the eigenvalues of A have strictly negative real parts.*

Note that stability is a property of an equilibrium point. This is particularly important to keep in mind for nonlinear systems, which may well have multiple equilibrium points, some of which may be asymptotically stable or unstable or any combination of stability properties. By contrast, linear systems cannot have isolated equilibrium points other than the origin. Consequently, in an abuse of terminology, it is not uncommon for the system itself to be referred to as "exponentially stable" or "unstable."

A matrix A is said to be *Hurwitz* if all the eigenvalues of A have strictly negative real parts. Therefore, based on Theorem 3.1 item 3, referring to the exponentially stable origin of a linear system (3.1) is often done by simply referring to a Hurwitz matrix A.

Moreover, note that Theorem 3.1 does not mention asymptotic stability and does not distinguish between local and global stability properties. The reason for this is that for linear systems asymptotic stability is equivalent to exponential stability and local (exponential) stability implies global (exponential) stability of the origin of the linear system (3.1). The former statement follows from the fact that solutions are combinations of exponential functions and hence the obtained bounds for an asymptotically stable origin will be exponential. The latter statement follows from the fact that a linear system with a locally exponentially stable origin can only have an isolated equilibrium point at the origin and hence if the origin is locally exponentially stable it is also globally exponentially stable.

3.1.2   Quadratic Lyapunov Functions

To analyze the stability properties of linear systems (3.1) through Lyapunov methods, we can rely on a special class of Lyapunov functions given by

$$V(x) = x^T P x,$$

i.e., quadratic Lyapunov functions described through *symmetric positive definite matrices* $P \in \mathcal{S}^n$. Here

$$\mathcal{S}^n = \{P \in \mathbb{R}^{n \times n} : P = P^T\}$$

denotes the vector space of real symmetric matrices.

A symmetric matrix $P \in \mathcal{S}^n$ is positive definite if

$$x^T P x > 0, \quad \forall x \neq 0.$$

Similarly, the matrix P is *positive semidefinite* if $x^T P x \geq 0$, *negative definite* if $x^T P x < 0$, or *negative semidefinite* if $x^T P x \leq 0$. The set of positive definite matrices is denoted by $\mathcal{S}^n_{>0}$. The validation that a symmetric matrix is positive definite can be checked through different criteria (see, for example, [64, Sections 7.1–7.2]).

Lemma 3.2. *The following are equivalent:*

1. *$P \in \mathcal{S}^n$ is positive definite;*
2. *All the eigenvalues of P are positive;*
3. *The determinants of all the upper left submatrices (the so-called leading principal minors) of P are positive;*
4. *There exists a nonsingular matrix $H \in \mathbb{R}^{n \times n}$ such that $P = H^T H$.*

In addition to Lemma 3.2 item 2, a positive definite matrix $P \in \mathcal{S}^n_{>0}$ satisfies

$$0 < \lambda_{\min} x^T x \leq x^T P x \leq \lambda_{\max} x^T x, \quad \forall \, x \neq 0, \tag{3.3}$$

where $\lambda_{\min}$ and $\lambda_{\max}$ denote the minimum and maximum eigenvalues of P, respectively. Note that the eigenvalues of symmetric matrices are real valued and thus the minimum and maximum eigenvalue are well defined. Also, particularly with reference to the notation in Theorem 2.17, note that $x^T x = |x|^2$.

Recall from the previous chapter that a characterization of positive definite functions is the existence of a lower bound given by a $\mathcal{K}$ function of the norm of the state (see (2.16)). We see that a positive definite matrix P corresponds to the quadratic function $x^T P x$ being a positive definite function where the desired lower bound $\alpha \in \mathcal{K}$ is $\alpha(s) = \lambda_{\min} s^2$ for $s \geq 0$.

Theorem 3.3. *For the linear system (3.1), the following are equivalent:*

1. *The origin is exponentially stable;*
2. *All eigenvalues of A have strictly negative real parts;*
3. *For every symmetric positive definite $Q \in \mathcal{S}^n_{>0}$ there exists a unique symmetric positive definite $P \in \mathcal{S}^n_{>0}$, satisfying*

$$A^T P + P A = -Q. \tag{3.4}$$

Proof. The equivalence of items 1 and 2 is simply part of Theorem 3.1, which leaves us to prove the equivalence of items 1 and 3.

'3 $\Rightarrow$ 1': For simplicity take $Q = I$ (the interested reader may consider the changes needed below for an arbitrary symmetric positive definite Q) and the Lyapunov function candidate $V(x) = x^T P x$. Then

$$x^T P x \leq \lambda_{\max} x^T x \quad \Rightarrow \quad -x^T x \leq -\frac{1}{\lambda_{\max}} x^T P x$$

and, applying the chain rule,

$$\begin{aligned} \tfrac{d}{dt} V(x) &= \dot{x}^T P x + x^T P \dot{x} = x^T A^T P x + x^T P A x \\ &= x^T (A^T P + P A) x = -x^T x \leq -\frac{1}{\lambda_{\max}} x^T P x \\ &= -\frac{1}{\lambda_{\max}} V(x). \end{aligned} \tag{3.5}$$

An alternate derivation using the previous (equivalent) gradient notation, keeping in mind that $P = P^T$, uses $\nabla V(x) = \left(x^T P\right)^T + P x = 2 P x$, which gives

$$\langle \nabla V(x), Ax \rangle = 2 x^T P A x = x^T P A x + \left(x^T P A x\right)^T = x^T P A x + x^T A^T P x.$$

Continuing from (3.5), the comparison principle (Lemma 2.14) yields

$$V(x(t)) \leq V(x(0)) \exp\left(-\frac{1}{\lambda_{\max}} t\right),$$

from which we can compute

$$\lambda_{\min} |x(t)|^2 \leq V(x(t)) \leq V(x(0)) \exp\left(-\frac{1}{\lambda_{\max}} t\right) \leq \lambda_{\max} |x(0)|^2 \exp\left(-\frac{1}{\lambda_{\max}} t\right),$$

and hence

$$|x(t)| \leq \sqrt{\frac{\lambda_{\max}}{\lambda_{\min}}} |x(0)| \exp\left(-\frac{1}{2\lambda_{\max}} t\right).$$

Referring to Definition 2.7 we see that with $M = \sqrt{\frac{\lambda_{\max}}{\lambda_{\min}}}$ and $\lambda = 1/(2\lambda_{\max})$, the origin is exponentially stable.

'1 $\Rightarrow$ 3': Given the symmetric positive definite matrix $Q \in \mathcal{S}_{>0}^n$, let

$$P = \int_0^\infty e^{A^T \tau} Q e^{A \tau} d\tau. \tag{3.6}$$

Note that the integral in (3.6) is well defined since $\|e^{A^T t} Q e^{At}\|$ converges to zero exponentially fast.

To see that P solves the Lyapunov equation, first note that

$$\tfrac{d}{dt}\left(e^{A^T t} Q e^{At}\right) = A^T e^{A^T t} Q e^{At} + e^{A^T t} Q e^{At} A.$$

Then we can directly compute

$$A^T P + PA = \int_0^\infty \left(A^T e^{A^T \tau} Q e^{A\tau} + e^{A^T \tau} Q e^{A\tau} A \right) d\tau = \int_0^\infty \frac{d}{d\tau} \left(e^{A^T \tau} Q e^{A\tau} \right) d\tau$$
$$= e^{A^T t} Q e^{At} \Big|_0^\infty = \left(\lim_{t \to \infty} e^{A^T t} Q e^{At} \right) - e^{A^T 0} Q e^{A0} = -Q.$$

It remains to show that P defined by (3.6) is symmetric, positive definite, and unique. That P is symmetric follows from the fact that $Q = Q^T$ since

$$P^T = \int_0^\infty \left(e^{A^T \tau} Q e^{A\tau} \right)^T d\tau = \int_0^\infty e^{A^T \tau} Q^T e^{A\tau} d\tau$$
$$= \int_0^\infty e^{A^T \tau} Q e^{A\tau} d\tau = P.$$

To show that P is positive definite, let $z \in \mathbb{R}^n$ and consider

$$z^T P z = \int_0^\infty z^T e^{A^T \tau} Q e^{A\tau} z \, d\tau.$$

Note that if $z \neq 0$ then $x(\tau) = e^{A\tau} z \neq 0$ and, since Q is positive definite,

$$z^T P z = \int_0^\infty x(\tau)^T Q x(\tau) d\tau > 0.$$

Additionally, if $z = 0$ then $x(\tau) = 0$, and so P is indeed positive definite.

Finally, that P is unique can be proved by contradiction and is left to the interested reader. $\square$

Recognizing that the matrices P and Q in (3.4) are used in constructing a Lyapunov function as seen in the proof above, (3.4) is referred to as the *Lyapunov equation*. Note that this provides a constructive method to find a Lyapunov function for linear systems. In fact, there is a MATLAB command that computes P given a positive definite symmetric matrix Q. See Exercise 3.3.

3.2   LINEARIZATION

We now return to the nonlinear system (1.1) given by

$$\dot{x} = f(x) \tag{3.7}$$

with $f(0) = 0$ and assume that f is continuously differentiable. Recall that, as argued for (1.9), we can shift any equilibrium point of interest to the origin so that, for the translated system, $f(0) = 0$.

Define the matrix A by the Jacobian of f evaluated at the origin,

$$A = \left[\frac{\partial f(x)}{\partial x} \right]_{x=0},$$

and let $f_1(x) = f(x) - Ax$. Note that

$$\lim_{|x| \to 0} \frac{|f_1(x)|}{|x|} = \lim_{|x| \to 0} \frac{|f(x) - Ax|}{|x|} = 0, \tag{3.8}$$

where the last equality follows from an application of L'Hôpital's rule.

A slightly different point of view is to take the Taylor expansion of f at 0, $f(x) = Ax + f_1(x)$, where all the higher order terms are collapsed into f_1.

The system

$$\dot{z}(t) = Az(t) \tag{3.9}$$

is called the linearization of (3.7) at the origin. Note that, in an abuse of notation, (3.9) is almost always written as $\dot{x}(t) = Ax(t)$, blurring the distinction between the actual state x of the nonlinear system (3.7) and its linear approximation (3.9).

Theorem 3.4. *Consider the nonlinear system (3.7) with continuously differentiable right-hand side f and its linearization (3.9). If the origin of the linear system (3.9) is globally exponentially stable then the origin of (3.7) is locally exponentially stable.*

Proof. Let the origin of (3.9) be globally exponentially stable and define $Q = I$. Since the origin is exponentially stable for (3.9), Theorem 3.3 provides a symmetric and positive definite P satisfying (3.4). Take $V(x) = x^T P x$. Then

$$\langle \nabla V(x), f(x) \rangle = -x^T x + 2x^T P f_1(x). \tag{3.10}$$

As before, denote the maximum eigenvalue of P by $\lambda_{\max}$. Choose $r > 0$ and $\rho < \frac{1}{2}$ such that, for all x satisfying $|x| \le r$,

$$|f_1(x)| \le \frac{\rho}{\lambda_{\max}} |x|. \tag{3.11}$$

That this can be done follows from (3.8). Then

$$|2x^T P f_1(x)| \le 2|Px| |f_1(x)| \le 2 \left(\lambda_{\max} |x| \right) \left(\frac{\rho}{\lambda_{\max}} |x| \right) = 2\rho x^T x.$$

Therefore, for $|x| \le r$,

$$\langle \nabla V(x), f(x) \rangle \le -x^T x + 2\rho x^T x = -(1 - 2\rho) x^T x$$
$$\le -\frac{1 - 2\rho}{\lambda_{\max}} V(x) = -cV(x),$$

where $c = \frac{1-2\rho}{\lambda_{\max}} > 0$ since $\rho < \frac{1}{2}$. We see that V satisfies all the assumptions of Theorem 2.17 and so the origin of (3.7) is locally exponentially stable. $\square$

We can also use the linearization to ascertain if the origin is unstable.

Theorem 3.5. *Consider the nonlinear system (3.7) with continuously differentiable right-hand side f and its linearization (3.9) and assume that the eigenvalues of A satisfy $\lambda_i + \lambda_j \neq 0$ for all i, j. The equilibrium 0 is unstable for (3.7) if A has at least one eigenvalue with positive real part.*

Here, we only prove a special case of the theorem which relies on the two following lemmas presented without a proof. For a proof of the general statement of Theorem 3.5 we refer to [86, Theorem 3.7].

Lemma 3.6 ([158, Lemma 5.4.35]). *The Lyapunov equation (3.4) has a unique (real symmetric) solution P for each (real symmetric) Q if and only if the eigenvalues of A satisfy $\lambda_i + \lambda_j \neq 0$ for all i, j.*

Lemma 3.7 ([158, Lemma 5.4.52]). *Suppose the eigenvalues of A satisfy $\lambda_i + \lambda_j \neq 0$ for all i, j. If Q is positive definite, and P solves the Lyapunov equation (3.4), then P has as many negative eigenvalues as there are eigenvalues of A with positive real part.*

Sketch of the proof of Theorem 3.5: Let $f(x) = Ax + f_1(x)$ satisfy (3.8). Take $Q = I$, $\widehat{P} = -P$, and $V(x) = x^T \widehat{P} x$. If A has at least one eigenvalue with positive real part then P has at least one negative eigenvalue and $\widehat{P}$ has at least one positive eigenvalue. Therefore, there exists an x_0 arbitrarily close to the origin such that $V(x_0) > 0$.

In order to apply Theorem 2.25 (Chetaev's theorem), it remains to show that there is a neighborhood of the origin where $\langle \nabla V(x), f(x) \rangle > 0$. This can be done using arguments similar to those of the proof of Theorem 3.4, where (3.10) can be shown to satisfy

$$\langle \nabla V(x), f(x) \rangle = x^T x + 2x^T \widehat{P} f_1(x) \geq cV(x)$$

on some neighborhood of the origin and for some $c > 0$. The details are left to Exercise 3.4. □

Note that if all eigenvalues of A have non-positive real part but A has any eigenvalues with zero real part, then the linearization is inconclusive.

Example 3.8. Consider the nonlinear system

$$\dot{x} = cx^3 \tag{3.12}$$

with parameter $c \in \mathbb{R}$. The function $V(x) = \frac{1}{2}x^2$ satisfies (2.17) and

$$\dot{V}(x) = \langle \nabla V(x), cx^2 \rangle = cx^4.$$

Thus, for $c < 0$, the origin of (3.12) is asymptotically stable according to Theorem 2.16 and for $c > 0$ the origin of (3.12) is unstable according to Theorem 2.24.

However, independently of the parameter c, the linearization of the system (3.12) around the origin is given by $\dot{z} = Az = 0 \cdot z$. Hence, it is impossible to conclude stability properties of the origin for the nonlinear system based on its linearization if the matrix A contains eigenvalues with zero real part.

Example 3.9. Consider the mass-spring system of Example 1.2 with a hardening spring given by $F_{sp} = k_0 y + k_1 y^3 = k_0 x_1 + k_1 x_1^3$ with $k_0, k_1 > 0$, which yields the state space system

$$\begin{aligned}
\dot{x}_1 &= x_2 \\
\dot{x}_2 &= \tfrac{1}{m}\left(-k_0 x_1 - k_1 x_1^3 - cx_2\right).
\end{aligned} \tag{3.13}$$

The origin is an equilibrium and the matrix defining the linear system is given by

$$A = \left[\frac{\partial f(x)}{\partial x}\right]_{x=0} = \left[\begin{array}{cc} 0 & 1 \\ -\frac{k_0}{m} - 3\frac{k_1}{m}x_1^2 & -\frac{c}{m} \end{array}\right]_{x=0} = \left[\begin{array}{cc} 0 & 1 \\ -\frac{k_0}{m} & -\frac{c}{m} \end{array}\right].$$

We can compute the eigenvalues for A as

$$0 = \det(\lambda I - A) = \lambda\left(\lambda + \frac{c}{m}\right) + \frac{k_0}{m} = \lambda^2 + \lambda\frac{c}{m} + \frac{k_0}{m},$$

from which we have $\lambda = -\frac{c}{2m} \pm \sqrt{\frac{c^2}{4m^2} - \frac{k_0}{m}}$. We can identify three distinct cases for the eigenvalues ($k_0 = \frac{c^2}{4}$, $k_0 < \frac{c^2}{4}$, and $k_0 > \frac{c^2}{4}$), all of which yield eigenvalues with negative real parts. Therefore, Theorem 3.3 tells us that the origin is exponentially stable for $\dot{z} = Az$ and Theorem 3.4 yields that the origin is exponentially stable for (3.13).

Example 3.10. Consider the pendulum of Example 1.4 with the origin shifted to the upright equilibrium:

$$\begin{aligned} \dot{x}_1 &= x_2 \\ \dot{x}_2 &= -\frac{g}{\ell}\sin(x_1 + \pi) - \frac{k}{m}x_2. \end{aligned} \qquad (3.14)$$

We compute the matrix describing the linearized system by

$$A = \left[\frac{\partial f(x)}{\partial x}\right]_{x=0} = \left[\begin{array}{cc} 0 & 1 \\ -\frac{g}{\ell}\cos(x_1 + \pi) & -\frac{k}{m} \end{array}\right]_{x=0} = \left[\begin{array}{cc} 0 & 1 \\ \frac{g}{\ell} & -\frac{k}{m} \end{array}\right].$$

The eigenvalues of A are given by

$$0 = \det(\lambda I - A) = \lambda\left(\lambda + \frac{k}{m}\right) - \frac{g}{\ell} = \lambda^2 + \lambda\frac{k}{m} - \frac{g}{\ell}$$

so that

$$\lambda = -\frac{k}{2m} \pm \sqrt{\left(\frac{k}{2m}\right)^2 + \frac{g}{\ell}},$$

which yields two real eigenvalues, where one eigenvalue is negative and the other is positive. Therefore, from Theorem 3.5, the origin, which is the upright equilibrium, is unstable.

Example 3.11. Consider the mass-spring-damper from Example 2.35:

$$\begin{aligned} \dot{x}_1 &= x_2 \\ \dot{x}_2 &= \frac{1}{m}\left(-k_0 x_1 - k_1 x_1^3 - bx_2|x_2|\right). \end{aligned} \qquad (3.15)$$

The linearized system is described by

$$A = \left[\frac{\partial f(x)}{\partial x}\right]_{x=0} = \left[\begin{array}{cc} 0 & 1 \\ -\frac{k_0}{m} - 3\frac{k_1}{m}x_1^2 & -2\frac{b}{m}x_2 \end{array}\right]_{x=0} = \left[\begin{array}{cc} 0 & 1 \\ -\frac{k_0}{m} & 0 \end{array}\right].$$

The eigenvalues are given by

$$0 = \det(\lambda I - A) = \lambda^2 + \frac{k_0}{m},$$

which implies $\lambda = \pm j\sqrt{k_0/m}$. Since the eigenvalues of A have zero real parts, the linearization tells us nothing about stability of the origin for (3.15).

In addition to being used to study stability properties of equilibria, the linearization of a nonlinear system can also be used to construct local Lyapunov functions. In particular, Theorem 3.4 and its proof imply the following result.

Corollary 3.12. *Consider the nonlinear system (3.7) with continuously differentiable right-hand side f and its linearization (3.9) with a locally and globally exponentially stable origin for (3.7) and (3.9), respectively. Let $P \in \mathcal{S}^n$ be the unique solution of the Lyapunov equation (3.4) for an arbitrary positive definite matrix $Q \in \mathcal{S}^n_{>0}$. Then $V(x) = x^T P x$ is a local Lyapunov function of the nonlinear system (3.7).*

This corollary shows that it is straighforward to compute local Lyapunov functions for nonlinear systems (3.7) with continuously differentiable right-hand side and with respect to an exponentially stable equilibrium. However, recalling Theorem 2.17, it is in general nontrivial to obtain the domain $\mathcal{D} \subset \mathbb{R}^n$ where the Lyapunov function satisfies the conditions of Theorem 2.17. Corollary 3.12 only guarantees the existence of a $c > 0$ such that the forward-invariant sublevel set $\{x \in \mathbb{R}^n : V(x) \leq c\}$ is contained in the region of attraction $\mathcal{R}_f(0)$. Note that V and c depend on the selection of the positive definite matrix $Q \in \mathcal{S}^n_{>0}$. The calculation of c is again far from being trivial as outlined in Section 2.4. In Section 3.4 we will present a numerical method to compute Lyapunov functions and estimate the region of attraction for a special class of systems.

3.3   TIME-VARYING SYSTEMS

Linear time-varying systems

$$\dot{x}(t) = A(t)x(t) \tag{3.16}$$

represent a special class of time-varying systems (1.2). It is important to note that if the matrix $A(t)$ is time-dependent and not constant, then Theorem 3.1 and Theorem 3.3 are not applicable. Even if all the eigenvalues of $A(t)$ have a negative real part for all $t \in \mathbb{R}_{\geq 0}$, (exponential) stability of the origin cannot be concluded.

Example 3.13. The matrix

$$A(t) = \begin{bmatrix} -1 + 1.5\cos^2(t) & 1 - 1.5\sin(t)\cos(t) \\ -1 - 1.5\sin(t)\cos(t) & -1 + 1.5\sin^2(t) \end{bmatrix} \tag{3.17}$$

has eigenvalues at $\lambda_{1,2} = -0.25 \pm j0.25\sqrt{7}$. However, the solution of $\dot{x}(t) = A(t)x(t)$ is given by

$$x(t) = \begin{bmatrix} e^{0.5t}\cos(t) & e^{-t}\sin(t) \\ -e^{0.5t}\sin(t) & e^{-t}\cos(t) \end{bmatrix} x(0), \tag{3.18}$$

which clearly has a component that exponentially diverges from zero.

For time-invariant systems, the linearization stability theorem (Theorem 3.4) relied on

$$\lim_{|x|\to 0} \frac{|f(x) - Ax|}{|x|} = 0,$$

which is always true when $f(x)$ is continuously differentiable and A is defined as the Jacobian of f at the origin. In particular, this property is used in equation (3.11) to define the neighborhood of the origin where the linearization behaves similarly to the original nonlinear system.

However, from (1.2) with

$$A(t) = \left[\frac{\partial f(t, x)}{\partial x}\right]_{x=0},$$

it is not necessarily true that

$$\lim_{|x|\to 0} \sup_{t\geq 0} \frac{|f(t, x) - A(t)x|}{|x|} = 0. \tag{3.19}$$

Consequently, in order to obtain a result similar to Theorem 3.4 for time-varying systems, it is necessary to assume that (3.19) holds.

Example 3.14. ([158, Chapter 5.5]) As an example we consider the nonlinear system

$$\dot{x} = f(t, x) = \begin{bmatrix} -x_1 + tx_2^2 \\ x_1 - x_2 \end{bmatrix} \tag{3.20}$$

with

$$\left[\frac{\partial f(t, x)}{\partial x}\right]_{x=0} x = A(t)x = \begin{bmatrix} -1 & 0 \\ 1 & -1 \end{bmatrix} x.$$

We see that

$$\lim_{|x|\to 0} \sup_{t\geq 0} \frac{|f(t, x) - A(t)x|}{|x|} \geq \lim_{|x_2|\to 0} \sup_{t\geq 0} \frac{|tx_2^2|}{|x_2|} \geq \lim_{x_2\to 0} \frac{|\frac{1}{x_2}x_2^2|}{|x_2|} = 1,$$

and thus (3.20) is a time-varying system with continuously differentable right-hand side which does not satisfy (3.19). Hence, to obtain a similar result to Theorem 3.4 for time-varying systems, condition (3.19) needs to be explicitly included in the assumptions, in contrast to condition (3.8) for autonomous systems.

The utility of a linearization is that it provides a reasonable approximation of the behavior of the original system. However, for this example the initial time plays a critical role and, in fact, for initial times different from zero, the linearization is not a good approximation of the nonlinear system, as can be seen in Figure 3.1 and Figure 3.2.

Theorem 3.15 ([158, Theorem 5.5.15]). *Consider the nonlinear time-varying system* (1.2) *and suppose that* $f(t, 0) = 0$ *for all* $t \geq t_0$ *and that* f *is locally Lipschitz continuous and continuously differentiable with respect to* x. *Assume that* (3.19) *holds and that* $A(\cdot)$ *is bounded. If the origin is an exponentially stable equilibrium for* $\dot{z}(t) = A(t)z(t)$, *then it is also an exponentially stable equilibrium of* (1.2).

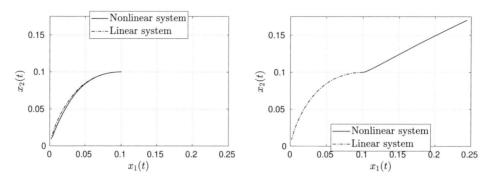

Figure 3.1: Solutions of the dynamics (3.20) and its linearization for $t \in [0, 4]$ (left) and $t \in [10, 14]$ (right) and initial value $x(t_0) = [0.1, 0.1]^T$.

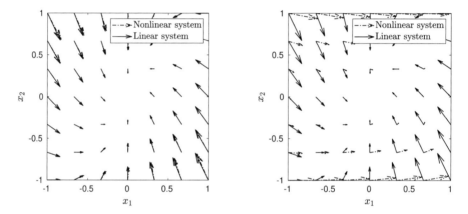

Figure 3.2: Phase portrait of the dynamics (3.20) and its linearization for $t_0 = 0.1$ (left) and $t_0 = 10$ (right).

3.4   NUMERICAL CALCULATION OF LYAPUNOV FUNCTIONS

While we can use Lyapunov functions to establish stability properties of equilibria, it is in general difficult to find a Lyapunov function for a given system. In this section we present a method to construct Lyapunov functions for systems $f : \mathbb{R}^n \to \mathbb{R}^n$,

$$\dot{x} = f(x), \tag{3.21}$$

where the right-hand side f is a polynomial function. In particular, we deviate from the main focus of linear systems in this section and discuss a slightly more general class of systems here. For linear systems, i.e., if $f(x)$ is a polynomial of degree 1, the solution of the Lyapunov equation (3.4) provides a quadratic Lyapunov function. Here, we present a method which can be applied to polynomials of higher degree. In particular we rewrite the conditions on V in Theorem 2.16 as a *semidefinite program*, which is a special form of a *convex optimization problem*, The conditions on V are phrased as *linear matrix inequalities* (LMIs), which can be solved through *semidefinite programming*. To this end, candidate Lyapunov functions are defined as *sum of squares* of polynomial functions.

In general it is difficult to validate if a function $W : \mathbb{R}^m \to \mathbb{R}$ satisfies $W(z) \geq 0$ for all $z \in \mathbb{R}^m$. However, if W is of the special form

$$W(z) = |Hz|^2 = z^T H^T H z$$

for a matrix $H \in \mathbb{R}^{m \times m}$, then $W(z) \geq 0$ for all $z \in \mathbb{R}^m$ follows from the positivity of the norm. Note that for every symmetric positive semidefinite matrix $P \in \mathcal{S}_{\geq 0}^m$ there exists $H \in \mathbb{R}^{m \times m}$ such that $P = H^T H$. We thus focus on candidate Lyapunov functions $W(z) = z^T P z$ with positive semidefinite matrix $P \in \mathcal{S}_{\geq 0}^m$.

Here $z : \mathbb{R}^n \to \mathbb{R}^m$, $m \in \mathbb{N}$, denotes monomial functions

$$z_j(x) = \prod_{i=1}^n x_i^{j_i}$$

for $j_i \in \mathbb{N}$, for all $i \in \{1, \ldots, n\}$ for all $j \in \{1, \ldots, m\}$. For example, $z : \mathbb{R}^2 \to \mathbb{R}^5$,

$$z(x) \doteq \begin{bmatrix} x_1 & x_2 & x_1^2 & x_2^2 & x_1 x_2 \end{bmatrix}^T, \tag{3.22}$$

captures the monomials of degree less than 3 of a two-dimensional system. Similarly, $y : \mathbb{R}^2 \to \mathbb{R}^9$,

$$y(x) \doteq \begin{bmatrix} x_1 & x_2 & x_1^2 & x_2^2 & x_1 x_2 & x_1^3 & x_2^3 & x_1^2 x_2 & x_1 x_2^2 \end{bmatrix}^T, \tag{3.23}$$

contains the monomials of degree less than 4.

This definition allows us to define candidate Lyapunov functions

$$V(x) = W(z(x)) = z(x)^T P z(x)$$

of arbitrary polynomial degree. Note that such functions, being quadratic in $z(x)$, will have polynomials that are twice the degree of the monomials in $z(x)$. For the purposes of constructing Lyapunov functions, we will use a weaker formulation of Theorem 2.16.

Theorem 3.16. *Consider* (3.21) *with* $f(0) = 0$, *a domain* $\mathcal{D} \subset \mathbb{R}^n$, *and a function* $\kappa : \mathbb{R}^n \to \mathbb{R}$ *such that* $\kappa(x) \leq 0$ *for all* $x \in \mathcal{D}$ *and* $\kappa(x) > 0$ *for all* $x \in \mathbb{R}^n \backslash \mathcal{D}$. *Additionally, suppose we have a continuously differentiable function* $V : \mathbb{R}^n \to \mathbb{R}$, $V(0) = 0$, $\alpha_1, \rho \in \mathcal{K}_\infty$, *and* $\delta_1, \delta_2 : \mathbb{R}^n \to \mathbb{R}_{\geq 0}$ *satisfying*

$$\alpha_1(|x|) - \delta_1(x)\kappa(x) \leq V(x) \tag{3.24}$$
$$\langle \nabla V(x), f(x) \rangle \leq -\rho(|x|) + \delta_2(x)\kappa(x) \tag{3.25}$$

for all $x \in \mathbb{R}^n$. *Then the origin is locally asymptotically stable. If, additionally,* $\mathcal{D} = \mathbb{R}^n$, *then the origin is globally asymptotically stable.*

For all $x \in \mathcal{D}$, conditions (3.24) and (3.25) satisfy

$$\alpha_1(|x|) \leq \alpha_1(|x|) - \delta_1(x)\kappa(x) \leq V(x)$$
$$\langle \nabla V(x), f(x) \rangle \leq -\rho(|x|) + \delta_2(x)\kappa(x) \leq -\rho(|x|),$$

and thus if the conditions of Theorem 3.16 are satisfied then so are the conditions of Theorem 2.16. To illustrate how semidefinite programming can be used to compute Lyapunov functions satisfying the assumptions of Theorem 3.16 we will focus on the dynamics

$$\begin{aligned}
\dot{x}_1 &= x_2 \\
\dot{x}_2 &= -x_1 - x_2 + cx_1^3
\end{aligned}$$

(3.26)

for $c \in \{0, -\frac{1}{4}, \frac{1}{4}\}$. In particular, we will show how the problem of finding a Lyapunov function can be translated into a finite-dimensional convex optimization problem.

Since the expressions become quite lengthy, we start with linear dynamics ($c = 0$), even though we have already seen how to construct Lyapunov functions for linear systems. For $c = -\frac{1}{4}$, the origin is the unique equilibrium of (3.26) and thus a global Lyapunov function will be constructed as a second example. For $c = \frac{1}{4}$, the dynamics (3.26) admit three equilibria $x_1 \in \{0, -2, 2\}$, $x_2 = 0$, and we will construct a local Lyapunov function in a neighborhood of the origin.

3.4.1 Linear Matrix Inequalities and Semidefinite Programming

The linear system $\dot{x} = Ax$ describing the dynamics (3.26) for $c = 0$ is given by

$$\begin{bmatrix} \dot{x}_1 \\ \dot{x}_2 \end{bmatrix} = \begin{bmatrix} 0 & 1 \\ -1 & -1 \end{bmatrix} \begin{bmatrix} x_1 \\ x_2 \end{bmatrix}.$$

(3.27)

In this case, solving the Lyapunov equation (3.4) with $Q = I$ leads to the positive definite matrix

$$P = \begin{bmatrix} \frac{3}{2} & \frac{1}{2} \\ \frac{1}{2} & 1 \end{bmatrix},$$

(3.28)

which implies exponential stability of the origin according to Theorem 3.3. Positive definiteness of P can be verified using Lemma 3.2, guaranteeing that $V(x) = x^T P x$ is a Lyapunov function.

While this is a straightforward approach for linear systems to compute a Lyapunov function, we will consider a different approach here to establish asymptotic stability of the origin and to obtain a Lyapunov function which is also applicable to a more general class of systems.

To this end, instead of solving the Lyapunov equation (3.4) we consider the conditions in Theorem 3.16 directly, i.e., we focus on the inequalities

$$\alpha_1(|x|) \leq V(x),$$

(3.29a)

$$\langle \nabla V(x), f(x) \rangle \leq -\rho(|x|).$$

(3.29b)

Since we want to find a global Lyapunov function to show global asymptotic stability, we set $\kappa(x) = 0$ for all $x \in \mathbb{R}^2$.

We fix $\varepsilon > 0$, select $\alpha_1(|x|) = \rho(|x|) = \varepsilon|x|^2$, and assume that the Lyapunov function V as well as the left-hand side of (3.29b) can be written as quadratic functions

$$V(x) = x^T P x, \qquad \langle \nabla V(x), f(x) \rangle = -x^T Q x,$$

for symmetric matrices $P, Q \in \mathcal{S}^2$ with unknown parameters (and in particular assume that we do not know P in (3.28)). With these assumptions, rearranging the conditions (3.29) yields

$$-x^T P x + \varepsilon x^T x \le 0, \qquad -x^T Q x + \varepsilon x^T x \le 0 \qquad (3.30)$$

for all $x \in \mathbb{R}^2$, and the entries

$$P = \begin{bmatrix} p_{11} & p_{12} \\ p_{12} & p_{22} \end{bmatrix} \quad \text{and} \quad Q = \begin{bmatrix} q_{11} & q_{12} \\ q_{12} & q_{22} \end{bmatrix}$$

need to be determined. Alternatively, instead of inequalities in $\mathbb{R}$, the inequalities (3.30) can be interpreted in terms of definiteness of matrices leading to so-called *linear matrix inequalities* (LMIs)

$$-P + \varepsilon I \le 0, \qquad -Q + \varepsilon I \le 0. \qquad (3.31)$$

However, the conditions (3.30), or equivalently (3.31), are incomplete, since they do not capture the connection of V and $\dot{V}$ yet. For linear systems, we obtain the condition

$$-x^T Q x = \langle \nabla V(x), f(x) \rangle = x^T (A^T P + P A) x \qquad (3.32)$$

recovering the Lyapunov equation (3.4). With the dynamics (3.27) this implies

$$-q_{11} x_1^2 - q_{22} x_2^2 - 2 q_{12} x_1 x_2 =$$
$$- 2 p_{12} x_1^2 + (2 p_{12} - 2 p_{22}) x_2^2 + (2 p_{11} - 2 p_{12} - 2 p_{22}) x_1 x_2$$

and thus the linear equations

$$q_{11} - 2 p_{12} = 0$$
$$q_{22} + 2 p_{12} - 2 p_{22} = 0 \qquad (3.33)$$
$$2 q_{12} + 2 p_{11} - 2 p_{12} - 2 p_{22} = 0$$

need to be satisfied.

These conditions together with the linear matrix inequalities (3.31) can be formulated as a *semidefinite program*

$$\min_{P, Q \in \mathcal{S}^2} \quad 1 \qquad (3.34a)$$

$$\text{subject to} \quad 0 \ge - P + \varepsilon I \qquad (3.34b)$$
$$0 \ge - Q + \varepsilon I \qquad (3.34c)$$
$$0 = q_{11} - 2 p_{12} \qquad (3.34d)$$
$$0 = q_{22} + 2 p_{12} - 2 p_{22} \qquad (3.34e)$$
$$0 = 2 q_{12} + 2 p_{11} - 2 p_{12} - 2 p_{22}. \qquad (3.34f)$$

Such problems can be solved efficiently when a solution exists.

Since we are only interested in a feasible solution of (3.34), the objective function (3.34a) in this example is not relevant and is simply set to 1. However, from the derivation it follows that every solution of (3.34) provides positive definite matrices P and Q, such that $V(x) = x^T P x$ is a Lyapunov function for the linear system (3.27). The inequality constraints (3.34b) and (3.34c) describe linear matrix inequalities while (3.34d)–(3.34f) are linear equality constraints in $\mathbb{R}$. The constraints are linear, since the unknowns appear as linear terms. For the selection $\varepsilon = 0.1$, the software package CVX [56], [57] (used as a toolbox in MATLAB), returns the solution

$$P = \begin{bmatrix} 6.6525 & 1.9515 \\ 1.9515 & 4.7614 \end{bmatrix}.$$

Here, we will not go into detail on how to solve linear matrix inequalities or semidefinite programs. Examples on how to solve linear matrix inequalities (and semidefinite programs) in MATLAB are given in the exercises with supporting material in Appendix B. Additional software and references are given in Section 3.7 and linear matrix inequalities will appear again in Chapter 8.

Note that the semidefinite program (3.34) can alternatively be written as

$$\min_{P \in \mathcal{S}^2} \; 1$$
$$\text{subject to} \quad 0 \geq -P + \varepsilon I \tag{3.35}$$
$$0 \geq (A^T P + PA) + \varepsilon I,$$

where we use the condition (3.32) directly and thus can eliminate the unknown Q. The consideration of Q will, however, be useful in the next section, when we discuss nonlinear dynamics.

3.4.2   Global Lyapunov Functions for Polynomial Systems

For linear systems $\dot{x} = Ax$ with A Hurwitz, we do not need to solve an optimization problem of the form (3.34) to obtain a positive definite matrix P. Instead, the Lyapunov equation (3.4) can be solved directly.

However, the approach discussed in Section 3.4.1 can also be applied to nonlinear systems with polynomial right-hand side. We now consider the system (3.26) with $c = -\frac{1}{4}$, construct a global Lyapunov function, and thus show that the origin is globally asymptotically stable.

We follow the same steps as in the previous section, but instead of a quadratic Lyapunov function we use a candidate Lyapunov function with higher order terms

$$V(x) = W(z(x)) = z(x)Pz(x), \tag{3.36}$$

where $z(x)$ is defined in (3.22) and $P \in \mathcal{S}^5$ is an unknown matrix

$$P = \begin{bmatrix} p_{11} & p_{12} & p_{13} & p_{14} & p_{15} \\ p_{12} & p_{22} & p_{23} & p_{24} & p_{25} \\ p_{13} & p_{23} & p_{33} & p_{34} & p_{35} \\ p_{14} & p_{24} & p_{34} & p_{44} & p_{45} \\ p_{15} & p_{25} & p_{35} & p_{45} & p_{55} \end{bmatrix}.$$

Note that additional higher order terms could be included in the definition of V and it is not clear, a priori, that the specific definition of V will lead to parameters such that V is a Lyapunov function. Through the definition (3.36), $V : \mathbb{R}^n \to \mathbb{R}_{\geq 0}$ is a positive definite mapping if P is a positive semidefinite matrix. Compared to the linear setting in the previous section, the candidate Lyapunov function is of degree 4 instead of degree 2.

Since we seek a global Lyapunov function, we again take $\kappa(x) = 0$ for all $x \in \mathbb{R}^n$ and set $\alpha_1(|x|) = \rho(|x|) = \varepsilon|x|^2$. Then, the condition (3.29a) can be written as the linear matrix inequality

$$-P + \begin{bmatrix} \varepsilon I & 0 \\ 0 & 0 \end{bmatrix} \leq 0,$$

where $I \in \mathbb{R}^{2 \times 2}$.

To compute conditions of the form (3.33), we calculate $\langle \nabla V(x), f(x) \rangle$ based on the definition of V. Tedious (but simple) calculations provide the expressions

$$\begin{aligned}
V(x) &= p_{11}x_1^2 + 2p_{13}x_1^3 + p_{22}x_2^2 + 2p_{24}x_2^3 + p_{33}x_1^4 + p_{44}x_2^4 \\
&\quad + (2p_{14} + 2p_{25})x_1x_2^2 + (2p_{15} + 2p_{23})x_1^2x_2 + 2p_{35}x_1^3x_2 \\
&\quad + 2p_{45}x_1x_2^3 + (2p_{34} + p_{55})x_1^2x_2^2 + 2p_{12}x_1x_2, \\
\nabla_{x_1}V(x) &= 2p_{11}x_1 + 2p_{12}x_2 + 6p_{13}x_1^2 + (2p_{14} + 2p_{25})x_2^2 + 4p_{33}x_1^3 + 2p_{45}x_2^3 \\
&\quad + 6p_{35}x_1^2x_2 + (4p_{34} + 2p_{55})x_1x_2^2 + (4p_{15} + 4p_{23})x_1x_2, \\
\nabla_{x_2}V(x) &= 2p_{12}x_1 + 2p_{22}x_2 + (2p_{15} + 2p_{23})x_1^2 + 6p_{24}x_2^2 + 2p_{35}x_1^3 \\
&\quad + 4p_{44}x_2^3 + (4p_{34} + 2p_{55})x_1^2x_2 + 6p_{45}x_1x_2^2 + (4p_{14} + 4p_{25})x_1x_2,
\end{aligned}$$

and

$$\begin{aligned}
\langle \nabla V(x), f(x) \rangle &= (2p_{12} - 2p_{22})x_2^2 - 2p_{12}x_1^2 + (2cp_{12} - 2p_{35})x_1^4 \qquad\qquad (3.37) \\
&\quad + (-2p_{15} - 2p_{23})x_1^3 + (2cp_{15} + 2cp_{23})x_1^5 + (2p_{14} - 6p_{24} + 2p_{25})x_2^3 \\
&\quad + 2cp_{35}x_1^6 + (-4p_{44} + 2p_{45})x_2^4 + (6p_{13} - 4p_{14} - 2p_{15} - 2p_{23} - 4p_{25})x_1^2x_2 \\
&\quad + (-4p_{14} + 4p_{15} + 4p_{23} - 6p_{24} - 4p_{25})x_1x_2^2 + (4cp_{14} + 4cp_{25})x_1^4x_2 \\
&\quad + (2cp_{22} + 4p_{33} - 4p_{34} - 2p_{35} - 2p_{55})x_1^3x_2 + (4cp_{34} + 2cp_{55})x_1^5x_2 \\
&\quad + (4p_{34} - 4p_{44} - 6p_{45} + 2p_{55})x_1x_2^3 + 6cp_{24}x_1^3x_2^2 + (2p_{11} - 2p_{12} - 2p_{22})x_1x_2 \\
&\quad + (-4p_{34} + 6p_{35} - 6p_{45} - 2p_{55})x_1^2x_2^2 + 4cp_{44}x_1^3x_2^3 + 6cp_{45}x_1^4x_2^2.
\end{aligned}$$

The decrease condition $\langle \nabla V(x), f(x) \rangle$ contains terms of degree 6. We thus assume that $\langle \nabla V(x), f(x) \rangle$ can be written as

$$\langle \nabla V(x), f(x) \rangle = -y(x)^T Q y(x), \qquad\qquad (3.38)$$

where y is defined in (3.23) and $Q \in \mathcal{S}^9$,

$$Q = \begin{bmatrix} q_{11} & \cdots & q_{19} \\ \vdots & \ddots & \vdots \\ q_{19} & \cdots & q_{99} \end{bmatrix}.$$

Condition (3.29b) is thus of the form

$$-Q + \begin{bmatrix} \varepsilon I & 0 \\ 0 & 0 \end{bmatrix} \leq 0$$

with $I \in \mathbb{R}^{2 \times 2}$. As a last step, we need to combine (3.37) and (3.38) to obtain linear equality constraints as in (3.33). In particular with

$$
\begin{aligned}
y^T Q y =\,& q_{11}x_1^2 + 2q_{13}x_1^3 + (2q_{16}+q_{33})x_1^4 + q_{22}x_2^2 + 2q_{24}x_2^3 + (2q_{27}+q_{44})x_2^4 \\
& + 2q_{36}x_1^5 + 2q_{47}x_2^5 + q_{66}x_1^6 + q_{77}x_2^6 + (2q_{14}+2q_{25})x_1x_2^2 \\
& + (2q_{15}+2q_{23})x_1^2x_2 + (2q_{17}+2q_{29}+2q_{45})x_1x_2^3 \\
& + (2q_{18}+2q_{26}+2q_{35})x_1^3x_2 + (2q_{38}+2q_{56})x_1^4x_2 + (2q_{49}+2q_{57})x_1x_2^4 \\
& + 2q_{68}x_1^5x_2 + 2q_{79}x_1x_2^5 + (2q_{19}+2q_{28}+2q_{34}+q_{55})x_1^2x_2^2 \\
& + (2q_{39}+2q_{46}+2q_{58})x_1^3x_2^2 + (2q_{37}+2q_{48}+2q_{59})x_1^2x_2^3 \\
& + (2q_{67}+2q_{89})x_1^3x_2^3 + (2q_{69}+q_{88})x_1^4x_2^2 + (2q_{78}+q_{99})x_1^2x_2^4 \\
& + 2q_{12}x_1x_2
\end{aligned}
$$

the equality constraints capturing Equation (3.38) are summarized in Table 3.1. Similarly to (3.34), the optimization problem

$$
\begin{array}{r|rcl}
x_1^2 & -2p_{12} & = & -q_{11} \\
x_1^3 & -2p_{15}-2p_{23} & = & -2q_{13} \\
x_1^4 & 2cp_{12}-2p_{35} & = & -(2q_{16}+q_{33}) \\
x_1^5 & 2cp_{15}+2cp_{23} & = & -2q_{36} \\
x_1^6 & 2cp_{35} & = & -q_{66} \\
x_2^2 & 2p_{12}-2p_{22} & = & -q_{22} \\
x_2^3 & 2p_{14}-6p_{24}+2p_{25} & = & -2q_{24} \\
x_2^4 & -4p_{44}+2p_{45} & = & -(2q_{27}+q_{44}) \\
x_2^5 & 0 & = & -2q_{47} \\
x_2^6 & 0 & = & -q_{77} \\
x_1x_2 & 2p_{11}-2p_{12}-2p_{22} & = & -2q_{12} \\
x_1x_2^2 & -4p_{14}+4p_{15}+4p_{23}-6p_{24}-4p_{25} & = & -(2q_{14}+2q_{25}) \\
x_1x_2^3 & 4p_{34}-4p_{44}-6p_{45}+2p_{55} & = & -(2q_{17}+2q_{29}+2q_{45}) \\
x_1x_2^4 & 0 & = & -(2q_{49}+2q_{57}) \\
x_1x_2^5 & 0 & = & -2q_{79} \\
x_1^2x_2 & 6p_{13}-4p_{14}-2p_{15}-2p_{23}-4p_{25} & = & -(2q_{15}+2q_{23}) \\
x_1^2x_2^2 & -4p_{34}+6p_{35}-6p_{45}-2p_{55} & = & -(2q_{19}+2q_{28}+2q_{34}+q_{55}) \\
x_1^2x_2^3 & 0 & = & -(2q_{37}+2q_{48}+2q_{59}) \\
x_1^2x_2^4 & 0 & = & -(2q_{78}+q_{99}) \\
x_1^3x_2 & 2cp_{22}+4p_{33}-4p_{34}-2p_{35}-2p_{55} & = & -(2q_{18}+2q_{26}+2q_{35}) \\
x_1^3x_2^2 & 6cp_{24} & = & -(2q_{39}+2q_{46}+2q_{58}) \\
x_1^3x_2^3 & 4cp_{44} & = & -(2q_{67}+2q_{89}) \\
x_1^4x_2 & 4cp_{14}+4cp_{25} & = & -(2q_{38}+2q_{56}) \\
x_1^4x_2^2 & 6cp_{45} & = & -(2q_{69}+q_{88}) \\
x_1^5x_2 & 4cp_{34}+2cp_{55} & = & -2q_{68}
\end{array}
$$

$$(3.39)$$

Table 3.1: Equality constraints induced through condition (3.38). The conditions are linear in the unknowns p_{ij} and q_{ij}.

$$\min_{P \in \mathcal{S}^5, \, Q \in \mathcal{S}^9} \quad 1$$

subject to   equality constraints (3.39)

$$0 \geq -P + \begin{bmatrix} \varepsilon I & 0 \\ 0 & 0 \end{bmatrix} \tag{3.40}$$

$$0 \geq -Q + \begin{bmatrix} \varepsilon I & 0 \\ 0 & 0 \end{bmatrix}$$

is obtained. If the semidefinite program is feasible, then according to Theorem 3.16, $V(x) = z(x)^T P z(x)$ is a global Lyapunov function and the origin of the dynamical system (3.26) is globally asymptotically stable.

For $\varepsilon = 0.1$ and the parameter $c = -\frac{1}{4}$, an implementation using CVX returns the matrix

$$P = \begin{bmatrix} 7.8740 & 3.2678 & 0.0000 & -0.0000 & -0.0000 \\ 3.2678 & 7.5878 & 0.0000 & -0.0000 & -0.0000 \\ 0.0000 & 0.0000 & 0.9667 & -0.0001 & 0.0003 \\ -0.0000 & -0.0000 & -0.0001 & 0.0000 & 0.0000 \\ -0.0000 & -0.0000 & 0.0003 & 0.0000 & 0.0002 \end{bmatrix},$$

where we additionally used lower and upper bounds of -10 and 10 on the entries of P and Q in the optimization problem (to ensure that the matrix entries are not unnecessarily large). If we only consider three digits, we obtain the Lyapunov function

$$V(x) = \begin{bmatrix} x_1 & x_2 & x_1^2 \end{bmatrix} \begin{bmatrix} 7.87 & 3.27 & 0 \\ 3.27 & 7.59 & 0 \\ 0 & 0 & 0.97 \end{bmatrix} \begin{bmatrix} x_1 \\ x_2 \\ x_1^2 \end{bmatrix}$$

$$= 7.87 x_1^2 + 7.59 x_2^2 + 6.54 x_1 x_2 + 0.96 x_1^4. \tag{3.41}$$

Thus, we are able to conclude global asymptotic stability from Theorem 3.16. Note that in the case of an infeasible optimization problem (3.40) one cannot conclude that the origin is unstable. To be able to conclude that, one would have to test the optimization problem (3.40) for all possible candidate Lyapunov functions, including nonpolynomial functions. Level sets of the Lyapunov function (3.41) are shown in Figure 3.3.

3.4.3   Local Lyapunov Functions for Polynomial Systems

If we consider the system (3.26) with $c = \frac{1}{4}$ then the dynamics admit the three equilibria $x_1 \in \{0, \pm 2\}$, $x_2 = 0$. Thus, the origin cannot be globally asymptotically stable. Hence, we will make use of the function κ and the additional degrees of freedom provided by δ_1 and δ_2 in Theorem 3.16 to investigate local stability properties of the origin.

We consider $\mathcal{D} = \mathcal{B}_1(0) = \{x \in \mathbb{R}^n : |x| < 1\}$ and the corresponding function

$$\kappa(x) = x^T x - 1$$

satisfying the assumption of Theorem 3.16. The functions δ_1 and δ_2 are defined as

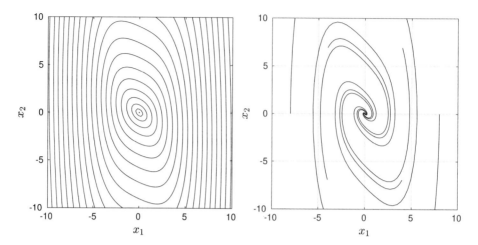

Figure 3.3: Level sets of the Lyapunov function (3.41) (on the left) and solutions of (3.26) for $c = -\frac{1}{4}$ (on the right).

unknown functions of the form

$$\delta_1(x) = z(x)^T D_{sm} z(x) \quad \text{and} \quad \delta_2(x) = z(x)^T E_{sm} z(x)$$

where

$$D_{sm} = \begin{bmatrix} d_{11} & \cdots & d_{15} \\ \vdots & \ddots & \vdots \\ d_{15} & \cdots & d_{55} \end{bmatrix} \quad \text{and} \quad E_{sm} = \begin{bmatrix} e_{11} & \cdots & e_{15} \\ \vdots & \ddots & \vdots \\ e_{15} & \cdots & e_{55} \end{bmatrix}$$

and z is defined in (3.22). With the help of y defined in (3.23), $\delta_1(x)\kappa(x)$ can be written as

$$\begin{aligned}
\delta_1(x)\kappa(x) &= z(x)^T D_{sm} z(x) \cdot (x^T x - 1) \\
&= z(x)^T D_{sm} z(x) x_1^2 + z(x)^T D_{sm} z(x) x_2^2 - z(x)^T D_{sm} z(x) \\
&= y(x)^T D_1 y(x) + y(x)^T D_2 y(x) - y(x)^T D_3 y(x) \\
&= y(x)^T D_{la} y(x).
\end{aligned} \tag{3.42}$$

Here $D_1, D_2, D_3, D_{la} \in \mathcal{S}^9$ are given by

$$D_1 = \begin{bmatrix} 0 & 0 & 0 & 0 & 0 & 0 & 0 & 0 & 0 \\ 0 & 0 & 0 & 0 & 0 & 0 & 0 & 0 & 0 \\ 0 & 0 & d_{11} & 0 & d_{12} & d_{13} & 0 & d_{15} & d_{14} \\ 0 & 0 & 0 & 0 & 0 & 0 & 0 & 0 & 0 \\ 0 & 0 & d_{12} & 0 & d_{22} & d_{23} & 0 & d_{25} & d_{24} \\ 0 & 0 & d_{13} & 0 & d_{23} & d_{33} & 0 & d_{35} & d_{34} \\ 0 & 0 & 0 & 0 & 0 & 0 & 0 & 0 & 0 \\ 0 & 0 & d_{15} & 0 & d_{25} & d_{35} & 0 & d_{55} & d_{45} \\ 0 & 0 & d_{14} & 0 & d_{24} & d_{34} & 0 & d_{45} & d_{44} \end{bmatrix}, \quad D_2 = \begin{bmatrix} 0 & 0 & 0 & 0 & 0 & 0 & 0 & 0 & 0 \\ 0 & 0 & 0 & 0 & 0 & 0 & 0 & 0 & 0 \\ 0 & 0 & 0 & 0 & 0 & 0 & 0 & 0 & 0 \\ 0 & 0 & 0 & d_{22} & d_{12} & 0 & d_{24} & d_{23} & d_{25} \\ 0 & 0 & 0 & d_{12} & d_{11} & 0 & d_{14} & d_{13} & d_{15} \\ 0 & 0 & 0 & 0 & 0 & 0 & 0 & 0 & 0 \\ 0 & 0 & 0 & d_{24} & d_{14} & 0 & d_{44} & d_{34} & d_{45} \\ 0 & 0 & 0 & d_{23} & d_{13} & 0 & d_{34} & d_{33} & d_{35} \\ 0 & 0 & 0 & d_{25} & d_{15} & 0 & d_{45} & d_{35} & d_{55} \end{bmatrix},$$

$$D_3 = \begin{bmatrix} D_{sm} & 0 \\ 0 & 0 \end{bmatrix},$$

and

$$D_{la} = \begin{bmatrix} -d_{11} & -d_{12} & -d_{13} & -d_{14} & -d_{15} & 0 & 0 & 0 & 0 \\ -d_{12} & -d_{22} & -d_{23} & -d_{24} & -d_{25} & 0 & 0 & 0 & 0 \\ -d_{13} & -d_{23} & d_{11}-d_{33} & -d_{34} & d_{12}-d_{35} & d_{13} & 0 & d_{15} & d_{14} \\ -d_{14} & -d_{24} & -d_{34} & d_{22}-d_{44} & d_{12}-d_{45} & 0 & d_{24} & d_{23} & d_{25} \\ -d_{15} & -d_{25} & d_{12}-d_{35} & d_{12}-d_{45} & d_{22}+d_{11}-d_{55} & d_{23} & d_{14} & d_{25}+d_{13} & d_{24}+d_{15} \\ 0 & 0 & d_{13} & 0 & d_{23} & d_{33} & 0 & d_{35} & d_{34} \\ 0 & 0 & 0 & d_{24} & d_{14} & 0 & d_{44} & d_{34} & d_{45} \\ 0 & 0 & d_{15} & d_{23} & d_{25}+d_{13} & d_{35} & d_{34} & d_{55}+d_{33} & d_{45}+d_{35} \\ 0 & 0 & d_{14} & d_{25} & d_{24}+d_{15} & d_{34} & d_{45} & d_{45}+d_{35} & d_{44}+d_{55} \end{bmatrix}. \tag{3.43}$$

In the same way, $\delta_2(x)\kappa(x)$ can be rewritten as

$$\delta_2(x)\kappa(x) = z(x)^T E_{sm} z(x) \cdot (x^T x - 1) = y(x)^T E_{la} y(x).$$

Thus we can extend the feasiblity problem (3.40) to

$$\min_{\substack{P \in \mathcal{S}^5, \ Q \in \mathcal{S}^9 \\ D_{sm}, E_{sm} \in \mathcal{S}^5 \\ \hat{D}, \hat{E} \in \mathcal{S}^9}} 1$$

subject to   equality constraints (3.39)

$$0 \geq \begin{bmatrix} -P & 0 \\ 0 & 0 \end{bmatrix} + \begin{bmatrix} \varepsilon I & 0 \\ 0 & 0 \end{bmatrix} - \hat{D}$$

$$0 \geq -Q + \begin{bmatrix} \varepsilon I & 0 \\ 0 & 0 \end{bmatrix} - \hat{E} \tag{3.44}$$

$$0 \geq -D_{sm}$$

$$0 \geq -E_{sm}$$

$$0 = \hat{D} - D_{la}$$

$$0 = \hat{E} - E_{la}$$

to compute a local Lyapunov function. If the optimization problem has a feasible solution, then $V(x) = z(x)^T P z(x)$ is a local Lyapunov function and the origin is locally asymptotically stable. Note that P does not need to be positive definite as a feasible solution of the optimization problem (3.44). However, the constraints in (3.44) ensure that $V(x) = z(x)^T P z(x) > 0$ for all $x \in \mathcal{D}\backslash\{0\}$.

For $\varepsilon = 0.1$ and the parameter $c = \frac{1}{4}$, an implementation using CVX returns the matrix (where we only report three digits)

$$P = \begin{bmatrix} 8.69 & 3.50 & 0 & 0 & 0 \\ 3.50 & 7.63 & 0 & 0 & 0 \\ 0 & 0 & 5.40 & 1.08 & 2.42 \\ 0 & 0 & 1.08 & 2.66 & 0.64 \\ 0 & 0 & 2.42 & 0.64 & 5.78 \end{bmatrix}, \tag{3.45}$$

which shows local asymptotic stability of the origin. Level sets of the function $V(x) = z(x)Pz(x)$ are shown in Figure 3.4.

In the same way as linearization can be used to construct local Lyapunov functions for nonlinear systems, the approach discussed in this section can be used to construct local Lyapunov functions for nonlinear systems considering higher order terms in the approximation of the right-hand side. Note, however, that we have presented a tool only to verify asymptotic stability. Instability cannot be concluded

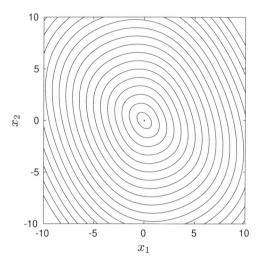

Figure 3.4: Level sets of the Lyapunov function $V(x) = z(x)Pz(x)$ defined through the matrix P in (3.45).

if a semidefinite program fails to find a feasible solution.

3.4.4   Estimation of the Region of Attraction

Even though $V(x)$ computed in the last section is a local Lyapunov function and satisfies the decrease condition (3.25) for all $x \in \mathcal{B}_1(0)$, it is not clear if $\mathcal{B}_1(0)$ is a subset of the region of attraction since $\mathcal{B}_1(0)$ does not need to be forward-invariant. Semidefinite programming can also be used to compute a forward-invariant sublevel set of the Lyapunov function $\{x \in \mathbb{R}^n : V(x) \leq c\}$, $c \in \mathbb{R}_{>0}$.

Theorem 3.17 ([5, Section 3.1]). *Consider* (3.21) *with* $f(0) = 0$, *a domain* $\mathcal{D} \subset \mathbb{R}^n$, *and a function* $\kappa : \mathbb{R}^n \to \mathbb{R}$ *such that* $\kappa(x) \leq 0$ *for all* $x \in \mathcal{D}$ *and* $\kappa(x) > 0$ *for all* $x \in \mathbb{R}^n \backslash \mathcal{D}$. *Additionally, let* $V : \mathbb{R}^n \to \mathbb{R}$ *be a Lyapunov function according to Theorem 3.16, and let* $k \in \mathbb{N}$, $\delta_3 : \mathbb{R}^n \to \mathbb{R}_{\geq 0}$, *and* $c \in \mathbb{R}_{>0}$. *If*

$$(V(\dot{x}) - c)|x|^{2k} - \delta_3(x)\kappa(x) \geq 0 \qquad (3.46)$$

holds for all $x \in \mathbb{R}^n$, *then the sublevel set* $\{x \in \mathbb{R}^n : V(x) \leq c\} \subset \mathcal{D}$ *is contained in the region of attraction.*

To indicate why condition (3.46) provides a sublevel set contained in $\mathcal{D}$ assume that (3.46) is satisfied but $\{x \in \mathbb{R}^n : V(x) \leq c\} \not\subseteq \mathcal{D}$. Let $x \in \{x \in \mathbb{R}^n : V(x) < c\}$ and $x \in \partial\mathcal{D}$ (which then needs to exist). Then $\kappa(x) = 0$ leads to a contradiction in (3.46).

Thus, if (3.46) can be written in terms of LMIs, the parameter c can be maximized. To illustrate this, we continue with the example from the previous section and solve the optimization problem

$$\max_{\substack{c \in \mathbb{R} \\ \delta_3 : \mathbb{R}^n \to \mathbb{R}_{\geq 0}}} \quad c$$

$$\text{subject to} \quad (V(x) - c)|x|^{2k} - \delta_3(x)\kappa(x) \geq 0 \qquad \forall\, x \in \mathbb{R}^n.$$

In particular, consider the case $k = 1$.

Observe that the product $V(x)|x|^2$ can be written as

$$V(x)|x|^2 = z(x)^T P z(x)|x|^2 = y(x)^T P_{la} y(x),$$

where P_{la} is defined as

$$P_{la} = \begin{bmatrix} 0 & 0 & 0 & 0 & 0 & 0 & 0 & 0 & 0 \\ 0 & 0 & 0 & 0 & 0 & 0 & 0 & 0 & 0 \\ 0 & 0 & p_{11} & 0 & p_{12} & p_{13} & 0 & p_{15} & p_{14} \\ 0 & 0 & 0 & p_{22} & p_{12} & 0 & p_{24} & p_{23} & p_{25} \\ 0 & 0 & p_{12} & p_{12} & p_{22} + p_{11} & p_{23} & p_{14} & p_{25} + p_{13} & p_{24} + p_{15} \\ 0 & 0 & p_{13} & 0 & p_{23} & p_{33} & 0 & p_{35} & p_{34} \\ 0 & 0 & 0 & p_{24} & p_{14} & 0 & p_{44} & p_{34} & p_{45} \\ 0 & 0 & p_{15} & p_{23} & p_{25} + p_{13} & p_{35} & p_{34} & p_{55} + p_{33} & p_{45} + p_{35} \\ 0 & 0 & p_{14} & p_{25} & p_{24} + p_{15} & p_{34} & p_{45} & p_{45} + p_{35} & p_{44} + p_{55} \end{bmatrix},$$

similarly to the definition of D_{la} in (3.43). The remaining term containing the unknown c can be written as

$$cx^T x = cy(x)^T \begin{bmatrix} I & 0 \\ 0 & 0 \end{bmatrix} y(x)$$

in terms of the variables $y(x)$. The product $\delta_3(x)\kappa(x)$ is of the same form as $\delta_1(x)\kappa(x)$ in (3.42). We thus obtain the semidefinite program

$$\max_{\substack{c \in \mathbb{R} \\ \hat{D} \in \mathcal{S}^9, D_{sm} \in \mathcal{S}^5}} \quad c$$

$$\text{subject to} \quad 0 \leq P_{la} - c \begin{bmatrix} I & 0 \\ 0 & 0 \end{bmatrix} - \hat{D} \tag{3.47}$$

$$0 \leq D_{sm}$$

$$0 = \hat{D} - D_{la}$$

maximizing the forward-invariant sublevel set through the parameter c.

For the Lyapunov function constructed in the previous section, using CVX returns the parameter $c^\star = 6.96$ as the optimal solution of (3.47). The level set $\{x \in \mathbb{R}^n : V(x) = c^\star\}$ is shown in Figure 3.5. Based on the phase portrait, in Figure 3.5 the estimate of the region of attraction is quite conservative. The estimate can be improved by changing the function κ and by increasing the parameter k in Theorem 3.17.

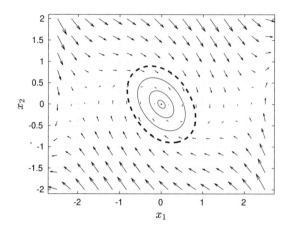

Figure 3.5: Visualization of the forward invariant level set $\{x \in \mathbb{R}^n : V(x) = c^\star\}$ (dashed line). The level set is contained in the region of interest $\mathcal{D} = \mathcal{B}_1(0)$. Additionally the phase portrait is shown.

3.5   SYSTEMS WITH INPUTS

Systems with external inputs (1.5) will be the primary focus of our efforts in what follows. Similarly to the linearization of system (3.7) in Section 3.2, we can linearize the right-hand side of nonlinear systems with external inputs

$$\dot{x} = f(x, u) \tag{3.48}$$

if the function $f : \mathbb{R}^n \times \mathbb{R}^m \to \mathbb{R}^n$ is continuously differentiable with respect to x and u.

Before we linearize (3.48) around an induced equilibrium pair $(x^e, u^e) \in \mathbb{R}^n \times \mathbb{R}^m$, $f(x^e, u^e) = 0$, remember that we can assume without loss of generality that $(x^e, u^e) = 0$ and $f(0, 0) = 0$ by applying the coordinate transformation $z = x - x^e$ and $v = u - u^e$.

Assuming now that $f(0, 0) = 0$, define the matrices

$$A = \left[\frac{\partial f}{\partial x}(x, u) \right]_{(x,u)=0} \quad \text{and} \quad B = \left[\frac{\partial f}{\partial u}(x, u) \right]_{(x,u)=0} . \tag{3.49}$$

Then the linearization of (3.48) is given by $\dot{z} = Az + Bu$. Here, as in (3.9), the variable z is used to highlight that the dynamics of the nonlinear system differ from its linearization.

In general, a linear system with input

$$\dot{x} = Ax + Bu \tag{3.50}$$

is defined through the pair (A, B), with $A \in \mathbb{R}^{n \times n}$ and $B \in \mathbb{R}^{n \times m}$. The linear system has the solution

$$x(t) = e^{At}x(0) + \int_0^t e^{A(t-\tau)}Bu(\tau)d\tau, \tag{3.51}$$

which can be computed analytically for a given $u : \mathbb{R}_{\geq 0} \to \mathbb{R}^m$ if the integral is sufficiently simple. If it is not clear from the context, we will use the notation

$$x(t; x_0, u) = e^{At}x_0 + \int_0^t e^{A(t-\tau)}Bu(\tau)d\tau \qquad (3.52)$$

to highlight the dependency on the initial state $x(0) = x_0$ and the dependency on the selection of the input u.

For linear systems with outputs, the linear system is accompanied by the linear output equation

$$y = Cx + Du \qquad (3.53)$$

for $y \in \mathbb{R}^p$, $C \in \mathbb{R}^{p \times n}$, and $D \in \mathbb{R}^{p \times m}$, which is a special case of (1.14). In many applications the *direct feedthrough* term D is not present.

A linear system (3.50) with output (3.53) is unambiguously defined through the quadruple (A, B, C, D) or the triple (A, B, C) if $D = 0$. Moreover, the pair (A, B) describes a linear system without output and the pair (A, C) captures the output behavior of a linear system without input.

Example 3.18. Recall the dynamics of the inverted pendulum on a cart (1.35) discussed in Section 1.2.1. The set of equilibria is given by

$$x \in \mathbb{R} \times \{k\pi \in \mathbb{R} : k \in \mathbb{Z}\} \times \{0\} \times \{0\} \subset \mathbb{R}^4, \qquad u \in \{0\},$$

as derived in (1.39). We focus on the upright position $x^{e_1} = [0, 0, 0, 0]^T$, and the downward position $x^{e_2} = [0, \pi, 0, 0]^T$ when the cart is at the origin and the input $u^e = 0$.

Using the notation introduced in (3.49), the pair (A_1, B_1) defining the linearization around (x^{e_1}, u^e) is given by

$$A_1 = \left[\frac{\partial f}{\partial x}(x, u) \right]_{(x,u)=(x^{e_1},0)} = \begin{bmatrix} 0 & 0 & 1 & 0 \\ 0 & 0 & 0 & 1 \\ 0 & \frac{g}{M\bar{J}-1} & -\frac{\bar{J}\bar{c}}{M\bar{J}-1} & -\frac{\bar{\gamma}}{M\bar{J}-1} \\ 0 & \frac{Mg}{M\bar{J}-1} & -\frac{\bar{c}}{M\bar{J}-1} & -\frac{M\bar{\gamma}}{M\bar{J}-1} \end{bmatrix},$$

$$B_1 = \left[\frac{\partial f}{\partial u}(x, u) \right]_{(x,u)=(x^{e_1},0)} = \begin{bmatrix} 0 \\ 0 \\ \frac{\bar{J}}{M\bar{J}-1} \\ \frac{1}{M\bar{J}-1} \end{bmatrix}.$$

In the same way, the pair (A_2, B_2) corresponding to the downward position

(x^{e_2}, u^e) is obtained:

$$
A_2 = \left[\frac{\partial f}{\partial x}(x, u) \right]_{(x,u)=(x^{e_2},0)} = \begin{bmatrix} 0 & 0 & 1 & 0 \\ 0 & 0 & 0 & 1 \\ 0 & -\frac{g}{M\bar{J}-1} & -\frac{\bar{J}\bar{c}}{M\bar{J}-1} & \frac{\bar{\gamma}}{M\bar{J}-1} \\ 0 & -\frac{Mg}{M\bar{J}-1} & \frac{\bar{c}}{M\bar{J}-1} & -\frac{M\bar{\gamma}}{M\bar{J}-1} \end{bmatrix},
$$

$$
B_2 = \left[\frac{\partial f}{\partial u}(x, u) \right]_{(x,u)=(x^{e_2},0)} = \begin{bmatrix} 0 \\ 0 \\ \frac{\bar{J}}{M\bar{J}-1} \\ -\frac{1}{M\bar{J}-1} \end{bmatrix}.
$$

3.5.1   Controllability and Observability

The intuitive properties of controllability (that we can drive the system to desirable states) and observability (that we can distinguish between two states by examining the system output) have nice algebraic characterizations for linear systems.

Definition 3.19 (Controllability). *Consider the linear system* (3.50) *uniquely defined through the pair* (A, B). *The linear system (or equivalently the pair* (A, B)*) is said to be controllable if for all* $x_1, x_2 \in \mathbb{R}^n$ *there exists* $T \in \mathbb{R}_{\geq 0}$ *and* $u : [0, T] \to \mathbb{R}^m$ *such that*

$$
x_2 = e^{AT} x_1 + \int_0^T e^{A(T-\tau)} Bu(\tau) d\tau. \tag{3.54}
$$

Definition 3.20 (Observability). *Consider the linear system* (3.1) *with output* (3.53) *defined through the pair* (A, C). *The linear system with output (or equivalently the pair* (A, C)*) is said to be observable if for all* $x_1, x_2 \in \mathbb{R}^n$, $x_1 \neq x_2$, *there exists* $T \in \mathbb{R}_{\geq 0}$ *such that*

$$
Ce^{AT} x_2 \neq Ce^{AT} x_1. \tag{3.55}
$$

Note that neither controllability nor observability depends on the matrix D. Controllability indicates that, by choosing an appropriate input $u : [0, T] \to \mathbb{R}^m$, any initial system state can be steered to any desired target system state. Observability indicates that a given state $x(0)$ can be uniquely determined by measuring the output signal $y(t) = Cx(t) = Ce^{At}x(0)$ over a given time window $t \in [0, T]$.

The triple (A, B, C) is called controllable and observable if the pair (A, B) is controllable and the pair (A, C) is observable. Controllability and observability are indeed independent concepts for linear systems. In Definition 3.19 the output matrix C is not present and, in Definition 3.20, equation (3.55) can be replaced by

$$
Ce^{AT} x_2 + C \int_0^T e^{A(T-\tau)} Bu(\tau) d\tau \neq Ce^{AT} x_1 + C \int_0^T e^{A(T-\tau)} Bu(\tau) d\tau \tag{3.56}
$$

to capture the impact of a possible input without changing the definition.

The following results present a variety of ways to check if a system is controllable.

Theorem 3.21 (Controllability [63, Theorem 12.1]). *Consider the linear system* (3.50) *defined through the pair* (A, B). *The linear system (or equivalently the pair*

(A, B)) *is controllable if and only if*

$$\text{rank}\left([B \ AB \ A^2B \ \cdots \ A^{n-1}B]\right) = n. \tag{3.57}$$

Theorem 3.22 (Popov-Belevitch-Hautus (PBH) test [63, Theorem 12.3]). *The linear system* (3.50) *is controllable if and only if*

$$\text{rank}\left([A - \lambda I \ \ B]\right) = n \tag{3.58}$$

for all $\lambda \in \mathbb{C}$.

Controllability and observability are frequently referred to as dual concepts for linear time-invariant systems defined by (3.1) and (3.53). Specifically, in the linear time-invariant setting, observability tests can be obtained from controllability tests for a transposed system as described following the next two results.

Theorem 3.23 (Observability [63, Theorem 15.7]). *Consider the linear system* (3.1) *with output* (3.53) *defined through the pair* (A, C). *The linear system with output (or equivalently the pair* (A, C)) *is observable if and only if*

$$\text{rank}\left(\begin{bmatrix} C \\ CA \\ CA^2 \\ \vdots \\ CA^{n-1} \end{bmatrix}\right) = n. \tag{3.59}$$

Theorem 3.24 (PBH test [63, Theorem 15.9]). *The linear system* (3.50) *with output* (3.53) *(or equivalently the pair* (A, C)) *is observable if and only if*

$$\text{rank}\left(\begin{bmatrix} A - \lambda I \\ C \end{bmatrix}\right) = n$$

for all $\lambda \in \mathbb{C}$.

Remark 3.25. The rank of a matrix has to be considered with caution. While mathematically the rank of the matrix

$$M_\varepsilon = \begin{bmatrix} 1 & 0 \\ 0 & \varepsilon \end{bmatrix}$$

is 2 for all $\varepsilon \neq 0$, it is difficult to justify that M_ε has rank 2 in a control application with arbitrarily small ε. It is generally good practice when relying on such rank conditions to verify that the singular values of the matrix in question are not very small.

From Theorem 3.21 and Theorem 3.23 it follows immediately that the pair (A, B) is controllable if and only if (A^T, B^T) is observable. Conversely, (A, C) is observable if and only if (A^T, C^T) is controllable.

Note that in Theorem 3.21 and Theorem 3.23, in contrast to Definition 3.19 and Definition 3.20, the time $T > 0$ does not play a role. Indeed, the length of the time interval $[0, T]$ does not influence the controllability and observability properties of a linear system. If a system is controllable or observable, then T in Definition 3.19

or Definition 3.20, respectively, can be chosen arbitrarily small.

3.5.2   Stabilizability and Detectability

Controllability describes the ability to go from any initial state $x_1 \in \mathbb{R}^n$ to any target state $x_2 \in \mathbb{R}^n$. In general, we will be interested in the particular target state $x_2 = 0$. In this case, we can weaken the assumptions on the matrix pair (A, B).

Definition 3.26 (Stabilizability). *Consider the linear system (3.50) defined through the pair (A, B). The linear system (or equivalently the pair (A, B)) is said to be stabilizable if for all $x \in \mathbb{R}^n$ there exists $u : \mathbb{R}_{\geq 0} \to \mathbb{R}^m$ such that*

$$|x(t; x, u)| \to 0 \qquad for \quad t \to \infty. \tag{3.60}$$

To obtain a result similar to Theorem 3.21 we consider a coordinate transformation of the linear system

$$\dot{x} = Ax + Bu, \qquad y = Cx + Du.$$

For an invertible matrix $T \in \mathbb{R}^{n \times n}$, we can write the linear system equivalently as

$$T\dot{x} = TAT^{-1}Tx + TBu,$$
$$y = CT^{-1}Tx + Du.$$

Using the notation $\tilde{x} = Tx$, $\tilde{A} = TAT^{-1}$, $\tilde{B} = TB$, and $\tilde{C} = CT^{-1}$, the compact representation

$$\dot{\tilde{x}} = \tilde{A}\tilde{x} + \tilde{B}u, \qquad y = \tilde{C}x + Du$$

with the same input-output behavior is derived. While the state x transforms to $x \mapsto \tilde{x} = Tx$, the matrix T does not change the convergence properties; i.e., $|x(t)| \to 0$ for $t \to \infty$ if and only if $|\tilde{x}(t)| \to 0$ for $t \to \infty$. (Specifically, note that A and $\tilde{A}$ have the same eigenvalues.) To obtain a result similar to Theorem 3.21 we will make use of a specific coordinate transformation.

Proposition 3.27 ([63, Theorem 13.2]). *Consider the pair (A, B). There exists an invertible matrix $T \in \mathbb{R}^{n \times n}$ such that*

$$TAT^{-1} = \begin{bmatrix} A_{11} & A_{12} \\ 0 & A_{22} \end{bmatrix} \qquad and \qquad TB = \begin{bmatrix} B_1 \\ 0 \end{bmatrix} \tag{3.61}$$

and the pair (A_{11}, B_1) is controllable.

Theorem 3.28. *Consider the pair (A, B) together with the coordinate transformation (3.61) where (A_{11}, B_1) is controllable. Then the pair (A, B) is stabilizable if and only if A_{22} is Hurwitz.*

The PBH test of Theorem 3.22 can be modified to test for stabilizability by limiting the values of $\lambda \in \mathbb{C}$ to $\lambda \in \overline{\mathbb{C}}_+$.

Theorem 3.29 ([63, Theorem 14.2]). *The linear system (3.50) is stabilizable if and only if*

$$\mathrm{rank}\,([A - \lambda I \quad B]) = n \tag{3.62}$$

for all $\lambda \in \overline{\mathbb{C}}_+$.

Finally, we observe that there is a Lyapunov test for stabilizability (compare with (3.4)).

Theorem 3.30 ([63, Theorem 14.3]). *The linear system* (3.50) *is stabilizable if and only if there exists* $P \in \mathcal{S}^n_{>0}$ *satisfying*

$$AP + PA^T - BB^T < 0. \tag{3.63}$$

In the same way that stabilizability is a weaker condition than controllability, *detectability* is a weaker property than observability. Observability of (A, C) implies that for each $x_0 \neq 0$ there exists a $t \geq 0$ such that

$$Cx(t; x_0, 0) \neq Cx(t; 0, 0) = 0,$$

i.e., x_0 can be distinguished from 0. If the pair (A, C) is not observable we can define the set

$$\mathcal{N} = \{x_0 \in \mathbb{R}^n : Cx(t; x_0, 0) = 0 \ \forall t \geq 0\} \tag{3.64}$$

of unobservable states.

Definition 3.31 (Detectability). *Consider the linear system* (3.1) *with output* (3.53) *defined through the pair* (A, C). *The linear system with output (or equivalently the pair* (A, C)*) is said to be detectable if for all* $x_0 \in \mathcal{N}$ *(where* $\mathcal{N}$ *is defined in* (3.64)*) the solution satisfies*

$$|x(t; x_0, 0)| \to 0 \quad \text{for} \quad t \to \infty. \tag{3.65}$$

Duality of controllability and observability for linear time-invariant systems immediately yields the following two results for detectable systems. The second result is discussed in [164, Chapter 10.4], for example.

Proposition 3.32 ([164, Corollary 5.3.14]). *Consider the pair* (A, C). *There exists an invertible matrix* $T \in \mathbb{R}^{n \times n}$ *such that*

$$TAT^{-1} = \begin{bmatrix} A_{11} & A_{12} \\ 0 & A_{22} \end{bmatrix} \quad \text{and} \quad CT^{-1} = \begin{bmatrix} 0 & C_2 \end{bmatrix} \tag{3.66}$$

and the pair (A_{22}, C_2) *is observable.*

Theorem 3.33 ([164, Corollary 5.3.19]). *Consider the pair* (A, C) *together with the coordinate transformation* (3.66) *where* (A_{22}, C_2) *is observable. Then the pair* (A, C) *is detectable if and only if* A_{11} *is Hurwitz.*

Theorem 3.34. *The linear system* (3.50) *with output* (3.53) *(or equivalently the pair* (A, C)*) is detectable if and only if*

$$\text{rank} \left(\begin{bmatrix} A - \lambda I \\ C \end{bmatrix} \right) = n$$

for all $\lambda \in \overline{\mathbb{C}}_+$.

Theorem 3.35. *The linear system* (3.50) *with output* (3.53) *(or equivalently the pair* (A, C)*) is detectable if and only if there exists* $P \in \mathcal{S}^n_{>0}$ *satisfying*

$$A^T P + PA - C^T C < 0.$$

We conclude this section with a result combining Propositions 3.27 and 3.32 to obtain what is usually referred to as the *Kalman decomposition*.

Proposition 3.36 (Kalman decomposition, [164, Theorem 5.4.1]). *Consider the linear system* (3.53) *with output* (3.50) *defined through* (A, B, C, D). *There exists an invertible matrix* $T \in \mathbb{R}^{n \times n}$ *such that*

$$TAT^{-1} = \begin{bmatrix} A_{11} & A_{12} & A_{13} & A_{14} \\ 0 & A_{22} & 0 & A_{24} \\ 0 & 0 & A_{33} & A_{34} \\ 0 & 0 & 0 & A_{44} \end{bmatrix}, \qquad TB = \begin{bmatrix} B_1 \\ B_2 \\ 0 \\ 0 \end{bmatrix},$$

$$CT^{-1} = \begin{bmatrix} 0 & C_2 & 0 & C_4 \end{bmatrix}$$

and such that

$$\left(\begin{bmatrix} A_{11} & A_{12} \\ 0 & A_{22} \end{bmatrix}, \begin{bmatrix} B_1 \\ B_2 \end{bmatrix} \right)$$

is controllable and

$$\left(\begin{bmatrix} A_{22} & A_{24} \\ 0 & A_{44} \end{bmatrix}, \begin{bmatrix} C_2 & C_4 \end{bmatrix} \right)$$

is observable.

3.5.3   Pole Placement

Theorem 3.3 states that the stability properties of a linear system $\dot{x} = Ax$ depend solely on the eigenvalues of the matrix A. If the eigenvalues of the matrix A are not contained in $\mathbb{C}_-$, i.e., A is not Hurwitz, then the origin is not asymptotically (exponentially) stable. A reasonable question is: If A is not Hurwitz, can we design a state feedback law $u = Kx$, $K \in \mathbb{R}^{m \times n}$ so that for (3.50) the closed-loop matrix $A + BK$ is Hurwitz?

This question can be answered in the affirmative through the concept of *pole placement* and depends on the controllability of the pair (A, B).

Theorem 3.37 (Pole Placement [10, Theorem 9.2]). *Consider the linear system* (3.50) *with input. Let* $\lambda_1, \ldots, \lambda_n \in \mathbb{C}$ *with* $\{\lambda_1, \ldots, \lambda_n\} = \{\bar{\lambda}_1, \ldots, \bar{\lambda}_n\}$. *If* (A, B) *is controllable, then there exists* $K \in \mathbb{R}^{m \times n}$ *such that* $\{\lambda_1, \ldots, \lambda_n\}$ *is the set of eigenvalues of the closed-loop matrix* $A + BK$.

Thus, by designing an appropriate state-dependent input $u = Kx$ it can be ensured that the origin of $\dot{x} = Ax + Bu$ is asymptotically stable if (A, B) is controllable, regardless of the eigenvalues of A. In particular, for given $\lambda_1, \ldots, \lambda_n$, the matrix K can be computed efficiently using MATLAB.

Example 3.38. Consider the linearization of the pendulum on a cart in the upright position derived in Example 3.18. For simplicity we assume that the constants are defined as

$$M = J = m = \ell = 1, \quad \gamma = c = 0.1, \quad \text{and} \quad g = 9.81,$$

which leads to the dynamics $\dot{x} = A_1 x + B_1 u$ defined by

$$A_1 = \begin{bmatrix} 0 & 0 & 1.0000 & 0 \\ 0 & 0 & 0 & 1.0000 \\ 0 & 3.2700 & -0.0667 & -0.0333 \\ 0 & 6.5400 & -0.0333 & -0.0667 \end{bmatrix} \quad \text{and} \quad B_1 = \begin{bmatrix} 0 \\ 0 \\ -0.6667 \\ 0.3333 \end{bmatrix}.$$

The eigenvalues of A_1 (obtained using eig.m in MATLAB) are given by

$$\{0, 2.5162, -2.5995, -0.05\},$$

so A_1 is not Hurwitz and the uncontrolled dynamics are unstable. With the feedback gain matrix

$$K = \begin{bmatrix} 7.3394 & -140.8354 & 15.4653 & -60.5307 \end{bmatrix}$$

the closed-loop matrix

$$A_{cl} = A + BK = \begin{bmatrix} 0 & 0 & 1.0000 & 0 \\ 0 & 0 & 0 & 1.0000 \\ 4.8930 & -90.6203 & 10.2436 & -40.3871 \\ 2.4465 & -40.4051 & 5.1218 & -20.2436 \end{bmatrix},$$

has the eigenvalues $\{-1, -2, -3, -4\}$ and consequently the origin of $\dot{x} = A_{cl}x$ is asymptotically stable. Here, the feedback gain matrix K was obtained using place.m in MATLAB.

We conclude this section with the observation that pole placement is only possible for static *state* feedback. The case of static *output* feedback is significantly more complicated. In other words, it is not always possible to design a feedback gain matrix $K \in \mathbb{R}^{m \times p}$ such that the closed-loop system

$$\dot{x} = A_{cl}x = (A + BKC)x \tag{3.67}$$

obtained from (3.50), (3.53) with $D = 0$, and $u = Ky$ has arbitrarily chosen eigenvalues.

Theorem 3.39. *For* (3.50), (3.53) *with* $D = 0$, *and* $u = Ky$, *if* $\text{trace}(A) > 0$ *and* $CB = 0$ *then there is no matrix gain* $K \in \mathbb{R}^{m \times p}$ *such that* $A + BKC$ *is Hurwitz.*

Proof. The proof follows from a few results regarding the trace of a matrix. First, since $CB = 0$, observe that

$$\text{trace}(BKC) = \text{trace}(CBK) = 0,$$

from which we obtain

$$\text{trace}(A + BKC) = \text{trace}(A) + \text{trace}(BKC) = \text{trace}(A) > 0.$$

Since the trace of a matrix is equal to the sum of its eigenvalues, this inequality implies that $A + BKC$ has at least one eigenvalue in the right half plane.   □

As a consequence, if A is not Hurwitz, it may not be possible to stabilize the origin using static output feedback. In other words, Theorem 3.37 indicates that, for a controllable pair (A, B), it is possible to arbitrarily place the poles of $A + BK$ by choice of K while Theorem 3.39 indicates that arbitrarily placing the poles of $A + BKC$ is not possible. However, as we discuss in Chapter 16, the problem of stabilization using outputs can be solved by employing *dynamic* output feedback.

3.6 EXERCISES

Exercise 3.1. Verify that the solution of $\dot{x} = Ax$ is indeed given by $x(t) = e^{At}x(0)$.

Exercise 3.2. Compute e^{Jt} for the third-order Jordan block

$$J = \begin{bmatrix} \lambda & 1 & 0 \\ 0 & \lambda & 1 \\ 0 & 0 & \lambda \end{bmatrix}.$$

Exercise 3.3. Solve the Lyapunov equation (3.4) for

$$A = \begin{bmatrix} -1 & -1 & 2 \\ 1 & -1 & 1 \\ -1 & 0 & -2 \end{bmatrix} \quad \text{and} \quad Q = \begin{bmatrix} 1 & 0 & 0 \\ 0 & 1 & 0 \\ 0 & 0 & 1 \end{bmatrix}$$

using `lyap.m`, in MATLAB. Verify the result. Based on the solution P, is A Hurwitz?
   Hint: Note that in MATLAB the Lyapunov equation is defined slightly differently.

Exercise 3.4. Complete the proof of Theorem 3.5.

Exercise 3.5. Show that (3.51) is the solution of (3.50).

Exercise 3.6. Recall the linearizations of the inverted pendulum on a cart from Example 3.18. Additionally assume that only the position and the angle can be measured, i.e.,

$$C = \begin{bmatrix} 1 & 0 & 0 & 0 \\ 0 & 1 & 0 & 0 \end{bmatrix}.$$

Show that the triple (A_1, B_1, C) is controllable and observable for the parameters $M = m = \ell = 1$, $J = c = \gamma = 0.1$, and $g = 9.81$.
   Hint: You can use the functions `ctrb.m` and `obsv.m` to compute the controllability and the observability matrices in MATLAB and you can use `rank.m` and `svd.m` to calculate the (numerical) rank of a matrix.

Exercise 3.7. Verify the statements in Example 3.13, i.e., show that the eigenvalues of (3.17) are indeed given by $\lambda = -0.25 \pm j0.25\sqrt{7}$ for all $t \in \mathbb{R}_{\geq 0}$ and show that (3.18) is indeed a solution of the time-varying linear system $\dot{x}(t) = A(t)x(t)$.

Exercise 3.8. (Compare this exercise with Exercise 2.5.) Consider the differential equations

$$\dot{v} = 0.1v, \qquad \dot{w} = 0, \qquad \dot{x} = -x^3, \qquad \dot{y} = -y,$$

which all have the origin as unique equilibrium.

1. Investigate the stability properties of the origin of the differential equations. In particular, what is the difference between instability, stability, asymptotic stability, and exponential stability?
2. Give Lyapunov functions (if possible) to manifest your findings with respect to the first question.
3. Visualize the solutions of the ordinary differential equations for the initial conditions -1 and 1 over the time interval $t \in [0, 10]$ in MATLAB using `ode45.m`.
4. Solve the differential equations analytically. (In MATLAB `syms`, `dsolve.m` and `diff.m` might be helpful.)

Exercise 3.9. Consider the linearization of the pendulum (discussed and introduced in Example 1.4) around the origin

$$\begin{bmatrix} \dot{x}_1 \\ \dot{x}_2 \end{bmatrix} = \begin{bmatrix} 0 & 1 \\ -\frac{g}{\ell} & -\frac{k}{m} \end{bmatrix} \begin{bmatrix} x_1 \\ x_2 \end{bmatrix} = Ax \qquad (3.68)$$

with parameters $g, \ell, m > 0$ and $k \geq 0$.

1. Investigate the stability properties of the origin of the linear system (3.68) depending on the parameters $g, \ell, m > 0$ and $k \geq 0$.
2. For $g = 9.81$, $k = 0.1$, and $\ell = m = 1$ compute a quadratic Lyapunov function $V(x) = x^T P x$ for the linear system by solving the Lyapunov equation

$$A^T P + PA = -Q, \qquad \text{for} \qquad Q = \begin{bmatrix} 1 & 0 \\ 0 & 1 \end{bmatrix}. \qquad (3.69)$$

Plot $(t, V(x(t)))$ for $t \in [0, 20]$ and initial value $x_0 = [3, 0]^T$. (In MATLAB you can use `lyap.m` to solve the Lyapunov equation. Note that the Lyapunov function in MATLAB is defined slightly different.)
3. According to Corollary 3.12, the quadratic function $V(x) = x^T P x$ is a local Lyapunov function (with respect to the origin) for the nonlinear system (1.11), i.e., there exists a $c > 0$ such that $V(x)$ is a Lyapunov function on the domain

$$\{x \in \mathbb{R}^2 : V(x) \leq c\}. \qquad (3.70)$$

Numerically find a $c > 0$ such that $V(x)$ is a Lyapunov function on the sublevel set defined in (3.70).
(*Additional task:* What is the biggest sublevel set you can find? The matrix Q in (3.69) can be replaced by any positive definite matrix. Can you find a Lyapunov function which is valid on a bigger sublevel set?)

Exercise 3.10. Consider the differential equation

$$\dot{x} = f(x) = \begin{bmatrix} \cos(x_2)\sin(x_1) + x_1^3 - 2x_1 \\ x_1 x_2^2 + x_1^2 + (1 + x_2)(2 + x_2) - \sin(x_2 + 2) \end{bmatrix}. \qquad (3.71)$$

We are interested in the stability properties of the equilibrium $x^e = [0, -2]^T$. (Note that the ordinary differential equation might have multiple equilibria.)

1. Use the coordinate transformation $y = x - x^e$ to define the ordinary differential equation $\dot{y} = \tilde{f}(y)$ with $y^e = [0, 0]^T$ as the equilibrium of interest.
2. Linearize the ordinary differential equation $\dot{y} = \tilde{f}(y)$ in the origin. (We denote the linearization by $\dot{z} = Az$.)
3. Investigate the stability properties of the equilibrium $x^e = [0, -2]^T$ of the ordinary differential equation (3.71).
4. Compute a global Lyapunov function $V(z) = z^T P z$, $P \in \mathcal{S}_{>0}^{2 \times 2}$, for the linear system $\dot{z} = Az$ with respect to the origin $z = [0, 0]^T$.
5. Show that V is a local Lyapunov function for the nonlinear system $\dot{y} = \tilde{f}(y)$ with respect to the origin $y = [0, 0]^T$ by numerically computing a sublevel set

$$L_c := \{y \in \mathbb{R}^2 | V(y) \leq c\}, \qquad c > 0,$$

where the decrease condition $\langle \nabla V(y), \tilde{f}(y) \rangle < 0$ is satisfied for all $y \in L_c \backslash \{0\}$.
6. Visualize $(y_1(t), y_2(t))$ and $(z_1(t), z_2(t))$ for

- $y(0) = z(0) = [0.75, -0.75]^T$ for $t \in [0, 20]$, and
- $y(0) = [1.5, -1.5]^T$ for $t \in [0, 0.5]$ and $z(0) = [1.5, -1.5]^T$ for $t \in [0, 20]$.

Exercise 3.11. (This exercise is equivalent to Exercise 3.10 but uses a different differential equation.) Consider the differential equation

$$\dot{x} = f(x) = \begin{bmatrix} \sin(x_2) + x_1 x_2^2 \\ x_1 - x_1^2 - x_2 \end{bmatrix}. \tag{3.72}$$

We are interested in the stability properties of the equilibrium $x^e = [1, 0]^T$. (Note that the ordinary differential equation has multiple equilibria.)

1. Use the coordinate transformation $y = x - x^e$ to define the ordinary differential equation $\dot{y} = \tilde{f}(y)$ with $y^e = [0, 0]^T$ as the equilibrium of interest.
2. Linearize the ordinary differential equation $\dot{y} = \tilde{f}(y)$ in the origin. (We denote the linearization by $\dot{z} = Az$.)
3. Investigate the stability properties of the equilibrium $x^e = [1, 0]^T$ of the ordinary differential equation (3.72).
4. Compute a global Lyapunov function $V(z) = z^T P z$, $P \in \mathcal{S}_{>0}^2$, for the linear system $\dot{z} = Az$ with respect to the origin $z = [0, 0]^T$.
5. Show that V is a local Lyapunov function for the nonlinear system $\dot{y} = \tilde{f}(y)$ with respect to the origin $y = [0, 0]^T$ by numerically computing a level set

$$L_c := \{y \in \mathbb{R}^2 | V(y) \leq c\}, \qquad c > 0,$$

where the decrease condition $\langle \nabla V(y), \tilde{f}(y) \rangle < 0$ is satisfied for all $y \in L_c \backslash \{0\}$.
6. Visualize $(y_1(t), y_2(t))$ and $(z_1(t), z_2(t))$ for

- $y(0) = z(0) = [0.4, -0.4]^T$ for $t \in [0, 20]$, and
- $y(0) = [1, -1]^T$ for $t \in [0, 0.3]$ and $z(0) = [1, -1]^T$ for $t \in [0, 20]$.

Exercise 3.12. In this exercise we use ideas discussed in Section 3.4 to calculate Lyapunov functions and estimates of regions of attractions. (For this exercise we recommend SOSTOOLS or CVX in MATLAB.)

1. Consider the nonlinear system

$$\dot{x} = f_g(x) = \begin{bmatrix} -x_2 - 1.5x_1^2 - 0.5x_1^3 \\ 3x_1 - x_2 \end{bmatrix}. \tag{3.73}$$

Compute a global Lyapunov function V_g with respect to the origin of the nonlinear system (3.73) using the ideas in Section 3.4.2. Assume that the function V_g can be written as a fourth order polynomial

$$P(x) = p_1 x_1^2 + p_2 x_2^2 + p_3 x_1 x_2 + p_4 x_1^3 + p_5 x_2^3 + p_6 x_1^2 x_2 + p_7 x_1 x_2^2 + p_8 x_1^4$$
$$+ p_9 x_2^4 + p_{10} x_1^3 x_2 + p_{11} x_1 x_2^3 + p_{12} x_1^2 x_2^2. \tag{3.74}$$

Visualize level sets of the function V_g and plot the phase portrait corresponding to $f_g(x)$.

2. Consider the nonlinear system

$$\dot{x} = f_l(x) = \begin{bmatrix} x_2 \\ -x_1 - x_2 + \frac{1}{3}x_1^3 \end{bmatrix} \tag{3.75}$$

with three equilibria $[0,0]^T$, $[-\sqrt{3},0]^T$ and $[\sqrt{3},0]^T$. Compute a local Lyapunov function V_l with respect to the origin of the nonlinear system (3.75) following the ideas in Section 3.4.3. Here, we define κ used in Section 3.4.3 as

$$\kappa(x) = x_1^2 + x_2^2 - 2 \tag{3.76}$$

and we assume that V_l, δ_1 and δ_2 can be written as polynomials of the form (3.74).

Visualize level sets of the function V_l and plot the phase portrait corresponding to $f_l(x)$.

3. We consider again the nonlinear system (3.75) together with the Lyapunov function V_l computed in part 2. In this exercise we compute an estimate of the region of attraction in terms of a sublevel set $\{x \in \mathbb{R}^2 : V_l(x) \leq c\}$ for $c > 0$. Solve the optimization problem

$$c^* = \max \quad c$$
$$\text{s.t.} \quad (V_l(x) - c)(x_1^2 + x_2^2) \geq \delta_3(x)\kappa(x) \tag{3.77}$$
$$\delta_3(x) \geq 0,$$

where V_l denotes the local Lyapunov function computed in 2., κ is defined in (3.76), and δ_3 is an unknown function of the form (3.74). Visualize the level set $\{x \in \mathbb{R}^2 : V_l(x) = c^*\}$ and the level set $\{x \in \mathbb{R}^2 : \kappa(x) = 0\}$.

Exercise 3.13. In this exercise we illustrate the linear quadratic regulator on the example of the dynamics of the inverted (double) pendulum shown in Figure 3.6. The dynamics are derived in [27], for example, using the Lagrange formalism:

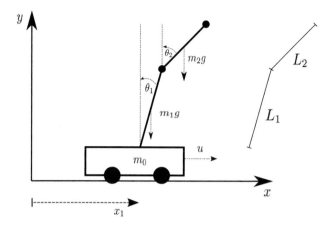

Figure 3.6: Visualization of the inverted double pendulum.

$$u = (m_0 + m_1 + m_2)\,\ddot{x}_1 + (m_1 l_1 + m_2 L_1)\cos(\theta_1)\ddot{\theta}_1 + m_2 l_2 \cos(\theta_2)\ddot{\theta}_2$$
$$\quad - (m_1 l_1 + m_2 L_1)\sin(\theta_1)\dot{\theta}_1^2 - m_2 l_2 \sin(\theta_2)\dot{\theta}_2^2$$
$$0 = (m_1 l_1 + m_2 L_1)\cos(\theta_1)\ddot{x}_1 + \left(m_1 l_1^2 + m_2 L_1^2 + I_1\right)\ddot{\theta}_1$$
$$\quad + m_2 L_1 l_2 \cos(\theta_1 - \theta_2)\ddot{\theta}_2 + m_2 L_1 l_2 \sin(\theta_1 - \theta_2)\dot{\theta}_2^2 \qquad (3.78)$$
$$\quad - (m_1 l_1 + m_2 L_1)\,g\sin\theta_1$$
$$0 = m_2 l_2 \cos(\theta_2)\ddot{x}_1 + m_2 L_1 l_2 \cos(\theta_1 - \theta_2)\ddot{\theta}_1$$
$$\quad + \left(m_2 l_2^2 + I_2\right)\ddot{\theta}_2 - m_2 L_1 l_2 \sin(\theta_1 - \theta_2)\dot{\theta}_1^2 - m_2 l_2 g\sin(\theta_2).$$

The parameters are defined in Table 3.2. We additionally assume that the mass

m_i	Mass of the pendulum/cart		Parameter	
L_i	Length of pendulum i		m_0	1.5 kg
l_i	Half the length of the ith pendulum		m_1	0.5 kg
x_1	Position of the cart		m_2	0.75 kg
θ_i	Angle of the ith pendulum		L_1	0.5 m
I_i	Moment of inertia of ith pendulum		L_2	0.75 m
g	Gravity constant ($g = 9.81\frac{m}{s^2}$)		l_i	$\frac{1}{2}L_i$
u	Control force		I_i	$\frac{1}{12}m_i L_i^2$

Table 3.2: Parameters of the inverted pendulum and parameter selection in this exercise.

of the pendulum is centered in the middle and the moment of inertia is given by $I_i = m_i L_i^2/12$. Friction is neglected in the model description (3.78). To simplify the presentation of the dynamics, we introduce the following constants:

$$d_1 = m_0 + m_1 + m_2, \qquad\qquad d_5 = m_2 L_1 l_2 = \tfrac{1}{2} m_2 L_1 L_2,$$

$$d_2 = m_1 l_1 + m_2 L_1 = \left(\tfrac{1}{2} m_1 + m_2\right) L_1, \qquad d_6 = m_2 l_2^2 + I_2 = \frac{1}{3} m_2 L_2^2,$$

$$d_3 = m_2 l_2 = \tfrac{1}{2} m_2 L_2, \qquad\qquad f_1 = \left(\tfrac{1}{2} m_1 + m_2\right) L_1 g,$$

$$d_4 = m_1 l_1^2 + m_2 L_1^2 + I_1 = \left(\tfrac{1}{3} m_1 + m_2\right) L_1^2, \quad f_2 = m_2 l_2 g = \tfrac{1}{2} m_2 L_2 g.$$

1. Rewrite (3.78) in the form of an ordinary differential equation of order 1, in the state $x = [x_1, \theta_1, \theta_2, \dot{x}_1, \dot{\theta}_1, \dot{\theta}_2]^T$ and with respect to the constants $d_1, \ldots, d_6$ and f_1, f_2.
2. Linearize the ordinary differential equation in the equilibrium $x = 0$.
   Hint: Linearize the equations (3.78). The linearization of (3.78) is equivalent to the linearization of the system represented as $\dot{x} = f(x, u)$.
3. Compute a linear state space representation of the form

$$\dot{z} = Az + Bu, \tag{3.79}$$

   capturing the linearization around the origin.
4. Give a representation of the nonlinear system dynamics of the form

$$\dot{x} = A(x)x + B(x)u + K(x). \tag{3.80}$$

5. Compute the linear system for the constants provided in Table 3.2.
6. Check if the origin of the linear system is asymptotically stable.
7. Is the pair (A, B) controllable?
8. Assume that only the position x_1 of the cart and the angle $\theta_1 + \theta_2$ can be measured. What does the corresponding matrix C (with $y(t) = Cx(t)$) look like? Is the pair (A, C) observable?
9. Use `care.m` (or `icare.m`) to compute a feedback law $u = Fx$ which globally stabilizes the origin of the linear system.
10. Simulate and visualize the closed-loop solutions of the linear and nonlinear systems for different initial values using the feedback $u = Fx$.

3.7   BIBLIOGRAPHICAL NOTES AND FURTHER READING

There are a multitude of excellent and thorough texts on linear systems theory. The primary source for results in this chapter is [63]. Additional excellent texts include [10], [34], and, particularly for time-varying linear systems, [133]. Though not strictly speaking a linear systems text, [164] contains many linear systems results viewed through a slightly different lens.

The constructive approach in Section 3.4 is directly applicable to dynamical systems (3.21) with polynomial right-hand side. In some cases also rational right-hand sides can be covered by the approach (see [122] for an example). It is important to note, however, that this approach is not guaranteed to provide a Lyapunov function even for systems with polynomial right-hand side.

There are differential equations with polynomial right-hand side and globally asymptotically stable origin where no global polynomial Lyapunov function exists.

This has been observed for the relatively simple dynamics

$$\dot{x}_1 = -x_1 + x_1 x_2, \qquad \dot{x}_2 = -x_1 \qquad (3.81)$$

in the paper [3], for example. While one might conjecture that (3.81) at least admits local quadratic Lyapunov functions, the authors in [2] additionally provide a (more complicated) example showing that even local polynomial Lyapunov functions do not need to exist for dynamical systems with polynomial right-hand side. For more detailed results on the numerical construction of polynomial Lyapunov functions we refer to [5] and references therein. The PhD thesis [117] largely initiated the sum of squares approach presented in Section 3.4. Extensions of the approach to the combined stability analysis and controller design also exist [121], [123].

For software packages developed for semidefinite programming we refer to CVX [56], [57], and SOSTOOLS [116], which can be used as toolboxes in MATLAB.

For a summary of results on static output feedback, see [151].

Chapter Four

Frequency Domain Analysis

.

Frequency domain analysis is a classical and powerful approach to the analysis of linear systems. These tools can be extended in a limited, though useful, way to nonlinear systems. Here, we provide a brief introduction to the frequency domain concepts necessary for subsequent developments in this text, in particular for our development of absolute stability results in Chapter 6.

4.1 FUNDAMENTAL RESULTS IN THE FREQUENCY DOMAIN

For simplicity, we consider single-input single-output linear systems

$$\dot{x}(t) = Ax(t) + bu(t), \qquad y(t) = cx(t) + du(t), \tag{4.1}$$

with $A \in \mathbb{R}^{n \times n}$, $b \in \mathbb{R}^{n \times 1}$, $c \in \mathbb{R}^{1 \times n}$, and $d \in \mathbb{R}$, and the system representation in the frequency domain

$$\hat{y}(s) = G(s)\hat{u}(s) \tag{4.2}$$

defined through the transfer function $G : \mathbb{C} \to \mathbb{C}$. In this chapter we ignore the initial value $x(0) \in \mathbb{R}$, since it is not captured in the representation (4.2), and generally assume that $x(0) = 0$ holds if not specified differently. We also deviate slightly from our previous notation and use lower case b, c, and d in (4.1) to highlight our focus on single-input single-output systems. Even though we are in general interested in systems without direct feedthrough (i.e., $d = 0$) in this book, we include the case $d \neq 0$ here to derive necessary tools for Chapter 6. The function G is a rational function, i.e., there exist polynomial functions $P, Q \in \mathbb{R}[s]$ (with coefficients in $\mathbb{R}$) such that the transfer function can be written as

$$G(s) = \frac{P(s)}{Q(s)}.$$

Additionally, without loss of generality we assume that P, Q are of minimal degree. We say that $P(s)$, $Q(s)$ have minimal degree if $Q(s) = 0$ for $s \in \mathbb{C}$ implies that $P(s) \neq 0$. In other words, $P(s)$ and $Q(s)$ do not have any zeros in common.

4.1.1 The Laplace Transform

The Laplace transform allows us to move between the time domain $t \in \mathbb{R}_{\geq 0}$ and the frequency domain $s \in \mathbb{C}$. We consider signals in the time domain $\psi : \mathbb{R}_{\geq 0} \to \mathbb{R}^m$, $m \in \mathbb{N}$, and the integral $\int_0^\infty \psi(t)e^{-st}\,dt$ for $s \in \mathbb{C}$, for which the integral is well defined, i.e.,

$$\int_0^\infty \psi(t)e^{-st} \, dt < \infty. \tag{4.3}$$

Definition 4.1 (Laplace transform). *Consider $\psi : \mathbb{R}_{\geq 0} \to \mathbb{R}^m$. For $s \in \mathcal{C} \subset \mathbb{C}$ for which (4.3) is satisfied, the Laplace transform $\hat{\psi} : \mathcal{C} \to \mathbb{C}^m$ of ψ is defined as*

$$\hat{\psi}(s) \doteq (\mathscr{L}\psi)(s) \doteq \int_0^\infty \psi(t)e^{-st} \, dt.$$

Example 4.2. Consider the constant signal $\psi(t) = 1$. Then for fixed $s \in \mathbb{C}$ we can compute the integral

$$\int_0^\infty e^{-st} \, dt = -\frac{1}{s}e^{-st}\Big|_0^\infty = \frac{1}{s},$$

which shows that the Laplace transform of ψ is given by $\hat{\psi}(s) = (\mathscr{L}\psi)(s) = \frac{1}{s}$.

Comprehensive lists of Laplace transforms for various signals ψ can be easily found online or in texts on signals and systems such as [115]. The Laplace transform maps a signal ψ from the time domain $t \in \mathbb{R}_{\geq 0}$ to $\hat{\psi}$ in the frequency domain $s \in \mathbb{C}$. To obtain a mapping from the frequency domain to the time domain, we use the inverse Laplace transform.

Definition 4.3 (Inverse Laplace transform). *Consider $\hat{\varphi} : \mathcal{C} \to \mathbb{C}^m$ and let $\alpha \in \mathbb{R}$ such that $\alpha + j\beta \in \mathcal{C} \subset \mathbb{C}$ for all $\beta \in \mathbb{R}$. Then the inverse Laplace transform $\varphi : \mathbb{R}_{\geq 0} \to \mathbb{R}^m$ of $\hat{\varphi}$ is defined as*

$$\varphi(t) \doteq (\mathscr{L}^{-1}\hat{\varphi})(t) \doteq \frac{1}{2\pi j} \int_{\alpha - j\infty}^{\alpha + j\infty} e^{st}\hat{\varphi}(s) \, ds = \frac{e^{\alpha t}}{2\pi j} \int_{-\infty}^{\infty} e^{jwt}\hat{\varphi}(\alpha + jw) \, dw. \tag{4.4}$$

In (4.4) we again tacitly assume that the integral is well defined, which is sufficient for the presentation of results in this book. Among other properties, the Laplace transform and its inverse satisfy the following properties, which we will make use of in the analysis of dynamical systems.

Proposition 4.4 (Properties of the Laplace transform). *Consider three signals $\varphi, \varphi_1, \varphi_2 : \mathbb{R}_{\geq 0} \to \mathbb{R}^m$ in the time domain and constants $a \in \mathbb{R}_{>0}$, $a_1, a_2 \in \mathbb{R}$. Then the Laplace transform and its inverse satisfy the following properties:*

$$\mathscr{L}^{-1}\mathscr{L}\varphi(t) = \varphi(t), \tag{4.5a}$$

$$\mathscr{L}(a_1\varphi + a_2\varphi_2)(s) = a_1\hat{\varphi}_1(s) + a_2\hat{\varphi}_2(s), \tag{4.5b}$$

$$\mathscr{L}(\varphi(a\cdot))(s) = \frac{1}{a}\hat{\varphi}\left(\frac{s}{a}\right), \tag{4.5c}$$

$$\mathscr{L}(\varphi(\cdot - a))(s) = e^{-sa}\hat{\varphi}(s), \tag{4.5d}$$

$$\mathscr{L}(\tfrac{d^k}{dt^k}\varphi)(s) = s^k\hat{\varphi}(s) - \sum_{j=1}^{k-1} s^{j-1}\tfrac{d^{k-1-j}}{dt^{k-1-j}}\varphi(0), \tag{4.5e}$$

$$\mathscr{L}\left(\int_0^\cdot \varphi(\tau) \, d\tau\right)(s) = \frac{1}{s}\hat{\varphi}(s). \tag{4.5f}$$

4.1.2   The Transfer Function

Using the properties given in Proposition 4.4 we can easily move between the linear system (4.1) in the time domain and its representation (4.2) in the frequency domain. The linearity of the Laplace transform (4.5b) implies that the linear terms in (4.1) can be considered separately. Using this fact together with (4.5e) implies

$$s\hat{x}(s) - x(0) = A\hat{x}(s) + b\hat{u}(s), \qquad \hat{y}(s) = c\hat{x}(s) + d\hat{u}(s). \qquad (4.6)$$

Under the assumption that $x(0) = 0$, we can rearrange the terms in (4.6) to eliminate $\hat{x}$ and to obtain

$$\hat{y}(s) = \left(c(sI - A)^{-1}b + d\right)\hat{u}(s),$$

from which we can identify the relationship between the input and output known as the transfer function

$$G(s) = \frac{\hat{y}(s)}{\hat{u}(s)} = c(sI - A)^{-1}b + d \qquad (4.7)$$

in (4.2).

Example 4.5. Consider the linear system

$$\dot{x}(t) = \begin{bmatrix} a_{11} & a_{12} \\ a_{21} & a_{22} \end{bmatrix} x(t) + \begin{bmatrix} b_1 \\ b_2 \end{bmatrix} u(t), \qquad y(t) = \begin{bmatrix} c_1 & c_1 \end{bmatrix} x(t) + du(t).$$

If $\det(sI - A) \neq 0$, we can use the formula (1.33) to derive the inverse of $sI - A$. Then the transfer function is given by

$$
\begin{aligned}
G(s) &= \frac{1}{\det(sI - A)} \begin{bmatrix} c_1 & c_2 \end{bmatrix} \begin{bmatrix} s - a_{22} & a_{12} \\ a_{21} & s - a_{11} \end{bmatrix} \begin{bmatrix} b_1 \\ b_2 \end{bmatrix} + d \\
&= \frac{s(c_1 b_1 + c_2 b_2) + c_1(a_{12}b_2 - b_1 a_{22}) + c_2(a_{21}b_1 - b_2 a_{11})}{s^2 - s(a_{11} + a_{22}) + a_{11}^2 + a_{22}^2 - a_{12}a_{21}} + d \\
&= \frac{s^2 d + s(c_1 b_1 + c_2 b_2 - d(a_{11} + a_{22}))}{s^2 - s(a_{11} + a_{22}) + a_{11}^2 + a_{22}^2 - a_{12}a_{21}} \\
&\quad + \frac{c_1(a_{12}b_2 - b_1 a_{22}) + c_2(a_{21}b_1 - b_2 a_{11}) + d(a_{11}^2 + a_{22}^2 - a_{12}a_{21})}{s^2 - s(a_{11} + a_{22}) + a_{11}^2 + a_{22}^2 - a_{12}a_{21}}.
\end{aligned}
$$

As this example shows, it is relatively easy to compute the transfer function for two-dimensional systems with $x \in \mathbb{R}^2$. For higher-dimensional systems the hand calculations quickly become tedious. The same is true for the inverse question, i.e., the calculation of matrices (A, b, c, d) for a given transfer function $G(s)$.

Definition 4.6 (Realization). *Consider a transfer function $G(s)$ and assume that (4.7) is satisfied for (A, b, c, d). Then $G(s)$ is called realizable and the quadruple (A, b, c, d) is called a realization of $G(s)$.*

With software packages such as MATLAB it is straightforward to compute $G(s)$ from a quadruple (A, b, c, d) and vice versa. We refer to Exercise 4.2 for a computation of transfer functions using MATLAB. We conclude this section with three results answering the question of the existence of realizations for given transfer functions, characterizing minimal realizations, and describing the connection between the eigenvalues of A and the poles of the transfer function.

Theorem 4.7 (Realizable transfer functions [10, Theorem 8.5]). *Consider a transfer function $G(s) = \frac{P(s)}{Q(s)}$, $P, Q \in \mathbb{R}[s]$. The transfer function $G(s)$ is realizable if and only if it is proper, i.e., $\deg(P) \leq \deg(Q)$.*

When $G(s) = \frac{P(s)}{Q(s)}$ is strictly proper, i.e., $\deg(P) < \deg(Q)$, the transfer function is realizable through a triple (A, b, c). Here, $\deg(\cdot)$ denotes the degree of a polynomial. It is not only possible to show that a realization of a proper transfer function exists, one can additionally construct a realization such that the dimension of the matrix $A \in \mathbb{R}^{n \times n}$ is minimal. A realization of $G(s) = c(sI - A)^{-1}b + d$ is minimal if there does not exist $(\tilde{A}, \tilde{b}, \tilde{c}, \tilde{d})$ with $G(s) = \tilde{c}(sI - \tilde{A})^{-1}\tilde{b} + \tilde{d}$ and $\dim(\tilde{A}) < \dim(A)$.

Theorem 4.8 (Minimal realization, [10, Thm. 8.9]). *The quadruple (A, b, c, d) is a minimal realization of $G(s) = c(sI - A)^{-1}b + d$ if and only if (A, b) is controllable and (A, c) is observable.*

Note that while the dimension of a minimal realization is unique and defined, the minimal realization itself is not unique. This follows from Exercise 4.2 and can be understood from the coordinate transformation discussed in Section 3.5.2.

Theorem 4.9 (Uncontrollable and unobservable modes [9, Theorem 5.3.11]). *Let (A, b, c, d) be a realization of $G(s) = \frac{P(s)}{Q(s)}$. If $\lambda \in \mathbb{C}$ is a pole of G, i.e., $Q(\lambda) = 0$, then λ is an eigenvalue of A. Conversely, let λ be an eigenvalue of A such that $G(\lambda) \neq 0$; then λ is an uncontrollable mode of (A, b) or an unobservable mode of (A, c).*

In the above, uncontrollable or unobservable modes refer to eigenvalues of the uncontrollable or unobservable components of the system, for example, as made explicit in the Kalman decomposition of Proposition 3.36.

From Theorem 4.8 and Theorem 4.9 it in particular follows that for a minimal realization (A, b, c, d) of $G(s)$, the eigenvalues of A coincide with the poles of $G(s)$.

4.1.3 The $\mathcal{L}_2$-, $\mathcal{L}_\infty$- and $\mathcal{H}_\infty$-norm

To be able to quantify the size of signals in the time domain and in the frequency domain, we introduce the $\mathcal{L}_2$-norm, $\mathcal{L}_\infty$-norm, and the $\mathcal{H}_\infty$-norm. Consider a function $\psi : [0, t) \to \mathbb{R}^n$, $n \in \mathbb{N}$, for $t \in \mathbb{R}_{\geq 0} \cup \{\infty\}$. For functions ψ for which the integral $(\int_0^t |\psi(\tau)|^2 \, d\tau)^{\frac{1}{2}} < \infty$ is well defined (i.e., finite), we define the norm

$$\|\psi\|_{\mathcal{L}_2[0,t)} \doteq \left(\int_0^t |\psi(\tau)|^2 \, d\tau \right)^{\frac{1}{2}}. \tag{4.8}$$

For essentially bounded functions ψ we additionally consider the norm

$$\|\psi\|_{\mathcal{L}_\infty[0,t)} \doteq \operatorname{ess\,sup}_{\tau \in [0,t)} |\psi(\tau)| \tag{4.9}$$

$$\doteq \inf\{\eta \in \mathbb{R}_{\geq 0} : |\psi(t)| \leq \eta \text{ for almost all } \tau \in [0, t)\}.$$

To simplify the notation we will omit the time interval in the norm and simply write $\|\cdot\|_{\mathcal{L}_2}$ and $\|\cdot\|_{\mathcal{L}_\infty}$, in general. Note that (4.8) and (4.9) combine two norms. In particular, $\|\cdot\|_{\mathcal{L}_2}$, $\|\cdot\|_{\mathcal{L}_\infty}$ define a norm of a function $\psi(\cdot) : \mathbb{R}_{\geq 0} \to \mathbb{R}^n$ and $|\cdot|$ denotes a vector norm of a vector $\psi(t) \in \mathbb{R}^n$ for a fixed $t \in \mathbb{R}_{\geq 0}$. In general, $|\cdot|$ denotes the Euclidean norm $|\psi(t)| = \sqrt{\sum_{i=1}^n \psi_i(t)^2}$ if not specified differently.

In the frequency domain, we will make use of the $\mathcal{H}_\infty$-norm of a function $\hat{\psi} : \mathbb{C} \to \mathbb{C}^n$, which is defined as

$$\|\hat{\psi}\|_\infty = \sup_{\omega \in \mathbb{R}} |\hat{\psi}(j\omega)|. \tag{4.10}$$

The $\mathcal{H}_\infty$-norm will be implicitly computed in Section 4.2.3 as a byproduct of developing the *Bode plot* of a given transfer function. Note that for a complex number $\eta \in \mathbb{C}^n$ the complex conjugate needs to be considered in the computation of the Euclidean norm $|\eta| = \sqrt{\overline{\eta}^T \eta}$.

The final result of this section relates the $\mathcal{L}_2$-norm of a time domain signal with its Laplace transform.

Proposition 4.10 (Parseval's theorem, [115, Sec. 4.3.7]). *Consider a signal $\psi : \mathbb{R}_{\geq 0} \to \mathbb{R}^n$ in the time domain satisfying $\|\psi\|_{\mathcal{L}_2[0,\infty)} < \infty$ and its Laplace transform $\hat{\psi} : \mathbb{C} \to \mathbb{C}^n$. Then Parseval's relation*

$$\int_0^\infty |\psi(\tau)|^2 \, d\tau = \frac{1}{2\pi} \int_{-\infty}^\infty |\hat{\psi}(j\omega)|^2 \, d\omega \tag{4.11}$$

is satisfied.

The implications of (4.11) are for example useful to derive upper bounds on the $\mathcal{L}_2$-norm of a state trajectory $x(\cdot)$ of perturbed systems (see Section 7.4).

Remark 4.11. As in [115], Parseval's relation is usually introduced using the Fourier transform. In this case the left-hand side of (4.11) is replaced by $\int_{-\infty}^\infty |\psi(\tau)|^2 \, d\tau$. Since we are interested in $\psi : \mathbb{R}_{\geq 0} \to \mathbb{R}^n$, Proposition 4.10 is a simplification of the general result where we tacitly assume that $\psi(t) = 0$ for all $t < 0$. With respect to the Fourier transform used in [115], the Fourier transform is a special case of the Laplace transform here, where $s \in \mathbb{C}$ is restricted to $s = j\omega$, $\omega \in \mathbb{R}$.

4.2 STABILITY ANALYSIS IN THE FREQUENCY DOMAIN

While stability properties of the origin of the linear system (4.1) can be deduced from the eigenvalues of the matrix A, in the frequency domain the same properties can be deduced from the poles of the transfer function G in (4.2). In this section we discuss additional criteria investigating stability in the frequency domain.

4.2.1 Bounded-Input, Bounded-Output Stability

Definition 4.12 (Bounded-input, bounded-output stability). *The linear system (4.1) is called bounded-input, bounded-output (BIBO) stable if $\|u\|_{\mathcal{L}_\infty} < \infty$ implies $\|y\|_{\mathcal{L}_\infty} < \infty$.*

It can be shown that the condition in Definition 4.12 is equivalent to the existence of a constant $\eta \in \mathbb{R}_{>0}$ such that

$$\|y\|_{\mathcal{L}_\infty} \leq \eta \|u\|_{\mathcal{L}_\infty}$$

for all $u : \mathbb{R}_{\geq 0} \to \mathbb{R}^m$. Furthermore, one can show that (4.1) is BIBO stable if and only if

$$\int_0^\infty |ce^{A\tau}b|\, d\tau < \infty.$$

This result immediately implies the following result relating exponential/asymptotic stability for linear systems to BIBO stability.

Corollary 4.13. *Assume that the origin of the linear system (4.1) with zero-input is exponentially/asymptotically stable. Then (4.1) is BIBO stable.*

It also follows easily that the converse result is not true; namely, BIBO stability does not imply exponential stability of the origin for (4.1). Consider for example the case $c = 0$ (and $d = 0$). In this case $y(t) \doteq 0$ for all inputs $u(\cdot)$, implying BIBO stability independently of the matrices A and b. However, from Theorem 4.9 we obtain a converse result.

Lemma 4.14. *Consider the transfer function $G(s)$ and an arbitrary realization (A, b, c, d). Then the system (4.2) and the corresponding system (4.1) is BIBO stable if and only if all poles of $G(s)$ are in $\mathbb{C}_-$.*

4.2.2   System Interconnections in the Frequency Domain

It is relatively straightforward to analyze intereonnected linear systems using transfer function representations in the frequency domain. Consider for example two systems,

$$\hat{y}_1(s) = G_1(s)\hat{u}_1(s),$$
$$\hat{y}_2(s) = G_2(s)\hat{u}_2(s),$$

in a cascade interconnection $\hat{u}_2(s) = \hat{y}_1(s)$ as visualized in Figure 4.1. Through

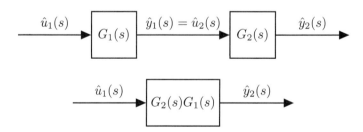

Figure 4.1: Cascade interconnection.

the representation in the frequency domain, the overall input-output behavior can be studied by looking at the transfer functions $G_1(s)$ and $G_2(s)$ individually, or by considering the overall transfer function, which is simply given by the product $G_2(s)G_1(s)$. Stability properties of cascaded nonlinear systems are studied in Chapter 7 using the concept of *Input-to-State Stability*.

Since BIBO stability of the system (4.2), for $G(s) = c(sI - A)^{-1}b + d$ with a minimal realization (A, b, c, d) only depends on the eigenvalues of the matrix A, BIBO stability can be considered as open-loop stability. To obtain closed-loop stability using an appropriate feedback law, we rewrite the input

$$u(t) = v(t) - ky(t), \tag{4.12}$$

where $k \in \mathbb{R}$ defines a feedback gain and $v : \mathbb{R}_{\geq 0} \to \mathbb{R}$ defines a new input (see Figure 4.2 below). The Laplace transform of (4.12) is given by

$$\hat{u}(s) = \hat{v}(s) - k\hat{y}(s). \tag{4.13}$$

Replacing $\hat{u}(s)$ in the linear system (4.2) leads to

$$\hat{y}(s) = G(s)(\hat{v}(s) - k\hat{y}(s))$$

and, after rearranging terms, provides the closed-loop expression

$$\hat{y}(s) = \frac{G(s)}{1 + G(s)k}\hat{v}(s). \tag{4.14}$$

The feedback interconnection is visualized in Figure 4.2. BIBO stability of the

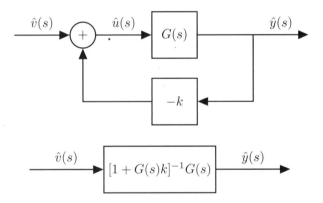

Figure 4.2: Feedback interconnection.

feedback interconnection (4.14) can be guaranteed by selecting the feedback gain k such that the closed-loop transfer function $(1 + G(s)k)^{-1}G(s)$ only has poles in the open left halfplane $\mathbb{C}_-$ according to Lemma 4.14. Note that here, a state feedback instead of a feedback defined based on the output y is covered by the special case $y(t) = Ix(t)$ (in the multi-input multi-output setting).

Example 4.15. Consider the linear system in the frequency domain defined through the transfer function

$$G(s) = \frac{1}{s^2 + 0.1s - 9.81}.$$

The poles of the transfer function are given by $\lambda_1 = -3.18$ and $\lambda_2 = 3.08$, which shows that $G(s)$ is unstable since $\lambda_2 \notin \mathbb{C}_-$. The closed-loop transfer function of the feedback interconnection (4.14) for $k \in \mathbb{R}$ is given by

$$[1 + G(s)k]^{-1}G(s) = \frac{\frac{1}{s^2+0.1s-9.81}}{1 + \frac{k}{s^2+0.1s-9.81}} = \frac{1}{s^2 + 0.1s - 9.81 + k}. \tag{4.15}$$

For the selection $k = 10$, the roots of the transfer function (4.15) are given by $-0.05 \pm 0.433j \in \mathbb{C}_-$, which shows that the feedback interconnection is BIBO stable.

In the next two sections we discuss two graphical methods for the selection of $k \in \mathbb{R}$.

4.2.3   The Bode Plot

The *Bode plot* is a classical tool to investigate the correlation between the input $u(t)$ and the output $y(t)$ of the system

$$\hat{y}(s) = G(s)\hat{u}(s).$$

In particular the output $y(t)$ for the complex input

$$u(t) = e^{j\omega t} = \cos(\omega t) + j\sin(\omega t), \qquad t \geq 0 \tag{4.16}$$

for $\omega \in \mathbb{R}$ is investigated. For a BIBO stable system (4.2) it can be shown that the output $y(t)$ converges to the steady-state solution

$$y_{\text{ss}}(t) = Me^{j(\omega t+\varphi)} = M\cos(\omega t + \varphi) + jM\sin(\omega t + \varphi) \tag{4.17}$$

(for $t \to \infty$), where

$$M = |G(j\omega)| \quad \text{and} \quad \varphi = \varphi(\omega) = \arctan_2(\text{Im}(G(j\omega)), \text{Re}(G(j\omega))) \tag{4.18}$$

define the *gain* and the *phase* of the transfer function, respectively. Recall that the function $\arctan_2 : \mathbb{R}^2 \backslash \{0\} \to (-\pi, \pi]$ was introduced in (1.48).

For the real input $u(t) = \sin(\omega t)$, $t \geq 0$, the solution converges to the real output $y_{\text{ss}}(t) = M\sin(\omega t + \varphi)$ (for $t \to \infty$). Thus, the gain $|G(j\omega)|$ captures the amplification of the input signal at the output while φ can be interpreted as a phase shift or delay. Observe that we have previously mentioned the maximal gain, i.e., the maximal amplification of a signal, in the definition of the $\mathcal{H}_\infty$-norm (4.10).

Having defined the gain and the phase of a transfer function $G(s)$, the Bode plot is defined as plots of $|G(j\omega)|$ and $\varphi(\omega)$ over $\omega \in \mathbb{R}$ on a $\log_{10}/\log_{10}$-scale and a $\log_{10}$/linear-scale, respectively. Rather than a comprehensive development of the Bode plot, we discuss some fundamental features based on the example of a single-input, single-output system and refer to the references provided in Section 4.4 for detailed discussions.

We consider the inverted pendulum modelled by (1.45) and the parameters defined in Table 3.1. Instead of stabilizing the equilibrium at the upright position we are interested in the angle $\theta(t)$ for inputs of the form (4.16) in a neighborhood around the stable equilibrium $[\theta, \dot{\theta}]^T = [\pi, 0]^T$. The linearization at the stable equilibrium is given by the linear system

$$\begin{bmatrix} \dot{x}_1 \\ \dot{x}_2 \end{bmatrix} = \begin{bmatrix} 0 & 1 \\ -\frac{mg\ell}{J+m\ell^2} & -\frac{\gamma}{J+m\ell^2} \end{bmatrix} \begin{bmatrix} x_1 \\ x_2 \end{bmatrix} + \begin{bmatrix} 0 \\ \frac{\ell}{J+m\ell^2} \end{bmatrix} u, \qquad y = \begin{bmatrix} 1 & 0 \end{bmatrix} x.$$

For simplicity, we use the constants $m = \ell = 1$, $J = 0$, $g = 9.81$, and $\gamma = 0.1$, which leads to the transfer function

$$G(s) = \frac{P(s)}{Q(s)} = \frac{1}{s^2 + 0.1s + 9.81}. \tag{4.19}$$

The Bode plot of the transfer function is shown in Figure 4.3 and motivates the selection of a $\log_{10}$-scale on the horizontal axis in order to illustrate the system behavior over a wide range of frequencies. The gain and the phase of the transfer

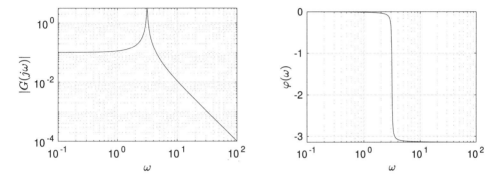

Figure 4.3: Magnitude and phase for the linearized pendulum.

function (4.19) can be described through three domains. For small ω, the gain $|G(j\omega)|$ behaves almost linearly in a $\log_{10}/\log_{10}$-scale before it peaks for $\omega \approx 3.13$ caused by the pair of complex poles $-0.05 \pm 3.13j$ of the transfer function $G(s)$. For large ω, after the peak, $|G(j\omega)|$ decreases linearly in the $\log_{10}/\log_{10}$-scale. Similarly, the phase is almost constant for small and large ω and switches nearly instantaneously from $\varphi = 0$ to $\varphi = -\pi$ at $\omega \approx 3.13$.

From the Bode plot, the input-output behavior for $u(t) = \sin(\omega t)$ can be read off immediately. In Figure 4.4, the pair $y(t)$ and $u(t) = \sin(2t)$ is visualized as an example. As expected, we observe that $y(t)$ converges to the steady-state solution (4.17) for large t (Figure 4.4 left). Once t is large enough we see that $y(t)$ is amplified by the gain $|G(2j)|$ (obtained from Figure 4.3 left) and the input is in phase with the output (as expected from Figure 4.3 right).

In Figure 4.5 additionally the input-output behavior for $\omega = 3.13$ (left) and $\omega = 4$ (right) at steady state is visualized. Here we can particularly observe the amplification of the input as well as the shift in the phase, which can be interpreted as a delay in the input-output correlation.

To sum up, for a given transfer function, the Bode plot captures the input output behavior of the linear system. Conversely, if the transfer function is not known, it can be approximated by experimentally driving the system with sinusoidal inputs of varying frequencies and measuring the output gain and phase as in Figures 4.4 and 4.5.

We conclude this section by illustrating how a sketch of the Bode plot can be

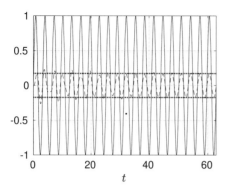
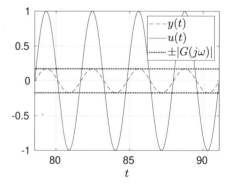

Figure 4.4: Input-output behavior for $u(t) = \sin(2t)$.

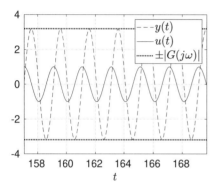
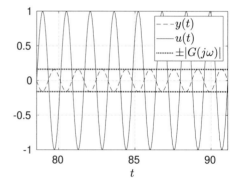

Figure 4.5: Input-output behavior for $u(t) = \sin(3.13t)$ (left) and $u(t) = \sin(4t)$ (right).

obtained based on the example of the transfer function

$$G(s) = \frac{P(s)}{Q(s)} = \frac{s + 10}{s^2 + 101s + 100}. \tag{4.20}$$

As pointed out, the peak in the Bode plot in Figure 4.3 is due to the complex poles of the transfer function. If the numerator and the denominator of the transfer function P and Q only have real zeros $p_i, q_j \in \mathbb{R}$, for $i \in \mathbb{N}_{d_P}$, $j \in \mathbb{N}_{d_Q}$ where d_P and d_Q denote the degree of P and Q, respectively, then $G(s)$ can be written as

$$G(s) = c \frac{\prod_{i=1}^{d_P}(\frac{s}{p_i} - 1)}{\prod_{j=1}^{d_Q}(\frac{s}{q_j} - 1)}, \tag{4.21}$$

for a constant $c \in \mathbb{R}$. (The case of complex conjugate pairs introduces additional cases which are not discussed for simplicity here, but which are covered in [12], for example.) With the representation (4.21), we can use the properties of the logarithm to write the gain (4.18) as a sum

$$\log_{10}|G(j\omega)| = \log_{10}(|c|) + \sum_{i=1}^{d_P}\log_{10}\left(\left|\frac{j\omega}{p_i} - 1\right|\right) - \sum_{i=1}^{d_Q}\log_{10}\left(\left|\frac{j\omega}{q_i} - 1\right|\right) \quad (4.22)$$

instead of a product. The individual terms in (4.21) can be interpreted as a cascade interconnection as in Figure 4.1. Then Equation (4.22) implies that we can analyze the gain of the subsystems individually and the gain of the cascade interconnection is obtained as the sum of the gains due to each term.

Observe that for small values $\omega \in \mathbb{R}_{>0}$ the terms in the sum (4.22) satisfy

$$\log_{10}\left(\left|\frac{j\omega}{\kappa} - 1\right|\right) \approx \log_{10}(1) = 0 \quad (4.23a)$$

for all $\kappa \in \mathbb{R}$. On the other hand, for large $\omega \in \mathbb{R}_{>0}$ the estimate

$$\log_{10}\left(\left|\frac{j\omega}{\kappa} - 1\right|\right) \approx \log_{10}(\omega) - \log_{10}(|\kappa|) \quad (4.23b)$$

holds for all $\kappa \in \mathbb{R}$. Thus, excluding the constant term $\log_{10}(\kappa)$, the individual parts in the sum (4.22) can be approximated by a constant zero for small ω and by a linear function in the $\log_{10}/\log_{10}$-scale for large ω, where the slope of the linear function is independent of the values of p_i or q_i. The change from a constant to a linear function can be taken approximately at the value of p_i or q_i. Finally, we observe that the slope of the linear function will be positive for terms corresponding to zeros (i.e., p_i) and negative for terms corresponding to poles (i.e., q_i).

For the phase we can make similar observations. The phase satisfies the equation

$$\varphi = \arctan_2(0, c) + \sum_{i=1}^{d_P}\arctan_2\left(\frac{\omega}{p_i}, -1\right) - \sum_{i=1}^{d_Q}\arctan_2\left(\frac{\omega}{q_i}, -1\right). \quad (4.24)$$

Moreover, it holds that

$$\begin{array}{llll}
\arctan_2(\frac{\omega}{\kappa}, -1) & \rightarrow & \pi & \text{for } \omega \rightarrow 0, \quad \kappa > 0, \\
\arctan_2(\frac{\omega}{\kappa}, -1) & \rightarrow & -\pi & \text{for } \omega \rightarrow 0, \quad \kappa < 0, \\
\arctan_2(\frac{\omega}{\kappa}, -1) & \rightarrow & \frac{\pi}{2} & \text{for } \omega \rightarrow \infty, \quad \kappa > 0, \\
\arctan_2(\frac{\omega}{\kappa}, -1) & \rightarrow & -\frac{\pi}{2} & \text{for } \omega \rightarrow \infty, \quad \kappa < 0, \\
\arctan_2(0, \kappa) & = & \pi & \text{for} \quad \kappa > 0, \\
\arctan_2(0, \kappa) & = & 0 & \text{for} \quad \kappa < 0,
\end{array} \quad (4.25)$$

which allows us to approximate the individual terms in (4.24) through piecewise constant functions for a wide range of $\omega \in \mathbb{R}$.

Returning to the transfer function (4.20), this can be written as

$$G(s) = \frac{s + 10}{(s + 1)(s + 100)} = \frac{-10}{(-1)(-100)}\frac{\frac{s}{-10} - 1}{\left(\frac{s}{-1} - 1\right)\left(\frac{s}{-100} - 1\right)}$$

$$= -\frac{1}{10}\frac{\frac{s}{-10} - 1}{\left(\frac{s}{-1} - 1\right)\left(\frac{s}{-100} - 1\right)}.$$

Consequently, we can explicitly write the gain and the phase in the form (4.22) and

(4.24),

$$\log_{10}(|G(j\omega)|) = \log_{10}(|\tfrac{1}{10}|) + \log_{10}(|\tfrac{j\omega}{-10} - 1|)$$
$$- \log_{10}(|\tfrac{j\omega}{-1} - 1|) - \log_{10}(|\tfrac{j\omega}{-100} - 1|),$$
$$\varphi = \arctan_2(0, -0.1) + \arctan_2(\tfrac{\omega}{-10}, -1)$$
$$- \arctan_2(\tfrac{\omega}{-1}, -1) - \arctan_2(\tfrac{\omega}{-100}, -1).$$

From the discussion based on (4.23) and (4.25), an approximation of the gain and the phase of the individual terms is shown in Figure 4.6. The actual Bode plot is

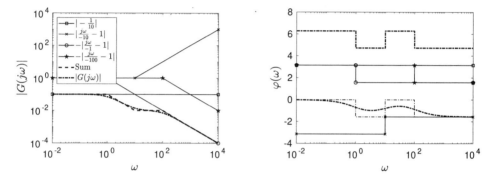

Figure 4.6: Sketch of the Bode plot for the transfer function (4.20) based on the individual terms.

obtained by smoothing the sum of the individual gains and phases in Figure 4.6.

Remark 4.16. The sum of the individual phases φ does not need to satisfy $\varphi \in (-\pi, \pi]$ but can be shifted to this interval as is done in Figure 4.6 on the right.

4.2.4   The Nyquist Criterion

The Bode plot visualizes the input-output behavior of dynamical systems in the frequency domain. As a second graphical method we study the *Nyquist criterion* to draw conclusions about systems represented through the feedback interconnected systems as shown in Figure 4.2. We consider the single-input single-output feedback interconnection

$$\hat{y}(s) = G_{cl}(s)\hat{u}(s) = \frac{G_{ol}(s)}{1 + G_{ol}(s)k}\hat{u}(s) \tag{4.26}$$

capturing (4.14). Here G_{ol} denotes an *open-loop* transfer function and G_{cl} denotes the corresponding *closed-loop* transfer function of a feedback interconnected system.

   The Nyquist plot for a given transfer function is a graphical representation of the transfer function evaluated along a closed contour, which we denote by Γ, in $\mathbb{C}$ that traverses the imaginary axis and a semicircle of infinite radius, as shown in Figure 4.7. In other words, the contour Γ encloses the closed right-half plane in $\mathbb{C}$. Note that, as long as the transfer function of interest is proper (i.e., has either the same number of zeros and poles or fewer zeros than poles), the arc at infinity plays essentially no role in the evaluation.

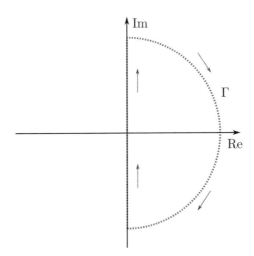

Figure 4.7: Closed contour Γ representing a semicircle of infinite radius.

As per Lemma 4.14, system (4.26) is BIBO stable if and only if G_{cl} does not have any poles in $\overline{\mathbb{C}}_+$. From (4.26), we see that the zeros of $1 + G_{ol}(s)k$ are the poles of $G_{cl}(s)$. It is clear that for stability of the closed-loop system we require that $1 + G_{ol}(j\omega)k \neq 0$ or, equivalently, $G_{ol}(j\omega) \neq -1/k$. Furthermore, we require that $1 + G_{ol}(s)k$ have no zeros in the closed right-half complex plane.

In order to relate this latter requirement to the Nyquist plot, we will use Cauchy's Argument Principle, which relates a particular contour integral of a meromorphic function $g : \mathbb{C} \to \mathbb{C}$, the number of zeros and poles of $g(\cdot)$ within the contour, and the winding number as

$$w_n = -\frac{1}{2\pi j} \oint_\Gamma \frac{g'(s)}{g(s)} ds = Z - P, \qquad (4.27)$$

where w_n is the winding number and Z and P are the number of zeros and poles, respectively, of $g(\cdot)$ contained within the contour Γ. Using the right-hand equality, BIBO stability requires $Z = 0$ so that

$$P = \frac{1}{2\pi j} \oint_\Gamma \frac{G'_{ol}(s)G_{ol}(s)k}{1 + G_{ol}(s)G_{ol}(s)k} ds. \qquad (4.28)$$

Observing that the poles of $G_{ol}(s)$ and $1 + G_{ol}(s)k$ are the same, we can determine P immediately given the open-loop transfer function.

Turning now to the contour integral in (4.28), we will make two successive changes of variables. First, take $T_1(s) = 1 + G_{ol}(s)k$. Then $dT_1(s) = G'_{ol}(s)kds$ and

$$-\frac{1}{2\pi j} \oint_\Gamma \frac{G'_{ol}(s)k}{1 + G_{ol}(s)k} ds = -\frac{1}{2\pi j} \oint_{T_1(\Gamma)} \frac{dT_1}{T_1}.$$

Next, take $T_2(T_1) = (T_1 - 1)/k$ so that

$$-\frac{1}{2\pi j} \oint_\Gamma \frac{G'_{ol}(s)k}{1 + G_{ol}(s)k} ds = -\frac{1}{2\pi j} \oint_{T_2(T_1(\Gamma))} \frac{dT_2}{T_2 + \frac{1}{k}}.$$

Finally, observe that $T_2(T_1(\Gamma)) = \frac{1 + G_{ol}(\Gamma)k - 1}{k} = G_{ol}(\Gamma)$, allowing us to write

$$w_n = -\frac{1}{2\pi j} \oint_\Gamma \frac{G'_{ol}(s)k}{1 + G_{ol}(s)k} ds = -\frac{1}{2\pi j} \oint_{G_{ol}(\Gamma)} \frac{dT_2}{T_2 + \frac{1}{k}}.$$

Note that the latter contour is exactly the Nyquist plot and hence w_n is the number of times that $G_{ol}(\Gamma)$ encircles the point $-1/k$. In other words, the relation (4.28) requires that the number of poles of the open-loop transfer function match the number of encirclements of the point $-1/k$ by $G_{ol}(\Gamma)$.

Theorem 4.17 (Nyquist Criterion). *Consider the single-input single-output closed-loop system* (4.26). *Let $P \in \mathbb{N}$ denote the number of poles of G_{ol} in $\mathbb{C}^+$. Moreover, assume that G_{ol} does not have any poles in $j\mathbb{R}$. Then the system* (4.26) *is BIBO stable if and only if $G_{ol}(jw)$, $w \in [-\infty, \infty]$, encircles $-1/k \in \mathbb{C}$ exactly $-P$ times clockwise.*

As an example of the Nyquist criterion we consider again the pendulum (1.45) but this time intend to stabilize the pendulum in the upright position $[\theta, \dot\theta]^T = [0, 0]^T$. In this case the linearization at the equilibrium is given by the single-input single-output system

$$\begin{bmatrix} \dot{x}_1 \\ \dot{x}_2 \end{bmatrix} = \begin{bmatrix} 0 & 1 \\ \frac{mg\ell}{J+m\ell^2} & -\frac{\gamma}{J+m\ell^2} \end{bmatrix} \begin{bmatrix} x_1 \\ x_2 \end{bmatrix} + \begin{bmatrix} 0 \\ \frac{\ell}{J+m\ell^2} \end{bmatrix} u,$$
$$y = \begin{bmatrix} 1 & 0 \end{bmatrix} x \tag{4.29}$$

with the transfer function

$$G_{ol}(s) = \frac{1}{s^2 + 0.1s - 9.81} \tag{4.30}$$

for $m = \ell = 1$, $J = 0$, $g = 9.81$, and $\gamma = 0.1$, which we have already analyzed in Example 4.15. From Example 4.15 we additionally know that

$$G_{cl}(s) = \frac{1}{s^2 + 0.1s - 9.81 + k}. \tag{4.31}$$

The roots of G_{ol} are given by $\lambda_1 = -3.18$ and $\lambda_2 = 3.08$ and thus $P = 1$ in Theorem 4.17. In Figure 4.8 the Nyquist plot of G_{ol} and $-\frac{1}{k}$ are visualized for different values of k. For $k < 9.81$ the graph of $G_{ol}(j\omega)$ encircles the point $-\frac{1}{k}$ zero times and for $k > 9.81$ the graph encircles the point $-\frac{1}{k}$ exactly -1 time clockwise. Thus from Theorem 4.17 BIBO stability of the closed-loop can be concluded for $k > 9.81$.

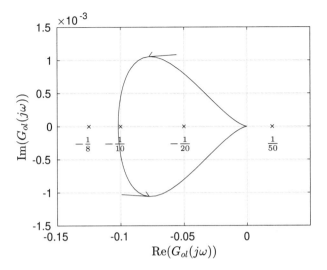

Figure 4.8: Nyquist plot of (4.30) and different $k \in \{8, 10, 20, -50\}$.

4.3 EXERCISES

Exercise 4.1. Consider the third-order differential equation

$$\dddot{y} + \ddot{y} + \dot{y} = u. \tag{4.32}$$

1. Rewrite the third-order differential equation into a first-order linear system

$$\dot{x} = Ax + bu, \qquad y = cx.$$

2. Compute the transfer function $G(s) = c(sI - A)^{-1}b$.
3. Compute the transfer function by computing the Laplace transform of (4.32) directly.

Exercise 4.2. Consider the linear system defined through the triple (A, b, c) defined as

$$A = \begin{bmatrix} 3 & 0 & 0 & 0 \\ 3 & \frac{3}{2} & -\frac{3}{2} & 1 \\ 1 & -\frac{1}{2} & \frac{5}{2} & -1 \\ 2 & -\frac{1}{2} & -\frac{3}{2} & 3 \end{bmatrix}, \quad b = \begin{bmatrix} 1 \\ 1 \\ 1 \\ 1 \end{bmatrix}, \quad c = \begin{bmatrix} 1 & -1 & 1 & -1 \end{bmatrix}.$$

Use MATLAB and `ss2tf.m` to compute a transfer function $G(s)$. Use `tf2ss.m` to compute a minimal realization $(\tilde{A}, \tilde{b}, \tilde{c})$. Show that there exists a matrix $T \in \mathbb{R}^{4 \times 4}$ such that $A = T\tilde{A}T^{-1}$.

Exercise 4.3. Consider the transfer function

$$G_{ol}(s) = \frac{s^2 + 2s - 1}{-s^3 - 2s^2 + 3s - 8}$$

Use Theorem 4.17 to derive a $k \in \mathbb{R}$ such that (4.26) is BIBO stable. Verify that (4.26) is BIBO stable by calculating the poles of $G_{cl}(s)$. Visualize the Bode diagram of $G_{cl}(s)$.

Hint: In MATLAB the Bode diagram and the Nyquist plot are generated through `bode.m` and `nyquist.m`, respectively.

Exercise 4.4. Compute the transfer function of the differential equation

$$u = k_p y + k_d \dot{y} + k_i \int_0^{\cdot} y$$

corresponding to a PID controller (i.e., under the assumption that initial conditions are zero compute a relation $\hat{u}(s) = G(s)\hat{y}(s)$).

4.4 BIBLIOGRAPHICAL NOTES AND FURTHER READING

While the state space methods of previous chapters have their origins in the Russian literature from the late 1800s, frequency domain analysis dates to the 1930s and work done at Bell Laboratories designing and analyzing amplifiers for long distance telephony [112, 24, 26] (these papers, and other important papers in the field of control, can be found with contextual commentary in [16]).

Frequency domain methods have been part of the standard curriculum for control system design for forty-plus years and detailed development of Nyquist and Bode plots can be found in most undergraduate control texts, such as [45, 51, 114]. Excellent presentations of these topics can also be found in [12] and [164].

Numerous texts covering Laplace transforms are available, including [115] and [61]. Tables of Laplace transforms and properties are also readily available on the Web.

Chapter Five

Discrete Time Systems

So far we have focused on systems modeled by ordinary differential equations which evolve in *continuous time*; that is, states are functions of time $t \in \mathbb{R}_{\geq 0}$. Such models can be used for a wide range of systems and will be the primary focus of this text. However, modern feedback controllers are often implemented digitally via some form of digital computing element. In this case, a model which evolves in *discrete time* can be useful; i.e., states and inputs are functions of time $k \in \mathbb{N}$.

For many systems and feedback designs this distinction is not terribly important. If states are sampled and control actions are computed very quickly, then the system likely behaves much like a continuous time system and the discrete time nature of the controller can be essentially treated as a continuous time element. However, this is not always the case and sometimes it is necessary to explicitly deal with discrete time systems. Indeed, when it comes to real-time numerical optimal control such as model predictive control presented in Chapter 15, the discrete time setting is crucial.

5.1 DISCRETE TIME SYSTEMS—FUNDAMENTALS

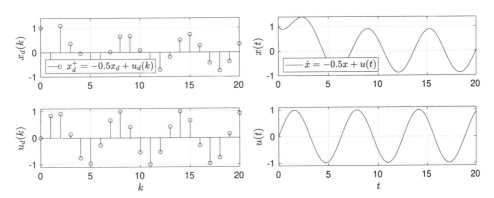

Figure 5.1: Evolution of a discrete time system (left) and a continuous time system (right) in $k \in \mathbb{N}$ and $t \in \mathbb{R}_{\geq 0}$, respectively.

The *discrete time* counterpart to the continuous time system with output (1.5), (1.13) is defined as

$$x_d(k+1) = F(x_d(k), u_d(k)), \quad x_d(0) = x_{d,0} \in \mathbb{R}^n \tag{5.1}$$

$$y_d(k) = H(x_d(k), u_d(k)), \tag{5.2}$$

with right-hand side $F : \mathbb{R}^n \times \mathbb{R}^m \to \mathbb{R}^n$, $H : \mathbb{R}^n \times \mathbb{R}^m \to \mathbb{R}^p$. Instead of the continuous time $t \in \mathbb{R}_{\geq 0}$, (5.1)–(5.2) evolves in discrete time $k \in \mathbb{N}$. In particular, $x_d : \mathbb{N} \to \mathbb{R}^n$, $u_d : \mathbb{N} \to \mathbb{R}^m$, and $y_d : \mathbb{N} \to \mathbb{R}^p$, where we use the subscript $\cdot_d$ to highlight quantities evolving in discrete time as opposed to continuous time.

While (5.1) is time invariant, *time-varying discrete time systems*

$$x_d(k+1) = F(k, x_d(k)), \quad x_d(k_0) = x_{d,0} \in \mathbb{R}^n, \ k \geq k_0 \geq 0, \tag{5.3}$$

where $F : \mathbb{N} \times \mathbb{R}^n \to \mathbb{R}^n$, can be defined in the same way as their continuous time counterparts. In this chapter we will concentrate on time-invariant discrete time systems with input as in (5.1), and without input as given by

$$x_d(k+1) = F(x_d(k)), \quad x_d(0) = x_{d,0} \in \mathbb{R}^n, \tag{5.4}$$

where $F : \mathbb{R}^n \to \mathbb{R}^n$. Instead of *ordinary differential equations*, discrete time systems are described by *difference equations*. Here,

$$x_d^+ = F(x_d, u_d)$$

is a common shorthand notation for difference equations capturing the dynamics (5.1). The evolution of a discrete time system together with the evolution of a continuous time system is visualized in Figure 5.1.

Many results and definitions for continuous time systems can be directly carried over to discrete time systems in an obvious way. However, after studying this chapter, the reader should be aware of slight differences between ordinary differential equations and difference equations.

Definition 5.1 (Equilibrium). *The point $x_d^e \in \mathbb{R}^n$ is called an equilibrium of the system (5.4) if $x_d^e = F(x_d^e)$ is satisfied. If $x_d^e = F(k, x_d^e)$ for all $k \in \mathbb{N}$ then x^e is an equilibrium of the time-varying system (5.3). Finally, the pair $(x_d^e, u_d^e) \in \mathbb{R}^n \times \mathbb{R}^m$ is called an equilibrium pair of the system (5.1) if $x_d^e = F(x_d^e, u_d^e)$ holds.*

Thus, while a constant solution in continuous time satisfies $\dot{x} = 0$, a constant solution in discrete time satisfies $x_d^+ = F(x_d)$. Nevertheless, equilibria (and equilibrium pairs) in discrete time can be shifted to the origin in the same way as in the continuous time case (see Definition 1.3 and the subsequent discussion).

5.2 SAMPLING: FROM CONTINUOUS TO DISCRETE TIME

While the laws of motion in general lead to continuous time systems, computers operate in discrete time. Even though plants are described through differential equations, they are usually numerically solved as difference equations and thus continuous time systems frequently are approximated by a discrete time representation.

Recall the definition of the derivative of a continuously differentiable function:

$$\tfrac{d}{dt} x(t) = \lim_{\Delta \to 0} \frac{x(t + \Delta) - x(t)}{\Delta}.$$

For small $\Delta > 0$, the difference quotient

$$\tfrac{d}{dt}x(t) \approx \frac{x(t+\Delta) - x(t)}{\Delta} \tag{5.5}$$

can be used as an approximation of the derivative. Using the definition of the right-hand side of the continuous time system

$$\dot{x}(t) = f(x(t), u(t)) \tag{5.6}$$

the difference quotient satisfies

$$\frac{x(t+\Delta) - x(t)}{\Delta} \approx f(x(t), u(t)).$$

Multiplying by $\Delta > 0$ and adding $x(t)$ on both sides allows us to write

$$x(t+\Delta) \approx x(t) + \Delta f(x(t), u(t)), \tag{5.7}$$

which is close to the definition of the discrete time system (5.1). Indeed, by identifying t with $k \cdot \Delta$ and

$$F(x_d, u_d) \doteq x_d + \Delta f(x_d, u_d), \tag{5.8}$$

the continuous time system (5.6) can be approximated by the discrete time system

$$x_d^+ = x_d(k+1) = F(x_d(k), u_d(k)).$$

The approximation (5.7) is known as *Euler's method*.

For continuous time systems the state $x(\cdot)$ and the input $u(\cdot)$ are functions $x : \mathbb{R}_{\geq 0} \to \mathbb{R}^n$ and $u : \mathbb{R}_{\geq 0} \to \mathbb{R}^m$. In discrete time, $x_d(\cdot)$ and $u_d(\cdot)$ are functions $x_d : \mathbb{N} \to \mathbb{R}^n$ and $u_d : \mathbb{N} \to \mathbb{R}^m$. One way to close the gap between continuous time and discrete time is to use a *zero-order hold*, i.e., to restrict x and u to piecewise constant functions

$$x_d(k) = x(\Delta k) = x(t + \Delta k),$$
$$u_d(k) = u(\Delta k) = u(t + \Delta k)$$

for all $k \in \mathbb{N}$, for all $t \in [0, \Delta)$, and $\Delta > 0$. A piecewise constant input satisfying $u(\Delta k) = u(t + \Delta k)$, $k \in \mathbb{N}$, for all $t \in [0, \Delta)$ is called a *sample-and-hold* input with *sampling rate* $\Delta > 0$. If a digital controller is used to control a plant, it is common to apply a piecewise constant sample-and-hold input to a continuous time system. In Figure 5.2 solutions of a dynamical system with respect to a sample-and-hold input (with a large sampling time $\Delta = 1$ for illustration) and with respect to a continuous input are shown.

5.2.1 Discretization of Linear Systems

The general time discretization for nonlinear systems can also be applied to linear systems

$$\dot{x}(t) = Ax(t) + Bu(t) \tag{5.9}$$

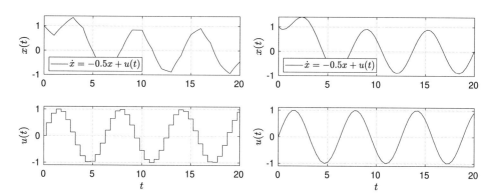

Figure 5.2: Solution of a continuous time system with a sample-and-hold input with sampling time $\Delta = 1$ (left) and with a continuous input (right).

with $A \in \mathbb{R}^{n \times n}$ and $B \in \mathbb{R}^{n \times m}$. Here, (5.7) takes the form

$$x(t + \Delta) \approx x(t) + \Delta(Ax(t) + Bu(t)) = (I + \Delta A)x(t) + \Delta Bu(t),$$

which leads us to consider the linear discrete time system

$$x_d(k + 1) = A_d x_d(k) + B_d u_d(k), \qquad x_d^+ = A_d x_d + B_d u_d, \qquad (5.10)$$

where A_d and B_d are defined as

$$A_d \doteq (I + \Delta A) \qquad \text{and} \qquad B_d \doteq \Delta B. \qquad (5.11)$$

As in the general case, (5.10) is an approximation of the linear continuous time system (5.9).

For linear systems we can consider an alternative discretization which is exact (i.e., not an approximation) under the assumption that $u(\cdot)$ is a piecewise constant (sample-and-hold) function. Recalling (3.51), the solution of the linear system (5.9) is given by

$$x(t + \Delta) = e^{A_c \Delta} x(t) + \int_0^\Delta e^{A_c(\Delta - \tau)} B_c u(t + \tau) d\tau.$$

If $u(\cdot)$ is constant on the interval $\tau \in [t, t+\Delta)$, i.e., $u(t+\tau) = u(t)$ for all $\tau \in [0, \Delta)$, the solution satisfies

$$x(t + \Delta) = e^{A_c \Delta} x(t) + \int_0^\Delta e^{A_c(\Delta - \tau)} d\tau B_c u(t).$$

We can define the matrices

$$A_{de} \doteq e^{A_c \Delta} \qquad \text{and} \qquad B_{de} \doteq \int_0^\Delta e^{A_c(\Delta - \tau)} d\tau B_c \qquad (5.12)$$

to obtain a second linear discrete time system

$$x_d(k + 1) = A_{de} x_d(k) + B_{de} u_d(k), \qquad x_d^+ = A_{de} x_d + B_{de} u_d. \qquad (5.13)$$

In contrast to the approximate discretization (5.10) which satisfies

$$x(k\Delta) \approx x_d(k), \qquad \text{for all } k \in \mathbb{N},$$

the discretization (5.13) satisfies

$$x(k\Delta) = x_d(k), \qquad \text{for all } k \in \mathbb{N},$$

if $u(t + \Delta k) = u(\Delta k) = u_d(k)$ for all $t \in [0, \Delta)$, for all $k \in \mathbb{N}$.

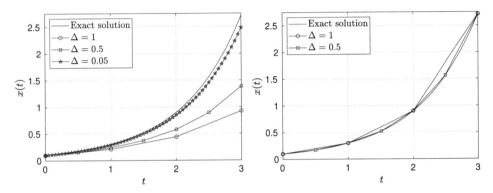

Figure 5.3: Approximation of the solution of $\dot{x} = 1.1x$ using (5.8) (left) and (5.13) (right) for different Δ.

Figure 5.3 shows the differences between the discretization (5.8) and (5.13) based on the example of a linear system without input. Additionally, it illustrates the impact of the sampling time Δ, which is important for (5.8) but not for (5.13). Specifically, as the sampling time decreases, the approximate discretization (5.8) approaches the underlying continuous time solution. By contrast, the discretization (5.13) exactly matches the underlying continuous time solution at the discrete time steps.

5.2.2 Higher Order Discretization Schemes

The Euler method derived in Equation (5.7) is a simple and straightforward discretization which is sufficient in many applications if a small sampling time $\Delta > 0$ is selected. However, if necessary, also more accurate discrete time approximations of nonlinear continuous time systems can be derived. Consider the continuous time system

$$\dot{x}(t) = f(x(t), u(t)) \tag{5.14}$$

where $f : \mathbb{R}^n \times \mathbb{R}^m \to \mathbb{R}^n$ is $r + 1$, $r \in \mathbb{N}$, times continuously differentiable. Since

$$\frac{d^{i+1}}{dt^{i+1}} x(t) = \frac{d^i}{dt^i} f(x(t), u(t)), \qquad i = 1, \ldots, r,$$

also the solution $x(t)$ is $r+1$ times continuously differentiable if the input signal is sufficiently smooth. Using a Taylor approximation, $x(t)$ can be written as

$$x(t+\Delta) = x(t) + \dot{x}(t)\Delta + \frac{1}{2!}\ddot{x}(t)\Delta^2 + \cdots + \frac{1}{r!}\frac{d^r}{dt^r}x(t)\Delta^r + R_r(\Delta)$$

and the remainder term is given by

$$R_r(\Delta) = \frac{1}{(r+1)!}\frac{d^{r+1}}{dt^{r+1}}x(\tau)\Delta^{r+1}$$

for $\tau \in [t, t+\Delta]$.

By ignoring the remainder in the case $r = 1$ and by identifying $\frac{d}{dt}x(t) = f(x(t))$, equation (5.7) is obtained. Since in this case the remainder depends on Δ^2, it can be expected that the error between the true state and the approximation converges at least quadratically, i.e.,

$$\frac{1}{\Delta^2}R_1(\Delta) = \frac{1}{2}\frac{d^2}{dt^2}x(\tau(\Delta)) \to c \in \mathbb{R},$$

$\tau(\Delta) \in [t, t+\Delta]$, for $\Delta \to 0$.

To obtain a faster convergence, also higher order terms of the Taylor polynomial can be taken into account. The remainder of the approximation

$$x(t+\Delta) \approx x(t) + \dot{x}(t)\Delta + \frac{1}{2}\ddot{x}(t)\Delta^2 \tag{5.15}$$

depends on Δ^3 instead of Δ^2, for example. Using the identity (5.14) together with its time derivative, i.e.,

$$\begin{aligned}
\ddot{x} = \frac{df}{dt}(x,u) &= \frac{\partial}{\partial x}f(x,u)\dot{x} + \frac{\partial}{\partial u}f(x,u)\dot{u}, \\
&= \frac{\partial}{\partial x}f(x,u)f(x,u) + \frac{\partial}{\partial u}f(x,u)\dot{u},
\end{aligned} \tag{5.16}$$

allows us to rewrite (5.15)

$$\begin{aligned}
x(t+\Delta) \approx {}&x(t) + \Delta f(x(t), u(t)) \\
&+ \frac{\Delta^2}{2}\left(\frac{\partial}{\partial x}f(x(t),u(t))f(x(t),u(t)) + \frac{\partial}{\partial u}f(x(t),u(t))\dot{u}(t)\right).
\end{aligned}$$

In this way, approximations of arbitrary order can be derived by considering more and more terms of the Taylor polynomial.

If we additionally assume that the input is constant $u(t+\delta) = u_d \in \mathbb{R}^m$ for all $\delta \in [0, \Delta)$, as we have done in the previous section, then the update of $x(t+\Delta)$ simplifies to

$$x(t+\Delta) \approx x(t) + \Delta f(x(t), u_d) + \frac{\Delta^2}{2}\frac{\partial}{\partial x}f(x(t), u_d)f(x(t), u_d). \tag{5.17}$$

To avoid the necessity to derive the derivative of f in (5.16), a Taylor expansion of f can be used instead of the derivative. In particular, it holds that

$$f(x+\Delta\dot{x}, u_d) = f(x, u_d) + \frac{\partial f}{\partial x}(x, u_d)\dot{x}\Delta + \frac{1}{2}\frac{d^2f}{d\Delta^2}(x+\delta\dot{x}, u_d)\Delta^2$$

for a $\delta \in [0, \Delta]$. Rearranging the terms, an expression for the time derivative of f is obtained:

$$\Delta\frac{\partial f}{\partial x}(x, u_d)f(x, u_d) = f(x+\Delta f(x, u_d), u_d) - f(x, u_d) - \frac{1}{2}\frac{d^2f}{d\Delta^2}(x+\delta\dot{x}, u_d)\Delta^2.$$

If we use this expression in the approximation (5.17) it holds that

$$x(t + \Delta) \approx x(t) + \Delta f(x(t), u_d) + \tfrac{\Delta}{2} \left(f(x(t) + \Delta f(x(t), u_d), u_d) - f(x(t), u_d) \right)$$
$$+ \tfrac{\Delta}{2} \left(-\tfrac{1}{2} \tfrac{d^2 f}{d\Delta^2} (x(t) + \delta \dot{x}, u_d) \Delta^2 \right)$$
$$= x(t) + \tfrac{\Delta}{2} f(x(t), u_d) + \tfrac{\Delta}{2} f(x(t) + \Delta f(x(t), u_d), u_d)$$
$$- \tfrac{1}{4} \tfrac{d^2 f}{d\Delta^2} (x(t) + \delta \dot{x}, u_d) \Delta^3. \tag{5.18}$$

Since we have already neglected terms of order Δ^3 in the approximation (5.17), we can also neglect the last term in the expression (5.18) without changing the order of accuracy of the approximation, which leaves us with the update formula

$$x(t + \Delta) \approx x(t) + \tfrac{\Delta}{2} f(x(t), u_d) + \tfrac{\Delta}{2} f(x(t) + \Delta f(x(t), u_d), u_d). \tag{5.19}$$

Following the notation in (5.8), we can define the function

$$F(x_d, u_d) = x_d + \tfrac{\Delta}{2} f(x_d, u_d) + \tfrac{\Delta}{2} f(x_d + \Delta f(x_d, u_d), u_d) \tag{5.20}$$

to obtain the discrete time system $x_d^+ = F(x_d, u_d)$ corresponding to this discretization scheme.

The iterative method (5.19) is known as *Heun's method*. The right-hand side of (5.19) is independent of the time derivative of f. Compared with (5.7) the update in (5.19) depends not only on $x(t)$ but also on $x(t) + \Delta \dot{x}(t)$, or more precisely, an approximation of the state at time $x(t + \Delta)$.

Figure 5.4 compares the Euler method (5.8) with Heun's method (5.20) on the example of a linear system without input. While Heun's method clearly outperforms

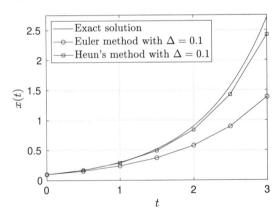

Figure 5.4: Comparison of the discretization method (5.8) and (5.20) applied to the dynamics $\dot{x} = 1.1x$.

the Euler discretization if the same sampling time is used, one has to keep in mind that the right-hand side of (5.20) is more complicated than the right-hand side in (5.8).

The Euler method and Heun's method are representatives of the wider class of *Runge-Kutta methods* used for the numerical solution of ordinary differential equations

$$\dot{x} = g(t, x). \tag{5.21}$$

Runge-Kutta methods define $x(t + \Delta)$ based on $x(t)$ through the formula,

$$x(t + \Delta) = x(t) + \Delta \sum_{i=1}^{s} b_i k_i \tag{5.22a}$$

where

$$
\begin{aligned}
k_1 &= g(t, x(t)) \\
k_2 &= g(t + c_2 \Delta, x + \Delta(a_{21} k_1)) \\
k_3 &= g(t + c_3 \Delta, x + \Delta(a_{31} k_1 + a_{32} k_2)) \\
&\vdots \\
k_s &= g(t + c_s \Delta, x + \Delta(a_{s1} k_1 + a_{s2} k_2 + \cdots + a_{s(s-1)} k(s))),
\end{aligned} \tag{5.22b}
$$

$s \in \mathbb{N}$ denotes the *stage*, and $a_{ij}, b_\ell, c_i \in \mathbb{R}$, $1 \leq j < i \leq s$, $1 \leq \ell \leq s$ are given parameters. A particular method is usually summarized in a so-called *Butcher tableau*:

$$
\begin{array}{c|ccccc}
0 & & & & & \\
c_2 & a_{21} & & & & \\
c_3 & a_{31} & a_{32} & & & \\
\vdots & \vdots & & \ddots & & \\
c_s & a_{s1} & a_{s2} & \cdots & a_{s(s-1)} & \\
\hline
& b_1 & b_2 & \cdots & b_{s-1} & b_s
\end{array}
$$

The parameters c_i are only necessary in the case of time-varying systems (5.21).

The Butcher tableaus corresponding to the Euler method and Heun's method are defined as

$$
\begin{array}{c|c}
0 & \\
\hline
& 1
\end{array}
\quad \text{and} \quad
\begin{array}{c|cc}
0 & & \\
1 & 1 & \\
\hline
& \frac{1}{2} & \frac{1}{2}
\end{array}
\tag{5.23}
$$

respectively.

In terms of (5.22) the update $x(t + \Delta)$ in (5.19) is calculated in three steps:

$$
\begin{aligned}
k_1 &= f(x(t), u_d), \\
k_2 &= f(x(t) + \Delta k_1, u_d), \\
x(t + \Delta) &= x(t) + \Delta \left(\tfrac{1}{2} k_1 + \tfrac{1}{2} k_2 \right).
\end{aligned}
$$

The function `ode23.m`, a numerical solver for ordinary differential equations in

MATLAB, uses the Runge-Kutta methods encoded in the Butcher tableaus

$$
\begin{array}{c|cccc}
0 \\
\frac{1}{2} & \frac{1}{2} \\
\frac{3}{4} & 0 & \frac{3}{4} \\
\hline
& \frac{2}{9} & \frac{1}{3} & \frac{4}{9}
\end{array}
\qquad \text{and} \qquad
\begin{array}{c|cccc}
0 \\
\frac{1}{2} & \frac{1}{2} \\
\frac{3}{4} & 0 & \frac{3}{4} \\
1 & \frac{2}{9} & \frac{1}{3} & \frac{4}{9} \\
\hline
& \frac{7}{24} & \frac{1}{4} & \frac{1}{3} & \frac{1}{8}
\end{array}
$$

One scheme is used to approximate the state $x(t+\Delta)$. The second scheme is needed to approximate the error introduced through the discretization based on which the step size Δ is selected. Similarly, `ode45.m` uses the schemes encoded in the tableau

$$
\begin{array}{c|ccccccc}
0 \\
\frac{1}{5} & \frac{1}{5} \\
\frac{3}{10} & \frac{3}{40} & \frac{9}{40} \\
\frac{4}{5} & \frac{44}{45} & -\frac{56}{15} & \frac{32}{9} \\
\frac{8}{9} & \frac{19372}{6561} & -\frac{25360}{2187} & \frac{64448}{6561} & -\frac{212}{729} \\
1 & \frac{9017}{3168} & -\frac{355}{33} & \frac{46732}{5247} & \frac{49}{176} & -\frac{5103}{18656} \\
1 & \frac{35}{384} & 0 & \frac{500}{1113} & \frac{125}{192} & -\frac{2187}{6784} & \frac{11}{84} \\
\hline
& \frac{35}{384} & 0 & \frac{500}{1113} & \frac{125}{192} & -\frac{2187}{6784} & \frac{11}{84} & 0
\end{array}
\tag{5.24}
$$

and and also the tableau (5.24) where the last row is replaced by

$$
\begin{array}{c|ccccccc}
& \frac{5179}{57600} & 0 & \frac{7571}{16695} & \frac{393}{640} & -\frac{92097}{339200} & \frac{187}{2100} & \frac{1}{40}
\end{array}
$$

The ordinary differential equation (5.21) covers the case of the nonlinear system (5.14) in the case of piecewise constant sample-and-hold input signals where $u(t + \delta) = u_d \in \mathbb{R}^m$ for all $\delta \in [0, \Delta)$ and

$$
g(t, x(t)) = f(x(t), u_d).
$$

5.3   STABILITY NOTIONS

The different stability properties studied in the context of continuous time systems in Section 2.1 are essentially the same in the discrete time setting, with the exception of exponential stability, as indicated below.

Note that the subscript $\cdot_d$ was useful in the previous section to differentiate between continuous time quantities and their (sometimes approximate) discretizations. In the remainder of this chapter we drop the subscript $\cdot_d$ if the system is not directly related to a continuous time system.

Definition 5.2. *Consider the origin of the discrete time system (5.4).*

1. *(Stability) The origin is* Lyapunov stable *(or simply* stable*) if, for any $\varepsilon > 0$ there exists $\delta > 0$ (possibly dependent on ε) such that if $|x(0)| \le \delta$ then, for all $k \ge 0$,*

$$
|x(k)| \le \varepsilon.
\tag{5.25}
$$

2. *(Instability) The origin is* unstable *for system (5.4) if it is not stable.*
3. *(Attractivity) The origin is* attractive *if there exists $\delta > 0$ such that if $|x(0)| < \delta$*

then

$$\lim_{k \to \infty} x(k) = 0.$$

4. (Asymptotic stability) The origin is asymptotically stable *if it is both stable and attractive.*

As in the continuous time setting, stability properties can alternatively be expressed through comparison functions. For example, the origin of (5.4) is globally asymptotically stable, or alternatively $\mathcal{KL}$-stable, if there exists $\beta \in \mathcal{KL}$ such that

$$|x(k)| \leq \beta(|x(0)|, k), \qquad \forall\, k \in \mathbb{N},$$

is satisfied for all $x(0) \in \mathbb{R}^n$.

Definition 5.3. *Consider the origin of the discrete time system* (5.4). *If there exist $M > 0$ and $\gamma \in (0, 1)$ such that for each $x(0) \in \mathbb{R}^n$ the inequality*

$$|x(k)| \leq M|x(0)|\gamma^k, \qquad \forall\, k \in \mathbb{N},$$

is satisfied, then the origin is globally exponentially stable.

Note the difference between the exponential decay in discrete time, which is described by the decaying sequence γ^k, $\gamma \in (0, 1)$, $k \in \mathbb{N}$, and the continuous time exponential decay described by $e^{-\lambda t}$, for $\lambda, t > 0$ in Definition 2.7.

5.3.1   Lyapunov Characterizations

The idea of Lyapunov's second method, as described in Section 2.3, can be applied in discrete time in a natural way, either by replacing the time derivative with a finite difference or by directly interpreting the idea that the abstract energy embodied in the Lyapunov function should not increase under the influence of the difference equation. The following definitions take the former view, but both ideas lead to the same decrease condition.

As in Section 2.3, we consider a domain $\mathcal{D} \subset \mathbb{R}^n$ where we assume $\mathcal{D}$ is open and contains the origin.

Theorem 5.4 (Lyapunov stability theorem [1, Theorem 5.9.1]). *Given* (5.4) *with $f(0) = 0$, suppose there exists a continuous function $V : \mathcal{D} \to \mathbb{R}_{\geq 0}$ satisfying $V(0) = 0$ and a function $\alpha_1 \in \mathcal{K}$ such that, for all $x \in \mathcal{D}$,*

$$\alpha_1(|x|) \leq V(x) \tag{5.26}$$

and

$$V(f(x)) - V(x) \leq 0. \tag{5.27}$$

Then the origin is stable.

As described above, we note that (5.27) can be arrived at by replacing the derivative in (2.18) with a finite difference. Alternatively, the idea that the Lyapunov function should not increase along solutions can be easily seen to correspond to the requirement that

$$V(x^+) = V(f(x)) \leq V(x),$$

which is just another way of writing (5.27). The decrease conditions below can be rewritten in a similar manner.

Note that differentiability of V is not required in discrete time since the decrease condition (5.27) depends only on the difference. In fact, continuity is not even required. Nonetheless, continuity of V is a desirable property. For example, it provides a certain robustness of the stability property. Hence, in our statements here we make the assumption that the Lyapunov function is continuous.

Theorem 5.5 (Lyapunov asymptotic stability theorem [1, Theorem 5.9.2]). *Given (5.4) with $f(0) = 0$, suppose there exists a continuous function $V : \mathcal{D} \to \mathbb{R}_{\geq 0}$ satisfying $V(0) = 0$, a function $\alpha_1 \in \mathcal{K}$, and a positive definite function $\rho \in \mathcal{P}$ satisfying $\rho(s) < s$ for all $s > 0$, such that, for all $x \in \mathcal{D}$, (5.26) holds and*

$$V(f(x)) - V(x) \leq -\rho(V(x)). \tag{5.28}$$

Then the origin is asymptotically stable.

Theorem 5.6 (Lyapunov exponential stability theorem). *Given (5.4) with $f(0) = 0$, suppose there exists a continuous function $V : \mathcal{D} \to \mathbb{R}_{\geq 0}$ and constants $\lambda_1, \lambda_2 > 0$, $p \geq 1$, and $c \in (0,1)$ such that, for all $x \in \mathcal{D}$*

$$\lambda_1 |x|^p \leq V(x) \leq \lambda_2 |x|^p \tag{5.29}$$

and

$$V(f(x)) - V(x) \leq -cV(x). \tag{5.30}$$

Then the origin is exponentially stable.

Proof. First rewrite (5.30) as

$$V(f(x)) \leq (1 - c)V(x) \tag{5.31}$$

and define $\gamma \doteq (1 - c)^{1/p}$. Note that $c \in (0,1)$ implies $\gamma \in (0,1)$. Then (5.31) implies that

$$V(x(k)) \leq \gamma^{kp} V(x(0))$$

and the bounds (5.29) imply

$$|x(k)| \leq \left(\frac{\lambda_2}{\lambda_1}\right)^{1/p} |x(0)| \gamma^k.$$

$\square$

For time-varying discrete time systems (5.3), the available results largely follow as above. We limit ourselves to quoting one such result.

Theorem 5.7 ([108, Theorem 6.3.1(c)]). *Given the time-varying system (5.3) with*

$f(k, 0) = 0$ for all $k \geq k_0 \geq 0$, if there exist a function $V : \mathbb{N} \times \mathbb{R}^n \to \mathbb{R}_{\geq 0}$ and functions $\alpha_1, \alpha_2 \in \mathcal{K}_\infty$ and $\rho \in \mathcal{P}$ such that, for all $x \in \mathbb{R}^n$ and $k \geq k_0 \geq 0$,

$$\alpha_1(|x|) \leq V(k, x) \leq \alpha_2(|x|) \qquad and \tag{5.32}$$
$$V(k + 1, f(k, x)) - V(k, x) \leq -\rho(|x|) \tag{5.33}$$

then the origin is uniformly globally asymptotically stable.

5.3.2   Linear Systems

Consider the discrete time linear system

$$x^+ = Ax, \qquad x(0) \in \mathbb{R}^n. \tag{5.34}$$

Theorem 5.8. *For the linear system* (5.34), *the following properties are equivalent:*

1. *The origin $x^e = 0$ is exponentially stable;*
2. *The eigenvalues $\lambda_1, \ldots, \lambda_n \in \mathbb{C}$ of A satisfy $|\lambda_i| < 1$ for all $i = 1, \ldots, n$; and*
3. *For every symmetric positive definite matrix $Q \in \mathcal{S}^n_{>0}$ there exists a unique symmetric positive definite matrix $P \in \mathcal{S}^n_{>0}$ satisfying the* discrete time Lyapunov equation

$$A^T P A - P = -Q. \tag{5.35}$$

Proof. Note that the solution of the linear discrete time system is given by

$$x(k) = A^k x(0) \qquad \forall k \in \mathbb{N}.$$

Assume $|\lambda_i| < 1$ is satisfied for all eigenvalues $\lambda_1, \ldots, \lambda_n \in \mathbb{C}$ of A. Thus,

$$|x(k)| \leq |A^k x(0)| \leq \max_{i=1,\ldots,n} |\lambda_i^k| \cdot |x(0)| \leq \max_{i=1,\ldots,n} |\lambda_i|^k \cdot |x(0)|$$

for all $k \in \mathbb{N}$, which indicates that item 2. implies item 1. Similarly, if there exists $i \in \{1, \ldots, n\}$ with $|\lambda_i| \geq 1$, it follows that the origin is not exponentially stable, i.e., item 1. implies item 2.

To show that item 3. implies item 1., assume that $P \in \mathcal{S}^n_{>0}$ satisfies the Lyapunov equation (5.35) for $Q = I$. Then the function $V(x) = x^T P x$ satisfies

$$V(x^+) - V(x) = x^T A^T P A x - x^T P x = x^T (A^T P A - P)x = -x^T x,$$

from which exponential stability can be concluded. Conversely, to show that item 1. implies item 3., P can be defined as $P = \sum_{k=0}^\infty (A^k)^T Q A^k$ where the sum replaces the integral compared to the continuous time setting (compare with equation (3.6)). $\qquad \square$

A matrix A which satisfies $|\lambda_i| < 1$ for all $i = 1, \ldots, n$ is called a *Schur matrix*. Schur matrices characterize exponential stability of the origin for linear time-invariant discrete time systems in the same way that Hurwitz matrices characterize exponential stability of the origin for linear time-invariant continuous time systems.

Inferring stability of the origin for a nonlinear system from its linearization is the same in discrete time as it is in continuous time (c.f. Theorem 3.4).

Theorem 5.9 ([63, Theorem 8.7]). *Consider the discrete time system* (5.4) *with* $0 = F(0)$. *If the origin of the linear system* $z^+ = Az$ *with*

$$A = \left[\frac{\partial F}{\partial x}(x)\right]_{x=0}$$

is globally exponentially stable, then the origin of (5.4) *is locally exponentially stable.*

5.3.3 Stability Preservation of Discretized Systems

We have seen how to verify stability properties of equilibria of continuous time systems and of discrete time systems. Additionally, we have seen how to obtain a discrete time approximation of a continuous time system. Thus a natural question is whether or not stability properties of equilibria of continuous time systems are preserved in their discrete time counterparts. Unfortunately, the answer to this question is, in general, no.

The discretization scheme discussed in Section 5.2.1 and specified through equation (5.13) passes the properties of the continuous time system on to its discrete time counterpart for all $\Delta > 0$ since $x(k\Delta) = x_d(k)$ is satisfied for all $k \in \mathbb{N}$.

For general nonlinear systems, however, the preservation of stability properties depends on the discretization scheme and on the sampling rate $\Delta > 0$. We illustrate this fact on the simple (linear) continuous time system

$$\dot{x} = \lambda x \tag{5.36}$$

for $\lambda \in \mathbb{R}$ together with the Euler method (5.8), i.e., the corresponding discrete time dynamics are defined through the difference equation

$$x^+ = x + \Delta\lambda x = (1 + \Delta\lambda)x. \tag{5.37}$$

The origin of (5.36) is exponentially stable if and only if $\lambda < 0$ while the origin of (5.37) is exponentially stable if and only if $|1 + \Delta\lambda| < 1$.

Let $\lambda < 0$. Then the condition $|1 + \Delta\lambda| < 1$ is equivalent to

$$1 + \Delta\lambda < 1 \qquad \text{and} \qquad -1 - \Delta\lambda < 1,$$

which can be rewritten as upper and lower bounds on the sampling rate:

$$0 < \Delta < -\tfrac{2}{\lambda}. \tag{5.38}$$

While the lower bound of (5.38) is trivially satisfied, the upper bound is restrictive, and, in particular, Δ needs to be sufficiently small.

Surprisingly, if $\lambda \to 0$, i.e., convergence to the equilibrium is slow, the condition is not restrictive, while in the case $\lambda \ll 0$, i.e., fast convergence to the equilibrium, the condition is restrictive. For example, in the case that $\lambda = -1000$, Δ needs to satisfy $\Delta < 0.002$ just to ensure that the origin of the discrete time system (5.37) is stable. For $\lambda \ll 0$ the ordinary differential equation is called *stiff*. In general stiff ordinary differential equations are more difficult to solve numerically due to their restriction on the sampling rate.

For $\lambda > 0$ the origin of (5.36) is unstable, leading to the condition $|1 + \Delta\lambda| > 1$

if this property is supposed to be inherited by the discrete time system (5.37). The condition $|1 + \Delta\lambda| > 1$ implies that $\Delta > 0$ or $\Delta < -\frac{2}{\lambda}$ needs to be satisfied by the sampling rate. Thus, instability of the origin of (5.37) is preserved under arbitrary sampling rates $\Delta > 0$ through the Euler discretization. This is, however, not guaranteed for general continuous time systems and general discretization schemes where an unstable equilibrium might become stable in a corresponding discrete time system.

The preservation of properties of the continuous time system in its discrete time approximation is investigated in sections on *stability* in the literature on the *numerical solution of ordinary differential equations*. In this context, stability is not to be understood as a property of an equilibrium but as a property of a discretization scheme. While this topic is out of the scope of this book, the reader should have understood from the presentation so far that properties of a continuous time system are not inherited automatically by its discrete time counterpart obtained through a numerical discretization. For further reading we refer to the references in Section 5.6 at the end of this chapter.

5.4   CONTROLLABILITY AND OBSERVABILITY

The definitions and the tests for controllability and observability for continuous time systems and for discrete time systems are essentially the same. However, as with stability, there are slight differences. Consider the linear system

$$
\begin{aligned}
x^+ &= Ax + Bu, \\
y &= Cx + Du.
\end{aligned}
\tag{5.39}
$$

Definition 5.10 (Controllability). *Consider the linear system* (5.39) *uniquely defined through the pair* (A, B) *(i.e., ignoring the output equation). The linear system (or equivalently the pair* (A, B)*) is said to be controllable if, for all* $x_1, x_2 \in \mathbb{R}^n$, *there exists* $K \in \mathbb{N}$ *and* $u : \mathbb{N}_0 \to \mathbb{R}^m$ *such that*

$$
x_2 = A^K x_1 + \sum_{i=1}^{K} A^{K-i} Bu(i - 1).
\tag{5.40}
$$

Definition 5.11 (Observability). *Consider the linear system* (5.39) *defined through the pair* (A, C) *(i.e., ignoring the input matrix). The linear system with output (or equivalently the pair* (A, C)*) is said to be observable if, for all* $x_1, x_2 \in \mathbb{R}^n$, $x_1 \neq x_2$, *there exists* $K \in N$ *such that*

$$
CA^K x_2 \neq CA^K x_1.
\tag{5.41}
$$

With these definitions Theorems 3.21–3.24 can be used to verify controllability and/or observability of (5.39).

In Section 3.5.1 on continuous time systems, we have pointed out that the time interval $[0, T]$, is not essential and $T > 0$ in the Definitions 3.19 and 3.20 can be selected arbitrarily small. This is however not the case in the discrete time setting where, depending on the dynamics of the controllable/observable system, $K = n$ time steps might be necessary to ensure that (5.40) or (5.41) is satisfied. As an example consider the pairs (A, B) and (A^T, B^T) defined through the matrices

$$A = \begin{bmatrix} 0 & 1 & 0 \\ 0 & 0 & 1 \\ 0 & 0 & 0 \end{bmatrix}, \qquad B = \begin{bmatrix} 0 \\ 0 \\ 1 \end{bmatrix}.$$

With the Theorems 3.21–3.24 it is straightforward to verify that (A, B) is controllable and (A^T, B^T) is observable.

Consider the states $x_1 = [0, 0, 1]^T$ and $x_2 = [0, 0, 0]^T$. Then it holds that

$$Ax_1 = \begin{bmatrix} 0 \\ 1 \\ 0 \end{bmatrix}, \qquad A^2 x_1 = \begin{bmatrix} 1 \\ 0 \\ 0 \end{bmatrix} \qquad \text{and} \qquad A^3 x_1 = \begin{bmatrix} 0 \\ 0 \\ 0 \end{bmatrix}.$$

Hence, without input, the origin is reached in $K = n = 3$ steps and $x_2 = A^3 x_1$ holds. Due to the vector B, which is only unequal to zero in the last entry, x_1 cannot be steered to the origin in fewer steps.

Similarly, for $x_1 = [1, 0, 0]^T$ and $x_2 = [0, 0, 0]^T$ we ask the reader to verify that $K = n = 3$ discrete time steps are necessary until (5.41) is satisfied for the pair (A^T, B^T).

In the preceding section, Section 5.3.3, we have highlighted that stability properties might be lost when a continuous time system is discretized and when the sampling time Δ is not chosen with care. Similarly, controllability may be lost, which we illustrate on an example.

Consider the continuous time system $\dot{x} = A_c x + B_c u$ defined through the matrices

$$A_c = \begin{bmatrix} 0 & 1 \\ -1 & 0 \end{bmatrix} \qquad \text{and} \qquad B_c = \begin{bmatrix} 0 \\ 1 \end{bmatrix}.$$

With Theorem 3.21 it is again straightforward to verify that the pair (A_c, B_c) is controllable. Moreover, using the exact discretization scheme (5.12) it holds that

$$A_{de}(\Delta) = e^{A_c \Delta} = \begin{bmatrix} \cos(\Delta) & \sin(\Delta) \\ -\sin(\Delta) & \cos(\Delta) \end{bmatrix}$$

(obtained by using symbolic variables and `expm.m` in MATLAB, for example) and

$$\begin{aligned} B_{de}(\Delta) &= \int_0^\Delta e^{A_c(\Delta - \tau)} d\tau B_c \\ &= \begin{bmatrix} -\sin(\Delta - \tau) & \cos(\Delta - \tau) \\ -\cos(\Delta - \tau) & -\sin(\Delta - \tau) \end{bmatrix} \Bigg|_0^\Delta \cdot \begin{bmatrix} 0 \\ 1 \end{bmatrix} \\ &= \begin{bmatrix} \sin(\Delta) & 1 - \cos(\Delta) \\ -1 + \cos(\Delta) & \sin(\Delta) \end{bmatrix} \cdot \begin{bmatrix} 0 \\ 1 \end{bmatrix} = \begin{bmatrix} 1 - \cos(\Delta) \\ \sin(\Delta) \end{bmatrix}. \end{aligned}$$

This implies that in the case $\Delta = 2\pi \ell$, $\ell \in \mathbb{N}$, $B_{de}(\Delta)$ satisfies

$$B_{de}(\Delta) = \begin{bmatrix} 0 \\ 0 \end{bmatrix},$$

and thus the discrete time system represented through $(A_{de}(2\pi \ell), B_{de}(2\pi \ell))$, $\ell \in \mathbb{N}$, is not controllable.

In general, the following result can be derived.

Lemma 5.12 ([144, Lemma 3.4.1]). *Consider the pair (A, B) and let (A_{de}, B_{de}) be defined through (5.12) for $\Delta > 0$. The pair (A_{de}, B_{de}) is controllable if and only if $(e^{A\Delta}, B)$ is controllable and A has no eigenvalues of the form $\frac{2}{\Delta}\pi\ell$, $\ell \in \mathbb{N}$.*

For discrete time systems defined through (5.11) controllability cannot be lost. In particular, for (A, B) controllable it holds that $(A_d, B_d) = (I + \Delta A, \Delta B)$ is controllable for all $\Delta > 0$ (see Exercice 5.4).

5.5   EXERCISES

Exercise 5.1. Compute the equilibria of the difference equation

$$x_1^+ = 2x_1 + 1$$
$$x_2^+ = 3x_1 + x_2^2 + 3x_2.$$

Exercise 5.2. Consider the linear system

$$x^+ = Ax = \begin{bmatrix} -0.26 & -0.68 & -0.24 \\ -0.07 & 0.71 & -0.61 \\ 0.96 & 0.28 & -0.14 \end{bmatrix} x.$$

Show that the matrix A is a Schur matrix. Compute a quadratic Lyapunov function $V(x) = x^T P x$, i.e., compute a positive definite matrix P such that the decrease condition $V(x^+) - V(x) < 0$ is satisfied for all $x \neq 0$.

Hint: In MATLAB the function `dlyap.m` might be helpful.

Exercise 5.3. Consider the time-varying linear system

$$\dot{x}(t) = A(t)x(t), \tag{5.42}$$

where

$$A(t) = \begin{bmatrix} -1 + 1.5\cos^2(t) & 1 - 1.5\sin(t)\cos(t) \\ -1 - 1.5\sin(t)\cos(t) & -1 + 1.5\sin^2(t) \end{bmatrix} \tag{5.43}$$

and with solution

$$x(t) = \begin{bmatrix} e^{0.5t}\cos(t) & e^{-t}\sin(t) \\ -e^{0.5t}\sin(t) & e^{-t}\cos(t) \end{bmatrix} x(0). \tag{5.44}$$

Visualize the solution (5.44) in MATLAB for the initial condition $x_0 = [1, 1]^T$ for $t \in [0, 1]$.

Compare the solution with the numerical solutions obtained through the Euler and the Heun methods characterized through the Butcher tableaus (5.23) for $\Delta = 0.1$ and $\Delta = 0.01$.

Exercise 5.4. Let the pair (A, B) be controllable. Use Theorem 3.21 to show that $(A_d, B_d) = (I + \Delta A, \Delta B)$ is controllable for all $\Delta > 0$.

Hint: The multiplication of a column of a matrix by a nonzero constant does not change the rank of the matrix. Moreover, subtracting one column of a matrix from a different column does not change the rank of a matrix.

5.6   BIBLIOGRAPHICAL NOTES AND FURTHER READING

As is somewhat obvious from the results in this chapter, many results for continuous time systems have direct analogues for discrete time systems, though sometimes certain details or required proof techniques are different. Some texts present continuous/discrete results together, for example using a notation such as $\dot{x}/x^+ = \dots$ to indicate either continuous or discrete time systems; see [63]. Alternatively, a more general, mathematical system representation can be used that allows one to take the time variable as either $t \in \mathbb{R}$ or $t \in \mathbb{Z}$; see [59] or [144].

Though not covered here, frequency domain results for linear discrete time systems (using the z-transform in place of the Laplace transform) have been developed and are available; see, for example, [113].

An increasingly common modelling framework combines continuous and discrete time elements, for example to capture effects such as controller resets, switching between different controllers depending on certain conditions, or to capture other discontinuous phenomena. Such models are referred to as *hybrid dynamical systems*; see, for example, [54].

For discretization and numerical solution of ordinary differential equations, as well as stability of discretization schemes, we refer to the books [14], [60], [136], [148], and [163].

Controllability with respect to sampling is discussed in detail in [144, Section 3.4].

Chapter Six

Absolute Stability

This chapter pulls together material from the linear systems in Chapter 3 and the frequency domain tools and techniques in Chapter 4 to solve a commonly encountered, nonlinear, implementation issue. The techniques and calculations developed in this chapter will be applied, with appropriate modifications and extensions, in developing the antiwindup techniques in Chapter 8 as well as some of the observer designs in Part III.

6.1   A COMMONLY IGNORED DESIGN ISSUE

In Section 4.2 we studied the stability properties of linear systems

$$\dot{x} = Ax + bu, \qquad y = cx, \tag{6.1}$$

$A \in \mathbb{R}^{n \times n}$, $b \in \mathbb{R}^{n \times 1}$, $c \in \mathbb{R}^{1 \times n}$, in the frequency domain. As an important application, we have investigated the feedback interconnection

$$\dot{x} = (A - bkc)x \tag{6.2}$$

in Figure 6.1 (shown in Figure 4.2 with a frequency domain representation). More

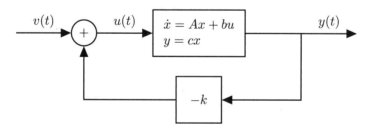

Figure 6.1: Idealized feedback interconnection.

realistically in many applications is the feedback interconnection shown in Figure 6.2, where $\psi : \mathbb{R} \to \mathbb{R}$ denotes a nonlinear function which impacts the input u and thus the stability properties of the closed-loop dynamics 6.2.

Example 6.1. Consider the inverted pendulum (1.45) and in particular its lineariza-

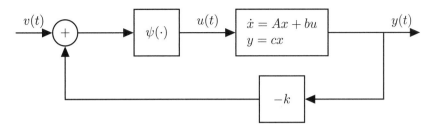

Figure 6.2: Realistic feedback interconnection with nonlinearity in the input.

tion in the upright position (4.29) described through the dynamics

$$\begin{bmatrix} \dot{x}_1 \\ \dot{x}_2 \end{bmatrix} = \begin{bmatrix} 0 & 1 \\ \frac{mg\ell}{J+m\ell^2} & -\frac{\gamma}{J+m\ell^2} \end{bmatrix} \begin{bmatrix} x_1 \\ x_2 \end{bmatrix} + \begin{bmatrix} 0 \\ \frac{\ell}{J+m\ell^2} \end{bmatrix} u, \qquad y = \begin{bmatrix} 1 & 0 \end{bmatrix} x.$$

If we again fix the parameters $m = \ell = 1$, $J = 0$, $g = 9.81$, and $\gamma = 0.1$ for simplicity, the system is defined through the matrices

$$A = \begin{bmatrix} 0 & 1 \\ 9.81 & -0.1 \end{bmatrix}, \qquad b = \begin{bmatrix} 0 \\ 1 \end{bmatrix}, \qquad \text{and} \qquad c = \begin{bmatrix} 1 & 0 \end{bmatrix},$$

and the feedback gain $k = 10$ ensures that the closed-loop matrix $A - bkc$ is Hurwitz.

However, any motor used to drive the cart has limited power, limiting the achievable input values $u \in [u_{lb}, u_{ub}]$, $u_{lb}, u_{ub} \in \mathbb{R}$. Hence, the input needs to be adjusted to

$$\psi(e) = \begin{cases} u_{lb}, & \text{for } e \leq u_{lb}, \\ e, & \text{for } u_{lb} \leq e \leq u_{ub}, \\ u_{ub}, & \text{for } e \geq u_{ub}, \end{cases} \tag{6.3}$$

to be implementable. Here, $e(t) = v(t) - ky(t)$ denotes the error variable. The definition of the input u in (6.3) describes a particular form of the nonlinearity $\psi(\cdot)$ known as *saturation*. While we have seen techniques to verify the stability properties of the origin for $\dot{x} = (A - bkc)x$ we require new tools in order to characterize the stability properties of $\dot{x} = Ax - b\psi(kcx)$.

The *saturation* described in this example is a common problem in control theory. The corresponding block diagram is shown in Figure 6.3. The saturation in (6.3)

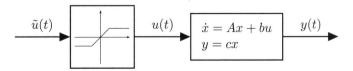

Figure 6.3: Input saturation.

depends on the upper and lower bound. The *saturation function* sat : $\mathbb{R} \to [-1, 1]$

is defined as

$$\mathrm{sat}(y) = \begin{cases} -1, & \text{for } y \leq -1, \\ y, & \text{for } -1 \leq y \leq 1, \\ 1, & \text{for } y \geq 1. \end{cases} \tag{6.4}$$

From this normalized function a specific saturation as in (6.3) can be obtained through an appropriate scaling and translation.

Example 6.2. ([152, Sec. 2.1.2]) As a second example we consider the *servo-valve* shown in Figure 6.4. Observe that raising the spool allows an inflow of pressure

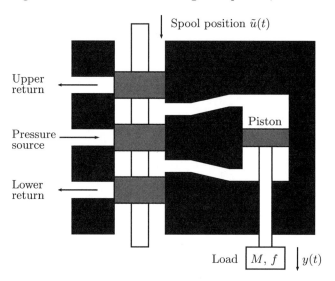

Figure 6.4: Servo-valve with deadzone.

(e.g., hydraulic fluid) from the pressure source while simultaneously allowing a pressure drop via the upper return so that the piston will rise. However, the spool and the various openings overlap and, near the configuration shown in Figure 6.4, there are a range of spool positions where the fluid flow is zero. This has clear benefits in a physically realized valve since it prevents leakage. However, from a control perspective it introduces a nonlinearity in the input. In particular, if the position of the spool $\tilde{u}$ defines a linear input of the system and the load position y denotes the output, then the block diagram shown in Figure 6.5 is obtained. The

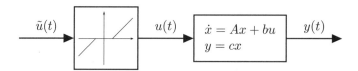

Figure 6.5: Deadzone.

linear dynamics are defined through the matrices

$$A = \begin{bmatrix} 0 & 1 \\ 0 & -\frac{B}{M} \end{bmatrix}, \quad b = \begin{bmatrix} 0 \\ 1 \end{bmatrix}, \quad c = \begin{bmatrix} \frac{K}{M} & 0 \end{bmatrix},$$

where

$$K = a \frac{\frac{\partial g}{\partial x}}{\frac{\partial g}{\partial p}} \quad \text{and} \quad B = f + \frac{a^2}{\frac{\partial g}{\partial p}}.$$

Here, $g(x, p)$ denotes the flow, a the area of the piston, p the pressure, and f the viscous friction.

The phenomenon described in Example 6.2 and illustrated in Figure 6.5 is called a *deadzone* and can be described through the function $dz : \mathbb{R} \to \mathbb{R}$,

$$dz(y) = \begin{cases} y + 1, & \text{for } y \le -1, \\ 0, & \text{for } -1 \le y \le 1, \\ y - 1, & \text{for } y \ge 1. \end{cases} \tag{6.5}$$

While we know how to analyze the stability properties of a linear system with a linear feedback interconnection $u(y) = ky$, it is not clear which properties can be preserved in the presence of a nonlinear (possibly time-dependent) actuator in the feedback interconnection $u = \psi(t, y)$, $\psi : \mathbb{R}_{\ge 0} \times \mathbb{R} \to \mathbb{R}$. This problem of non-ideal actuators served as the motivation for the Soviet academician Anatoly Lur'e to study the block diagram visualized in Figure 6.6. Such a system is frequently referred to as a *Lur'e system* and the question of whether or not the origin is asymptotically stable is sometimes referred to as the *Lur'e problem*. The block

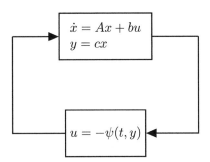

Figure 6.6: Lur'e problem.

diagram in Figure 6.6 generalizes the feedback interconnection in Figure 6.2, where we assume that the reference signal $v(t)$ is set to zero for simplicity. In contrast to Figure 6.2, the nonlinearity ψ can be time-dependent and can (additionally) be a nonlinearity in the output y instead of the input u.

For simplicity we consider single-input single-output systems (6.1) in this chapter. However, many of the results presented are extendable to multi-input multi-output systems. The nonlinearities above, and many others of interest, satisfy a similar property called a *sector condition*.

Definition 6.3 (Sector condition). *Let $\alpha, \beta \in \mathbb{R}$, $\alpha < \beta$, and $\Omega \subset \mathbb{R}$. A nonlinearity $\psi : \mathbb{R}_{\geq 0} \times \mathbb{R} \to \mathbb{R}$ satisfies a sector condition if*

$$\alpha y^2 \leq y\psi(t, y) \leq \beta y^2 \tag{6.6}$$

for all $t \geq 0$ and for all $y \in \Omega$. For $\Omega = \mathbb{R}$ we say that the sector condition is satisfied globally.

The sector condition (6.6) is visualized in Figure 6.7. In Exercise 6.1, the reader is asked to verify that the quadratic inequalities in (6.6) indeed correspond to the linear bounds shown in Figure 6.7. We have already seen two special classes

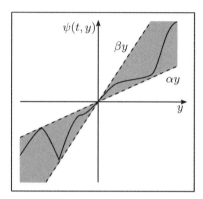

Figure 6.7: Visualization of the sector condition defined through $\alpha < \beta$.

of nonlinearities. The saturation (6.4) and the deadzone (6.5) are shown again together with the *sign* (or *signum*) function, sign : $\mathbb{R} \to \mathbb{R}$,

$$\mathrm{sign}(y) = \begin{cases} -1, & \text{for } y < 0, \\ 0, & \text{for } y = 0, \\ 1, & \text{for } y > 0, \end{cases} \tag{6.7}$$

in Figure 6.8, representing a third type of nonlinearity. While the saturation and the

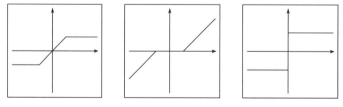

Figure 6.8: Examples of nonlinearities $y \mapsto \psi(y)$. In particular, from left to right: saturation, deadzone, and sign function.

deadzone satisfy the sector condition (6.6) for appropriate $\alpha, \beta \in \mathbb{R}$ (for all $\Omega \subset \mathbb{R}$), the sign function cannot be placed within a sector (for all $\Omega \subset \mathbb{R}$ containing the origin). For the sign function one would need to allow $\beta = \infty$. Nevertheless, the sector condition covers a wide range of nonlinearities based on $\alpha, \beta \in \mathbb{R}$.

Definition 6.4 (Absolute stability). *Let $\alpha, \beta \in \mathbb{R}$, $\alpha < \beta$, and $\Omega \subset \mathbb{R}$. The Lur'e system*

$$\dot{x} = Ax - b\psi(t, y), \qquad y = cx, \qquad (6.8)$$

is called absolutely stable (with respect to α, β, Ω) if the origin of (6.8) is asymptotically stable for all $\psi : \mathbb{R}_{\geq 0} \times \mathbb{R} \to \mathbb{R}$ satisfying the sector condition (6.6) for all $t \geq 0$ and for all $y_0 \in \Omega$.

While we have described how various nonlinearities such as saturations and deadzones might arise in feedback loops, the absolute stability formulation holding for all possible nonlinearities satisfying a sector condition perhaps requires some further motivation. Supposing first that the nonlinearity is time-invariant, having a feedback design that accommodates all nonlinearities in a sector allows for imperfections in mass manufacturing processes and yet guarantees that as long as certain tolerances are met the feedback designs will still work as expected. Consider, for example, mass producing the servo-valve in Figure 6.4, where all of the pieces may vary slightly due to imperfections in the machining process.

A different problem can arise during the operation of a system, whereby parameters might drift over time. For example, run continuously, a DC motor may heat up and the proportionality constant from current to torque can drift. Again, a desirable property of a feedback design would be that it accommodates such drift. One way to model this is via a time-varying nonlinearity.

6.2   HISTORICAL PERSPECTIVE ON THE LUR'E PROBLEM

An intuitively appealing possible answer to the question of absolute stability is to assume the origin is asymptotically stable for all linear feedbacks in the sector. Indeed, this was one of the first conjectured solutions to this problem.

Conjecture 6.5 (Aizerman's conjecture (1949)). *Let $\alpha, \beta \in \mathbb{R}$, $\alpha < \beta$, and suppose the origin of the linear system (6.1) is globally asymptotically stable for all linear feedbacks*

$$u = -\psi(y) = -ky, \quad k \in [\alpha, \beta]. \qquad (6.9)$$

Then the origin is globally asymptotically stable for all nonlinear feedbacks in the sector

$$\alpha \leq \frac{\psi(y)}{y} \leq \beta, \quad y \neq 0.$$

Unfortunately, this simple idea does not work and multiple counterexamples have been developed (see, e.g., [30]). Eight years later, Aizerman's Conjecture was further refined as follows:

Conjecture 6.6 (Kalman's Conjecture (1957)). *Let $\alpha, \beta \in \mathbb{R}$, $\alpha < \beta$, and suppose the origin of the linear system (6.1) is globally asymptotically stable for all linear feedbacks (6.9). Then the origin is globally asymptotically stable for all nonlinear* *feedbacks belonging to the incremental sector*

$$\alpha \leq \frac{\partial}{\partial y}\psi(y) \leq \beta. \tag{6.10}$$

However, this conjecture also turned out to be incorrect (see again [30] for a counterexample).

In fact, obtaining a useful solution to this problem involved a somewhat surprising detour into the frequency domain.

Definition 6.7 (Positive real). *A transfer function $H(s)$ is positive real if*

$$\mathrm{Re}(H(s)) \geq 0 \quad \text{for all } s \in \overline{\mathbb{C}}_+.$$

The transfer function is strictly positive real (SPR) if $H(s - \varepsilon)$ is positive real for some $\varepsilon > 0$.

Note that the strictly positive real definition above is equivalent to the requirement that

$$\mathrm{Re}(H(s)) > 0 \quad \text{for all } s \in \overline{\mathbb{C}}_+.$$

Example 6.8. Consider $H(s) = \frac{1}{s}$. Then

$$\mathrm{Re}(H(s)) = \mathrm{Re}\left(\frac{1}{\sigma + j\omega}\right) = \mathrm{Re}\left(\frac{\sigma - j\omega}{\sigma^2 + \omega^2}\right) = \frac{\sigma}{\sigma^2 + \omega^2} \geq 0$$

for $\sigma \geq 0$. So $H(s)$ is positive real.

Example 6.9. Consider $H(s) = \frac{1}{s+1}$. Then

$$\mathrm{Re}(H(s)) = \mathrm{Re}\left(\frac{1}{\sigma + j\omega + 1}\right) = \mathrm{Re}\left(\frac{\sigma + 1 - j\omega}{(\sigma+1)^2 + \omega^2}\right) = \frac{\sigma + 1}{\sigma^2 + \omega^2} > 0$$

for $\sigma \geq 0$. So $H(s)$ is strictly positive real.

A useful characterization of strict positive realness is given in the following lemma.

Lemma 6.10 ([139, Theorem 4.10]). *The transfer function $H(s)$ is strictly positive real if and only if $H(s)$ is Hurwitz (i.e., has all its poles in the open left-half complex plane) and*

$$\mathrm{Re}(H(j\omega)) > 0, \quad \text{for all } \omega \in \mathbb{R}. \tag{6.11}$$

The interested reader is invited to verify that $H(s) = \frac{s}{s+1}$ is Hurwitz but does not satisfy (6.11), while $H(s) = \frac{s-2}{s-1}$ satisfies (6.11) but is not Hurwitz.

Finally, we present a fundamental result that ties together the Lur'e Problem across the frequency and time domains. The following lemma is frequently referred to as the *Kalman-Yakubovich-Popov lemma* or the *strictly positive real (SPR) lemma*.

Lemma 6.11 (Kalman-Yakubovich-Popov lemma [158, Theorem 5.6.13]). *Let*

$$H(s) = c(sI - A)^{-1}b + d \tag{6.12}$$

be a transfer function where $A \in \mathbb{R}^{n \times n}$ is Hurwitz, (A, b) is controllable, and (A, c) is observable. Then $H(s)$ is strictly positive real if and only if there exist a positive definite symmetric matrix $P \in \mathbb{R}^{n \times n}$, a (row) vector $L \in \mathbb{R}^{1 \times n}$, a number $w \in \mathbb{R}$, and positive constant $\varepsilon > 0$ such that

$$A^T P + PA = -L^T L - \varepsilon P, \tag{6.13a}$$

$$Pb = c^T - L^T w, \tag{6.13b}$$

$$w^2 = 2d. \tag{6.13c}$$

6.3   SUFFICIENT CONDITIONS FOR ABSOLUTE STABILITY

In this section we present three sufficient conditions, under various assumptions, for absolute stability. As previously mentioned, we restrict our attention to single-input single-output systems for two reasons. The first is that it significantly simplifies the exposition. The second is that the graphical conditions we develop (i.e., the circle and Popov criteria) only work in the single-input single-output case.

Theorem 6.12 (Absolute stability). *Assume that A is Hurwitz, (A, b) is controllable, (A, c) is observable, $H(s)$ in (6.12) is strictly positive real, and the nonlinearity $\psi(t, y)$ is in the sector $[0, \infty)$. Then the origin of the Lur'e system (6.8) is globally exponentially stable (i.e., the system is absolutely stable).*

Proof. Since all of the assumptions of Lemma 6.11 are satisfied, we can solve the matrix equations (6.13). Define the function $V(x) = x^T P x$. Then we can calculate

$$
\begin{aligned}
\dot{V}(x) &= \dot{x}^T P x + x^T P \dot{x} \\
&= (x^T A^T - b^T \psi(t, y)) P x + x^T P (Ax - b\psi(t, y)) \\
&= x^T (A^T P + PA) x - 2\psi(t, y) b^T P x.
\end{aligned} \tag{6.14}
$$

Observe that, using equation (6.13b) and the output equation,

$$
\begin{aligned}
\psi(t, y) b^T P x &= \psi(t, y)(c - Lw)x \\
&= \psi(t, y)(y - du - Lwx) \\
&= \psi(t, y)y + d\psi^2(t, y) - \psi(t, y)Lwx
\end{aligned} \tag{6.15}
$$

is satisfied. Continuing from (6.14), and noting that the sector condition assumption implies $0 \leq y\psi(t, y)$, we see that

$$
\begin{aligned}
\dot{V}(x) &= x^T (A^T P + PA)x - 2(y\psi(t, y) + \psi^2(t, y)d - \psi(t, y)Lwx) \\
&\leq x^T (A^T P + PA)x - 2(\psi^2(t, y)d - \psi(t, y)Lwx) \\
&= -\varepsilon x^T P x - x^T L^T L x - \psi^2(t, y)w^2 + 2\psi(t, y)Lwx \\
&= -\varepsilon x^T P x - (Lx - \psi(t, y)w)^2 \leq -\varepsilon x^T P x,
\end{aligned} \tag{6.16}
$$

where the second-to-last equality uses (6.13a) and (6.13c). With the final equality, appealing to Theorem 2.17 we see that the origin is globally exponentially stable. $\square$

Unfortunately, Theorem 6.12 only applies to linear systems (A, b, c, d) with A Hurwitz. However, using a loop transformation it may be possible to "borrow" a stabilizing factor from the sector condition in order to apply Theorem 6.12 when A is not Hurwitz. Specifically, consider the loop transformation illustrated in Figure 6.9. Consider the Lur'e system

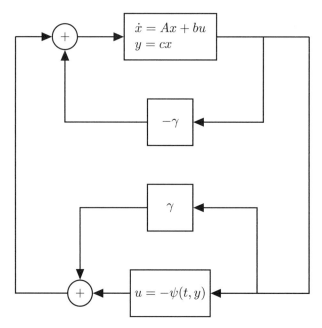

Figure 6.9: Loop transformation of the nonlinearity.

$$\dot{x} = Ax - b\psi(t, y), \qquad y = cx,$$

with (A, b) controllable and (A, c) observable but A not necessarily Hurwitz. If ψ satisfies the sector condition (6.6) for $\alpha, \beta \in \mathbb{R}$, then $\hat{\psi}(t, y) = \psi(t, y) - \gamma y$ satisfies the sector condition

$$\hat{\alpha} y^2 \le y\hat{\psi}(t, y) \le \hat{\beta} y^2 \tag{6.17}$$

for $\hat{\alpha} = \alpha - \gamma$ and $\hat{\beta} = \beta - \gamma$ for $\gamma \in \mathbb{R}$. Let $\gamma \in \mathbb{R}$ be defined such that the closed-loop matrix $\hat{A} = A - \gamma bc$ is Hurwitz. The closed-loop dynamics for the loop transformation depicted in Figure 6.9 can be written as

$$\begin{aligned}
\dot{x} &= Ax + b(-\gamma cx + \gamma cx - \psi(t, y)) \\
&= (A - \gamma bc) - b(\psi(t, y) - \gamma y) \\
&= \hat{A}x - b\hat{\psi}(t, y), \tag{6.18}
\end{aligned}$$

which satisfies the assumptions of Theorem 6.12 as long as $\hat{\alpha} > 0$.

6.3.1 Circle Criterion

We now pull together several previous results, including the Nyquist plot, to develop a graphical sufficient condition for absolute stability. This *circle criterion* extends Theorem 6.12 by making use of the above loop transformation (6.18) as well as a sector upper bound. Specifically, we choose $\gamma = \alpha$ in the loop transformation so that the sector nonlinearity $\hat{\psi}(t, y)$ satisfies

$$0 \leq y\hat{\psi}(t, y) \leq \hat{\beta}y^2 \tag{6.19}$$

where $\hat{\beta} \doteq \beta - \alpha$.

Before presenting the circle criterion, we develop a preliminary lemma whose proof is similar to that of Theorem 6.12.

Lemma 6.13. *Consider the Lur'e system* (6.8) *defined through* (A, b, c) *where* (A, b) *is controllable and* (A, c) *is observable, and with transfer function* $G(s) = c(sI - A)^{-1}b$. *The system* (6.8) *is absolutely stable according to Definition 6.4 (and with respect to* $\alpha, \beta \in \mathbb{R}$, $\alpha < \beta$, $\Omega = \mathbb{R}$) *if* $\hat{A} = A - \alpha bc$ *is Hurwitz, and*

$$H(s) = \frac{1 + \beta G(s)}{1 + \alpha G(s)} \tag{6.20}$$

is strictly positive real; i.e., if

$$\mathrm{Re}(H(s)) = \mathrm{Re}\left(\frac{1 + \beta G(s)}{1 + \alpha G(s)}\right) > 0 \quad \text{for all} \quad s \in \overline{\mathbb{C}}_+. \tag{6.21}$$

Proof. We start with the loop transformed Lur'e system

$$\dot{x} = (A - \alpha bc)x - b(\psi(t, y) - \alpha y) \doteq \hat{A}x - b\hat{\psi}(t, y)$$

using the loop transformation (6.18).

Consider the equivalent closed-loop systems shown in Figure 6.10. For the right

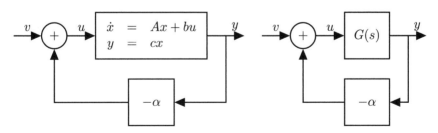

Figure 6.10: Closed-loop system in the time domain and in the frequency domain.

hand, frequency domain representation, it holds that

$$y(s) = G(s)u(s) = G(s)v(s) - \alpha G(s)y(s),$$

which implies

$$y(s) = \frac{G(s)}{1 + \alpha G(s)}v(s).$$

On the other hand, for the left hand, time domain representation, we may write

$$\dot{x} = Ax + bu = Ax + b(v - \alpha cx) = (A - \alpha bc)x + bv = \hat{A}x + bv.$$

Computing the transfer function of the time domain representation thus yields

$$c(sI - \hat{A})^{-1}b = \frac{G(s)}{1 + \alpha G(s)}.$$

Observe that (6.20) can be rewritten in the form

$$
\begin{aligned}
H(s) &= \frac{1 + \beta G(s)}{1 + \alpha G(s)} = \frac{1 + \alpha G(s) + (\beta - \alpha)G(s)}{1 + \alpha G(s)} = 1 + \hat{\beta}\frac{G(s)}{1 + \alpha G(s)} \\
&= 1 + \hat{\beta}c(sI - \hat{A})^{-1}b.
\end{aligned}
\tag{6.22}
$$

Since we have assumed $H(s)$ is strictly positive real, Lemma 6.11 implies we can solve the matrix equations

$$
\begin{aligned}
\hat{A}^T P + P\hat{A} &= -L^T L - \varepsilon P \\
Pb &= \hat{\beta}c^T - L^T w \\
w^2 &= 2.
\end{aligned}
$$

The sector condition (6.19) implies that, when $\hat{\psi}(t, y) \geq 0$, the inequality $\hat{\psi}(t, y) - \hat{\beta}y \leq 0$ is satisfied, which then ensures that

$$-2\hat{\psi}(t, y)(\hat{\psi}(t, y) - \hat{\beta}y) \geq 0. \tag{6.23}$$

Conversely, if $\hat{\psi}(t, y) \leq 0$ then $\hat{\psi}(t, y) - \beta\hat{\psi}(t, y) \geq 0$ and thus again (6.23) holds. Hence, a nonlinearity satisfying the sector condition (6.19) necessarily satisfies (6.23).

To show absolute stability (i.e., asymptotic stability of the origin) we define the Lyapunov function candidate $V(x) = x^T Px$. Then

$$
\begin{aligned}
\dot{V}(x) &= x^T(\hat{A}^T P + P\hat{A})x - 2x^T Pb\hat{\psi}(t, y) \\
&\leq x^T(\hat{A}^T P + P\hat{A})x - 2x^T Pb\hat{\psi}(t, y) - 2\hat{\psi}(t, y)(\hat{\psi}(t, y) - \hat{\beta}y) \\
&= -\varepsilon x^T Px - x^T L^T Lx - 2x^T Pb\hat{\psi}(t, y) + 2\hat{\psi}(t, y)\hat{\beta}cx - 2\hat{\psi}(t, y)^2 \\
&= -\varepsilon x^T Px - (Lx)^2 - 2x^T(Pb - \hat{\beta}c^T)\hat{\psi}(t, y) - 2\hat{\psi}(t, y)^2 \\
&= -\varepsilon x^T Px - (Lx)^2 + 2x^T(\sqrt{2}L^T)\hat{\psi}(t, y) - 2\hat{\psi}(t, y)^2 \\
&= -\varepsilon x^T Px - (Lx - \sqrt{2}\hat{\psi}(t, y))^2 \leq -\varepsilon x^T Px,
\end{aligned}
$$

which concludes the proof.   □

Before we proceed, we define a disc in the complex plane based on a center $\sigma : \mathbb{R}\backslash\{0\} \times \mathbb{R}_{>0} \to \mathbb{R}$ and a radius $r : \mathbb{R}\backslash\{0\} \times \mathbb{R}_{>0} \to \mathbb{R}$. In particular, for $\alpha \neq 0$ and $\beta > 0$ we define

$$\sigma(\alpha, \beta) = \frac{1}{2}\left(\frac{1}{\alpha} + \frac{1}{\beta}\right) \qquad \text{and} \qquad r(\alpha, \beta) = \frac{\text{sign}(\alpha)}{2}\left(\frac{1}{\alpha} - \frac{1}{\beta}\right).$$

Then, the disc $D(\cdot, \cdot)$ is defined as

$$D(\alpha, \beta) = \left\{ \begin{array}{ll} \{x \in \mathbb{C} : x = -\frac{1}{\beta} + j\omega, \omega \in \mathbb{R}\}, & \text{if } \alpha = 0 < \beta, \\ \{x \in \mathbb{C} : |x - \sigma(\alpha, \beta)| = r(\alpha, \beta)\}, & \text{if } 0 < \alpha < \beta, \\ \{x \in \mathbb{C} : |x - \sigma(\alpha, \beta)| = r(\alpha, \beta)\}, & \text{if } \alpha < 0 < \beta. \end{array} \right. \tag{6.24}$$

Indeed, for $\alpha \neq 0$, $D(\alpha, \beta) \subset \mathbb{C}$ defines a disc centered around $\sigma(\alpha, \beta)$ with radius $r(\alpha, \beta)$. In the special case where $\alpha = 0$, $D(0, \beta) \subset \mathbb{C}$ defines a vertical line. The three cases are shown in Figure 6.11.

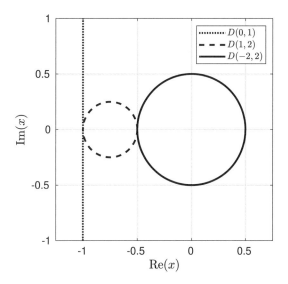

Figure 6.11: Visualization of the disc $D(\alpha, \beta)$ defined in (6.24) for the three different cases.

In order to arrive at the circle criterion below, we consider three cases depending on the sign of α, and relate these three cases to the Nyquist plot of $G(j\omega)$.

First case ($\alpha = 0 < \beta$): In this case, we see that (6.20) reduces to the requirement that

$$H(s) = 1 + \beta G(s) \tag{6.25}$$

is strictly positive real. Appealing to the characterization of strict positive realness in Lemma 6.10, we see that $H(s)$ will have the same poles as $G(s)$ and so a requirement for strict positive realness is that $G(s)$ be Hurwitz. Furthermore, we require that

$$\text{Re}(1 + \beta G(j\omega)) > 0, \quad \text{for all } \omega \in \mathbb{R}, \tag{6.26}$$

which, since $\beta > 0$, is equivalent to

$$\text{Re}(G(j\omega)) > -\frac{1}{\beta}, \quad \text{for all } \omega \in \mathbb{R}. \tag{6.27}$$

Note that this corresponds to the Nyquist plot of $G(j\omega)$ being to the right of the vertical line in the complex plane through $-1/\beta$.

Second case $(0 < \alpha < \beta)$: Similarly to the first case, with $0 < \alpha < \beta$ we see that

$$\text{Re}\left(\frac{1 + \beta G(j\omega)}{1 + \alpha G(j\omega)}\right) > 0 \quad \text{for all } \omega \in \mathbb{R}$$

is equivalent to

$$\text{Re}\left(\frac{\frac{1}{\beta} + G(j\omega)}{\frac{1}{\alpha} + G(j\omega)}\right) > 0 \quad \text{for all } \omega \in \mathbb{R}. \tag{6.28}$$

Observe that for a fixed $\hat{\omega} \in \mathbb{R}$, $G(j\hat{\omega})$ is simply a complex number, as are the other quantities in (6.28). Let $q = G(j\hat{\omega})$ and choose the polar coordinates $r_1, r_2 > 0$ and $\theta_1, \theta_2 \in (-\pi, \pi)$ so that

$$\frac{1}{\beta} + q = r_1 e^{j\theta_1} \quad \text{and} \quad \frac{1}{\alpha} + q = r_2 e^{j\theta_2}. \tag{6.29}$$

Then we can rewrite (6.28) as

$$\begin{aligned}
\text{Re}\left(\frac{\frac{1}{\beta} + G(j\omega)}{\frac{1}{\alpha} + G(j\omega)}\right) &= \text{Re}\left(\frac{r_1 e^{j\theta_1}}{r_2 e^{j\theta_2}}\right) = \text{Re}\left(\frac{r_1}{r_2} e^{j(\theta_1 - \theta_2)}\right) \\
&= \frac{r_1}{r_2} \text{Re}(\cos(\theta_1 - \theta_2) + j\sin(\theta_1 - \theta_2) \\
&= \frac{r_1}{r_2} \cos(\theta_1 - \theta_2),
\end{aligned} \tag{6.30}$$

which is greater than zero when $|\theta_1 - \theta_2| < \frac{\pi}{2}$.

We can relate the geometry of these points and the disc $D(\alpha, \beta)$, as shown in Figure 6.12. It is clear (and can be rigorously shown using, for example, the law of cosines) that $|\theta_1 - \theta_2| < \frac{\pi}{2}$ corresponds to the point q lying outside the disc $D(\alpha, \beta)$. Since $q = G(j\hat{\omega})$ is any point on the Nyquist plot of $G(j\omega)$, we see that (6.28) corresponds to the Nyquist plot lying entirely outside of the disc $D(\alpha, \beta)$.

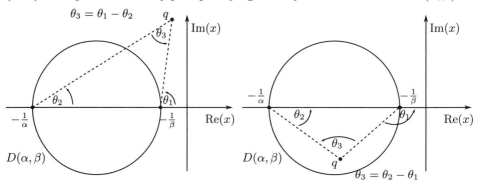

Figure 6.12: Angle condition.

Recall that, in addition to (6.28), to conclude strict positive realness from Lemma 6.10 requires that $H(s)$ be Hurwitz. Note that the poles of $H(s)$ are given by $1 + \alpha G(j\omega)$, and so in order to apply Theorem 4.17 we require that the Nyquist plot encircle the point $-1/\alpha$ as many times as there are right-half plane poles of $G(s)$. Since we additionally require that the Nyquist plot not enter the disc $D(\alpha, \beta)$, we can combine these two requirements by insisting that the Nyquist plot not enter the disc and encircle the entire disc as many times as there are right-half plane poles of $G(s)$.

Third case ($\alpha < 0 < \beta$): Since $\alpha < 0 < \beta$, the first step of the previous case yields that

$$\text{Re}\left(\frac{1 + \beta G(j\omega)}{1 + \alpha G(j\omega)}\right) > 0 \quad \text{for all } \omega \in \mathbb{R}$$

is equivalent to

$$\text{Re}\left(\frac{\frac{1}{\beta} + G(j\omega)}{\frac{1}{\alpha} + G(j\omega)}\right) < 0 \quad \text{for all } \omega \in \mathbb{R}, \tag{6.31}$$

where the inequality changes direction due to the multiplication by $\alpha/\beta < 0$. We can largely repeat the analysis of the second case above, where now we want (6.30) less than zero and hence require $|\theta_1 - \theta_2| > \frac{\pi}{2}$. Examining Figure 6.12, we see that this corresponds to q, and hence the Nyquist plot, lying entirely within the disc $D(\alpha, \beta)$. Furthermore, since the Nyquist plot cannot leave the disc, it cannot encircle the point $-1/\alpha$, and hence cannot have any right-half plane poles. In other words, $G(s)$ must be Hurwitz.

Summarizing these three cases, we have thus proved the circle criterion as stated below.

Theorem 6.14 (Circle criterion). *Suppose (A, b, c) is a minimal realization of $G(s)$ and $\psi(t, y)$ satisfies the sector condition (6.6) globally. Then the system is absolutely stable if:*

1. *$\alpha = 0 < \beta$, the Nyquist plot is to the right of the line $\text{Re}(s) = -\frac{1}{\beta}$ (i.e., to the right of $D(0, \beta)$), and $G(s)$ is Hurwitz;*
2. *$0 < \alpha < \beta$, the Nyquist plot does not enter the disk $D(\alpha, \beta)$, and encircles it in the counterclockwise direction as many times, N, as there are right-half plane poles of $G(s)$; or*
3. *$\alpha < 0 < \beta$, the Nyquist plot lies in the interior of the disk $D(\alpha, \beta)$, and $G(s)$ is Hurwitz.*

Example 6.15. Consider the transfer function $G(s) = \frac{1}{s+1}$ which has a single pole at $s = -1$. Since $G(s)$ is Hurwitz, all three items of the circle criterion are potentially applicable. The three cases of the circle criterion are illustrated in Figure 6.13, which shows the Nyquist plot and the disc $D(\alpha, \beta)$ for different values of α and β. For $\alpha = 0$ and $\beta = 10$ the Nyquist plot is to the right of the line $D(0, 10)$ and thus $G(s)$ is absolutely stable according to Theorem 6.14, item 1. In particular, $G(s)$ is asymptotically stable for all nonlinearities $\psi(t, y)$ satisfying the sector condition (6.6) with $\alpha = 0$ and $\beta = 10$.

From Figure 6.13 we can additionally see that the Nyquist plot is outside the disc $D(1, 10)$ and encircles it zero times and thus absolute stability follows from Theorem 6.14, item 2. Finally, $D(-0.9, 10)$ encircles the Nyquist plot which shows absolute stability according to Theorem 6.14, item 3.

Here, the best estimate, i.e., the largest sector in terms of the sector condition 6.6, is obtained through Theorem 6.14, item 3, which covers the other two cases.

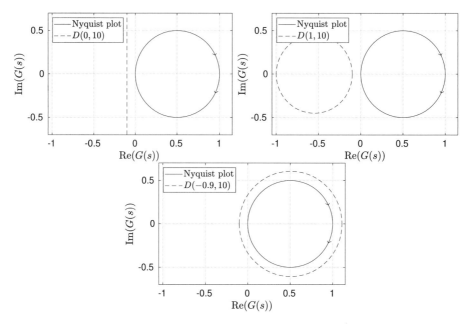

Figure 6.13: Circle criterion for $G(s) = \frac{1}{s+1}$.

We invite the reader to check that β can be chosen arbitrarily large and also that the lower bound can be improved by increasing the width of the sector.

Example 6.16. Consider the transfer function $G(s) = \frac{1}{s-1}$ with one pole in the right-half plane. Since $G(s)$ is not Hurwitz, only item 2 of Theorem 6.14 might be applicable. As shown in Figure 6.14, the disc $D(1.01, 100)$ encircles the Nyquist plot

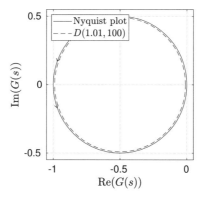

Figure 6.14: Circle criterion for $G(s) = \frac{1}{s-1}$.

exactly once in the counterclockwise direction, and hence we can conclude absolute stability for the sector defined by $\alpha = 1.01$ and $\beta = 100$.

Example 6.17. Consider the transfer function

$$G(s) = \frac{s+1}{s^2 - 2s + 2} = \frac{s+1}{(s-1+j)(s-1-j)}$$

with two poles in the right-half plane. Again, only item 2 of Theorem 6.14 might be applicable. In Figure 6.15 we see that the Nyquist plot encircles the disc $D(2.5, 20)$

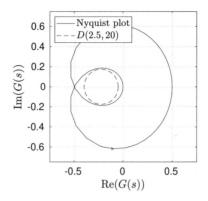

Figure 6.15: Circle criterion for $G(s) = \frac{s+1}{s^2-2s+2}$.

twice in the counterclockwise direction and thus absolute stability follows.

Example 6.18. Consider the transfer function

$$G(s) = \frac{s+1}{s^2 + 2s + 2} = \frac{s+1}{(s+1+j)(s+1-j)}.$$

Since $G(s)$ is Hurwitz we can potentially apply all three items of Theorem 6.14. From Theorem 6.14, item 3, and Figure 6.16 we obtain absolute stability for the pairs $(\alpha, \beta) = (-1.5, 3.5)$ and $(\alpha, \beta) = (-1, 100)$ but not for $(\alpha, \beta) = (-1.5, 100)$. From Theorem 6.14, item 1, we observe that β can be selected arbitrarily large if α is set to zero.

Example 6.19. As a last example, consider the transfer function

$$G(s) = \frac{2}{s^2 - 2s + 2}. \tag{6.32}$$

With the poles in the right-half plane (at $1 \pm j$), we are constrained to use the second item in the circle criterion. In addition to the Nyquist plot not entering the disc $D(\alpha, \beta)$, we also require that the Nyquist plot encircles the disc as many times as there are right-half plane poles of $G(s)$. Hence, the Nyquist plot needs to encircle the disc twice, and it is clear this cannot happen.

Note, however, that this does *not* imply that no sector exists such that the closed loop has an asymptotically stable equilibrium. Indeed, the circle criterion is a *sufficient* condition; i.e., *if* the condition is satisfied, *then* the system is absolutely stable. This statement tells us nothing about what happens when the condition is *not* satisfied.

However, the system (6.32) is not stabilizable via static output feedback. This can be seen directly from the root locus plot, where it is clear that no static output feedback (i.e., no $u = ky$) can result in a stable system since the closed-loop poles will never be in the right-half plane.

Alternatively, recall that Theorem 3.39 shows that (6.1) cannot be stabilized via the static output feedback $y = Kx$ when $CB = 0$ and $\text{trace}(A) > 0$. The above

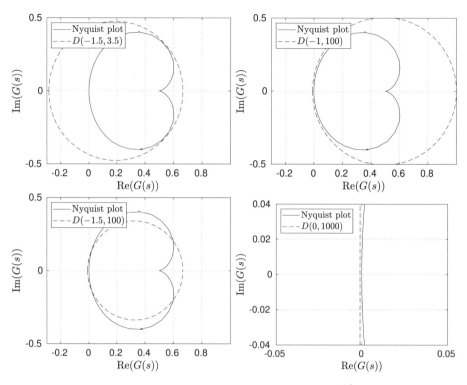

Figure 6.16: Circle criterion for $G(s) = \frac{s+1}{s^2+2s+2}$.

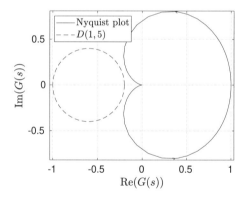

Figure 6.17: Circle criterion for $G(s) = \frac{2}{s^2-2s+2}$. Since $G(s)$ is not Hurwitz the application of the circle criterion 6.14 is restricted to item 2. For any $0 < \alpha < \beta$ the disc $D(\alpha, \beta)$ is to the left of the imaginary axis.

transfer function can be converted to state space form (e.g., via the MATLAB function `tf2ss.m`), yielding matrices

$$A = \begin{bmatrix} 2 & -1 \\ 1 & 0 \end{bmatrix}, \qquad b = \begin{bmatrix} 1 \\ 0 \end{bmatrix}, \qquad c = \begin{bmatrix} 0 & 1 \end{bmatrix},$$

from which we see that, in fact, $cb = 0$ and $\mathrm{trace}(A) = 2$.

6.3.2   Popov Criterion

The circle criterion provides a sufficient condition for absolute stability. The *Popov criterion* provides a second sufficient condition which we discuss next. For the Popov criterion we assume that A is Hurwitz and restrict the sector condition to $\alpha = 0 < \beta$. Additionally, we assume that the nonlinearity is memoryless, i.e., $\psi : \mathbb{R} \to \mathbb{R}$ is a function of y and not a function of (t, y). Recall that a nonlinearity ψ in the sector $0 \le y\psi(y) \le \beta y^2$ satisfies (6.23).

Theorem 6.20 (Popov criterion). *Suppose A is Hurwitz, (A, b) is controllable, (A, c) is observable, and $\psi(y)$ satisfies the sector condition*

$$0 \le y\psi(y) \le \beta y^2 \tag{6.33}$$

for all $y \in \mathbb{R}$. Then the Lur'e system with $G(s) = c(sI - A)^{-1}b$ is absolutely stable if there is an $\eta \ge 0$ with $-\frac{1}{\eta}$ not an eigenvalue of A such that

$$H(s) = 1 + (1 + \eta s)\beta G(s)$$

is strictly positive real.

If ψ only satisfies the sector condition (6.33) for $y \in \Omega \subset \mathbb{R}$, then the system is absolutely stable with a finite domain.

Proof. We assume that there exists an $\eta \ge 0$ such that $-\frac{1}{\eta}$ is not an eigenvalue of A and $H(s)$ is strictly positive real. Then, we can rewrite $H(s)$ in the form

$$
\begin{aligned}
H(s) &= 1 + (1 + \eta s)\beta G(s) \\
&= 1 + (1 + \eta s)\beta c(sI - A)^{-1}b \\
&= 1 + c\left(\beta I + \eta\beta A + \eta\beta sI - \eta\beta A\right)(sI - A)^{-1}b \\
&= 1 + c\left(\beta I + \eta\beta A\right)(sI - A)^{-1}b + \eta\beta c(sI - A)(sI - A)^{-1}b \\
&= 1 + \eta\beta cb + c\left(\beta I + \eta\beta A\right)(sI - A)^{-1}b \\
&= d + \hat{c}(sI - A)^{-1}b,
\end{aligned}
$$

where we have used the definition $d = 1 + \eta\beta cb$ and $\hat{c} = c\left(\beta I + \eta\beta A\right)$ in the last equation. Note that the eigenvalue condition on $-\frac{1}{\eta}$ in particular ensures that $\hat{c} \ne 0$ whenever $c \ne 0$. Since $H(s)$ is strictly positive real by assumption, there exist $P > 0$, L, ε, and w satisfying (6.13).

To prove absolute stability we consider the candidate Lyapunov function

$$V(x) = x^T P x + 2\eta\beta \int_0^y \psi(r)dr. \tag{6.34}$$

Then we can derive the estimate

$$
\begin{aligned}
\dot{V}(x) &= x^T(A^TP + PA)x - 2x^TPb\psi(y) + 2\eta\beta\psi(y)c\dot{x} \\
&= x^T(A^TP + PA)x - 2x^TPb\psi(y) + 2\eta\beta\psi(y)c(Ax - b\psi(y)) \\
&\leq x^T(A^TP + PA)x - 2x^TPb\psi(y) \\
&\quad + 2\eta\beta\psi(y)c(Ax - b\psi(y)) - 2\psi(y)(\psi(y) - \beta y) \\
&= x^T(A^TP + PA)x - 2x^T\left(Pb - \eta\beta A^Tc^T - \beta c^T\right)\psi(y) \\
&\quad - 2\left(\eta\beta cb + 1\right)\psi(y)^2 \\
&= -\varepsilon x^T(A^TP + PA)x - 2x^T\left(Pb - \hat{c}^T\right)\psi(y) - 2d\psi(y)^2.
\end{aligned}
$$

Here, the inequality follows from (6.23).

Using the solution of the equations from Lemma 6.11 we can further rewrite the estimate

$$
\begin{aligned}
\dot{V}(x) &\leq -\varepsilon x^TPx - x^TL^TLx - 2x^T\left(-L^Tw\right)\psi(y) - w^2\psi(y)^2 \\
&= -\varepsilon x^TPx - (x^TL^T)^2 + 2(x^TL^T)w\psi(y) - w^2\psi(y)^2 \\
&= -\varepsilon x^TPx - (x^TL^T - w\psi(y))^2 \\
&\leq -\varepsilon x^TPx.
\end{aligned}
$$

Thus V is a Lyapunov function which shows absolute stability of the system and completes the proof.   □

To develop a graphical technique similar to that of the circle criterion, Lemma 6.10 states that $H(s)$ in Theorem 6.20 is strictly positive real if and only if $H(s)$ is Hurwitz and

$$
\mathrm{Re}(1 + (1 + j\eta\omega)\beta G(j\omega)) > 0 \quad \text{for all } \omega \in \mathbb{R}. \tag{6.35}
$$

The Hurwitz requirement is satisfied since $H(s)$ will have the same poles as $G(s)$ since $-1/\eta$ is not an eigenvalue of A.

For (6.35), $G(j\omega) \in \mathbb{C}$ can be written as $\gamma + j\delta = G(j\omega)$ for $\gamma, \delta \in \mathbb{R}$ for all $\omega \in \mathbb{R}$. Then

$$
\begin{aligned}
\mathrm{Re}(1 + \beta G(j\omega) + j\eta\omega\beta G(j\omega)) &= \mathrm{Re}(1 + \beta(\gamma + j\delta) + j\eta\omega\beta(\gamma + j\delta)) \\
&= \mathrm{Re}(1 + \beta\gamma - \eta\omega\beta\delta + j(\beta\delta + \eta\omega\beta\gamma)) \\
&= 1 + \beta\gamma - \eta\omega\beta\delta \\
&= 1 + \beta\,\mathrm{Re}(G(j\omega)) - \eta\omega\beta\,\mathrm{Im}(G(j\omega)).
\end{aligned}
$$

Hence, (6.35) is equivalent to

$$
\tfrac{1}{\beta} + \mathrm{Re}(G(j\omega)) - \eta\omega\,\mathrm{Im}(G(j\omega)) > 0.
$$

On a plot of $\mathrm{Re}(G(j\omega))$ versus $\omega\,\mathrm{Im}(G(j\omega))$, the above inequality defines a half space on the right side of the line through $-\frac{1}{\beta}$ of slope $\frac{1}{\eta}$. Similarly to how we defined a disc $D(\cdot, \cdot)$ depending on α, β, we define the line

$$
L(\beta, \eta) = \{x \in \mathbb{C} : x = (-\tfrac{1}{\beta} + j\tfrac{1}{\eta})w,\ w \in \mathbb{R}\} \tag{6.36}
$$

depending on $\beta > 0$ and $\eta > 0$. We refer to a plot of $\omega \operatorname{Im}(G(j\omega))$ versus $\operatorname{Re}(G(j\omega))$ including the line $L(\beta, \eta)$ as a *Popov plot*.

Example 6.21. Consider the transfer function $G(s) = \frac{1}{s+1}$. Since the pole of $G(s)$ is -1, the Popov criterion is applicable. Absolute stability can be concluded from the Popov plot in Figure 6.18 for the sector defined through $\alpha = 0$ and $\beta = 100$ and the parameter η defined as $\eta = 0.5$. It is clear from the Popov plot that, in fact, β can be chosen arbitrarily large.

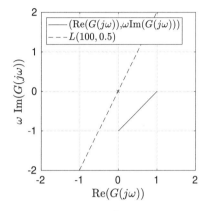

Figure 6.18: Popov criterion for $G(s) = \frac{1}{s+1}$. Absolute stability follows from the fact that $(\operatorname{Re}(G(j\omega)), \omega \operatorname{Im}(G(j\omega))$ is to the right of the line $L(100, 0.5)$.

Example 6.22. Consider the transfer function

$$G(s) = \frac{1}{s^2 + s + 1}$$

with two poles in the open left-half plane. Absolute stability is obtained through the Popov criterion with $\beta = 10$ and $\eta = 0.5$ (Figure 6.19, left) or $\beta = 1000$ and $\eta = 1$ (Figure 6.19, right), for example.

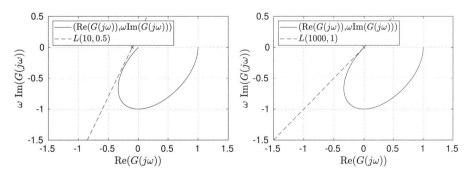

Figure 6.19: Popov criterion for $G(s) = \frac{1}{s^2 + s + 1}$.

6.3.3   Circle versus Popov Criterion

Even though we have only discussed the single-input single-output case here, the circle criterion easily extends to multi-input multi-output systems. The Popov criterion can be extended, but appears to require more structure in the interconnection of the input-output behavior and types of nonlinearities that can be accommodated. The circle criterion allows for time-varying nonlinearities while the Popov criterion requires that the nonlinearities be time-invariant.

The circle criterion yields nice checkable conditions for $\alpha \neq 0$. This stems from the fact that we arrived at investigating the strict positive realness of the ratio

$$\frac{1 + \beta G(s)}{1 + \alpha G(s)}.$$

However, the derivation of the Popov criterion did not provide such a nice expression due to the presence of the $(1 + \eta s)$ factor. Hence, it is generally reasonable to assume $G(s)$ is Hurwitz and take $\alpha = 0$ in the Popov criterion. In this latter case, we observe that if we take $\eta = 0$, we obtain the circle criterion since, with $\alpha = 0$, the circle criterion requires that $1 + \beta G(s)$ be strictly positive real. Hence, as shown in Example 6.22, the freedom to choose $\eta \geq 0$ can provide a less conservative result. The assumption $G(s)$ Hurwitz and $\alpha = 0$ can be accomplished through an appropriate loop transform as in Figure 6.9.

6.4   EXERCISES

Exercise 6.1. Confirm that the quadratic inequalities (6.6) capture the linear sector bounds visualized in Figure 6.7.

Exercise 6.2. Show that the two systems defined through the transfer functions $H(s) = \frac{s}{s+1}$ and $H(s) = \frac{s-2}{s-1}$ are not strictly positive real. In particular, show that $H(s) = \frac{s}{s+1}$ is Hurwitz but does not satisfy (6.11), while $H(s) = \frac{s-2}{s-1}$ satisfies (6.11) but is not Hurwitz.

Exercise 6.3. Consider the transfer function

$$H(s) = c(sI - A)^{-1}b + d$$

defined through

$$A = \begin{bmatrix} 0 & 1 & 1 \\ -3 & -3 & -1 \\ -2 & 2 & 0 \end{bmatrix}, \quad b = \begin{bmatrix} 1 \\ 2 \\ 0 \end{bmatrix}, \quad c = \begin{bmatrix} 0 \\ 2 \\ 1 \end{bmatrix}^T, \quad d = 2.$$

Verify that the $H(s)$ satisfies the assumptions of Lemma 6.11. Compute a solution of the Equations (6.13).

Hint: The calculation of w is straightforward. With w known, the unknown L in (6.13b) can be eliminated through (6.13a). Afterward, for fixed $\varepsilon > 0$, through several transformations, (6.13b) can be reformulated as an algebraic Riccati equation which can be solved in MATLAB through `icare.m` (or `care.m`), for example.

Exercise 6.4. Consider the two linear systems

$$\dot{x}_1 = \begin{bmatrix} -2 & -3 & -4 \\ 1 & 0 & 0 \\ 0 & 1 & 0 \end{bmatrix} x_1 + \begin{bmatrix} 1 \\ 0 \\ 0 \end{bmatrix} u_1, \quad y_1 = \begin{bmatrix} 1 & 2 & 3 \end{bmatrix} x_1 \tag{6.37}$$

$$\dot{x}_2 = \begin{bmatrix} 0 & -1 & 1 \\ 1 & 0 & 0 \\ 0 & 1 & 0 \end{bmatrix} x_2 + \begin{bmatrix} 1 \\ 0 \\ 0 \end{bmatrix} u_2, \quad y_2 = \begin{bmatrix} 1 & 2 & 1 \end{bmatrix} x_2. \tag{6.38}$$

1. Compute transfer functions $G_1(s)$ and $G_2(s)$ of the two linear systems.
2. If applicable, use the three items of the circle criterion to derive sector conditions for the two systems to derive absolute stability properties. In particular, apply every item of the circle criterion or give an explanation why it is not applicable.
3. If applicable, use the Popov criterion to derive sector conditions for the systems or give an explanation why the Popov criterion is not applicable.

Hint: The following functions can be used in MATLAB to visualize the circle criterion (`circle_plot.m`), the Popov criterion (`popov_plot.m`), and the sector condition (`sector_plot.m`).

```
1  function circle_plot(alpha,beta,c_colour)
2
3      if alpha < 0
4          r = (-(1/alpha) + (1/beta))/2;
5          x = linspace(-1/beta,-1/alpha,500);
6          y1 = sqrt(r^2 - (x + 1/beta - r).^2);
7          y2 = -sqrt(r^2 - (x + 1/beta - r).^2);
8          plot(x,y1,c_colour,'LineWidth',2)
9          plot(x,y2,c_colour,'LineWidth',2)
10     elseif alpha > 0
11         r = ((1/alpha) - (1/beta))/2;
12         x = linspace(-1/beta,-1/alpha,500);
13         y1 = sqrt(r^2 - (x + 1/alpha - r).^2);
14         y2 = -sqrt(r^2 - (x + 1/alpha - r).^2);
15         plot(x,y1,c_colour,'LineWidth',2)
16         plot(x,y2,c_colour,'LineWidth',2)
17     else
18         plot([-1/beta -1/beta],[-10 10],c_colour,'LineWidth',2);
19     end
20
21 end
```

```
1  function popov_plot(G,eta,beta)
2      w = 0:0.01:10000;
3      G1 = squeeze(freqresp(G,w));
4      plot(real(G1),w'.*imag(G1),'k','LineWidth',2)
5      hold on
6
7      s = 10*[-1 1];
8      plot(s,eta*s + eta/beta,'r','LineWidth',2)
9      plot(-1/beta,0,'rx','MarkerSize',10,'LineWidth',3)
10     grid on
11     box on
12     title('Popov Plot')
13     xlabel('Re\{G(j\omega)\}')
14     ylabel('\omega Im\{G(j\omega)\}')
15     axis equal
16     xlim([-2 2])
```

```
17    hold off
18 end
```

```
 1 function sector_plot(alpha,beta)
 2     s = -10:0.1:10;
 3     plot(s,alpha*s,'k','LineWidth',3)
 4     hold on
 5     plot(s,beta*s,'k','LineWidth',3)
 6     fill([s fliplr(s)], [beta*s fliplr(alpha*s)], 'r');
 7     grid on
 8   . box on
 9     hold off
10     axis([-10 10 -10 10])
11 end
```

Exercise 6.5. (This exercise discusses [86, Example 10.5].) Consider the feedback connection of the Lur'e problem (6.8) where the linear system is given by

$$\dot{x} = Ax + bu = \begin{bmatrix} 0 & 1 \\ 1 & 0 \end{bmatrix} x + \begin{bmatrix} 0 \\ 1 \end{bmatrix} u \qquad (6.39\text{a})$$

$$y = Cx \quad = \begin{bmatrix} 2 & 1 \end{bmatrix} x \qquad (6.39\text{b})$$

and the nonlinear element is defined through the saturation $\psi(y) = \text{sat}(y)$.

1. Investigate stability properties of the origin of the linear system (6.39) and compute the transfer function.
2. Is it possible to conclude absolute stability (on an infinite domain) using the circle criterion? (With explanation!)
3. Conclude absolute stability on a finite domain using the circle criterion. (For example, the parameters $\alpha = 0.7$ and $\beta = 1.1$ can be used to conclude absolute stability on a finite domain. Additionally, the MATLAB functions in Exercise 6.4 might be helpful.)

We consider the loop transformation

$$u(y) = -\alpha y - \psi(y) + \alpha y = -\alpha y - \psi_T(y) \qquad (6.40)$$

which leads to the dynamics

$$\dot{x} = \hat{A}x - b\psi_T(y) = \begin{bmatrix} 0 & 1 \\ 1 - 2\alpha & -\alpha \end{bmatrix} x - \begin{bmatrix} 0 \\ 1 \end{bmatrix} \psi_T(y) \qquad (6.41\text{a})$$

$$y = Cx = \begin{bmatrix} 2 & 1 \end{bmatrix} x. \qquad (6.41\text{b})$$

The new disturbance ψ_T of the closed loop satisfies the sector condition for the sector defined by $\alpha_T = 0$ and $\beta_T = \beta - \alpha$ on the domain $y \in [-\frac{1}{\alpha}, \frac{1}{\alpha}]$.

4. Show that the system (6.41) is absolutely stable with respect to the nonlinearity ψ_T on the domain $y \in [-\frac{1}{\alpha}, \frac{1}{\alpha}]$.

With the loop transformation (6.40), the transfer function of the closed-loop system is defined as $H(s) = 1 + \beta_T c(sI - \hat{A})^{-1} b$.

5. Compute a quadratic Lyapunov function $V(x) = x^T P x$ with respect to the origin of the system (6.41) by solving the equations of the Kalman-Yakubovich-Popov

lemma (Lemma 6.11).

6. Estimate the region of attraction in form of a level set $\Omega_c = \{x \in \mathbb{R}^2 | V(x) \leq c\}$ (and based on the value α).

6.5 BIBLIOGRAPHICAL NOTES AND FURTHER READING

The absolute stability problem first appeared in a paper by Lur'e and Postnikov [101], which also proposed the form of a Lyapunov function with the integral of the nonlinearity as in equation (6.34). The Lur'e problem is also discussed, for example, in [158, Chapter 5.6].

Questions of intellectual priority, relative importance, and Cold War tensions have led to various orderings in the names when referencing Lemma 6.11. We leave the interested reader to form their own opinion on this matter, with the relevant references being Popov [119] and Yakubovitch [166] (both of which appear in [16] with historical commentary). These together with the Kalman Decomposition Theorem [78] (included here as Proposition 3.36) led to the formulation of the lemma [79].

The connection to positive real functions was recognized by Popov [120] and Brockett [32], which in turn led to the formulation of the circle criterion by Zames [169]. Again, many of these papers, with commentary, can be found in [16].

The servo valve of Example 6.2 is taken from [152, Sec. 2.1.2].

The loop transformation of Figure 6.9 is perhaps the simplest possible example of a more general technique, using so-called *multipliers*, to simplify the absolute stability problem in order to accommodate a wider range of situations. See, for example, [38] for a description. The multiplier approach, and hence the results in this chapter, can be embedded in the more general framework of *integral quadratic constraints (IQCs)* [107]. These more general approaches lose the graphical techniques described in this chapter, which were particularly useful before significant computational power was available. However, both multipliers and IQCs can be cast as linear matrix inequality (LMI) problems (such as those discussed in Section 3.4), and hence are amenable to modern computational techniques.

Chapter Seven

Input-to-State Stability

The absolute stability problem of the previous chapter can be viewed as a first attempt to deal with the issue of *robustness*, which might be framed as an ability for idealized designs to cope with uncertainties, imperfections, or other factors that are not included in the mathematical model used for design purposes. This chapter continues this thread in a way that accommodates a natural extension of the Lyapunov function introduced in Chapter 2. This notion of *input-to-state stability* (ISS) is a cornerstone of the antiwindup designs in the subsequent chapter and, in its extension of Lyapunov functions to allow for inputs, also foreshadows a similar development of control Lyapunov functions in Chapter 9.

7.1 MOTIVATION AND DEFINITION

In this chapter, we develop an analysis framework for *robust* stability, i.e., stability in the presence of external disturbances. As a motivation, consider the linear system

$$\dot{x} = Ax + Ew, \quad x(0) = x_0 \in \mathbb{R}^n, \tag{7.1}$$

with $A \in \mathbb{R}^{n \times n}$, $E \in \mathbb{R}^{n \times m}$, A Hurwitz, and where we have used w, rather than u, to explicitly indicate that the input is a disturbance or uncontrolled signal. We can calculate a bound on the norm of $x(t)$, $t \in \mathbb{R}_{\geq 0}$, directly from the solution

$$x(t) = e^{At}x(0) + \int_0^t e^{A(t-\tau)}Ew(\tau)d\tau$$

as

$$|x(t)| \leq \left|e^{At}x(0)\right| + \left|\int_0^t e^{A(t-\tau)}Ew(\tau)d\tau\right|$$

$$\leq \left\|e^{At}\right\||x(0)| + \int_0^t \left\|e^{A(t-\tau)}\right\|\|E\||w(\tau)|d\tau$$

$$\leq \left\|e^{At}\right\||x(0)| + \left(\|E\|\int_0^\infty \left\|e^{A\tau}\right\|d\tau\right)\operatorname*{ess\,sup}_{\tau \geq 0}|w(\tau)|.$$

Here, $\|\cdot\|$ denotes the *spectral norm* for matrices, in contrast to the *Euclidean norm* $|\cdot|$ for vectors (see Appendix A.1). Note that the integral in the last line is finite since A is Hurwitz. Using the definition $\gamma = \|E\|\int_0^\infty \left\|e^{A\tau}\right\|d\tau$ for fixed

$t \in \mathbb{R}_{\geq 0}$ allows us to simplify the estimate to

$$|x(t)| \leq \|e^{At}\| \, |x(0)| + \gamma \|w\|_{\mathcal{L}_\infty}. \tag{7.2}$$

We note that this bound essentially consists of two components. The first term is a transient bound that shows the decaying effect of the initial state $x(0)$, while the second term is related to the worst-case or largest input disturbance, w, that impacts the system. This leads us to propose the following *input-to-state stability* (ISS) property for the nonlinear system

$$\dot{x} = f(x, w), \quad x(0) = x_0 \in \mathbb{R}^n \tag{7.3}$$

with $w : \mathbb{R}_{\geq 0} \to \mathbb{R}^m$. We denote the set of allowable input functions by

$$\mathcal{W} = \{w : \mathbb{R}_{\geq 0} \to \mathbb{R}^m | \ w \text{ essentially bounded}\}.$$

Definition 7.1 (Input-to-state stability). *System* (7.3) *is said to be* input-to-state stable (ISS) *if there exist* $\beta \in \mathcal{KL}$ *and* $\gamma \in \mathcal{K}$ *such that solutions satisfy*

$$|x(t)| \leq \beta(|x(0)|, t) + \gamma\left(\|w\|_{\mathcal{L}_\infty}\right) \tag{7.4}$$

for all $x \in \mathbb{R}^n$, $w \in \mathcal{W}$, *and* $t \geq 0$.

The function $\gamma \in \mathcal{K}$ is referred to as the *ISS-gain* of (7.3) while $\beta \in \mathcal{KL}$ is referred to as the *transient bound*.

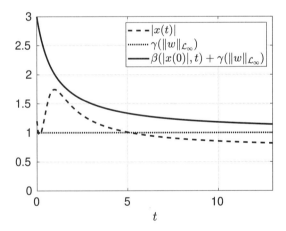

Figure 7.1: Illustration of the ISS property through the bound (7.4).

Another commonly used and equivalent definition for ISS replaces the bound (7.4) with

$$|x(t)| \leq \max\left\{\hat{\beta}(|x(0)|, t), \ \hat{\gamma}\left(\|w\|_{\mathcal{L}_\infty}\right)\right\} \tag{7.5}$$

for appropriate functions $\hat{\beta} \in \mathcal{KL}$ and $\hat{\gamma} \in \mathcal{K}$. The equivalence of these formulations follows from

$$a + b \leq \max\{2a, 2b\} \leq 2a + 2b, \quad \forall \, a, b \in \mathbb{R}_{\geq 0}.$$

In other words, we can replace (7.4) with (7.5) and vice versa, being careful to modify the ISS-gain and transient bound functions as appropriate.

Example 7.2. Based on the estimate (7.2), if A is Hurwitz we see that (7.1) is ISS with the transient bound and ISS-gain given by

$$\beta(s,t) \doteq s\|e^{At}\| \quad \text{and} \quad \gamma(s) \doteq \left(\|E\| \int_0^\infty \|e^{A\tau}\| \, d\tau\right) s,$$

respectively. In particular, observe that the ISS-gain in this case is linear and the transient bound is given by the product of the identity and an exponentially decaying function of time.

We observe that the derivations above indicate that A Hurwitz is sufficient to guarantee that (7.1) is ISS. In other words, if with the input w equal to zero the origin is globally asymptotically stable for (7.1), then (7.1) is ISS. This is not the case for nonlinear systems.

Example 7.3. Consider the one-dimensional bilinear system

$$\dot{x} = -x + xw. \tag{7.6}$$

This system is 0-input globally asymptotically stable since $w = 0$ implies $\dot{x} = -x$ and so $x(t) = x(0)e^{-t}$. Here, 0-input global asymptotic stability (0-GAS) refers to the fact that the origin of the system $\dot{x} = -x$, which represents (7.6) for $w = 0$, is globally asymptotically stable. However, with the constant and bounded input $w = 2$, the system becomes $\dot{x} = x$ and so $x(t) = x(0)e^t$. Consequently, it is impossible to find functions $\beta \in \mathcal{KL}$ and $\gamma \in \mathcal{K}$ such that

$$|x(t)| = |x(0)|e^t \leq \beta(|x(0)|, t) + \gamma(2).$$

7.2   LYAPUNOV CHARACTERIZATIONS

As with proving asymptotic stability of an equilibrium by examining solutions of the relevant differential equation, directly demonstrating that a system is ISS based on Definition 7.1 is difficult. Fortunately, as in the asymptotic stability case, we can use a modified Lyapunov function to prove ISS.

Theorem 7.4 (ISS-Lyapunov function [147, Theorem 1]). *The system (7.3) is input-to-state stable if and only if there exist a continuously differentiable function* $V : \mathbb{R}^n \to \mathbb{R}_{\geq 0}$ *and* $\alpha_1, \alpha_2, \alpha_3, \chi \in \mathcal{K}_\infty$ *such that for all* $x \in \mathbb{R}^n$ *and all* $w \in \mathbb{R}^m$

$$\alpha_1(|x|) \leq V(x) \leq \alpha_2(|x|) \tag{7.7}$$

$$|x| \geq \chi(|w|) \quad \Rightarrow \quad \langle \nabla V(x), f(x, w) \rangle \leq -\alpha_3(|x|). \tag{7.8}$$

We refer to a function satisfying (7.7)–(7.8) as an *ISS-Lyapunov function*. *Sketch of the Proof:* To show that the existence of an ISS-Lyapunov function implies ISS, it is possible to show that the set

$$S_w = \{x \in \mathbb{R}^n : |x| \leq \chi(|w|)\} \tag{7.9}$$

is forward invariant; i.e., once solutions enter the set S_w they remain there for all future time. Importantly, the "size" of this set is dependent only on $|w|$ scaled via $\chi \in \mathcal{K}_\infty$. Outside the set S_w, the decrease condition (7.8) holds and we can use the comparison principle (Lemma 2.13) to obtain a transient bound $\beta \in \mathcal{KL}$. Combining the set S_w and the transient bound thus obtained yields an estimate

$$|x(t)| \leq \max\left\{\beta(|x(0)|, t), \gamma\left(\|w\|_{\mathcal{L}_\infty}\right)\right\}. \tag{7.10}$$

The converse direction, namely that ISS implies the existence of an ISS-Lyapunov function, is significantly more difficult. We refer to the references in Section 7.8 for a proof.   □

As indicated in the sketch of the proof, the set (7.9) is forward-invariant. Since additionally $\dot{V}(x(t)) < 0$ for all $x \notin S_w$ according to (7.8), the set S_w is asymptotically stable.

The decrease condition (7.8) can be shown to be equivalent to

$$\langle \nabla V(x), f(x, w) \rangle \leq -\alpha_3(|x|) + \sigma(|w|), \tag{7.11}$$

where $\sigma \in \mathcal{K}_\infty$. This can be strengthened even further to

$$\langle \nabla V(x), f(x, w) \rangle \leq -V(x) + \sigma(|w|) \tag{7.12}$$

or, in alignment with (7.8),

$$|x| \geq \chi(|w|) \quad \Rightarrow \quad \langle \nabla V(x), f(x, w) \rangle \leq -V(x). \tag{7.13}$$

However, it is important to note that the *same* function V will generally not satisfy all of these conditions.

For a Lyapunov function in dissipation-form (7.11), a forward invariant set S_w (see (7.9)) can be defined through the condition

$$-\alpha_3(|x|) + \sigma(|w|) \leq -\varepsilon\alpha_3(|x|)$$

for $\varepsilon \in (0, 1)$ fixed but arbitrary, for example. Rearranging terms implies $|x| \geq \alpha_3^{-1}(\frac{1}{1-\varepsilon}\sigma(|w|))$ and through the definition $\chi(\cdot) = \alpha_3^{-1}(\frac{1}{1-\varepsilon}\sigma(\cdot))$ the condition (7.8) is recovered where $\alpha_3(\cdot)$ on the right-hand side needs to be replaced by $\varepsilon\alpha_3(\cdot)$.

In the dissipative systems literature, the V of (7.11) is referred to as a *storage function* and (α_3, σ) is referred to as a *supply pair* (see the references in Section 7.8). The terminology is motivated by energy considerations whereby V is a measure of the energy stored in the system and σ and α_3 characterize how much energy has been supplied to the system and how much has been dissipated.

A function satisfying (7.7)–(7.8) is called an *implication-form ISS-Lyapunov function* (due to the implication in (7.8)) while a function satisfying (7.7) and (7.13) is called an *exponential implication-form ISS-Lyapunov function* since, when the state is sufficiently large when compared to the input, the ISS-Lyapunov function decreases at an exponential rate. A function satisfying (7.7) and (7.11) (or (7.12)) is called an *(exponential) dissipation-form ISS-Lyapunov function*. Since all of these formulations are equivalent, if the form of the decrease is unimportant or obvious from context, we simply refer to such a function as an *ISS-Lyapunov function*.

Example 7.5. Consider the simple system

$$\dot{x} = -x - x^3 + xw, \quad x(0) = x_0 \in \mathbb{R}. \tag{7.14}$$

The candidate ISS-Lyapunov function $V(x) = \frac{1}{2}x^2$ satisfies

$$\langle \nabla V(x), f(x, w) \rangle = \langle x, -x - x^3 + xw \rangle = -x^2 - x^4 + x^2 w$$
$$\leq -x^2 - x^4 + \frac{1}{2}x^4 + \frac{1}{2}w^2 = -x^2 - \frac{1}{2}x^4 + \frac{1}{2}w^2,$$

where the inequality follows from Lemma A.4 (Young's inequality). Therefore, with $\alpha(s) \doteq s^2 + \frac{1}{2}s^4$ and $\sigma(s) \doteq \frac{1}{2}s^2$, we have

$$\dot{V}(x) \leq -\alpha(|x|) + \sigma(|w|),$$

so V is an ISS-Lyapunov function and (7.14) is ISS.

Observe that with respect to the discussion at the beginning of the chapter, (7.14) is ISS while (7.6) is not ISS. This is the case even though the systems differ only in a nonlinear term and the linearization of both systems at the origin is the same.

Example 7.6. Consider the second-order system

$$\begin{aligned}\dot{x}_1 &= -x_1 + w \\ \dot{x}_2 &= -x_2^3 + x_1 x_2\end{aligned} \tag{7.15}$$

and the candidate ISS-Lyapunov function

$$V(x) = \frac{1}{2}x_1^2 + \frac{1}{2}x_2^2.$$

Then

$$\langle \nabla V(x), f(x, w) \rangle = \left\langle \begin{bmatrix} x_1 \\ x_2 \end{bmatrix}, \begin{bmatrix} -x_1 + w \\ -x_2^3 + x_1 x_2 \end{bmatrix} \right\rangle$$
$$= -x_1^2 + x_1 w - x_2^4 + x_2^2 x_1$$
$$\leq -x_1^2 + \frac{1}{4}x_1^2 + w^2 - x_2^4 + \frac{1}{2}x_2^4 + \frac{1}{2}x_1^2$$
$$= -\frac{1}{4}x_1^2 - \frac{1}{2}x_2^4 + w^2.$$

Here, the estimate is again obtained using Lemma A.4 (Young's inequality) applied to the terms $x_1 w$ and $x_2^2 x_1$. With

$$\alpha(s) \doteq \begin{cases} \frac{1}{8}s^4, & s \leq 1 \\ \frac{1}{8}s^2, & s > 1 \end{cases} \quad \text{and} \quad \sigma(s) \doteq s^2$$

we have $\dot{V}(x) \leq -\alpha(|x|) + \sigma(|w|)$ and hence (7.15) is ISS.

7.3   SYSTEM INTERCONNECTIONS

ISS extends the concept of asymptotic stability to additionally account for the effect of external disturbances in a natural way. This provides a methodology for the modular analysis of systems, assuming that interconnections of ISS systems are well behaved. In other words, when it is possible to decompose a given system into smaller subsystems, a natural approach to analyzing the behavior of the overall system is to first analyze the individual subsystems and then to investigate how these (hopefully well-behaved) subsystems behave when interconnected in the overall system.

In this section, we will investigate the interconnection of two systems

$$\dot{x}_1 = f_1(x_1, w_1), \tag{7.16a}$$
$$\dot{x}_2 = f_2(x_2, w_2), \tag{7.16b}$$

in two different configurations. First, we will examine systems connected in cascade; i.e., when the state of one system acts as the input to the other system (see Figure 7.2). Second, we will examine systems connected in feedback with external disturbances; i.e., the state of one system plus an external input act as the input to the other system, and vice versa (see Figure 7.4). We emphasize that in this section, the states x_1 and x_2 as well as the disturbances w_1 and w_2 can be multi-dimensional and do not denote one-dimensional variables of the overall system. Specifically, for $n_1, n_2, m_1, m_2 \in \mathbb{N}$, we consider $x_1 \in \mathbb{R}^{n_1}$, $x_2 \in \mathbb{R}^{n_2}$, and with essentially bounded inputs taking values in $\mathbb{R}^{m_1}$ and $\mathbb{R}^{m_2}$.

7.3.1   Cascade Connections

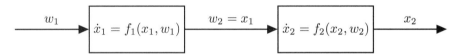

Figure 7.2: Cascade interconnection.

The cascade interconnection shown in Figure 7.2 corresponds to the interconnection condition $w_2 = x_1$; that is, the state of the first system is the input to the second system. Note that necessarily the dimensions must match; i.e., $m_2 = n_1$. The question that we pose is: if both systems are ISS, is the interconnected system also ISS? The answer is a simple "yes." However, the proof relies on the following nontrivial result, for which we first need to introduce the *big $\mathcal{O}$ notation*.

Definition 7.7 (Big $\mathcal{O}$ notation). *Consider two positive functions $\rho_1, \rho_2 \in \mathcal{P}$ and let $c \in \mathbb{R}_{\geq 0} \cup \{\infty\}$. We say that $\rho_1(s) = \mathcal{O}[\rho_2(s)]$ as $s \to c$ if and only if*

$$\limsup_{s \to c} \left| \frac{\rho_1(s)}{\rho_2(s)} \right| < \infty.$$

As an example consider the functions $\alpha_1, \alpha_2 \in \mathcal{K}_\infty$ defined as

$$\alpha_1(s) = 4s^2 \quad \text{and} \quad \alpha_2(s) = \begin{cases} s^2, & s \leq 1, \\ s^4, & s > 1. \end{cases}$$

Then, the calculations

$$\limsup_{s \to 0} \left| \frac{\alpha_1(s)}{\alpha_2(s)} \right| = \limsup_{s \to 0} \left| \frac{4s^2}{s^2} \right| = \limsup_{s \to 0} 4 = 4 < \infty$$

show that $\alpha_1(s) = \mathcal{O}[\alpha_2(s)]$ as $s \to 0$, and similar calculations show that $\alpha_1(s) = \mathcal{O}[\alpha_2(s)]$ as $s \to \infty$. Convince yourself that the converse statements, namely $\alpha_2(s) = \mathcal{O}[\alpha_1(s)]$ as $s \to c$, $c \in \{0, \infty\}$, may be true but do not need to be satisfied in general.

Theorem 7.8 (Changing supply pairs [146, Theorems 1, 2]). *Consider* (7.16), *with* $[x_1^T, x_2^T]^T \in \mathbb{R}^n$, *and with the cascade interconnection* $w_2 = x_1$. *Assume that* $V : \mathbb{R}^n \to \mathbb{R}_{\geq 0}$ *and* $\sigma, \alpha_3 \in \mathcal{K}_\infty$ *satisfy* (7.11).

1. *Suppose that* $\tilde{\sigma} \in \mathcal{K}_\infty$ *satisfies* $\sigma(r) = \mathcal{O}[\tilde{\sigma}(r)]$ *as* $r \to \infty$. *Then there exists* $\tilde{\alpha}_3 \in \mathcal{K}_\infty$ *so that* $(\tilde{\sigma}, \tilde{\alpha}_3)$ *satisfy* (7.11) *for some* $\tilde{V} : \mathbb{R}^n \to \mathbb{R}_{\geq 0}$.
2. *Suppose that* $\tilde{\alpha}_3 \in \mathcal{K}_\infty$ *satisfies* $\tilde{\alpha}_3(r) = \mathcal{O}[\alpha_3(r)]$ *as* $r \to 0$. *Then there exists a* $\tilde{\sigma} \in \mathcal{K}_\infty$ *so that* $(\tilde{\sigma}, \tilde{\alpha}_3)$ *satisfies* (7.11) *for some* $\tilde{V} : \mathbb{R}^n \to \mathbb{R}_{\geq 0}$.

The above theorem essentially says that, if we have an ISS-Lyapunov function, we can find a new ISS-Lyapunov function where we can freely choose the gain $\tilde{\sigma}$ for small arguments (item 1) or we can freely choose the decrease rate $\tilde{\alpha}_3$ for large arguments (item 2). Theorem 7.8 states this in the opposite direction, using the big $\mathcal{O}$ notation to constrain the new gain $\tilde{\sigma}$ to behave similarly to the original gain σ for large arguments (item 1) or to constrain the new decrease rate $\tilde{\alpha}$ to behave similarly to the original decrease rate α for small arguments; i.e.,

$$\limsup_{s \to \infty} \left| \frac{\sigma(s)}{\tilde{\sigma}(s)} \right| < \infty, \quad \text{and} \quad \limsup_{s \to 0} \left| \frac{\tilde{\alpha}_3(s)}{\alpha_3(s)} \right| < \infty.$$

Theorem 7.8 enables a straightforward proof for the following theorem.

Theorem 7.9 (ISS Cascade). *Consider the system* (7.16) *with* $x = [x_1^T, x_2^T]^T \in \mathbb{R}^n$ *where* $n = n_1 + n_2$ *and* $m_2 = n_1$. *If each of the subsystems* (7.16a) *and* (7.16b) *are ISS, then* (7.16) *with* $w_2 = x_1$ *is ISS with* w_1 *as input and* x *as state*.

Proof. Since both subsystems are ISS, there exist ISS-Lyapunov functions (by Theorem 7.4 and (7.11)) satisfying

$$\dot{V}_1(x_1) \leq -\alpha_{3,1}(|x_1|) + \sigma_1(|w_1|),$$
$$\dot{V}_2(x_2) \leq -\alpha_{3,2}(|x_2|) + \sigma_2(|w_2|).$$

The basic idea is now to "match" $\alpha_{3,1}$ and σ_2 using Theorem 7.8. In particular, we can construct a function $\varphi \in \mathcal{K}_\infty$ such that

$$\varphi(s) = \begin{cases} O[\alpha_{3,1}(s)], & \text{as } s \to 0 \\ O[2\sigma_2(s)], & \text{as } s \to \infty \end{cases} \tag{7.17}$$

and apply Theorem 7.8 to obtain $\widetilde{V}_1, \widetilde{V}_2, \tilde{\sigma}_1, \tilde{\alpha}_{3,2}$ satisfying

$$\dot{\widetilde{V}}_1(x_1) \le -\varphi(|x_1|) + \tilde{\sigma}_1(|w_1|), \tag{7.18a}$$

$$\dot{\widetilde{V}}_2(x_2) \le -\tilde{\alpha}_{3,2}(|x_2|) + \tfrac{1}{2}\varphi(|w_2|). \tag{7.18b}$$

Taking $V(x) = \widetilde{V}_1(x_1) + \widetilde{V}_2(x_2)$, and using the interconnection condition $w_2 = x_1$ we have

$$\dot{V}(x) = \dot{\widetilde{V}}_1(x_1) + \dot{\widetilde{V}}_2(x_2) \le -\tfrac{1}{2}\varphi(|x_1|) - \tilde{\alpha}_{3,2}(|x_2|) + \tilde{\sigma}_1(|w_1|). \tag{7.19}$$

Additionally we define the function

$$\alpha_3(s) \doteq \min\{\tfrac{1}{2}\varphi(\tfrac{1}{2}s), \tilde{\alpha}_{3,2}(\tfrac{1}{2}s)\},$$

which satisfies $\alpha_3 \in \mathcal{K}_\infty$, since $\varphi, \tilde{\alpha}_{3,2} \in \mathcal{K}_\infty$, and

$$\alpha_3(|x|) \le \alpha_3(|x_1| + |x_2|) \le \alpha_3(2|x_1|) + \alpha_3(2|x_2|) \le \tfrac{1}{2}\varphi(|x_1|) + \tilde{\alpha}_{3,2}(|x_2|)$$

(see [81, Lemma 10] for the triangle inequality for comparison functions). Thus, we can rewrite the estimate (7.19) in the form

$$\dot{V}(x) \le -\alpha_3(|x|) + \tilde{\sigma}_1(|w_1|),$$

and therefore V is an ISS-Lyapunov function for the system (7.16). $\qquad\square$

It is important to note that the application of Theorem 7.9 does not rely on finding the matched ISS-Lyapunov functions used in the proof. In other words, once we have established that each of the systems (7.16a) and (7.16b) is ISS, we immediately have that the cascade connection is also ISS.

However, in the subsequent section, we will make use of such "matched" ISS-Lyapunov functions as in (7.18) and it is worth noting that the factor of $\frac{1}{2}$ in (7.18b) is somewhat arbitrary. Indeed, the important outcome is the negative definiteness in the state in (7.19). This can be achieved with any scalar value less than 1 and the choice of $\frac{1}{2}$ is for simplicity of exposition only.

Example 7.10. Consider the second-order system

$$\begin{aligned} \dot{x}_1 &= -x_1 + w_1 \\ \dot{x}_2 &= -x_2^3 + x_2 w_2, \end{aligned} \tag{7.20}$$

which represents the dynamics (7.15) with cascade interconnection $w_2 = x_1$. The two Lyapunov functions $V_1(x_1) = \frac{1}{2}x_1^2$ and $V_2(x_2) = \frac{1}{2}x_2^2$ together with the estimates

$$\begin{aligned} \dot{V}_1(x_1) &= -x_1^2 + x_1 w_1 \le -x_1^2 + \tfrac{1}{2}x_1^2 + \tfrac{1}{2}w_1^2 = -\tfrac{1}{2}x_1^2 + \tfrac{1}{2}w_1^2 \\ \dot{V}_2(x_2) &= -x_2^4 + x_2^2 w_2 \le -x_2^4 + \tfrac{1}{2}x_2^4 + \tfrac{1}{2}w_2^2 = -\tfrac{1}{2}x_2^4 + \tfrac{1}{2}w_2^2 \end{aligned} \tag{7.21}$$

show ISS of the two subsystems. Therefore, Theorem 7.9 guarantees ISS of the cascade system (7.20) with interconnection $x_1 = w_2$, which we already concluded in Example 7.6 by constructing an ISS-Lyapunov function $V(x)$.

Note that, as the input and state dimensions match, and since both systems are ISS, we can swap the order of the cascade interconnection (i.e., $x_2 = w_1$), and the resulting system is again ISS.

7.3.2   Feedback Interconnections

Consider now a feedback interconnection of (7.16) as shown in Figure 7.3, where the interconnections are given by $w_2 = x_1$ and $w_1 = k(x_2)$ where $k : \mathbb{R}^{n_2} \to \mathbb{R}^{m_1}$ is a continuous function. We pose a similar question as in the previous section: if we connect two ISS systems in this way is the resulting system well behaved? In particular, we might consider "well behaved" to mean the origin is asymptotically stable.

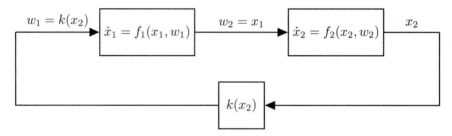

Figure 7.3: Feedback interconnection.

Assume that we have matched ISS-Lyapunov functions as in (7.18), except replace the factor of 2 in (7.17) with $1/\varepsilon$, where $\varepsilon \in (0,1)$; i.e.,

$$\varphi(s) = \begin{cases} \mathcal{O}[\alpha_{3,1}(s)], & \text{as } s \to 0 \\ \mathcal{O}\left[\frac{1}{\varepsilon}\sigma_2(s)\right], & \text{as } s \to \infty \end{cases}$$

so that

$$\dot{V}_1(x_1) \leq -\varphi(|x_1|) + \sigma_1(|w_1|) \tag{7.22a}$$

$$\dot{V}_2(x_2) \leq -\alpha_{3,2}(|x_2|) + \varepsilon\varphi(|w_2|), \tag{7.22b}$$

where, to simplify the notation, we have dropped the tilde markers. Take $V(x) = V_1(x_1) + V_2(x_2)$ as a candidate Lyapunov function. Then, with the feedback interconnection conditions in Figure 7.3,

$$\dot{V}(x) = \dot{V}_1(x_1) + \dot{V}_2(x_2) \leq -\varphi(|x_1|) + \sigma_1(|k(x_2)|) - \alpha_{3,2}(|x_2|) + \varepsilon\varphi(|x_1|)$$
$$= -(1-\varepsilon)\varphi(|x_1|) - \alpha_{3,2}(|x_2|) + \sigma_1(|k(x_2)|). \tag{7.23}$$

If the right-hand side of (7.23) is negative definite, then V is a Lyapunov function and we can conclude asymptotic stability of the origin. This will be the case if, for some $\bar{\varepsilon} \in (0,1)$,

$$\sigma_1(|k(x_2)|) \leq (1-\bar{\varepsilon})\alpha_{3,2}(|x_2|).$$

We have therefore proved the following theorem.

Theorem 7.11 (ISS small-gain). *Consider system (7.16) with $w_2 = x_1 \in \mathbb{R}^{n_1}$, $x_2 \in \mathbb{R}^{n_2}$ and $w_1 \in \mathbb{R}^{m_1}$. Suppose the subsystems (7.16a) and (7.16b) are connected in feedback as in Figure 7.3 and we have matched ISS-Lyapunov functions as in (7.22). If the nonlinear function $k : \mathbb{R}^{n_1} \to \mathbb{R}^{m_2}$ satisfies*

$$|k(x_2)| \leq \sigma_1^{-1}\left((1 - \bar{\varepsilon})\alpha_{3,2}(|x_2|)\right) \tag{7.24}$$

for some $\bar{\varepsilon} \in (0, 1)$, then the origin of (7.16) is asymptotically stable for the closed-loop system.

Theorem 7.11 is an example of what is called a *small-gain theorem* and (7.24) is referred to as a *small-gain condition*. Intuitively, small-gain theorems place limits on the loop-gain of a feedback system so that signals are not amplified as they traverse the feedback loop.

Without relying on matched ISS-Lyapunov functions, we can similarly show the following result.

Theorem 7.12. *Consider system (7.16) with $w_2, x_1 \in \mathbb{R}^{n_1}$ and $w_1, x_2 \in \mathbb{R}^{n_2}$. Suppose the subsystems (7.16a) and (7.16b) are connected in feedback by $w_1 = x_2$ and $w_2 = x_1$. If each of the systems is ISS with ISS-Lyapunov functions*

$$\begin{aligned}
\dot{V}_1(x_1) &\leq -\alpha_{3,1}(V_1(x_1)) + \sigma_1(V_2(x_2)) \\
\dot{V}_2(x_2) &\leq -\alpha_{3,2}(V_2(x_2)) + \sigma_2(V_1(x_1))
\end{aligned} \tag{7.25}$$

(and $\alpha_{3,1}, \alpha_{3,2}, \sigma_1, \sigma_2 \in \mathcal{K}_\infty$) and if, for all $s \geq 0$,

$$\begin{aligned}
\alpha_{3,1}^{-1} \circ \sigma_2(s) &< s \\
\alpha_{3,2}^{-1} \circ \sigma_1(s) &< s
\end{aligned} \tag{7.26}$$

then the origin of (7.16) is asymptotically stable for the closed-loop system.

The proof of the theorem is left to the reader in Exercise 7.3. It is important to note that small-gain theorems generally present *sufficient* conditions. In other words, the fact that the small-gain conditions do not hold does *not* imply that the origin is not asymptotically stable.

Example 7.13. In this example we illustrate the application of Theorem 7.11. We consider again the dynamics (7.20). Additionally we consider the functions $V_1(x_1) = \frac{\varepsilon}{2}x_1^2$ for $\varepsilon \in (0, 1)$ and $V_2(x_2) = \frac{1}{2}x_2^2$. For $\alpha_{3,1}(s) = \frac{\varepsilon}{2}s^4$, $\sigma_2(s) = \frac{1}{2}s^2$, and $\varphi(s) = \frac{1}{2}s^2$, similar calculations to (7.21) show that the inequalities

$$\dot{V}_2(x_2) \leq -\varphi(|x_2|) + \sigma_2(|w_2|) \tag{7.27a}$$

$$\dot{V}_1(x_1) \leq -\alpha_{3,1}(|x_1|) + \varepsilon\varphi(|w_1|) \tag{7.27b}$$

are satisfied. Thus, for the feedback interconnection $w_2 = k(x_1)$, $w_1 = x_2$, the two ISS-Lyapunov functions V_1, V_2 are matched as in (7.22). We can use Theorem 7.11 to verify asymptotic stability of the origin for any feedback satisfying (7.24). In particular, with $\sigma_2^{-1}(s) = \sqrt{2s}$ we obtain the condition

$$|k(x_1)| \leq \sqrt{2(1 - \bar{\varepsilon})\frac{\varepsilon}{2}|x_1|^4} = \sqrt{(1 - \bar{\varepsilon})\varepsilon}x_1^2. \tag{7.28}$$

Since in this example the selection of $\varepsilon \in (0,1)$ in the ISS-Lyapunov function as well as $\bar{\varepsilon} \in (0,1)$ in Theorem 7.11 are arbitrary, all of the feedback functions

$$k(x_1) = \tfrac{1}{2}x_1^2, \quad k(x_1) = \tfrac{1}{2}\operatorname{sign}(x_1)x_1^2 \quad \text{or} \quad k(x_1) = \tfrac{1}{2}\operatorname{sat}(x_1^2)$$

satisfy the condition (7.28) for $\varepsilon = \bar{\varepsilon} = \tfrac{1}{2}$. Therefore asymptotic stability of the origin of the feedback interconnected system using any of these is guaranteed through Theorem 7.11.

Note that if we change the feedback interconnection to $w_1 = k(x_2)$, $w_2 = x_1$, then the Lyapunov functions V_1 and V_2 are not appropriately matched and thus Theorem 7.11 cannot be applied directly. However, for a different selection of V_1 and V_2 as well as φ, a stabilizing feedback can be constructed (see Exercise 7.2).

Remark 7.14. Note the similarity between Theorem 7.11 and the absolute stability results of Chapter 6. In particular, Theorems 6.12, 6.14, and 6.20 require some assumptions on the system (e.g., strict positive realness of a transfer function) and guarantee asymptotic stability for all nonlinearities satisfying a sector bound. In Theorem 7.11, the subsystems must be ISS and then the nonlinearity needs to satisfy the bound (7.24).

Before we continue with the next topic we show how Theorem 7.12 can be applied.

Example 7.15. Consider the dynamical system

$$\begin{aligned}
\dot{x}_1 &= -x_1^3 + x_1 w_1, \\
\dot{x}_2 &= -x_2 + \tfrac{1}{2}w_2^2.
\end{aligned} \tag{7.29}$$

The functions $V_1(x_1) = \tfrac{1}{2}x_1^2$ and $V_2(x_2) = \tfrac{1}{2}x_2^2$ satisfy the estimates

$$\begin{aligned}
\dot{V}_1(x_1) &= -x_1^4 + x_1^2 w_1 \le -x_1^4 + \tfrac{1}{2}x_1^4 + \tfrac{1}{2}w_1^2 = -2V_1(x_1)^2 + V_2(w_1), \\
\dot{V}_2(x_2) &= -x_2^2 + \tfrac{1}{2}x_2 w_2^2 \le -x_2^2 + \tfrac{1}{4}x_2^2 + \tfrac{1}{4}w_2^4 = -\tfrac{3}{2}V_2(x_2) + V_1(w_1)^2.
\end{aligned}$$

With the definitions $\alpha_{3,1}(s) = 2s^2$, $\sigma_1(s) = \tfrac{1}{2}s$, $\alpha_{3,2}(s) = \tfrac{3}{2}s$, and $\sigma_2(s) = s^2$, the inequalities (7.25) are satisfied. It is not difficult to show that (7.26) holds and therefore the origin of the system (7.29) is ISS according to Theorem 7.12.

7.4   INTEGRAL-TO-INTEGRAL ESTIMATES AND $\mathcal{L}_2$-GAIN

We can derive an alternate ISS estimate by starting from a dissipation-form ISS-Lyapunov function (7.11) and integrating both sides. In other words, from

$$\tfrac{d}{dt}V(x(t)) = \langle \nabla V(x(t)), f(x(t), w(t)) \rangle \le -\alpha_3(|x(t)|) + \sigma(|w(t)|)$$

we compute

$$V(x(t)) - V(x(0)) \le -\int_0^t \alpha_3(|x(\tau)|)d\tau + \int_0^t \sigma(|w(\tau)|)d\tau.$$

Rearranging terms and using the upper bound $V(x) \leq \alpha_2(|x|)$ from (7.7) and the fact that V is positive definite, we have

$$\int_0^t \alpha_3(|x(\tau)|)d\tau \leq \int_0^t \alpha_3(|x(\tau)|)d\tau + V(x(t)) \leq V(x(0)) + \int_0^t \sigma(|w(\tau)|)d\tau$$

$$\leq \alpha_2(|x(0)|) + \int_0^t \sigma(|w(\tau)|)d\tau. \tag{7.30}$$

Since ISS implies the existence of an ISS-Lyapunov function (Theorem 7.4), we see that ISS also implies an estimate of the form (7.30). With an additional technical assumption, it is possible to show that (7.30) also implies ISS. In other words, the ISS estimate (7.4) is equivalent to (7.30). This observation is summarized in the next result, where forward completeness means that solutions of (7.3) are defined for all inputs in $\mathcal{W}$ for all forward time $t_{\geq 0}$.

Lemma 7.16 ([143, Theorem 1]). *Consider the nonlinear system (7.3). If the system is ISS according to Definition 7.1, then there exist $\alpha_2, \alpha_3, \sigma \in \mathcal{K}_\infty$ such that (7.30) is satisfied for all $t \geq 0$. Conversely, if (7.3) is forward complete and satisfies (7.30) for $\alpha_2, \alpha_3, \sigma \in \mathcal{K}_\infty$ for all $t \geq 0$, then the system is ISS according to Definition 7.1.*

Consider the linear system (7.1), where we have assumed that A is Hurwitz. By Theorem 3.3 we can find a solution P to the Lyapunov equation

$$A^T P + PA = -2I.$$

For the function $V(x) = x^T Px$ we observe that the time derivative can be upper bounded through the estimate

$$\dot{V}(x) = \dot{x}^T Px + x^T P\dot{x} = x^T A^T Px + w^T E^T Px + x^T PAx + x^T PEw$$
$$= -2x^T x + 2x^T PEw \leq -2x^T x + 2|x|\,|w|\,\|PE\| \tag{7.31}$$
$$\leq -2x^T x + x^T x + \|PE\|^2 w^T w = -x^T x + \|PE\|^2 w^T w,$$

where we have used the Cauchy-Schwarz inequality and Young's inequality (see Lemmas A.2 and A.4) to obtain the inequalities. As above, we can integrate both sides of this inequality and rearrange terms to obtain

$$\int_0^t |x(\tau)|^2 d\tau \leq \lambda_{\max}|x(0)|^2 + \|PE\|^2 \int_0^t |w(\tau)|^2 d\tau. \tag{7.32}$$

As before, $\lambda_{\max} \in \mathbb{R}_{>0}$ denotes the largest eigenvalue of the symmetric matrix P. With respect to the estimate (7.30) we can identify the functions $\alpha_3(s) = s^2$, $\alpha_2(s) = \lambda_{\max}s^2$, and $\sigma(s) = \|PE\|^2 s^2$ as a general selection for linear systems (7.1) with A Hurwitz. Using the notation from Chapter 4 we can alternatively write the estimate (7.32) using the $\mathcal{L}_2$-norm

$$\|x\|_{\mathcal{L}_2}^2 \leq \lambda_{\max}|x(0)|^2 + \gamma^2 \|w\|_{\mathcal{L}_2}^2,$$

where $\gamma = \|PE\|$. This motivates the following definition.

Definition 7.17 ($\mathcal{L}_2$-stability). *System (7.3) is said to be $\mathcal{L}_2$-stable or to have* finite $\mathcal{L}_2$-gain *if there exist constants* $\kappa, \gamma > 0$ *so that*

$$\|x\|_{\mathcal{L}_2}^2 \leq \kappa |x(0)|^2 + \gamma^2 \|w\|_{\mathcal{L}_2}^2 \tag{7.33}$$

for all $w \in \mathcal{W}$.

Note that, in the literature, it is common to assume $x(0) = 0$ and hence the above definition is frequently written simply as

$$\|x\|_{\mathcal{L}_2}^2 \leq \gamma^2 \|w\|_{\mathcal{L}_2}^2.$$

For linear systems, we can use Parseval's theorem, Proposition 4.10, to make an interesting connection between the Bode Plot of the transfer function introduced in Section 4.2.3 and input-output $\mathcal{L}_2$-stability. Consider the single-input single-output system

$$\dot{x} = Ax + Ew, \qquad y = Cx$$

and its representation in the frequency domain

$$\hat{y}(s) = G(s)\hat{w}(s), \qquad G(s) = C(sI - A)^{-1}E.$$

Then Proposition 4.10 yields

$$\|y\|_{\mathcal{L}_2}^2 = \int_0^\infty |y(\tau)|^2 \, d\tau = \frac{1}{2\pi} \int_{-\infty}^\infty |\hat{y}(jw)|^2 \, dw,$$

which can be further rewritten by using the system representation in the frequency domain and the definition of the $\mathcal{H}_\infty$-norm (see (4.10))

$$\begin{aligned}
\|y\|_{\mathcal{L}_2}^2 &\leq \frac{1}{2\pi} \int_{-\infty}^\infty |G(jw)|^2 |\hat{w}(jw)|^2 \, d\omega \\
&\leq \operatorname{ess\,sup}_\omega |G(j\omega)|^2 \frac{1}{2\pi} \int_{-\infty}^\infty |\hat{w}(jw)|^2 \, d\omega \\
&= \|G\|_\infty^2 \int_0^\infty |w(\tau)|^2 \, d\tau \\
&= \|G\|_\infty^2 \|w\|_{\mathcal{L}_2}^2. \tag{7.34}
\end{aligned}$$

Therefore, with $\gamma = \|G\|_\infty$ we see that the $\mathcal{L}_2$-gain of a linear system is precisely the peak magnitude of the transfer function and can be read off directly from the Bode Plot.

Remark 7.18. Note that the estimate also holds for multi-input multi-output systems. However, in this case the definition of the $\mathcal{H}_\infty$-norm for multi-input multi-output systems has to be used.

7.4.1   System interconnections

Observe that Lemma 7.16 and (7.33) imply that $\mathcal{L}_2$-stability implies ISS. This follows from the fact that the functions $\alpha_3(s) = s^2$, $\alpha_2(s) = \kappa s^2$, and $\sigma(s) = \gamma s^2$ are a special case of the more general relationship (7.30). However, consideration of this particular special case leads to some relatively straightforward and easily applicable results.

For simplicity, in this section we will assume $x(0) = 0$. Note that the $\mathcal{L}_2$-norm satisfies the inequality

$$\|v_1 + v_2\|_{\mathcal{L}_2}^2 \le \|v_1\|_{\mathcal{L}_2}^2 + \|v_2\|_{\mathcal{L}_2}^2, \tag{7.35}$$

for $v_1, v_2 : [0, t) \to \mathbb{R}^n$, which follows immediately from the definition of the $\mathcal{L}_2$-norm.

We have already seen that the cascade interconnection of two ISS systems is again ISS (Theorem 7.9). It is therefore immediate that the cascade interconnection of two $\mathcal{L}_2$-stable systems is ISS. However, it is not difficult to demonstrate the following theorem.

Theorem 7.19 ($\mathcal{L}_2$-stable cascade). *Consider system (7.16) with $[x_1^T, x_2^T]^T \in \mathbb{R}^n$ where $n = n_1 + n_2$ and $m_2 = n_1$. If each of the subsystems (7.16a) and (7.16b) is $\mathcal{L}_2$-stable, then (7.16) with $w_2 = x_1$ is $\mathcal{L}_2$-stable.*

The proof follows directly from Definition 7.17 and is left to the reader as Exercise 7.5.

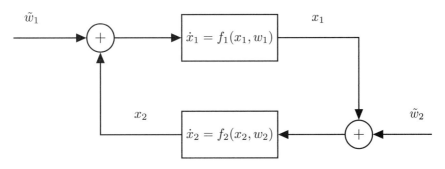

Figure 7.4: Feedback interconnection with external inputs.

For the feedback interconnection in Figure 7.4 it is possible to derive the following simple small-gain theorem.

Theorem 7.20 ($\mathcal{L}_2$ small-gain). *Consider the system (7.16) with $[x_1^T, x_2^T]^T \in \mathbb{R}^n$ where $n = n_1 + n_2$, $m_1 = n_2$, and $m_2 = n_1$. If each of the subsystems (7.16a) and (7.16b) are $\mathcal{L}_2$-stable with gains $\gamma_1, \gamma_2 > 0$, then (7.16) with $w_1 = x_2 + \tilde{w}_1$ and $w_2 = x_1 + \tilde{w}_2$ is $\mathcal{L}_2$-stable if $\gamma_1 \gamma_2 < 1$.*

Proof. Using (7.33) and (7.35), we first bound the $\mathcal{L}_2$-norm of x_1 by

$$\begin{aligned}
\|x_1\|_{\mathcal{L}_2}^2 &\le \gamma_1^2 \|w_1\|_{\mathcal{L}_2}^2 = \gamma_1^2 \|\tilde{w}_1 + x_2\|_{\mathcal{L}_2}^2 \\
&\le \gamma_1^2 \|\tilde{w}_1\|_{\mathcal{L}_2}^2 + \gamma_1^2 \|x_2\|_{\mathcal{L}_2}^2 \\
&\le \gamma_1^2 \|\tilde{w}_1\|_{\mathcal{L}_2}^2 + \gamma_1^2 \gamma_2^2 \|w_2\|_{\mathcal{L}_2}^2 = \gamma_1^2 \|\tilde{w}_1\|_{\mathcal{L}_2}^2 + \gamma_1^2 \gamma_2^2 \|\tilde{w}_2 + x_1\|_{\mathcal{L}_2}^2 \\
&\le \gamma_1^2 \|\tilde{w}_1\|_{\mathcal{L}_2}^2 + \gamma_1^2 \gamma_2^2 \|\tilde{w}_2\|_{\mathcal{L}_2}^2 + \gamma_1^2 \gamma_2^2 \|x_1\|_{\mathcal{L}_2}^2,
\end{aligned}$$

which implies

$$\|x_1\|_{\mathcal{L}_2}^2(1 - \gamma_1^2\gamma_2^2) \leq \gamma_1^2\|\tilde{w}_1\|_{\mathcal{L}_2}^2 + \gamma_1^2\gamma_2^2\|\tilde{w}_2\|_{\mathcal{L}_2}^2.$$

If $\gamma_1\gamma_2 < 1$ then $\gamma_1^2\gamma_2^2 < 1$ and the estimate

$$\|x_1\|_{\mathcal{L}_2}^2 \leq \frac{1}{1 - \gamma_1^2\gamma_2^2}\left(\gamma_1^2\|\tilde{w}_1\|_{\mathcal{L}_2}^2 + \gamma_1^2\gamma_2^2\|\tilde{w}_2\|_{\mathcal{L}_2}^2\right) \tag{7.36}$$

is satisfied. We can repeat the same arguments to obtain a similar upper bound on the $\mathcal{L}_2$-norm of x_2.

The bounds on x_1 and x_2 can be combined as

$$\begin{aligned}\|x\|_{\mathcal{L}_2}^2 &= \|x_1\|_{\mathcal{L}_2}^2 + \|x_2\|_{\mathcal{L}_2}^2 \\ &\leq \frac{1}{1 - \gamma_1^2\gamma_2^2}\left(\gamma_1^2\|\tilde{w}_1\|_{\mathcal{L}_2}^2 + \gamma_2^2\|\tilde{w}_2\|_{\mathcal{L}_2}^2 + \gamma_1^2\gamma_2^2(\|\tilde{w}_1\|_{\mathcal{L}_2}^2 + \|\tilde{w}_2\|_{\mathcal{L}_2}^2)\right)\end{aligned}$$

and hence the feedback interconnection in Figure 7.4 is $\mathcal{L}_2$-stable. □

The reader should take the time to compare the small-gain conditions presented in Theorem 7.12 and Theorem 7.20, noting the relative simplicity of the condition $\gamma_1\gamma_2 < 1$.

7.5 INTEGRAL ISS AND NONLINEAR $\mathcal{L}_2$-GAIN

As we observed in Example 7.3 by considering the constant input $w(t) = 2$, the bilinear system

$$\dot{x} = -x + xw \tag{7.37}$$

is not ISS. However, if we consider the integrable input $w(t) = e^{-t}$ then the solution of (7.37) is given by

$$x(t) = x(0)\exp(e^{-t} - 1). \tag{7.38}$$

So, while a constant input causes the solution $x(t)$ to diverge, what we might consider a well-behaved input—in the sense of the input being integrable—leads to reasonable behavior of the solution—in the sense of the solution being bounded. This leads to the following concept of *integral* input to state stability.

Definition 7.21 (Integral input-to-state stability). *System* (7.3) *is said to be integral input-to-state stable (iISS) if there exist* $\beta \in \mathcal{KL}$ *and* $\alpha, \gamma \in \mathcal{K}_\infty$ *such that solutions satisfy*

$$\alpha(|x(t)|) \leq \beta(|x(0)|, t) + \int_0^t \gamma\left(|w(\tau)|\right)d\tau \tag{7.39}$$

for all $x \in \mathbb{R}^n$, $w \in \mathcal{W}$, *and* $t \geq 0$.

Like ISS, iISS is not the same as 0-GAS. As an example consider the system

$$\dot{x} = -x + (x^2 + 1)w. \tag{7.40}$$

Since $\dot{x} = -x$ for $w = 0$, it follows immediately that (7.40) is 0-GAS. However, for $w(t) = (2t+2)^{-1/2}$ and $x(0) = \sqrt{2}$ one can verify that the solution of (7.40) satisfies $x(t) = \sqrt{2t+2}$. In other words, even in the presence of inputs that converge to zero the state may diverge.

Moreover, the constant input $w(t) = 1$ leads to a finite escape time. The interested reader can verify that the solution includes a tangent function of time. It is clear then that the estimate (7.39) is violated for any choice of $\alpha, \gamma \in \mathcal{K}_\infty$ and (7.40) is not iISS. This example will occur again in Section 9.2 when we discuss robust feedback designs.

Theorem 7.22 (iISS-Lyapunov function [8, Theorem 1]). *The system (7.3) is integral input-to-state stable (iISS) if and only if there exist a continuously differentiable function $V : \mathbb{R}^n \to \mathbb{R}_{\geq 0}$ and $\alpha_1, \alpha_2, \sigma \in \mathcal{K}_\infty$ and $\alpha_3 \in \mathcal{P}$ such that for all $x \in \mathbb{R}^n$ and all $w \in \mathbb{R}^m$*

$$\alpha_1(|x|) \leq V(x) \leq \alpha_2(|x|) \tag{7.41}$$
$$\langle \nabla V(x), f(x, w) \rangle \leq -\alpha_3(|x|) + \sigma(|w|). \tag{7.42}$$

Note that the iISS-Lyapunov function decrease (7.42) is almost the same as the ISS-Lyapunov function decrease (7.11), with only a change in the required property for α_3. In particular, for ISS we require $\alpha_3 \in \mathcal{K}_\infty$, while for iISS we only require $\alpha_3 \in \mathcal{P}$. Since $\mathcal{K}_\infty \subset \mathcal{P}$, we see that ISS implies iISS. However, the bilinear system (7.37) provides an example of a system that is iISS but not ISS.

Without providing details, similarly to the derivations in Section 7.4 it is possible to obtain an integral-to-integral estimate that is equivalent to the iISS definition (7.39). Similarly to how (7.30) leads to the finite $\mathcal{L}_2$-gain property (7.33) as a special but important case, this equivalent integral-to-integral estimate leads to the special case of finite nonlinear $\mathcal{L}_2$-gain.

Definition 7.23. *System (7.3) is said to have* finite nonlinear $\mathcal{L}_2$-gain *if there exist functions $\kappa, \gamma \in \mathcal{K}_\infty$ so that*

$$\|x\|_{\mathcal{L}_2}^2 \leq \kappa(|x(0)|) + \gamma\left(\|w\|_{\mathcal{L}_2}^2\right) \tag{7.43}$$

for all $w \in \mathcal{W}$.

Finally, the following theorem provides explicit bounds for the important class of bilinear systems.

Theorem 7.24 ([46, Proposition 1]). *Fix $k > 0$. Given the bilinear system*

$$\dot{x} = Ax + \sum_{i=1}^{m} w_i A_i x + Bw \tag{7.44}$$

with $A, A_i \in \mathbb{R}^{n \times n}$, $B \in \mathbb{R}^{n \times m}$, *and* A *Hurwitz, choose* $\epsilon_0 > 0$ *and* $\epsilon \in \mathbb{R}^m_{>0}$ *so that*

$$\eta = 2\|A\| - \epsilon_0^2 - \sum_{i=1}^{m} \|A_i\| \epsilon_i^2 > 0.$$

Let

$$\rho_\epsilon = \max_{i \in \{1, \ldots, m\}} \left(\frac{\|A_i\|}{\epsilon_i^2} \right).$$

Then solutions of (7.44) *satisfy* (7.43) *with*

$$\kappa(s) = \frac{1}{\eta} s^2 + \frac{1}{2k^2 \eta} s^4$$

$$\gamma(s) = \frac{1}{\eta} \left(\frac{k^2}{2} (\exp(\rho_\epsilon s) - 1)^2 + \frac{\|B\|^2}{\epsilon_0^2} \exp(\rho_\epsilon s) s \right).$$

We conclude this section by observing that integral ISS is not as useful as ISS when considering modular analysis of interconnected systems. For example, in contrast to the ISS case, the cascade interconnection of two iISS systems is not necessarily iISS and a necessary requirement for asymptotic stability of the origin for the feedback interconnection of two iISS systems is that one of them must, in fact, be ISS. These topics continue to be the subject of active research and some references are included in the bibliographical notes in Section 7.8.

7.6 DISSIPATIVITY AND PASSIVITY

ISS is defined in Definition 7.1 based on a particular desired behavior of system trajectories. However, given the equivalence of Theorem 7.4, we could have alternatively defined an ISS system as one possessing an ISS-Lyapunov function. The properties of dissipativity and passivity take this view.

Consider the nonlinear system (7.3) with an output given by

$$\dot{x} = f(x, w), \quad y = h(x, w), \quad x \in \mathbb{R}^n, w \in \mathbb{R}^m, y \in \mathbb{R}^p. \tag{7.45}$$

Definition 7.25 (Dissipativity). *System* (7.45) *is said to be* dissipative *if there exists a continuously differentiable function* $S : \mathbb{R}^n \to \mathbb{R}_{\geq 0}$, *which we call a* storage function, *and a continuous function* $s : \mathbb{R}^m \times \mathbb{R}^p \to \mathbb{R}$, *which we call a* supply rate, *satisfying*

$$S(x(t)) \leq S(x(0)) + \int_0^t s(w(\tau), y(\tau)) d\tau. \tag{7.46}$$

The notion of dissipativity is closely tied to energy considerations and the capacity for systems to store supplied energy. The intuition behind (7.46) is that the energy stored in the system at time $t \geq 0$, given by $S(x(t))$, must be less than the energy initially stored in the system, given by $S(x(0))$, plus the energy *supplied* to the system between time 0 and time t.

Frequently, the differential form of (7.46) is easier to work with; i.e.,

$$\frac{d}{dt}S(x) = \dot{S}(x) \leq s(w, y). \tag{7.47}$$

Remark 7.26. Observe that ISS is a special case of dissipativity. Specifically, choose the output function $h(x, w) = x$ and the supply rate

$$s(w, y) = -\alpha_3(|y|) + \sigma(|w|)$$

from (7.11). Then the ISS-Lyapunov function is the required storage function, i.e., $S(x) = V(x)$.

We can see that ISS implies 0-GAS. However, to infer 0-GAS from dissipativity requires some extra assumptions. The primary one is an assumption of zero-state detectability, which is that if $w(t) = 0$, then $y(t) = 0$ implies that $x(t) \to 0$ as $t \to \infty$. By additionally assuming a positive definite and radially unbounded storage function, and a supply rate satisfying both $s(0, y) \leq 0$ and $s(0, y) = 0$ implies $y = 0$, the interested reader should be able to conclude 0-GAS from dissipativity.

An important sub-class of dissipative systems is the class of passive systems. Note the requirement that the input and output have the same dimension.

Definition 7.27 (Passive systems). *System* (7.45), *with* $m = p$, *is said to be passive if it is dissipative with supply rate* $s(w, y) = \langle w, y \rangle$. *Furthermore, the system is said to be input strictly passive if the supply rate satisfies* $s(w, y) = \langle w, y \rangle - \delta |w|^2$ *for some* $\delta > 0$, *and output strictly passive if* $s(w, y) = \langle w, y \rangle - \varepsilon |y|^2$ *for some* $\varepsilon > 0$.

Example 7.28. Consider the mass-spring system described in Example 1.2 and take as the output the mass velocity. For ease of reference, we repeat the system equations

$$\dot{x}_1(t) = x_2(t) \tag{7.48a}$$
$$\dot{x}_2(t) = -\frac{k}{m}x_1(t) - \frac{c}{m}x_2(t) + \frac{1}{m}u(t) \tag{7.48b}$$
$$y = x_2 \tag{7.48c}$$

as well as the total energy

$$E(x_1, x_2) = E(x) = \tfrac{1}{2}kx_1^2 + \tfrac{1}{2}mx_2^2.$$

Relabeling $S = E$ and computing the time derivative yields

$$\dot{S} = -cy^2 + uy.$$

Therefore, we see that the mass-spring system is output strictly passive and, recalling that the input is the external force applied to the mass, the supply rate is the mechanical power, i.e., force times velocity.

7.7   EXERCISES

Exercise 7.1. Demonstrate an ISS-Lyapunov function for the linear system (7.1) when A is Hurwitz.

Hint: Use the Lyapunov equation and Young's inequality (Lemma A.4).

Exercise 7.2. Consider the dynamical system (7.20). Construct a stabilizing feedback interconnection $w_1 = k(x_2)$, $w_2 = x_1$ using Theorem 7.11.

Hint: Use $V_1(x_1) = x_1^4$ and $V_2(x_2) = \frac{1}{2}x_2^2$ as candidates for the matched ISS-Lyapunov functions.

Exercise 7.3. Prove Theorem 7.12.

Exercise 7.4. Show that the inequalities (7.8) and (7.11) used to characterize ISS-Lyapunov functions are equivalent. Additionally show that (7.12) and (7.13) imply (7.8) and (7.11).

Exercise 7.5. Prove Theorem 7.19.

Exercise 7.6. Consider the systems $\dot{x}_1 = f_1(x_1, w_1)$ and $\dot{x}_2 = f_2(x_2, w_2)$ defined through

$$f_1(x_1, w_1) = -x_1 + x_1^2 - x_1^3 - 2x_1^5 - \frac{1}{2}x_1 w_1^3 + x_1^3 w_1^2$$

$$f_2(x_2, w_2) = Ax_2 + Bw_2 = \begin{bmatrix} -2 & -3 \\ 0 & -1 \end{bmatrix} x_2 + \begin{bmatrix} 1 \\ 1 \end{bmatrix} w_2.$$

(i) Investigate if the cascade interconnection in Figure 7.2, i.e.,

$$\begin{bmatrix} \dot{x}_2 \\ \dot{x}_1 \end{bmatrix} = \begin{bmatrix} f_2(x_2, x_1) \\ f_1(x_1, w_1) \end{bmatrix},$$

is input-to-state stable.

(ii) Define a feedback law $w_1 = k(x_2)$ with the property $|k(x_2)| \to \infty$ for $|x_2| \to \infty$ such that the origin of the feedback interconnection

$$\begin{bmatrix} \dot{x}_2 \\ \dot{x}_1 \end{bmatrix} = \begin{bmatrix} f_2(x_2, x_1) \\ f_1(x_1, k(x_2)) \end{bmatrix}$$

is globally asymptotically stable.

Exercise 7.7. Consider the linear system

$$\dot{x} = Ax + Bu, \qquad x(0) \in \mathbb{R}^n,$$

with solution

$$x(t) = e^{At}x(0) + \int_0^t e^{A\tau}Bu(\tau)d\tau.$$

We assume that A is Hurwitz, and for simplicity we assume that the matrix A is diagonalizable, i.e., A can be written as $A = TDT^{-1}$, where D is a diagonal matrix with the eigenvalues of A on the diagonal. Additionally, we denote the largest real part of the eigenvalues of A by $\lambda_{\max}$.

(i) Compute $M, \gamma \in \mathbb{R}$ such that $|x(t)| \le Me^{\lambda_{\max}t}|x(0)| + \gamma\|u\|_\infty$ is satisfied for all $t \in \mathbb{R}_{\ge 0}$.

Consider the linear system defined through the matrices

$$A = \begin{bmatrix} -2 & -3 & -4 \\ 1 & -1 & 0 \\ 0 & 1 & -1 \end{bmatrix} \quad \text{and} \quad B = \begin{bmatrix} 1 & 2 \\ 3 & 4 \\ 1 & 4 \end{bmatrix}.$$

(ii) For $|\cdot| = |\cdot|_2$ compute M and γ.

(iii) For $|\cdot| = |\cdot|_\infty$ compute M and γ.

7.8 BIBLIOGRAPHICAL NOTES AND FURTHER READING

ISS was first introduced by Sontag in [141], with the Lyapunov characterization of Theorem 7.4 provided in [147, Theorem 1] (which additionally provides a characterization of ISS in terms of robust stability). Integral ISS was introduced in [143].

The simple interconnections described here (cascade and feedback) can be extended to the consideration of many subsystems rather than just two. Small-gain theorems for such large-scale systems typically come in two flavors: so-called *cyclic* small-gain theorems that explicitly enumerate all possible loops in the overall system (see [98]) and those that rely on algebraic conditions on particular monotone operators (see [43], [132]).

Numerical calculation of (i)ISS transient and gain bounds has been considered in [66], [170], and [171].

While interconnections of iISS systems may not behave well, it has been recently shown that an intermediate class of systems, called *strongly iISS*, behaves fairly well. A system that is strongly iISS is iISS and also ISS with respect to small inputs. The various relations between ISS, iISS, strong iISS, and linear and nonlinear $\mathcal{L}_2$-gain are investigated in [83], [84], where the latter contains simple examples for systems in each class. Notably, the bilinear system (7.44) is strongly iISS.

An excellent and comprehensive text on dissipative systems is [109].

The use of $\mathcal{L}_2$ norms and the associated small-gain theorem predate ISS by some twenty years and is a cornerstone of what has sometimes been referred to as the *input-output approach*, which brought certain mathematical tools from functional analysis to bear on problems in feedback systems. This was pioneered by Zames [169] and Sandberg [134] (the former can be found in [16]), and an excellent early text is [44]. A more recent text that also provides an excellent introduction to the important class of *Port-Hamiltonian Systems* is [157].

Part II

Controller Design

Chapter Eight

LMI-Based Controller and Antiwindup Designs

The available design tools for linear systems are very powerful and generally well understood. As a consequence, a common design methodology is to pursue a linear design that provides the desired performance when the system is operating near its desired operating point. However, disturbances can push the system far away from this point and in some cases returning to the region where the controller and plant work as designed can take an unacceptably long time. Techniques to reduce this recovery time, while maintaining the designed local performance, go under the heading of *antiwindup* techniques.

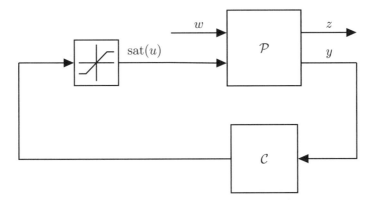

Figure 8.1: Feedback loop with saturated input.

Specifically, in this chapter we analyze the stability properties of the closed-loop dynamics shown in Figure 8.1, where the plant is modelled as a linear system

$$\mathcal{P}: \begin{cases} \dot{x}_p &= A_p x_p + B_p \operatorname{sat}(u) + B_w w \\ y &= C_{p,y} x_p + D_{p,y} w \\ z &= C_{p,z} x_p + D_{p,z} w, \end{cases} \tag{8.1}$$

with a possible saturation nonlinearity in the input. Additionally, the controller is defined by the linear system

$$\mathcal{C}: \begin{cases} \dot{x}_c &= A_c x_c + B_c y \\ u &= C_c x_c + D_{c,y} y, \end{cases} \tag{8.2}$$

where the feedback law u is defined based on the measured output of the plant y.

The various dimensions of the system are given by

$$x_p \in \mathbb{R}^{n_p}, x_c \in \mathbb{R}^{n_c}, y \in \mathbb{R}^{n_y}, z \in \mathbb{R}^{n_z}, u \in \mathbb{R}^{n_u}, w \in \mathbb{R}^{n_w} \qquad (8.3)$$

for $n_p, n_c, n_y, n_z, n_u, n_w \in \mathbb{N}$, from which the dimensions of the matrices can be deduced. Through the definition of the overall state $x \doteq [x_p^T, x_c^T]^T \in \mathbb{R}^n$ and the matrices

$$\left[\begin{array}{c|c|c} A & B & E \\ \hline C & D & F \\ \hline K & L & G \end{array}\right] = \left[\begin{array}{cc|c|c} A_p + B_p D_{c,y} C_{p,y} & B_p C_c & -B_p & B_p D_{c,y} D_{p,y} + B_w \\ B_c C_{p,y} & A_c & 0 & B_c D_{p,y} \\ \hline C_{p,z} & 0 & 0 & D_{p,z} \\ \hline D_{c,y} C_{p,y} & C_c & 0 & D_{c,y} D_{p,y} \end{array}\right]$$

$$(8.4)$$

the feedback loop in Figure 8.1 can be compactly represented through the dynamics

$$\begin{aligned}
\dot{x} &= Ax + Bq + Ew \\
z &= Cx + Dq + Fw \\
u &= Kx + Lq + Gw \\
q &= u - \mathrm{sat}(u).
\end{aligned} \qquad (8.5)$$

Note that (8.5) is more general than the motivational example in Figure 8.1 since it contains the additional matrices D and L, for example, which are not present in the feedback loop in Figure 8.1. The system is nonlinear due to the saturation in the input u. Moreover, if the matrix L is unequal to zero, note that u is only implicitly defined in (8.5) as the solution of the nonlinear *algebraic loop* or *algebraic equation*

$$u = L(u - \mathrm{sat}(u)) + \mu, \qquad (8.6)$$

where $\mu = Kx + Gw \in \mathbb{R}^{n_u}$. Existence and uniqueness of u satisfying (8.6) is not automatically guaranteed and solving the algebraic loop for a given $\mu \in \mathbb{R}^{n_u}$ is in general nontrivial. To ensure that the solution of (8.5) is well defined in terms of existence and uniqueness, we make use of the following definition.

Definition 8.1 (Well-posed algebraic loop). *For $L \in \mathbb{R}^{n_u \times n_u}$, the algebraic equation (8.6) is well posed if (8.6) admits a unique solution for all $\mu \in \mathbb{R}^{n_u}$ and if $\mu \mapsto u(\mu)$ is Lipschitz continuous.*

A sufficient condition for well-posedness is given in the following lemma.

Lemma 8.2 ([168, Section 3.4.2]). *Consider the algebraic loop (8.6) for $L \in \mathbb{R}^{n_u \times n_u}$ and $u, \mu \in \mathbb{R}^{n_u}$. If there exists a positive definite matrix $W \in \mathcal{S}_{>0}^{n_u}$ satisfying the matrix inequality*

$$\tfrac{1}{\|W\|} \left(L^T W + W L - 2W \right) < 0, \qquad (8.7)$$

then the algebraic loop (8.6) is well posed.

Note that the factor $\frac{1}{\|W\|}$ is not necessary in (8.7), but it indicates how far the inequality is from being violated and thus provides a certain robustness margin with respect to well-posedness.

As mentioned above, our main focus in this chapter will be the analysis of the stability properties of (8.1) and (8.2) represented through the compact dynamics (8.5). Our main tool will be *linear matrix inequalities* (LMIs), which we have already seen in Section 3.4 in the context of the numerical calculation of Lyapunov functions containing higher order terms.

8.1 $\mathcal{L}_2$-GAIN OPTIMIZATION FOR LINEAR SYSTEMS

Recall from Chapter 3 that exponential stability of the origin for the linear system $\dot{x} = Ax$ is equivalent to being able to solve the Lyapunov equation; i.e., for a given matrix $Q > 0$ there exists a matrix $P > 0$ satisfying

$$A^T P + PA = -Q. \tag{8.8}$$

Furthermore, recall that the fundamental idea of the Lyapunov equation is that $V(x) = x^T P x$ is a Lyapunov function since

$$\dot{V}(x) = x^T (A^T P + PA)x = -x^T Qx < 0, \qquad x \neq 0.$$

However, for the purposes of the Lyapunov argument above, it is the sign-definiteness of the inequality that is important, not the equality of the Lyapunov equation. This suggests that if, given A, we could solve the LMIs

$$\begin{aligned} 0 &< P \\ A^T P + PA &< 0, \end{aligned} \tag{8.9}$$

then we would have a Lyapunov function. The advantage of LMI (8.9) over the Lyapunov Equation (8.8) is the additional degree of freedom by not fixing Q to a specific positive definite matrix. This degree of freedom will be used in the following to indirectly find "optimal" matrices Q.

Leveraging the convexity of (8.9), and recalling the results of Section 3.4, rather than solving the Lyapunov Equation (8.8), we can solve the following convex optimization problem:

$$\begin{aligned} \min_{P,\,k} \quad & k \\ \text{subject to} \quad 0 &< k \\ 0 &< P - \alpha I \\ 0 &> P - (k + \alpha)I \\ 0 &> A^T P + PA. \end{aligned} \tag{8.10}$$

Here, for $\alpha > 0$, the second constraint is simply to avoid the optimizer returning an arbitrarily small P and the third constraint is chosen to avoid an arbitrarily large P. If the numerical conditioning of the particular optimization problem is not an issue, it may be reasonable to take $\alpha = 0$. From a solution P, an optimal Q with respect to the optimization problem (8.10) is then uniquely defined through the Lyapunov equation (8.8).

There are many tools available to solve such optimization problems. In Appendix B.3 and Appendix B.4 we indicate how this can be done using CVX [56], [57] or SOSTOOLS [116] in MATLAB. An alternative toolbox in MATLAB is YALMIP [99]. Since in optimization problems with strict inequality constraints the optimal value is in general not attained, tools such as CVX are not able to handle strict inequality constraints. In this case we can approximate the LMI (and all LMIs introduced in the following) through

$$\min_{P,\,k} \quad k$$

$$\begin{aligned}
\text{subject to} \quad & 0 \;\le\; k \\
& 0 \;\le\; P - \alpha I - \varepsilon I \\
& 0 \;\ge\; P - (k + \alpha)I + \varepsilon I \\
& 0 \;\ge\; A^T P + P A + \varepsilon I,
\end{aligned} \tag{8.11}$$

where $\varepsilon > 0$ denotes an arbitrarily small parameter.

8.1.1 Asymptotic Stability and $\mathcal{L}_2$-Gain Optimization

Before we derive results for the general system (8.5) we start with the simplified linear system with (disturbance) inputs given by

$$\begin{aligned}
\dot{x} &= Ax + Ew \\
z &= Cx + Fw.
\end{aligned} \tag{8.12}$$

Here $x \in \mathbb{R}^n$, $w \in \mathbb{R}^m$ and $z \in \mathbb{R}^p$ for $n, m, p \in \mathbb{N}$. We call a system with disturbance inputs *0-input Globally Asymptotically Stable (0-GAS)* if the origin is asymptotically stable for $w = 0$. In case of the linear system (8.12), 0-GAS is equivalent to A being Hurwitz. Note that with respect to [168], which is the main reference in this chapter for further reading, 0-GAS is referred to as *internal stability*.

Recall the estimate (7.31) from Chapter 7 that, if A is Hurwitz and $P > 0$ solves the Lyapunov Equation (8.8) with $Q = -2I$, then taking $V(x) = x^T P x$ yields

$$\dot{V}(x) \le -x^T x + \gamma^2 w^T w, \tag{8.13}$$

where $\gamma = \|PE\|$. Rearranging terms in (8.13), integrating both sides of the inequality, and assuming $x(0) = 0$ yields the $\mathcal{L}_2$ stability bound

$$\|x\|^2_{\mathcal{L}_2[0,t)} \le \int_0^t x(\tau)^T x(\tau) d\tau + V(x(t))$$

$$\le \gamma^2 \int_0^t w(\tau)^T w(\tau) d\tau = \gamma^2 \|w\|^2_{\mathcal{L}_2[0,t)}. \tag{8.14}$$

We propose a slight modification of (8.13) that accounts for the output equation in (8.12). Suppose we can find a $P > 0$ so that $V(x) = x^T P x$ and

$$\dot{V}(x) = x^T (A^T P + P A)x + 2x^T P E w < -\gamma \left(\tfrac{1}{\gamma^2} z^T z - w^T w \right) \tag{8.15}$$

for all $(x, w) \neq 0$. We can show that this guarantees asymptotic stability and an $\mathcal{L}_2$-gain bound of $\gamma > 0$ from input w to output z. For the former, with $w = 0$, we see that $\dot{V}(x) < 0$ for all $x \neq 0$ and thus asymptotic stability is guaranteed. For the latter, similarly to the calculation in (8.14), by integrating both sides of (8.15) and assuming $x(0) = 0$, we get

$$\frac{1}{\gamma} \int_0^t z^T(\tau)z(\tau)d\tau + V(x(t)) \leq \gamma \int_0^t w^T(\tau)w(\tau)d\tau,$$

which yields

$$\|z\|_{\mathcal{L}_2[0,t)} \leq \gamma \|w\|_{\mathcal{L}_2[0,t)}. \tag{8.16}$$

As in the previous stability problem, we can write (8.15) as an LMI. Using the output equation and factoring out x and w, we get

$$\begin{bmatrix} x \\ w \end{bmatrix}^T \left(\begin{bmatrix} A^T P + PA & PE \\ E^T P & -\gamma I \end{bmatrix} + \frac{1}{\gamma} \begin{bmatrix} C^T \\ F^T \end{bmatrix} \begin{bmatrix} C & F \end{bmatrix} \right) \begin{bmatrix} x \\ w \end{bmatrix} < 0$$

for all $(x, w) \neq 0$ or just the matrix equations

$$0 < P,$$

$$\begin{bmatrix} A^T P + PA & PE \\ E^T P & -\gamma I \end{bmatrix} + \frac{1}{\gamma} \begin{bmatrix} C^T \\ F^T \end{bmatrix} \begin{bmatrix} C & F \end{bmatrix} < 0. \tag{8.17}$$

If $\gamma > 0$ is fixed, the above is just a feasibility problem; i.e., is it possible to find a matrix P so that the above holds? However, we might be interested in how small γ can be. In other words, we may wish to consider γ as an additional decision variable rather than something given a priori. This poses a problem since, due to the presence of the $\frac{1}{\gamma}$ term, the above is not a *linear* matrix inequality. Using the *Schur complement* will allow us to circumvent this problem.

Lemma 8.3 (Schur complement, [64, Theorem 7.7.6]). *Let $Q \in \mathcal{S}^r$ and $R \in \mathcal{S}^q$ for $r, q \in \mathbb{N}$ and let $S \in \mathbb{R}^{r \times q}$. Then the matrix condition*

$$\begin{bmatrix} Q & S \\ S^T & R \end{bmatrix} < 0$$

is equivalent to the matrix conditions

$$R < 0$$
$$Q - SR^{-1}S^T < 0.$$

To apply Lemma 8.3, take $R = -\gamma$, $S = \begin{bmatrix} C & F \end{bmatrix}$, and Q as the leftmost matrix in (8.17). Then, by Lemma 8.3, the expression (8.17) is equivalent to

$$\left[\begin{array}{cc|c} A^T P + PA & PE & C^T \\ E^T P & -\gamma I & F^T \\ \hline C & F & -\gamma I \end{array} \right] < 0, \tag{8.18}$$

which is linear in the matrix decision variables P and γ. Therefore, to obtain an estimate of the smallest gain bound γ, we can solve the optimization problem

$$\min_{P,\,\gamma}\quad \gamma$$

$$\text{subject to}\quad 0 < P \quad\text{symmetric}$$

$$0 < \gamma \tag{8.19}$$

$$0 > \left[\begin{array}{cc|c} A^T P + PA & PE & C^T \\ E^T P & -\gamma I & F^T \\ \hline C & F & -\gamma I \end{array}\right].$$

Remark 8.4. Note that Lemma 8.3 contains an *if and only if* statement. Hence, (8.18) implies (8.17) and $-\gamma < 0$. Therefore, $0 < \gamma$ is not necessary in the optimization problem (8.19) and the information $\gamma \in \mathbb{R}_{>0}$ is implicitly contained in (8.18). Nevertheless, to make the condition explicit, $0 < \gamma$ is added in (8.19). Similarly, note that $0 < P$ in (8.19) implies that P is symmetric since definiteness is only defined for $P \in \mathcal{S}^{n_u}$. Thus also the statement P *symmetric* in (8.19) is not necessary. However, in an implementation P needs to be specified as a symmetric matrix to reduce the number of unknowns in the optimization problem.

Observe that the symmetric matrix in (8.18) contains quite a bit of redundant information. As a way to reduce the redundant information and simplify expressions, we introduce the notation

$$\text{He } X = X + X^T. \tag{8.20}$$

This allows us to compactly rewrite (8.18) as

$$\text{He}\left[\begin{array}{cc|c} PA & PE & 0 \\ 0 & -\frac{\gamma}{2}I & 0 \\ \hline C & F & -\frac{\gamma}{2}I \end{array}\right] < 0. \tag{8.21}$$

There are two things worth noting in (8.21). First, general convention is that the block matrices without transposes are retained. Hence, we used PE above the diagonal and C and F below the diagonal. However, this is merely a commonly used convention and we could as easily have used C^T and F^T above the diagonal, which would yield an upper triangular matrix. The second thing to observe is the factor of $\frac{1}{2}$ on the two lower diagonal elements, as this is frequently overlooked.

Example 8.5. Consider the system

$$\dot{x} = \begin{bmatrix} -1 & -2 & 2 \\ 1 & -2 & 1 \\ 3 & -2 & -2 \end{bmatrix} x + \begin{bmatrix} 2 & -2 \\ 1 & 3 \\ 3 & -2 \end{bmatrix} w$$

$$z = \begin{bmatrix} 1 & 0 & 0 \end{bmatrix} x + \begin{bmatrix} -3 & 2 \end{bmatrix} w.$$

Solving the Lyapunov Equation (8.8) for $Q = -2I$ yields the positive definite matrix

$$P = \begin{bmatrix} 3.10 & -3.20 & 1.77 \\ -3.20 & 5.50 & -1.80 \\ 1.77 & -1.80 & 1.37 \end{bmatrix}.$$

Young's inequality (Lemma A.4) together with the estimate (8.14) leads to the bound

$$\|z\|_{\mathcal{L}_2[0,t)}^2 = \|Cx + Fw\|_{\mathcal{L}_2[0,t)}^2$$

$$= \left(\int_0^t x^T CCx + 2x^T C^T Fw + w^T F^T Fw \, d\tau \right)^2$$

$$\leq \left(\int_0^t 2x^T CCx + 2w^T F^T Fw \, d\tau \right)^2$$

$$\leq \left(\int_0^t 2\lambda_{\max}(C^T C)x^T x + 2\lambda_{\max}(F^T F)w^T w \, d\tau \right)^2$$

$$= 2\lambda_{\max}(C^T C)\|x\|_{\mathcal{L}_2[0,t)}^2 + 2\lambda_{\max}(F^T F)\|w\|_{\mathcal{L}_2[0,t)}^2$$

$$\leq 2\left(\lambda_{\max}(C^T C)\|PE\|^2 + \lambda_{\max}(F^T F) \right) \|w\|_{\mathcal{L}_2[0,t)}^2$$

$$= 2711 \cdot \|w\|_{\mathcal{L}_2[0,t)}^2.$$

Hence, the estimate provides the bound $\gamma = 52.07$ on the input-output behavior (8.16).

Alternatively, $\gamma = 7.43$ is returned as the solution of the optimization problem (8.19). A corresponding matrix P such that the estimate (8.15) is satisfied is given by

$$P = \begin{bmatrix} 4.38 & -0.22 & -4.12 \\ -0.22 & 0.32 & -0.02 \\ -4.12 & -0.02 & 4.18 \end{bmatrix}.$$

Here, γ and P are obtained using CVX. The strict inequalities are approximated by inequalities as in (8.11) using the parameter $\varepsilon = 0.001$.

8.1.2   Feedback Synthesis

Suppose now we have a control input for which we wish to design a static state feedback; i.e., consider

$$\begin{aligned} \dot{x} &= Ax + Bu + Ew \\ z &= Cx + Du + Fw \\ u &= Kx. \end{aligned} \qquad (8.22)$$

Written in closed-loop form, this becomes

$$\begin{aligned} \dot{x} &= (A + BK)x + Ew \\ z &= (C + DK)x + Fw. \end{aligned}$$

We propose the same Lyapunov function for input-output $\mathcal{L}_2$-gain with 0-GAS but for the above closed-loop system; i.e., $P > 0$, $V(x) = x^T Px$, and

$$\begin{aligned} \dot{V}(x) &= x^T \left((A + BK)^T P + P(A + BK) \right) x + 2x^T PEw \\ &< -\gamma \left(\tfrac{1}{\gamma^2} z^T z - w^T w \right) \end{aligned} \qquad (8.23)$$

for all $(x, w) \neq 0$.

Before moving to the development of the full matrix inequality, we first focus just on the term guaranteeing 0-GAS given by

$$(A + BK)^T P + P(A + BK) = A^T P + PA + K^T B^T P + PBK < 0.$$

Here, both K and P are matrix decision variables and they appear in the product terms PBK. In other words, this term is not a *linear* matrix inequality. In order to circumvent this difficulty, define $\Lambda = P^{-1}$ and multiply the above expression from both the left and the right by Λ to obtain

$$\Lambda A^T P \Lambda + \Lambda P A \Lambda + \Lambda K^T B^T P \Lambda + \Lambda PBK\Lambda$$
$$= \Lambda A^T + A\Lambda + \Lambda(BK)^T + BK\Lambda = \text{He}\,(A\Lambda + BK\Lambda).$$

This still does not yield a *linear* matrix inequality since we still have a product between the matrix decision variables K and Λ. However, without the B matrix between them, we can define a matrix variable $X = K\Lambda$ so that the above term becomes $\text{He}\,(A\Lambda + BX)$, which is indeed linear in the matrix decision variables Λ and X.

With these ideas at hand, we now return to (8.23), which follows from (8.21) in an obvious way; namely,

$$\text{He} \left[\begin{array}{cc|c} P(A+BK) & PE & 0 \\ 0 & -\frac{\gamma}{2}I & 0 \\ \hline C+DK & F & -\frac{\gamma}{2}I \end{array} \right] < 0. \tag{8.24}$$

We multiply both sides of the inequality from the left and the right by a block diagonal matrix containing $\Lambda = P^{-1}$ in the first block and the identity in the remaining blocks (i.e., $\text{diag}(\Lambda, I, I)$) to get

$$\left[\begin{array}{ccc} \Lambda & 0 & 0 \\ 0 & I & 0 \\ 0 & 0 & I \end{array} \right] \text{He} \left[\begin{array}{cc|c} P(A+BK) & PE & 0 \\ 0 & -\frac{\gamma}{2}I & 0 \\ \hline C+DK & F & -\frac{\gamma}{2}I \end{array} \right] \left[\begin{array}{ccc} \Lambda & 0 & 0 \\ 0 & I & 0 \\ 0 & 0 & I \end{array} \right]$$
$$= \text{He} \left[\begin{array}{cc|c} A\Lambda+BK\Lambda & E & 0 \\ 0 & -\frac{\gamma}{2}I & 0 \\ \hline C\Lambda+DK\Lambda & F & -\frac{\gamma}{2}I \end{array} \right] < 0.$$

As above, with $X = K\Lambda$ this yields the LMI

$$\text{He} \left[\begin{array}{cc|c} A\Lambda+BX & E & 0 \\ 0 & -\frac{\gamma}{2}I & 0 \\ \hline C\Lambda+DX & F & -\frac{\gamma}{2}I \end{array} \right] < 0.$$

We can thus solve the optimization problem

$$\min_{\Lambda, X, \gamma} \quad \gamma$$

$$\text{subject to} \quad 0 < \Lambda \quad \text{symmetric}$$

$$0 < \gamma \tag{8.25}$$

$$0 > \text{He} \left[\begin{array}{cc|c} (A\Lambda + BX) & E & 0 \\ 0 & -\frac{\gamma}{2}I & 0 \\ \hline C\Lambda + DX & F & -\frac{\gamma}{2}I \end{array} \right]$$

to obtain a Lyapunov function $V(x) = x^T \Lambda^{-1} x$ and a feedback gain matrix $K = X\Lambda^{-1}$ such that γ in (8.16) is minimal.

Example 8.6. We consider the dynamics

$$\dot{x} = \left[\begin{array}{ccc} 1 & 2 & -2 \\ -1 & 2 & -1 \\ -3 & 2 & 2 \end{array} \right] x + \left[\begin{array}{c} 3 \\ 2 \\ -1 \end{array} \right] u + \left[\begin{array}{cc} 2 & -2 \\ 1 & 3 \\ 3 & -2 \end{array} \right] w$$

$$z = \left[\begin{array}{ccc} 1 & 0 & 0 \end{array} \right] x + \left[\begin{array}{cc} -3 & 2 \end{array} \right] w$$

as an example of (8.22).

As a solution of the optimization problem we obtain the $\mathcal{L}_2$-gain $\gamma = 8.1910$ as well as the feedback gain

$$K = \left[\begin{array}{ccc} -7.32 & 6.64 & 4.62 \end{array} \right]. \tag{8.26}$$

The matrix A has three eigenvalues in $\mathbb{C}_+$ while K ensures 0-GAS, i.e., the eigenvalues of $A + BK$ are in the open left half-plane $\mathbb{C}_-$. The Lyapunov function $V(x) = x^T \Lambda x = x^T P x$ is defined through

$$P = \left[\begin{array}{ccc} 1.18 & -1.19 & -0.62 \\ -1.19 & 1.33 & 0.71 \\ -0.62 & 0.71 & 0.40 \end{array} \right].$$

Here, the solution is obtained through CVX using the parameter $\varepsilon = 0.001$ and the solution may depend on the solver.

8.2 SYSTEMS WITH SATURATION

We now turn to the analysis of linear systems with input saturation. In particular, as discussed in Section 6.1 it is more realistic to assume that a closed-loop system is of the form

$$\dot{x} = Ax + B \operatorname{sat}(u)$$
$$u = Kx \tag{8.27}$$

or the more general form (8.5) motivated at the beginning of the chapter. The saturation operator has been formally introduced in Chapter 6 in the context of absolute stability. Here, we recall the definition specifically in terms of the input

$$\mathrm{sat}(u) \doteq \begin{cases} -1, & u < -1 \\ u, & -1 \leq u \leq 1 \\ 1, & 1 < u. \end{cases}$$

More generally, an arbitrary saturation level, $\bar{u} > 0$ can be chosen so that

$$\mathrm{sat}_{\bar{u}}(u) \doteq \begin{cases} -\bar{u}, & u < -\bar{u} \\ u, & -\bar{u} \leq u \leq \bar{u} \\ \bar{u}, & \bar{u} < u. \end{cases}$$

We could further allow non-symmetric saturations by taking different upper and lower limits. However, the notation becomes burdensome. In what follows, we will generally suppress the limit $\bar{u}$. In addition to the saturation, we have considered deadzone nonlinearities in Chapter 6. For a given saturation function, the corresponding deadzone is uniquely defined through the identities

$$\mathrm{dz}(u) = u - \mathrm{sat}(u) \qquad \text{and} \qquad \mathrm{dz}_{\bar{u}}(u) = u - \mathrm{sat}_{\bar{u}}(u),$$

and we have used the variable $q = \mathrm{dz}(u)$ in (8.5) to denote the deadzone. The saturation function and the deadzone are visualized in Figure 8.2.

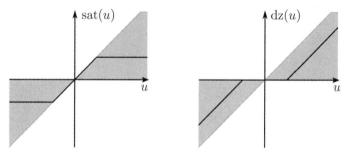

Figure 8.2: Deadzone and saturation nonlinearities and their corresponding sectors.

Dealing with the saturation in (8.5) or (8.27) directly is difficult. However, as we saw with the topic of absolute stability, sector-bounded nonlinearities have a convenient characterization that fits nicely in the framework of linear systems and quadratic functions. In particular, with $u \in \mathbb{R}$ and $q = u - \mathrm{sat}(u)$, the deadzone nonlinearity satisfies

$$\mathrm{dz}(u)\,\mathrm{sat}(u) \geq 0 \qquad \text{or equivalently} \qquad q(u - q) \geq 0.$$

This fact can be understood immediately from the visualization in Figure 8.2 which illustrates that $\mathrm{sign}(\mathrm{dz}(u)) = \mathrm{sign}(\mathrm{sat}(u))$ or equivalently

$$\mathrm{sign}(q) = \mathrm{sign}(u - q).$$

Furthermore, it is clear that multiplying by a number $w > 0$ does not change the inequality; i.e.,

$$wq(u - q) \geq 0. \tag{8.28}$$

In the case of multiple inputs $(u \in \mathbb{R}^{n_u})$ we refer to saturations as *decentralized* if each input has its own saturation function, possibly with different saturation levels $\bar{u}_i$ on the ith input. A vector version of (8.28) is then given by

$$\text{dz}(u)^T W \, \text{sat}(u) \geq 0 \qquad \text{or equivalently} \qquad q^T W(u - q) \geq 0, \qquad (8.29)$$

where $W \in \mathbb{R}^{n_u \times n_u}$, $W > 0$, is diagonal.

8.2.1 LMI-Based Saturated Linear State Feedback Design

Before we focus on (8.5), we design a feedback law $u = Kx$ stabilizing the dynamics (8.27). Using the notation $q = u - \text{sat}(u)$, (8.27) can be rewritten as

$$\dot{x} = Ax + B \, \text{sat}(Kx) = (A + BK)x - B \, \text{dz}(Kx). \qquad (8.30)$$

We start with the assumption of a quadratic Lyapunov function $V(x) = x^T P x$ with $P > 0$. A desirable property of the closed loop would be for the time derivative of the Lyapunov function to be negative definite *despite* the presence of the nonlinearity. One way of phrasing this would be that the sector condition (8.29) implies a negative definite derivative of the Lyapunov function. Formally, then, for $(x, q) = (x, \text{dz}(Kx)) \neq 0$ we want

$$q^T W(u - q) = (\text{dz}(Kx))^T W(Kx - \text{dz}(Kx)) \geq 0 \quad \Rightarrow \quad \dot{V}(x) < 0 \qquad (8.31)$$

with

$$\begin{aligned} \dot{V}(x) &= x^T((A + BK)^T P + P(A + BK))x - 2x^T PBq \\ &= x^T(A^T P + PA + K^T B^T P + PBK)x - 2x^T PB \, \text{dz}(Kx), \end{aligned}$$

where the diagonal matrix $W \in \mathcal{S}^{n_u}_{>0}$, as well as $P \in \mathcal{S}^{n_x}_{>0}$ and $K \in \mathbb{R}^{n_x \times n_u}$, are unknown design parameters.

In order to construct useful linear matrix inequalities as done in the previous section, we will need the following result, known as the S-lemma or S-procedure.

Lemma 8.7 (S-lemma, [28, Section 2.6.3]). *Let $M_0, M_1 \in \mathcal{S}^r$ and $r \in \mathbb{N}$, and suppose there exists $\zeta^* \in \mathbb{R}^r$ such that $(\zeta^*)^T M_1 \zeta^* > 0$. Then the following statements are equivalent:*

1. There exists $\tau > 0$ such that $M_0 - \tau M_1 > 0$.
2. For all $\zeta \neq 0$ such that $\zeta^T M_1 \zeta \geq 0$ it holds that $\zeta^T M_0 \zeta > 0$.

With Lemma 8.7, if

$$x^T(A^T P + PA + K^T B^T P + PBK)x - 2x^T PBq + 2q^T W(Kx - q) < 0 \qquad (8.32)$$

for all $(x, q) \neq 0$, then (8.31) is satisfied (this is 1. $\Rightarrow$ 2. in Lemma 8.7, with the factor τ being absorbed into the diagonal matrix W). It is thus sufficient to check if (8.32) holds to ensure that the implication (8.31) is true, or, equivalently, to design W, P, and K so that

$$\text{He} \begin{bmatrix} PA + PBK & -PB \\ WK & -W \end{bmatrix} \qquad (8.33)$$

is negative definite. By multiplying (8.33) from the left and from the right with $\text{diag}(P^{-1}, W^{-1})$ the LMI

$$\text{He} \begin{bmatrix} A\Lambda + BX & -BD \\ X & D \end{bmatrix} = \text{He} \begin{bmatrix} AP^{-1} + BKP^{-1} & -BW^{-1} \\ KP^{-1} & W^{-1} \end{bmatrix} \qquad (8.34)$$

$$= \begin{bmatrix} P^{-1} & 0 \\ 0 & W^{-1} \end{bmatrix} \text{He} \begin{bmatrix} PA + PBK & -PB \\ WK & -W \end{bmatrix} \begin{bmatrix} P^{-1} & 0 \\ 0 & W^{-1} \end{bmatrix}$$

in the unknowns $\Lambda = P^{-1}$, $X = KP^{-1}$, $D = W^{-1}$ is obtained, from which P, K, and W can be recovered. Consequently, if

$$\text{He} \begin{bmatrix} A\Lambda + BX & -BD \\ X & -D \end{bmatrix} < 0, \qquad \Lambda > 0, \qquad D > 0 \text{ diagonal} \qquad (8.35)$$

has a solution, then the bounded input $\text{sat}(Kx)$ with $K = XP$ globally asymptotically stabilizes the origin of (8.30).

Unfortunately, condition (8.35) is very restrictive. Indeed, if (8.35) has a solution, then (8.34) being negative definite implies that

$$0 > \begin{bmatrix} I & -B \\ 0 & I \end{bmatrix} \text{He} \begin{bmatrix} A\Lambda + BX & -BD \\ X & -D \end{bmatrix} \begin{bmatrix} I & 0 \\ -B^T & I \end{bmatrix} = \text{He} \begin{bmatrix} A\Lambda & BD \\ X & -D \end{bmatrix}.$$

Using the Schur complement (Lemma 8.3), the inequality implies that $A\Lambda + \Lambda A^T < 0$. With the additional information that Λ is positive definite, this implies that the matrix A is Hurwitz and thus the origin of (8.27) is asymptotically stable with $K = 0$.

Here, the problem is that we are trying to globally stabilize the origin of a system with a bounded input, which is only possible if the matrix A does not have any eigenvalues with positive real part. Local approaches overcoming this issue are discussed in Section 8.3.

8.2.2   Global Asymptotic Stability Analysis

We now proceed with (8.5), but assume that the external input w is set to zero. For given matrices K and L such that the algebraic loop (8.6) is well posed, i.e., the assumptions of Definition 8.1 are satisfied, we are looking for a quadratic Lyapunov function $V(x) = x^T P x$ with $P > 0$ establishing global asymptotic stability of the origin of (8.5). As in Section 8.2.1, a desirable property in this case would be for the time derivative of the Lyapunov function to be negative definite *despite* the presence of the nonlinearity. Using the sector condition (8.29), this can be phrased as

$$q^T W(u - q) \geq 0 \quad \Rightarrow \quad \dot{V}(x) = x^T(A^T P + PA)x + 2x^T PBq < 0, \qquad (8.36)$$

which needs to be satisfied for all $(x, q) \neq 0$. With Lemma 8.7, if

$$x^T(A^T P + PA)x + 2x^T PBq + 2q^T W(u - q) < 0 \qquad (8.37)$$

for all $(x, q) \neq 0$, then (8.36) is satisfied. Again, the factor τ is absorbed into the diagonal matrix W. Substituting the expression for the input u into (8.37) we obtain

$$x^T(A^TP + PA)x + 2x^TPBq + 2q^TW(Kx - Lq - q) < 0, \qquad (8.38)$$

for all $(x, q) \neq 0$, from which we can write the linear matrix inequality

$$\text{He} \begin{bmatrix} PA & PB \\ WK & -W + WL \end{bmatrix} < 0. \qquad (8.39)$$

Therefore, a sufficient condition for (8.36) (and thus global asymptotic stability of the origin) is given by feasibility of (8.39) in the unknowns $P \in \mathcal{S}_{>0}^n$ and $W \in \mathcal{S}_{>0}^{n_u}$ diagonal.

Note that the assumption on the well-posedness of the algebraic loop made at the beginning of this section is not necessary in the derivations. Feasibility of (8.39) together with the Schur complement (Lemma 8.3) guarantee that (8.7) holds and thus well-posedness according to Definition 8.1 is implied (see Lemma 8.2). To introduce a robustness margin with respect to well-posedness, an additional degree of freedom can be added in (8.39). In particular, let $\nu \in (0, 1]$ and observe that

$$\begin{aligned} x^T(A^TP + PA)x &+ 2x^TPBq + 2q^TW(Kx - Lq - q) \\ &= x^T(A^TP + PA)x + 2x^TPBq + 2q^TW(Kx - Lq) - 2q^TWq \\ &\leq x^T(A^TP + PA)x + 2x^TPBq + 2q^TW(Kx - Lq) - 2\nu q^TWq. \end{aligned} \qquad (8.40)$$

Therefore, if the right-hand side of (8.40) is less than zero, then so is the left-hand side. Consequently, if we can find a solution to the LMI

$$\text{He} \begin{bmatrix} PA & PB \\ WK & -\nu W + WL \end{bmatrix} < 0, \qquad (8.41)$$

this guarantees that (8.36) holds.

Again, if (8.41) holds, the Schur complement (Lemma 8.3) implies that

$$L^TW + WL - 2\nu W < 0.$$

Thus, inequality (8.7) guaranteeing well-posedness of the algebraic loop is satisfied with an additional margin

$$\tfrac{1}{\|W\|}\left(L^TW + WL - 2W\right) < -2(1 - \nu)\tfrac{W}{\|W\|} < 0,$$

depending on the selection of $\nu \in (0, 1)$. While (8.41) is more restrictive in terms of the existence of solutions for smaller values of $\nu \in (0, 1]$, robustness with respect to well-posedness of the algebraic loop increases if smaller values ν are selected.

In line with the arguments that conclude Section 8.2.1, global stability properties and feasibility of (8.39) pose strong conditions on the system (8.1). Strategies extending the ideas here to the case of local stability are discussed in Section 8.3.1.

8.2.3 $\mathcal{L}_2$-Stability and $\mathcal{L}_2$-Gain Optimization

We can approach the $\mathcal{L}_2$-stability problem in the same way as the asymptotic stability problem. In particular, we want to minimize the $\mathcal{L}_2$-gain γ in the presence of sector-bounded nonlinearities. As before, for $(x, q) \neq 0$, we can phrase this as

$$q^T W(u - q) \geq 0 \quad \Rightarrow$$
$$x^T(A^T P + PA)x + 2x^T P(Bq + Ew) < -\gamma \left(\tfrac{1}{\gamma^2} z^T z - w^T w \right). \quad (8.42)$$

Since (8.42) is not obviously amenable to computation, we instead examine the inequality

$$x^T(A^T P + PA)x + 2x^T P(Bq + Ew) + 2q^T W(u - q) + \gamma \left(\tfrac{1}{\gamma^2} z^T z - w^T w \right) < 0 \quad (8.43)$$

for all $(x, q) \neq 0$, and, if this holds, use the S-lemma (Lemma 8.7) to conclude (8.42). Substituting the expression for u into (8.43) yields

$$x^T(A^T P + PA)x + 2x^T P(Bq + Ew)$$
$$+ 2q^T W(Gw + Kx + Lq - q) + \tfrac{1}{\gamma} z^T z - \gamma w^T w < 0 \quad (8.44)$$

for all $(x, q) \neq 0$. We can factor this inequality as

$$\begin{bmatrix} x \\ q \\ w \end{bmatrix}^T \left(\begin{bmatrix} A^T P + PA & PB + K^T W & PE \\ B^T P + WK & -2W + WL + L^T W & WG \\ E^T P & G^T W & -\gamma I \end{bmatrix} \right.$$
$$\left. + \frac{1}{\gamma} \begin{bmatrix} C^T \\ D^T \\ F^T \end{bmatrix} [C \ D \ F] \right) \begin{bmatrix} x \\ q \\ w \end{bmatrix} < 0. \quad (8.45)$$

Using the Schur complement (Lemma 8.3), inequality (8.45) can be written as

$$\text{He} \left[\begin{array}{ccc|c} PA & PB & PE & 0 \\ WK & -W + WL & WG & 0 \\ 0 & 0 & -\frac{\gamma}{2}I & 0 \\ \hline C & D & F & -\frac{\gamma}{2}I \end{array} \right] < 0. \quad (8.46)$$

We summarize the result of this section in the optimization problem

$$\min_{P, W, \gamma} \quad \gamma$$
$$\text{subject to} \quad 0 < P \quad \text{symmetric}$$
$$\qquad\qquad\quad 0 < W \quad \text{diagonal}$$
$$\qquad\qquad\quad 0 < \gamma \qquad\qquad\qquad\qquad\qquad\qquad (8.47)$$
$$0 > \text{He} \left[\begin{array}{ccc|c} PA & PB & PE & 0 \\ WK & -\nu W + WL & WG & 0 \\ 0 & 0 & -\frac{\gamma}{2}I & 0 \\ \hline C & D & F & -\frac{\gamma}{2}I \end{array} \right]$$

in the unknowns $P > 0$, $W > 0$, and $\gamma > 0$, for a fixed parameter $\nu \in (0, 1]$. Here, ν can be used to add additional flexibility as discussed in Section 8.2.2. If the LMI

(8.47) is feasible, then the system (8.5) is $\mathcal{L}_2$-stable with $\mathcal{L}_2$-gain γ.

Example 8.8. We extend the dynamics of Example 8.5 to include the deadzone nonlinearity

$$\dot{x} = \begin{bmatrix} -1 & -2 & 2 \\ 1 & -2 & 1 \\ 3 & -2 & -2 \end{bmatrix} x + \begin{bmatrix} 3 \\ 2 \\ -1 \end{bmatrix} q + \begin{bmatrix} 2 & -2 \\ 1 & 3 \\ 3 & -2 \end{bmatrix} w$$

$$z = \begin{bmatrix} 1 & 0 & 0 \end{bmatrix} x + \begin{bmatrix} -3 & 2 \end{bmatrix} w$$

$$u = \begin{bmatrix} -1 & -2 & 1 \end{bmatrix} x + \begin{bmatrix} 2 & -3 \end{bmatrix} w$$

$$q = u - \text{sat}(u).$$

As a solution of the optimization problem (8.47) the $\mathcal{L}_2$-gain $\gamma = 7.8607$ is returned.

8.3 REGIONAL ANALYSIS

In the presence of the saturation nonlinearity in (8.5), many systems will fail to be globally asymptotically stable in the case $w = 0$ or $\mathcal{L}_2$-stable in the case with disturbances. Therefore, the analysis in the previous section will not be possible, with the failure manifesting itself as infeasibility of the LMIs.

Fortunately, we can encode a regional restriction into the sector condition (8.29), which allows us to pursue local rather than global results. This is done in the following lemma.

Lemma 8.9. *Let $u \in \mathbb{R}$ and $q = u - \text{sat}(u)$. For an arbitrary row vector $H \in \mathbb{R}^{1 \times n}$, for all $x \in \mathbb{R}^n$ such that $\text{sat}(Hx) = Hx$, the sector condition*

$$(u - q + Hx)q \geq 0 \tag{8.48}$$

holds.

Proof. Let $x \in \mathbb{R}^n$ such that $\text{sat}(Hx) = Hx$ is satisfied. First, using $q = u - \text{sat}(u)$, rewrite the left-hand side of (8.48) as

$$\begin{aligned} (u - q + Hx)q &= (u - u + \text{sat}(u) + Hx)(u - \text{sat}(u)) \\ &= (\text{sat}(u) + Hx)(u - \text{sat}(u)) \\ &= (\text{sat}(u) + \text{sat}(Hx))(u - \text{sat}(u)), \end{aligned} \tag{8.49}$$

where the last equality follows from $\text{sat}(Hx) = Hx$ for $x \in \mathbb{R}^n$ under consideration. Now, consider two cases:

1. $u = \text{sat}(u)$: It is immediate that (8.49) is equal to zero and, hence, (8.48) is satisfied.
2. $u \neq \text{sat}(u)$: Observe that both terms in (8.49) will always have the same sign or be zero. Specifically, if $u > \text{sat}(u)$ then $\text{sat}(u) \geq \text{sat}(Hx)$, if $u = \text{sat}(u)$ the right term in (8.49) vanishes, and if $u < \text{sat}(u)$ then $\text{sat}(u) \leq \text{sat}(Hx)$. Consequently, (8.48) is satisfied.

□

In the above result it is important to observe that the saturation used in the condition $\mathrm{sat}(Hx) = Hx$ needs to be the same saturation impacting the input u. In other words, for a given saturation level $\bar{u}_0 > 0$ on the input u, the above condition needs to be $\mathrm{sat}_{\bar{u}_0}(Hx) = Hx$. Specifically, this is needed in the argument of item 2 above.

Similarly to (8.29), a vector version of (8.48) is given by

$$(u - q + Hx)^T Wq \geq 0 \tag{8.50}$$

for all $x \in \mathbb{R}^n$ such that $\mathrm{sat}(Hx) = Hx$. Here, we require $W > 0$ diagonal and $H \in \mathbb{R}^{n_u \times n}$, where $n_u \in \mathbb{N}$ denotes the dimension of the input. Additionally, the saturations are assumed to be decentralized with the saturation level on the ith input (given by $\bar{u}_i$) matching the saturation level used on the ith row of H. A visualization of the domain satisfying $\mathrm{sat}(Hx) = Hx$ for a two-dimensional example is provided in Figure 8.3.

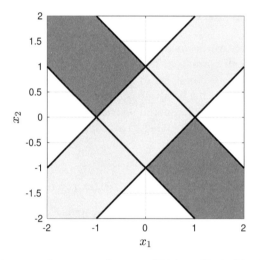

Figure 8.3: Visualization of sectors where $\mathrm{sat}(Hx) = Hx$ holds. Here the dark and light shaded domains correspond to $H_d = [1, 1]$ and $H_\ell = [1, -1]$, respectively. The square centered around the origin represents the area where both conditions are satisfied, i.e., $\mathrm{sat}(Hx) = Hx$ holds.

8.3.1   Local Asymptotic Stability

As we did when investigating stability properties in relation to linearized systems, we can use Lyapunov functions to estimate the region of attraction. Specifically, we can restrict attention to those $x \in \mathbb{R}^n$ such that $V(x) = x^T P x \leq 1$ and then check if $\dot{V}(x) < 0$ for all such $x \neq 0$.

If we can ensure that

$$\tfrac{1}{\bar{u}_i^2} x^T H_i^T H_i x < x^T P x, \qquad \forall\, x \neq 0 \tag{8.51}$$

for each $i = 1, \ldots, n_u$, then on the sublevel set $\{x \in \mathbb{R}^{n_u} \mid x^T P x \leq 1\}$ condition (8.51) implies $\mathrm{sat}_{\bar{u}_i}(H_i x) = H_i x$ for all $i \in \{1, \ldots, n_u\}$. Rearranging the inequality and using the Schur complement (Lemma 8.3) yields

$$0 < \begin{bmatrix} P & H_i^T \\ H_i & \bar{u}_i^2 \end{bmatrix}, \quad i = 1, \ldots, n_u, \tag{8.52}$$

which is a linear matrix inequality in the free variables H_i and P.

To obtain an LMI whose feasibility guarantees local asymptotic stability we can proceed as in Section 8.2.2 with the only difference being that the inequality $q^T W(u - q) \geq 0$ needs to be replaced by (8.50). Here, the S-lemma (Lemma 8.7) leads to the inequality

$$x^T(A^T P + PA)x + 2x^T PBq + 2q^T W(Kx - Lq - q + Hx) < 0 \quad \forall\, (x, q) \neq 0$$

and the corresponding matrix inequality representation

$$\text{He} \begin{bmatrix} PA & PB \\ WK + WH & -W + WL \end{bmatrix} < 0. \tag{8.53}$$

Note that the matrix is not linear in the unknowns P, W, and H due to the product WH. Additionally, since the unknown H also appears in the LMI (8.52), we cannot simply introduce a new unknown WH.

As before, we define the matrices

$$\begin{bmatrix} P^{-1} & 0 \\ 0 & 1 \end{bmatrix} \quad \text{and} \quad \begin{bmatrix} P^{-1} & 0 \\ 0 & W^{-1} \end{bmatrix} \tag{8.54}$$

and multiply (8.52) and (8.53) from the left and the right. Additionally, we define a new set of unknowns $\Lambda_1 = P^{-1}$, $\Lambda_2 = W^{-1}$, and $\Gamma = HP^{-1}$, and the definition of Γ satisfies

$$\begin{bmatrix} \Gamma_1 \\ \vdots \\ \Gamma_{n_u} \end{bmatrix} = \Gamma = HP^{-1} = \begin{bmatrix} H_1 P^{-1} \\ \vdots \\ H_{n_u} P^{-1} \end{bmatrix}.$$

With these definitions, it holds that

$$0 < \begin{bmatrix} P^{-1} & 0 \\ 0 & 1 \end{bmatrix} \begin{bmatrix} P & H_i^T \\ H_i & \bar{u}_i^2 \end{bmatrix} \begin{bmatrix} P^{-1} & 0 \\ 0 & 1 \end{bmatrix}$$

$$= \begin{bmatrix} P^{-1} & P^{-1} H_i^T \\ H_i P^{-1} & \bar{u}_i^2 \end{bmatrix} = \begin{bmatrix} \Lambda_1 & \Gamma_i^T \\ \Gamma_i & \bar{u}_i^2 \end{bmatrix}$$

for all $i \in \{1, \ldots, n_u\}$, and

$$0 > \begin{bmatrix} P^{-1} & 0 \\ 0 & W^{-1} \end{bmatrix} \text{He} \begin{bmatrix} PA & PB \\ WK + WH & -W + WL \end{bmatrix} \begin{bmatrix} P^{-1} & 0 \\ 0 & W^{-1} \end{bmatrix}$$

$$= \text{He} \begin{bmatrix} AP^{-1} & BW^{-1} \\ KP^{-1} + HP^{-1} & -W^{-1} + LW^{-1} \end{bmatrix}$$

$$= \text{He} \begin{bmatrix} A\Lambda_1 & B\Lambda_2 \\ K\Lambda_1 + \Gamma & -\Lambda_2 + L\Lambda_2 \end{bmatrix} \tag{8.55}$$

and the new inequalities are linear in the unknowns Λ_1, λ_2, Γ.

We summarize the results of this section in the following optimization problem. If

$$\min_{\Lambda_1,\Lambda_2,\Gamma,k} \quad k$$

$$\text{subject to} \quad 0 < \Lambda_1 \quad \text{symmetric}$$

$$0 < \Lambda_2 \quad \text{diagonal}$$

$$0 < k$$

$$0 < kI - \Lambda_1 \tag{8.56}$$

$$0 < \text{He} \begin{bmatrix} \frac{1}{2}\Lambda_1 & 0 \\ \Gamma_i & \frac{1}{2}\bar{u}_i^2 \end{bmatrix} \quad i = 1,\ldots,n_u$$

$$0 > \text{He} \begin{bmatrix} A\Lambda_1 & B\Lambda_2 \\ K\Lambda_1 + \Gamma & -\Lambda_2 + L\Lambda_2 \end{bmatrix}$$

admits a solution, then system (8.5) is locally asymptotically stable. In particular, for $w = 0$, the sublevel set $\{x \in \mathbb{R}^n : x^T P x \leq 1\}$ is contained in the region of attraction of the asymptotically stable origin of the nonlinear system (8.5). Recall that P is obtained through the coordinate transformation $P = \Lambda_1^{-1}$. The variable $k \in \mathbb{R}_{>0}$ and the inequalities involving k are not necessary. Through $0 < kI - \Lambda_1$ (which is equivalent to $\frac{1}{k}I < P$) and the objective function, the smallest eigenvalue of P is maximized. In the same way as in (8.41), the parameter $\nu \in (0,1)$ can be included in the LMI (8.55).

Example 8.10. We consider the saturated version of the dynamics in Example 8.6; i.e., consider

$$\dot{x} = Ax + B \operatorname{sat}(u) + Ew$$
$$z = Cx + D \operatorname{sat}(u) + Fw \tag{8.57}$$
$$u = Kx.$$

Using the definition of the deadzone $q = u - \operatorname{sat}(u)$ the dynamics (8.27) can be rewritten in the form

$$\dot{x} = Ax + BKx - Bq + Ew = (A + BK)x - Bq + Ew$$
$$z = Cx + DBx - Dq + Fw = (C + DB)x - Dq + Fw$$
$$u = Kx \tag{8.58}$$
$$q = u - \operatorname{sat}(u).$$

By identifying the closed-loop matrices $A + BK$ and $C + DK$ as well as $-B$ and $-D$ with their counterparts in (8.5), an analysis of asymptotic stability and $\mathcal{L}_2$-stability properties of (8.5) allows us to analyze the dynamics (8.57) with saturated input. Continuing with the feedback gain K from Example 8.6 derived for the unconstrained system, the closed-loop dynamics

$$\dot{x} = \begin{bmatrix} -20.93 & 21.92 & 11.83 \\ -15.62 & 15.28 & 8.22 \\ 4.31 & -4.64 & -2.61 \end{bmatrix} x + \begin{bmatrix} -3 \\ -2 \\ 1 \end{bmatrix} q + \begin{bmatrix} 2 & -2 \\ 1 & 3 \\ 3 & -2 \end{bmatrix} w$$

$$z = \begin{bmatrix} 1 & 0 & 0 \end{bmatrix} x + \begin{bmatrix} -3 & 2 \end{bmatrix} w$$

$$u = \begin{bmatrix} -7.31 & 6.64 & 4.61 \end{bmatrix} x + \begin{bmatrix} 2 & -3 \end{bmatrix} w$$

$$q = u - \text{sat}(u)$$

(8.59)

of the form (8.27) are obtained. While the system without saturation is asymptotically stable, the system with saturated control input is not globally asymptotically stable.

From the optimization problem (8.56) we obtain the Lyapunov function $V(x) = x^T P x$ where

$$P = \begin{bmatrix} 21.93 & -22.11 & -10.78 \\ -22.11 & 26.01 & 12.83 \\ -10.78 & 12.83 & 9.43 \end{bmatrix}$$

and the vector $H = [2.75, -1.98, -2.06]$. For this setting, the relevant level sets, i.e., $|\text{sat}(Hx)| = 1$ and $V(x) = 1$, are visualized in Figure 8.4.

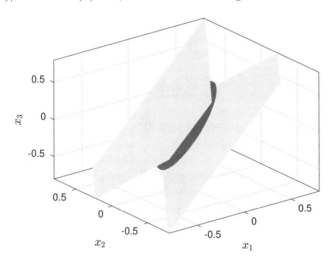

Figure 8.4: Visualization of the sector $|\text{sat}(Hx)| = 1$ (light shaded) together with the level set $V(x) = 1$ corresponding to Example 8.10. The sublevel set $\{x \in \mathbb{R}^3 : V(x) \leq 1\}$ contained in the sector $|\text{sat}(Hx)| \leq 1$ provides an estimate of the region of attraction of the origin of the closed-loop system with bounded input.

8.3.2 $\mathcal{L}_2$-Stability and $\mathcal{L}_2$-Gain Optimization

In this section we derive an LMI whose feasibility guarantees local $\mathcal{L}_2$-stability for the nonlinear system (8.5) building upon the ideas of Section 8.2.3 and Section 8.3.1. Here, we are not only going to restrict the domain of the states $x \in \{x \in \mathbb{R}^n : x^T P x \leq 1\}$ but additionally restrict the disturbances $\|w\|_{\mathcal{L}_2} \leq s$ for a positive value $s \in \mathbb{R}_{>0}$.

For a given parameter $s > 0$, we thus seek to find a Lyapunov function $V(x) = x^T P x$ and an $\mathcal{L}_2$-gain $\gamma > 0$ such that

$$\|z\|_{\mathcal{L}_2} \leq \gamma \|w\|_{\mathcal{L}_2} \qquad \forall \, \|w\|_{\mathcal{L}_2} \leq s \tag{8.60}$$

(and $x(0) = 0$). Instead of condition (8.15) the result will be derived based on condition

$$\dot{V}(x(t)) \leq w^T w \tag{8.61}$$

and, instead of considering the domain $\{x \in \mathbb{R}^n : V(x) \leq 1\}$ (cf. Section 8.3.1) we consider the domain $\{x \in \mathbb{R}^n : V(x) \leq s^2\}$ for $s > 0$ here. In particular, observe that the condition (8.61) together with the assumption $\|w\|_{\mathcal{L}_2} \leq s$ (and $x(0) = 0$) implies that

$$x(t)^T P x(t) = V(x(t)) - V(x(0)) = \int_0^t w(\tau)^T w(\tau) \, d\tau = \|w\|_{\mathcal{L}_2[0,t)}^2 \leq s^2$$

for all $t \geq 0$. Thus, when starting from $x(0) = 0$, $x(t)$ stays in the forward-invariant set $\{x \in \mathbb{R}^n : V(x) \leq s^2\}$ for all $t \geq 0$ and the condition can be enforced as in (8.51) through the LMIs

$$0 < \begin{bmatrix} P & H_i^T \\ H_i & \frac{\bar{u}_i^2}{s^2} \end{bmatrix}, \quad i = 1, \dots, n_u. \tag{8.62}$$

Note that (8.51) is written in terms of $s = 1$. Thus, (8.51) needs to be multiplied by s^2, which explains the term $\frac{1}{s^2}$ in (8.62).

As in Section 8.2.3 we derive P based on a variation of condition (8.42). In particular, for all $(x, q) \neq 0$ we use the condition

$$(u - q + Hx)^T W q \geq 0 \quad \Rightarrow$$
$$x^T (A^T P + PA)x + 2x^T P(Bq + Ew) < -\tfrac{1}{\gamma^2} z^T z + w^T w \tag{8.63}$$

where the local sector condition (8.50) is used and the right-hand side in (8.42) is changed in accordance with the considerations in (8.61). Continuing as in Section 8.2.3 we can proceed with the inequality

$$x^T (A^T P + PA)x + 2x^T P(Bq + Ew)$$
$$+ 2q^T W(Gw + Kx + Lq - q) + \tfrac{1}{\gamma^2} z^T z - w^T w < 0 \qquad \forall \, (x, q) \neq 0 \tag{8.64}$$

and conclude that (8.64) implies (8.63) using the S-lemma (Lemma 8.7).

In matrix notation, after applying the Schur complement (Lemma 8.3), inequality (8.64) translates to the matrix inequality

$$\text{He} \begin{bmatrix} PA & PB & PE & 0 \\ WH + WK & -W + WL & WG & 0 \\ 0 & 0 & -\tfrac{1}{2}I & 0 \\ \hline C & D & F & -\tfrac{\gamma^2}{2}I \end{bmatrix} < 0. \tag{8.65}$$

With the matrices

$$
\begin{bmatrix} P^{-1} & 0 \\ 0 & 1 \end{bmatrix}
\qquad \text{and} \qquad
\begin{bmatrix} P^{-1} & 0 & 0 & 0 \\ 0 & W^{-1} & 0 & 0 \\ 0 & 0 & I & 0 \\ 0 & 0 & 0 & I \end{bmatrix},
\tag{8.66}
$$

which extend the matrices defined in (8.54), left and right multiplication of (8.62) and (8.65) lead to the LMIs

$$
0 < \begin{bmatrix} \Lambda_1 & \Gamma_i^T \\ \Gamma_i & \frac{\bar{u}_i^2}{s^2} \end{bmatrix}
\qquad \forall\, i \in \{1, \dots, n_u\}
$$

and

$$
\text{He} \left[\begin{array}{ccc|c}
A\Lambda_1 & B\Lambda_2 & E & 0 \\
\Gamma + K\Lambda_1 & -\Lambda_2 + L\Lambda_2 & G & 0 \\
0 & 0 & -\frac{1}{2}I & 0 \\ \hline
C\Lambda_1 & D\Lambda_2 & F & -\frac{\gamma^2}{2}I
\end{array} \right] < 0,
$$

where $\Lambda_1 = P^{-1}$, $\Lambda_2 = W^{-1}$, and $\Gamma = HP^{-1}$.

We summarize the results of this section in the following optimization problem. If

$$
\min_{\Lambda_1, \Lambda_2, \Gamma, \delta} \quad \delta
$$

$$
\begin{aligned}
\text{subject to} \quad & 0 < \Lambda_1 \quad \text{symmetric} \\
& 0 < \Lambda_2 \quad \text{diagonal} \\
& 0 < \delta
\end{aligned}
\tag{8.67}
$$

$$
0 < \text{He} \begin{bmatrix} \frac{1}{2}\Lambda_1 & 0 \\ \Gamma_i & \frac{\bar{u}_i^2}{2s^2} \end{bmatrix} \qquad i = 1, \dots, n_u
$$

$$
0 > \text{He} \left[\begin{array}{ccc|c}
A\Lambda_1 & B\Lambda_2 & E & 0 \\
\Gamma + K\Lambda_1 & -\Lambda_2 + L\Lambda_2 & G & 0 \\
0 & 0 & -\frac{1}{2}I & 0 \\ \hline
C\Lambda_1 & D\Lambda_2 & F & -\frac{\delta}{2}I
\end{array} \right]
$$

admits a solution for fixed $s > 0$, then (8.5) satisfies the local $\mathcal{L}_2$-bound (8.60) for $\gamma = \sqrt{\delta}$. Additionally, from the derivations in Section 8.3.1 it follows that the origin of (8.5) is locally asymptotically stable for all $x \in \mathbb{R}^n$ contained in the sublevel set $\{x \in \mathbb{R}^n : x^T P x \leq s^2\}$.

Example 8.11. We consider again the dynamics (8.59) and investigate the local $\mathcal{L}_2$-stability properties. In particular, we solve the optimization problem (8.67) for different values s. The corresponding minimal $\mathcal{L}_2$-gain γ is visualized in Figure 8.5. For $s \geq 0.066$ the optimization problem is infeasible (8.67).

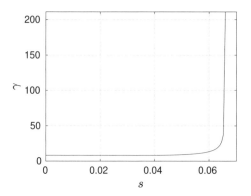

Figure 8.5: Optimal $\mathcal{L}_2$-gain γ (with respect to s) corresponding to the dynamical system (8.59).

8.4 ANTIWINDUP SYNTHESIS

In this section we recall the feedback loop in Figure 8.1 and the corresponding plant and controller dynamics (8.1)–(8.2) motivated at the beginning of this chapter. To improve the performance of the controller (8.2), in particular with respect to saturated inputs, we modify the controller dynamics

$$\mathcal{C} : \begin{cases} \dot{x}_c & = A_c x_c + B_c y + D_{aw,1} q \\ u & = C_c x_c + D_{c,y} y + D_{aw,2} q, \end{cases} \qquad (8.68)$$

where the degree of freedom in the antiwindup injection terms $D_{aw,1}$ and $D_{aw,2}$ are to be designed to improve the closed-loop performance. The structure of this antiwindup scheme is shown in Figure 8.6. With the dimensions defined in (8.3), we see that the injection matrices will satisfy $D_{aw,1} \in \mathbb{R}^{n_c \times n_u}$ and $D_{aw,2} \in \mathbb{R}^{n_u \times n_u}$.

Using the representation (8.4) the modified system capturing the dynamics in Figure 8.6 is given by

$$\begin{aligned} \dot{x} & = Ax + (B + B_{aw} D_{aw}) q + Ew \\ z & = Cx + Dq + Fw \\ u & = Kx + (L + L_{aw} D_{aw}) q + Gw \\ q & = u - \text{sat}(u), \end{aligned} \qquad (8.69)$$

which is a slight modification of (8.5) to include the antiwindup injection terms. As in (8.5) we consider the more general case where possibly $D \neq 0$ and $L \neq 0$ even though according to (8.4) the matrices L and D vanish in the dynamics (8.1), (8.68). Here, the additional matrices D_{aw}, B_{aw}, and L_{aw} are defined as

$$D_{aw} = \begin{bmatrix} D_{aw,1} \\ D_{aw,2} \end{bmatrix}, \quad B_{aw} = \begin{bmatrix} 0 & B_p \\ I_{n_c} & 0 \end{bmatrix}, \quad \text{and} \quad L_{aw} = [0 \ I_{n_u}].$$

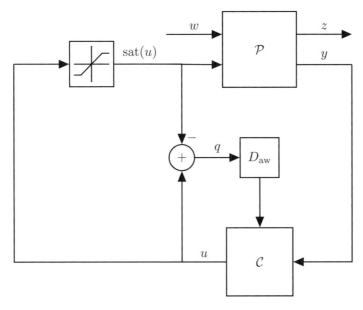

Figure 8.6: Static antiwindup architecture.

8.4.1 Global Antiwindup Synthesis

Following the derivations of Section 8.2.3 we arrive at the matrix inequality

$$
\text{He} \left[\begin{array}{ccc|c}
PA & P\left(B + B_{\text{aw}}D_{\text{aw}}\right) & PE & 0 \\
WK & -W + WL + WL_{\text{aw}}D_{\text{aw}} & WG & 0 \\
0 & 0 & -\frac{\gamma}{2}I & 0 \\
\hline
C & D & F & -\frac{\gamma}{2}I
\end{array} \right] < 0 \qquad (8.70)
$$

with matrix decision variables given by P, W, D_{aw}, and γ.

The inequality (8.70) is equivalent to (8.46), where the dynamics (8.5) is replaced by (8.69). However, in contrast to (8.46), we note that this is not a *linear* matrix inequality. As we did for the feedback synthesis problem, and as we did in Section 8.3.2, we can pre- and post-multiply by the right matrix in (8.66):

$$
\text{He} \left[\begin{array}{ccc|c}
AP^{-1} & \left(B + B_{\text{aw}}D_{\text{aw}}\right)W^{-1} & E & 0 \\
KP^{-1} & -W^{-1} + LW^{-1} + L_{\text{aw}}D_{\text{aw}}W^{-1} & G & 0 \\
0 & 0 & -\frac{\gamma}{2}I & 0 \\
\hline
CP^{-1} & D & F & -\frac{\gamma}{2}I
\end{array} \right] < 0.
$$

We see that this is still not linear in the matrix decision variables D_{aw} and W^{-1}. Hence, we define $X = D_{\text{aw}}W^{-1}$, $\Lambda_1 = P^{-1}$, and $\Lambda_2 = W^{-1}$ to obtain

$$
\text{He} \left[\begin{array}{ccc|c}
A\Lambda_1 & B\Lambda_2 + B_{\text{aw}}X & E & 0 \\
K\Lambda_1 & -\Lambda_2 + L\Lambda_2 + L_{\text{aw}}X & G & 0 \\
0 & 0 & -\frac{\gamma}{2}I & 0 \\
\hline
C\Lambda_1 & D & F & -\frac{\gamma}{2}I
\end{array} \right] < 0, \qquad (8.71)
$$

which is linear in the matrix decision variables Λ_1, Λ_2, X, and γ.

To obtain the antiwindup injection terms, we can thus solve the optimization problem

$$\min_{\Lambda_1,\Lambda_2,X,\gamma} \quad \gamma$$

$$\begin{aligned}
\text{subject to} \quad & 0 < \Lambda_1 \quad \text{symmetric} \\
& 0 < \Lambda_2 \quad \text{diagonal} \\
& 0 < \gamma \\[4pt]
& 0 > \text{He} \begin{bmatrix} A\Lambda_1 & B\Lambda_2 + B_{\text{aw}}X & E & 0 \\ K\Lambda_1 & -\Lambda_2 + L\Lambda_2 + L_{\text{aw}}X & G & 0 \\ 0 & 0 & -\frac{\gamma}{2}I & 0 \\ \hline C\Lambda_1 & D & F & -\frac{\gamma}{2}I \end{bmatrix}.
\end{aligned}$$
$$(8.72)$$

If the optimization problem is feasible, D_{aw} is given by $D_{\text{aw}} = X\Lambda_2^{-1} = XW$.

8.4.2   Well-Posedness of the Control Law

As in (8.7), observe that the controller (8.68) imposes an algebraic condition on $D_{\text{aw},2}$. Specifically, the controller output equation has u on both sides of the equation. Therefore,

$$u = C_c x_c + D_{c,y} y + D_{\text{aw},2}(u - \text{sat}(u))$$

needs to be well posed as per to Definition 8.1 to ensure that $u(x)$ is well defined and in particular Lipschitz continuous. In the single-input case, $D_{\text{aw},2}$ is just a number, in which case we need to be able to solve

$$u - D_{\text{aw},2}(u - \text{sat}(u)) = (1 - D_{\text{aw},2})u + D_{\text{aw},2}\,\text{sat}(u) = v \qquad (8.73)$$

for any $v \in \mathbb{R}$.

As argued in Section 8.2.2, introducing an additional parameter $\nu \in (0,1]$ in (8.71) may fix the problem of well-posedness of the feedback law.

In other words, if the optimization problem (8.72) returns a $D_{\text{aw},2}$ that does not guarantee well-posedness of (8.72), then the LMI (8.71) can be replaced by

$$\text{He} \begin{bmatrix} A\Lambda_1 & B\Lambda_2 + B_{\text{aw}}X & E & 0 \\ K\Lambda_1 & -\nu\Lambda_2 + L\Lambda_2 + L_{\text{aw}}X & G & 0 \\ 0 & 0 & -\frac{\gamma}{2}I & 0 \\ \hline C\Lambda_1 & D & F & -\frac{\gamma}{2}I \end{bmatrix} < 0$$

in (8.72), with $\nu \in (0,1]$ fixed, which may provide an implementable $D_{\text{aw},2}$.

Example 8.12. We consider the plant (8.1) together with the controller dynamics (8.68) defined through the matrices

$$\left[\begin{array}{c|c|c} A_p & B_p & B_w \\ \hline C_{p,y} & & D_{p,y} \\ \hline C_{p,z} & & D_{p,z} \end{array} \right] = \left[\begin{array}{cc|c|c} -0.2 & -0.2 & 0.6 & 3 \\ 1 & 0 & 0.4 & 3 \\ \hline -0.4 & -0.9 & & 0 \\ \hline -0.4 & -0.9 & & 0 \end{array} \right]$$

and

$$\left[\begin{array}{c|c} A_c & B_c \\ \hline C_c & D_{c,y} \end{array} \right] = \left[\begin{array}{c|c} 0 & 1 \\ \hline 2 & 2 \end{array} \right].$$

By solving the optimization problem (8.72) the antiwindup injection terms $D_{aw,1} = -127.30$ and $D_{aw,2} = 0.45$ are obtained.

Figure 8.7 shows the closed-loop response of the system for different scenarios and with respect to a disturbance

$$w(t) = \left\{ \begin{array}{ll} 1, & \text{if } t \leq 1 \\ 0, & \text{if } t > 1 \end{array} \right.$$

driving the states away from the origin. The best performance in terms of driving

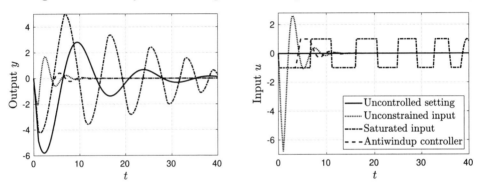

Figure 8.7: Closed-loop solution using the static antiwindup scheme.

the output y back to the origin is achieved through the unconstrained controller. Interestingly, the saturated controller, i.e., the control law in the case that the antiwindup injection terms $D_{aw,1}$ and $D_{aw,2}$ are set to zero, performs worse than the uncontrolled closed-loop system. The antiwindup controller improves the saturated controller significantly.

To obtain the antiwindup controller satisfying the algebraic condition

$$u = C_c x_c + D_{c,y} y + D_{aw,2}(u - \text{sat}(u))$$

we define

$$u = \left\{ \begin{array}{ll} \frac{C_c x_c + D_{c,y} y + D_{aw,2}}{1 - D_{aw,2}}, & \text{if } C_c x_c + D_{c,y} y < -1 \\ C_c x_c + D_{c,y} y, & \text{if } C_c x_c + D_{c,y} y \in [-1, 1] \\ \frac{C_c x_c + D_{c,y} y - D_{aw,2}}{1 - D_{aw,2}}, & \text{if } C_c x_c + D_{c,y} y > 1. \end{array} \right.$$

Then q is defined as $q = u - \text{sat}(u)$ and $\text{sat}(u) = u - q$ is provided as an input to the plant. Note that this explicit formulation of the solution of the algebraic equation is only possible since the input u is one-dimensional.

8.4.3   Regional Antiwindup Synthesis

The above antiwindup structure can provide a surprising improvement in the performance of a closed-loop system. However, its applicability is fairly limited as the achieved global result significantly restricts the plants that can be considered. At a minimum, the plant (8.1) must have the origin as an exponentially stable equilibrium. Even this may not be sufficient though, as the presence of the saturation may make it impossible to find an appropriate Lyapunov function.

A much more widely applicable approach is to combine the regional analysis of Section 8.3 with the antiwindup synthesis of Section 8.4.1. In particular, we can start with the LMIs (8.62) encoding the sector condition and the bound on w, and with the matrix inequality (8.65), where we additionally include the antiwindup injection terms

$$
\text{He} \left[
\begin{array}{ccc|c}
PA & \dot{P}(B + B_{\text{aw}}D_{\text{aw}}) & PE & 0 \\
WH + WK & -\dot{W} + WL + WL_{\text{aw}}D_{\text{aw}} & WG & 0 \\
0 & 0 & -\frac{1}{2}I & 0 \\
\hline
C & D & F & -\frac{\gamma^2}{2}I
\end{array}
\right] < 0. \qquad (8.74)
$$

As in Section 8.3 and Section 8.4.1, multiplication with the matrices (8.66) together with the coordinate transformation $X = D_{\text{aw}}W^{-1}$, $\Lambda_1 = P^{-1}$, $\Lambda_2 = W^{-1}$, and $\Gamma = HP^{-1}$ leads to the LMIs

$$
0 < \left[
\begin{array}{cc}
\Lambda_1 & \Gamma_i^T \\
\Gamma_i & \frac{\bar{u}_i^2}{s^2}
\end{array}
\right] \qquad \forall\, i \in \{1, \ldots, n_u\}
$$

and

$$
\text{He} \left[
\begin{array}{ccc|c}
A\Lambda_1 & B\Lambda_2 + B_{\text{aw}}X & E & 0 \\
K\Lambda_1 & -\Lambda_2 + L\Lambda_2 + L_{\text{aw}}X & G & 0 \\
0 & 0 & -\frac{1}{2}I & 0 \\
\hline
C\Lambda_1 & D & F & -\frac{\gamma^2}{2}I
\end{array}
\right] < 0, \qquad (8.75)
$$

which we can use to formulate the following optimization problem. For fixed $s > 0$, the local equivalent to (8.72) can be formulated as

$$
\min_{\Lambda_1, \Lambda_2, \Gamma, X, \delta} \quad \delta
$$

$$
\text{subject to} \quad
\begin{aligned}
&0 < \Lambda_1 \quad \text{symmetric} \\
&0 < \Lambda_2 \quad \text{diagonal} \\
&0 < \delta
\end{aligned} \qquad (8.76)
$$

$$
0 < \text{He} \left[
\begin{array}{cc}
\frac{1}{2}\Lambda_1 & 0 \\
\Gamma_i & \frac{\bar{u}_i^2}{2s^2}
\end{array}
\right] \qquad \forall\, i \in \{1, \ldots, n_u\}
$$

$$
0 > \text{He} \left[
\begin{array}{ccc|c}
A\Lambda_1 & B\Lambda_2 + B_{\text{aw}}X & E & 0 \\
K\Lambda_1 & -\Lambda_2 + L\Lambda_2 + L_{\text{aw}}X & G & 0 \\
0 & 0 & -\frac{1}{2}I & 0 \\
\hline
C\Lambda_1 & D & F & -\frac{\delta}{2}I
\end{array}
\right].
$$

If the optimization problem is feasible, we obtain D_{aw} from $D_{\text{aw}} = X\Lambda_2^{-1} = XW$, the $\mathcal{L}_2$-gain is given by $\gamma = \sqrt{\delta}$, and the relevant sublevel set is defined through $\{x \in \mathbb{R}^n : x^T P x \leq s^2\}$, where $P = \Lambda^{-1}$. As argued in Section 8.4.2 on the basis of the global antiwindup optimization problem (8.72), the parameter $\nu \in (0, 1)$ can be incorporated in (8.76) using the same arguments.

8.5 EXERCISES

Exercise 8.1. Consider the system discussed in Example 8.5. Solve LMI (8.19) to confirm that the optimal γ is given by $\gamma = 7.43$.

Exercise 8.2. Consider the system discussed in Example 8.6. Compute an optimal γ, a stabilizing feedback law K, and a Lyapunov function V as in Example 8.6.

Exercise 8.3. Implement and solve the LMI discussed in Example 8.10.

Exercise 8.4. Combine the ideas discussed in Section 8.2.1 and Section 8.3 to derive an LMI whose solution provides a locally stabilizing feedback law of the origin of the saturated dynamics (8.27).

Exercise 8.5. Reproduce Example 8.12. In particular, consider the plant

$$\dot{x}_p = A_p x_p + B_p(u - q) + B_w w,$$
$$y = C_{p,y} x_p + D_{p,y} w,$$

and the controller (without the z-dynamics)

$$\dot{x}_c = A_c x_c + B_c y + D_{aw,1} q,$$
$$u = C_c x_c + D_{c,y} y + D_{aw,2} q,$$

defined through

$$\left[\begin{array}{c|c|c} A_p & B_p & B_w \\ \hline C_{p,y} & & D_{p,y} \end{array}\right] = \left[\begin{array}{cc|c|c} -0.2 & -0.2 & 1 & 3 \\ 1 & 0 & 0.5 & 3 \\ \hline -0.4 & -0.9 & & 0 \end{array}\right]$$

$$\left[\begin{array}{c|c} A_c & B_c \\ \hline C_c & D_{c,y} \end{array}\right] = \left[\begin{array}{c|c} 0 & 1 \\ \hline 2 & 2 \end{array}\right].$$

Additionally, we assume that $x_p(0) = [0, 0]^T$, $x_c(0) = 0$,

$$w(t) = \begin{cases} 1 & \text{for } t \le 1 \\ 0 & \text{for } t > 1, \end{cases}$$

and we are interested in the input $(t, u(t))$ as well as the output $(t, y(t))$ for $t \in [0, 40]$.

(i) (Uncontrolled dynamics) Simulate and visualize $(t, u(t))$ and $(t, y(t))$ for the setting $u(t) = 0$ and $q(t) = 0$.
(ii) (Unconstrained dynamics) Simulate and visualize $(t, u(t))$ and $(t, y(t))$ for and $q(t) = 0$.
(iii) (Constrained dynamics without antiwindup controller) Simulate and visualize $(t, u(t))$ and $(t, y(t))$ for $q = u - \text{sat}(u)$, $D_{aw,1} = 0$ and $D_{aw,2} = 0$.
(iv) (Antiwindup controller) Compute $D_{aw,1}$ and $D_{aw,2}$ by solving the optimization problem (8.72). Simulate and visualize $(t, u(t))$ and $(t, y(t))$ for $q = u - \text{sat}(u)$.

Exercise 8.6. Consider the linear system

$$\dot{x} = Ax + Bu + Ew$$
$$z = Cx + Fw$$
$$u = Kx$$

where

$$A = \begin{bmatrix} 3 & 0 & 0 & -2 \\ -1 & -1 & 2 & -2 \\ 2 & -1 & 1 & 1 \\ -1 & 1 & 1 & 2 \end{bmatrix}, \quad B = \begin{bmatrix} -2 & 1 \\ 1 & -1 \\ 0 & 1 \\ -2 & 2 \end{bmatrix}, \quad E = \begin{bmatrix} 2 & -2 \\ 1 & 3 \\ 3 & -2 \\ -2 & 2 \end{bmatrix},$$
$$C = \begin{bmatrix} 1 & 0 & 0 & 0 \end{bmatrix}, \quad F = \begin{bmatrix} -3 & 2 \end{bmatrix}.$$

Compute an optimal feedback gain matrix K and a corresponding $\mathcal{L}_2$-gain γ such that γ in $\|z\|_{\mathcal{L}_2} \le \gamma \|w\|_{\mathcal{L}_2}$ is minimal.

8.6   BIBLIOGRAPHICAL NOTES AND FURTHER READING

The term *antiwindup* stems from the classical behavior of a PI controller in the presence of input saturation. In particular, input saturation can lead to a lengthy, and frequently large, error. This error then is integrated by the PI controller and, as a consequence, the controller state (i.e., the integrator state) can grow very large. This phenomenon is referred to as *integrator windup* and techniques to cope with this are referred to as *antiwindup techniques*. Most classical undergraduate texts, such as [45], [51], and [114], will include at least one simple scheme for addressing this issue.

The approach presented in this chapter is based on the book by Zaccarian and Teel [168]. Whereas classical antiwindup techniques focus on the undesirable behavior of the controller state, Zaccarian and Teel asked a different question, which is: once the input has saturated, what is the best possible overall system behavior we can achieve? This led to a series of designs whereby presumably desirable local behavior is unmodified and behavior in the presence of saturation is compensated via an optimally designed controller augmentation (for example, the augmentation shown in Figure 8.6). The book [168] contains designs for a wide variety of plants satisfying different commonly encountered assumptions.

As with the design presented in this chapter, the designs in [168] rely heavily on optimization techniques involving Linear Matrix Inequalities. The standard reference for LMIs is [28]. For more information on well-posedness of algebraic loops, we refer to [25].

Chapter Nine

Control Lyapunov Functions

In discussing absolute stability and input-to-state stability, inputs have generally not been available to us as designers but, rather, have been the source of undesirable behavior. We now turn to the question of design; that is, how can we *select* inputs, or design feedback control laws, so as to obtain desirable system behavior? To this end, compared to the previous chapters we make a minor notational change and consider nonlinear systems of the form

$$\dot{x} = f(x, u), \tag{9.1}$$

$f : \mathbb{R}^n \times \mathbb{R}^m \to \mathbb{R}^n$, where u denotes a control input as distinct from the previously considered external or disturbance input which we had labelled as w.

Our general goal in this chapter is to leverage Lyapunov's second method for the purpose of designing feedback control laws that asymptotically stabilize the origin. For this purpose, we will use a *control Lyapunov function (CLF)*, namely a continuously differentiable and positive definite $V : \mathbb{R}^n \to \mathbb{R}_{\geq 0}$ such that for each state x we can find a control input u to enforce a satisfactory decrease condition

$$\tfrac{d}{dt} V(x(t)) = \langle \nabla V(x), f(x, u) \rangle < 0, \qquad \forall \, x \neq 0. \tag{9.2}$$

In particular, our goal is to define the input as a function of the state so that the origin is asymptotically stable. In other words, we wish to find a function $u = k(x)$ so that the function $V(x)$ satisfies

$$\tfrac{d}{dt} V(x(t)) = \langle \nabla V(x), f(x, k(x)) \rangle < 0, \qquad \forall \, x \neq 0$$

and is therefore a Lyapunov function.

Definition 9.1 (Control Lyapunov function (CLF))**.** *Consider the nonlinear system* (9.1)*. A continuously differentiable function* $V : \mathbb{R}^n \to \mathbb{R}_{\geq 0}$ *is called control Lyapunov function for* (9.1) *if there exist* $\alpha_1, \alpha_2 \in \mathcal{K}_\infty$ *such that*

$$\alpha_1(|x|) \leq V(x) \leq \alpha_2(|x|), \qquad \forall \, x \in \mathbb{R}^n,$$

and for all $x \in \mathbb{R}^n \backslash \{0\}$ *there exists* $u \in \mathbb{R}^m$ *such that*

$$\langle \nabla V(x), f(x, u) \rangle < 0. \tag{9.3}$$

9.1   CONTROL AFFINE SYSTEMS

In this chapter, we will largely focus on a class of systems that are affine in the control input u; i.e.,

$$\dot{x} = f(x) + g(x)u, \qquad (9.4)$$

where $f, g : \mathbb{R}^n \to \mathbb{R}^n$ satisfy our usual regularity conditions (e.g., locally Lipschitz). For simplicity, we will focus on single-input systems, $u \in \mathbb{R}$, though the results in this chapter generalize in a straightforward manner to multi-input systems. Without loss of generality, we will assume $f(0) = 0$.

When discussing Lyapunov methods for input affine systems, it is convenient to use what is called the *Lie derivative* notation

$$\langle \nabla V(x), f(x) \rangle = L_f V(x). \qquad (9.5)$$

Importantly for our purposes, the inner product is distributive over addition and homogeneous with respect to scalar multiplication; that is, for scalars $a_1, a_2 \in \mathbb{R}$ and vectors $v, v_1, v_2 \in \mathbb{R}^n$,

$$\langle v, a_1 v_1 + a_2 v_2 \rangle = a_1 \langle v, v_1 \rangle + a_2 \langle v, v_2 \rangle.$$

Using the Lie derivative notation (9.5), the decrease condition (9.2) for a control affine system (9.4) can be written as

$$\dot{V}(x) = \langle \nabla V(x), f(x) + g(x)u \rangle = L_f V(x) + L_g V(x)u < 0, \quad \forall\, x \neq 0. \qquad (9.6)$$

For such control affine systems, one common definition of a control Lyapunov function is a continuously differentiable and positive definite function V satisfying

$$L_f V(x) < 0 \quad \text{for all} \ \ x \in \mathbb{R}^n \backslash \{0\} \ \ \text{such that} \ \ L_g V(x) = 0. \qquad (9.7)$$

In other words, when the derivative in (9.6) cannot be made negative via the control input u (because the term $L_g V(x)$ is zero), the term $L_f V(x)$ provides the required negative derivative.

9.2   ISS REDESIGN VIA $L_g V$ DAMPING

In Chapter 7, we investigated system robustness in the form of input-to-state stability (ISS). As the following example shows, feedback designs that provide asymptotic stability of the origin are not necessarily ISS with respect to disturbances.

Example 9.2. Consider the system

$$\dot{x} = x + (x^2 + 1)u.$$

One possible feedback stabilizer is given by

$$u = k(x) = -\frac{2x}{x^2 + 1},$$

which results in the (linear) closed-loop system $\dot{x} = -x$.

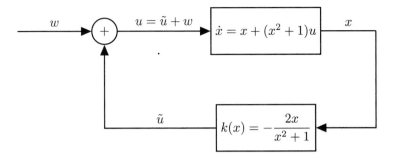

Figure 9.1: Feedback with disturbance.

However, consider the feedback control subject to an additive disturbance as shown in Figure 9.1; i.e., where $u = k(x) + w$. Then the closed-loop system is given by

$$\dot{x} = -x + (x^2 + 1)w.$$

As discussed in Section 7.5, this system is clearly not ISS. In fact, with $w(t) = 1$ for all $t \geq 0$ the system exhibits finite escape time.

Suppose $V(x)$ is a Lyapunov function for the closed-loop system in the absence of the disturbance w so that V satisfies

$$\dot{V}(x) = L_f V(x) + L_g V(x)k(x).$$

Then a candidate ISS-Lyapunov function V for the closed loop system would have a decrease condition given by

$$\dot{V}(x) = L_f V(x) + L_g V(x)k(x) + L_g V(x)w.$$

The usual trick of applying Young's inequality (Lemma A.4) yields

$$\dot{V}(x) = L_f V(x) + L_g V(x)k(x) + \tfrac{1}{2}\left(L_g V(x)\right)^2 + \tfrac{1}{2}w^2,$$

from which we see that if the feedback controller had an additional term in the form of $-L_g V(x)$, this would dominate the $\tfrac{1}{2}\left(L_g V(x)\right)^2$ term and $V(x)$ would indeed be an ISS-Lyapunov function.

Returning to the example system, first take $V_1(x) = \tfrac{1}{2}x^2$ and augment the feedback controller as

$$u = k_1(x) = k(x) - L_g V_1(x) = -\frac{2x}{x^2 + 1} - x(x^2 + 1).$$

Then, in the presence of the additive disturbance w the closed-loop system is

$$\dot{x} = -x - x(x^2 + 1)^2 + (x^2 + 1)w \doteq f_1(x, w).$$

The following computation shows that this new closed-loop system is now ISS:

$$\langle \nabla V_1(x), f_1(x, w) \rangle = -x^2 - x^2(x^2 + 1)^2 + x(x^2 + 1)w$$
$$\leq -x^2 - \tfrac{1}{2}x^2(x^2 + 1)^2 + \tfrac{1}{2}w^2.$$

Alternatively, consider the closed-loop Lyapunov function given by

$$V_2(x) = \tfrac{1}{2}\log(1 + x^2)$$

and augment the feedback controller as

$$u = k_2(x) = k(x) - L_g V_2(x) = -\frac{2x}{x^2 + 1} - x.$$

Again considering the additive disturbance, the closed loop is

$$\dot{x} = -x - x(x^2 + 1) + (x^2 + 1)w \doteq f_2(x, w).$$

Similarly, V_2 is an ISS-Lyapunov function, as shown by

$$\langle \nabla V_2(x), f_2(x, w) \rangle = -\frac{x^2}{1 + x^2} - x^2 + xw$$

$$\leq -\frac{x^2}{1 + x^2} - \tfrac{1}{2}x^2 + \tfrac{1}{2}w^2.$$

The above example suggests the following ISS redesign procedure. Suppose we have constructed a stabilizing state feedback $u = k(x)$ for (9.4); i.e., the origin is asymptotically stable for

$$\dot{x} = f(x) + g(x)k(x).$$

Suppose further that we have a Lyapunov function for the closed-loop system; i.e., a function $V : \mathbb{R}^n \to \mathbb{R}_{\geq 0}$ so that

$$L_f V(x) + L_g V(x)k(x) \leq -\alpha(|x|), \tag{9.8}$$

where $\alpha \in \mathcal{K}_\infty$. (That such a Lyapunov function always exists follows from Theorem 2.31.) The augmented feedback

$$u = k(x) - L_g V(x)$$

yields a closed-loop system that is ISS with respect to an additive disturbance. In other words, the system

$$\dot{x} = f(x) + g(x)\left(k(x) - L_g V(x)\right) + g(x)w$$

is ISS. Indeed, V is an ISS-Lyapunov function since

$$\langle \nabla V(x), f(x) + g(x)\left(k(x) - L_g V(x)\right) + g(x)w \rangle$$
$$= L_f V(x) + L_g V(x)k(x) - \left(L_g V(x)\right)^2 + L_g V(x)w$$
$$\leq L_f V(x) + L_g V(x)k(x) - \tfrac{1}{2}\left(L_g V(x)\right)^2 + \tfrac{1}{2}w^2$$
$$\leq -\alpha(|x|) + \tfrac{1}{2}w^2. \tag{9.9}$$

Such a feedback stabilizer is also sometimes referred to as a nonlinear damping or Jurdjevic-Quinn controller.

9.3   SONTAG'S UNIVERSAL FORMULA

Given a control affine system (9.4) and a control Lyapunov function (9.7), the function

$$
k(x) = \begin{cases} -\left(\kappa + \dfrac{L_f V(x) + \sqrt{L_f V(x)^2 + L_g V(x)^4}}{L_g V(x)^2} \right) L_g V(x), & L_g V(x) \neq 0 \\ 0, & L_g V(x) = 0, \end{cases} \tag{9.10}
$$

where $\kappa \geq 0$, provides a feedback law that yields an asymptotically stable origin. This feedback inherits the regularity property of the control Lyapunov function everywhere except at the origin. At the origin, the feedback is continuous if the control Lyapunov function satisfies a small control property whereby states close to the origin only require small-in-norm inputs to guarantee the decrease property (9.6).

We can see how (9.10) provides a stabilizing feedback by examining the formula in two steps. First, consider $\kappa = 0$. Then we see that, in the time derivative of the (closed-loop) control Lyapunov function, (9.10) cancels the $L_f V$ term and replaces it with a negative definite quantity as

$$
\begin{aligned}
\dot{V}(x) &= L_f V(x) + L_g V(x) k(x) \\
&= L_f V(x) - L_g V(x) \left(\frac{L_f V(x) + \sqrt{L_f V(x)^2 + L_g V(x)^4}}{L_g V(x)^2} \right) L_g V(x) \\
&= L_f V(x) - L_f V(x) - \sqrt{L_f V(x)^2 + L_g V(x)^4} = -\sqrt{L_f V(x)^2 + L_g V(x)^4}.
\end{aligned}
$$

Furthermore, with $\kappa > 0$, we see that this merely adds a nonlinear damping term, $-\kappa(L_g V(x))^2$, as in the ISS redesign procedure above, which implies that the feedback (9.10) guarantees ISS of the closed-loop system. Developed by Eduardo Sontag, (9.10) is known as Sontag's Universal Formula.

9.4   BACKSTEPPING

In the previous two sections we have assumed we have a control Lyapunov function available. This is reasonable for first order systems, but may present a difficulty for higher order systems. In this section as well as the subsequent section, we investigate systems with additional structure in order to derive recursive methods of constructing control Lyapunov functions.

First, we consider systems in *strict feedback form*. Mathematically, such systems take the form

$$
\begin{aligned}
\dot{x}_1 &= f_1(x_1, x_2) \\
\dot{x}_2 &= f_2(x_1, x_2, x_3) \\
&\ \ \vdots \\
\dot{x}_{n-1} &= f_{n-1}(x_1, x_2, \ldots, x_{n-1}, x_n) \\
\dot{x}_n &= f_n(x_1, x_2, \ldots, x_n, u).
\end{aligned} \tag{9.11}
$$

Figure 9.2 shows a third-order system in a strict feedback form. This form is

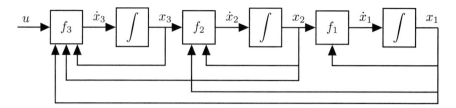

Figure 9.2: Third-order strict feedback system.

referred to as a *feedback* structure due to the fact that, with the exception of a single state feeding forward into each function, states are fed back to the various functions. For example, in Figure 9.2, x_3 feeds into f_2 but is not fed forward into f_1.

We introduce the basic idea of backstepping via a small example. Consider the system

$$\dot{x} = x^3 + x\xi$$
$$\dot{\xi} = u. \tag{9.12}$$

Step 1: Define Virtual Control. Suppose, for the moment, that ξ is a control input for the x-subsystem; that is, consider ξ as a virtual control for x. Then a simple choice for a feedback stabilizer is to take

$$\xi = k(x) = -2x^2,$$

which would result in a closed-loop system $\dot{x} = -x^3$ with an asymptotically stable origin. A simple Lyapunov function for this closed-loop system is given by $V(x) = \frac{1}{2}x^2$. Adding and subtracting the term $xk(x)$ on the right-hand side of (9.12), we can rewrite the x-subsystem as

$$\dot{x} = x^3 + xk(x) - xk(x) + x\xi = -x^3 + x(\xi + 2x^2).$$

Step 2: Define an Error Variable. Of course ξ is a state, not a control input. However, if we can drive the difference between ξ and the function $k(x)$ to zero, we might hope that ξ could still essentially behave as a control input. Therefore, let us define an error variable

$$z = \xi - k(x) = \xi + 2x^2, \tag{9.13}$$

which we may rewrite as $\xi = z - 2x^2$. We calculate

$$\dot{z} = \dot{\xi} - \widehat{k(x)} = u - \frac{\partial}{\partial x}k(x)\dot{x} = u + 4x(x^3 + x\xi) = u + 4x(-x^3 + xz)$$
$$= u - 4x^4 + 4x^2 z.$$

and rewrite (9.12) in the (x, z) coordinates:

$$\dot{x} = -x^3 + xz$$
$$\dot{z} = u - 4x^4 + 4x^2 z. \tag{9.14}$$

Step 3: Construct a Control Lyapunov Function. In order to construct a control Lyapunov function, we augment the Lyapunov function for the x-subsystem with the square of the error variable; i.e.,

$$V_a(x, z) = V(x) + \tfrac{1}{2} z^2 = \tfrac{1}{2} x^2 + \tfrac{1}{2} z^2. \tag{9.15}$$

To see that this is a control Lyapunov function, we compute the time derivative

$$\dot{V}_a(x, z) = -x^4 + x^2 z + z(u - 4x^4 + 4x^2 z)$$
$$= -x^4 + z(u + x^2 - 4x^4 + 4x^2 z).$$

From the above expression it is clear that we can choose u to make the time derivative of V_a negative definite. Alternatively, we note that $L_g V_a(x, z) = z$ and when $L_g V_a(x, z) = 0$, $L_f V_a(x, z) = -x^4 < 0$.

Step 4: Construct a feedback stabilizer. At this stage, the most obvious feedback stabilizer is given by

$$u = k_1(x, z) = -x^2 + 4x^4 - 4x^2 z - z, \tag{9.16}$$

which cancels all the cross-terms in the Lyapunov function and leaves

$$\dot{V}_a(x, z) = -x^4 - z^2. \tag{9.17}$$

However, any feedback stabilizer that results in $V_a(x, z)$ being a Lyapunov function is sufficient, for example, using Sontag's formula (9.10). The closed-loop solution of (9.12) using the input (9.16) is visualized in Figure 9.3. Algorithm 1 summarizes

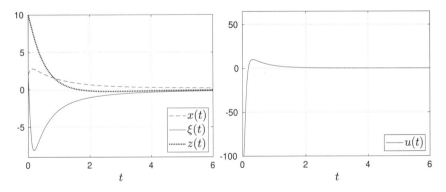

Figure 9.3: Closed-loop solution (9.12), error dynamics (9.13), and corresponding input (9.16).

the steps of this section for higher order systems where additionally the definition of error dynamics

$$
\begin{bmatrix}
\dot{z}_0 \\
\dot{z}_1 \\
\vdots \\
\dot{z}_i
\end{bmatrix}
=
\begin{bmatrix}
\tilde{f}_1(z_0, k_1(z_0)) \\
\tilde{f}_2(z_0, z_1, k_2(z_0, z_1)) \\
\vdots \\
\tilde{f}_{i+1}(z_0, z_1, \ldots, z_{i-1}, x_{i+1})
\end{bmatrix}
\tag{9.18}
$$

for $i = 1, \ldots, n$ is used.

Algorithm 1: Backstepping

Input: System in strict feedback form (9.11). Define $z_0 = x_1$, $x_{n+1} = u$, $\tilde{f}_1 = f_1$, and $V_0 = 0$.

Output: Stabilizing feedback law u.

For $i = 1, 2, \ldots, n$

1. Consider the error dynamics (9.18) and the virtual control $x_{i+1} = k_i(z_0, \ldots, z_{i-1})$.
2. Define k_i in such a way that the origin of the error dynamics (9.18) is asymptotically stable and define $\tilde{V}_i(z_0, \ldots, z_{i-1})$ so that

$$
V_i(z_0, \ldots, z_{i-1}) \doteq V_{i-1}(z_0, \ldots, z_{i-2}) + \tilde{V}_i(z_0, \ldots, z_{i-1})
$$

   is a Lyapunov function for (9.18).
3. If $i \neq n$, define the error dynamics $z_i = x_{i+1} - k_i(z_0, \ldots, z_{i-1})$ with

$$
\dot{z}_i = \dot{x}_{i+1} - \tfrac{d}{dt}k_i(z_0, \ldots, z_{i-1}) = f_{i+1}(x_1, \ldots, x_{i+1}) - \tfrac{d}{dt}k_i(z_0, \ldots, z_{i-1})
$$
$$
\doteq \tilde{f}_{i+1}(z_0, \ldots, z_{i-1}, x_{i+1}).
$$

4. If $i = n$ return the input

$$
u(x_1, \ldots, x_n) \doteq k_n(z_0, \ldots, z_{n-1})
$$

   and the control Lyapunov function $V(x_1, \ldots, x_n) \doteq V_n(z_0, \ldots, z_{n-1})$.

9.4.1   Avoiding Cancellations

In some cases, including the above, it may be possible to simplify the virtual control by avoiding cancellations. In order to present a more general procedure, consider

$$
\dot{x} = f(x) + g(x)\xi
$$
$$
\dot{\xi} = u.
$$

With the virtual stabilizing feedback $\xi = k(x)$ and the error variable $z = \xi - k(x)$ this yields

$$
\dot{x} = f(x) + g(x)k(x) + g(x)z
$$
$$
\dot{z} = u - \tfrac{\partial}{\partial x}k(x)\dot{x}.
$$

The feedback control derived in the previous section is given by

$$u(x, z) = -L_g V(x) + \tfrac{\partial}{\partial x} k(x) \left(f(x) + g(x)k(x) + g(x)z \right) - z. \qquad (9.19)$$

Rather than taking the augmented Lyapunov function (9.15), consider a candidate control Lyapunov function given by

$$V_a(x, z) = V(x) + \tfrac{1}{2} z^2 + W(x), \qquad (9.20)$$

where $W(x)$ satisfies

$$L_f W(x) + L_g W(x)k(x) < 0 \qquad (9.21)$$

for all $x \neq 0$ and

$$\langle \nabla W(x), g(x) \rangle = \left. \tfrac{\partial}{\partial x} k(x) \dot{x} \right|_{z=0} = \tfrac{\partial}{\partial x} k(x)(f(x) + g(x)k(x)). \qquad (9.22)$$

In other words, W is a Lyapunov function for the closed-loop system if the virtual control was the real control, (9.21), and also satisfies a particular constraint on the term $L_g W$ given by (9.22).

Computing the time derivative of (9.20) we see that

$$\begin{aligned}
\dot{V}_a(x, z) &= L_f V(x) + L_g V(x)k(x) + L_g V(x)z + L_f W(x) + L_g W(x)k(x) \\
&\quad + L_g W(x)z + z \left(u - \tfrac{\partial k}{\partial x}(x)\dot{x} \right) \\
&= L_f V(x) + L_g V(x)k(x) + L_f W(x) + L_g W(x)k(x) \\
&\quad + z \left(u + L_g V(x) - \tfrac{\partial k}{\partial x}(x) \left(f(x) + g(x)k(x) - g(x)z \right) + L_g W(x) \right) \\
&= L_f V(x) + L_g V(x)k(x) + L_f W(x) + L_g W(x)k(x) + z \left(u + L_g V(x) \right) \\
&\quad + \tfrac{\partial k}{\partial x}(x)g(x)z - \tfrac{\partial k}{\partial x}(x) \left(f(x) + g(x)k(x) \right) + L_g W(x) \Big).
\end{aligned}$$

Using (9.22), we see that the last terms above cancel, leaving

$$\begin{aligned}
\dot{V}_a(x, z) &= L_f V(x) + L_g V(x)k(x) + L_f W(x) + L_g W(x)k(x) \\
&\quad + z \left(u + L_g V(x) + \tfrac{\partial k}{\partial x}(x)g(x)z \right),
\end{aligned}$$

which yields the feedback stabilizer

$$u(x, z) = -L_g V(x) - \tfrac{\partial k}{\partial x}(x)g(x)z - z.$$

This feedback is simpler than (9.19) but at the expense of a more complicated control Lyapunov function (9.20).

In the example above, we had $f(x) = x^3$, $g(x) = x$, and $k(x) = -2x^2$. Therefore, the constraint (9.22) is

$$\langle \nabla W(x), x \rangle = \tfrac{\partial W}{\partial x}(x)x = (-4x)(x^3 - 2x^3) = 4x^4.$$

By inspection, this yields $W(x) = x^4$, and it is straightforward to check (9.21) since

$$L_f W(x) + L_g W(x)k(x) = 4x^6 - 8x^6 = -4x^6.$$

Therefore, a control Lyapunov function for the above example is

$$V_a(x, z) = \tfrac{1}{2}x^2 + x^4 + \tfrac{1}{2}z^2 \tag{9.23}$$

and a stabilizing feedback is

$$u(x, z) = -x^2 - 4x^2 z - z. \tag{9.24}$$

Not only is this simpler than (9.16) in the sense that it has fewer mathematical operations, the term it avoids is the highest order term $4x^4$, meaning the feedback (9.24) will require less control effort, especially far away from the origin. Similarly to Figure 9.3, Figure 9.4 shows the closed-loop solution of (9.12), but with the feedback stabilizer (9.24) instead of (9.16).

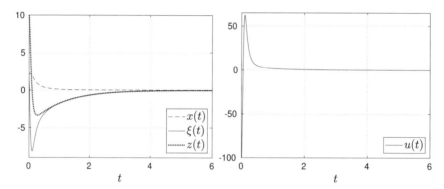

Figure 9.4: Closed-loop solution (9.12), error dynamics (9.13), and corresponding input (9.24).

9.4.2   Exact Backstepping and a High-Gain Alternative

Consider now (9.14), but with an additional integrator

$$\begin{aligned}
\dot{x} &= x^3 + x\xi_1 \\
\dot{\xi}_1 &= \xi_2 \\
\dot{\xi}_2 &= u.
\end{aligned} \tag{9.25}$$

Take the first virtual control as $\xi_1 = k_1(x) = -2x^2$ and define the error variable

$$z_1 = \xi_1 - k_1(x) = \xi_1 + 2x^2.$$

Then we can rewrite the top two subsystems as

$$\begin{aligned}
\dot{x} &= -x^3 + xz_1 \\
\dot{z}_1 &= \xi_2 - \tfrac{\partial k_1}{\partial x}(x)\left(-x^3 + xz_1\right) = \xi_2 - 4x^4 + 4x^2 z_1.
\end{aligned}$$

As in the previous section in (9.23), we take the control Lyapunov function

$$V(x, z_1) = \tfrac{1}{2}x^2 + x^4 + \tfrac{1}{2}z_1^2 \tag{9.26}$$

and the virtual control

$$\xi_2 = k_2(x, z_1) = -x^2 - 4x^2 z_1 - z_1.$$

Define the error variable $z_2 = \xi_2 - k_2(x, z_1)$ so that

$$\dot{x} = -x^3 + x z_1$$
$$\dot{z}_1 = z_2 + k_2(x, z_1) - 4x^4 + 4x^2 z_1 = -z_1 + z_2 - x^2 - 4x^4$$
$$\dot{z}_2 = u - \overbrace{\dot{k_2(x, z_1)}}.$$

Note that this has become a very complicated expression as

$$\overbrace{\dot{k_2(x, z_1)}} = (-8x z_1 - 2x)\dot{x} + (-4x^2 - 1)\dot{z}_1$$
$$= (-8x z_1 - 2x)(-x^3 + x z_1) + (-4x^2 - 1)(z_2 - x^2 - z_1 - 4x^4).$$

The computations at this point become a bit onerous, but using again the Lyapunov function

$$V(x, z_1, z_2) = \tfrac{1}{2}x^2 + x^4 + \tfrac{1}{2}z_1^2 + \tfrac{1}{2}z_2^2 \tag{9.27}$$

(extending (9.26) by an additional term) would lead naturally to the control

$$u = -z_1 - z_2 + \overbrace{\dot{k_2(x, z_1)}}. \tag{9.28}$$

We refer to this process as *exact backstepping*, where we obtain a CLF at each recursive step as well as a globally stabilizing feedback.

As an alternative design at the first step of backstepping, rather than canceling the $\dot{k}_1$ term as above, we can dominate it via a linear term. In other words, consider the virtual control

$$\xi_2 = -\kappa z_1,$$

where we will take $\kappa > 0$ sufficiently large. Still with $k_1(x) = -2x^2$ and $z_1 = \xi_1 - k_1(x)$, and for the moment assuming ξ_2 is the actual input, we have

$$\dot{x} = -x^3 + x z_1$$
$$\dot{z}_1 = -\kappa z_1 - 4x^4 + 4x^2 z_1.$$

Using the Lyapunov function (9.26) we have

$$\dot{V}(x, z_1) = -x^4 + x^2 z_1 - 4x^6 + 4x^4 z_1 - \kappa z_1^2 - 4x^4 z_1 + 4x^2 z_1^2$$
$$\leq -x^4 - 4x^6 + \tfrac{1}{2}x^4 + \tfrac{1}{2}z_1^2 - \kappa z_1^2 + 4x^2 z_1^2$$
$$= -\tfrac{1}{2}x^4 - 4x^6 - z_1^2\left(\kappa - \tfrac{1}{2} - 4x^2\right).$$

Therefore, if

$$\kappa > \tfrac{1}{2} + 4x^2$$

or, equivalently,

$$x^2 < \frac{\kappa - \frac{1}{2}}{4} \tag{9.29}$$

then the origin is locally asymptotically stable with an estimate of the region of attraction given by the largest level set of $V(x, z_1)$ such that (9.29) holds for all x in that level set. Notice that by taking κ sufficiently large, we can make the region in (9.29) as large as we wish. As opposed to the previous exact backstepping, we refer to this process as *approximate* or *high-gain* backstepping.

The subsequent step of backstepping is relatively simple. Let $z_2 = \xi_2 + \kappa z_1$ so that

$$\dot{x} = -x^3 + x z_1$$
$$\dot{z}_1 = -\kappa z_1 - 4x^4 - 4x^2 z_1 + z_2$$
$$\dot{z}_2 = u + \kappa \left(-\kappa z_1 - 4x^4 - 4x^2 z_1 + z_2 \right).$$

If we again use a dominating linear term $u = -\kappa z_2$ we can back substitute the error variables to see that the feedback control thus constructed is

$$u = -\kappa \left(\xi_2 + \kappa(\xi_1 + 2x^2) \right). \tag{9.30}$$

Note that we do not need to use the same value of κ at both steps. We have only done so here to keep the notation simple.

Theorem 9.3 (High-gain backstepping [135, Proposition 6.3]). *Consider the system*

$$\dot{x} = f(x) + g(x)\xi_1$$
$$\dot{\xi}_1 = \xi_2$$
$$\vdots$$
$$\dot{\xi}_n = u$$

in strict feedback form, let $\kappa \in \mathbb{R}_{>0}$ be a design parameter, and assume there exists a feedback stabilizer $\xi_1 = k(x)$ and an associated control Lyapunov function $V(x)$. Let

$$p(\lambda) = \lambda^n + a_{n-1}\lambda^{n-1} + \cdots + a_1\lambda + a_0$$

be an arbitrary Hurwitz polynomial. Then the feedback

$$u = -\kappa \left(a_{n-1}\xi_n + \kappa \left(a_{n-2}\xi_{n-1} + \kappa \left(\cdots + \kappa(a_1\xi_2 + \kappa a_0(\xi_1 - k(x))) \cdots \right) \right) \right)$$

achieves semiglobal stabilization of $[x^T, \xi^T]^T = 0$.

Here, the term semiglobal refers to the fact that we have a design parameter, κ, which can be tuned to make the region of attraction for the origin as large as we wish.

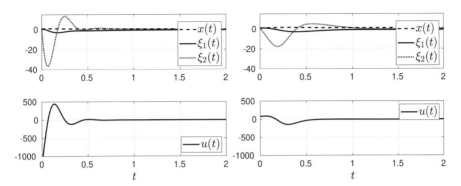

Figure 9.5: Closed-loop solution (9.25) and corresponding input u. On the left, high-gain backstepping (9.30) with $\kappa = 20$, and exact backstepping (9.28) on the right.

9.4.3   Convergence Structure

We conclude our discussion on backstepping by looking at the convergence behavior of the closed-loop solution for both exact and high-gain backstepping. Recall the dynamics (9.12), defining the two-dimensional system

$$\dot{x} = x^3 + x\xi$$
$$\dot{\xi} = u$$

for which we have derived the error dynamics

$$z = \xi - k(x) = \xi + 2x^2$$

(see (9.13)), and using exact backstepping we have obtained the feedback law

$$u(x, \xi) = -x^2 + 4x^4 - 4x^2(\xi + 2x^2) - (\xi + 2x^2) \tag{9.31a}$$

(see (9.16)). Alternatively, for $\kappa > 0$, with Theorem 9.3 we can design the control law

$$u(x, \xi) = -\kappa^2(\xi + 2x^2) = -\kappa^2 z \tag{9.31b}$$

using high-gain backstepping. Here, the Hurwitz polynomial is defined as $p(\lambda) = \lambda + 1$. Figure 9.6 shows the closed-loop solution and the error dynamics for exact backstepping as well as for different values of κ for high-gain backstepping. Additionally, the set $\mathcal{Z} := \{[x, \xi]^T \in \mathbb{R}^2 : 0 = \xi - 2x^2\}$ where the z is zero is shown in Figure 9.6, left. We observe that for larger κ, the solution rapidly converges to the set $\mathcal{Z}$ and hence the error dynamics go to zero. After $\mathcal{Z}$ is reached, the solution slides along the surface toward the origin $[x, \xi]^T = 0$.

   This is a typical property of the high-gain backstepping control law, where the selection of κ impacts the speed of convergence driving the error dynamics to zero.

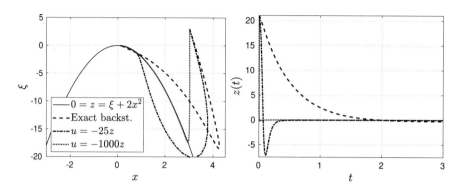

Figure 9.6: Closed-loop solution (left) and error dynamics (right) for different feedback laws (9.31).

9.5 FORWARDING

Instead of the strict feedback structure of (9.11) illustrated in Figure 9.2, we now consider systems in *strict feedforward form*. Mathematically, such systems are described by

$$
\begin{aligned}
\dot{x}_1 &= f_1(x_2, x_3, \ldots, x_n, u) \\
\dot{x}_2 &= f_2(x_3, x_4, \ldots, x_n, u) \\
&\;\;\vdots \\
\dot{x}_{n-1} &= f_{n-1}(x_n, u) \\
\dot{x}_n &= f_n(u)
\end{aligned}
\tag{9.32}
$$

and a block diagram for a third-order strict feedforward system is shown in Figure 9.7. Notice that, in contrast to Figure 9.2, states and the input are fed forward in Figure 9.7.

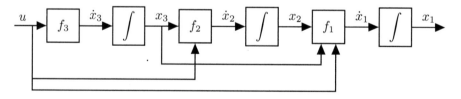

Figure 9.7: Third-order strict feedforward system.

In order to introduce the basic idea of forwarding, consider a system of the form

$$
\begin{aligned}
\dot{z} &= h(x) + \ell(x)u \\
\dot{x} &= f(x) + g(x)u
\end{aligned}
\tag{9.33}
$$

shown in Figure 9.8, and suppose the origin is globally asymptotically stable for $\dot{x} = f(x)$ and that we have an appropriate Lyapunov function; i.e., a continuously differentiable positive definite $V : \mathbb{R}^n \to \mathbb{R}_{\geq 0}$ that satisfies $L_f V(x) < 0$ for all

$x \neq 0$. Furthermore, suppose we can find a solution $\mathcal{M}(x)$ to the partial differential equation

$$L_f \mathcal{M}(x) = \langle \nabla \mathcal{M}(x), f(x) \rangle = h(x), \quad \mathcal{M}(0) = 0. \tag{9.34}$$

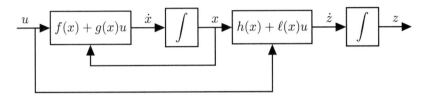

Figure 9.8: Representation of the dynamics (9.33).

If we are able to find a solution to (9.34) satisfying $\ell(0) - L_g \mathcal{M}(0) \neq 0$ then a control Lyapunov function for the overall system is given by

$$W(x, z) = V(x) + \tfrac{1}{2} \left(z - \mathcal{M}(x) \right)^2. \tag{9.35}$$

Indeed, computing the time derivative of W yields

$$\begin{aligned}
\dot{W}(x, z) &= L_f V(x) + L_g V(x)u + (z - \mathcal{M}(x))(\dot{z} - L_f \mathcal{M}(x) - L_g \mathcal{M}(x)u) \\
&= L_f V(x) + L_g V(x)u + (z - \mathcal{M}(x))(h(x) + \ell(x)u - L_f \mathcal{M}(x) - L_g \mathcal{M}(x)u) \\
&= L_f V(x) + L_g V(x)u + (z - \mathcal{M}(x))(\ell(x)u - L_g \mathcal{M}(x)u) \\
&= L_f V(x) + u \left(L_g V(x) + (z - \mathcal{M}(x))(\ell(x) - L_g \mathcal{M}(x)) \right).
\end{aligned} \tag{9.36}$$

Note that the condition $\ell(0) - L_g \mathcal{M}(0) \neq 0$ is required to guarantee a decrease in z.

One choice for a feedback control is then

$$u = -\kappa \left(L_g V(x) + (z - \mathcal{M}(x))(\ell(x) - L_g \mathcal{M}(x)) \right), \tag{9.37}$$

where $\kappa > 0$ is a design parameter.

Before we give an example of forwarding we summarize the derivations so far in the following theorem.

Theorem 9.4. *Consider the system* (9.33) *and let* $V : \mathbb{R}^n \to \mathbb{R}_{\geq 0}$ *be a continuously differentiable positive definite Lyapunov function for* $\dot{x} = f(x)$. *Suppose there exists a solution* $\mathcal{M} : \mathbb{R}^n \to \mathbb{R}^n$ *to the partial differential equation* (9.34) *such that* $\ell(0) - L_g \mathcal{M}(0) \neq 0$. *Then* (9.35) *is a control Lyapunov function of* (9.33) *and* (9.37) *is a globally asymptotically stabilizing feedback law.*

Example 9.5. Consider

$$\begin{aligned}
\dot{z} &= x - x^2 u \\
\dot{x} &= u.
\end{aligned} \tag{9.38}$$

This system is not immediately in the standard form above, but is easily modified by defining the initial feedback $u = -x + v$, where v is now the input to be designed for the system

$$\dot{z} = x - x^2(-x + v) = x + x^3 - x^2 v$$
$$\dot{x} = -x + v.$$

Here, we see that $h(x) = x + x^3$, $f(x) = -x$, $\ell(x) = -x^2$, and the obvious Lyapunov function for $\dot{x} = -x$ is $V(x) = \frac{1}{2}x^2$. Following the above, we would like to find a function $\mathcal{M}(x)$ satisfying $\mathcal{M}(0) = 0$,

$$h(x) = \frac{\partial \mathcal{M}(x)}{\partial x} f(x),$$

and $\ell(0) - L_g \mathcal{M}(0) \neq 0$; that is,

$$x + x^3 = \frac{\partial \mathcal{M}(x)}{\partial x}(-x).$$

It is clear that $\mathcal{M}(x) = -\frac{1}{3}x^3 - x$ solves this equation and the condition

$$\ell(0) - L_g \mathcal{M}(0) = -1 \neq 0$$

is satisfied.

Therefore, a control Lyapunov function is given by

$$W(x, z) = \frac{1}{2}x^2 + \frac{1}{2}\left(z + x + \frac{1}{3}x^3\right)^2.$$

Indeed,

$$\dot{W}(x, z) = -x^2 + xv + \left(z + x + \frac{1}{3}x^3\right)\left(\dot{z} + \dot{x} + x^2\dot{x}\right)$$
$$= -x^2 + xv + \left(z + x + \frac{1}{3}x^3\right)\left(x + x^3 - x^2v - x + v - x^3 + x^2v\right)$$
$$= -x^2 + xv + \left(z + x + \frac{1}{3}x^3\right)v$$
$$= -x^2 + \left(z + 2x + \frac{1}{3}x^3\right)v.$$

We choose the feedback stabilizer

$$v = -\left(z + 2x + \frac{1}{3}x^3\right) \tag{9.39}$$

and hence the control for the original system (9.38) is given by

$$u = -x + v = -z - 3x + \frac{1}{3}x^3. \tag{9.40}$$

The corresponding closed-loop solution using the feedback law (9.40) is shown in Figure 9.9.

Before we move on, note that as with backstepping, forwarding can be applied recursively. To make this clear, suppose now the system of interest is

$$\dot{z}_2 = h_2(x, z_1) + \ell_2(x, z_1)u$$
$$\dot{z}_1 = h_1(x) + \ell_1(x)u$$
$$\dot{x} = f(x) + g(x)u,$$

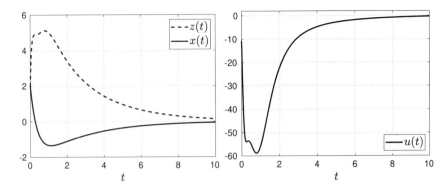

Figure 9.9: Closed-loop solution of the dynamics (9.38) with control (9.40).

which is simply (9.33) with the addition of the state z_2. The process described above constructs a control Lyapunov function for the (x, z_1) system. However, a straightforward relabelling of variables whereby x is redefined as (x, z_1) and z is defined as z_2 allows us to apply the forwarding procedure again.

9.5.1 Forwarding mod $L_g V$

It is not always straightforward to solve the forwarding PDE (9.34). Here, we briefly present a technique that can provide some extra flexibility when trying to solve (9.34).

We consider the system

$$\begin{aligned} \dot{z} &= h(x) \\ \dot{x} &= f(x) + g(x)u, \end{aligned} \qquad (9.41)$$

where $x \in \mathbb{R}^n$, $z \in \mathbb{R}^m$ and which represents a special class of (9.33) with $\ell(x) = 0$. In this case the function W in (9.35) satisfies

$$\dot{W}(x, z) = L_f V(x) + (L_g V(x) - (z - \mathcal{M}(x)) L_g \mathcal{M}(x)) u.$$

Theorem 9.6 ([125, Proposition 1]). *Consider the system* (9.41) *and assume that it satisfies the following assumptions.*

1. *There exists a positive definite function* $V : \mathbb{R}^n \to \mathbb{R}_{\geq 0}$ *such that* $L_f V(x) < 0$ *for all* $x \neq 0$.
2. *There exist a continuous function* $k : \mathbb{R}^n \to \mathbb{R}$ *and a continuously differentiable function* $\mathcal{M} : \mathbb{R}^n \to \mathbb{R}^m$ *with the properties*

$$L_f \mathcal{M}(x) = h(x) + k(x) L_g V(x), \qquad \mathcal{M}(0) = 0, \qquad (9.42)$$

$$L_f V(x) - \frac{k(x) L_g V(x)^2}{L_g \mathcal{M}(x)} < 0 \quad \forall\, x \in \{x \in \mathbb{R}^n : L_g \mathcal{M}(x) \neq 0,\ x \neq 0\},$$

$$L_g \mathcal{M}(0) \neq 0.$$

Then $W(x, z) = V(x) + \frac{1}{2}(z - \mathcal{M}(x))^2$ *is a control Lyapunov function of* (9.41) *which satisfies the small control property.*

Compared to (9.34), the partial differential equation (9.42) contains the additional term $k(x)L_gV(x)$, which may be easier to solve due to the free function $k(x)$.

With a control Lyapunov function $W(x, z)$ from Theorem 9.6, a feedback u can be defined based on the following conditions:

1. If $k(x) = 0$ we define

$$u(x, z) = -(L_gV(x) - (z - \mathcal{M}(x))L_g\mathcal{M}(x)).$$

2. If $k(x) \neq 0$ and $L_g\mathcal{M}(x) = 0$ or $k(x)L_g\mathcal{M}(x) > 0$ we define

$$u(x, z) = k(x)(z - \mathcal{M}(x)).$$

3. If $k(x)L_g\mathcal{M}(x) < 0$, we define

$$u(x, z) = -\frac{k(x)L_gV(x)}{L_g\mathcal{M}(x)} - (L_gV(x) - (z - \mathcal{M}(x))L_g\mathcal{M}(x))^2. \tag{9.43}$$

Then $u(x, z)$ asymptotically stabilizes the origin.

Example 9.7. Consider the dynamics

$$\dot{z} = 2x_2^2 - 3x_2^3$$
$$\dot{x} = \begin{bmatrix} -1 & 0 \\ -3 & -2 \end{bmatrix} x + \begin{bmatrix} 1 \\ 1 \end{bmatrix} v. \tag{9.44}$$

A Lyapunov function for the unforced linear part of the system is

$$V(x) = \frac{1}{2}x^T \begin{bmatrix} 5 & -1 \\ -1 & 1 \end{bmatrix} x. \tag{9.45}$$

Using this function V in the partial differential equation (9.42) yields

$$\tfrac{\partial \mathcal{M}}{\partial x_1}(x)(-x_1) + \tfrac{\partial \mathcal{M}}{\partial x_2}(x)(-3x_1 - 2x_2) = 2x_2^2 - 3x_2^3 + k(x)(5x_1 - x_2 - x_1 + x_2)$$

which we can rewrite as

$$-\left(\tfrac{\partial \mathcal{M}}{\partial x_1}(x) + 3\tfrac{\partial \mathcal{M}}{\partial x_2}(x)\right)x_1 - 2\tfrac{\partial \mathcal{M}}{\partial x_2}(x)x_2 = 2x_2^2 - 3x_2^3 + 4x_1k(x).$$

From this expression we can conclude that the unknown functions $\mathcal{M}$ and k need to satisfy

$$\tfrac{\partial \mathcal{M}}{\partial x_2}(x) = -x_2 + \tfrac{3}{2}x_2^2$$
$$\tfrac{\partial \mathcal{M}}{\partial x_1}(x) + 3\tfrac{\partial \mathcal{M}}{\partial x_2}(x) = -4k(x) \tag{9.46}$$

and it is straightforward to verify that

$$\mathcal{M}(x) = -\tfrac{1}{2}x_2^2 + \tfrac{1}{2}x_2^3 \qquad \text{and} \qquad k(x) = \tfrac{3}{4}x_2 - \tfrac{9}{8}x_2^2$$

satisfies these conditions.

However, this choice of $\mathcal{M}$ does not satisfy $L_g\mathcal{M}(0) \neq 0$ as required by Theorem 9.6. Examining the forwarding PDE (9.46), there is an easy fix in the form of

$$\mathcal{M}(x) = -\tfrac{1}{2}x_2^2 + \tfrac{1}{2}x_2^3 + x_1 \qquad \text{and} \qquad k(x) = \tfrac{3}{4}x_2 - \tfrac{9}{8}x_2^2 - \tfrac{1}{4}.$$

From Theorem 9.6, a control Lyapunov function is thus given by

$$W(x,z) = V(x) + \tfrac{1}{2}(z - \mathcal{M}(x))^2.$$

With the above choices, it is straightforward to verify that $k(x)L_g\mathcal{M}(x) < 0$ for all $x \in \mathbb{R}^2$ and thus a stabilizing feedback law is given by (9.43).

As a final observation on this example, we can try to use the standard forwarding presented in Theorem 9.4. This requires us to solve (9.46) with $k(x) = 0$. However, some straightforward calculations show that this is not possible.

9.5.2   Convergence Structure

To analyze the convergence structure induced by a forwarding design for (9.33), we examine the time derivative of the function W derived in (9.36), given by

$$\dot{W}(x,z) = L_f V(x) + u\left(L_g V(x) + (z - \mathcal{M}(x))\left(\ell(x) - L_g\mathcal{M}(x)\right)\right)$$

together with the feedback

$$u = -\left(L_g V(x) + (z - \mathcal{M}(x))\left(\ell(x) - L_g\mathcal{M}(x)\right)\right)$$

defined in (9.37).

In addition, as above we assume that $x = 0$ is asymptotically stable for $\dot{x} = f(x)$; i.e., $L_f V(x) < 0$ for all $x \neq 0$ and $u = 0$.

Rearranging the terms in (9.37) shows that $u(x,z) = 0$ is satisfied on the *forwarding manifold*

$$\left\{ (x,z) \in \mathbb{R}^{n+m} : z = \mathcal{M}(x) + \tfrac{L_g V(x)}{\ell(x) - L_g\mathcal{M}(x)} \right\}. \tag{9.47}$$

The feedback u thus drives the system to the forwarding manifold, while the condition $L_f V(x) < 0$ for all $x \neq 0$ guarantees convergence to the origin once (x,z) is close to the forwarding manifold where the influence of u vanishes.

We illustrate this behavior by continuing with Example 9.5.

Example 9.8. In Example 9.5, using $\mathcal{M}(x) = -\tfrac{1}{3}x^3 - x$, we derived the feedback law $v(x,z) = (-z + 2x + \tfrac{1}{3}x^3)$ for the system

$$\dot{z} = x + x^3 - x^2 v, \qquad \dot{x} = -x + v. \tag{9.48}$$

The forwarding manifold, where v vanishes is defined based on the condition

$$z = -2x - \tfrac{1}{3}x^3.$$

The forwarding manifold is shown in Figure 9.10 together with a closed-loop solution. As expected, solutions converge to the forwarding manifold (9.47) and, on the forwarding manifold, the system behaves as $\dot{x} = f(x) = -x$ so that solutions converge to the origin.

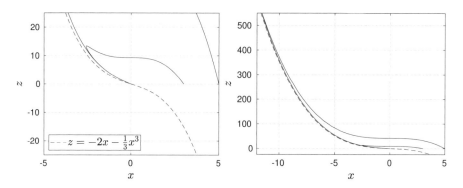

Figure 9.10: The forwarding manifold. Solutions converge to the forwarding manifold before they approach the origin.

9.5.3   Saturated Control

We maintain the assumption that the origin is asymptotically stable for $\dot{x} = f(x)$. Figure 9.10 shows that the solution can make a large detour before converging to the origin. In particular, the solution with initial condition

$$[x(0), z(0)]^T = [5, 0]^T$$

generates a relatively large value for the z-coordinate. To avoid this behavior observe that

$$u = -c \cdot \text{sat}\left(\frac{1}{c}\left(L_g V(x) + (z - \mathcal{M}(x))\left(\ell(x) - L_g \mathcal{M}(x)\right)\right)\right) \qquad (9.49)$$

guarantees $\dot{W}(x, z) < 0$, $(x, z) \neq 0$, for all values of $c > 0$. Compared to (9.37), the control law (9.49) has the advantage that it is bounded by c by design, i.e., $u \in [-c, c]$, but this feedback still ensures that the origin $(x, z) = 0$ is globally asymptotically stabilized.

However, note that if the origin of $\dot{x} = f(x)$ has been stabilized using a transformation involving the input, then (9.49) does not guarantee that the original input is bounded.

Example 9.9. Figure 9.11 shows closed-loop solutions of the system (9.48) using the saturated feedback law (9.49) for $c \in \{1, 2\}$. While $v(x, z)$ is bounded by c, the original input $u = -x + v$ defined in Example 9.5 is clearly unbounded due to its dependence on x. Still, the presence of the saturation avoids the large state detour seen in Figure 9.10. However, avoiding this detour does not imply that the solutions converge faster to the origin, as can be seen in the bottom plots in Figure 9.11.

9.6   STABILIZABILITY AND CONTROL LYAPUNOV
FUNCTIONS

In this section we return to the definition of CLFs in Definition 9.1 and investigate their existence for a given system (9.1). Since we intend to stabilize the origin of the nonlinear system (9.1) we need to assume that an input $u : \mathbb{R}_{\geq 0} \to \mathbb{R}^m$ exists

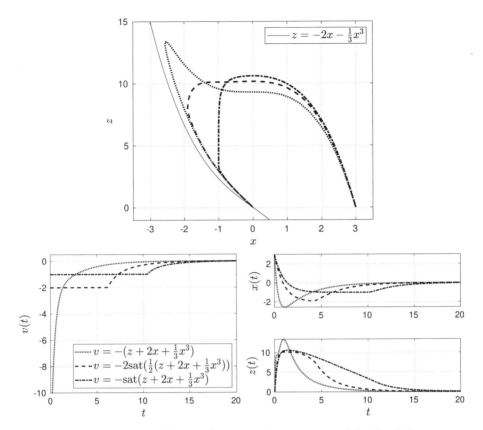

Figure 9.11: Closed-loop solutions using a saturated feedback law.

which renders the origin of the closed-loop system asymptotically stable.

Definition 9.10 (Asymptotic stabilizability). *Consider the nonlinear system* (9.1). *The origin is called (locally)* asymptotically stabilizable *if there exist* $\varepsilon > 0$ *and* $\beta \in \mathcal{KL}$ *such that for all* $x_0 = x(0) \in \mathcal{B}_\varepsilon$ *there exists a bounded and continuous function* $u_{x_0} : \mathbb{R}_{\geq 0} \to \mathbb{R}^m$ *such that*

$$|x(t; x_0, u_{x_0})| \leq \beta(|x_0|, t) \qquad \forall\, t \geq 0. \tag{9.50}$$

If ε *can be chosen arbitrarily large, then the origin is called* globally asymptotically stabilizable.

In (9.50), we highlight the dependency of the solution $x(t)$ on the selection of the initial value x_0 and on the input u_{x_0}. Definition 9.10 generalizes Definition 3.26 in Section 3.5.2 for linear systems. If the origin of (9.1) is asymptotically stabilizable according to Definition 9.10 then we know that it is possible to drive $x_0 \in \mathcal{B}_\varepsilon$ asymptotically to the origin by selecting an appropriate input function $u_{x_0} : \mathbb{R}_{\geq 0} \to \mathbb{R}^m$ depending on the initial state. Since (9.1) is in general only a model of a plant, we are interested in feedback laws $u = u(x)$, to compensate for unmodelled disturbances. Thus, instead of a time-dependent function, $u(t)$, $t \in \mathbb{R}_{\geq 0}$, we are interested in state-dependent feedback laws $u(x) = k(x)$, $k : \mathbb{R}^n \to \mathbb{R}^m$.

Definition 9.11 (Lipschitz continuous feedback stabilizability). *Consider the non-linear system* (9.1). *The origin is called (locally) Lipschitz continuous feedback stabilizable if there exist* $\varepsilon > 0$, $\beta \in \mathcal{KL}$, *and a Lipschitz continuous feedback law* $k : \mathbb{R}^n \to \mathbb{R}^m$ *such that*

$$|x(t; x_0, k(x))| \leq \beta(|x_0|, t) \qquad \forall\, t \geq 0, \quad \forall\, x_0 \in \mathcal{B}_\varepsilon.$$

If ε *can be chosen arbitrarily large, then the origin is called globally Lipschitz continuous feedback stabilizable.*

It follows immediately that Lipschitz continuous feedback stabilizability implies asymptotic stabilizability. In the case of linear systems in Section 3.5.2 we did not distinguish between Lipschitz continuous feedback stabilizability and asymptotic stabilizability. This is due to the fact that we have discussed a constructive method to design linear (and thus, in particular, Lipschitz continuous) stabilizing feedback laws for stabilizable linear systems in Section 3.5.3. With a stabilizing linear feedback law at hand, we know how to construct a quadratic CLF $V(x) = x^T P x$ for linear systems by solving the Lyapunov Equation (3.4).

For nonlinear systems (9.1), on the other hand, the following questions still need to be addressed:

- Are there systems (9.1) with an asymptotically stabilizable, but not Lipschitz continuous feedback stabilizable, origin?
- Does asymptotic stabilizability or Lipschitz continuous feedback stabilizability imply the existence of a control Lyapunov function according to Definition 9.1?

9.6.1 Existence of Lipschitz Continuous Feedback Laws

The question whether Definition 9.10 and Definition 9.11 are equivalent can simply be answered through an appropriate counterexample. Here, we use dynamics known as Artstein's circles to derive such an example. A second famous counterexample can be derived based on the dynamics of the Brockett integrator (1.56).

Example 9.12. Consider the two-dimensional system

$$\begin{aligned} \dot{x}_1 &= \left(x_1^2 - x_2^2\right) u \\ \dot{x}_2 &= 2x_1 x_2 u \end{aligned} \tag{9.51}$$

with one dimensional input $u \in \mathbb{R}$ and known as Artstein's circles [11]. Figure 9.12 shows solutions of (9.51) for different inital conditions (left) and the phase portrait (right) with respect to constant input signals $u = -1$ and $u = 1$, respectively. From the illustration it can be observed that the solutions with $x_0 \in \mathbb{R}^2 \backslash (\mathbb{R} \times \{0\})$ evolve on circles with center $c = c(x_0) \in \mathbb{R}^2$. More precisely, every solution of (9.51) with $x_0 \in \mathbb{R}^2 \backslash (\mathbb{R} \times \{0\})$ satisfies

$$|x(t; x_0, u(t)) - c(x_0)| = |x_0 - c(x_0)|, \qquad \forall\, t \geq 0 \tag{9.52}$$

independently of the input $u : \mathbb{R} \to \mathbb{R}$. While we only provide an intuitive explanation based on numerical simulations here, a rigorous analytical derivation of this property can be found in [145], for example.

The only impact the input has is to change the orientation of the solutions.

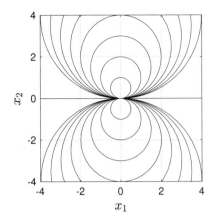 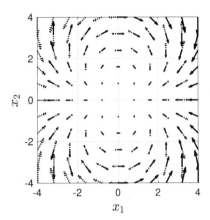

Figure 9.12: On the left, solutions of the dynamics (9.51) for different initial conditions and $u \in \{-1, 1\}$. On the right, phase portrait of (9.51) for $u = 1$ (dotted arrows) and $u = -1$.

Namely, for $x_2(0) > 0$ and $u > 0$, the solution follows its circle counterclockwise, $u = 0$ implies $\dot{x} = 0$, leading to an equilibrium, and for $x_2(0) > 0$ and $u < 0$ the solution follows its circle clockwise. Finally, for $x_2(0) < 0$ the roles of $u > 0$ and $u < 0$ are switched. Solutions starting on the x_1-axis, i.e., $x_0 \in \mathbb{R} \times \{0\}$, satisfy $x(t; x_0, u(t)) \in \mathbb{R} \times \{0\}$ for all $t \geq 0$ and for all $u : \mathbb{R}_{\geq 0} \to \mathbb{R}$.

For all $x_0 \in \mathbb{R}^2$ the state-dependent input

$$k(x) = \begin{cases} |x|, & x_1 < 0 \\ |x|, & x_1 = 0 \\ -|x|, & x_1 > 0 \end{cases} \tag{9.53}$$

guarantees $x(t; x_0, k(x)) \to 0$ for $t \to \infty$ and thus shows that (9.51) is asymptotically stabilizable according to Definition 9.10. Note that the choice of feedback (9.53) on the x_2-axis is essentially arbitrary and either $k(x) = |x|$ or $k(x) = -|x|$ ensures convergence to the origin. However, it is important to note that a choice to move left or right along a particular circle is necessary.

The feedback law (9.53) does not guarantee Lipschitz continuous feedback stabilizabilty according to Definition 9.11 since (9.53) is discontinuous. In fact, a locally Lipschitz continuous feedback law cannot exist, not even locally for an arbitrarily small neighborhood around the origin. To see this, consider for the sake of contradiction that there exists a neighborhood $\mathcal{B}_\delta$ and a Lipschitz continuous feedback law $k(\cdot)$ ensuring that the origin of the corresponding closed-loop system is asymptotically stable. Consider $x_m = [-\varepsilon, 0]^T$ and $x_p = [\varepsilon, 0]^T$ for $\varepsilon > 0$ arbitrarily small. Since k is Lipschitz continuous and according to the observations above, there exist neighborhoods $\mathcal{B}_{\varepsilon_m}(x_m)$ and $\mathcal{B}_{\varepsilon_p}(x_p)$, $\varepsilon_m, \varepsilon_p > 0$, such that $u(x) > 0$ for all $\mathcal{B}_{\varepsilon_m}(x_m)$ and $u(x) < 0$ for all $\mathcal{B}_{\varepsilon_p}(x_p)$. Let $x_0 \in \mathcal{B}_{\varepsilon_p}(x_p)$ be selected such that

$$\{x \in \mathbb{R}^2 : |x - c(x_0)| = |x_0 - c(x_0)|\} \subset \mathcal{B}_\delta(0)$$

and

$$\{x \in \mathbb{R}^2 : |x - c(x_0)| = |x_0 - c(x_0)|\} \cap \mathcal{B}_{\varepsilon_m}(x_m) \neq \emptyset.$$

Then the Lipschitz continuity of $u(\cdot)$ together with the different signs implies that there exists

$$\tilde{x}_0 \in \{x \in \mathbb{R}^2 : |x - c(x_0)| = |x_0 - c(x_0)|\} \setminus \{0\}$$

such that $k(\tilde{x}_0) = 0$. Thus the corresponding solution $x(t, \tilde{x}_0, k) = \tilde{x}_0$ is constant, which contradicts the assumption on asymptotic stability of the origin of the closed-loop system.

This example illustrates the difference between asymptotic stabilizability and Lipschitz continuous feedback stabilizability. In general, it is difficult to decide whether a system is Lipschitz continuous feedback stabilizable or not. We will give one sufficient and one necessary condition for Lipschitz continuous feedback stabilizability here.

Theorem 9.13. *Consider the dynamics* (9.1) *and its linearization* (3.9) *around the origin represented through the pair* (A, B). *If* (A, B) *is stabilizable, then* (9.1) *is Lipschitz continuous feedback stabilizable.*

Proof. Since the linearization defined through (3.49) is stabilizable we can apply the results from Section 3.5.3 to construct a feedback gain matrix $K \in \mathbb{R}^{m \times n}$ such that the origin of the linear system $\dot{x} = Ax + BKx$ is exponentially stable and $u = Kx$ is a Lipschitz continuous feedback law.
    Since

$$\left[\frac{\partial f}{\partial x}(x, Kx)\right]_{(x,u)=(0,0)} = \left[\frac{\partial f}{\partial x}(x, u)\right]_{(x,u)=(0,0)} + \left[\frac{\partial f}{\partial u}(x, u)\right]_{(x,u)=(0,0)} \cdot K$$
$$= A + BK$$

it follows from Theorem 3.4 that the origin of $\dot{x} = f(x, Kx)$ is locally exponentially stable. Thus $u(x) = Kx$ is a Lipschitz continuous stabilizing feedback law for (9.1) which completes the proof. $\square$

Theorem 9.14 (Brockett condition, [15, Theorem 2.16] or [33]). *Consider the dynamics* (9.1) *and assume there exists a Lipschitz continuous feedback law* $u : \mathbb{R}^n \to \mathbb{R}^m$, $u(0) = 0$, *such that the origin of the closed-loop dynamics* $\dot{x} = f(x, u(x))$ *is asymptotically stable.*
    Then, for each $\varepsilon > 0$ *there exists* $\delta > 0$ *such that for all* $y \in \mathcal{B}_\delta$ *there exist* $x \in \mathcal{B}_\varepsilon$ *and* $u \in \mathcal{B}_\varepsilon$ *with* $y = f(x, u)$.

While Theorem 9.13 and Theorem 9.14 provide a sufficient and a necessary condition for Lipschitz continuous feedback stabilizability it is important to note that there is a gap between these results, i.e., there are systems for which neither the assumptions of Theorem 9.13 nor the assumptions of Theorem 9.14 are satisfied. We refer to Exercise 9.7 addressing this fact based on the dynamics (9.51).
    A second well-known example of a system which is asymptotically stabilizable but not Lipschitz continuous feedback stabilizable is the *Brockett integrator* (1.55) introduced in Section 1.2.2.

Example 9.15. Recall the dynamics of the *Brockett integrator* (1.55) describing the dynamics of a mobile robot, $\dot{x} = f(x, u)$,

$$\dot{x}_1 = u_1, \tag{9.54a}$$
$$\dot{x}_2 = u_2, \tag{9.54b}$$
$$\dot{x}_3 = x_1 u_2 - x_2 u_1. \tag{9.54c}$$

Asymptotic feedback stabilizability is verified in Exercise 9.8 through a Lyapunov argument.

Here, we show that (9.54) does not admit a stabilizing Lipschitz continuous feedback law. Consider the vector $y = [0, 0, \delta]^T \in \mathcal{B}_\delta$ for $\delta > 0$ arbitrarily small. According to Theorem 9.14, the origin of (9.54) can only be Lipschitz continuous feedback stabilizable if there exist $x \in \mathbb{R}^3$ and $u \in \mathbb{R}^2$ such that $y = f(x, u)$. However, (9.54a) and (9.54b) imply $u_1 = u_2 = 0$ and thus $\delta \neq f_3(x, u) = 0$ for all $x \in \mathbb{R}^3$ according to (9.54c). Hence, the necessary condition of Theorem 9.14 for Lipschitz continuous feedback stabilizability is not satisfied.

Even though less obvious, the same conclusions can be drawn for the mobile robot represented through the equivalent dynamics (1.46). For $x_3 \in [0, \pi]$ we can equivalently express the dynamics (1.46) as

$$\dot{x}_1 = u_1 \cos(x_3), \qquad \dot{x}_2 = u_1 \sqrt{1 - \cos^2(x_3)}, \qquad \dot{x}_3 = u_2,$$

which follows from the identity $1 = \sin^2(x_3) + \cos^2(x_3)$, for all $x_3 \in \mathbb{R}$. Thus, for $0 < \delta < 1$, we can conclude that the vector $y = [\delta, 2\sqrt{1 - \delta^2}, 0]^T$ satisfies $y \neq f(x, u)$ for all $(x, u) \in \mathbb{R}^3 \times \mathbb{R}^2$.

For the dynamics (1.46) it is straightforward to verify that the origin is asymptotically feedback stabilizable according to Definition 9.10.

9.6.2   Nonsmooth Control Lyapunov Functions

We now focus on the second question: the existence of CLFs according to Definition 9.1. In Equation (9.10) in Section 9.3 we have seen an explicit formula to define a feedback law $k(x)$ based on a given CLF. From the right-hand side in (9.10) it follows that the feedback law inherits continuity properties of the CLF V.

In Example 9.12 we have seen that the dynamics (9.51) cannot be stabilized through a continuous feedback law even though the definition of a stabilizing (discontinuous) input in (9.53) is straightforward.

This indicates that the assumption on the existence of a continuously differentiable CLF according to Definition 9.1 might be too strong for asymptotic stabilizability. Indeed, it can be shown that the dynamical system (9.51) does not admit a continuously differentiable CLF, [15, Theorem 2.19].

In order to arrive at an equivalence between asymptotic controllability and Lyapunov-like functions, the definiton of control Lyapunov functions have been extended based on the (upper right) *Dini derivative*.

Definition 9.16 (Dini derivative). *Consider a Lipschitz continuous function V : $\mathbb{R}^n \to \mathbb{R}$. The (upper right) Dini derivative of V at $x \in \mathbb{R}^n$ in direction $w \in \mathbb{R}^n$ is* *defined as*

$$DV(x;w) = \limsup_{h \searrow 0} \frac{1}{h} \left(V(x + hw) - V(x) \right).$$

The Dini derivative is a generalization of the directional derivative. In particular, if $V : \mathbb{R}^n \to \mathbb{R}$ is continuously differentiable on a neighborhood around $x \in \mathbb{R}^n$ then the identity

$$DV(x;w) = \langle \nabla V(x), w \rangle$$

is satisfied for all $w \in \mathbb{R}^n$. Additionally, it follows from *Rademacher's Theorem* (see [47, Theorem 3.2], for example) that a Lipschitz continuous function is continuously differentiable almost everywhere.

Definition 9.17 (Nonsmooth control Lyapunov function). *Consider the nonlinear system (9.1) and let $\alpha_1, \alpha_2 \in \mathcal{K}_\infty$. A Lipschitz continuous function $V : \mathbb{R}^n \to \mathbb{R}_{\geq 0}$ is called control Lyapunov function for (9.1) if*

$$\alpha_1(|x|) \leq V(x) \leq \alpha_2(|x|), \qquad \forall\, x \in \mathbb{R}^n,$$

and for all $x \in \mathbb{R}^n \backslash \{0\}$ there exists $u \in \mathbb{R}^m$ such that

$$DV(x; f(x,u)) < 0. \tag{9.55}$$

Definition (9.17) relaxes the differentiability assumption of the function V. From a control perspective it is usually desirable to ensure that $u(x) \to 0$ for $x \to 0$. If $0 = f(0,0)$, i.e., $(x^e, u^e) = (0,0)$ is an equilibrium pair, and $k(\cdot)$ is a stabilizing Lipschitz continuous feedback law, then $k(x) \to 0$ for $x \to 0$ is automatically satisfied. This so-called *small control property* is however not necessarily satisfied if the stabilizing input is not Lipschitz continuous.

If for all $\varepsilon > 0$ there exists a $\delta > 0$ such that whenever $x \in \mathcal{B}_\delta$ there exists a $u \in \mathcal{B}_\varepsilon$ such that (9.55) is satisfied, then we say that V has the small control property.

Theorem 9.18 (Existence of control Lyapunov functions, [129, 140]). *Consider the dynamical system (9.1). The origin of (9.1) is asymptotically stabilizable according to Definition 9.10 if and only if there exists a Lipschitz continuous CLF according to Definition 9.17.*

Example 9.19. As an example of a nonsmooth CLF according to Definition 9.17 we consider the dynamics (9.51) together with the function

$$V(x) = \sqrt{4x_1^2 + 3x_2^2} - |x_1| \tag{9.56}$$

discussed in [15, Example 2.14], for example.

To show that V is a nonsmooth CLF we start by establishing a lower and an upper bound $\alpha_1, \alpha_2 \in \mathcal{K}_\infty$. The estimate $4x_1^4 + 4x_1^2 x_2^2 + x_2^4 \geq 4x_1^4 + 3x_1^2 x_2^2$ implies the chain of inequalities

$$19x_1^2 + 11x_2^2 \geq 8(2x_1^2 + x_2^2) \geq 8\sqrt{4x_1^4 + 3x_1^2 x_2^2}.$$

Using this bound, it holds that

$$4 \left(4x_1^2 + 3x_2^2 - 2\sqrt{4x_1^2 + 3x_2^2}|x_1| + x_1^2 \right)$$
$$= 16x_1^2 + 12x_2^2 - 8\sqrt{4x_1^2 + 3x_2^2}|x_1| + 4x_1^2 \geq x_1^2 + x_2^2.$$

Taking the square root and dividing by 2 provides the lower bound

$$V(x) = \sqrt{4x_1^2 + 3x_2^2} - |x_1| \geq \tfrac{1}{2}|x| \doteq \alpha_1(|x|).$$

Conversely, an upper bound is obtained through

$$V(x) \leq \sqrt{4x_1^2 + 3x_2^2} \leq 2\sqrt{x_1^2 + x_2^2} \doteq \alpha_2(|x|).$$

For $x_1 \neq 0$, the function V is continuously differentiable and the gradient is given by

$$\nabla V(x) = \frac{1}{\sqrt{4x_1^2 + 3x_2^2}} \begin{bmatrix} 4x_1 \\ 3x_2 \end{bmatrix} + \begin{bmatrix} \text{sign}(x_1) \\ 0 \end{bmatrix}.$$

To compute the Dini derivative of V we first consider the function $g(x_1) = |x_1|$. In the domain $\mathbb{R}\backslash\{0\}$ the function g is continuously differentiable with $\nabla g(x_1) = \text{sign}(x_1)$. Thus, the Dini derivative of g in direction $w_1 \in \mathbb{R}\backslash\{0\}$ is given by

$$Dg(x_1; w_1) = \langle \nabla g(x_1), w_1 \rangle = \text{sign}(x_1)w_1.$$

At the origin $x_1 = 0$ the Dini derivative can be calculated as

$$Dg(0; w_1) = \limsup_{h \searrow 0} \tfrac{1}{h}(|0 + hw_1| - |0|) = |w_1|.$$

For V, we observe that the function $\sqrt{4x_1^2 + 3x_2^2}$ is continuously differentiable everywhere except for the origin. The Dini derivative of V can thus be calculated by considering different domains. For $x \in \mathbb{R}^2$ and $x_1 \neq 0$ it holds that

$$DV(x; w) = \langle \nabla V(x), w \rangle = \frac{4x_1 w_1 + 3x_2 w_2}{\sqrt{4x_1^2 + 3x_2^2}} + \text{sign}(x_1)w_1.$$

For $x \in \mathbb{R}^2$ with $x_1 = 0$ and $x_2 \neq 0$ it holds that

$$DV(x; w) = \frac{4x_1 w_1 + 3x_2 w_2}{\sqrt{4x_1^2 + 3x_2^2}} + |w_1|.$$

At the origin $x = 0$ the Dini derivative is given by

$$DV(0; w) = \limsup_{h \searrow 0} \frac{1}{h} \left(\sqrt{4h^2 w_1^2 + 3h^2 w_2^2} - |hw_1| \right) = \sqrt{4w_1^2 + 3w_2^2} - |w_1|.$$

We leave it to the reader to verify that for all $x \in \mathbb{R}^2\backslash\{0\}$ there exists a $u \in$

$[-1, 1]$ such that

$$DV(x; f(x, u)) < 0, \qquad (9.57)$$

which implies that V is indeed a CLF according to Definition 9.17. In particular, u defined in (9.53) satisfies the decrease condition (9.57).

On the nonsmooth domain of the function V, i.e., on the x_2-axis, any input $u \neq 0$ leads to a decrease. The function V is visualized in Figure 9.13.

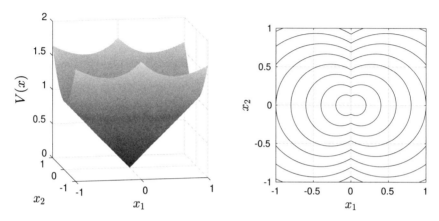

Figure 9.13: Nonsmooth CLF (9.56) for the dynamics (9.51). The function V is nonsmooth on the x_2-axis.

9.6.3   Robustness and Discontinuous Feedback Laws

We conclude this section by pointing out robustness issues induced through discontinuous feedback laws based on the example of Artstein's circles (9.51).

As a general setting we consider the perturbed closed-loop dynamics

$$\dot{x}(t) = f(x(t), k(x(t)) + e(t)) + d(t), \qquad (9.58)$$

where $k : \mathbb{R}^n \to \mathbb{R}^m$ denotes a (possibly discontinuous) feedback law and $e : \mathbb{R}_{\geq 0} \to \mathbb{R}^m$ and $d : \mathbb{R}_{\geq 0} \to \mathbb{R}^n$ denote time-dependent disturbances. For $e = 0$ and $d = 0$ the closed-loop dynamics of the original system (9.1) are recovered.

Theorem 9.20 ([145, Theorem F]). *There exists a feedback $k : \mathbb{R}^n \to \mathbb{R}^n$ which asymptotically stabilizes (9.58) if and only if there exists a continuously differentiable CLF according to Definition 9.1 for the unperturbed system (9.1).*

We have seen that the dynamics (9.51) cannot be stabilized through a continuous feedback law and a continuously differentiable CLF does not exist. Thus, according to Theorem 9.20 Artstein's circles cannot be robustly stabilized through a feedback law $k(x)$. We illustrate this fact through a particular perturbation. Consider the perturbed dynamics

$$\begin{aligned}
\dot{x}_1 &= \left(x_1^2 - x_2^2\right) k(x) \\
\dot{x}_2 &= 2x_1 x_2 k(x) + d_2(x_1),
\end{aligned} \qquad (9.59)$$

where $e(t) = 0$, $d_1(t) = 0$, and $d_2(x_1(t))$ is possibly unequal to zero. For a given feedback law $k(x)$ we define

$$d_2(x) = \begin{cases} \varepsilon & \text{if } x_1 x_2 k(x) \leq 0, \\ 0 & \text{if } x_1 x_2 k(x) = 0, \\ -\varepsilon & \text{if } x_1 x_2 k(x) \geq 0. \end{cases}$$

With these definitions, solutions of (9.59) starting close to the x_2-axis, i.e., x_1 small, do not converge to the origin. In particular, the definition of the disturbance $d_2(x)$ is such that the solution cannot escape a neighborhood around the x_2-axis and will stay there indefinitely.

While one might argue that for a specific application it is unlikely that the disturbance cancels the effect of the control law, it hightlights an intrinsic problem of asymptotically stabilizable systems which cannot be stabilized through Lipschitz continuous feedback laws. Ways to overcome the robustness issues include time-dependent feedback laws (including sample and hold) or feedback laws incorporating memory through a hybrid control structure [145], [126], [53].

9.7   EXERCISES

Exercise 9.1. Consider the system

$$\dot{x} = f_1(x, u) = 2x^2 + (x^4 + 1)u. \tag{9.60}$$

A feedback law which is stabilizing the origin is given by

$$u = k_1(x) = \frac{-2x^2 - x}{x^4 + 1}.$$

However, the perturbed dynamics

$$\dot{x} = f_2(x, u, w) = 2x^2 + (x^4 + 1)(u + w) \tag{9.61}$$

are not robust with respect perturbations $w \in \mathbb{R}$.

(i) Visualize the closed-loop solution $x(t)$ of the dynamics $\dot{x} = f_1(x, k_1(x))$ as well as $k_1(x(t))$ with respect to the initial condition $x(0) = 1$. (In particular, plot $(t, x(t))$ and $(t, k_1(x(t)))$ for $t \in [0, 5]$.)

(ii) Visualize the closed-loop solution $x(t)$ of the dynamics $\dot{x} = f_2(x, k_1(x), w)$ as well as $k_1(x(t))$ with respect to the initial condition $x(0) = 0.5$ and the constant perturbation $w = 1$. (In particular, plot $(t, x(t))$ and $(t, k_1(x(t)))$ for $t \in [0, 0.98]$.)

(iii) Use the ISS redesign to define a feedback law $u = k_2(x)$ such that (9.61) is ISS. Visualize the closed-loop solution $x(t)$ of the dynamics $\dot{x} = f_2(x, k_2(x), w)$ as well as $k_2(x(t))$ with respect to the initial condition $x(0) = 1$ and the constant perturbation $w = 1$. (In particular, plot $(t, x(t))$ and $(t, k_2(x(t)))$ for $t \in [0, 5]$.)

Exercise 9.2. Consider the system

$$\begin{aligned} \dot{x}_1 &= x_1^3 + x_1 x_2, \\ \dot{x}_2 &= u. \end{aligned} \tag{9.62}$$

A control Lyapunov function of the system is given by

$$V(x) = \tfrac{1}{2}x_1^2 + \tfrac{1}{2}x_2^2 + 2x_2x_1^2 + 2x_1^4.$$

Use Sontag's universal formula (with $\kappa = 1$) to compute a stabilizing feedback law $u = k(x)$.

Visualize the closed-loop solution $x(t)$ of the dynamics (9.62) (with $u = k(x)$), and visualize $k(x(t))$ and $V(x(t))$ with respect to the initial condition $x(0) = [1, 1]^T$. (In particular, plot $(t, x(t))$, $(t, k(x(t)))$ and $(t, V(x(t)))$ for $t \in [0, 10]$.)

Exercise 9.3. Consider the nonlinear system

$$\begin{bmatrix} \dot{x}_1 \\ \dot{x}_2 \\ \dot{x}_3 \end{bmatrix} = \begin{bmatrix} x_1 + 3x_2 \\ (1 + x_1)x_2 + x_3 \\ u + x_1 \end{bmatrix} \tag{9.63}$$

with an equilibrium at the origin. Use backstepping to compute a feedback law $u = k(x)$ which globally stabilizes the origin of the system (9.63).

Exercise 9.4. Consider the system of Example 9.5, having applied the initial feedback transformation; i.e.,

$$\dot{z} = x + x^3 - x^2 v$$
$$\dot{x} = -x + v.$$

Linearize the system about the origin and apply the forwarding technique to the linear system. Show that the obvious linear control in this case, as well as the saturation of this linear control, yield a closed-loop system with a globally asymptotically stable origin. Provide a result for a more general class of systems.

Hint: Consider systems where the origin is globally asymptotically stable for $\dot{x} = f(x)$ and where the linearized system is stabilizable.

Exercise 9.5. Consider the nonlinear system

$$\begin{bmatrix} \dot{x}_1 \\ \dot{x}_2 \\ \dot{x}_3 \end{bmatrix} = \begin{bmatrix} x_1^2 + x_2 \\ x_3 \\ u \end{bmatrix} \tag{9.64}$$

with an equilibrium at the origin.

(i) Use backstepping to compute a feedback law $u = k(x)$ which globally stabilizes the origin of the system (9.63). Additionally, compute a Lyapunov function of the closed-loop system.

(ii) Implement the closed-loop dynamics (in MATLAB) and visualize $(t, x(t))$ as well as $(t, k(x(t)))$ for different initial values.

Exercise 9.6. Consider the system $\dot{x} = f(x, u)$ given by

$$\dot{x}_1 = x_1^2 + x_2$$
$$\dot{x}_2 = x_3 \tag{9.65}$$
$$\dot{x}_3 = u.$$

Use high-gain backstepping with $\kappa = 10$ and $\kappa = 20$ to compute two feedback laws $u = k_{10}(x)$ and $u = k_{20}(x)$ to stabilize the origin.

- Visualize the closed-loop solution $x(t)$ of the dynamics $\dot{x} = f(x, k_{10}(x))$ and $k_{10}(x(t))$ with respect to the initial condition $x(0) = [1, 1, 1]^T$. (In particular, plot $(t, x(t))$ and $(t, k_{10}(x(t)))$ for $t \in [0, 3]$.)
- Visualize the closed-loop solution $x(t)$ of the dynamics $\dot{x} = f(x, k_{20}(x))$ and $k_{20}(x(t))$ with respect to the initial condition $x(0) = [1, 1, 1]^T$. (In particular, plot $(t, x(t))$ and $(t, k_{20}(x(t)))$ for $t \in [0, 3]$.)

Exercise 9.7. Show that the system dynamics (9.51) satisfy neither the assumptions of Theorem 9.13 nor the assumptions of Theorem 9.14. In particular show that

- the linearization around the origin is not stabilizable,
- for each $\varepsilon > 0$ there exists a $\delta > 0$ such that for all $y \in \mathcal{B}_\delta$ there exist $x \in \mathcal{B}_\varepsilon$ and $u \in \mathcal{B}_\varepsilon$ with $y = f(x, u)$.

Hint: It might be easier to consider the ∞-norm instead of the 2-norm in this exercise, i.e., consider $\mathcal{B}_r = \{z \in \mathbb{R}^n : |z|_\infty < r\}$ instead of $\mathcal{B}_r = \{z \in \mathbb{R}^n : |z| < r\}$ for $r > 0$. Since norms are equivalent in $\mathbb{R}^n$, the selection of a particular norm does not change the statement of Theorem 9.14.

Exercise 9.8. Show that the function

$$V(x) = x_1^2 + x_2^2 + 2x_3^2 - 2|x_3|\sqrt{x_1^2 + x_2^2}$$

is a nonsmooth CLF for the dynamics (9.54).

9.8   BIBLIOGRAPHICAL NOTES AND FURTHER READING

The fundamental idea of designing feedback controllers to achieve a decrease in a Lyapunov function can be found in works in the Russian literature as far back as the 1940s. Interesting early texts include [94] and [95].

The nonlinear damping, or Jurdjevic-Quinn, control first appeared in [76]. Sontag's Universal Formula first appeared in [142].

Key advanced texts for the backstepping and forwarding material in this chapter (and other topics as well) are [93] and [135]. Work that brings together CLF designs and ISS notions as robust CLFs can be found in [52]. An excellent summary can be found in [89] and an interesting view of the developments as they were happening is contained in [90].

When applied recursively, the saturated control technique in Section 9.5.3 leads to so-called nested saturations which are critical to overcoming certain pathological behaviors that can occur in feedforward systems. These designs have their origin in [153].

Chapter Ten

Sliding Mode Control

In this chapter we discuss the technique of *sliding mode control*, a robust control method, taking the effect of unmodeled unknown disturbances into account. In particular, we consider systems of the form

$$\dot{x} = f(x, u, \delta(t, x))$$

with state $x \in \mathbb{R}^n$, input $u \in \mathbb{R}^m$, and potentially time- and state-dependent unknown disturbance $\delta : \mathbb{R}_{\geq 0} \times \mathbb{R}^n \to \mathbb{R}^n$. While we primarily focus on the state x, we will also consider systems with output

$$y = h(x)$$

to study the input-output behavior in the context of *output tracking*.

However, before we turn to the actual topic of this chapter we need to introduce the concept of *finite-time stability*.

10.1   FINITE-TIME STABILITY

Consider the dynamics $f : \mathbb{R}^n \to \mathbb{R}^n$,

$$\dot{x} = f(x), \qquad x(0) = x_0 \in \mathbb{R}^n, \tag{10.1}$$

without input. As usual, without loss of generality, we assume that $f(0) = 0$, and thus $x^e = 0$ is an equilibrium of (10.1).

Definition 10.1 (Finite-time stability). *Consider the system (10.1) with $f(0) = 0$. The origin is said to be finite-time stable if there exist an open neighborhood $\mathcal{D} \subset \mathbb{R}^n$ of the origin and a function $T : \mathcal{D} \backslash \{0\} \to (0, \infty)$, called the settling-time function, such that the following statements hold:*

- *(Stability) For every $\varepsilon > 0$ there exists a $\delta > 0$ such that, for every $x(0) = x_0 \in \mathcal{B}_\delta \cap \mathcal{D} \backslash \{0\}$, $x(t) \in \mathcal{B}_\varepsilon$ for all $t \in [0, T(x_0))$.*
- *(Finite-time convergence) For every $x(0) = x_0 \in \mathcal{D} \backslash \{0\}$, $x(\cdot)$ is defined on $[0, T(x_0))$, $x(t) \in \mathcal{D} \backslash \{0\}$ for all $t \in [0, T(x_0))$, and $x(t) \to 0$ for $t \to T(x_0)$.*

The origin is said to be globally finite-time stable if it is finite-time stable with $\mathcal{D} = \mathbb{R}^n$.

To be able to analyze finite-time stability we need to consider systems (10.1) where the right-hand side $f(x)$ is not Lipschitz continuous, and thus uniqueness of solutions is, in general, not guaranteed.

Example 10.2. Consider the one-dimensional differential equation

$$\dot{x} = f(x) = -\sqrt[3]{x^2} \tag{10.2}$$

with unique equilibrium at the origin. The function $f(x)$ is Lipschitz continuous for all $x \neq 0$. However, at the origin $f(x)$ is not Lipschitz continuous and thus a unique solution can only be guaranteed if $x(t) \neq 0$.

We can verify that

$$x(t) = -\tfrac{1}{27}(t - 3\operatorname{sign}(x(0))\sqrt[3]{|x(0)|})^3$$

satisfies the differential equation for all $x \in \mathbb{R}$. However, for $x(0) > 0$

$$x(t) = \begin{cases} -\tfrac{1}{27}(t - 3\sqrt[3]{|x(0)|})^3, & \text{if } t \le 3\sqrt[3]{|x(0)|} \\ 0, & \text{if } t \ge 3\sqrt[3]{|x(0)|} \end{cases}$$

is also a solution. These two different solutions are shown in Figure 10.1.

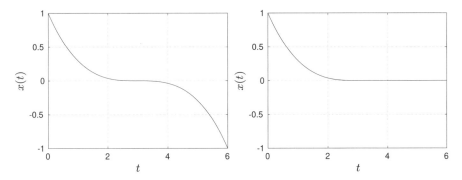

Figure 10.1: Two solutions of (10.2). Since the right-hand side is not Lipschitz continuous, solutions of the ordinary differential equation are not unique.

Example 10.3. As a second example, consider

$$\dot{x} = f(x) = -\operatorname{sign}(x)\sqrt[3]{x^2}. \tag{10.3}$$

In contrast to Example 10.2, the ordinary differential equation (10.3) has a unique solution defined as

$$x(t) = \begin{cases} -\tfrac{1}{27}\operatorname{sign}(x(0))(t - 3\sqrt[3]{|x(0)|})^3 & \text{if } t \le 3\sqrt[3]{|x(0)|} \\ 0 & \text{if } t \ge 3\sqrt[3]{|x(0)|}. \end{cases} \tag{10.4}$$

Once the equilibrium is reached, the inequalities

$$-\operatorname{sign}(x)\sqrt[3]{x^2} < 0 \text{ for all } x > 0 \quad \text{and}$$
$$-\operatorname{sign}(x)\sqrt[3]{x^2} > 0 \text{ for all } x < 0$$

ensure that the origin is attractive. Moreover, (10.4) implies that the origin is globally finite-time stable according to Definition 10.1. The settling-time function is given by $T(x) = 3\sqrt[3]{|x|}$. The solution (10.4) and the settling-time function are visualized in Figure 10.2.

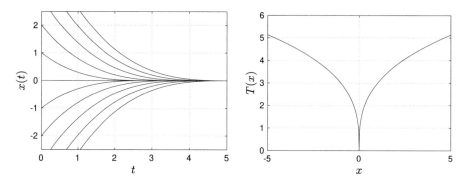

Figure 10.2: Solutions of the ordinary differential equation from different initial conditions on the left and the settling-time function on the right.

Finite-time stability of system (10.1) can be characterized by an appropriate Lyapunov function.

Theorem 10.4 (Lyapunov function for finite-time stability). *Consider system (10.1) with $f(0) = 0$. Assume there exist a continuous function $V : \mathbb{R}^n \to \mathbb{R}_{\geq 0}$ that is continuously differentiable on $\mathbb{R}^n \backslash \{0\}$, $\alpha_1, \alpha_2 \in \mathcal{K}_\infty$ and a constant $\kappa > 0$ such that*

$$\alpha_1(|x|) \leq V(x) \leq \alpha_2(|x|), \tag{10.5}$$

$$\dot{V}(x) = \langle \nabla V(x), f(x) \rangle \leq -\kappa\sqrt{V(x)} \tag{10.6}$$

for all $x \neq 0$. Then the origin is globally finite-time stable.

Moreover, the settling-time $T(x) : \mathbb{R}^n \to \mathbb{R}_{\geq 0}$ is upper bounded by

$$T(x) \leq \tfrac{2}{\kappa}\sqrt{\alpha_2(|x|)}.$$

Proof. Using a comparison principle, similar to that of Lemma 2.13, implies

$$\int_0^t \frac{dV(x(t))}{\sqrt{V(x(t))}} \leq -\int_0^t \kappa\, dt,$$

which allows us to calculate

$$\sqrt{V(x(t))} \leq \sqrt{V(x(0))} - \frac{\kappa t}{2}.$$

With the bounds (10.5), we obtain the inequality

$$|x(t)| \leq \alpha_1^{-1}\left(\left(\sqrt{\alpha_2(|x(0)|)} - \frac{\kappa t}{2}\right)^2\right) \tag{10.7}$$

and thus finite-time convergence follows. In particular, the condition

$$\sqrt{\alpha_2(|x(0)|)} - \frac{\kappa t}{2} = 0$$

implies $|x(T)| \leq 0$ for all $T \geq \frac{2}{\kappa}\sqrt{\alpha_2(|x(0)|)}$ and thus not only shows finite-time convergence but also provides the upper bound on the settling-time.   □

The expression in (10.7) is a bit clumsy. However, for the commonly used quadratic Lyapunov function, the bounds (10.5) become

$$a_1|x|^2 \leq V(x) \leq a_2|x|^2 \tag{10.8}$$

for some $a_1, a_2 > 0$. The bound (10.7) then simplifies to

$$|x(t)| \leq \tfrac{1}{\sqrt{a_1}}\left(\sqrt{a_2}|x(0)| - \frac{\kappa t}{2}\right) \tag{10.9}$$

and we see that $|x(t)| = 0$ no later than

$$T(|x(0)|) = |x(0)|\frac{2\sqrt{a_2}}{\kappa}. \tag{10.10}$$

Example 10.5. We consider again the ordinary differential equation (10.3) defined as $\dot{x} = f(x) = -\operatorname{sign}(x)\sqrt[3]{x^2}$. Additionally, consider the function $V(x) = \sqrt[3]{x^2}$, which is continuously differentiable for all $x \neq 0$. The definition of V allows us to define the $\mathcal{K}_\infty$ functions $\alpha_1(s) = \alpha_2(s) = \sqrt[3]{x^2}$.

For all $x \neq 0$, the time derivative of V with respect to the dynamics (10.3) satisfies

$$\dot{V}(x) = \langle \nabla V(x), -\operatorname{sign}(x)\sqrt[3]{x^2}\rangle = \tfrac{2}{3}\operatorname{sign}(x)|x|^{-\frac{1}{3}}(-\operatorname{sign}(x)|x|^{\frac{2}{3}})$$
$$= -\tfrac{2}{3}|x|^{\frac{1}{3}} = -\tfrac{2}{3}\sqrt{V(x)},$$

which shows that V is a Lyapunov function that guarantees finite-time stability by Theorem 10.4. For the settling-time we obtain the bound

$$T(x) \leq \tfrac{2}{\kappa}\sqrt{\alpha_2(|x|)} = \frac{2}{\tfrac{2}{3}}\sqrt{|x|^{\frac{2}{3}}} = 3\sqrt[3]{|x|},$$

which coincides with the settling-time derived in Example 10.3.

10.2   BASIC SLIDING MODE CONTROL

In this section we introduce the main features of sliding mode control based on the two dimensional system

$$\begin{aligned}\dot{x} &= x^3 + z,\\ \dot{z} &= u + \delta(t, x, z).\end{aligned} \tag{10.11}$$

Here, $\delta : \mathbb{R}_{\geq 0} \times \mathbb{R}^2 \to \mathbb{R}$ denotes an unknown disturbance. The only assumption we make about this disturbance is that there exists a constant $L_\delta \in \mathbb{R}_{>0}$ such that

$|\delta(t, x, z)| \leq L_\delta$ for all $(t, x, z) \in \mathbb{R}_{\geq 0} \times \mathbb{R}^2$. Thus, we assume that δ is bounded but we do not make any assumptions about the continuity of δ.

Suppose we would like the x-subsystem to behave as $\dot{x} = -x$, i.e., our goal is to exponentially stabilize the origin of the x-subsystem for all bounded disturbances. In particular, our desired behavior implies $\dot{x} + x = 0$ and substituting the $\dot{x}$ dynamics from (10.11) yields

$$x^3 + z + x = 0. \tag{10.12}$$

Define

$$\sigma = x^3 + z + x \tag{10.13}$$

as a new state and consider $V(\sigma) = \frac{1}{2}\sigma^2$. Then

$$\dot{V}(\sigma) = \sigma\dot{\sigma} = \sigma\left(3x^2\dot{x} + \dot{z} + \dot{x}\right)$$
$$= \sigma\left(3x^5 + 3x^2 z + u + \delta(t, x, z) + x^3 + z\right).$$

To cancel the known terms (remembering that δ is unknown) we define

$$u = v - 3x^5 - 3x^2 z - x^3 - z$$

so that

$$\dot{V}(\sigma) = \sigma\left(v + \delta(t, x, z)\right)$$

and v is a new input to be defined.

Selecting $v = -\rho\,\text{sign}(\sigma)$, $\rho > 0$, provides the estimate

$$\dot{V}(\sigma) = \sigma\left(-\rho\,\text{sign}(\sigma) + \delta(t, x, z)\right) = -\rho|\sigma| + \sigma\delta(t, x, z)$$
$$\leq -\rho|\sigma| + L_\delta|\sigma| = -(\rho - L_\delta)|\sigma|.$$

Finally, with $\rho = L_\delta + \frac{\kappa}{\sqrt{2}}$, $\kappa > 0$, we have

$$\dot{V}(\sigma) \leq -\frac{\kappa}{\sqrt{2}}|\sigma| = -\alpha\sqrt{V(\sigma)},$$

which, from Theorem 10.4 implies $\sigma(t) \to 0$ in finite time. Note that the control

$$u = -\left(L_\delta + \frac{\kappa}{\sqrt{2}}\right)\text{sign}\left(x^3 + z + x\right) - 3x^5 - 3x^2 z - x^3 - z \tag{10.14}$$

is independent of the term $\delta(t, x, z)$. This is crucial because we have assumed we do not know anything about δ other than a bounding constant L_δ.

In Figure 10.3 and Figure 10.4 closed-loop solutions of the system (10.11) using the control law (10.14) are shown. For the simulations the disturbance is defined as $\delta(t, x, z) = \sin(x)$ and $\delta(t, x, z) = \text{sign}(\cos(2t)\sin(2t))$ (see Figure 10.3, left and right, respectively). Since both disturbances are bounded by 1, we use $L_\delta = 1$ for the simulations. Additionally, the parameter influencing the finite-time convergence is taken as $\kappa = 2$. From Figure 10.3 and Figure 10.4 we observe that σ converges to zero in finite time before (x, z) asymptotically approach the origin while keeping σ equal to zero. Since the ordinary differential equation is solved numerically, σ

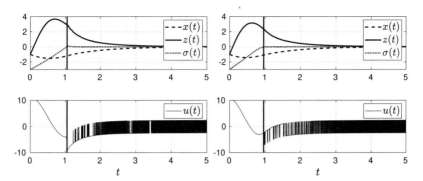

Figure 10.3: Closed-loop solution of (10.11) using the sliding mode control law (10.14) with $L_\delta = 1$ and $\kappa = 2$. On the left, the disturbance $\delta(t, x, z) = \sin(t)$ is used for the simulation, while $\delta(t, x, z) = \text{sign}(\cos(2t) \sin(2t))$ is used for the simulations on the right.

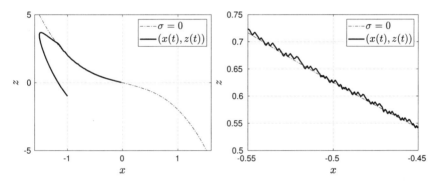

Figure 10.4: Closed-loop solution of (10.11) using the control law (10.14) with $L_\delta = 1$ and $\kappa = 2$ and with respect to $\delta(t, x, z) = \sin(t)$. Additionally, the set $\sigma = x^3 + z + x = 0$ is shown.

is not exactly zero (see Figure 10.4, right). While the solutions in Figure 10.3 differ slightly, the different choices of disturbance $\delta(t, x, z)$ are not obvious in the simulations.

Remark 10.6. Note that a similar convergence structure as in Figure 10.4 has been observed in Section 9.4.3 and Section 9.5.2 for high-gain backstepping and forwarding. First, the solution of the dynamical system converges to a lower-dimensional manifold, and as a second step the solution converges to the origin while staying on the manifold (see Figures 9.6 and 9.10).

10.2.1   Terminology

The example discussed so far contains the main ideas and covers the main properties of sliding mode control.

The new state σ defined in (10.13) is called the *sliding variable*. The set $\sigma = 0$, or more precisely

$$\{(x, z) \in \mathbb{R}^2 : \sigma(x, z) = 0, \ (x, z) \in \mathbb{R}^2\},$$

is called the *sliding surface*. The sliding variable, and thus implicitly the sliding surface, is defined such that the origin of the x-subsystem is exponentially stable if $\sigma(t) = 0$, $t \in \mathbb{R}_{\geq 0}$, is satisfied. While we have defined σ based on the condition $\dot{x} = -x$ in (10.12), a definition based on $\dot{x} = -2x$ or $\dot{x} = -x^3$ would have led to similar closed-loop properties via different feedback laws.

Using the results on finite-time convergence in Section 10.1, a control law u is derived such that states converge to the sliding surface in finite time. This initial convergence of $\sigma(t) \to 0$ is called the *reaching phase*.

On the sliding surface, condition (10.12) ensures that the dynamics behave like $\dot{x} = -x$. This is called the *sliding phase* and guarantees asymptotic stability of the origin for the overall closed-loop system. In Figure 10.3 the separation of the reaching phase and the sliding phase are indicated by a vertical line.

10.2.2   Chattering and Chattering Avoidance

The control law u derived in (10.14) is discontinuous due to the definition of

$$v = \rho \operatorname{sign}(\sigma),$$

which switches between ρ and $-\rho$ depending on the sign of the sliding variable σ. This leads to *chattering* of the control law u which can be observed in Figure 10.3 and which results from the sliding variable being slightly off the sliding surface (see Figure 10.4).

To obtain a control law with better regularity properties, the discontinuous feedback law can be approximated through continuous or smooth functions. In particular, the continuous saturation function $\operatorname{sat}_\varepsilon : \mathbb{R} \to [-1, 1]$,

$$\operatorname{sat}_\varepsilon(\sigma) = \operatorname{sat}(\tfrac{\sigma}{\varepsilon}) = \begin{cases} 1, & \tfrac{\sigma}{\varepsilon} \geq 1 \\ \tfrac{\sigma}{\varepsilon}, & -1 \leq \tfrac{\sigma}{\varepsilon} \leq 1 \\ -1, & \tfrac{\sigma}{\varepsilon} \leq -1, \end{cases} \tag{10.15}$$

can be used to approximate the *sign function*. For $\varepsilon > 0$ arbitrarily small, $\operatorname{sat}_\varepsilon(\sigma)$ approximates $\operatorname{sign}(\sigma)$ arbitrarily well. Note that we have already seen the sign function and the sat function in Chapter 6.

Alternatively the smooth *sigmoid function* $\operatorname{sig}_\varepsilon : \mathbb{R} \to [-1, 1]$, given by

$$\operatorname{sig}_\varepsilon(\sigma) = \frac{1 - e^{-\sigma\varepsilon}}{1 + e^{-\sigma/\varepsilon}}, \tag{10.16}$$

can be used to obtain a smooth approximation of the discontinuous feedback law (10.14). The functions (10.15) and (10.16) are shown in Figure 10.5 for different values of $\varepsilon > 0$.

The closed-loop solution of (10.11) using the approximated sliding mode control law

$$u_\varepsilon = -\left(L_\delta + \tfrac{\kappa}{\sqrt{2}}\right) \operatorname{sat}_\varepsilon\left(x^3 + z + x\right) - 3x^5 - 3x^2 z - x^3 - z \tag{10.17}$$

is shown in Figures 10.6 and 10.7 for different values of $\varepsilon > 0$. For the simulation, we again use the function $\delta(t, x, z) = \sin(t)$ as an "unknown" disturbance. Additionally, L_δ and κ are set to $L_\delta = 1$ and $\kappa = 2$.

Due to the approximation of the sign function through (10.15), only convergence

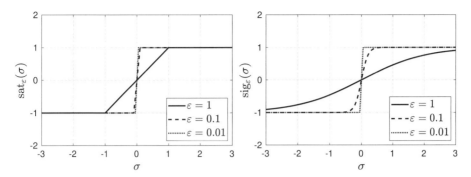

Figure 10.5: Approximation of the sign function through a continuous saturation (left) and a smooth sigmoid function (right).

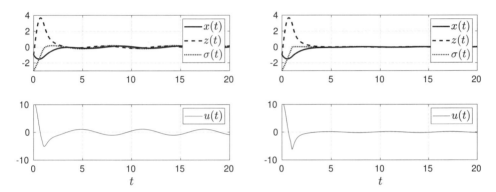

Figure 10.6: Closed-loop solution of (10.11) using the control law (10.17) for $\varepsilon = 0.5$ (left) and $\varepsilon = 0.1$ (right). For the simulations, the disturbance is set to $\delta(t, x, z) = \sin(t)$ and the parameters are defined as $L_\delta = 1$ and $\kappa = 2$.

to a neighborhood around the origin is guaranteed. Nevertheless, for $\varepsilon \to 0$, the neighborhood becomes arbitrarily small. With respect to the discussed example, a value of $\varepsilon = 0.1$ already leads to a small neighborhood around the origin and a significantly better behaved control law (see Figures 10.6 and 10.7).

10.3 A MORE GENERAL STRUCTURE

The ideas discussed in the preceding section based on the example of the dynamics (10.11) can be easily extended to a more general setting. Here, we consider

$$\dot{x} = f_1(x, z) \tag{10.18a}$$
$$\dot{z} = f_2(x, z) + g(x, z)(u + \delta(t, x, z)), \tag{10.18b}$$

where $f_1 : \mathbb{R}^n \times \mathbb{R} \to \mathbb{R}^n$, $f_2 : \mathbb{R}^n \times \mathbb{R} \to \mathbb{R}$, $g : \mathbb{R}^n \times \mathbb{R} \to \mathbb{R}$, and $\delta : \mathbb{R}_{\geq 0} \times \mathbb{R}^n \times \mathbb{R} \to \mathbb{R}$. Note that the x-dynamics are n-dimensional while the z-dynamics are one-dimensional. The function δ again models unknown disturbances. We assume that $f_1(0, 0) = 0$, i.e., the origin is an equilibrium of the x-dynamics. For the system (10.18) we not only assume that δ is bounded, i.e., we assume that $|\delta(t, x, z)| \leq$

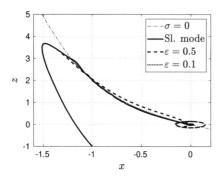
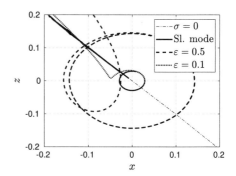

Figure 10.7: Comparison of the closed-loop solution (10.11) using the sliding mode controller (10.14) and its approximation (10.17) with $\varepsilon = 0.5$ and $\varepsilon = 0.1$. Additionally, the definitions $\delta(t, x, z) = \sin(t)$, $L_\delta = 1$ and $\kappa = 2$ are used.

L_δ for all $(t, x, z) \in \mathbb{R}_{\geq 0} \times \mathbb{R}^{n+1}$, but additionally assume that $g(x, z) \neq 0$ and $|g(x, z)| \leq L_g$ is satisfied for all $(x, z) \in \mathbb{R}^{n+1}$ and a given bound $L_g > 0$. Since the disturbance only appears in the z-dynamics together with the input u, the disturbance is referred to as a *matched disturbance*; i.e., the disturbance is matched by the input.

To obtain a sliding mode control law, we follow similar steps as before and as a first step design a virtual control law $z = k(x)$ that asymptotically stabilizes the origin of the x-dynamics (10.18a). Hence, we assume that the origin of the system $\dot{x} = f_1(x, k(x))$ is asymptotically stable.

As argued in Section 9.4 where we discussed backstepping, unfortunately z is a state and not an input. Thus, we consider the error variable, or in this context, the sliding variable

$$\sigma = z - k(x)$$

whose time derivative is given by

$$\dot{\sigma} = \dot{z} - \tfrac{dk}{dt}(x)$$
$$= f_2(x, z) + g(x, z)(u + \delta(t, x, z)) - \tfrac{\partial k}{\partial x}(x)f_1(x, z). \qquad (10.19)$$

Now, we can proceed in the same way as in the preceding section. We define the candidate control Lyapunov function

$$V(\sigma) = \tfrac{1}{2}\sigma^2,$$

which satisfies

$$\dot{V}(\sigma) = \sigma\left(f_2(x, z) + g(x, z)(u + \delta(t, x, z)) - \tfrac{\partial k}{\partial x}f_1(x, z)\right). \qquad (10.20)$$

To cancel the known terms we choose

$$u = \frac{1}{g(x, z)}\left(-f_2(x, z) + \frac{\partial k}{\partial x}(x)f_1(x, z) + v\right)$$

with new input v. Note that u is well defined for all $(x, z) \in \mathbb{R}^{n+1}$ due to the assumption that $g(x, z) \neq 0$ for all $(x, z) \in \mathbb{R}^{n+1}$.

Substituting the expression for u in (10.20) simplifies the time derivative

$$\dot{V}(\sigma) = \sigma(v + g(x, z)\delta(t, x, z))$$

and the expression is independent of the functions f_1 and f_2. Finally, selecting the new input as

$$v = -\left(\frac{\kappa}{\sqrt{2}} + L_g L_\delta\right) \text{sign}(\sigma)$$

and making use of the assumption that δ and g are bounded we obtain the estimate

$$\dot{V}(\sigma) \leq \sigma v + |\sigma| L_g L_\delta = -\frac{\kappa}{\sqrt{2}} |\sigma| = -\kappa \sqrt{V(\sigma)},$$

which shows that V is a finite-time Lyapunov function and, by Theorem 10.4, finite-time convergence $\sigma(t) \to 0$ can be concluded.

We summarize the result derived so far in the following theorem.

Theorem 10.7. *Consider the dynamics* (10.18) *with* $f_1(0, 0) = 0$. *Assume that* $g(x, z) \neq 0$ *and there exists a constant* $L_g > 0$ *such that* $|g(x, z)| \leq L_g$ *for all* $(x, z) \in \mathbb{R}^{n+1}$. *Additionally assume that* $k(x)$ *is defined such that the origin of* $\dot{x} = f(x, k(x))$ *is asymptotically stable with* $k(0) = 0$.

Then for all disturbances δ *satisfying the condition*

$$|\delta(t, x, z)| \leq L_\delta \qquad \forall\, (t, x, z) \in \mathbb{R}_{\geq 0} \times \mathbb{R}^{n+1}$$

for some $L_\delta > 0$, *the feedback law*

$$u = \frac{1}{g(x, z)}\left(-f_2(x, z) + \frac{\partial k}{\partial x}(x)f_1(x, z)\right) - \left(\frac{\kappa}{\sqrt{2}} + L_g L_\delta\right)\frac{\text{sign}(z - k(x))}{g(x, z)} \tag{10.21}$$

asymptotically stabilizes the origin of the system (10.18) *for all* $\kappa > 0$. *Additionally, the sliding surface* $\sigma = z - k(x) = 0$ *is reached no later than*

$$T(\sigma(0)) = T(z(0) - k(x(0))) = \frac{1}{\sqrt{2}\kappa}|z(0) - k(x(0))|.$$

10.4 ESTIMATING THE DISTURBANCE

To obtain the sliding mode control (10.21) which asymptotically stabilizes the origin for the system (10.18) we introduced the sliding variable σ. Through appropriate Lyapunov arguments we ensured that σ reaches the sliding surface, defined by $\sigma = 0$, in finite time. In theory, i.e, if an arbitrarily fast sampling is possible, the control law (10.21) guarantees that $\sigma(T) = 0$ implies $\sigma(t) = 0$ for all $t \geq T$. As we have seen in the examples in Section 10.2, in a numerical simulation this will generally not be satisfied.

Returning to the sliding variable dynamics given in (10.19), rather than dominating the unknown disturbance, consider the unimplementable control defined by

$$u_{eq} = \frac{1}{g(x,z)}\left(-f_2(x,z) + \frac{\partial k}{\partial x}(x)f_1(x,z)\right) - \delta(t,x,z), \qquad (10.22)$$

which is called the *equivalent control*. Note that this control is not implementable because, in contrast to the control laws (10.11) and (10.18), it depends on the unknown function δ. However, assuming we are on the sliding surface where $\dot{\sigma} = 0$, it follows that the equivalent control guarantees $\sigma(t) = 0$ for all $t \geq T$ if $\sigma(T) = 0$ without the chattering effects of (10.21) (assuming that δ is sufficiently smooth).

Nevertheless, if δ is a smooth function, the equivalent control is a smooth average of the chattering control (10.21) and an estimate of δ can be obtained by applying a *low-pass filter* to the chattering input

$$v = -\left(\frac{\kappa}{\sqrt{2}} + L_g L_\delta\right)\operatorname{sign}(\sigma).$$

In particular, we can augment the dynamics (10.18) by adding an additional state $\xi \in \mathbb{R}$ leading to the dynamical system

$$\dot{x} = f_1(x,z) \qquad\qquad\qquad\qquad (10.23\text{a})$$
$$\dot{z} = f_2(x,z) + g(x,z)(\hat{u}_{eq} + \delta(t,x,z)) \qquad\qquad (10.23\text{b})$$
$$\dot{\xi} = -\tfrac{1}{\tau}\xi + \tfrac{1}{\tau}\operatorname{sign}(z - k(x)), \qquad\qquad (10.23\text{c})$$

where $\tau > 0$ is a small parameter and the approximated equivalent control is defined as

$$\hat{u}_{eq} = \frac{1}{g(x,z)}\left(-f_2(x,z) + \frac{\partial k}{\partial x}(x)f_1(x,z)\right)' - \left(\frac{\kappa}{\sqrt{2}} + L_g L_\delta\right)\frac{\xi}{g(x,z)}. \qquad (10.24)$$

Here, we have replaced $\operatorname{sign}(\sigma)$ by the augmented state variable ξ.

Remark 10.8. The dynamics (10.23c) represent a *low-pass filter*. To see this, consider the one-dimensional system

$$\dot{x} = -\tfrac{1}{\tau}x + \tfrac{1}{\tau}u$$
$$y = x$$

and its representation in the frequency domain

$$\hat{y}(s) = (s + \tfrac{1}{\tau})^{-1}\tfrac{1}{\tau}\hat{u}(s) = \frac{\tfrac{1}{\tau}}{s + \tfrac{1}{\tau}}\hat{u}(s).$$

For $\tau > 0$ small we observe from the transfer function

$$G(s) = \frac{\tfrac{1}{\tau}}{s + \tfrac{1}{\tau}}$$

that for low frequencies the system approximately satisfies $\hat{y}(s) \approx \hat{u}(s)$ and for high frequencies it holds that $\hat{y}(s) \approx 0$.

With these considerations, the control law (10.24) provides an alternative to the sliding mode control law (10.21) which can be obtained by considering the augmented dynamics (10.23) for $\tau > 0$ small. The control law (10.24) is an approximation of the equivalent control (10.22). We illustrate the properties of the control law on our initial example (10.11).

Example 10.9. We consider the dynamics (10.11) and augment it with a low-pass filter

$$
\begin{aligned}
\dot{x} &= x^3 + z, \\
\dot{z} &= u + \delta(t, x, z) \\
\dot{\xi} &= -\tfrac{1}{\tau}\xi + \tfrac{1}{\tau}\operatorname{sign}(z - k(x))
\end{aligned}
\tag{10.25}
$$

for $\tau > 0$, small.

To stabilize the x-dynamics using the result from Theorem 10.7 we define the virtual control law $z = k(x) = -x^3 - x$, which ensures that the origin of the dynamics $\dot{x} = x^3 - x^3 - x = -x$ is exponentially stable.

Theorem 10.7 provides the sliding mode control law (10.21) defined as

$$
\begin{aligned}
u &= (-3x^2 - 1)(x^3 + z) - \left(\tfrac{\kappa}{\sqrt{2}} + L_\delta\right)\operatorname{sign}(z + (x^3 + x)) \\
&= -3x^5 - x^3 - 3x^2 z - z - \left(\tfrac{\kappa}{\sqrt{2}} + L_\delta\right)\operatorname{sign}(z + x^3 + x),
\end{aligned}
\tag{10.26}
$$

where we have identified the functions $f_1(x, z) = -x^3 + z$, $f_2(x, z) = 0$, and $g(x, z) = 1$ (and thus $L_g = 1$). Note that the control law coincides with the control law derived in (10.14). The approximated equivalent control (10.24) is thus defined as

$$
\dot{\hat{u}}_{eq} = -3x^5 - x^3 - 3x^2 z - z - \left(\tfrac{\kappa}{\sqrt{2}} + L_\delta\right)\xi.
\tag{10.27}
$$

In Figure 10.8 the closed-loop solution of (10.25) using the approximated equivalent control (10.27) is shown. For the simulation, again the disturbance $\delta(t, x, z) =$

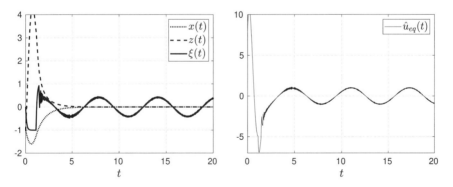

Figure 10.8: Closed-loop solution of (10.25) using the approximated equivalent control (10.27) derived through a low-pass filter.

$\sin(t)$ is used. Additionally the parameters are set to $L_\delta = 1$, $\kappa = 2$, and $\tau = 0.1$.

Instead of using a low-pass filter to replace the discontinuous sliding mode control law (10.21) with the approximated equivalent control law (10.24), the low-pass filter can be used to estimate the disturbance. In particular, if we compare the equivalent control law (10.22) and the control law (10.24), we observe that an approximation $\hat{\delta} : \mathbb{R}_{\geq 0} \times \mathbb{R}^{n+1} \to \mathbb{R}$ of the disturbance $\delta(t, x, z)$ is given by

$$\hat{\delta}(t, x, z) = \left(\frac{\kappa}{\sqrt{2}} + L_g L_\delta \right) \frac{\xi}{g(x, z)}. \tag{10.28}$$

The accuracy of the approximation is again discussed based on the example (10.25).

Example 10.10. Consider the dynamics (10.25) together with the sliding mode control law (10.26). Figure 10.9 shows the estimated disturbance

$$\hat{\delta}(t, x, z) = (\sqrt{2} + 1)\xi \tag{10.29}$$

defined through equation (10.28), together with the actual disturbance $\delta(t, x, z)$. Here the parameters are again defined as $L_\delta = 1$, $\kappa = 2$, and $g(x, z) = 1$ yields $L_g = 1$. The parameter defining the low-pass filter is set to $\tau = 0.05$. Since the

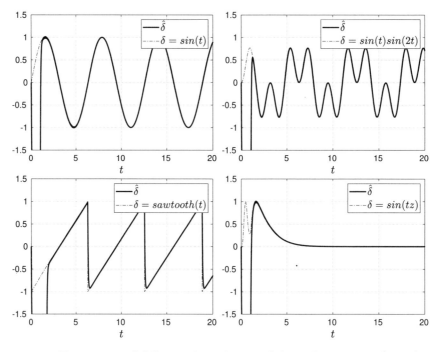

Figure 10.9: Estimation of different disturbances δ through Equation (10.29) using a low-pass filter.

derivation above relied on being on the sliding surface $\sigma = 0$, a good estimation of the disturbance can only be expected during the sliding phase. During the sliding phase $\hat{\delta}$ and δ are almost indistinguishable in Figure 10.9, even in the case where δ is discontinuous.

10.5   OUTPUT TRACKING

So far we have designed a sliding mode control law which asymptotically stabilizes the origin for a system subject to unknown bounded disturbances. In this section we extend these ideas to output tracking on the example of the dynamics (10.11).

In particular, we construct a sliding mode control law which ensures that the output $y : \mathbb{R}_{\geq 0} \to \mathbb{R}$ defined through the dynamics

$$y = x, \qquad \dot{x} = x^3 + z, \qquad \dot{z} = u + \delta(t, x, z) \tag{10.30}$$

follows a given time-dependent, twice continuously differentiable reference signal $y_r : \mathbb{R}_{\geq 0} \to \mathbb{R}$.

The output tracking condition that $y(t) \to y_r(t)$ for $t \to \infty$ can equivalently be stated in terms of the error dynamics

$$e(t) = y_r(t) - y(t),$$

and the requirement that $e(t) \to 0$ for $t \to \infty$. Similarly to condition (10.12), which ensures that x converges to the origin exponentially fast, the definition

$$\dot{e} = -e$$

by design provides exponential convergence of the error dynamics.

Following the same steps as in the preceding sections, we define the sliding variable $\sigma = \dot{e} + e$. Hence, σ can be written as

$$\sigma = \dot{e} + e = \dot{y}_r - \dot{y} + y_r - y = \dot{y}_r + y_r - x^3 - z - x.$$

The time derivative of the sliding variable is given by

$$\begin{aligned}\dot{\sigma} &= \ddot{y}_r + \dot{y}_r - 3x^2\dot{x} - \dot{z} - \dot{x} \\ &= \ddot{y}_r + \dot{y}_r - 3x^5 - 3x^2 z - u - \delta(t, x, z) - x^3 - z,\end{aligned}$$

which together with the candidate Lyapunov function $V(\sigma) = \frac{1}{2}\sigma^2$ leads to the expression

$$\dot{V}(\sigma) = \dot{\sigma}\sigma = \sigma(\ddot{y}_r + \dot{y}_r - 3x^5 - 3x^2 z - u - \delta(t, x, z) - x^3 - z). \tag{10.31}$$

To simplify the expression, we define the input to cancel the known terms and include a new input v as

$$u = -3x^5 - 3x^2 z - x^3 - z - v. \tag{10.32}$$

Assuming we do not know the reference $y_r(t)$ to be tracked a priori, just that it is twice differentiable, define a new unknown disturbance as

$$\psi(t, x, z) = \ddot{y}_r + \dot{y}_r - \delta(t, x, z). \tag{10.33}$$

With these definitions the time derivative of the candidate Lyapunov function (10.31) reads

$$\dot{V}(\sigma) = \sigma(\psi(t, x, z) + v).$$

Under the assumption that $\psi(t, x, z)$ is bounded, i.e., $|\psi(t, x, z)| \leq L_\psi$, for $L_\psi > 0$, we can define v as

$$v = -(L_\psi + \tfrac{\kappa}{\sqrt{2}}) \operatorname{sign}(\sigma) \tag{10.34}$$

for an arbitrary parameter $\kappa > 0$. Then $\dot{V}(\sigma)$ satisfies the estimate

$$\dot{V}(\sigma) = \sigma(\psi(t, x, z) - (L_\psi + \tfrac{\kappa}{\sqrt{2}}) \operatorname{sign}(\sigma))$$

$$\leq |\sigma| L_\psi - |\sigma| \left(L_\psi + \frac{\kappa}{\sqrt{2}} \right)$$

$$= -\kappa \sqrt{V(x)},$$

which shows finite-time convergence of the error variable by Theorem 10.4.

It is important to note that L_ψ needs to be defined based on the unknown disturbance δ and some knowledge regarding bounds of the first two derivatives of the reference signal y_r.

Example 10.11. Consider the system with output given by (10.30), together with the reference signal

$$y_r(t) = \begin{cases} 0.8 \sin(2t) & \text{for } t < 8, \\ 1.2 \sin(4t) & \text{for } t \geq 8. \end{cases} \tag{10.35}$$

The reference signal is twice continuously differentiable for all $t \neq 8$. For the disturbance $\delta(t, x, z) = \sin(t)$ the new disturbance ψ defined in (10.33) satisfies $|\psi(t, x, z)| \leq 25$ for all $t \neq 8$.

Figure 10.10 shows the output y of the closed-loop system (10.30) using the control law given by (10.32) and (10.34), together with the reference signal $y_r(t)$. On the left, L_ψ is defined as $L_\psi = 25$ and on the right $L_\psi = 1$ is used. Additionally

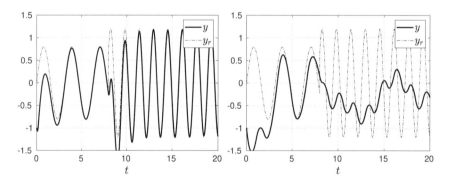

Figure 10.10: Sliding mode reference tracking of y_r defined in (10.35). On the left, L_ψ is defined as $L_\psi = 25$ while $L_\psi = 1$ is used on the right.

the parameter κ is set to $\kappa = 2$. For $L_\psi = 25$, good tracking performance can be observed (see Figure 10.10, left). At time $t = 8$, where $|\psi(t, x, z)| \leq 25$ is not satisfied, the error increases before y exponentially converges to y_r again. However, in Figure 10.10, right, we observe that we cannot select the constant L_ψ based on the disturbance δ only and expect good tracking performance.

10.6   EXERCISES

Exercise 10.1. Consider the nonlinear system

$$\dot{x} = x\sin(x) + z,$$
$$\dot{z} = u + \delta(x, z, t). \tag{10.36}$$

1. Define a stabilizing sliding mode control law u asymptotically stabilizing the origin of (10.36).
2. Approximate the discontinuous control law u by a continuous control law u_ε using $\mathrm{sign}(x) \approx \mathrm{sat}(\frac{x}{\varepsilon})$, $\varepsilon > 0$.
3. Write MATLAB functions capturing the dynamics of the closed-loop system (10.36) using the feedback laws defined in 1. and 2. Simulate/visualize the closed-loop dynamics for the disturbances

$$\delta_a(x, z, t) = 0, \qquad \delta_b(x, z, t) = \sin(t), \quad \text{and} \quad \delta_c(x, z, t) = \cos(x \cdot z),$$

the initial value $x = z = 1$, $t \in [0, 10]$, and the parameters $\varepsilon = 0.1$, $\alpha = 1$, and $L = 5$. In the plots, visualize $(x(t), z(t))$ and the feedback $(t, u(t))$. (You can additionally visualize the sliding surface using `fimplicit.m`.)
4. Use the method discussed in Section 10.4 to estimate the disturbances δ_b and δ_c for the simulations in 3. using the additional parameter $\tau = 0.1$. Visualize $(t, \delta(x, z, t))$ as well as the estimated disturbances $(t, \hat{\delta}(x, z, t))$.
5. We extend the dynamical system (10.36) by considering the output

$$y = x.$$

Use output tracking to track (or try to track) the reference signals

$$y_a^{\mathrm{ref}}(t, x, z) = \cos(t),$$
$$y_b^{\mathrm{ref}}(t, x, z) = \begin{cases} 1 & \text{if } t \in [2k, 2k+1), \ k \in \mathbb{N}, \\ -1 & \text{if } t \in [2k+1, 2k+2), \ k \in \mathbb{N}, \end{cases}$$

for the setting defined in 3. but only with respect to δ_a. Visualize $(t, y(t))$ as well as $(t, y^{\mathrm{ref}}(t, x, z))$.

Exercise 10.2. Consider the nonlinear system

$$\dot{x} = x^2 + 3x - \tfrac{1}{2}z$$
$$\dot{z} = -2u + \delta(x, z, t) \tag{10.37}$$

with unknown disturbance δ. For the following numerical simulations, use the disturbance

$$\delta(x, z, t) = \cos(t) + 2\sin(3xz)$$

as an illustration. Additionally, for the specific questions below, use the initial condition $[x(0), z(0)] = [1, 1]^T$ and consider the time interval $t \in [0, 10]$. For κ, ε, and τ defined in this chapter we use the parameters $\kappa = 2$, $\varepsilon = 0.5$, and $\tau = 0.02$. Moreover, the parameter L needs to be defined based on δ. (You may wish to experiment with different values, and even different disturbances δ, to get a sense of how these things change the performance.)

1. Define a stabilizing sliding mode control law u_{sm} asymptotically stabilizing the origin of (10.37). Additionally,

   - visualize the closed-loop solution $(x(t), z(t))$ together with the sliding surface $\sigma = 0$;
   - visualize $(t, x(t))$, $(t, z(t))$, and $(t, \sigma(t))$; and
   - visualize $(t, u_{sm}(t))$.

2. Approximate the discontinuous control law u_{sm} through a continuous control law u_ε, i.e., use $\text{sign}(\sigma) \approx \text{sat}(\frac{\sigma}{\varepsilon})$ to approximate u_{sm}. Simulate the dynamical system (10.37) using the feedback law u_ε and

   - visualize $(t, x(t))$, $(t, z(t))$, and $(t, \sigma(t))$;
   - visualize $(t, u_\varepsilon(t))$.

3. Use a low-pass filter as discussed in Section 10.4 to stabilize the origin of (10.37) with an approximated equivalent control law $\hat{u}_{eq}$. (Initialize ξ through $\xi(0) = 0$.)

   - Visualize $(t, x(t))$, $(t, z(t))$, and $(t, \xi(t))$; and
   - visualize $(t, \hat{u}_{eq}(t))$.

4. Use a low-pass filter as discussed in Section 10.4 to approximate the disturbance $\delta(x, z, t)$.

   - Visualize the approximated disturbance $\hat{\delta}(x, z, t)$ together with $\delta(x, z, t)$ (i.e., visualize $(t, \hat{\delta})$ and (t, δ)).

5. Consider the output $y = x$ and the reference signal

$$y_r(t) = \cos(3t).$$

   Use the ideas discussed in Section 10.5 to track the reference signal $y_r(t)$.

   - Visualize $(t, y(t))$ and $(t, y_r(t))$.
   - Justify your selection of the parameter L.

Hint: The chattering behavior of the sliding mode controller leads to a very short sampling rate of the ode-solver (have a look at the size of your solution in MATLAB) and it may take MATLAB a while to compute a solution. While you are testing your code it might thus be beneficial to reduce the time span $[0, 10]$ to a smaller time interval. Additionally, there is a difference between `tspan=[0 10]` and `tspan=0:0.1:10` in the definition of the time span in the ode-solver. (What is the difference?) Use the definition `tspan=[0 10]` in your visualizations.

10.7   BIBLIOGRAPHICAL NOTES AND FURTHER READING

The switching between the reaching and sliding phases caused the idea of sliding mode control to frequently be referred to as an approach under the heading of *variable structure systems*. Specifically, the two phases were classified as different "structures" and hence the system has variable structures. One of the first descriptions of this idea in the English language literature is the survey [155].

Modern references for finite-time and fixed-time stability include [21] and [118].

An important advancement to cope with the chattering effect of sliding mode control was the development of *higher order sliding mode (HOSM)* control. These ideas and others are fully explored in the text [137], where, additionally, the ideas are applied to the problem of state estimation. We provide a brief introduction to sliding mode observers in Section 18.2.

Chapter Eleven

Adaptive Control

In this chapter we discuss *adaptive controller designs*, that is, control methods that are applicable to parameter-dependent systems

$$\dot{x} = f(x, u, \theta), \qquad x \in \mathbb{R}^n, \ u \in \mathbb{R}^m, \ \theta \in \mathbb{R}^q, \tag{11.1}$$

with unknown parameter θ. In particular, despite the presence of the parameter θ, which captures the fact that the plant dynamics are not perfectly known, control laws are derived, which render the origin asymptotically stable. In contrast to sliding mode controller designs discussed in the preceding chapter, the adaptive controller designs covered here are not compensating for the unknown parameters or the disturbances with respect to the mismatch of the model and the plant by using larger inputs. Instead, the adaptive control laws u depend on the state x and parameter estimates $\hat{\theta}$. Thus, adaptive controller designs stabilize the origin of (11.1) while simultaneously estimating the unknown parameter θ.

Here, we focus on adaptive controller designs based on Lyapunov arguments. Alternative methods can be found in references given in the bibliographical notes.

11.1 MOTIVATING EXAMPLES AND CHALLENGES

As a motivating example, we focus on the nonlinear plant

$$\dot{x} = \theta \phi(x) + u \tag{11.2}$$

with state $x \in \mathbb{R}$, known function $\phi : \mathbb{R} \to \mathbb{R}$, and unknown parameter $\theta \in \mathbb{R}$. If an upper bound on the term $|\theta\phi(x)|$ is known, i.e., $|\theta\phi(x)| \leq L$ for all $x \in \mathbb{R}$ and $L \in \mathbb{R}$ selected appropriately, then (11.2) defines simplified dynamics of the form (10.11) discussed in the preceding chapter. Hence, a sliding mode controller can be derived to stabilize the origin, for example. This is possible despite the unknown term $\theta\phi(x)$ and is achieved solely by relying on the bound L. However, if $\phi(x) = x$, the term $|\theta\phi(x)|$ is unbounded and the dynamics (11.2) reduce to the linear system

$$\dot{x} = \theta x + u \tag{11.3}$$

with unknown gain θ. In this case, the sliding mode controller design in Chapter 10 is at best applicable locally, i.e., for x sufficiently small.

To systematically overcome this issue, we ask the following question: What can be achieved in terms of stability properties of the origin of the linear system (11.3) depending on the controller structure? To provide an answer, we consider linear and nonlinear control laws as well as static and dynamic controllers.

11.1.1 Limitations of Static Feedback Laws

Focusing on the one-dimensional linear dynamics (11.3), a controller design which globally stabilizes the origin is straightforward if an upper bound $|\theta| \leq L_\theta \in \mathbb{R}$ on the unknown parameter is known. Selecting a linear control law

$$u = -kx$$

with gain $k > L_\theta$ globally asymptotically stabilizes the origin since the closed-loop dynamics are given by $\dot{x} = -(k - \theta)x$ with $k - \theta > 0$.

Conversely, if an upper bound on θ is not known, a linear stabilizing feedback law which guarantees asymptotic stability of the origin does not exist. Even worse, if $k - \theta < 0$, all solutions with initial condition $x(0) \in \mathbb{R}\backslash\{0\}$ satisfy $|x(t)| \to \infty$, and thus, the origin is not even locally stable.

What happens if we consider a nonlinear controller

$$u = -k_1 x - k_2 x^3 \tag{11.4}$$

with controller gains $k_1, k_2 \in \mathbb{R}_{>0}$ instead? The control law leads to the closed-loop dynamics

$$\dot{x} = (\theta - k_1)x - k_2 x^3 = \left[(\theta - k_1) - k_2 x^2\right] x. \tag{11.5}$$

As a first observation, for $\theta \leq k_1$ the closed-loop dynamics exhibits a unique real-valued equilibrium at the origin $x^e = 0$, while (11.5) has three equilibria $x^e \in \left\{0, \pm\sqrt{\frac{\theta - k_1}{k_2}}\right\}$ for $\theta > k_1$. To analyze stability properties of the closed-loop system we consider the candidate Lyapunov function $V(x) = \frac{1}{2}x^2$ which satisfies

$$\dot{V}(x) = (\theta - k_1)x^2 - k_2 x^4.$$

Thus, as in the case of the linear feedback law, asymptotic stability of the origin can be concluded from Theorem (2.16) if an upper bound on θ is known and k_1 is selected accordingly so that $\theta - k_1$ is negative.

In the case that an upper bound on θ is not known, Young's inequality (Lemma A.4) leads to the estimate

$$\begin{aligned}
\dot{V}(x) &= -k_1 x^2 - k_2 x^4 + \theta x^2 \\
&\leq -k_1 x^2 - k_2 x^4 + \tfrac{1}{2}x^4 + \tfrac{1}{2}\theta^2 = -k_1 x^2 - (k_2 - \tfrac{1}{2})x^4 + \tfrac{1}{2}\theta^2,
\end{aligned}$$

which shows that V is an ISS-Lyapunov function (see Theorem 7.4 and Equation (7.11)) for $k_1 > 0$ and $k_2 > \frac{1}{2}$ if θ is interpreted as an unknown disturbance.

Even if an upper bound on $|\theta|$ is not known, ISS implies boundedness of solutions. Even more, through the definitions $\alpha_3(|x|) = k_1|x|^2$ and $\sigma(|\theta|) = \frac{1}{2}|\theta|^2$ as well as $\chi(\theta) = \alpha_3^{-1}(2\sigma(|\theta|)) = \sqrt{\frac{1}{k_1}}|\theta|$ it holds that all solutions converge to the set

$$S_\theta = \left\{x \in \mathbb{R} \,\middle|\, |x| \leq \sqrt{\tfrac{1}{k_1}}|\theta|\right\}$$

according to the discussion after Theorem 7.4. (Here, we have chosen $\varepsilon = \frac{1}{2}$ to convert the decrease condition (7.11) of the ISS-Lyapunov function in dissipation form to the decrease condition (7.8) in implication form, for simplicity). Thus, not only is it possible to guarantee boundedness of solutions using a nonlinear feedback law, by increasing the gain k_1 we can make the neighborhood around the origin to which solutions necessarily converge arbitrarily small. This is achieved despite $|\theta|$ being potentially arbitrarily large but constant.

Note that a similar derivation (see Exercise 11.1) together with the controller $u = -k_1 x - k_2 x \phi(x)^2$ can be used to derive the same qualitative behavior for the nonlinear system dynamics (11.2), i.e., the derivations do not rely on the fact that (11.3) is a linear parameter-dependent system.

Despite its advantages in terms of boundedness and convergence of solutions to S_θ, the static nonlinear control law (11.4) is not able to globally asymptotically stabilize the origin of (11.5). This follows immediately from the fact that the closed-loop dynamics have multiple equilibria.

To achieve global asymptotic stability of the origin, we consider a dynamic controller

$$
\begin{aligned}
u &= -k_1 x - \xi x \\
\dot{\xi} &= x^2
\end{aligned}
\tag{11.6}
$$

with gain $k_1 \in \mathbb{R}_{>0}$. In this case, the closed-loop system (11.3) is given by

$$
\left[\begin{array}{c} \dot{x} \\ \hline \dot{\xi} \end{array} \right] = \left[\begin{array}{c} \theta x - k_1 x - \xi x \\ \hline x^2 \end{array} \right],
\tag{11.7}
$$

where the line separates the plant states from the controller states.

Equivalently, in terms of the error variable $\hat{\theta} = \xi - \theta$ we observe that the closed-loop dynamics (11.7) are of the form

$$
\left[\begin{array}{c} \dot{x} \\ \hline \dot{\hat{\theta}} \end{array} \right] = \left[\begin{array}{c} -\hat{\theta} x - k_1 x \\ \hline x^2 \end{array} \right],
$$

since θ is constant by assumption, i.e., $\dot{\hat{\theta}} = \dot{\xi}$. This interpretation reveals the role of the state ξ of the dynamic controller (11.6) as an estimate of the unknown parameter θ. The adaptive control law attempts to simultaneously asymptotically stabilize the origin of the plant while estimating the unknown parameters, i.e., while driving the error $\hat{\theta}(t) = \xi(t) - \theta$ to zero.

Taking the candidate Lyapunov function $V(x, \hat{\theta}) = \frac{1}{2} x^2 + \frac{1}{2} \hat{\theta}^2$, we have

$$
\dot{V}(x, \hat{\theta}) = (-(\xi - \theta) x - k_1 x) x + (\xi - \theta) x^2 = -k_1 x^2,
$$

which shows that the origin of the augmented dynamics (11.7) is globally stable (see Theorem 2.15). Moreover, the LaSalle-Yoshizawa theorem (Theorem 2.27) guarantees global asymptotic stability of the origin of the state of the plant (11.3), i.e., $x(t) \to 0$ for $t \to \infty$ for all initial conditions $x(0) \in \mathbb{R}$, $\xi(0) \in \mathbb{R}$, independently of the unknown parameter θ. Convergence of the parameter estimate $\xi(t) \to \theta$ for $t \to \infty$ is however not guaranteed through Theorem 2.27 and it will in general not hold. Nevertheless, this does not present a problem with respect to our original goal to stabilize the origin of (11.3).

Note again that a similar dynamic controller can be used to stabilize the origin of the nonlinear dynamics (11.2) (see Exercise 11.2).

11.1.2   Estimation-Based Controller Designs

In the preceding section, we have focused on the state x and not on the parameter θ. Alternatively, one can first focus on the parameter θ, try to find an estimate of $\hat{\theta}$ which converges to the true value of the parameter θ, and then define a control law u based on θ which asymptotically stabilizes the origin of (11.2), under the assumption that the mismatch $\hat{\theta}(t) - \theta$ is sufficiently small.

We illustrate a corresponding adaptive controller design again based on the example of the linear dynamics (11.3). We assume that the signal $x(t)$ is available to define a state feedback, but $\dot{x}(t)$ is not known. If $x(t)$, $\dot{x}(t)$, and $u(t)$ are known, θ can be computed explicitly via (11.3).

As a first step in the controller design, we filter both sides of (11.3) through the transfer function $G(s) = \frac{1}{s+1}$, leading to filtered versions of the signals

$$\hat{x}_f(s) = \frac{1}{s+1}\hat{x}(s), \qquad \hat{u}_f(s) = \frac{1}{s+1}\hat{u}(s) \tag{11.8}$$

in the frequency domain.

Remark 11.1. The filtered state $x_f(t)$ in (11.8) satisfies $(s+1)\hat{x}_f(s) = \hat{x}(s)$ and thus in the time domain can be simulated through the linear dynamics

$$\dot{x}_f(t) = -x_f(t) + x(t)$$

with state $x_f(t)$ and input $x(t)$. The same relation holds for the filtered input.

The Laplace transform of (11.3) in combination with the filter $G(s)$ satisfies

$$\frac{s}{s+1}\hat{x}(s) = \theta\frac{1}{s+1}\hat{x}(s) + \frac{1}{s+1}\hat{u}(s),$$

which is equivalent to

$$\begin{aligned} \hat{x}(s) &= \frac{s}{s+1}\hat{x}(s) + \frac{1}{s+1}\hat{x}(s) = \theta\frac{1}{s+1}\hat{x}(s) + \frac{1}{s+1}\hat{u}(s) + \frac{1}{s+1}\hat{x}(s) \\ &= \frac{\theta+1}{s+1}\hat{x}(s) + \frac{1}{s+1}\hat{u}(s) = (\theta+1)\hat{x}_f(s) + \hat{u}_f(s). \end{aligned}$$

In the time domain, we can thus express $x(t)$ through $x(t) = (\theta+1)x_f(t) + u_f(t)$.

Since θ is not known, we replace it by an estimate $\hat{\theta} \in \mathbb{R}$, leading to the estimate of the state

$$\chi(t) = (\hat{\theta}+1)x_f(t) + u_f(t).$$

The difference, or the error e between the true state x and the estimate of the state χ, satisfies

$$e(t) = x(t) - \chi(t) = (\theta - \hat{\theta})x_f(t).$$

For a time-dependent estimate $\hat{\theta} : \mathbb{R}_{\geq 0} \to \mathbb{R}$, the squared error $(e(\hat{\theta}))^2$ can be minimized by appropriately updating the parameter $\hat{\theta}$ over time. The gradient of the function $(e(\hat{\theta}))^2$ satisfies

$$\frac{d}{d\hat{\theta}}\left(e(\hat{\theta})^2\right) = \frac{d}{d\hat{\theta}}\left((\theta - \hat{\theta})^2 x_f^2\right) = -2(\theta - \hat{\theta})x_f^2 = -2ex_f.$$

Using a gradient descent method to minimize $(e(\hat{\theta}))^2$ thus yields

$$\dot{\hat{\theta}} = \gamma e x_f, \tag{11.9}$$

where $\gamma \in \mathbb{R}_{>0}$ denotes a positive constant.

Remark 11.2. Recall that the gradient is the direction of steepest ascent of a continuously differentiable function. This fact can for example be derived from the Taylor expansion. A decreasing direction is thus given by the negative direction of the gradient.

With the feedback law

$$u = -k_1 x - \hat{\theta}x \tag{11.10}$$

the dynamics (11.3) become

$$\dot{x} = \theta x - k_1 x - \hat{\theta}x = ([\theta - \hat{\theta}] - k_1)x$$

and the origin $x = 0$ is globally asymptotically stable for $k_1 > 0$ and if $\hat{\theta}$ approximates θ sufficiently well, i.e., if $\theta - \hat{\theta} - k_1 < 0$. Combining these derivations, the overall system is given by

$$\begin{bmatrix} \dot{x} \\ \dot{\hat{\theta}} \\ \dot{x}_f \\ \dot{u}_f \end{bmatrix} = \begin{bmatrix} \theta x + u \\ \gamma(x - \chi)x_f \\ -x_f + x \\ -u_f + u \end{bmatrix} = \begin{bmatrix} (\theta - \hat{\theta} - k_1)x \\ \gamma(x - (\hat{\theta} + 1)x_f - u_f)x_f \\ -x_f + x \\ -u_f - k_1 x - \hat{\theta}x \end{bmatrix}, \tag{11.11}$$

where we have again separated the state of the plant from the dynamics of the controller. The $(\hat{\theta}, x_f, u_f)$ dynamics is notably independent of θ. The gradient descent method guarantees convergence of the parameter $\hat{\theta}(t)$ for $t \to \infty$, while the selection of the controller gain $k_1 > 0$ ensures that $x(t) \to 0$ for $t \to \infty$.

Remark 11.3. Note that convergence of $\hat{\theta}(t) \to \theta$ is not guaranteed for (11.11) since the closed-loop dynamics exhibit infinitely many equilibrium points. In other words, while the gradient descent method guarantees convergence to an equilibrium, this equilibrium is not necessarily the true parameter value. In particular, for $x = x_f = 0$, the $\hat{\theta}$-dynamics reduces to $\dot{\hat{\theta}} = 0$.

Figure 11.1 shows the closed-loop solution of the dynamics (11.11) together with the feedback law (11.10). Here, the parameters are selected as $k_1 = 3$ and $\gamma = 2$, the unknown parameter is set to $\theta = 10$, and the initial condition is defined as $[x(0), \hat{\theta}(0), x_f(0), u_f(0)]^T = [1, 0, 0, 0]^T$ for the solution shown in Figure 11.1. The numerical simulations confirm the theoretical properties derived in this section.

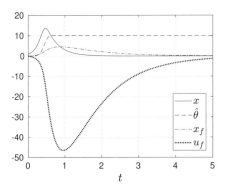
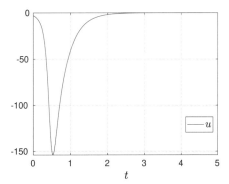

Figure 11.1:   Visualization of the solution of the closed-loop dynamics (11.11) together with the input (11.10).   The system is initialized through $[x(0), \hat{\theta}(0), x_f(0), u_f(0)]^T = [1, 0, 0, 0]^T$.   The parameters are defined as $k_1 = 3$, $\gamma = 2$ and the unknown parameter θ is set to $\theta = 10$ for the simulation.

Instead of the feedback law (11.9), a normalized update

$$\dot{\hat{\theta}} = \gamma \frac{1}{1 + x_f^2} e x_f = \gamma \frac{x_f^2}{1 + x_f^2} (\theta - \hat{\theta}) \tag{11.12}$$

is commonly used in the controller design, leading to the same asymptotic behavior (see Exercise 11.3).

For nonlinear dynamics, the controller selection described here is more subtle. To highlight the difficulties arising from nonlinear dynamics, consider (11.2) with $\phi(x) = x^2$, i.e.,

$$\dot{x} = \theta x^2 + u. \tag{11.13}$$

In this case, the feedback law

$$u = -k_1 x - \hat{\theta} x^2 \tag{11.14}$$

again implies that $\dot{x} = -k_1 x$ if $\hat{\theta} = \theta$ and thus the origin is globally asymptotically stable for $k_1 > 0$ and θ known. However, guaranteeing $\hat{\theta}(t) \to \theta$ for $t \to \infty$ sufficiently fast is not straightforward to ensure. To illustrate this fact, we consider again the Laplace transform of filtered signals leading to the equation

$$\hat{x}(s) = \frac{s}{s+1} \hat{x}(s) + \frac{1}{s+1} \hat{x}(s) = \theta \frac{1}{s+1} \hat{x}^2(s) + \frac{1}{s+1} \hat{u}(s) + \frac{1}{s+1} \hat{x}(s),$$

where $\hat{x}^2(s) = (\mathcal{L}(x^2))(s)$ denotes the Laplace transform of $x(t)^2$.

With the definitions

$$\hat{x}_{f_1}(s) = \frac{1}{s+1} \hat{x}(s), \qquad \hat{x}_{f_2}(s) = \frac{1}{s+1} \hat{x}^2(s), \qquad \hat{u}_f(s) = \frac{1}{s+1} \hat{u}(s), \tag{11.15}$$

we can thus express $x(t)$ through $x(t) = \theta x_{f_2}(t) + x_{f_1}(t) + u_f(t)$.

As in the linear setting, since θ is not known, an estimate $\hat{\theta} \in \mathbb{R}$ is necessary, leading to the estimate of the state

$$\chi(t) = \hat{\theta}x_{f_2}(t) + x_{f_1}(t) + u_f(t).$$

The difference, or the error e between the true state x and its estimate χ, satisfies

$$e(t) = x(t) - \chi(t) = (\theta - \hat{\theta})x_{f_2}(t).$$

As before, the gradient descent method yields

$$\dot{\hat{\theta}} = \gamma x_{f_2}^2 (\theta - \hat{\theta}) \qquad \text{or} \qquad \dot{\hat{\theta}} = \gamma \frac{x_{f_2}^2}{1 + x_{f_2}^2}(\theta - \hat{\theta}) \qquad (11.16)$$

in its unscaled and its scaled form, respectively.

Summarizing the derivations so far, using the scaled update (11.16), the overall control loop

$$\begin{bmatrix} \dot{x} \\ \dot{\hat{\theta}} \\ \dot{x}_{f_1} \\ \dot{x}_{f_2} \\ \dot{u}_f \end{bmatrix} = \begin{bmatrix} -k_1 x + (\theta - \hat{\theta})x^2 \\ \gamma \frac{x_{f_2}}{1 + x_{f_2}^2}(x - \hat{\theta}x_{f_2}(t) - x_{f_1}(t) - u_f(t)) \\ -x_{f_1} + x \\ -x_{f_2} + x^2 \\ -u_f - k_1 x - \hat{\theta}x^2 \end{bmatrix} \qquad (11.17)$$

has been derived for the nonlinear dynamics (11.13).

Returning to the scaled differential equation (11.16), the convergence of $\bar{\theta}(t) = \theta - \hat{\theta} \to 0$ for $t \to \infty$ is slower than the exponential convergence

$$\bar{\theta}(t) = e^{-\gamma t}\bar{\theta}(0)$$

since $\frac{x_{f_2}^2}{1 + x_{f_2}^2} < 1$. Nevertheless, using this most favorable estimate of the error of the parameter θ leads to the closed-loop dynamics

$$\dot{x} = e^{-\gamma t}\bar{\theta}(0)x^2 - k_1 x \qquad (11.18)$$

when combined with (11.13) and the feedback law (11.14). The solution of (11.18) is given by

$$x(t) = \frac{k_1 + \gamma}{\bar{\theta}(0)x(0)e^{-\gamma t} + \left[(k_1 + \gamma) - \bar{\theta}(0)x(0)\right]e^{kt}}x(0) \qquad (11.19)$$

(see Exercise 11.4). From the solution we can observe that for

$$0 < \bar{\theta}(0)x(0) \qquad \text{and} \qquad k + \gamma < \bar{\theta}(0)x(0)$$

there exists a $t \in \mathbb{R}_{>0}$ such that the denominator in (11.19) is zero, i.e., $|x(t)| \to \infty$ in finite time. Despite $\bar{\theta}(t) \to 0$, or equivalently $\hat{\theta}(t) \to \theta$, for $t \to \infty$, the convergence is too slow to ensure that $x(t)$ stays bounded in the transient behavior.

Closed-loop solutions of (11.17) are visualized in Figure 11.2. For the simulations the same parameters as in the linear setting in Figure 11.1 are used. Figure 11.2 confirms that the asymptotic behavior depends on the selection of the initial

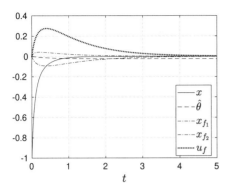 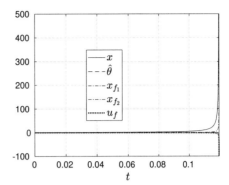

Figure 11.2: Visualization of the solution of the closed-loop dynamics (11.17) for different initial conditions. On the left, the dynamics are initialized through $[x(0), \hat{\theta}(0), x_f(0), u_f(0)]^T = [-1, 0, 0, 0]^T$, and on the right $[x(0), \hat{\theta}(0), x_f(0), u_f(0)]^T = [1, 0, 0, 0]^T$ is used.

condition and in the worst case finite escape time can be observed (Figure 11.2, right). In Figure 11.2, left, even though $x(t) \to 0$ for $t \to \infty$, the estimate $\hat{\theta}$ is converging but $\hat{\theta}(t) \to \theta$ for $t \to \infty$ is not satisfied.

11.2 MODEL REFERENCE ADAPTIVE CONTROL

Consider linear systems

$$\dot{x} = Ax + Bu \tag{11.20}$$

with unknown matrices $A \in \mathbb{R}^{n \times n}$, $B \in \mathbb{R}^{n \times m}$.

We intend to design a controller so that the unknown system behaves like

$$\dot{\tilde{x}} = \tilde{A}\tilde{x} + \tilde{B}u^e \tag{11.21}$$

where $\tilde{A} \in \mathbb{R}^{n \times n}$ and $\tilde{B} \in \mathbb{R}^{n \times m}$ are design parameters and $u^e \in \mathbb{R}^m$ is a constant reference value which defines the induced equilibrium of the system. In particular, if $\tilde{A}$ is Hurwitz, the equilibrium of the system

$$\tilde{x}^e = -\tilde{A}^{-1}\tilde{B}u^e \tag{11.22}$$

is globally asymptotically stable. System (11.21) is called the reference model.

As a control law, consider

$$u = M(\theta)u^e + L(\theta)x \tag{11.23}$$

for parameter-dependent matrices $M : \mathbb{R}^q \to \mathbb{R}^{m \times m}$, $L : \mathbb{R}^q \to \mathbb{R}^{m \times n}$, $q \in \mathbb{N}$, to be designed. Observe that the role of the parameter θ here is slightly different than in the previous section, where the assumption was that the plant to be controlled had one or more parameters which were unknown. Here, the unknown parameters to be estimated are part of the controller designed to make system (11.20) behave like the reference model (11.21).

We assume that the components $M_{ij}(\cdot)$, $i, j \in \{1, \ldots, m\}$ and $L_{ij}(\cdot)$, $i \in \{1, \ldots, m\}$, $j \in \{1, \ldots, n\}$, are affine functions. This implies that for all $i \in \{1, \ldots, m\}$ there exists $c \in \mathbb{R}^q$ and $d \in \mathbb{R}$ such that

$$M_{ij}(\theta) = c^T \theta + d.$$

The same representation is valid for the entries of $L(\cdot)$.

The closed-loop dynamics are given by

$$\begin{aligned}
\dot{x} &= Ax + B(M(\theta)u^e + L(\theta)x) \\
&= (A + BL(\theta))x + BM(\theta)u^e \\
&= A_{\mathrm{cl}}(\theta)x + B_{\mathrm{cl}}(\theta)u^e,
\end{aligned}$$

where the last equality follows by defining $A_{\mathrm{cl}}(\theta) = A + BL(\theta)$, $B_{\mathrm{cl}}(\theta) = BM(\theta)$. Comparing this system with the desired behavior (11.21) leads to the conditions

$$\begin{aligned}
A_{\mathrm{cl}}(\theta) = \tilde{A} &\quad \Longleftrightarrow \quad BL(\theta) = \tilde{A} - A, \\
B_{\mathrm{cl}}(\theta) = \tilde{B} &\quad \Longleftrightarrow \quad BM(\theta) = \tilde{B}.
\end{aligned} \tag{11.24}$$

In general, it is not clear that parameters $\theta \in \mathbb{R}^q$ exist such that (11.24) is satisfied. For the derivations we thus need the following *compatibility assumption*.

Assumption 11.4 (Compatibility Condition). *Let $A, \tilde{A} \in \mathbb{R}^{n \times n}$, $B, \tilde{B} \in \mathbb{R}^{n \times m}$, and $M : \mathbb{R}^q \to \mathbb{R}^{m \times m}$, $L : \mathbb{R}^q \to \mathbb{R}^{m \times n}$ for $q \in \mathbb{N}$. We say that the compatibility condition is satisfied if there exists $\theta \in \mathbb{R}^q$ such that (11.24) is satisfied.*

Example 11.5. Recall the mass-spring system from Example 1.2 described by the second-order differential equation

$$m\ddot{y} = u - c\dot{y} - ky. \tag{11.25}$$

Following the definitions in this chapter, we assume that the constants $m, c, k \in \mathbb{R}$ are unknown. With $x_1 = y$ and $x_2 = \dot{x}$, the system can equivalently be represented through the linear dynamics $\dot{x} = Ax + Bu$ with

$$A = \begin{bmatrix} 0 & 1 \\ -\frac{k}{m} & -\frac{c}{m} \end{bmatrix} = \begin{bmatrix} 0 & 1 \\ \mu_1 & \mu_2 \end{bmatrix} \quad \text{and} \quad B = \begin{bmatrix} 0 \\ \frac{1}{m} \end{bmatrix} = \begin{bmatrix} 0 \\ \mu_3 \end{bmatrix}$$

and unknown parameters $\mu_1, \mu_2, \mu_3 \in \mathbb{R}$. In particular, since the linear system is derived from a second-order differential equation, some entries of the matrices A and B are known.

We define the controller

$$u = M(\theta)u^e + L(\theta)x = \theta_1 u^e + \begin{bmatrix} \theta_2 & \theta_3 \end{bmatrix} x, \tag{11.26}$$

consistent with the dimension of the two-dimensional system and with three gains, $\theta_1, \theta_2, \theta_3 \in \mathbb{R}$.

To verify the compatibility conditions, we compute the closed-loop matrices in

(11.24):

$$A_{cl}(\theta) = \begin{bmatrix} 0 & 1 \\ \mu_1 & \mu_2 \end{bmatrix} + \begin{bmatrix} 0 \\ \mu_3 \end{bmatrix} \begin{bmatrix} \theta_2 & \theta_3 \end{bmatrix} = \begin{bmatrix} 0 & 1 \\ \mu_1 + \mu_3\theta_2 & \mu_2 + \mu_3\theta_3 \end{bmatrix},$$

$$B_{cl}(\theta) = \begin{bmatrix} 0 \\ \mu_3 \end{bmatrix} \theta_1 = \begin{bmatrix} 0 \\ \mu_3\theta_1 \end{bmatrix}.$$

At this point it is important to note that θ_1, θ_2, and θ_3 represent degrees of freedom which can be selected arbitrarily. Thus under the additional assumption that $m \neq 0$ (which is a reasonable assumption and necessary to ensure that the matrices A and B are well defined) matrices $\tilde{A}$, $\tilde{B}$ satisfy the compatibility condition in Assumption 11.4 if and only if

$$\tilde{A} = \begin{bmatrix} 0 & 1 \\ a_{21} & a_{22} \end{bmatrix}, \qquad \tilde{B} = \begin{bmatrix} 0 \\ b_2 \end{bmatrix}, \tag{11.27}$$

for $a_{21}, a_{22} \in \mathbb{R}$ and $b_2 \in \mathbb{R}\backslash\{0\}$.

The characteristic polynomial of $\tilde{A}$ is given by

$$p_{\tilde{A}}(\lambda) = \det(\lambda I - \tilde{A}) = \lambda(\lambda - a_{22}) - a_{21} = \lambda^2 - a_{22}\lambda - a_{21}$$

and thus the eigenvalues of $\tilde{A}$ satisfy

$$\lambda_1 = \frac{1}{2}\left(a_{22} + \sqrt{a_{22}^2 + 4a_{21}}\right), \qquad \lambda_2 = \frac{1}{2}\left(a_{22} - \sqrt{a_{22}^2 + 4a_{21}}\right).$$

Equivalently, to ensure that the matrix $\tilde{A}$ is Hurwitz, desired eigenvalues $\lambda_1, \lambda_2 \in \mathbb{C}_-$ can be used to define a_{21}, a_{22} through the solution of the linear equation

$$\begin{bmatrix} 1 & \lambda_1 \\ 1 & \lambda_2 \end{bmatrix} \begin{bmatrix} a_{21} \\ a_{22} \end{bmatrix} = \begin{bmatrix} \lambda_2^2 \\ \lambda_1^2 \end{bmatrix}. \tag{11.28}$$

Using the formula (1.33), for $\tilde{A}$ Hurwitz, the induced equilibrium defined in (11.22) satisfies

$$x^e = -\frac{1}{-a_{21}} \begin{bmatrix} a_{22} & -1 \\ -a_{21} & 0 \end{bmatrix} \begin{bmatrix} 0 \\ b_2 \end{bmatrix} u^e = \begin{bmatrix} -\frac{b_2}{a_{21}} \\ 0 \end{bmatrix} u^e, \tag{11.29}$$

i.e., by appropriately selecting a constant reference input u^e, we can stabilize the center of the mass of the mass-spring system at an arbitrary position x^e.

Before we can continue with the example, we need an adaptive controller defining the functions $M(\cdot)$, $L(\cdot)$. We consider the error variable

$$e = x - \tilde{x},$$

which satisfies

$$\begin{aligned} \dot{e} = \dot{x} - \dot{\tilde{x}} &= Ax + Bu - \tilde{A}\tilde{x} - \tilde{B}u^e \\ &= (A + BL(\theta))x + BM(\theta)u^e - \tilde{A}\tilde{x} - \tilde{B}u^e \\ &= \tilde{A}(x - \tilde{x}) + (A_{cl}(\theta) - \tilde{A})x + (B_{cl}(\theta) - \tilde{B})u^e. \end{aligned}$$

Let Assumption 11.4 be satisfied and let θ^e be such that $BL(\theta^e) = \tilde{A} - A$ and $BM(\theta^e) = \tilde{B}$. Then we can further rewrite the error dynamics in the form

$$\dot{e} = \tilde{A}e + (A_{\mathrm{cl}}(\theta) - A_{\mathrm{cl}}(\theta^e))x + (B_{\mathrm{cl}}(\theta) - B_{\mathrm{cl}}(\theta^e))u^e,$$

which reduces to $\dot{e} = \tilde{A}e$ in the case that $\theta = \theta^e$.

Since the entries of $M(\cdot)$ and $L(\cdot)$ are affine in θ, the entries of $A_{\mathrm{cl}}(\cdot) = A + BL(\cdot)$ and $B_{\mathrm{cl}}(\cdot) = BM(\cdot)$ are affine in θ. Hence, there exists a function $\Psi : \mathbb{R}^n \times \mathbb{R}^m \to \mathbb{R}^{n \times q}$ such that

$$\Psi(x, u^e)(\theta - \theta^e) = (A_{\mathrm{cl}}(\theta) - A_{\mathrm{cl}}(\theta^e))x + (B_{\mathrm{cl}}(\theta) - B_{\mathrm{cl}}(\theta^e))u^e, \qquad (11.30)$$

which allows us to express the error dynamics as

$$\dot{e} = \tilde{A}e + \Psi(x, u^e)(\theta - \theta^e). \qquad (11.31)$$

Example 11.6. We continue with Example 11.5 to illustrate the definition of the function $\Psi(x, u^e)$. To derive Ψ, we calculate

$$
\begin{aligned}
A_{\mathrm{cl}}(\theta)x + B_{\mathrm{cl}}(\theta)u^e &= \begin{bmatrix} 0 & 1 \\ \mu_1 + \mu_3\theta_2 & \mu_2 + \mu_3\theta_3 \end{bmatrix} \begin{bmatrix} x_1 \\ x_2 \end{bmatrix} + \begin{bmatrix} 0 \\ \mu_3\theta_1 \end{bmatrix} u^e \\
&= \begin{bmatrix} x_2 \\ \mu_1 x_1 + \mu_3\theta_2 x_1 + \mu_2 x_2 + \mu_3\theta_3 x_2 + \mu_3\theta_1 u^e \end{bmatrix} \\
&= \begin{bmatrix} x_2 \\ \mu_1 x_1 + \mu_2 x_2 \end{bmatrix} + \begin{bmatrix} 0 & 0 & 0 \\ \mu_3 u^e & \mu_3 x_1 & \mu_3 x_2 \end{bmatrix} \begin{bmatrix} \theta_1 \\ \theta_2 \\ \theta_3 \end{bmatrix}.
\end{aligned}
$$

Replacing θ with θ^e in the expression above, we can take the difference on the right-hand side of (11.30), which leads to the definition

$$\Psi(x, u^e) = \begin{bmatrix} 0 & 0 & 0 \\ \mu_3 u^e & \mu_3 x_1 & \mu_3 x_2 \end{bmatrix}. \qquad (11.32)$$

Theorem 11.7. *Consider the system (11.20) with unknown matrices A, B together with the reference system (11.21) where $\tilde{A}$ is Hurwitz. Let Assumption 11.4 be satisfied and let Ψ be defined according to (11.30). Let $\gamma \in \mathbb{R}_{>0}$, $Q \in \mathcal{S}_{>0}^n$ and let $P \in \mathcal{S}_{>0}^n$ denote the solution of the Lyapunov equation*

$$\tilde{A}^T P + P\tilde{A} = -Q$$

(which exists according to Theorem 3.3).

Then the dynamic controller

$$
\begin{aligned}
u &= M(\theta)u^e + L(\theta)x \\
\dot{\tilde{x}} &= \tilde{A}\tilde{x} + \tilde{B}u^e \\
\dot{\theta} &= -\gamma\Psi(x, u^e)^T P(x - \tilde{x})
\end{aligned}
$$

ensures that the solutions of the closed-loop system

$$
\begin{bmatrix} \dot{x} \\ \dot{\tilde{x}} \\ \dot{\theta} \end{bmatrix} = \begin{bmatrix} (A + BL(\theta))x + BM(\theta)u^e \\ \tilde{A}\tilde{x} + \tilde{B}u^e \\ -\gamma\Psi(x, u^e)^T P(x - \tilde{x}) \end{bmatrix}
\tag{11.33}
$$

are globally bounded. Moreover, it holds that $x(t) \to -\tilde{A}^{-1}\tilde{B}u^e$ *for* $t \to \infty$ *for all initial conditions* $(x_0, \tilde{x}_0, \theta_0) \in \mathbb{R}^n \times \mathbb{R}^n \times \mathbb{R}^q$.

Proof. We first note that the $\tilde{x}$-dynamics in the closed-loop dynamics (11.33) are independent of the other states. Moreover, since $\tilde{A}$ is Hurwitz by assumption, the equilibrium $-\tilde{A}^{-1}\tilde{B}u^e$ of the subsystem $\dot{\tilde{x}} = \tilde{A}\tilde{x} + \tilde{B}u^e$ is globally asymptotically stable.

Using the error variable $e = x - \tilde{x}$, the error dynamics satisfies

$$
\begin{bmatrix} \dot{e} \\ \dot{\theta} \end{bmatrix} = \begin{bmatrix} \tilde{A}e + \Psi(x, u^e)(\theta - \theta^e) \\ -\gamma\Psi(x, u^e)^T Pe \end{bmatrix},
\tag{11.34}
$$

which follows from (11.33) and (11.31).

We consider the radially unbounded function $V : \mathbb{R}^n \times \mathbb{R}^q \to \mathbb{R}_{\geq 0}$,

$$
V(e, \bar{\theta}) = \tfrac{1}{2}\left(\gamma e^T Pe + \bar{\theta}^T \bar{\theta}\right),
$$

with $\bar{\theta} = \theta - \theta^e$ and $\dot{\bar{\theta}} = \dot{\theta}$, which satisfies

$$
\begin{aligned}
\dot{V}(e, \bar{\theta}) &= \gamma e^T P\dot{e} + \bar{\theta}^T \dot{\bar{\theta}} \\
&= \gamma e^T P\left(\tilde{A}e + \Psi(x, u^e)\bar{\theta}\right) - \gamma\bar{\theta}^T\Psi(x, u^e)^T Pe \\
&= \tfrac{\gamma}{2}\left(e^T P\tilde{A}e + e^T \tilde{A}^T Pe\right) \\
&= -\tfrac{\gamma}{2}e^T Qe \leq 0
\end{aligned}
\tag{11.35}
$$

for all $(e, \theta) \in \mathbb{R}^n \times \mathbb{R}^q$. Thus, V is a Lyapunov function according to Theorem 2.15 and the equilibrium $(0, \theta^e) \in \mathbb{R}^n \times \mathbb{R}^q$ of the system (11.34) is stable. In particular, the error $e(t)$ and $\theta(t)$ are bounded for all $t \in \mathbb{R}_{\geq 0}$. Due to the convergence $\tilde{x}(t) \to 0$ and the boundedness of $e(t)$ it additionally follows that $x(t)$ is bounded, which is necessary in the decrease condition (11.35) to ensure that $\Psi(x(t), u^e)$ is well defined for all $t \in \mathbb{R}_{\geq 0}$.

The convergence $e(t) \to 0$ for $t \to \infty$ follows from LaSalle-Yoshizawa Theorem 2.27. $\square$

As given in Theorem 11.7, the controller in the closed-loop dynamics (11.34) might not be implementable since Ψ might depend on unknown parameters as in (11.32) in Example 11.6. For the dynamics in Example 11.6, the Lyapunov function used in the proof of Theorem 11.7 can be adapted to obtain an implementable feedback law.

Example 11.8. We continue with Example 11.6 and demonstrate the adaptive controller design discussed in this section. We take the system parameters $m = 5$, $k = 1$, and $c = 0.2$ to define the matrices A and B. These parameters are used for the simulation, but are not used in the controller design and are assumed to be unknown. We only assume that the sign of $\mu_3 = \frac{1}{m} > 0$ is known, which is reasonable since m denotes the mass of the mass-spring system. For the reference system, we place the poles of the matrix $\tilde{A}$ in (11.27) at $\lambda_{1,2} = -1 \pm 0.5j$. Then, according to (11.28), the parameters of $\tilde{A}$ need to be selected as $a_{21} = -1.25$ and $a_{22} = -2$. To simplify the relation between u^e and x^e we define $b_2 = -a_{21} = 1.15$, which implies that $x^e = [u^e, 0]^T$ according to (11.29). Moreover, the parameters $\gamma = 10$ and $Q = 10 \cdot \mathrm{diag}(1, 1)$ are used. The selection of Q leads to the matrix

$$P = \begin{bmatrix} 13.625 & 4 \\ 4 & 4.5 \end{bmatrix}$$

in Theorem 11.7 (obtained through `lyap.m` in MATLAB). To define the parameter update $\dot{\bar{\theta}}$ we consider the candidate Lyapunov function

$$V(e, \bar{\theta}) = \gamma e^T P e + \mu_3 \bar{\theta}^T \bar{\theta}$$

whose derivative satisfies

$$\dot{V}(e, \bar{\theta}) = \gamma e^T P \dot{e} + \mu_3 \bar{\theta}^T \dot{\bar{\theta}} = \gamma e^T P \left(\tilde{A} e + \Psi(x, u^e) \bar{\theta} \right) + \mu_3 \dot{\bar{\theta}}^T \bar{\theta}$$

$$= \gamma e^T P \tilde{A} e + \mu_3 \left(\frac{\gamma}{\mu_3} e^T P \Psi(x, u^e) + \dot{\bar{\theta}}^T \right) \bar{\theta}.$$

Thus, the selection

$$\dot{\bar{\theta}} = -\frac{\gamma}{\mu_3} \Psi(x, u^e)^T P e = -\begin{bmatrix} 0 & u^e \\ 0 & x_1 \\ 0 & x_2 \end{bmatrix} \begin{bmatrix} 13.625 & 4 \\ 4 & 4.5 \end{bmatrix} e \qquad (11.36)$$

guarantees $\dot{V}(e, \bar{\theta}) = -\frac{\gamma}{2} e^T Q e \leq 0$ and the dynamics (11.36) only contain known terms.

The closed-loop solution of the adaptive controller combining (11.33) and (11.36) together with the input (11.26) are shown in Figure 11.3. For the simulation, the state of the closed-loop system (11.33) is initialized at zero. The reference value u^e is selected as $u^e = 1$, which implies that the induced equilibrium is defined as $x^e = [1, 0]^T$ according to (11.29). From Figure 11.3 we observe convergence $x(t) \to x^e$, $\tilde{x}(t) \to x^e$, and $u(t) \to u^e$ for $t \to \infty$, as expected.

Additionally, observe that the states θ converge. However, for the simulation presented here, $A_{\mathrm{cl}}(\theta(t))$ and $B_{\mathrm{cl}}(\theta(t))$ do not converge to $\tilde{A}$ and $\tilde{B}$, respectively, for $t \to \infty$ (see Figure 11.4). This is consistent with Theorem 11.7, which only guarantees convergence of $e(t) = x(t) - \tilde{x}(t) \to 0$ for $t \to \infty$. The asymptotic behavior of θ depends on the initial condition and the selected reference input u^e (see Exercise 11.5).

11.3   ADAPTIVE CONTROL FOR NONLINEAR SYSTEMS

In this section we discuss backstepping for systems with unknown constant parameters, i.e., we derive extensions of the controller design in Chapter 9.4. While *adaptive backstepping*, discussed in Section 11.3.1, leads to overparameterization, that is, multiple estimates of the same unknown parameter are needed in the dynamic

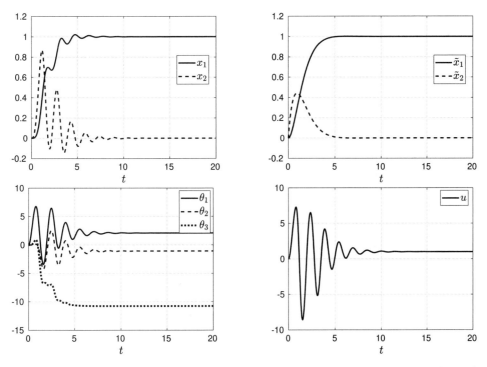

Figure 11.3: Closed-loop solution of the model reference adaptive controller (11.33) together with the input (11.26).

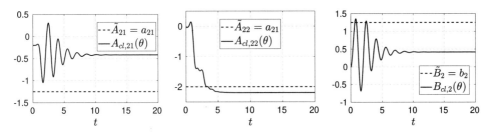

Figure 11.4: Evolution of the time-varying matrix entries of $A_{cl}(\theta)$ and $B_{cl}(\theta)$ compared to the entries in the reference matrices $\tilde{A}$ and $\tilde{B}$.

controller, *tuning functions*, discussed in Section 11.3.2, overcome this problem. The controller designs in this section are illustrated on the example of a *single link manipulator with flexible joint* in Section 11.3.3.

11.3.1 Adaptive Backstepping

In this section we extend the ideas of backstepping discussed in Chapter 9.4 to systems with unknown parameters. To introduce the main idea, we consider a system of the form

$$\dot{x} = f(x) + F(x)\theta + g(x)u, \tag{11.37}$$

where $x \in \mathbb{R}^n$ is the state, $\theta \in \mathbb{R}^q$ is a vector of unknown constant parameters, $u \in \mathbb{R}$ is the control input, and $x^e = 0$ is an equilibrium. We assume that we know a dynamic controller (such as one of those presented in the previous two sections),

$$
\begin{aligned}
u &= \alpha(x, \hat{\theta}), \\
\dot{\hat{\theta}} &= T(x, \hat{\theta}),
\end{aligned}
\tag{11.38}
$$

with parameter estimate $\hat{\theta} \in \mathbb{R}^q$ so that the equilibrium $[0, \theta]^T \in \mathbb{R}^{n+q}$ of the closed-loop dynamics

$$
\begin{bmatrix} \dot{x} \\ \dot{\hat{\theta}} \end{bmatrix} = \begin{bmatrix} f(x) + F(x)\theta + g(x)\alpha(x, \hat{\theta}) \\ T(x, \hat{\theta}) \end{bmatrix}
\tag{11.39}
$$

is globally stable and $\lim_{t \to \infty} x(t) = 0$. To characterize these stability properties, we additionally assume that we know functions

$$
V(\cdot, \cdot) : \mathbb{R}^{n+q} \to \mathbb{R}_{\geq 0} \qquad \text{and} \qquad W(\cdot, \cdot) : \mathbb{R}^{n+q} \to \mathbb{R}_{\geq 0}
\tag{11.40}
$$

satisfying the assumptions of Theorem 2.27, $W(x, \hat{\theta}) > 0$ for all $x \neq 0$, and in particular the decrease condition

$$
\frac{\partial V}{\partial x}(x, \hat{\theta}) \left[f(x) + F(x)\theta + g(x)\alpha(x, \hat{\theta}) \right] + \frac{\partial V}{\partial \hat{\theta}}(x, \hat{\theta}) T(x, \hat{\theta}) \leq -W(x, \hat{\theta}).
\tag{11.41}
$$

As in Chapter 9.4, we are now going to make use of these assumptions to asymptotically stabilize the origin of the augmented dynamics

$$
\begin{aligned}
\dot{x} &= f(x) + F(x)\theta + g(x)\xi, \\
\dot{\xi} &= u
\end{aligned}
\tag{11.42}
$$

with virtual input ξ. Accordingly, we define the error variable

$$
z = \xi - \alpha(x, \hat{\theta})
$$

whose derivative satisfies

$$
\begin{aligned}
\dot{z} &= \dot{\xi} - \tfrac{d}{dt}\alpha(x, \hat{\theta}) \\
&= u - \frac{\partial \alpha}{\partial x}(x, \hat{\theta})\dot{x} - \frac{\partial \alpha}{\partial \hat{\theta}}(x, \hat{\theta})\dot{\hat{\theta}} \\
&= u - \frac{\partial \alpha}{\partial x}(x, \hat{\theta}) \left[f(x) + F(x)\theta + g(x)\xi \right] - \frac{\partial \alpha}{\partial \hat{\theta}}(x, \hat{\theta}) T(x, \hat{\theta}) \\
&= u - \frac{\partial \alpha}{\partial x}(x, \hat{\theta}) \left[f(x) + F(x)\theta + g(x)(\alpha(x, \hat{\theta}) + z) \right] - \frac{\partial \alpha}{\partial \hat{\theta}}(x, \hat{\theta}) T(x, \hat{\theta}).
\end{aligned}
$$

In the variables x and z, the dynamics (11.42) thus read

$$
\begin{aligned}
\dot{x} &= f(x) + F(x)\theta + g(x)(\alpha(x, \hat{\theta}) + z), \\
\dot{z} &= u - \frac{\partial \alpha}{\partial x}(x, \hat{\theta}) \left[f(x) + F(x)\theta + g(x)(\alpha(x, \hat{\theta}) + z) \right] - \frac{\partial \alpha}{\partial \hat{\theta}}(x, \hat{\theta}) T(x, \hat{\theta}).
\end{aligned}
$$

As a next step, we consider a new estimate of $\hat{\theta}_s$ of the the unknown parameters θ and we augment the Lyapunov function V in (11.40) with additional terms

$$
\begin{aligned}
V_a(x, \xi, \hat{\theta}, \hat{\theta}_s) &= V(x, \hat{\theta}) + \tfrac{1}{2}(\xi - \alpha(x, \hat{\theta}))^T (\xi - \alpha(x, \hat{\theta})) \\
&\quad + \tfrac{1}{2}(\theta - \hat{\theta}_s)^T \Gamma^{-1}(\theta - \hat{\theta}_s),
\end{aligned}
\tag{11.43}
$$

where $\Gamma \in \mathcal{S}_{>0}^q$. The time derivative of the augmented function satisfies

$$
\begin{aligned}
\dot{V}_a(x, \xi, \hat{\theta}, \hat{\theta}_s) &= \dot{V}(x, \hat{\theta}) + (\xi - \alpha(x, \hat{\theta}))^T(\dot{\xi} - \dot{\alpha}(x, \hat{\theta})) - (\theta - \hat{\theta}_s)^T \Gamma^{-1}\dot{\hat{\theta}}_s \\
&= \tfrac{\partial V}{\partial x}(x, \hat{\theta}) \left[f(x) + F(x)\theta + g(x)\alpha(x, \hat{\theta}) + g(x)z \right] + \tfrac{\partial V}{\partial \hat{\theta}}(x, \hat{\theta})T(x, \hat{\theta}) \\
&\quad + \left(u - \tfrac{\partial \alpha}{\partial x}(x, \hat{\theta}) \left[f(x) + F(x)\theta + g(x)(\alpha(x, \hat{\theta}) + z) \right] - \tfrac{\partial \alpha}{\partial \hat{\theta}}(x, \hat{\theta})T(x, \hat{\theta}) \right) z \\
&\quad - (\theta - \hat{\theta}_s)^T \Gamma^{-1}\dot{\hat{\theta}}_s,
\end{aligned}
$$

where we have not specified the input u and the dynamics of the estimate $\hat{\theta}_s$ yet. With (11.41), this expression simplifies to

$$
\begin{aligned}
\dot{V}_a &\leq -W + \left(\tfrac{\partial V}{\partial x}g + u - \tfrac{\partial \alpha}{\partial x} \left[f + F(\theta - \hat{\theta}_s + \hat{\theta}_s) + g(\alpha + z) \right] - \tfrac{\partial \alpha}{\partial \hat{\theta}}T \right) z \\
&\quad - (\theta - \hat{\theta}_s)^T \Gamma^{-1}\dot{\hat{\theta}}_s \\
&= -W + \left(u + \tfrac{\partial V}{\partial x}g - \tfrac{\partial \alpha}{\partial x} \left[f + F\hat{\theta}_s + g(\alpha + z) \right] - \tfrac{\partial \alpha}{\partial \hat{\theta}}T \right) z \\
&\quad - \left(\tfrac{\partial \alpha}{\partial x}Fz + \dot{\hat{\theta}}_s^T \Gamma^{-1} \right)(\theta - \hat{\theta}_s),
\end{aligned}
$$

where we have dropped the arguments of the function to shorten the expressions.

To ensure that $\dot{V}_a(x, \xi, \hat{\theta}, \hat{\theta}_s) < 0$ for all $x \neq 0$ we can thus define

$$
\begin{aligned}
u &= -c(\xi - \alpha(x, \hat{\theta})) + \tfrac{\partial \alpha}{\partial x}(x, \hat{\theta}) \left[f(x) + F(x)\hat{\theta}_s + g(x)\xi \right] \\
&\quad + \tfrac{\partial \alpha}{\partial \hat{\theta}}(x, \hat{\theta})T(x, \hat{\theta}) - \tfrac{\partial V}{\partial x}(x, \hat{\theta})g(x) \\
\dot{\hat{\theta}} &= T(x, \hat{\theta}) \\
\dot{\hat{\theta}}_s &= -\Gamma \left[\tfrac{\partial \alpha}{\partial x}(x, \hat{\theta})F(x) \right]^T (\xi - \alpha(x, \hat{\theta})),
\end{aligned}
\tag{11.44}
$$

for $c > 0$, which implies that

$$
\dot{V}_a(x, \xi, \hat{\theta}, \hat{\theta}_s) \leq -W(x, \hat{\theta}) - c(\xi - \alpha(x, \hat{\theta}))^2.
\tag{11.45}
$$

(Note that $\dot{\hat{\theta}} = T(x, \hat{\theta})$ has already been used in the derivation of $\dot{z}$, but is repeated here to summarize the dynamic controller.) Thus, with the application of Theorem 2.27, we have derived the following result.

Theorem 11.9 (Adaptive backstepping, [93, Lemma 3.2]). *Assume that the dynamic controller (11.38) globally asymptotically stabilizes the origin of the system (11.37) and assume that the functions V and W in (11.40) satisfy (11.41). Let $c > 0$ and $\Gamma \in \mathcal{S}_{>0}^q$.*

Then, the adaptive controller (11.44) guarantees global boundedness of $x(\cdot)$, $\xi(\cdot)$,

$\hat{\theta}(\cdot)$, $\hat{\theta}_s(\cdot)$. *Moreover, the function V_a defined in (11.43) satisfies (11.45) and thus,*

$$\lim_{t \to 0} W(x(t), \hat{\theta}(t)) = 0 \qquad and \qquad \lim_{t \to \infty} \xi(t) - \alpha(x(t), \hat{\theta}(t)) = 0.$$

Theorem 11.9 relies on the assumption that the functions α and T in (11.38) are known. Exercise 11.6 discusses a definition of α and T in the case that the state x, the unknown parameter θ, and the input u are one-dimensional.

Theorem 11.9 can be applied iteratively, similarly to the derivations in Chapter 9.4 on backstepping. For systems in *parametric strict-feedback form*,

$$\begin{aligned}
\dot{x}_1 &= x_2 + \phi_1(x_1)^T \theta \\
\dot{x}_2 &= x_3 + \phi_2(x_1, x_2)^T \theta \\
&\vdots \\
\dot{x}_{n-1} &= x_n + \phi_{n-1}(x_1, \ldots, x_{n-1})^T \theta \\
\dot{x}_n &= \beta(x)u + \phi_n(x)^T \theta,
\end{aligned} \tag{11.46}$$

where $\beta(x) \neq 0$ for all $x \in \mathbb{R}^n$, the following general result can be derived. Here $\theta \in \mathbb{R}^q$ again denotes the vector of unknown parameters and $\beta : \mathbb{R}^n \to \mathbb{R}^q$, $\phi_i : \mathbb{R}^i \to \mathbb{R}^q$, $i = 1, \ldots, n$, represent known functions. With the definitions

$$A = \begin{bmatrix} 0 & 1 & & \\ \vdots & \ddots & \ddots & \\ \vdots & & \ddots & 1 \\ 0 & \cdots & \cdots & 0 \end{bmatrix}, \quad F(x) = \begin{bmatrix} \phi_1(x_1)^T \\ \phi_2(x_1, x_2)^T \\ \vdots \\ \phi_n(x)^T \end{bmatrix}, \quad g(x) = \begin{bmatrix} 0 \\ 0 \\ \vdots \\ \beta(x) \end{bmatrix} \tag{11.47}$$

the dynamics can be compactly written as

$$\dot{x} = Ax + F(x)\theta + g(x)u.$$

Theorem 11.10 (Parametric Strict-Feedback Systems, [93, Theorem 3.5]). *Let $c_i > 0$ for $i \in \{1, \ldots, n\}$. For the system (11.46) with $\beta(x) \neq 0$ for all $x \in \mathbb{R}^n$, consider the adaptive controller*

$$u = \frac{1}{\beta(x)} \alpha_n(x, \hat{\theta}_1, \ldots, \hat{\theta}_n)$$

$$\dot{\hat{\theta}}_i = \Gamma \left(\phi_i(x_1, \ldots, x_i) - \sum_{j=1}^{i-1} \frac{\partial \alpha_{i-1}}{\partial x_j} \phi_j(x_1, \ldots x_j) \right) z_i, \quad i = 1, \ldots, n, \tag{11.48}$$

where $\hat{\theta}_i \in \mathbb{R}^q$ are multiple estimates of θ, $\Gamma \in \mathcal{S}^q_{>0}$ is the adaptation gain matrix, and the variables z_i and the stabilizing functions

$$\alpha_i = \alpha_i(x_1, \ldots, x_i, \hat{\theta}_1, \ldots, \hat{\theta}_i), \qquad \alpha_i : \mathbb{R}^{i+i \cdot q} \to \mathbb{R}, \qquad i = 1, \ldots, n,$$

are defined by the following recursive expressions (and $z_0 \equiv 0$, $\alpha_0 \equiv 0$ for notational *convenience):*

$$z_i = x_i - \alpha_{i-1}(x_1, \ldots, x_i, \hat{\theta}_1, \ldots, \hat{\theta}_i)$$

$$\alpha_i = -c_i z_i - z_{i-1} - \left(\phi_i - \sum_{j=1}^{i-1} \frac{\partial \alpha_{i-1}}{\partial x_j} \phi_j \right)^T \hat{\theta}_i$$

$$+ \sum_{j=1}^{i-1} \left(\frac{\partial \alpha_{i-1}}{\partial x_j} x_{j+1} + \frac{\partial \alpha_{i-1}}{\partial \hat{\theta}_j} \Gamma \left(\phi_j - \sum_{k=1}^{j-1} \frac{\partial \alpha_{j-1}}{\partial x_k} \phi_k \right) z_j \right).$$

This adaptive controller guarantees global boundedness of $x(\cdot)$, $\hat{\theta}_1(\cdot)$, ..., $\hat{\theta}_n(\cdot)$, and $x_1(t) \to 0$, $x_i(t) \to x_i^e$ for $i = 2, \ldots, n$ for $t \to \infty$, where

$$x_i^e = -\theta^T \phi_{i-1}(0, x_2^e, \ldots, x_{i-1}^e), \qquad i = 2, \ldots, n.$$

An example of the adaptive controller design in Theorem 11.10 is discussed in Section 11.3.3. The result in Theorem 11.10 requires n estimates of the unknown parameter vector $\theta \in \mathbb{R}^q$. This property is known as *overparameterization*. A controller requiring only one parameter estimate is derived in the next section.

11.3.2 Tuning Function Designs

Adaptive backstepping, discussed in the preceding section, led to the general adaptive controller design for systems in parametric strict feedback form summarized in Theorem 11.10. The corresponding dynamic controller in (11.48) consists of the input u and n copies of the estimate of the unknown parameter vector θ. In this section an adaptive backstepping controller design based on *tuning functions* is presented to reduce the need for multiple copies of the same parameter vector. In addition, we extend the previous design to address the reference tracking problem.

We consider again the dynamics (11.46) with output

$$y(t) = x_1(t) \tag{11.49}$$

and intend to design an adaptive reference tracking controller

$$y(t) \to y_r(t), \qquad t \to \infty, \tag{11.50}$$

where $y_r : \mathbb{R}_{\geq 0} \to \mathbb{R}$ denotes an arbitrary given reference signal. Additionally, we assume that $y_r(\cdot)$ is n times continuously differentiable and the derivatives $y_r^{(i)} = \frac{d^i}{dt^i} y_r : \mathbb{R}_{\geq 0} \to \mathbb{R}$ are known.

For the recursive derivation of the adaptive controller using tuning functions we use the notation

$$\tilde{x}_i = [x_1, \ldots, x_i]^T, \qquad \tilde{y}_i = [y_r, \ldots, y_r^{(i-1)}]^T, \qquad i = 1, \ldots, n. \tag{11.51}$$

Additionally, for $i = 1, \ldots, n$, we derive the following functions (initialized at $z_0 \equiv 0$, $\alpha_0 \equiv 0$, $\tau_0 \equiv 0$).

- Error dynamics:

$$z_i = x_i - y_r^{(i-1)} - \alpha_{i-1}. \tag{11.52}$$

- Regressor functions:

$$w_i(\tilde{x}_i, \hat{\theta}, \tilde{y}_i) = \phi_i - \sum_{k=1}^{i-1} \frac{\partial \alpha_{i-1}}{\partial x_k} \phi_k. \tag{11.53}$$

- Tuning functions:

$$\tau_i(\tilde{x}_i, \hat{\theta}, \tilde{y}_i) = \tau_{i-1} + w_i z_i. \tag{11.54}$$

- Virtual control laws:

$$\alpha_i(\tilde{x}_i, \hat{\theta}, \tilde{y}_i) = -z_{i-1} - c_i z_i - w_i^T \hat{\theta} + \sum_{k=1}^{i-1} \left(\frac{\partial \alpha_{i-1}}{\partial x_k} x_{k+1} + \frac{\partial \alpha_{i-1}}{\partial y_r^{(k-1)}} y_r^{(k)} \right)$$
$$- \kappa_i w_i^T w_i z_i + \frac{\partial \alpha_{i-1}}{\partial \hat{\theta}} \Gamma \tau_i + \sum_{k=2}^{i-1} \frac{\partial \alpha_{k-1}}{\partial \hat{\theta}} \Gamma w_i z_k, \tag{11.55}$$

for parameters $c_i \in \mathbb{R}_{>0}$, $\kappa_i \in \mathbb{R}_{\geq 0}$, and a positive definite matrix $\Gamma \in \mathcal{S}_{>0}^q$, and where $\hat{\theta} \in \mathbb{R}^q$ denotes the estimate of the unknown parameter vector θ.

- Lyapunov functions:

$$V_i(z_1, \ldots, z_i, \bar{\theta}) = V_{i-1}(z_1, \ldots, z_{i-1}, \bar{\theta}) + \tfrac{1}{2} z_i^2, \tag{11.56}$$

where $\bar{\theta} = \theta - \hat{\theta}$ and $V_0(\bar{\theta}) = \tfrac{1}{2} \bar{\theta}^T \Gamma^{-1} \bar{\theta}$.

Step 1: The first error dynamics and the first tuning function. To derive the first elements in (11.52)–(11.56), we introduce the error variables

$$z_1 = y - y_r = x_1 - y_r, \tag{11.57}$$
$$z_2 = x_2 - \dot{y}_r - \alpha_1, \tag{11.58}$$

with function α_1 to be designed and (11.57) motivated by (11.49) and (11.50). If we can guarantee that $z_1(t) \to 0$ for $t \to \infty$, tracking of the reference signal is achieved. The z_1-dynamics satisfies

$$\dot{z}_1 = \dot{y} - \dot{y}_r = \dot{x}_1 - \dot{y}_r = x_2 + \phi_1^T \theta - \dot{y}_r = z_2 + \alpha_1 + \phi_1^T \theta, \tag{11.59}$$

which follows from the system dynamics (11.46) and Equation (11.58).

With the definition of the first virtual control law

$$\alpha_1(x_1, \hat{\theta}, y_r) = -c_1 z_1 - \phi_1^T \hat{\theta} = -c_1(x_1 - y_r) - \phi_1^T \hat{\theta} \tag{11.60}$$

(see Equation (11.55)) for $c_1 > 0$, $\kappa_1 \geq 0$ and a parameter estimate $\hat{\theta} \in \mathbb{R}^q$, (11.59) reads

$$\dot{z}_1 = -c_1 z_1 + z_2 + \phi_1^T (\theta - \hat{\theta}) - \kappa_1 |\phi_1|_2^2 z_1. \tag{11.61}$$

In the case that the error $\bar{\theta} = \theta - \hat{\theta}$ is zero, Equation (11.61) reduces to $\dot{z}_1 = -c_1 z_1 - \kappa_1 |\phi_1|_2^2 z_1 + z_2$. Thus, to ensure that $\lim_{t \to 0} z_1(t) = 0$, we need to guarantee $\lim_{t \to 0} z_2(t) = 0$, which is addressed in the next step of the recursive controller design.

Before we move to the next step, consider the candidate Lyapunov function

$$V_1(z_1, \bar{\theta}) = \frac{1}{2} z_1^2 + \frac{1}{2} \bar{\theta}^T \Gamma^{-1} \bar{\theta}$$

(see Equation (11.55)) for a given positive definite matrix $\Gamma \in \mathcal{S}_{>0}^q$. Since $\dot{\bar{\theta}} = -\dot{\hat{\theta}}$, The derivative of V_1 satisfies

$$\dot{V}_1(z_1, \bar{\theta}) = z_1(-c_1 z_1 - \kappa_1 |\phi_1|_2^2 z_1 + z_2 + \phi_1^T \bar{\theta}) - \bar{\theta}^T \Gamma^{-1} \dot{\hat{\theta}}$$

$$= z_1(-c_1 z_1 - \kappa_1 |\phi_1|_2^2 z_1 + z_2) - \bar{\theta}^T (\Gamma^{-1} \dot{\hat{\theta}} - \phi_1 z_1).$$

We define the first regressor function

$$w_1(x_1, \hat{\theta}, y_r) = \phi_1$$

(see Equation (11.53)) and the first tuning function

$$\tau_1(x_1, \hat{\theta}, y_r) = \phi_1 z_1 = w_1 z_1$$

(see Equation (11.54)), which leads to

$$\dot{V}_1(z_1, \bar{\theta}) = -c_1 z_1^2 - \kappa_1 |w_1|_2^2 z_1^2 + z_1 z_2 - \bar{\theta}^T (\Gamma^{-1} \dot{\hat{\theta}} - \tau_1). \tag{11.62}$$

Through this definition, note that the update law for the parameter estimate $\dot{\hat{\theta}} = \Gamma \tau_1$, which we have not fixed yet, ensures that the term containing the unknown parameter θ in (11.62) disappears.

Step 2: The second error dynamics and the second tuning function. We continue with the second equation in (11.46), consider x_3 as the next virtual control variable, and introduce the error variable

$$z_3 = x_3 - \ddot{y}_r - \alpha_2 \tag{11.63}$$

for α_2 yet to be defined. Using this expression, the derivative of z_2 in (11.58) satisfies

$$\dot{z}_2 = \dot{x}_2 - \ddot{y}_r - \dot{\alpha}_1(x_1, \hat{\theta}, y_r) = x_3 + \phi_2^T \theta - \ddot{y}_r - \dot{\alpha}_1(x_1, \hat{\theta}, y_r)$$

$$= z_3 + \alpha_2 + \phi_2^T \theta - \dot{\alpha}_1(x_1, \hat{\theta}, y_r).$$

With $\alpha_1(x_1(t), \hat{\theta}(t), y_r(t))$ in (11.60) it holds that

$$\frac{d}{dt} \alpha_1(x_1, \hat{\theta}, y_r) = \frac{\partial \alpha_1}{\partial x_1} \dot{x}_1 + \frac{\partial \alpha_1}{\partial \hat{\theta}} \dot{\hat{\theta}} + \frac{\partial \alpha_1}{\partial y_r} \dot{y}_r$$

$$= \frac{\partial \alpha_1}{\partial x_1} (x_2 + \phi_1^T \theta) + \frac{\partial \alpha_1}{\partial y_r} \dot{y}_r + \frac{\partial \alpha_1}{\partial \hat{\theta}} \dot{\hat{\theta}},$$

where we have additionally used the system dynamics (11.46). Thus, the error dynamics $\dot{z}_2$ become

$$\dot{z}_2 = z_3 + \alpha_2 - \frac{\partial \alpha_1}{\partial x_1} x_2 + \left(\phi_2 - \frac{\partial \alpha_1}{\partial x_1} \phi_1 \right)^T \theta - \frac{\partial \alpha_1}{\partial y_r} \dot{y}_r - \frac{\partial \alpha_1}{\partial \hat{\theta}} \dot{\hat{\theta}}. \qquad (11.64)$$

As in Step 1, the function α_2 is selected in such a way that terms in the error dynamics z_2 are canceled and such that the origin of the error dynamics $[z_1, z_2]^T$ is globally asymptotically stable under the assumption that $\hat{\theta} = \theta$ and $z_3 = 0$.
We select α_2 as

$$\alpha_2(\tilde{x}_2, \hat{\theta}, \tilde{y}_2) = -z_1 - c_2 z_2 + \frac{\partial \alpha_1}{\partial x_1} x_2 - \left(\phi_2 - \frac{\partial \alpha_1}{\partial x_1} \phi_1 \right)^T \hat{\theta} + \frac{\partial \alpha_1}{\partial y_r} \dot{y}_r + \frac{\partial \alpha_1}{\partial \hat{\theta}} \Gamma \tau_2$$

$$- \kappa_2 \left| \phi_2 - \frac{\partial \alpha_1}{\partial x_1} \phi_1 \right|_2^2 z_2$$

(see Equation (11.55)). Here, $c_2 > 0$, $\kappa_2 \geq 0$ are again constants and τ_2 is the second tuning function, defined as

$$\tau_2(\tilde{x}_2, \hat{\theta}, \tilde{y}_2) = \tau_1 + \left(\phi_2 - \frac{\partial \alpha_1}{\partial x_1} \phi_1 \right) z_2 = \tau_1 + w_2 z_2$$

(see (11.54)), where we have additionally used the definition

$$w_2(\tilde{x}_2, \hat{\theta}, \tilde{y}_2) = \phi_2 - \frac{\partial \alpha_1}{\partial x_1} \phi_1$$

(see Equation (11.53)). With α_2 and τ_2, Equation (11.64) becomes

$$\dot{z}_2 = z_3 - z_1 - c_2 z_2 + w_2^T (\theta - \hat{\theta}) + \frac{\partial \alpha_1}{\partial \hat{\theta}} \left(\Gamma \tau_2 - \dot{\hat{\theta}} \right) - \kappa_2 |w_2|_2^2 z_2.$$

As in step 1, we define α_2 based on the candidate Lyapunov function

$$V_2(z_1, z_2, \bar{\theta}) = V_1(z_1, \bar{\theta}) + \frac{1}{2} z_2^2.$$

The derivative of V satisfies

$$\dot{V}_2(z_1, z_2, \bar{\theta}) = -c_1 z_1^2 - \kappa_1 |w_1|_2^2 z_1^2 + z_1 z_2 - \bar{\theta}^T \left(\Gamma^{-1} \dot{\hat{\theta}} - \tau_1 \right)$$

$$+ z_2 \left(-z_1 - c_2 z_2 + z_3 + w_2^T \bar{\theta} + \frac{\partial \alpha_1}{\partial \hat{\theta}} \left(\Gamma \tau_2 - \dot{\hat{\theta}} \right) - \kappa_2 |w_2|_2^2 z_2 \right)$$

$$= -c_1 z_1^2 - c_2 z_2^2 - \kappa_1 |w_1|_2^2 z_1^2 - \kappa_2 |w_2|_2^2 z_2^2 + z_2 z_3$$

$$+ \bar{\theta}^T \left(\tau_1 + w_2 z_2 - \Gamma^{-1} \dot{\hat{\theta}} \right) + z_2 \frac{\partial \alpha_1}{\partial \hat{\theta}} \left(\Gamma \tau_2 - \dot{\hat{\theta}} \right)$$

$$= -c_1 z_1^2 - c_2 z_2^2 - \kappa_1 |w_1|_2^2 z_1^2 - \kappa_2 |w_2|_2^2 z_2^2 + z_2 z_3$$

$$+ \bar{\theta}^T \left(\tau_2 - \Gamma^{-1} \dot{\hat{\theta}} \right) + z_2 \frac{\partial \alpha_1}{\partial \hat{\theta}} \left(\Gamma \tau_2 - \dot{\hat{\theta}} \right).$$

Here, if the parameter update law is defined as

$$\dot{\hat{\theta}} = \Gamma\tau_2,$$

and if $z_3 = 0$, then the function V_2 satisfies $\dot{V}_2 = -c_1 z_1^2 - c_2 z_2^2 - \kappa_1|w_1|_2^2 z_1^2 - \kappa_2|w_2|_2^2 z_2^2$. Since in general $z_3 \neq 0$, the error dynamics z_3 are addressed in the next step.

Step 3: The third error dynamics and the third tuning function. We perform one last step of the recursive controller design. Alternatively, Step 3 can be written as Step i to obtain the more general form.

We continue with the third equation of the dynamical system (11.46) and introduce the error dynamics

$$z_4 = x_4 - y_r^{(3)} - \alpha_3, \tag{11.65}$$

and note that the derivative of α_2 satisfies

$$\dot{\alpha}_2 = \frac{\partial\alpha_2}{\partial x_1}\dot{x}_1 + \frac{\partial\alpha_2}{\partial x_2}\dot{x}_2 + \frac{\partial\alpha_2}{\partial\hat{\theta}}\dot{\hat{\theta}} + \frac{\partial\alpha_2}{\partial y_r}\dot{y}_r + \frac{\partial\alpha_2}{\partial\dot{y}_r}\ddot{y}_r$$

$$= \frac{\partial\alpha_2}{\partial x_1}x_2 + \frac{\partial\alpha_2}{\partial x_2}x_3 + \left(\frac{\partial\alpha_2}{\partial x_1}\phi_1 + \frac{\partial\alpha_2}{\partial x_2}\phi_2\right)^T\theta + \frac{\partial\alpha_2}{\partial\hat{\theta}}\dot{\hat{\theta}} + \frac{\partial\alpha_2}{\partial y_r}\dot{y}_r + \frac{\partial\alpha_2}{\partial\dot{y}_r}\ddot{y}_r.$$

Thus, the derivative of z_3 in (11.63) satisfies

$$\dot{z}_3 = \dot{x}_3 - y_r^{(3)} - \dot{\alpha}_2$$

$$= x_4 + \phi_3^T\theta - y_r^{(3)} - \dot{\alpha}_2$$

$$= z_4 + \alpha_3 + \phi_3^T\theta - \dot{\alpha}_2$$

$$= z_4 + \alpha_3 - \frac{\partial\alpha_2}{\partial x_1}x_2 - \frac{\partial\alpha_2}{\partial x_2}x_3 + \left(\phi_3 - \frac{\partial\alpha_2}{\partial x_1}\phi_1 - \frac{\partial\alpha_2}{\partial x_2}\phi_2\right)^T\theta$$

$$\quad - \frac{\partial\alpha_2}{\partial y_r}\dot{y}_r - \frac{\partial\alpha_2}{\partial\dot{y}_r}\ddot{y}_r - \frac{\partial\alpha_2}{\partial\hat{\theta}}\dot{\hat{\theta}}$$

$$= z_4 + \alpha_3 - \frac{\partial\alpha_2}{\partial x_1}x_2 - \frac{\partial\alpha_2}{\partial x_2}x_3 + w_3^T\theta - \frac{\partial\alpha_2}{\partial y_r}\dot{y}_r - \frac{\partial\alpha_2}{\partial\dot{y}_r}\ddot{y}_r - \frac{\partial\alpha_2}{\partial\hat{\theta}}\dot{\hat{\theta}},$$

where the last equality follows from the definition of

$$w_3 = \phi_3 - \frac{\partial\alpha_2}{\partial x_1}\phi_1 - \frac{\partial\alpha_2}{\partial x_2}\phi_2$$

(see Equation (11.53)). We continue as before, fix $c_3 > 0$, $\kappa_3 \geq 0$, and define α_3 and the third tuning function τ_3 as

$$\alpha_3(\tilde{x}_3, \hat{\theta}, \tilde{y}_3) = -z_2 - c_3 z_3 + \frac{\partial\alpha_2}{\partial x_1}x_2 + \frac{\partial\alpha_2}{\partial x_2}x_3 - w_3^T\hat{\theta}$$

$$\quad + \frac{\partial\alpha_2}{\partial y_r}\dot{y}_r + \frac{\partial\alpha_2}{\partial\dot{y}_r}\ddot{y}_r + \frac{\partial\alpha_2}{\partial\hat{\theta}}\Gamma\tau_3 + z_2\frac{\partial\alpha_1}{\partial\hat{\theta}}\Gamma w_3 - \kappa_3|w_3|_2^2 z_3$$

$$\tau_3 = \tau_2 + w_3 z_3$$

(see Equation (11.54)). With this definition $\dot{z}_3$ reduces to

$$\dot{z}_3 = z_4 - z_2 - c_3 z_3 + w_3^T(\hat{\theta} - \theta) + \frac{\partial \alpha_2}{\partial \hat{\theta}}\left(\Gamma \tau_3 - \dot{\hat{\theta}}\right) + z_2 \frac{\partial \alpha_1}{\partial \hat{\theta}}\Gamma w_3 - \kappa_3 |w_3|_2^2 z_3.$$

Define the candidate Lyapunov function

$$V_3(z_1, z_2, z_3, \bar{\theta}) = V_2(z_1, z_2, \bar{\theta}) + \frac{1}{2}z_3^2,$$

which satisfies

$$\dot{V}_3 = -c_1 z_1^2 - c_2 z_2^2 - c_3 z_3^2 - \kappa_1 |w_1|_2^2 z_1^2 - \kappa_2 |w_2|_2^2 z_2^2 - \kappa_3 |w_3|_2^2 z_3^2 + z_3 z_4$$
$$+ \bar{\theta}^T\left(\tau_2 - \Gamma^{-1}\dot{\hat{\theta}}\right) + z_2\frac{\partial \alpha_1}{\partial \hat{\theta}}\left(\Gamma \tau_2 - \dot{\hat{\theta}}\right)$$
$$+ z_3 w_3^T\bar{\theta} + z_3\frac{\partial \alpha_2}{\partial \hat{\theta}}\left(\Gamma \tau_3 - \dot{\hat{\theta}}\right) + z_3 z_2 \frac{\partial \alpha_1}{\partial \hat{\theta}}\Gamma w_3.$$

With $\tau_3 = \tau_2 + w_3 z_3$, or equivalently $\tau_2 = \tau_3 - w_3 z_3$, this expression simplifies to

$$\dot{V}_3 = -c_1 z_1^2 - c_2 z_2^2 - c_3 z_3^2 - \kappa_1 |w_1|_2^2 z_1^2 - \kappa_2 |w_2|_2^2 z_2^2 - \kappa_3 |w_3|_2^2 z_3^2 + z_3 z_4$$
$$- \bar{\theta}^T w_3 z_3 + \bar{\theta}^T\left(\tau_3 - \Gamma^{-1}\dot{\hat{\theta}}\right) - z_2 \frac{\partial \alpha_1}{\partial \hat{\theta}}\Gamma w_3 z_3$$
$$+ z_2\frac{\partial \alpha_1}{\partial \hat{\theta}}\left(\Gamma \tau_3 - \dot{\hat{\theta}}\right) + z_3 w_3^T\bar{\theta} + z_3\frac{\partial \alpha_2}{\partial \hat{\theta}}\left(\Gamma \tau_3 - \dot{\hat{\theta}}\right) + z_3 z_2 \frac{\partial \alpha_1}{\partial \hat{\theta}}\Gamma w_3$$
$$= \sum_{i=1}^{3} -c_1 z_1^2 - \kappa_i |w_i|_2^2 z_i^2 + z_3 z_4 + \bar{\theta}^T\left(\tau_3 - \Gamma^{-1}\dot{\hat{\theta}}\right)$$
$$+ z_2 \frac{\partial \alpha_1}{\partial \hat{\theta}}\left(\Gamma \tau_3 - \dot{\hat{\theta}}\right) + z_3 \frac{\partial \alpha_2}{\partial \hat{\theta}}\left(\Gamma \tau_3 - \dot{\hat{\theta}}\right)$$
$$= \sum_{i=1}^{3} -c_i z_i^2 + z_3 z_4 + \bar{\theta}^T\left(\tau_3 - \Gamma^{-1}\dot{\hat{\theta}}\right) + \left(z_2\frac{\partial \alpha_1}{\partial \hat{\theta}} + z_3\frac{\partial \alpha_2}{\partial \hat{\theta}}\right)\left(\Gamma \tau_3 - \dot{\hat{\theta}}\right).$$

Again, $\dot{V}_3$ reduces to $\dot{V}_3 = -\sum_{i=1}^{3} c_1 z_1^2 + \kappa_i |w_i|_2^2 z_i^2$ if $z_4 = 0$ and if $\dot{\hat{\theta}}$ is defined as $\dot{\hat{\theta}} = \Gamma \tau_3$.

Step 3 can be performed iteratively until the nth equation in (11.46) is reached. We complete the controller design by selecting the input u.

Final step: Definition of u. As a last step of the recursive controller design, we focus on the nth equation in (11.46) and, following the derivations in Step 3, we assume that $n = 4$. The general case follows the same ideas.

We take the derivative of the error dynamics (11.65), which satisfies

$$\dot{z}_4 = \dot{x}_4 - y_r^{(4)} - \dot{\alpha}_3$$
$$= \beta(x)u + \phi_4(x)^T\theta - y_r^{(4)} - \sum_{k=1}^{3}\frac{\partial \alpha_3}{\partial x_i}\dot{x}_i - \frac{\partial \alpha_3}{\partial \hat{\theta}}\dot{\hat{\theta}} - \sum_{k=1}^{3}\frac{\partial \alpha_3}{\partial y_r^{(i-1)}}y_r^{(i)}$$
$$= \beta u + \phi_4^T\theta - y_r^{(4)} - \sum_{k=1}^{3}\frac{\partial \alpha_3}{\partial x_i}\phi_i^T\theta - \frac{\partial \alpha_3}{\partial \hat{\theta}}\dot{\hat{\theta}} - \sum_{k=1}^{3}\left(\frac{\partial \alpha_3}{\partial x_i}x_{i+1} + \frac{\partial \alpha_3}{\partial y_r^{(i-1)}}y_r^{(i)}\right).$$

As before, we define the last virtual controller

$$\alpha_4 = -z_3 - c_4 z_4 - w_4^T \hat{\theta} + \sum_{k=1}^{3} \left(\frac{\partial \alpha_3}{\partial x_k} x_{k+1} + \frac{\partial \alpha_3}{\partial y_r^{(k-1)}} y_r^{(k)} \right)$$

$$+ \frac{\partial \alpha_3}{\partial \hat{\theta}} \Gamma \tau_4 + \sum_{k=2}^{3} \frac{\partial \alpha_{k-1}}{\partial \hat{\theta}} \Gamma w_4 z_k - \kappa_4 |w_4|_2^2 z_4,$$

where $c_4 > 0$, $\kappa_4 \geq 0$, and

$$w_4 = \phi_4 - \sum_{j=1}^{3} \frac{\partial \alpha_3}{\partial x_j} \phi_j, \qquad \tau_4 = \tau_3 + w_4 z_4.$$

Finally, under the assumption that $\beta(x) \neq 0$ for all $x \in \mathbb{R}^n$, the input u is defined as

$$u = \frac{1}{\beta(x)} \left(\alpha_4(\tilde{x}_4, \hat{\theta}, \tilde{y}_4) + y_r^{(4)} \right).$$

Combining these definitions in $\dot{z}_4$, the error dynamics satisfy

$$\dot{z}_4 = -z_3 - c_4 z_4 + w_4^T \bar{\theta} + \frac{\partial \alpha_3}{\partial \hat{\theta}} \left(\Gamma \tau_4 - \dot{\hat{\theta}} \right) + \sum_{k=2}^{3} \frac{\partial \alpha_{k-1}}{\partial \hat{\theta}} \Gamma w_4 z_k - \kappa_4 |w_4|_2^2 z_4.$$

With the Lyapunov function

$$V_4(z_1, z_2, z_3, z_4, \bar{\theta}) = V_4(z_1, z_2, z_3, \bar{\theta}) + \frac{1}{2} z_4^2,$$

the update of the parameter estimate $\dot{\hat{\theta}} = \Gamma \tau_4$ and $\tau_3 = \tau_4 - w_4 z_4$, the decrease condition

$$\dot{V}_4 = \dot{V}_3 + z_4 \dot{z}_4$$

$$= \sum_{i=1}^{3} -c_i z_i^2 - \kappa_i |w_i|_2^2 z_i^2 + z_3 z_4 + \bar{\theta}^T \left(\tau_3 - \Gamma^{-1} \dot{\hat{\theta}} \right) + \left(z_2 \frac{\partial \alpha_1}{\partial \hat{\theta}} + z_3 \frac{\partial \alpha_2}{\partial \hat{\theta}} \right) \left(\Gamma \tau_3 - \dot{\hat{\theta}} \right)$$

$$+ z_4 \left(-z_3 - c_4 z_4 + w_4^T \bar{\theta} + \frac{\partial \alpha_3}{\partial \hat{\theta}} \left(\Gamma \tau_4 - \dot{\hat{\theta}} \right) + \sum_{k=2}^{3} \frac{\partial \alpha_{k-1}}{\partial \hat{\theta}} \Gamma w_4 z_k - \kappa_4 |w_4|_2^2 z_4 \right)$$

$$= \sum_{i=1}^{4} -c_i z_i^2 - \kappa_i |w_i|_2^2 z_i^2 + \bar{\theta}^T \left(-w_4 z_4 \right) + \left(z_2 \frac{\partial \alpha_1}{\partial \hat{\theta}} + z_3 \frac{\partial \alpha_2}{\partial \hat{\theta}} \right) \left(-\Gamma w_4 z_4 \right)$$

$$+ z_4 w_4^T \bar{\theta} + z_4 \sum_{k=2}^{3} \frac{\partial \alpha_{k-1}}{\partial \hat{\theta}} \Gamma w_4 z_k$$

$$= \sum_{i=1}^{4} -c_i z_i^2 - \kappa_i |w_i|_2^2 z_i^2 \leq 0$$

is satisfied, which allows the application of the LaSalle-Yoshizawa Theorem 2.27. We have thus sketched the proof of the following result.

Theorem 11.11 ([93, Theorem 4.14]). *Let $\Gamma \in \mathcal{S}_{>0}^q$ and $c_i \in \mathbb{R}_{>0}$ for $i = 1, \ldots, n$. Consider the plant dynamics (11.46) with $\beta(x) \neq 0$ for all $x \in \mathbb{R}^n$, the notations and definitions in (11.51)–(11.56), and the controller*

$$u = \frac{1}{\beta(x)} \left(\alpha_n(x, \hat{\theta}, \tilde{y}_n) + y_r^{(n)} \right) \tag{11.66a}$$

with corresponding parameter update

$$\dot{\hat{\theta}} = \Gamma \tau_n(x, \hat{\theta}, \tilde{y}_n). \tag{11.66b}$$

Then the closed-loop system (11.46), (11.11) satisfies the following properties.

- *The augmented error dynamics $[z^T, \bar{\theta}^T]^T$ has a globally uniformly stable equilibrium at the origin $z^e = 0$, $\bar{\theta}^e = 0$.*
- *The error dynamics satisfies $\lim_{t \to \infty} z(t) = 0$.*
- *The Lyapunov function $V(z, \bar{\theta}) = \frac{1}{2} z^T z + \frac{1}{2} \bar{\theta}^T \Gamma^{-1} \bar{\theta}$ satisfies*

$$\dot{V}(z, \bar{\theta}) \leq -c_0 z^T z, \qquad c_0 = \min\{c_1, \ldots, c_n\}. \tag{11.67}$$

- *Global asymptotic tracking*

$$\lim_{t \to \infty} (y(t) - y_r(t)) = 0$$

is achieved.
- *If $\lim_{t \to \infty} y_r^{(i)}(t) = 0$, $i = 0, 1, \ldots, n-1$, and $F(0) = 0$ where F is defined in (11.47), then $\lim_{t \to \infty} x(t) = 0$.*

As can be seen from (11.67) and from the derivations leading to Theorem 11.11, the terms $-\kappa_i |w_i|_2^2 z_i$, $i \in \{1, \ldots, n\}$, with $\kappa_i \geq 0$ in the controller are not necessary to show the assertions in Theorem 11.11. However, $-\kappa_i |w_i|_2^2 z_i$ with $\kappa_i > 0$ adds a nonlinear damping term to the controller, as in the discussion in Section 9.2 on improving the robustness properties of the overall closed-loop dynamics.

Note that the role of the term $\dfrac{\partial \alpha_{i-1}}{\partial \hat{\theta}} \Gamma \tau_i$ in (11.55) is to exactly cancel the tuning function τ_i in the z_i dynamics. An alternative virtual control law that dominates, rather than cancels, the tuning function at each recursive step, and which consequently leads to lower order nonlinearities, can be found in [93, Remark 4.10].

The results so far do not allow for unknown parameters in the linear part of the dynamics (11.46) or in controller gain, i.e., the case

$$\dot{x}_i = b_i x_{i+1} + \phi_i(x_1, \ldots, x_i)^T \theta, \qquad i = 1, \ldots, n-1,$$
$$\dot{x}_n = b_n \beta(x) u + \phi_n(x)^T \theta \tag{11.68}$$

with unknown parameters $b_1, \ldots, b_n$ is not covered. Nevertheless, extensions exist under the assumption that the signs of the parameters b_i, $i \in \{1, \ldots, n\}$ are known [93, Section 4.5.1]. We conclude this section by reporting one particular extension where only the parameter b_n is present in (11.68).

Lemma 11.12 ([93, Section 4.5.1]). *Let the assumptions of Theorem 11.11 be satisfied and assume that the last equation in the system dynamics (11.46) is replaced* *by*

$$\dot{x}_n = b\beta(x)u + \phi_n(x_1, \ldots, x_n)^T\theta \tag{11.69}$$

with unknown parameter $b \in \mathbb{R}\backslash\{0\}$ so that only the sign of b is known. Let

$$u = \frac{\hat{\varrho}}{\beta(x)}\left(\alpha_n(x, \hat{\theta}, \tilde{y}_n) + y_r^{(n)}\right)$$

$$\dot{\hat{\varrho}} = -\gamma\,\mathrm{sign}(b)(\alpha_n(x, \hat{\theta}, \tilde{y}_n) + y_r^{(n)})z_n$$

for a positive constant $\gamma > 0$ and

$$\dot{\hat{\theta}} = \Gamma\tau_n(x, \hat{\theta}, \tilde{y}_n).$$

Then $\hat{\varrho}$ is an estimate of $\frac{1}{b}$ and the assertions of Theorem 11.11 are satisfied with the Lyapunov function being replaced by $V(z, \bar{\theta}, \frac{1}{b} - \hat{\varrho}) = \frac{1}{2}z^Tz + \frac{1}{2}\bar{\theta}^T\Gamma^{-1}\bar{\theta} + \frac{|b|}{2\gamma}(\frac{1}{b} - \hat{\varrho})^2.$

11.3.3 Application: Single Link Manipulator with Flexible Joint

We conclude this chapter by discussing the second-order differential equation

$$\begin{aligned} I\ddot{q}_1 + MgL\sin(q_1) + k(q_1 - q_2) &= 0 \\ J\ddot{q}_2 - k(q_1 - q_2) &= u, \end{aligned} \tag{11.70}$$

which models a single link manipulator with flexible joint (see [139], [105], for example). Here $M, g, L, k, I, J > 0$ denote positive unknown constants. In particular, k specifies the stiffness of the spring and I and J denote the moment of inertia of the link and the actuator, respectively. The physical system is shown in Figure 11.5.

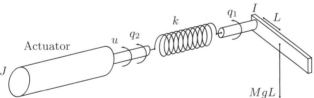

Figure 11.5: Single link manipulator with a flexible joint.

Through the coordinate transformation $x = [q_1, \dot{q}_1, q_2, \dot{q}_2]^T$ the second-order system (11.70) can be written in terms of the first-order system

$$\dot{x} = \begin{bmatrix} x_2 \\ -\frac{MgL}{I}\sin(x_1) - \frac{k}{I}(x_1 - x_3) \\ x_4 \\ \frac{k}{J}(x_1 - x_3) \end{bmatrix} + \begin{bmatrix} 0 \\ 0 \\ 0 \\ \frac{1}{J} \end{bmatrix}u. \tag{11.71}$$

The dynamics (11.71) are not of the form (11.46) yet. We thus consider the coordinate transformation

$$
\begin{bmatrix} \chi_1 \\ \chi_2 \\ \chi_3 \\ \chi_4 \end{bmatrix} = \begin{bmatrix} x_1 \\ x_2 \\ \frac{k}{I} x_3 \\ \frac{k}{I} x_4 \end{bmatrix} \qquad \Longleftrightarrow \qquad \begin{bmatrix} x_1 \\ x_2 \\ x_3 \\ x_4 \end{bmatrix} = \begin{bmatrix} \chi_1 \\ \chi_2 \\ \frac{I}{k} \chi_3 \\ \frac{I}{k} \chi_4 \end{bmatrix}.
$$

In the χ-coordinates, the dynamics (11.71) satisfy

$$
\begin{bmatrix} \dot{\chi}_1 \\ \dot{\chi}_2 \\ \frac{I}{k}\dot{\chi}_3 \\ \frac{I}{k}\dot{\chi}_4 \end{bmatrix} = \begin{bmatrix} \chi_2 \\ -\frac{MgL}{I}\sin(\chi_1) - \frac{k}{I}(\chi_1 - \frac{I}{k}\chi_3) \\ \frac{I}{k}\chi_4 \\ \frac{k}{J}(\chi_1 - \frac{I}{k}\chi_3) \end{bmatrix} + \begin{bmatrix} 0 \\ 0 \\ 0 \\ \frac{1}{J} \end{bmatrix} u,
$$

i.e., the dynamics of the form (11.46) are obtained:

$$
\begin{bmatrix} \dot{\chi}_1 \\ \dot{\chi}_2 \\ \dot{\chi}_3 \\ \dot{\chi}_4 \end{bmatrix} = \begin{bmatrix} \chi_2 \\ \chi_3 \\ \chi_4 \\ \frac{k}{JI} u \end{bmatrix} + \begin{bmatrix} 0 \\ -\frac{MgL}{I}\sin(\chi_1) - \frac{k}{I}\chi_1 \\ 0 \\ \frac{k^2}{JI}\chi_1 - \frac{k}{J}\chi_3 \end{bmatrix}
$$

$$
= \begin{bmatrix} \chi_2 \\ \chi_3 \\ \chi_4 \\ \frac{k}{JI} u \end{bmatrix} + \begin{bmatrix} 0 & 0 & 0 & 0 \\ -\sin(\chi_1) & -\chi_1 & 0 & 0 \\ 0 & 0 & 0 & 0 \\ 0 & 0 & \chi_1 & -\chi_3 \end{bmatrix} \begin{bmatrix} \frac{MgL}{I} \\ \frac{k}{I} \\ \frac{k^2}{JI} \\ \frac{k}{J} \end{bmatrix}.
$$

$$(11.72)$$

Accordingly,

$$
\phi_1(\chi_1) = \begin{bmatrix} 0 \\ 0 \\ 0 \\ 0 \end{bmatrix}, \qquad \phi_2(\chi_1, \chi_2) = \begin{bmatrix} -\sin(\chi_1) \\ -\chi_1 \\ 0 \\ 0 \end{bmatrix},
$$

$$
\phi_3(\chi_1, \ldots, \chi_3) = \begin{bmatrix} 0 \\ 0 \\ 0 \\ 0 \end{bmatrix}, \qquad \phi_4(\chi) = \begin{bmatrix} 0 \\ 0 \\ \chi_1 \\ -\chi_3 \end{bmatrix},
$$

and the vector of unknowns is given by

$$
\theta = \begin{bmatrix} \frac{MgL}{I} & \frac{k}{I} & \frac{k^2}{JI} & \frac{k}{J} \end{bmatrix}^T, \qquad b = \frac{k}{JI},
$$

where b corresponds to the representation in (11.69). For the simulations we use the unknown parameters $M = 1$, $g = 9.81$, $l = 1$, $k = 5$, $I = 1$, and $J = 0.2$, and thus θ and b are defined as

$$
\theta = \begin{bmatrix} 9.81 & 5 & 125 & 25 \end{bmatrix}^T, \qquad b = 25.
$$

Figure 11.6 shows the uncontrolled dynamics (11.71), i.e., $u(t) = 0$, initialized at

$$
\chi_0 = \begin{bmatrix} 1 & 0 & 0 & 0 \end{bmatrix}^T.
$$

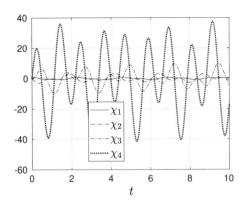

Figure 11.6: Solution of the uncontrolled dynamics (11.71), i.e., $u(t) = 0$, initialized at $\chi_0 = [1, 0, 0, 0]^T$.

In the remainder of this section, we investigate the controller designs in Theorem 11.10, Theorem 11.11, and Lemma 11.12.

Remark 11.13. Note that the controllers in Theorem 11.10, Theorem 11.11, and Lemma 11.12 rely on the knowledge of the states χ in (11.72) in the controller design. However, for the system (11.71) at best the states x can be measured. We ignore this fact here and focus on the numerical calculations of the control laws instead, i.e., we assume that χ is known.

For the application of Theorem 11.10, we first assume that b is known, i.e., $\beta(x) = b$. We select the parameters $c_1 = c_2 = c_3 = c_4 = 2$ and Γ is defined as the identity matrix $\Gamma = I$. The closed-loop solution together with the input are visualized in Figure 11.7. Here, the system dynamics and the controller dynamics

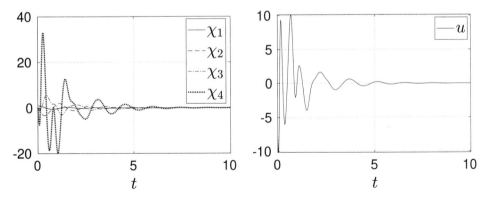

Figure 11.7: Closed-loop solution and feedback law of the controller design in Theorem 11.10 initialized at $\chi(0) = [1, 0, 0, 0]^T$ and the unknown parameters through zero.

are initialized at

$$\chi(0) = \begin{bmatrix} 1 \\ 0 \\ 0 \\ 0 \end{bmatrix}, \qquad [\hat{\theta}_1(0), \hat{\theta}_2(0), \hat{\theta}_3(0), \hat{\theta}_4(0)] = \begin{bmatrix} 0 & 0 & 0 & 0 \\ 0 & 0 & 0 & 0 \\ 0 & 0 & 0 & 0 \\ 0 & 0 & 0 & 0 \end{bmatrix},$$

respectively. The evolution of the parameter estimates is shown in Figure 11.8. The dynamics of the parameter estimates are given by

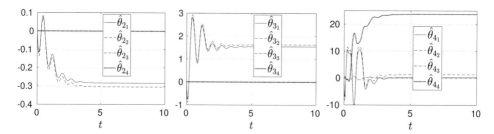

Figure 11.8: Evolution of the parameter estimates corresponding to the solution in Figure 11.7. The parameter estimates are converging but not to the correct parameter values. The parameter estimates $\hat{\theta}_{1_i}$, $i = 1, \ldots, 4$, satisfy $\hat{\theta}_{1_i}(t) = 0$ for all $t \in [0, 10]$ and are thus omitted.

$$\left[\dot{\hat{\theta}}_1, \dot{\hat{\theta}}_2, \dot{\hat{\theta}}_3, \dot{\hat{\theta}}_4\right] = \begin{bmatrix} 0 & -\sin(\chi_1)(2\chi_1 + \chi_2) & -4\sin(\chi_1)\sigma_3 & \sin(\chi_1)\sigma_2\sigma_1 \\ 0 & -\chi_1(2\chi_1 + \chi_2) & -4\chi_1\sigma_3 & \chi_1\sigma_2\sigma_1 \\ 0 & 0 & 0 & \chi_1\sigma_1 \\ 0 & 0 & 0 & -\chi_3\sigma_1 \end{bmatrix},$$

where

$$\sigma_1 = 12\chi_1 + 9\chi_2 + 6\chi_3 + \chi_4 - 2\hat{\theta}_{2_2}\chi_1 - 4\hat{\theta}_{3_2}\chi_1 - \chi_2\left(\hat{\theta}_{2_2} + \hat{\theta}_{2_1}\cos(\chi_1) - 5\right)$$
$$+ \left(\chi_1^2 + \sin^2(\chi_1)\right)(2\chi_1 + \chi_2) - 2\hat{\theta}_{2_1}\sin(\chi_1) - 4\hat{\theta}_{3_1}\sin(\chi_1)$$
$$\sigma_2 = -\chi_1^2 + \cos^2(\chi_1) + \hat{\theta}_{2_1}\cos(\chi_1) + \hat{\theta}_{2_2} - 15$$
$$\sigma_3 = 5\chi_1 + 4\chi_2 + \chi_3 - \hat{\theta}_{2_2}\chi_1 - \hat{\theta}_{2_1}\sin(\chi_1).$$

The control law u in (11.48) is defined as

$$
\begin{aligned}
u = \tfrac{1}{25}\Big(& 44\chi_1\cos\left(2\chi_1\right) - 79\chi_2 - \tfrac{71\chi_3}{2} - 8\chi_4 - 73\chi_1 + 35\chi_2\cos\left(2\chi_1\right) \\
& + \tfrac{17\chi_3\cos(2\chi_1)}{2} - 4\hat{\theta}_{2_1}\sin\left(3\chi_1\right) - \tfrac{\hat{\theta}_{4_1}\sin(3\chi_1)}{4} + 13\hat{\theta}_{2_2}\chi_1 + 4\hat{\theta}_{2_2}\chi_2 + 8\hat{\theta}_{3_2}\chi_1 \\
& + \hat{\theta}_{2_2}\chi_3 + 4\hat{\theta}_{3_2}\chi_2 + \tfrac{29\hat{\theta}_{4_2}\chi_1}{2} - \hat{\theta}_{4_3}\chi_1 + \hat{\theta}_{4_4}\chi_3 - \tfrac{3\chi_2^2\sin(2\chi_1)}{2} + 16\hat{\theta}_{2_2}\chi_1^3 \\
& + \hat{\theta}_{4_2}\chi_1^3 - 3\chi_1\chi_2^2 - 76\chi_1^2\chi_2 - 17\chi_1^2\chi_3 + 17\hat{\theta}_{2_1}\sin\left(\chi_1\right) + 8\hat{\theta}_{3_1}\sin\left(\chi_1\right) \\
& + \tfrac{59\hat{\theta}_{4_1}\sin(\chi_1)}{4} - 88\chi_1^3 + 4\hat{\theta}_{2_1}\chi_2\cos\left(\chi_1\right) + \hat{\theta}_{2_1}\chi_3\cos\left(\chi_1\right) + 4\hat{\theta}_{3_1}\chi_2\cos\left(\chi_1\right) \\
& - \hat{\theta}_{4_1}\hat{\theta}_{2_2}\sin\left(\chi_1\right) - 8\hat{\theta}_{2_2}\chi_1\cos\left(2\chi_1\right) - \tfrac{\hat{\theta}_{4_2}\chi_1\cos(2\chi_1)}{2} - \tfrac{\hat{\theta}_{2_1}\hat{\theta}_{4_1}\sin\left(2\chi_1\right)}{2} \\
& + 16\hat{\theta}_{2_1}\chi_1^2\sin\left(\chi_1\right) - \hat{\theta}_{2_2}\chi_2^2\sin\left(\chi_1\right) + \hat{\theta}_{4_1}\chi_1^2\sin\left(\chi_1\right) - 3\chi_1\chi_2\sin\left(2\chi_1\right) \\
& - \hat{\theta}_{2_2}\hat{\theta}_{4_2}\chi_1 - \hat{\theta}_{2_1}\hat{\theta}_{4_2}\chi_1\cos\left(\chi_1\right)\Big)
\end{aligned}
$$

and is obtained through symbolic calculations in MATLAB. The dynamics of the
parameter estimate and of the controller quickly become complicated and almost
impossible to calculate by hand. The same statement is true for the controller
design relying on tuning functions.

For the controller design relying on tuning functions outlined in Theorem 11.11,
we select the parameters $c_1 = c_2 = c_3 = c_4 = 2$, $\kappa_1 = \kappa_2 = \kappa_3 = \kappa_4 = 0.1$, and
$\Gamma = 0.5I$. As a reference signal, the constant function $y_r(t) = 1$ is used. The
closed-loop solution initialized at

$$
\chi(0) = \begin{bmatrix} 0 & 0 & 0 & 0 \end{bmatrix}^T, \qquad \hat{\theta}(0) = \begin{bmatrix} 0 & 0 & 0 & 0 \end{bmatrix}^T,
$$

and the feedback law defined in Theorem 11.11, are shown in Figure 11.9. Addi-

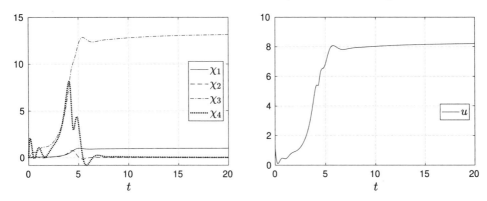

Figure 11.9: Closed-loop solution and control input defined in Theorem 11.11 initialized at the origin. The corresponding evolution of the parameter estimates is
shown in Figure 11.10.

tionally, the evolution of the parameter estimates and the decrease of the Lyapunov
function are shown in Figure 11.10.

Since the expressions of the control law as well the parameter update law are
even more complicated than in the adaptive backstepping case, the control law using
tuning functions is not reported here, but was again obtained through symbolic

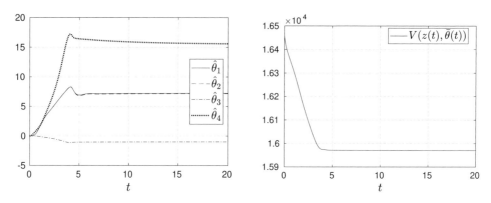

Figure 11.10: Evolution of the parameter estimates (left) and the decrease of the Lyapunov function (right) corresponding to the solution visualized in Figure 11.9.

calculations in MATLAB.

As a last example we additionally assume that the parameter b is unknown, i.e., we consider (11.69) with unknown b and $\beta(x) = 1$. For the controller design in Lemma 11.12 we select the parameters $c_1 = c_2 = c_3 = c_4 = 2$, $\kappa_1 = \kappa_2 = \kappa_3 = \kappa_4 = 0.1$, $\gamma = 0.5$, and $\Gamma = 0.5I$ and we intend to track the reference signal $y_r(t) = 0.5\sin(t)$. The closed-loop solution of (11.72) and the evolution of the dynamic controller are shown in Figure 11.11 and Figure 11.12. The input u and

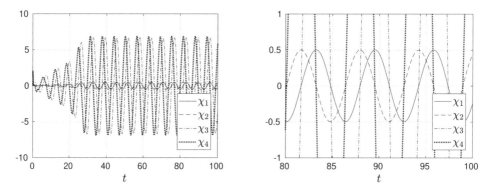

Figure 11.11: Closed-loop solution of the dynamical system (11.11), where $\chi_1(t) = y(t)$ is tracking the reference signal $y_r(t) = 0.5\sin(t)$.

the decrease of the Lyapunov function are shown in Figure 11.13.

The solution visualized in Figure 11.11 is initialized at

$$\chi(0) = \begin{bmatrix} 0 & 0 & 0 & 0 \end{bmatrix}, \qquad \hat{\theta}(0) = \begin{bmatrix} 0 & 0 & 0 & 0 \end{bmatrix}, \qquad \hat{\varrho}(0) = 0.$$

Again, the definition of the control law and the parameter update are not reported here. Since the control laws discussed in this section are very complicated, they can be very sensitive with respect to the parameter selection.

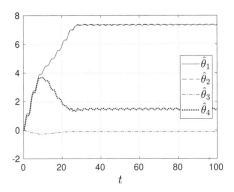

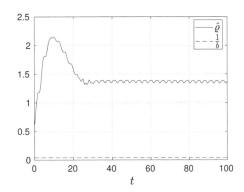

Figure 11.12: Evolution of the parameter estimates corresponding to the solution in Figure 11.11.

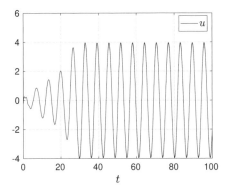

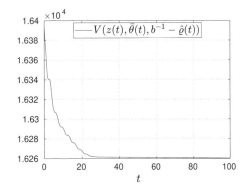

Figure 11.13: Input u (left) and the decrease of the Lyapunov function (right) corresponding to the solution in Figure 11.11.

11.4   EXERCISES

Exercise 11.1. Consider the dynamics (11.2) together with the control law $u = -k_1 x - k_2 x \phi(x)^2$, $k_1, k_2 \in \mathbb{R}_{>0}$. Follow the derivation in Section (11.1) to derive controller gains k_1, k_2 depending on upper bounds on δ so that all solutions of (11.2) are bounded and so that all solutions converge to the set $\{x \in \mathbb{R} : |x| \le 1\}$.

Exercise 11.2. Consider the dynamics (11.2). Show that the dynamic controller

$$u = -k_1 x - \xi \phi(x)$$
$$\dot{\xi} = x \phi(x)$$

with gain $k_1 \in \mathbb{R}_{>0}$ globally asymptotically stabilizes the origin of the system (11.2) independently of the unknown constant parameter $\theta \in \mathbb{R}$.

Exercise 11.3. Show that the gradient descent update rule (11.9) and its scaled version (11.12) lead to a similar closed-loop behavior, i.e., reproduce the results visualized in Figure 11.1 using (11.12) instead of (11.9) in the closed-loop dynamics.

Exercise 11.4. Consider the differential equation (11.18). Show that for $k_1, \gamma \in \mathbb{R}_{\geq 0}$ the solution $x(t)$ satisfies (11.19).

Exercise 11.5. Implement the controller in Theorem 11.7 for model reference adaptive systems to reproduce the results in Example 11.8. Verify numerically, that the asymptotic behavior of states θ depends on the initial condition $[x_0^T, \bar{x}_0^T, \theta_0^T]$ and the selection of the reference input u^e.

Exercise 11.6. Consider the dynamical system (11.37) with $x, \theta, u \in \mathbb{R}$ and $g(x) \neq 0$ for all $x \in \mathbb{R}$ together with the dynamic controller (11.38) defined as

$$u = \alpha(x, \hat{\theta}) = \frac{1}{g(x)} \left(-f(x) - F(x)\hat{\theta} - cx \right)$$

$$u = T(x, \hat{\theta}) = \frac{1}{\gamma} x F(x)$$

for $c, \gamma > 0$ constant. Use the function

$$V(x, \bar{\theta}) = \frac{1}{2} x^2 + \frac{1}{2} \gamma \bar{\theta}^2$$

with $\bar{\theta} = \theta - \hat{\theta}$ to show that the dynamic controller guarantees $x(t) \to \infty$ for $t \to 0$ for all initial conditions $x(0), \hat{\theta}(0) \in \mathbb{R}$.

Exercise 11.7. Implement the control laws discussed in Theorem 11.10, Theorem 11.11, and Lemma 11.12 on the example of the flexible link (11.70). To simplify the calculations and expressions, you can assume that only some of the parameters are unknown. In MATLAB you can use symbolic variables (`syms`) and symbolic differentiation (`diff.m`) to avoid tedious calculations by hand. Moreover, a symbolic expression can be converted to a MATLAB function through the command `matlabFunction.m`.

11.5   BIBLIOGRAPHICAL NOTES AND FURTHER READING

The motivating examples and challenges in Section 11.1 follow the presentation in [93, Chapters 1.1.3, 1.2.1, and 3.1].

The material on model reference adaptive control in Section 11.2 follows the presentation in [13, Chapters 5.1 and 5.5]. While we focus on derivations based on Lyapunov arguments here, [13] additionally discusses the *MIT rule*, a gradient-based controller design used in the original works on model reference adaptive control.

The material discussed in Section 11.3 can be found in [93, Chapters 3 and 4] and in [162, 172], for example.

The derivations in this chapter are solely focused on state feedback controllers. The more realistic case of output feedback controllers is not discussed here. Extensions to output feedback controllers extending the discussion in Section 11.3, for example, can be found in [93].

A certain amount of care is needed when applying adaptive control laws in the presence of uncertainty. Relatively simple examples exist where, in the presence of disturbances or unmodeled dynamics, the parameter estimator or the state may diverge. See [67, Chapter 8] for examples and potential algorithm modifications.

Chapter Twelve

Introduction to Differential Geometric Methods

In this chapter we introduce a slightly different set of ideas and techniques for non-linear systems analysis and design based on the mathematical tools of *differential geometry*. These tools allow us to develop nonlinear extensions of several concepts from linear systems theory. We consider systems

$$\dot{x} = f(x) + g(x)u \tag{12.1a}$$
$$y = h(x) \tag{12.1b}$$

with state $x \in \mathbb{R}^n$, input $u \in \mathbb{R}$, output $y \in \mathbb{R}$, and $f, g : \mathbb{R}^n \to \mathbb{R}^n$, $h : \mathbb{R}^n \to \mathbb{R}$. In particular, we restrict our attention to single-input single-output systems here. Additionally, we make the assumption that f, g, h are arbitrarily often continuously differentiable and $f(0) = 0$ holds.

The results discussed in this chapter rely on coordinate transformations $z = \Phi(x)$, where $\Phi : \mathbb{R}^n \to \mathbb{R}^n$, and we assume that Φ^{-1} is well defined. Systematic coordinate transformations will be used to rewrite nonlinear systems in equivalent linear or at least partially linear systems for which a controller design is significantly simplified. In particular, we will investigate how to construct Φ such that the coordinate transformation $z = \Phi(x)$ describes the dynamics

$$
\begin{bmatrix} \dot{z}_1 \\ \vdots \\ \dot{z}_{r-1} \\ \dot{z}_r \end{bmatrix} = \begin{bmatrix} z_2 \\ \vdots \\ z_r \\ \alpha(z) + \beta(z)u \end{bmatrix}
$$
$$
\begin{bmatrix} \dot{z}_{r+1} \\ \vdots \\ \dot{z}_n \end{bmatrix} = \gamma(z) \tag{12.2}
$$
$$
y = z_1,
$$

where $r \in \{1, \ldots, n\}$ and $\alpha, \beta : \mathbb{R}^n \to \mathbb{R}$, $\gamma : \mathbb{R}^n \to \mathbb{R}^{n-r}$. If we additionally consider a coordinate transformation in the input $v = \alpha(z) + \beta(z)u$ (under the assumption $\beta(z) \neq 0$), a linear controller (i.e., linear in v) driving the output y asymptotically to the origin can be easily defined using methods from linear systems theory. Whether this input is meaningful in the original coordinates

$$
u = \frac{1}{\beta(\Phi(x))}\left(v - \alpha(\Phi(x))\right)
$$

depends on the behavior of the coordinates $z_{r+1}, \ldots, z_n$. This procedure is called *feedback linearization.* If $r = n$, the function γ is not present and an *input-to-state linearization* of the system is achieved. In the case $r < n$, only the relationship between the input and the output is linear, i.e, (12.2) represents an *input-to-output linearization.*

12.1   INTRODUCTORY EXAMPLES

We begin this chapter by looking at several examples illustrating different linear-style designs and highlighting some of the ideas introduced in detail in the following sections.

Example 12.1. Consider the nonlinear system

$$
\begin{aligned}
\dot{x} &= x^3 + u \\
y &= x.
\end{aligned}
\tag{12.3}
$$

If we linearize this system about the origin, we get $\dot{x} = u$ and a natural stabilizer is then given by $u = -kx$ ($k > 0$). However, this yields the closed-loop system $\dot{x} = x(x^2 - k)$ which has three equilibria given by

$$
x^e \in \{0, \pm\sqrt{k}\}.
$$

Note that the origin is locally asymptotically stable, and we can make the region of attraction arbitrarily large by choosing the feedback gain k arbitrarily large.

Alternatively, consider the nonlinear feedback $u = -x^3 + v$ where v is an input still to be designed. This yields the initial closed-loop system $\dot{x} = v$, and a natural feedback is then $v = -kx$. This yields a closed-loop system of $\dot{x} = -kx$ and the origin is globally asymptotically stable for all choices of the feedback gain k.

In this simple example we have seen that a coordinate transformation in the input u allows us to equivalently rewrite the nonlinear system in terms of a linear system, and from Chapter 3 we know how to define stabilizing controllers for linear systems. The example additionally illustrates that even though the linearization around the origin and the coordinate transformation may provide similar results, globally the coordinate transformation provides a stronger result.

The next example shows that coordinate transformations can also be performed in the states. However, the result is the same, i.e., a nonlinear system is equivalently rewritten through linear dynamics.

Example 12.2. Consider the second-order system

$$
\begin{aligned}
\dot{x}_1 &= x_2 + x_1^2 \\
\dot{x}_2 &= -2x_1^3 - 2x_1x_2 + u \\
y &= x_1,
\end{aligned}
\tag{12.4}
$$

where y denotes an output. Take the change of coordinates $z_1 = x_1$ and $z_2 = x_2 + x_1^2$ (we will describe later how to arrive at this change of coordinates). In the new coordinates,

$$\dot{z}_1 = z_2$$
$$\dot{z}_2 = \dot{x}_2 + 2x_1\dot{x}_1 = u \tag{12.5}$$
$$y = z_1,$$

where we observe that, in the z coordinates, the system is linear. A feedback that globally exponentially stabilizes the origin is given by

$$u = -k_1 z_1 - k_2 z_2$$
$$= -k_1 x_1 - k_2 \left(x_2 + x_1^2 \right)$$

for all choices of gains $k_1, k_2 > 0$. This can be easily verified by checking that the closed-loop matrix defining the linear system (12.5) is Hurwitz.

Here, the coordinate transformation again allows us to stabilize and analyze a linear system instead of a nonlinear system. If the origin of the linear system (12.5) is asymptotically stable it is straightforward to draw conclusions about the stability properties of the corresponding equilibrium of the nonlinear system (12.4) based on the relation between the x and z coordinates. Additionally, for the input-output behavior of the system it is not important if the dynamics are written in terms of the x-coordinates or the z-coordinates.

Example 12.3. As a final example, consider the third-order nonlinear system

$$\dot{x}_1 = x_2$$
$$\dot{x}_2 = x_3^3 + u$$
$$\dot{x}_3 = x_1 + x_3^3 \tag{12.6}$$
$$y = x_1,$$

where again y is an output. Choose the change of coordinates

$$z_1 = x_3$$
$$z_2 = x_1 + x_3^3$$
$$z_3 = x_2 + 3x_1 x_3^2 + 3x_3^5.$$

With the initial feedback

$$u = -x_3^3 - 3x_2 x_3^2 - 6x_1^2 x_3 - 21x_1 x_3^4 - 15x_3^7 + v,$$

where v is an input to be designed, we obtain linear states but a nonlinear output

$$\dot{z}_1 = z_2$$
$$\dot{z}_2 = z_3$$
$$\dot{z}_3 = v \tag{12.7}$$
$$y = z_2 - z_1^3.$$

On the other hand, choosing the initial feedback

$$u = -x_3^3 + v$$

yields a linear input-output relationship from v to y,

$$\begin{aligned}
\dot{x}_1 &= x_2 \\
\dot{x}_2 &= v \\
\dot{x}_3 &= x_1 + x_3^3 \\
y &= x_1,
\end{aligned} \tag{12.8}$$

but the "internal" x_3 dynamics, which are not visible through the output, are nonlinear.

Here, we are only partially able to linearize the dynamics (12.6). For the linear dynamics (12.7), v can be defined such that the origin $z = 0$ is asymptotically stable, which implies that the output y asymptotically converges to zero. Similarly, for the representation (12.8) a controller guaranteeing $y(t) \to 0$ for $t \to \infty$ can be defined using pole placement, for example. Due to $\dot{x}_3 = x_1 + x_3^3$ it is however not immediately clear if the origin of the closed-loop system is asymptotically stable.

In the following sections, the ideas applied in these examples are generalized and made more precise.

12.2   ZERO DYNAMICS AND RELATIVE DEGREE

We begin our more formal discussions with two concepts that are closely tied to linear systems theory.

Definition 12.4 (Zero dynamics). *The* zero dynamics *of (12.1) are the internal dynamics when the output is kept at zero by the input u.*

Example 12.5. Consider again the system of Example 12.3. The zero dynamics are defined by the condition $y = x_1 \equiv 0$, which implies that $\dot{x}_3 = x_3^3$.

Note that the zero dynamics are the same in the z coordinates where the condition $y = z_2 - z_1^3 \equiv 0$ implies $\dot{z}_1 = z_1^3$.

Example 12.6. Consider the linear system defined by the transfer function

$$H(s) = \frac{s - 1}{(s + 2)(s + 3)}. \tag{12.9}$$

A minimal state space realization of this transfer function (see Section 4.1.2) is given by

$$\begin{aligned}
\dot{x}_1 &= -5x_1 - 6x_2 + u \\
\dot{x}_2 &= x_1 \\
y &= x_1 - x_2.
\end{aligned}$$

Hence, from Definition 12.4, the zero dynamics are given by $y \equiv 0$, which implies $\dot{x}_2 = x_2$.

Consider now the linear system

$$H(s) = \frac{s+1}{(s+2)(s+3)}, \tag{12.10}$$

which has a minimal state space realization given by

$$\begin{aligned}
\dot{x}_1 &= -5x_1 - 6x_2 + u \\
\dot{x}_2 &= x_1 \\
y &= x_1 + x_2.
\end{aligned} \tag{12.11}$$

The condition $y \equiv 0$ then implies that the zero dynamics are given by $\dot{x}_2 = -x_2$.

The above example illustrates an important distinction. For linear systems, zeros in the right-half complex plane give rise to zero dynamics where the origin is an unstable equilibrium, whereas zeros in the left-half complex plane give rise to zero dynamics where the origin is an exponentially stable equilibrium point. Such systems are referred to as *non-minimum phase* and *minimum phase*, respectively. This terminology is also used for nonlinear systems where systems with unstable zero dynamics are called non-minimum phase while systems with asymptotically stable zero dynamics are referred to as minimum phase. Hence, we see that the system of Example 12.5 is non-minimum phase.

A second definition used for linear systems that can be extended to nonlinear systems is the *relative degree*. To make this definition precise we recall the Lie derivative notation introduced in equation (9.5),

$$L_f \lambda(x) = \frac{\partial \lambda}{\partial x}(x) \cdot f(x), \tag{12.12}$$

though here we generally consider $\lambda : \mathbb{R}^n \to \mathbb{R}^m$, $m \in \mathbb{N}$, and consequently the Jacobian $\frac{\partial \lambda}{\partial x}(x)$ is an $m \times n$ matrix, rather than a vector as was the case when we discussed control Lyapunov functions in Chapter 9. If $\lambda : \mathbb{R}^n \to \mathbb{R}^m$ then $\frac{\partial \lambda}{\partial x} : \mathbb{R}^n \to \mathbb{R}^{m \times n}$. In contrast to (9.5), where we have used the inner product $\langle \cdot, \cdot \rangle : \mathbb{R}^n \times \mathbb{R}^n \to \mathbb{R}$, we need to be careful with the notation in (12.12) if $m > 1$. However, in this chapter, we generally consider single-input systems so that $m = 1$, and hence we will continue to use the inner product notation. Since $L_f \lambda : \mathbb{R}^n \to \mathbb{R}^m$ if $\lambda : \mathbb{R}^n \to \mathbb{R}^m$ we can take repeated Lie derivatives. Some notation to simplify the necessary calculations is given by

$$L_g L_f \lambda(x) = \tfrac{\partial}{\partial x} \left(L_f \lambda(x) \right) g(x), \quad \text{and} \quad L_f^k \lambda(x) = \tfrac{\partial}{\partial x} \left(L_f^{k-1} \lambda(x) \right) f(x),$$

where, by convention, $L_f^0 \lambda(x) = \lambda(x)$.

Definition 12.7 (Relative degree). *The system* (12.1) *has relative degree* $r \in \mathbb{N}$ *at a point* $x^\circ \in \mathbb{R}^n$ *if*

(i) *the repeated Lie derivatives satisfy* $L_g L_f^k h(x) = 0$ *for all* x *in a neighborhood of* x° *and all* $k < r - 1$; *and*

(ii) *the repeated Lie derivative satisfies* $L_g L_f^{r-1} h(x^\circ) \neq 0$.

For linear systems described by a transfer function, the relative degree is defined as the difference between the degree of the denominator and numerator polynomials.

Example 12.8. We see that for (12.10), the difference in the degree between the denominator and numerator polynomials is 1. For the state space representation of (12.10) given by (12.11) we see that

$$h(x) = x_1 + x_2 \quad \text{and} \quad g(x) = \begin{bmatrix} 1 \\ 0 \end{bmatrix}.$$

The Lie derivative in Definition 12.7 satisfies

$$L_g h(x) = \left\langle \begin{bmatrix} 1 \\ 1 \end{bmatrix}, \begin{bmatrix} 1 \\ 0 \end{bmatrix} \right\rangle = 1.$$

Thus, Definition 12.7(ii) implies $r - 1 = 0$, i.e., the relative degree is $r = 1$, which is consistent with the difference in degree of the denominator and numerator polynomials of the transfer function.

Consider now the transfer function

$$H(s) = \frac{1}{(s+2)(s+3)}, \tag{12.13}$$

which has a relative degree of 2. A state space representation is given by

$$\dot{x}_1 = -5x_1 - 6x_2 + u$$
$$\dot{x}_2 = x_1$$
$$y = x_2,$$

so that $h(x) = x_2$ and $g(x) = [1, 0]^T$. Computing the first Lie derivative of Definition 12.7 (i.e., with $k = 0$ in item (i)) we see that

$$L_g h(x) = \left\langle \begin{bmatrix} 0 \\ 1 \end{bmatrix}, \begin{bmatrix} 1 \\ 0 \end{bmatrix} \right\rangle = 0.$$

In order to compute $L_g L_f h(x)$, we first compute $L_f h(x)$ as

$$L_f h(x) = \left\langle \begin{bmatrix} 0 \\ 1 \end{bmatrix}, \begin{bmatrix} -5x_1 - 6x_2 \\ x_1 \end{bmatrix} \right\rangle = x_1$$

and then, since $\nabla L_f h(x) = [1, 0]^T$, we can compute

$$L_g L_f h(x) = \left\langle \begin{bmatrix} 1 \\ 0 \end{bmatrix}, \begin{bmatrix} 1 \\ 0 \end{bmatrix} \right\rangle = 1.$$

Therefore, from Definition 12.7 we see that $r - 1 = 1$ or that the relative degree is $r = 2$. As expected, this is consistent with the difference in degree of the transfer function polynomials.

We can interpret the relative degree condition in Definition 12.7 as the number of times the output needs to be differentiated in order to obtain an expression that includes the input u. Computing the time derivative of the output in (12.1) yields

$$\dot{y} = \overbrace{h(x)} = \langle \nabla h(x), \dot{x} \rangle = \langle \nabla h(x), f(x) + g(x)u \rangle$$
$$= L_f h(x) + L_g h(x)u.$$

If $L_g h(x) \neq 0$ then the input appears in the first derivative and the relative degree is 1. However, if $L_g h(x) = 0$ then the input is not visible in $\dot{y}$. Taking the second time derivative of y leads to

$$\ddot{y} = \langle \nabla L_f h(x), \dot{x} \rangle = L_f^2 h(x) + L_g L_f h(x)u,$$

where if $L_g L_f h(x) = 0$ then the input does not appear in the second derivative and we continue with higher order derivatives. On the other hand, if $L_g L_f h(x) \neq 0$ then the input appears in the second derivative and the relative degree is 2. Intuitively, this interpretation gives rise to the following result.

Lemma 12.9 ([139, Lemma 6.5])**.** *The relative degree* $r \in \mathbb{N}$ *of a system* (12.1) *satisfies* $r \leq n$.

Example 12.10. Consider again the system from Example 12.2. Differentiating the output twice yields

$$\dot{y} = \dot{x}_1 = x_2 + x_1^2$$
$$\ddot{y} = \dot{x}_2 + 2x_1 \dot{x}_1 = -2x_1^3 - 2x_1 x_2 + u + 2x_1 x_2 + 2x_1^3 = u,$$

where the input appears in the second time derivative. Writing out the Lie derivatives (which is essentially the same thing), we first observe that $L_g h(x) = 0$ and then

$$L_f h(x) = \langle \nabla h(x), f(x) \rangle = \left\langle \begin{bmatrix} 1 \\ 0 \end{bmatrix}, \begin{bmatrix} x_2 + x_1^2 \\ -2x_1^3 - 2x_1 x_2 \end{bmatrix} \right\rangle = x_2 + x_1^2,$$
$$L_g L_f h(x) = \langle \nabla(L_f h(x)), g(x) \rangle = \left\langle \begin{bmatrix} 2x_1 \\ 1 \end{bmatrix}, \begin{bmatrix} 0 \\ 1 \end{bmatrix} \right\rangle = 1.$$

Therefore, we see that the relative degree is 2.

Example 12.11. Consider the system from Example 12.3. The reader should check that $L_g h(x) = 0$. Then computing the relevant Lie derivatives yields

$$L_f h(x) = \left\langle \begin{bmatrix} 1 \\ 0 \\ 0 \end{bmatrix}, \begin{bmatrix} x_2 \\ x_3^3 \\ x_1 + x_3^3 \end{bmatrix} \right\rangle = x_2 \quad \text{and}$$
$$L_g L_f h(x) = \left\langle \begin{bmatrix} 0 \\ 1 \\ 0 \end{bmatrix}, \begin{bmatrix} 0 \\ 1 \\ 0 \end{bmatrix} \right\rangle = 1.$$

Hence the relative degree is 2.

12.3   FEEDBACK LINEARIZATION

In this section we use the definitions of zero dynamics and relative degree to systematically derive a *feedback linearization* for systems of the form (12.1) based on coordinate transformations. A *nonlinear change of coordinates* is a smooth function $\Phi : \mathbb{R}^n \to \mathbb{R}^n$ with a smooth inverse Φ^{-1}. We write the component functions of a coordinate change as

$$z = \Phi(x) = \begin{bmatrix} \phi_1(x) \\ \phi_2(x) \\ \vdots \\ \phi_n(x) \end{bmatrix}.$$

Such a Φ is called a global diffeomorphism. We will also allow *local* diffeomorphisms where Φ and its inverse may only be defined and smooth around a given point in the state space. An easily checkable sufficient condition for a function $\Phi : \mathbb{R}^n \to \mathbb{R}^n$ to be a local diffeomorphism is given by the following proposition.

Proposition 12.12 ([139, Lemma 6.2]). *Consider a smooth function $\Phi : \mathbb{R}^n \to \mathbb{R}^n$. Suppose the Jacobian of Φ is nonsingular at a point $x^\circ \in \mathbb{R}^n$. Then, on a suitable open set containing x°, Φ defines a local diffeomorphism.*

In Section 3.5.2 we have seen a coordinate transformation for linear systems. Here, a diffeomorphism enables us to define a general coordinate transformation for nonlinear systems. Consider the transformation $z = \Phi(x)$ and thus $x = \Phi^{-1}(z)$. Then it holds that

$$\dot{z} = \dot{\Phi}(x) = \frac{\partial \Phi}{\partial x} \dot{x} = \frac{\partial \Phi}{\partial x} f(x) + \frac{\partial \Phi}{\partial x} g(x) u$$
$$= \frac{\partial \Phi}{\partial x} f(\Phi^{-1}(z)) + \frac{\partial \Phi}{\partial x} g(\Phi^{-1}(z)) u,$$

and the output can be written as

$$y = h(\Phi^{-1}(z)).$$

With the definition of a diffeomorphism we are finally in the position to make the goal of this chapter more precise. As stated at the beginning of the chapter we intend to use coordinate transformations to rewrite the nonlinear system (12.1) as a linear, or at least partially linear, system. It turns out that the change of coordinates we need to consider is given by

$$z = \Phi(x) = \begin{bmatrix} \phi_1(x) \\ \phi_2(x) \\ \vdots \\ \phi_r(x) \end{bmatrix} = \begin{bmatrix} h(x) \\ L_f h(x) \\ \vdots \\ L_f^{r-1} h(x) \end{bmatrix}. \tag{12.14}$$

If the relative degree $r \neq n$, we augment this change of coordinates with additional $n - r$ functions to obtain a mapping $\Phi : \mathbb{R}^n \to \mathbb{R}^n$.

Example 12.13. Consider the system of Example 12.2. The above change of coordinates is given by

$$z = \Phi(x) = \left[\begin{array}{c} h(x) \\ L_f h(x) \end{array} \right] = \left[\begin{array}{c} x_1 \\ x_2 + x_1^2 \end{array} \right].$$

Note that these are the coordinates used in Example 12.2, which yielded a linear system.

Example 12.14. We have previously seen that the system in Example 12.3 has relative degree 2 with the output $y = x_1$. If we consider instead the output $\bar{y} = \bar{h}(x) = x_3$, the system has relative degree 3. (The reader should check this assertion in Exercise 12.1.) We compute a change of coordinates as

$$z_1 = \bar{h}(x) = x_3,$$

$$z_2 = L_f \bar{h}(x) = \left\langle \left[\begin{array}{c} 0 \\ 0 \\ 1 \end{array} \right], \left[\begin{array}{c} x_2 \\ x_3^3 \\ x_1 + x_3^3 \end{array} \right] \right\rangle = x_1 + x_3^3,$$

$$z_3 = L_f^2 \bar{h}(x) = \left\langle \left[\begin{array}{c} 1 \\ 0 \\ 3x_3^2 \end{array} \right], \left[\begin{array}{c} x_2 \\ x_3^3 \\ x_1 + x_3^3 \end{array} \right] \right\rangle = x_2 + 3x_1 x_3^2 + 3x_3^5.$$

Therefore, in the z-coordinates, we have

$$\dot{z}_1 = z_2$$
$$\dot{z}_2 = z_3$$
$$\dot{z}_3 = L_f^3 \bar{h}(x) + L_g L_f^2 \bar{h}(x) u.$$

Since $L_g L_f^2 \bar{h}(x) = 1 \neq 0$, we can compute a feedback that yields a linear system as $u = -L_f^3 \bar{h}(x) + v$, where v is now an input to be designed. We see that this is precisely the feedback used in Example 12.3. However, with the output $\bar{y}$ we have obtained a linear input-output relationship with linear states.

The above examples and change of coordinates can be formalized into the following proposition.

Proposition 12.15 (Input-to-state and input-to-output linearization). *Consider the system* (12.1) *with relative degree* $r \in \mathbb{N}$ *at* $x^\circ \in \mathbb{R}^n$ *together with the coordinate transformation* (12.14).

- *If* $r < n$, *then there exist* $n - r$ *functions* $\phi_{r+1}, \ldots, \phi_n : \mathbb{R}^n \to \mathbb{R}$, *so that* $\Phi(x) = [\phi_1, \ldots, \phi_n]^T$ *has a nonsingular Jacobian at* x^0. *The value of these additional functions can be chosen arbitrarily at* x° *and can additionally be chosen to satisfy*

$$L_g \phi_i(x) = 0, \quad r + 1 \leq i \leq n.$$

- *For $r \leq n$, the coordinate transformation satisfies*

$$\dot{z}_1 = \langle \nabla \phi_1(x), \dot{x} \rangle = L_f h(x) + L_g h(x) u = L_f h(x) = z_2$$
$$\dot{z}_2 = \langle \nabla (L_f h(x)), \dot{x} \rangle = L_f^2 h(x) = z_3$$

$$\vdots$$

$$\dot{z}_{r-1} = \langle \nabla (L_f^{r-2} h(x)), \dot{x} \rangle = L_f^{r-1} h(x) = z_r$$
$$\dot{z}_r = \langle \nabla (L_f^{r-1} h(x)), \dot{x} \rangle = L_f^r h(x) + L_g L_f^{r-1} h(x) u,$$

and if $r < n$, the remaining coordinates $i \in \{r+1, \ldots, n\}$ satisfy

$$\dot{z}_i = \langle \nabla \phi_i(x), \dot{x} \rangle = L_f \phi_i(x) + L_g \phi_i(x) u = L_f \phi_i(x).$$

Since Φ is invertible (otherwise it would not define a change of coordinates) we have $x = \Phi^{-1}(z)$. We summarize the system in the z-coordinates,

$$\dot{z}_1 = z_2$$

$$\vdots$$

$$\dot{z}_{r-1} = z_r$$
$$\dot{z}_r = L_f^r h(\Phi^{-1}(z)) + L_g L_f^{r-1} h(\Phi^{-1}(z)) u$$
$$\dot{z}_{r+1} = L_f \phi_{r+1}(\Phi^{-1}(z)) \tag{12.15}$$

$$\vdots$$

$$\dot{z}_n = L_f \phi_n(\Phi^{-1}(z))$$
$$y = z_1,$$

providing an explicit representation of the functions α, β, and γ in (12.2). With the input

$$u = \frac{1}{L_g L_f^{r-1} h(x)} (-L_f^r h(x) + v),$$

the system (12.15) is linear in v.

Moreover, if $r = n$, then (12.15) is linear in the state z and the input v, i.e., (12.15) represents an input-to-state linearization of (12.1). In this case, we have found a feedback linearization and in particular can use linear control methods to define a control law v which stabilizes the origin $x = 0$ of the original dynamics.

If $r < n$, then the right-hand sides of $\dot{z}_{r+1}, \ldots, \dot{z}_n$ are not necessarily linear. However, with the transformed input v, the variables $z_{r+1}, \ldots, z_n$ do not affect the output $y = z_1$, and due to the linear dynamics of the remaining variables, (12.15) represents an input-to-output linearization of (12.1). Furthermore, in the case $r < n$, we see that maintaining $y = z_1 = 0$ requires $\dot{z}_i = 0$ for $i = 1, \ldots, r$, but places no restrictions on $\dot{z}_i$ for $i = r + 1, \ldots, n$. Therefore, we see that these last $n - r$ states are precisely the zero dynamics of the system.

In the following we focus on the case of input-to-state linearization. However, before we can proceed, we need to introduce some additional tools from differential geometry.

12.3.1   Nonlinear Controllability

Controllability for linear systems was introduced in Section 3.5.1 in Definition 3.19 as the property of being able to reach any arbitrary state from any initial state in finite time by selecting the input u appropriately. For nonlinear systems, controllability can be defined in the same way.

Definition 12.16 ((Complete) Controllability). *Consider the control affine system*

$$\dot{x} = f(x) + \sum_{j=1}^{m} g_j(x)u_j, \tag{12.16}$$

$f, g_j : \mathbb{R}^n \to \mathbb{R}^n$, $j \in \{1, \ldots, m\}$, $m \in \mathbb{N}$. The system is called (completely) controllable if for any two points $x_1, x_2 \in \mathbb{R}^n$ there exist a finite time $T \in \mathbb{R}_{\geq 0}$ and an input $u : [0, T] \to \mathbb{R}^m$ such that $x_2 = x(T; x_1, u)$, where $x(\cdot; x_1, u)$ denotes the solution of (12.16) uniquely defined through the initial condition x_1 and the input u.

In the case of linear systems we have seen an efficient way to verify controllability by checking the rank of the controllability matrix (3.57). For nonlinear systems, verifying controllability is not as straightforward. Instead, we present a weaker concept, local accessibility, that relies on *reachable sets* and that can be verified through a similar rank condition. For $x_0 \in \mathbb{R}^n$, $T \in \mathbb{R}_{\geq 0}$, and a neighborhood $\mathcal{O}$ of x_0 define the reachable sets

$$\mathcal{R}^{\mathcal{O}}(x_0, T) = \left\{ x_1 \in \mathbb{R}^n \,\middle|\, \begin{array}{l} x_1 = x(T; x_0, u), \quad x(t; x_0, u) \in \mathcal{O} \;\; \forall \, t \leq T \\ u : [0, T] \to \mathbb{R}^m \text{ measurable} \end{array} \right\} \tag{12.17}$$

$$\mathcal{R}_T^{\mathcal{O}}(x_0) = \bigcup_{\tau \leq T} \mathcal{R}^{\mathcal{O}}(x_0, \tau). \tag{12.18}$$

Recall that a neighborhood $\mathcal{O} \subset \mathbb{R}^n$ of a point x_0 is an open set with $x_0 \in \mathcal{O}$.

Definition 12.17 (Local Accessibility). *Consider the system (12.16). The system is said to be locally accessible from $x_0 \in \mathbb{R}^n$ if the interior of $\mathcal{R}_T^{\mathcal{O}}(x_0)$ defined in (12.18) is nonempty for all nonempty neighborhoods $\mathcal{O} \subset \mathbb{R}^n$ of x_0 for all $T > 0$. If this property is satisfied for all $x_0 \in \mathbb{R}^n$ then the system is locally accessible.*

While local accessibility does not allow moving from any arbitrary point x_0 to any arbitrary point x_1, accessibility guarantees that from a given point x_0 every point in a neighborhood around x_0 can be reached. Thus, *accessibility* is also known as *weak controllability* in the literature (see [144, Chapter 4], for example) and (complete) controllability as per Definition 12.16 implies local accessibility as per Definition 12.17.

Local accessibility is in general not easy to verify. To obtain a definition which can be verified through a similar rank condition as in the linear setting we need a stronger definition.

Definition 12.18 (Strong Accessibility). *Consider the system (12.16). The system is said to be locally strongly accessible from $x_0 \in \mathbb{R}^n$ if for any neighborhood $\mathcal{R}_T^{\mathcal{O}}(x_0)$ of x_0 the set $\mathcal{R}^{\mathcal{O}}(x_0, T)$ defined in (12.17) contains a nonempty set for any $T > 0$ sufficiently small. If this property is satisfied for all $x_0 \in \mathbb{R}^n$ then the system is locally strongly accessible.*

Local strong accessibility implies local accessibility. It can be verified through the concept of *Lie brackets*.

Definition 12.19 (Lie bracket). *The* Lie bracket *of two smooth vector fields $f, g :$ $\mathbb{R}^n \to \mathbb{R}^n$ is given by*

$$[f, g](x) = \frac{\partial g}{\partial x}(x)f(x) - \frac{\partial f}{\partial x}(x)g(x)$$

where $\frac{\partial g}{\partial x}(x)$ and $\frac{\partial f}{\partial x}(x)$ are the Jacobian matrices for g and f, respectively.

To facilitate repeated application, define $\mathrm{ad}_f^0 g(x) = g(x)$ and

$$\mathrm{ad}_f^k g(x) = [f, \mathrm{ad}_f^{k-1} g](x). \tag{12.19}$$

While we generally consider only single input systems in this chapter, for the purposes of illustration consider a two-input nonlinear system

$$\dot{x} = u_1 g_1(x) + u_2 g_2(x) \tag{12.20}$$

with smooth vector fields $g_1, g_2 : \mathbb{R}^n \to \mathbb{R}^n$ and $u_1, u_2 \in \mathbb{R}$. Then at a point x° we can obviously steer in any direction contained in

$$\Delta(x^\circ) = \mathrm{span}\{g_1(x^\circ), g_2(x^\circ)\}.$$

In other words, at the point x°, with $u_1 \neq 0$ and $u_2 = 0$ we can steer the system in the direction $g_1(x^\circ)$ while with $u_1 = 0$ and $u_2 \neq 0$ we can steer in the direction of $g_2(x^\circ)$. By taking both $u_1 \neq 0$ and $u_2 \neq 0$, we can steer in the direction of a linear combination of $g_1(x^\circ)$ and $g_2(x^\circ)$.

A model that fits this form is the mobile robot, mobile robot unicycle, or shopping cart with state $x = [\chi_1, \chi_2, \theta]^T$ introduced in (1.46):

$$\frac{d}{dt} \begin{bmatrix} \chi_1 \\ \chi_2 \\ \theta \end{bmatrix} = \underbrace{\begin{bmatrix} \sin\theta \\ \cos\theta \\ 0 \end{bmatrix}}_{g_1(x)} u_1 + \underbrace{\begin{bmatrix} 0 \\ 0 \\ 1 \end{bmatrix}}_{g_2(x)} u_2. \tag{12.21}$$

A corresponding visualization is provided in Chapter 1 in Figure 1.9.

The linearization at any point can easily be seen to be uncontrollable as the A matrix is the 3×3 zero matrix and the B matrix has rank 2. Intuitively though, unicycles and shopping carts can be steered to an arbitrary configuration of position and angle $[\chi_1, \chi_2, \theta]^T$. Still, it is clear that simply a linear combination of u_1 and u_2 as described above will not span the entire space.

The following proposition, whose proof is omitted, indicates how to consider local controllability or accessibility for nonlinear systems.

Proposition 12.20 ([110, Proposition 3.6]). *Consider* (12.20) *and a control signal* *given by*

$$u(t) = \begin{cases} [1, 0]^T, & t \in [0, h) \\ [0, 1]^T, & t \in [h, 2h) \\ [-1, 0]^T, & t \in [2h, 3h) \\ [0, -1]^T, & t \in [3h, 4h) \end{cases}$$

where $h > 0$. Then the solution satisfies

$$x(4h) = x_0 + h^2 [g_1, g_2](x_0) + \mathcal{O}[h^3].$$

In words, by switching the control inputs in a particular sequence we can move from an initial condition of x_0 to a point that is approximately given by the Lie bracket of the two input vector fields. This suggests that the Lie bracket might provide the required additional information to arrive at an understanding of controllability for nonlinear systems.

Returning to the unicycle system above, in order to demonstrate accessibility we need to be able to essentially access any direction in $\mathbb{R}^3$. We observe that g_1 and g_2 are linearly independent, so we only need a third direction in which we can steer. The above proposition suggests we can construct an input that allows us to move in the direction of the Lie bracket, which we calculate as

$$[g_1, g_2] = \frac{\partial g_2}{\partial x} g_1 - \frac{\partial g_1}{\partial x} g_2 = - \begin{bmatrix} 0 & 0 & \cos\theta \\ 0 & 0 & -\sin\theta \\ 0 & 0 & 0 \end{bmatrix} \begin{bmatrix} 0 \\ 0 \\ 1 \end{bmatrix} = \begin{bmatrix} -\cos\theta \\ \sin\theta \\ 0 \end{bmatrix}.$$

Therefore, we see that we can steer in any direction in $\mathbb{R}^3$ since

$$\mathrm{span}\,\{g_1, g_2, [g_1, g_2]\} = \mathbb{R}^3.$$

Example 12.21. Consider the linear system $\dot{x} = Ax + Bu$. Then the vector fields of $\dot{x} = f(x) + g(x)u$ are given by

$$f(x) = Ax \quad \text{and} \quad g(x) = B.$$

We can calculate the Lie brackets

$$\mathrm{ad}_f g(x) = [f, g] = \frac{\partial g}{\partial x} f(x) - \frac{\partial f}{\partial x} g(x) = 0 \cdot Ax - AB = -AB,$$
$$\mathrm{ad}_f^2 g(x) = [f, \mathrm{ad}_f g] = 0 \cdot Ax - AAB = -A^2 B,$$
$$\vdots$$
$$\mathrm{ad}_f^{n-1} g = [f, \mathrm{ad}_f^{n-2} g] = 0 \cdot Ax - AA^{n-2} B = -A^{n-1} B.$$

We then see that the repeated Lie brackets essentially provide the controllability matrix of Theorem 3.21,

$$\begin{bmatrix} g & \mathrm{ad}_f g & \mathrm{ad}_f^2 g & \cdots & \mathrm{ad}_f^{n-1} g \end{bmatrix} = -\begin{bmatrix} -B & AB & A^2 B & \cdots & A^{n-1} B \end{bmatrix},$$

where it is useful to note that the minus sign plays no role in determining the rank of the matrix.

For nonlinear systems, accessibility can be concluded from repeated Lie brackets in a similar way.

Lemma 12.22 ([144, Theorem 9]). *Consider the control affine system* (12.16) *with* $m = 1$. *If*

$$\text{span}\left\{g(x), \text{ad}_f g(x), \ldots, \text{ad}_f^{n-1} g(x)\right\} = \mathbb{R}^n, \tag{12.22}$$

then (12.16) *is strongly locally accessible from* x.

For multi-input systems, a similar result holds. However, for $m > 1$ additional Lie brackets have to be taken into account in the left-hand side of (12.22).

As indicated at the beginning of this section, local strong accessibility does not imply controllability in general. However, in the special case that $f(x) = 0$ for all $x \in \mathbb{R}^n$, controllability can be concluded.

Lemma 12.23 ([144, Corollary 4.3.12]). *Consider the control affine system* (12.16) *with* $f(x) = 0$ *for all* $x \in \mathbb{R}^n$. *If the system is locally strongly accessible according to Definition 12.18 then it is controllable according to Definition 12.16.*

The following results, focusing on single-input systems, only rely on condition (12.22).

12.3.2   Input-to-State Linearization

In this section we return to input-to-state linearization or feedback linearization of (12.1a). Here, in contrast to the result in Proposition 12.15, the function h defining the output in (12.1b) is considered as a degree of freedom.

The vector space that we discussed informally in the previous section is referred to as a distribution.

Definition 12.24 (Smooth distribution). *Given* $d \in \mathbb{N}$ *smooth* n-*dimensional vector fields* $f_1, \ldots, f_d : \mathbb{R}^n \to \mathbb{R}^n$, *we call*

$$\Delta(x) = \text{span}\left\{f_1(x), \ldots, f_d(x)\right\}$$

a smooth distribution.

Note that $\Delta(x)$ defines a vector space at each point x. The distribution Δ is called *nonsingular* if $\dim(\Delta(x)) = d$ for all x. A *regular point* of a distribution Δ is a point $x^\circ \in \mathbb{R}^n$ such that $\Delta(x^\circ)$ is nonsingular in a neighborhood of x°.

Definition 12.25 (Involutive distribution). *A distribution* Δ *is* involutive *if the Lie bracket* $[\tau_1, \tau_2] : \mathbb{R}^n \to \mathbb{R}^n$ *of any pair of vector fields* $\tau_1 : \mathbb{R}^n \to \mathbb{R}^n$ *and* $\tau_2 : \mathbb{R}^n \to \mathbb{R}^n$ *belonging to* Δ *is a vector field which belongs to* Δ; *i.e.,*

$$\tau_1, \tau_2 \in \Delta \quad \Rightarrow \quad [\tau_1, \tau_2] \in \Delta.$$

It is obvious that constant vector fields define an involutive distribution since the Lie bracket of two constant vectors is zero. It is also immediate that a set composed of a single vector field is involutive. A less trivial example of an involutive distribution is when, for the set of vector fields $\{f_1(x), \ldots, f_d(x)\}$,

$$\text{rank}\left(\left[f_1(x) \ \cdots \ f_d(x)\right]\right) = \text{rank}\left(\left[f_1(x) \ \cdots \ f_d(x) \ [f_i, f_j](x)\right]\right)$$

for all x and all $i, j \in \{1, \ldots, d\}$ is satisfied.

With the above definitions in hand, we can state the following result indicating when a control affine nonlinear system is input-to-state linearizable.

Theorem 12.26 (Input-to-state linearization, [139, Theorem 6.2]). *The single-input system* (12.1a) *is input-to-state linearizable if and only if in a neighborhood of the origin*

(i) $\text{span}\left\{g, \text{ad}_f g, \ldots, \text{ad}_f^{n-1} g\right\} = \mathbb{R}^n$; *and*

(ii) the distribution $\Delta = \text{span}\left\{g, \text{ad}_f g, \ldots, \text{ad}_f^{n-2} g\right\}$ *is involutive and of constant rank* $n - 1$.

Based on our previous discussion, we observe that item (i) is essentially the accessibility condition or a local nonlinear controllability condition in the neighborhood of the origin. Item (ii) is a technical condition related to the integrability of the system dynamics. Note that a controllable linear system satisfies the above conditions.

Corollary 12.27. *A planar nonlinear system* $\dot{x} = f(x) + g(x)u$, $x \in \mathbb{R}^2$, $f(0) = 0$, *is locally state feedback linearizable in a neighborhood of the origin if and only if its linear approximation at the origin*

$$\dot{x} = \frac{\partial f}{\partial x}(0)x + g(0)u$$

is controllable.

Corollary 12.28. *Systems in triangular form,*

$$\dot{x}_i = x_{i+1} + \varphi_i(x_1, \ldots, x_i), \quad 1 \leq i \leq n - 1$$
$$\dot{x}_n = \varphi_n(x_1, \ldots, x_n) + u,$$

where φ_i *are smooth and* $\varphi_i(0) = 0$, *are locally feedback linearizable.*

We conclude the discussion of this section with a systematic procedure summarized in Algorithm 2 to compute the input-to-state linearization of a system and two examples.

Example 12.29. Consider the system from Example 12.3, without output, i.e.,

$$f(x) = \begin{bmatrix} x_2 \\ x_3^2 \\ x_1 + x_3^3 \end{bmatrix}, \quad g(x) = \begin{bmatrix} 0 \\ 1 \\ 0 \end{bmatrix}.$$

Algorithm 2: Input-to-state linearization

Input: System $\dot{x} = f(x) + g(x)u$ with smooth $f, g : \mathbb{R}^n \to \mathbb{R}^n$.

Output: Coordinate transformations $z = \Phi(x)$ and $u = \alpha(x) + \beta(x)v$ such that $\dot{z}_i = z_{i+1}$ for all $i = 1, \ldots, n-1$ and $\dot{z}_n = v$; or verify that the system is not input-to-state linearizable.

Algorithm:

1. Construct the vector fields $g, \mathrm{ad}_f g, \ldots, \mathrm{ad}_f^{n-1} g$.
2. Check the controllability and involutivity conditions of Theorem 12.26.
3. If the conditions are satisfied, compute the output function $h(x)$ by solving

   - $L_{\mathrm{ad}_f^i g} = 0$ for $i = 0, \ldots, n-2$
   - $L_{\mathrm{ad}_f^{n-1} g} \neq 0$.

4. Construct a change of coordinates

   $$z = \Phi(x) = \left[h(x) \quad L_f h(x) \quad \cdots \quad L_f^{n-1} h(x) \right]^T \qquad (12.23)$$

   and the feedback linearizing input $u = \alpha(x) + \beta(x)v$ where

   $$\alpha(x) = -\frac{L_f^n h(x)}{L_g L_f^{n-1} h(x)}; \quad \beta(x) = \frac{1}{L_g L_f^{n-1} h(x)}. \qquad (12.24)$$

Following Algorithm 2, we compute the vector fields

$$\mathrm{ad}_f g(x) = -\frac{\partial f}{\partial x} g(x) = \begin{bmatrix} 0 & 1 & 0 \\ 0 & 0 & 3x_3^2 \\ 1 & 0 & 3x_3^2 \end{bmatrix} \begin{bmatrix} 0 \\ 1 \\ 0 \end{bmatrix} = \begin{bmatrix} 1 \\ 0 \\ 0 \end{bmatrix}$$

$$\mathrm{ad}_f^2 g(x) = -\frac{\partial f}{\partial x} \mathrm{ad}_f g(x) = \begin{bmatrix} 0 & 1 & 0 \\ 0 & 0 & 3x_3^2 \\ 1 & 0 & 3x_3^2 \end{bmatrix} \begin{bmatrix} 1 \\ 0 \\ 0 \end{bmatrix} = \begin{bmatrix} 0 \\ 0 \\ 1 \end{bmatrix}.$$

Then the nonlinear controllability condition

$$\mathrm{span} \left\{ g, \mathrm{ad}_f g, \mathrm{ad}_f^2 g \right\} = \mathbb{R}^3$$

is satisfied and the distribution

$$\Delta(x) = \mathrm{span} \left\{ g, \mathrm{ad}_f g \right\}$$

is involutive (since it consists of constant vectors) and is of constant rank 2.

The various Lie derivative conditions imply the output function $h(x)$ needs to satisfy certain constraints:

$$L_g h(x) = 0 \quad \Rightarrow \quad \frac{\partial h}{\partial x_2}(x) = 0$$

$$L_{\mathrm{ad}_f g} h(x) = 0 \quad \Rightarrow \quad \frac{\partial h}{\partial x_1}(x) = 0$$

$$L_{\mathrm{ad}_f^2 g} h(x) \neq 0 \quad \Rightarrow \quad \frac{\partial h}{\partial x_3}(x) \neq 0.$$

This leads to the natural choice of the output function $h(x) = x_3$. The required change of coordinates and feedback linearizing input are then straightforward.

Example 12.30. ([139, Example 6.10], [105]) We consider again the single link manipulator with flexible joint discussed in Section 11.3.3. For convenience, the second-order differential equation

$$I\ddot{q}_1 + MgL\sin(q_1) + k(q_1 - q_2) = 0 \tag{12.25}$$
$$J\ddot{q}_2 - k(q_1 - q_2) = u$$

and its first-order representation

$$\dot{x} = f(x) + g(x)u, \tag{12.26}$$

where $x = [q_1, \dot{q}_1, q_2, \dot{q}_2]^T$, and f and g are defined by

$$f(x) = \begin{bmatrix} x_2 \\ -\frac{MgL}{I}\sin(x_1) - \frac{k}{I}(x_1 - x_3) \\ x_4 \\ \frac{k}{J}(x_1 - x_3) \end{bmatrix} \quad \text{and} \quad g(x) = \begin{bmatrix} 0 \\ 0 \\ 0 \\ \frac{1}{J} \end{bmatrix},$$

are repeated here. In contrast to Section 11.3.3, we assume that the constants $M, g, L, k, I, J > 0$ are known.

With the definition of the functions f and g the Lie brackets satisfy

$$\begin{bmatrix} g & \mathrm{ad}_f g & \mathrm{ad}_f^2 g & \mathrm{ad}_f^3 g \end{bmatrix} = \begin{bmatrix} 0 & 0 & 0 & -\frac{k}{IJ} \\ 0 & 0 & \frac{k}{IJ} & 0 \\ 0 & -\frac{1}{J} & 0 & \frac{k}{J^2} \\ \frac{1}{J} & 0 & -\frac{k}{J^2} & 0 \end{bmatrix}. \tag{12.27}$$

Thus, $\mathrm{span}\{g, \mathrm{ad}_f g, \mathrm{ad}_f^2 g, \mathrm{ad}_f^3 g\} = \mathbb{R}^4$ (since the determinant of the matrix (12.27) is unequal to zero, for example) and the distribution

$$\Delta = \mathrm{span}\{g, \mathrm{ad}_f g, \mathrm{ad}_f^2 g\}$$

is involutive and the rank of Δ is 3. This can be seen immediately from the fact that the first three columns of (12.27) are linearly independent and constant (and it follows immediately that the Lie bracket of two constant functions is zero).

As a next step, we compute an input-to-state linearization of the system using Algorithm 2. Since the conditions of Theorem 12.26 are satisfied, we know already that the system is input-to-state linearizable.

According to step 4 in Algorithm 2, the coordinate transformations are given by (12.23) and (12.24), depending on the unknown function h.

Following step 3, the function $h(x)$ needs to satisfy

$$L_g h(x) = 0, \quad L_{\mathrm{ad}_f g} h(x) = 0, \quad L_{\mathrm{ad}_f^2 g} h(x) = 0, \quad L_{\mathrm{ad}_f^3 g} h(x) \neq 0. \tag{12.28}$$

Thus using the first two columns of (12.27) it holds that

$$\frac{\partial h(x)}{\partial x_4} = 0, \qquad \frac{\partial h(x)}{\partial x_3} = 0 \tag{12.29}$$

and the last two columns of (12.27) together with (12.29) imply

$$\frac{\partial h(x)}{\partial x_2} = 0, \qquad \frac{\partial h(x)}{\partial x_1} \neq 0. \tag{12.30}$$

Hence $h(x)$ can only depend on x_1 and we can define

$$h(x) = x_1,$$

for example. After $h(x)$ is fixed, the coordinate transformations Φ, α, and β are obtained through (12.28) and (12.29), respectively. In particular, the coordinate transformation satisfies

$$z_1(x) = \Phi_1(x) = x_1$$
$$z_2(x) = \Phi_2(x) = x_2$$
$$z_3(x) = \Phi_3(x) = -\frac{MgL}{I}\sin(x_1) - \frac{k}{I}(x_1 - x_3)$$
$$z_4(x) = \Phi_4(x) = -\frac{MgL}{I}x_2\cos(x_1) - \frac{k}{I}(x_2 - x_4)$$
$$\alpha(x) = -\frac{MgLJ}{k}\sin x_1 \left(x_2^2 + \frac{MgL}{I}\cos(x_1) + \frac{k}{I} \right)$$
$$\qquad\qquad - J(x_1 - x_3)\left(\frac{k}{I} + \frac{k}{J} + \frac{MgL}{I}\cos x_1 \right)$$
$$\beta(x) = \frac{IJ}{k}.$$

By construction, this leads to the linear dynamics

$$\begin{bmatrix} \dot{z}_1 \\ \dot{z}_2 \\ \dot{z}_3 \\ \dot{z}_4 \end{bmatrix} = \begin{bmatrix} 0 & 1 & 0 & 0 \\ 0 & 0 & 1 & 0 \\ 0 & 0 & 0 & 1 \\ 0 & 0 & 0 & 0 \end{bmatrix} \begin{bmatrix} z_1 \\ z_2 \\ z_3 \\ z_4 \end{bmatrix} + \begin{bmatrix} 0 \\ 0 \\ 0 \\ 1 \end{bmatrix} v. \tag{12.31}$$

For the linear system we can calculate an optimal stabilizing feedback law using Theorem 14.4, for example (though one could also use pole placement). For $Q = I \in R^{4\times 4}$ and $R = 1$ the optimal control $v = Kz$ with

$$K = \begin{bmatrix} -1.0000 & -3.0777 & -4.2361 & -3.0777 \end{bmatrix}$$

is obtained through MATLAB. Here the constants are set to 1 for simplicity.

In Figure 12.1 the closed-loop solution as well as the input with respect to the initial condition $x(0) = [1, 0, 0, 0]^T$ in the original x-coordinates and the transformed z-coordinates are visualized.

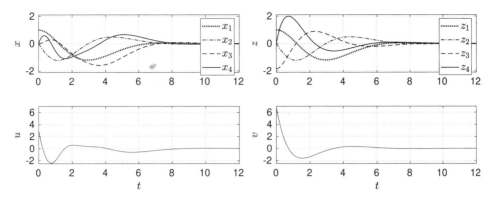

Figure 12.1: Closed-loop solution of (12.26) in the nonlinear and the linear coordinates (12.31).

12.4   EXERCISES

Exercise 12.1. Show that the system

$$\dot{x}_1 = x_2$$
$$\dot{x}_2 = x_3^3 + u$$
$$\dot{x}_3 = x_1 + x_3^3$$
$$y = x_3$$

$$(12.32)$$

has relative degree 3 (at the origin).

Exercise 12.2. Consider the dynamics

$$\dot{x}_1 = x_2,$$
$$\dot{x}_2 = x_3,$$
$$\dot{x}_3 = u,$$
$$\dot{x}_4 = x_1^2 + x_2^2 - x_3 - x_1^3 - x_4^2,$$
$$y = x_1.$$

Check whether the system is minimum or non-minimum phase.

Exercise 12.3. Consider the dynamics

$$\dot{x}_1 = x_1^2 + x_2^2 + u,$$
$$\dot{x}_2 = x_2^2 - x_1,$$
$$\dot{x}_3 = -x_2^2 + 2x_1,$$
$$y = x_1 + x_2 + x_3.$$

$$(12.33)$$

Compute the relative degree of the system and check whether the system is minimum or non-minimum phase.

Exercise 12.4. Consider the dynamics (12.33). Apply Proposition 12.15 to derive a coordinate transformation $z = \Phi(x)$ and a representation of the system dynamics in the z coordinates. Check whether the zero dynamics are asymptotically stable.

Exercise 12.5. (Nelson's car, [144, Example 4.3.12]) Consider the dynamics

$$
\begin{bmatrix} \dot{x}_1 \\ \dot{x}_2 \\ \dot{x}_3 \\ \dot{x}_4 \end{bmatrix} = \underbrace{\begin{bmatrix} \cos(x_3 + x_4) \\ \sin(x_3 + x_4) \\ \sin(x_4) \\ 0 \end{bmatrix}}_{g_1(x)} u_1 + \underbrace{\begin{bmatrix} 0 \\ 0 \\ 0 \\ 1 \end{bmatrix}}_{g_2(x)} u_2,
\tag{12.34}
$$

where x_1 and x_2 denote the position in the plane and $x_3 = \varphi$ and $x_4 = \theta$ describe angles of the axes of the car visualized in Figure 12.2. The dynamics (12.34) describe

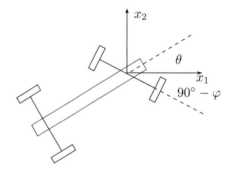

Figure 12.2: Visualization of Nelson's car in the plane.

an extension of (12.21) or (1.49), respectively.

1. Compute the Lie brackets $[g_1, g_2]$ and $[[g_1, g_2], g_1]$.
2. Show that $\operatorname{span}\{g_1, g_2, [g_1, g_2], [[g_1, g_2], g_1]\} = \mathbb{R}^4$ for all $x \in \mathbb{R}^4$ to conclude that the system is controllable.

Hint: Use MATLAB (or an equivalent program) to simplify the sine and cosine expressions.

Exercise 12.6. Prove Proposition 12.20.
    Hint: Derive and combine Taylor expansions of x, g_1, and g_2.

Exercise 12.7. Consider the dynamics

$$
\dot{x} = \begin{bmatrix} x_1 + x_3 + x_1^2 \\ x_1 + x_3 + x_2 x_1 + u \\ x_2 + 2x_3 + u \end{bmatrix}.
$$

Apply Algorithm 2 to construct an input-to-state linearization.

12.5   BIBLIOGRAPHICAL NOTES AND FURTHER READING

The increasing use of state space methods in the 1950s and 1960s led to key insights in linear systems, for example the notions of controllability and observability and the ability to consider linear subspaces related to these properties (e.g., the Kalman decomposition in Proposition 3.36). In the early 1970s it was recognized that the appropriate objects to consider for nonlinear systems were differentiable manifolds rather than subspaces, and the appropriate mathematical tools to extend the available results for linear systems to nonlinear systems were to be found in the area of differential geometry. A key first reference in this direction is [150].

The most widely used text for differential geometric control methods is [68]. Another excellent text is [35]. Proposition 12.15 follows from the results in [139, Sections 6.3, 6.4].

Chapter Thirteen

Output Regulation

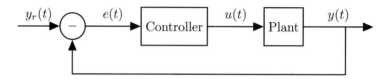

Figure 13.1: Output regulation control loop.

We now turn to the feedback loop of Figure 13.1 and consider the problem of *output regulation*. We consider perturbed systems with output,

$$\dot{x} = f(x, u, d),$$
$$y = h(x, u, d), \tag{13.1}$$

where $x \in \mathbb{R}^n$ denotes the state, $u \in \mathbb{R}^m$ denotes the input, $y \in \mathbb{R}^p$ denotes the output, and $d \in \mathbb{R}^q$ denotes an unknown perturbation or disturbance. Moreover, $f : \mathbb{R}^n \times \mathbb{R}^m \times \mathbb{R}^q \to \mathbb{R}^n$ and $h : \mathbb{R}^n \times \mathbb{R}^m \times \mathbb{R}^q \to \mathbb{R}^p$ are continuously differentiable by assumption. The goal of output regulation is a controller design u such that y asymptotically tracks a reference signal $y_r : \mathbb{R}_{\geq 0} \to \mathbb{R}^p$, regardless of the disturbance $d : \mathbb{R}_{\geq 0} \to \mathbb{R}^p$, and the control law is defined based on the error dynamics or tracking error

$$e(t) = y(t) - y_r(t). \tag{13.2}$$

We begin our analysis with linear systems, i.e., f and h in (13.1) are linear in the triple (x, u, d). In this context we will also make the *output regulation problem* more precise.

13.1 LINEAR OUTPUT REGULATION

We consider a special class of (13.1), namely the linear dynamics

$$\dot{x} = Ax + Bu + E_d d$$
$$y = Cx + Du + F_d d, \tag{13.3}$$

where the dimensions of the matrices follow from the definition of the functions f and h. In this case, the tracking error is defined as

$$e = Cx + Du + F_d d - y_r.$$

We assume furthermore that the references and disturbances come from a particular set of signals generated by

$$\dot{y}_r = A_r y_r \quad \text{and} \quad \dot{d} = A_d d,$$

where $A_r \in \mathbb{R}^{p \times p}$, $A_d \in \mathbb{R}^{q \times q}$.

We combine the reference and the disturbance in the vector $w = [y_r^T, d^T]^T$, which allows us to consider the dynamics

$$\dot{w} = A_1 w \tag{13.4}$$

called the *exosystem* and defined by

$$A_1 = \begin{bmatrix} A_r & 0 \\ 0 & A_d \end{bmatrix}.$$

With the additional definitions $E = \begin{bmatrix} 0 & E_d \end{bmatrix}$ and $F = \begin{bmatrix} -I & F_d \end{bmatrix}$, the output y in (13.3) can be replaced by the error variable

$$\begin{aligned} \dot{x} &= Ax + Bu + Ew, \\ e &= Cx + Du + Fw. \end{aligned} \tag{13.5}$$

Both asymptotic tracking and disturbance rejection correspond to driving e to zero, i.e., $\lim_{t \to \infty} e(t) = 0$. Additionally, to ensure that the states x are well behaved, we demand that $\lim_{t \to \infty} x(t) = 0$ in the case $w \equiv 0$. These two properties define the *output regulation problem*.

To solve the output regulation problem, we consider the overall dynamics, including the exosystem,

$$\begin{aligned} \dot{x} &= Ax + Bu + Ew \\ \dot{w} &= A_1 w \\ e &= Cx + Du + Fw, \end{aligned}$$

together with a static state feedback and disturbance/reference feedforward control

$$u = K_x x + K_w w. \tag{13.6}$$

Then the closed-loop system is given by

$$\begin{aligned} \dot{x}_c &= A_c x_c + B_c w \\ \dot{w} &= A_1 w \\ e &= C_c x_c + D_c w, \end{aligned} \tag{13.7}$$

where

$$A_c = A + BK_x, \qquad B_c = E + BK_w,$$
$$C_c = C + DK_x, \qquad D_c = F + DK_w.$$

Thus, the output regulation problem can be framed as trying to find feedback matrices $K_x \in \mathbb{R}^{m \times n}$ and $K_w \in \mathbb{R}^{m \times (p+q)}$ such that A_c is Hurwitz and

$$\lim_{t \to \infty} e(t) = \lim_{t \to \infty} (C_c x_c(t) + D_c w(t)) = 0 \qquad (13.8)$$

for all initial values $x_0 \in \mathbb{R}^n$, $w_0 \in \mathbb{R}^{p+q}$.

Lemma 13.1 ([65, Lemma 1.4.]). *Consider the closed-loop system* (13.7). *Assume A_c is Hurwitz and A_1 has no eigenvalues with negative real parts. Then* (13.8) *is satisfied if and only if there exists a unique matrix X_c satisfying*

$$X_c A_1 = A_c X_c + B_c, \qquad (13.9)$$
$$0 = C_c X_c + D_c. \qquad (13.10)$$

Proof. Equation (13.9) is known as the Sylvester Equation, a generalization of Lyapunov's Equation (3.4). Equation (13.9) has a unique solution if A_1 and A_c have no common eigenvalues [55, Lemma 7.1.5], which is satisfied due to the assumptions on A_c and A_1. Let $\bar{x} = x_c - X_c w$ so that

$$\dot{\bar{x}} = \dot{x}_c - X_c \dot{w} = A_c x_c + B_c w - X_c A_1 w = A_c x_c + B_c w - (A_c X_c + B_c) w$$
$$= A_c x_c - A_c X_c w = A_c (x_c - X_c w) = A_c \bar{x}.$$

Then $\lim_{t \to \infty} \bar{x}(t) = 0$ since A_c is Hurwitz.

- If $0 = C_c X_c + D_c$ then

$$\lim_{t \to \infty} e(t) = \lim_{t \to \infty} C_c \bar{x}(t) + (C_c X_c + D_c) w(t) = \lim_{t \to \infty} C_c \bar{x}(t) = 0.$$

- If (13.8) holds, then the limit condition implies $\lim_{t \to \infty} (C_c X_c + D_c) w(t) = 0$. Since $\lim_{t \to \infty} w(t) \neq 0$ (because A_1 has no eigenvalues with negative real part by assumption), necessarily $C_c X_c + D_c = 0$.

These items complete the proof. $\qquad \qquad \qquad \qquad \qquad \qquad \qquad \qquad \qquad \square$

With this result in mind, a static state feedback (13.6) can be computed in two steps:

1. Design a gain K_x so that $A_c = A + BK_x$ is Hurwitz.
2. Solve

$$X_c A_1 = (A + BK_x) X_c + BK_w + E$$
$$0 = (C + DK_x) X_c + DK_w + F \qquad (13.11)$$

for X_c and K_w.

A drawback of these two steps is that a redesign of K_x results in needing to re-solve the matrix equations (13.11). This can be avoided by considering the linear transformation

$$\begin{bmatrix} X \\ U \end{bmatrix} = \begin{bmatrix} I & 0 \\ K_x & I \end{bmatrix} \begin{bmatrix} X_c \\ K_w \end{bmatrix}$$

leading to the so-called *regulator equations,*

$$\begin{aligned} XA_1 &= AX + BU + E, \\ 0 &= CX + DU + F, \end{aligned} \tag{13.12}$$

in the unknowns $X \in \mathbb{R}^{n \times (p+q)}$ and $U \in \mathbb{R}^{m \times (p+q)}$. In these coordinates K_w satisfies $K_w = U - K_x X$ and, as a consequence, it is only necessary to solve the regulator equations once. We conclude this section with a necessary and sufficient condition on the solvability of the regulator Equations (13.12).

Theorem 13.2 (Regulator equations, [65, Theorem 1.9]). *Consider the regulator equations* (13.12) *defined through matrices* $A \in \mathbb{R}^{n \times n}$, $A_1 \in \mathbb{R}^{(p+q) \times (p+q)}$, $B \in \mathbb{R}^{n \times m}$, $C \in \mathbb{R}^{p \times m}$, $D \in \mathbb{R}^{p \times m}$, $E \in \mathbb{R}^{n \times (p+q)}$, *and* $F \in \mathbb{R}^{p \times (p+q)}$. *Then the equations* (13.12) *are solvable for any matrices* E *and* F *if and only if for all eigenvalues* λ *of* A_1,

$$\operatorname{rank}\left(\begin{bmatrix} A - \lambda I & B \\ C & D \end{bmatrix} \right) = n + p. \tag{13.13}$$

Example 13.3. Consider the system dynamics (13.5) defined by the matrices

$$A = \begin{bmatrix} 1 & 2 \\ 1 & -1 \end{bmatrix}, \quad B = \begin{bmatrix} 1 \\ 1 \end{bmatrix}, \quad C = \begin{bmatrix} 1 & 0 \end{bmatrix}, \quad D = 0$$

and

$$E = \begin{bmatrix} 0 & 1 & 1 \\ 0 & -1 & 1 \end{bmatrix}, \quad F = \begin{bmatrix} -1 & 0 & 0 \end{bmatrix}.$$

Note that the structure of F implies F_d is in fact zero, i.e., the disturbances do not directly affect the output but only act on the plant through E_d.

Assume that we wish to track constant references and that disturbances will be composed of sinusoids of a fixed frequency. This gives rise to an exosystem model (13.4) given by the matrix

$$A_1 = \begin{bmatrix} 0 & 0 & 0 \\ 0 & 0 & -1 \\ 0 & 1 & 0 \end{bmatrix}.$$

The eigenvalues of A_1 are given by $\lambda_1 = j$, $\lambda_2 = -j$, and $\lambda_3 = 0$. Thus, with respect to Theorem 13.2, the matrices

$$\begin{bmatrix} 1-j & 2 & 1 \\ 1 & -1-j & 1 \\ 1 & 0 & 0 \end{bmatrix}, \quad \begin{bmatrix} 1+j & 2 & 1 \\ 1 & -1+j & 1 \\ 1 & 0 & 0 \end{bmatrix}, \quad \begin{bmatrix} 1 & 2 & 1 \\ 1 & -1 & 1 \\ 1 & 0 & 0 \end{bmatrix}$$

need to have rank 3 to ensure that the equations (13.12) are solvable. It is straightforward to verify that the matrices have rank 3, either by calculating the determinants or by inspecting the columns of the matrices directly.

The regulator equations (13.12) can be solved through CVX [57], for example, (see Appendix B.3.2) and in this case return the solution

$$X = \begin{bmatrix} 1 & 0 & 0 \\ 0 & -0.6 & -0.2 \end{bmatrix}$$
$$U = \begin{bmatrix} -1 & 0.2 & -0.6 \end{bmatrix}.$$

A feedback gain matrix

$$K_x = \begin{bmatrix} -1.67 & -1.33 \end{bmatrix}$$

such that $A_c = A + BK_x$ is Hurwitz and has the eigenvalues -1 and -2 is derived in MATLAB through `place.m`. The matrices K_x, X, and U define the feedforward gain

$$K_w = U - K_x X = \begin{bmatrix} 0.67 & -0.6 & -0.87 \end{bmatrix},$$

and thus we have derived the static control law (13.6) for the given dynamics. The closed-loop solution of (13.7) with $x_0 = [1, 1]^T$ and $w_0 = [-2, 1, 1]^T$ using this control is shown in Figure 13.2. Observe that the output converges to the constant

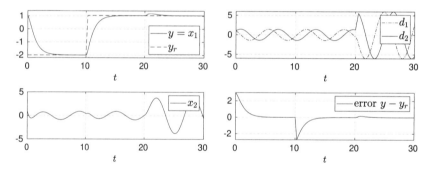

Figure 13.2: Closed-loop solution of (13.6)–(13.7) with $x_0 = [1, 1]^T$ and $w_0 = [-2, 1, 1]^T$.

reference of -2 prior to a new reference of 1 being commanded at $t = 10$, at which point y moves to track the new constant reference. At time $t = 20$ the amplitude of the disturbance changes, which, after a brief transient visible in the error, is rejected. Finally, note that the state x_2 does not converge, but this is perfectly fine as the stated goal of output regulation is for the output $y = x_1$ to track the reference y_r and for the disturbances d_1 and d_2 to not appear in the output. In other words, our stated goal says nothing about the behavior of states.

13.2   ROBUST LINEAR OUTPUT REGULATION

The results of Section 13.1 can be extended to (uncertain) linear systems

$$\begin{aligned}
\dot{x} &= A(\delta)x + B(\delta)u + E(\delta)w, \\
e &= C(\delta)x + D(\delta)u + F(\delta)w.
\end{aligned} \tag{13.14}$$

Here, for $\delta \in \mathbb{R}_{\geq 0}$,

$$A(\delta) = A + A_\delta$$

represents a nominal plant matrix A, as in (13.5), and an arbitrary perturbation A_δ with the property $\|A_\delta\| \leq \delta$. The other matrices $B(\delta)$, $C(\delta)$, $D(\delta)$, $E(\delta)$, and $F(\delta)$ are defined in the same way. For $\delta = 0$, the unperturbed dynamics (13.5) are recovered.

Recalling the arguments used in the preceding section, the same analysis could be applied here to solve the output regulation problem. However, the feedforward gain K_w in (13.6) depends on the solution of

$$\begin{aligned}
X A_1 &= A(\delta)X + B(\delta)U + E(\delta) \\
0 &= C(\delta)X + D(\delta)U + F(\delta)
\end{aligned} \tag{13.15}$$

and, if the perturbations vary, K_w would need to vary as well (recalling that we designed K_x in Step 1 before (13.11)). To provide a more reasonable solution to this problem, instead of the static state feedback (13.6) we consider a *dynamic state feedback*

$$\begin{aligned}
\dot{\xi} &= G_1\xi + G_2 e, \tag{13.16a} \\
u &= K_1 x + K_2 \xi \tag{13.16b}
\end{aligned}$$

consisting of a dynamic compensator (13.16a) with state $\xi \in \mathbb{R}^{n_\xi}$ and a linear feedback (13.16b). For the derivation of a dynamic state feedback we have a closer look at the exosystem (13.4) and its characteristic polynomial. To this end, we denote the characteristic polynomial of the exosystem by

$$\alpha_{A_1}(s) = \det(sI - A_1) = s^{n_m} + \alpha_1 s^{(n_m - 1)} + \cdots + \alpha_{(n_m - 1)}s + \alpha_{n_m}. \tag{13.17}$$

Then the matrices

$$\hat{G}_1 = \begin{bmatrix}
0 & 1 & \cdots & 0 \\
0 & 0 & \cdots & 0 \\
\vdots & \vdots & \ddots & \vdots \\
0 & 0 & \cdots & 1 \\
-\alpha_{n_m} & -\alpha_{(n_m - 1)} & \cdots & -\alpha_1
\end{bmatrix}, \qquad
\hat{G}_2 = \begin{bmatrix}
0 \\
0 \\
\vdots \\
0 \\
1
\end{bmatrix}$$

define a controllable pair $(\hat{G}_1, \hat{G}_2)$ and $\hat{G}_1$ has (13.17) as its characteristic polynomial. In other words, $\hat{G}_1$ and the exosystem A_1 have the same characteristic polynomial. We define G_1 and G_2 as block diagonal matrices with p copies of $\hat{G}_1$

and $\hat{G}_2$, respectively. In particular, we define the matrices

$$
G_1 = \begin{bmatrix} \hat{G}_1 & 0 & \cdots & 0 \\ 0 & \ddots & \ddots & \vdots \\ \vdots & \ddots & \ddots & 0 \\ 0 & \cdots & 0 & \hat{G}_1 \end{bmatrix} \quad \text{and} \quad G_2 = \begin{bmatrix} \hat{G}_2 & 0 & \cdots & 0 \\ 0 & \ddots & \ddots & \vdots \\ \vdots & \ddots & \ddots & 0 \\ 0 & \cdots & 0 & \hat{G}_2 \end{bmatrix}. \tag{13.18}
$$

Remember that p denotes the dimension of the error dynamics. Under these assumptions and definitions, (13.16a) is called a p copy *internal model* of the exosystem (13.4).

Before we proceed with the derivation of feedback gain matrices in (13.16b), we illustrate the constructions so far on an example.

Example 13.4. Let $p = 2$ and suppose the exosystem is given by

$$
\dot{w} = A_1 w = \begin{bmatrix} 0 & 1 \\ -1 & 0 \end{bmatrix} w.
$$

Note that A_1 has the eigenvalues $\pm j$ and the characteristic polynomial $s^2 + 1$, and thus A_1 satisfies the assumptions of Lemma 13.1. According to the construction so far we can define the matrices

$$
\hat{G}_1 = \begin{bmatrix} 0 & 1 \\ -1 & 0 \end{bmatrix}, \qquad \hat{G}_2 = \begin{bmatrix} 0 \\ 1 \end{bmatrix}
$$

and

$$
G_1 = \begin{bmatrix} 0 & 1 & 0 & 0 \\ -1 & 0 & 0 & 0 \\ 0 & 0 & 0 & 1 \\ 0 & 0 & -1 & 0 \end{bmatrix}, \qquad G_2 = \begin{bmatrix} 0 & 0 \\ 1 & 0 \\ 0 & 0 \\ 0 & 1 \end{bmatrix}.
$$

Note the structure of the G_1 and G_2, where each component of the error dynamics drives a copy of the exosystem. Specifically, $p = 2$ means $e = [e_1 \ e_2]^T \in \mathbb{R}^2$. With a slight abuse of notation, we write $\tilde{\xi}_1 = [\xi_1 \ \xi_2]^T \in \mathbb{R}^2$ and $\tilde{\xi}_2 = [\xi_3 \ \xi_4] \in \mathbb{R}^2$. Then the dynamic states of the controller (13.16a) are

$$
\dot{\tilde{\xi}}_1 = \hat{G}_1 \tilde{\xi}_1 + \hat{G}_2 e_1
$$
$$
\dot{\tilde{\xi}}_2 = \hat{G}_1 \tilde{\xi}_2 + \hat{G}_2 e_2.
$$

To proceed with the controller design, we consider the system composed of the (perturbed) system dynamics and the dynamic state feedback

$$
\begin{aligned}
\dot{x} &= A(\delta)x + B(\delta)u + E(\delta)w, \\
\dot{\xi} &= G_1 \xi + G_2 e, \\
\dot{w} &= A_1 w, \\
e &= C(\delta)x + D(\delta)u + F(\delta)w.
\end{aligned} \tag{13.19}
$$

With the definition $x_c = \begin{bmatrix} x^T, \xi^T \end{bmatrix}^T$ and $u = K_1 x + K_2 \xi$, with K_1 and K_2 to be determined, the system dynamics (13.19), can be expressed compactly as

$$\dot{x}_c = A_c(\delta)x_c + B_c(\delta)w$$
$$\dot{w} = A_1 w \qquad (13.20)$$
$$e = C_c(\delta)x_c + D_c(\delta)w,$$

where the closed-loop matrices are defined through the following identities

$$A_c(\delta) = \begin{bmatrix} A(\delta) + B(\delta)K_1 & B(\delta)K_2 \\ G_2(C(\delta) + D(\delta)K_1) & G_1 + G_2 D(\delta)K_2 \end{bmatrix}, \quad B_c(\delta) = \begin{bmatrix} E(\delta) \\ G_2 F(\delta) \end{bmatrix},$$

$$C_c(\delta) = \begin{bmatrix} C(\delta) + D(\delta)K_1 & D(\delta)K_2 \end{bmatrix}, \qquad\qquad D_c(\delta) = F(\delta).$$

This effectively brings us back to our original (non-robust) output regulation problem, where we now incorporate the dynamic state feedback (13.16), and where the existence of (K_1, K_2, G_1, G_2) corresponds to the solvability of the regulator equations

$$X_c A_1 = A_c(\delta)X_c + B_c(\delta),$$
$$0 = C_c(\delta)X_c + D_c(\delta). \qquad (13.21)$$

Since we have already fixed the matrices G_1 and G_2, we say that the dynamic state feedback (13.16) defined through (K_1, K_2) solves the *robust output regulation problem* if the matrix $A_c(0)$ is Hurwitz and if there exists a $\Delta > 0$ such that

$$\lim_{t\to\infty} e(t) = \lim_{t\to\infty} C_c(\delta)x_c(t) + D_c(\delta)w(t) = 0 \qquad (13.22)$$

for all $\delta \leq \Delta$.   .

Lemma 13.5 ([65, Lemma 1.26]). *Consider the dynamics (13.5) (or equivalently, the dynamics (13.14) for $\delta = 0$). Additionally, let the following properties be satisfied by assumption:*

- *The pair (A, B) is stabilizable.*
- *The matrix A_1 defining the exosystem (13.4) has no eigenvalues with negative real parts.*
- *Condition (13.13) is satisfied for all eigenvalues λ of A_1.*
- *The matrices (G_1, G_2) are defined according to (13.18).*

Then the pair

$$\left(\begin{bmatrix} A & 0 \\ G_2 C & G_1 \end{bmatrix}, \begin{bmatrix} B \\ G_2 D \end{bmatrix} \right)$$

is stabilizable.

This result shows that we can find matrices K_1 and K_2 (using pole placement discussed in Section 3.5.3, for example) such that $A_c(0)$ is Hurwitz. Under the assumptions of Lemma 13.5, the next result guarantees that the robust output regulation problem can be solved by a dynamic state feedback (13.16) and in particular guarantees the existence of K_1 and K_2 guaranteeing robust output regulation.

Theorem 13.6 (Robust output regulation, [65, Theorem 1.30]). *Consider the system* (13.14) *together with the exosystem* (13.4) *and the dynamic state feedback* (13.16), *and let* G_1, G_2 *be defined according to* (13.18). *Assume that* A_1 *has no eigenvalues with negative real parts and* (A, B) *is stabilizable.*

Then the following statements are equivalent.

1. *The condition* (13.13) *is satisfied for all eigenvalues* λ *of* A_1.
2. *The linear robust output regulation problem is solvable by a dynamic state feedback controller* (K_1, K_2, G_1, G_2).
3. *There exists* $\Delta > 0$ *such that for each* $\delta \leq \Delta$, *the regulator equations* (13.15) *have a solution* $(X(\delta), U(\delta))$.

The design of (G_1, G_2, K_1, K_2) in Theorem 13.6 is noticeably only dependent on the nominal dynamics (13.14) with $\delta = 0$. While Theorem 13.6 shows robustness with respect to uncertainty in the system dynamics, the robustness margin $\Delta > 0$ cannot be deduced from the result.

We have observed that the static control $u = K_x + K_w w$ of (13.6) includes a feedforward component, and the regulator equations (13.11) provide a way, in particular, to calculate the feedforward gain K_w. By contrast, (13.16) is a feedback problem which, under appropriate assumptions such as those here, can be solved for example by pole placement, as described in the examples below.

The result of Theorem 13.6 is effectively what is referred to as the *internal model principle*, which refers to the fact that in order to track a reference or reject a disturbance, the controller needs to incorporate a model of the reference or disturbance. This can be seen in the above development, where a model of the exosystem is embedded in the controller explicitly via the matrix $\hat{G}_1$, which has the same characteristic polynomial as the exosystem. In turn, the exosystem is a model of the references and disturbances for which perfect tracking can be achieved.

Example 13.7. (PI Control) Consider the one-dimensional linear system given by

$$\dot{x} = x + u.$$

Following the above development, we propose a dynamic compensator and control as

$$\dot{\xi} = G_1 \xi + G_2 e$$
$$u = K_1 x + K_2 \xi.$$

As a first step, we fix a nominal stabilizing controller $K_1 = -2$ and rewrite the plant-controller dynamics as

$$\dot{x} = -x + \tilde{u}$$
$$\dot{\xi} = G_1 \xi + G_2 e$$
$$u = -2x + K_2 \xi = -2x + \tilde{u}.$$

Suppose now that we wish to robustly track a constant reference, which we denote by $y_r \in \mathbb{R}$. The corresponding exosystem is simply

$$\dot{w} = 0,$$

which implies

$$G_1 = 0, \quad G_2 = 1.$$

This then leads to the overall system dynamics

$$\dot{x} = x + u$$
$$\dot{\xi} = e = x - y_r$$
$$u = -2x + K_2\xi,$$

where $K_2 > 0$ is a parameter we can tune to achieve a desired closed-loop performance. Note that K_1 could also be used as a tuneable parameter subject to $K_1 < -1$ to ensure asymptotic stability.

By adding and subtracting the tracking reference and making the integration specific, we can rewrite the control u purely in terms of the error and the reference as

$$\dot{x} = x + u$$
$$u = K_1 e + K_2 \int_0^t e(\tau)d\tau + K_1 y_r.$$

We immediately recognize the above as a classical PI (proportional-integral) controller, which, from classical frequency domain control design methods, we know can track a constant reference with an error that converges to zero. This then provides an illustration of the internal model principle whereby the controller contains a model of the exosystem.

Example 13.8. We return to the plant and exosystem presented in Example 13.3.

First, we consider the lack of robustness in the model by using the controller designed for A when the controlled plant is actually

$$A(\delta) = \begin{bmatrix} 1 & 2.1 \\ 1.1 & -1 \end{bmatrix}.$$

In Figure 13.3 we show the response limited to just trying to regulate the output to the constant -2 while rejecting the sinusoidal disturbances. Note that neither goal is accomplished. This is the same task achieved in the first 10 seconds in Figure 13.2 when the plant is known exactly.

Following the described design procedure, the characteristic polynomial of A_1 is given by $\alpha_{A_1}(s) = s^3 + s$, and hence, noting that $p = 1$,

$$G_1 = \begin{bmatrix} 0 & 1 & 0 \\ 0 & 0 & 1 \\ 0 & -1 & 0 \end{bmatrix}, \quad \text{and} \quad G_2 = \begin{bmatrix} 0 \\ 0 \\ 1 \end{bmatrix}.$$

With the disturbance w set to zero, the combined plant and internal model can be written as

$$\dot{z} = \begin{bmatrix} A & 0 \\ G_2 C & G_1 \end{bmatrix} z + \begin{bmatrix} B \\ 0 \end{bmatrix} u,$$

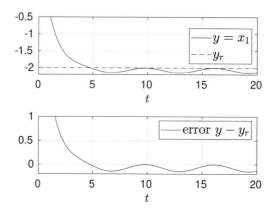

Figure 13.3: Closed-loop solution of (13.6)–(13.7) with $x_0 = [1,1]^T$, $w_0 = [-2,1,1]^T$, and plant $A(\delta)$.

where $z = [x, \xi]^T$ and the 0 matrices have appropriate dimensions. Choosing the desired closed-loop poles as $\{-1, -1.5, -2, -2.5, -3\}$ and using `place.m` in MATLAB yields the feedback gains $K_1 = [8\ 2]$ and $K_2 = [7.5\ 12.25\ 16.75]$.

Figure 13.4 shows the same simulations as Figure 13.2, except that the nominal plant A is replaced with the perturbed plant $A(\delta)$. Observe that, even though we designed the controller based on the nominal plant A, with the perturbed plant $A(\delta)$, we still track the reference signal y_r and reject the sinusoidal disturbances (not shown here, but taken as the same as in Figure 13.2). While the tracking and disturbance rejection goals are achieved, we do observe a degradation in the transient response from Figure 13.2 to Figure 13.4.

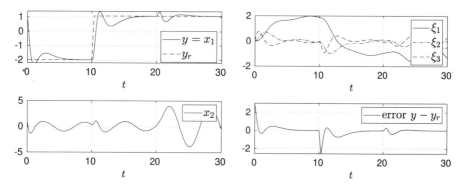

Figure 13.4: Closed-loop solution of (13.14) and (13.16) with $x_0 = [1,1]^T$, $w_0 = [-2,1,1]^T$, and plant $A(\delta)$.

13.3 NONLINEAR OUTPUT REGULATION

We now return to the nonlinear system (13.1) introduced at the beginning of the chapter. We again combine the reference and the disturbance in a single variable

$w = [y_r^T, d^T]^T$, but compared to (13.4) allow nonlinear dynamics for the exosystem

$$\dot{w} = a_1(w), \tag{13.23}$$

where $a_1 : \mathbb{R}^{p+q} \to \mathbb{R}^{p+q}$ is continuously differentiable and $a_1(0) = 0$. Moreover, we assume that the origin $w = 0$ is stable but not asymptotically stable. (This property is also known as *neutral stability*.)

· Similarly to the linear setting, with (13.23), the nonlinear system (13.1) and the error dynamics (13.2) can be combined in the overall system

$$\begin{aligned}
\dot{x} &= F(x, u, w) \\
\dot{w} &= a_1(w) \\
e &= H(x, u, w),
\end{aligned} \tag{13.24}$$

where the definitions of F and H follow from f and h, i.e.,

$$F(x, u, w) = f(x, u, d), \qquad H(x, u, w) = h(x, u, d) - y_r,$$

and where $w = [y_r^T, d^T]^T$ as per the definition before (13.4).

With this setting, the (local) *nonlinear output regulation problem* is to find a controller $\kappa(x, w)$, $\kappa : \mathbb{R}^n \times \mathbb{R}^{p+q} \to \mathbb{R}^p$ such that

- the origin is locally asymptotically stable for $\dot{x} = F(x, \kappa(x, 0), 0)$; and
- in some neighborhood $\mathcal{N} \subset \mathbb{R}^n \times \mathbb{R}^{p+q}$ of the origin,

$$\lim_{t \to \infty} H(x(t), \kappa(x(t), w(t)), w(t)) = 0$$

is satisfied for all initial conditions $(x_0, w_0) \in \mathcal{N}$.

To provide a solution to the (local) nonlinear output regulation problem we linearize the system at the origin,

$$A = \left[\frac{\partial F}{\partial x}(x, u, w)\right]_{(0,0,0)}, \qquad B = \left[\frac{\partial F}{\partial u}(x, u, w)\right]_{(0,0,0)} \tag{13.25}$$

(see Sections 3.2 and 3.5), and reduce the nonlinear setting to the linear setting discussed earlier.

Theorem 13.9 ([65, Lemma 3.6]). *Consider the system* (13.24) *with linearization* (13.25) *and let the origin of the exosystem be stable but not asymptotically stable. Then, the nonlinear output regulation problem is solvable if and only if the pair* (A, B) *is stabilizable and there exist continuously differentiable functions* $x = \pi(w)$ *and* $u = c(w)$ *defined in an open neighborhood around the origin with* $\pi(0) = 0$ *and* $c(0) = 0$, *satisfying the conditions*

$$\begin{aligned}
\frac{\partial \pi}{\partial w} a_1(w) &= F(\pi(w), c(w), w), \\
0 &= H(\pi(w), c(w), w).
\end{aligned} \tag{13.26}$$

If the pair (A, B) is stabilizable, then a feedback law κ satisfying $\kappa(x, 0) = Kx$ where $A + BK$ is Hurwitz renders the origin of the ordinary differential equation

$\dot{x} = F(x, \kappa(x, 0), 0)$ locally asymptotically stable. Thus, with the definitions of $c(x)$ and $\pi(w)$ in Theorem 13.9, the control law $\kappa(x, w) = c(w) + K(x - \pi(w))$ solves the nonlinear output regulation problem. Note that the control law locally also satisfies $u = \kappa(x, w) = \kappa(\pi(w), w) = c(w)$ under these conditions.

To illustrate Theorem 13.9 and in particular the partial differential equation (13.26), we return to linear systems to show that it extends the results we have seen so far.

Example 13.10. Consider the linear system

$$\begin{aligned}
\dot{x} &= F(x, u, w) = Ax + Bu + Ew, \\
\dot{w} &= a_1(w) &&= A_1 w, \\
e &= H(x, u, w) = Cx + Du + Fw,
\end{aligned}$$

leading to the partial differential equation (13.26) of the form

$$\begin{aligned}
\frac{\partial \pi}{\partial w} A_1 w &= A\pi(w) + Bc(w) + Ew, \\
0 &= C\pi(w) + Dc(w) + Fw.
\end{aligned}$$

We conjecture that $\pi(\cdot)$ and $c(\cdot)$ can be defined as linear functions in this context, i.e., we assume that $\pi(w) = Xw$ and $c(w) = Uw$ for unknown matrices X, U of appropriate dimension. Thus $\frac{\partial \pi}{\partial w}(w) = X$, and the partial differential equation reduces to the equations

$$\begin{aligned}
XA_1 w &= AXw + BUw + Ew, \\
0 &= CXw + DUw + Fw.
\end{aligned} \tag{13.27}$$

Since these equations need to be satisfied for all w (in a neighborhood around the origin), (13.27) recovers the regulator equations (13.12) and Theorem 13.9 is indeed an extension of the linear setting.

Based on Theorem 13.9 a solution of the partial differential equation (13.26) needs to be derived to solve the nonlinear output regulation problem. Since this is in general difficult or intractable, we focus on a specific form of (13.1). Specifically, we consider the control-affine single-input single-output case, i.e., $x \in \mathbb{R}^n$, $u, y \in \mathbb{R}$, and

$$\begin{aligned}
\dot{x} &= f(x) + g(x)u, \\
y &= h(x).
\end{aligned} \tag{13.28a}$$

We denote the exosystem and the reference signal by

$$\dot{w} = a_1(w) \qquad \text{and} \qquad \bar{y}_r = -p(w), \tag{13.28b}$$

respectively; i.e., we only consider reference signals $\bar{y}_r \in \mathbb{R}$ and no disturbances but still assume $w \in \mathbb{R}^{p+q}$.

We assume that the system has relative degree $r \leq n$ at the origin, according to Definition 12.7, and we use Proposition 12.15 to rewrite the system in the form

$$\dot{z}_1 = z_2,$$

$$\vdots$$

$$\dot{z}_{r-1} = z_r, \tag{13.29}$$
$$\dot{z}_r = \alpha(z, \eta) + \beta(z, \eta)u,$$
$$\dot{\eta} = q(z, \eta),$$
$$e = z_1 + p(w).$$

Moreover, we decompose $\pi(\cdot)$ in Theorem 13.9 as

$$\pi(w) = \begin{bmatrix} k_1(w) \\ \vdots \\ k_r(w) \\ \lambda(w) \end{bmatrix}, \tag{13.30}$$

where $\lambda : \mathbb{R}^{n_w} \to \mathbb{R}^{n-r}$ and, for each $i = 1, \ldots r$, $k_i : \mathbb{R}^{n_w} \to \mathbb{R}$. With the notation

$$[\dot{z}, \dot{\eta}] = F([z, \eta], u, w)$$
$$\dot{w} = a_1(w)$$
$$e = H([z, \eta], u, w)$$

representing the overall system (13.24), the partial differential equation (13.26) translates into the conditions

$$\left.\begin{array}{rcl} \frac{\partial k_1(w)}{\partial w} a_1(w) &=& k_2(w) \\ &\vdots& \\ \frac{\partial k_{r-1}(w)}{\partial w} a_1(w) &=& k_r(w) \end{array}\right\} \tag{13.31a}$$

$$\left.\frac{\partial k_r(w)}{\partial w} a_1(w) = \alpha(k(w), \lambda(w)) + \beta(k(w), \lambda(w))c(w)\right\} \tag{13.31b}$$

$$\left.\frac{\partial \lambda(w)}{\partial w} a_1(w) = q(k(w), \lambda(w))\right\} \tag{13.31c}$$

and

$$0 = k_1(w) + p(w). \tag{13.31d}$$

Thus, $k_1(w) = -p(w)$ can be obtained from (13.31d) and the conditions (13.31a) allow an iterative calculation of

$$k_i(w) = -L_{a_1}^{i-1} p(w)$$

for $i = 2, \ldots, r$ (recall the previously used Lie derivative notation in equation (9.5), for example). When $k(w)$ is known, the only unknown in (13.31c) is the function

$\lambda(w)$. Lastly, through (13.31b) the function $c(w)$ is given by

$$c(w) = - \left(\frac{L_{a_1}^r p(w) + \alpha(k(w), \lambda(w))}{\beta(k(w), \lambda(w))} \right).$$

Therefore, the nonlinear output regulation problem corresponding to (13.28) is solvable if and only if the pair (A, B), i.e., the linearization in the $[z, \eta]$-coordinates, is stabilizable and there exists $\lambda : \mathbb{R}^{n_w} \to \mathbb{R}^{n-r}$ satisfying (13.31c).

We conclude this chapter with an example.

Example 13.11. ([68, Example 8.3.5]) Consider the nonlinear dynamics

$$\begin{aligned}
\dot{z}_1 &= z_2 \\
\dot{z}_2 &= u \\
\dot{\eta} &= \eta + z_1 + z_2^2 \\
e &= z_1 + p(w)
\end{aligned} \tag{13.32}$$

in the form of (13.29), with relative degree $r = 2$, and exosystem

$$\dot{w} = \begin{bmatrix} w_2 \\ -w_1 \end{bmatrix} \doteq a_1(w), \qquad p(w) = -w_1. \tag{13.33}$$

The solution of the differential equation $\dot{w} = a_1(w)$ is given by

$$\begin{aligned}
w_1(t) &= c_1 \sin(t) + c_2 \cos(t), \\
w_2(t) &= c_1 \cos(t) - c_2 \sin(t),
\end{aligned}$$

where $c_1, c_2 \in \mathbb{R}$ are defined through the initial conditions. Thus, through the definition of the exosystem, reference signals of the form

$$\bar{y}_r(t) = c_1 \sin(t) + c_2 \cos(t) \tag{13.34}$$

can be tracked. In this example we use output regulation to compute a feedback law

$$u = \kappa(x, w) = c(w) + K(x - \pi(w))$$

such that $\lim_{t \to \infty} e(t) = 0$, i.e., $x_1(t) = z_1(t)$ converges to the reference signal. With $x = [x_1, x_2, x_3]^T = [z_1, z_2, \eta]^T$ the function F in (13.24) is defined as

$$F(x, u, w) = \begin{bmatrix} x_2 \\ u \\ x_3 + x_1 + x_2^2 \end{bmatrix}.$$

Thus, the matrices A and B describing the linearization (13.25) at the origin are given by

$$A = \begin{bmatrix} 0 & 1 & 0 \\ 0 & 0 & 0 \\ 1 & 0 & 1 \end{bmatrix} \quad \text{and} \quad B = \begin{bmatrix} 0 \\ 1 \\ 0 \end{bmatrix}.$$

With

$$K = [-18 \quad -7 \quad -24]$$

(obtained through `place.m` in MATLAB) the closed-loop matrix $A + BK$ has the eigenvalues -1, -2, and -3 and thus is Hurwitz.

As a next step, we calculate $\pi(w)$ of the form (13.30) and $c(w)$ from conditions (13.31a)–(13.31d), i.e.,

$$\frac{\partial k_1(w)}{\partial w} a_1(w) = k_2(w),$$

$$\frac{\partial k_2(w)}{\partial w} a_1(w) = \alpha(k(w), \lambda(w)) + \beta(k(w)), \lambda(w))c(w),$$

$$\frac{\partial \lambda}{\partial w} a_1(w) = q(k(w), \lambda(w)),$$

$$0 = k_1(w) + p(w).$$

Using the definition of the particular system (13.32) and the exosystem (13.33), these conditions simplify to

$$\frac{\partial k_1(w)}{\partial w_1} w_2 - \frac{\partial k_1(w)}{\partial w_2} w_1 = k_2(w), \tag{13.35a}$$

$$\frac{\partial k_2(w)}{\partial w_1} w_2 - \frac{\partial k_2(w)}{\partial w_2} w_1 = c(w), \tag{13.35b}$$

$$\frac{\partial \lambda(w)}{\partial w_1} w_2 - \frac{\partial \lambda(w)}{\partial w_2} w_1 = \lambda(w) + k_1(w) + k_2(w)^2, \tag{13.35c}$$

$$0 = k_1(w) - w_1. \tag{13.35d}$$

From (13.35d), k_1 is defined as $k_1(w) = w_1$ and thus

$$w_2 - 0 = k_2(w)$$

needs to be satisfied according to (13.35a). Using this result in (13.35b) implies that $c(w)$ is defined as $c(w) = -w_1$. For the calculation of $\lambda : \mathbb{R}^2 \to \mathbb{R}$, we make the assumption that

$$\lambda(w) = \gamma_1 w_1 + \gamma_2 w_2 + \gamma_{11} w_1^2 + \gamma_{12} w_1 w_2 + \gamma_{22} w_2^2$$

for unknown constants $\gamma_1, \gamma_2, \gamma_{11}, \gamma_{12}, \gamma_{22} \in \mathbb{R}$. Then, (13.35c) leads to the equation

$$\gamma_1 w_2 + 2\gamma_{11} w_1 w_2 + \gamma_{12} w_2^2 - \left(\gamma_2 w_1 + \gamma_{12} w_1^2 + 2\gamma_{22} w_2 w_1\right)$$

$$= -\gamma_2 w_1 + \gamma_1 w_2 - \gamma_{12} w_1^2 + 2(\gamma_{11} - \gamma_{22})w_1 w_2 + \gamma_{12} w_2^2$$

$$= \gamma_1 w_1 + \gamma_2 w_2 + \gamma_{11} w_1^2 + \gamma_{12} w_1 w_2 + \gamma_{22} w_2^2 + w_1 + w_2^2$$

$$= (\gamma_1 + 1)w_1 + \gamma_2 w_2 + \gamma_{11} w_1^2 + \gamma_{12} w_1 w_2 + (\gamma_{22} + 1)w_2^2.$$

Comparing the coefficients reduces this expression to the linear conditions

$$-\gamma_2 = \gamma_1 + 1,$$
$$\gamma_1 = \gamma_2,$$
$$-\gamma_{12} = \gamma_{11},$$
$$2\gamma_{11} - 2\gamma_{22} = \gamma_{12},$$
$$\gamma_{12} = \gamma_{22} + 1,$$

which can be solved in MATLAB, or by inspecting the conditions directly. Here, we obtain

$$\gamma_1 = \gamma_2 = -\tfrac{1}{2}, \qquad \gamma_{11} = -\tfrac{2}{5}, \qquad \gamma_{12} = \tfrac{2}{5}, \qquad \text{and} \qquad \gamma_{22} = -\tfrac{3}{5},$$

and hence

$$\lambda(w) = -\frac{1}{2}w_1 - \frac{1}{2}w_2 - \frac{2}{5}w_1^2 + \frac{2}{5}w_1 w_2 - \frac{3}{5}w_2^2.$$

Finally, we can define the feedback law

$$\kappa(x, w) = c(w) - K(x - \pi(w)) = -w_1 + [18 \quad 7 \quad 24] \begin{bmatrix} x_1 - w_1 \\ x_2 - w_2 \\ x_3 - \lambda(w) \end{bmatrix}, \qquad (13.36)$$

which solves the output regulation problem. In Figure 13.5 the tracking performance of the control law (13.36) for the nonlinear system is visualized. For the

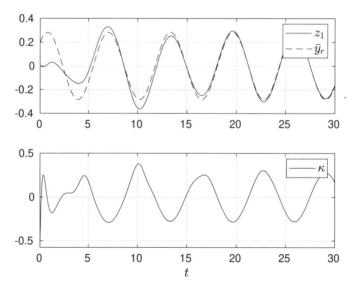

Figure 13.5: Closed-loop solution of (13.32) using the feedback law (13.36).

reference signal (13.34) the constants are set to $c_1 = c_2 = 0.2$ in the simulation.

13.4   EXERCISES

Exercise 13.1. Repeat the calculations in Example 13.3 to compute a static state feedback $u = K_x x + K_w w$ for the system defined through the matrices

$$A = \begin{bmatrix} 1 & -1 & 3 \\ 2 & -1 & 1 \\ 0 & 1 & 0 \end{bmatrix}, \quad B = \begin{bmatrix} 1 \\ 1 \\ 0 \end{bmatrix}, \quad C = \begin{bmatrix} 1 & 0 & 1 \end{bmatrix}, \quad D = 0$$

and

$$E = \begin{bmatrix} 0 & 1 & 1 \\ 0 & -1 & 1 \\ 0 & 0 & 1 \end{bmatrix}, \quad F = \begin{bmatrix} -1 & 0 & 0 \end{bmatrix}.$$

Here, define K_x such that A_c has the eigenvalues -1, -2, and -3 and initialize your simulations with $x = [1, 1, 1]^T$ and $w = [-2, 1, 1]^T$.

13.5   BIBLIOGRAPHICAL NOTES AND FURTHER READING

The internal model principle, and its necessity in the context of robust linear output regulation, was initially published in [50] (see also [165]). Results for the nonlinear setting have been collected in [36], [65], and [68]. The references [36] and [65] additionally provide summaries of the linear setting. Here we have largely followed the development in [65, Chapters 1 and 3] and [68, Chapter 8].

The controllers discussed in this chapter are referred to as the *full information output regulation problem* since a controller based on the state x is defined. A full theory is available for *error feedback*, rather than state feedback, based on the output y, but is not discussed here for simplicity.

Practical applications of output regulation can be found in [69]. This reference also incorporates ideas from adaptive control in order to adapt the internal model.

Chapter Fourteen

Optimal Control

In previous chapters we discussed methods to design stabilizing feedback laws. In this chapter we introduce *optimal feedback laws*, where optimality is to be understood with respect to a certain *performance criterion*. While we start with the continuous time setting, *optimal control* in both the continuous time setting as well as in the discrete time setting is discussed in this chapter.

14.1 OPTIMAL CONTROL—CONTINUOUS TIME SETTING

As before, we consider the continuous time system

$$\dot{x}(t) = f(x(t), u(t)), \quad x(0) = x_0 \in \mathbb{R}^n \tag{14.1}$$

with locally Lipschitz continuous right-hand side $f : \mathbb{R}^n \times \mathbb{R}^m \to \mathbb{R}^n$, state $x \in \mathbb{R}^n$, and input $u \in \mathbb{R}^m$. The set of inputs and the set of solutions of (14.1) is denoted by

$$\begin{aligned}
\mathcal{U} &= \{u(\cdot) : \mathbb{R}_{\geq 0} \to \mathbb{R}^m | \ u(\cdot) \text{ measurable}\}, \\
\mathcal{X} &= \{x(\cdot) : \mathbb{R}_{\geq 0} \to \mathbb{R}^n | \ x(\cdot) \text{ is absolutely continuous}\}.
\end{aligned} \tag{14.2}$$

We say that $(x(\cdot), u(\cdot)) \in \mathcal{X} \times \mathcal{U}$ is a *solution pair* if it satisfies (14.1) for almost all $t \in \mathbb{R}_{\geq 0}$. The condition *for almost all $t \in \mathbb{R}_{\geq 0}$* instead of *for all $t \in \mathbb{R}_{\geq 0}$* allows a larger class of solutions $x(\cdot)$. In particular, it is sufficient if $x(\cdot)$ is continuously differentiable for almost all $t \geq 0$. This furthermore allows us to consider functions $u(\cdot)$ which are not necessarily continuous but piecewise continuous, for example. Under appropriate regularity conditions on the function f, a solution $x(\cdot) \in \mathcal{X}$ is uniquely defined through $x_0 \in \mathbb{R}^n$ and $u(\cdot) \in \mathcal{U}$.

If the initial condition is important and not clear from the context, $x(\cdot; x_0) \in \mathcal{X}$ and $u(\cdot; x_0) \in \mathcal{U}$ is used to highlight the initial condition of $x(\cdot)$ and $u(\cdot)$, respectively.

For a solution pair $(x(\cdot), u(\cdot)) \in \mathcal{X} \times \mathcal{U}$ we define the *cost functional* $J : \mathbb{R}^n \times \mathcal{U} \to \mathbb{R} \cup \{\pm\infty\}$ as

$$J(x_0, u(\cdot)) = \int_0^\infty \ell(x(\tau), u(\tau)) d\tau. \tag{14.3}$$

Here, the cost functional is defined through the *running cost* $\ell : \mathbb{R}^n \times \mathbb{R}^m \to \mathbb{R}$. Alternatively, the cost functional J is referred to as the *performance criterion*. We further define $V : \mathbb{R}^n \to \mathbb{R}_{\geq 0}$ by

$$V(x_0) = \min_{u(\cdot) \in \mathcal{U}} J(x_0, u(\cdot)) \tag{14.4}$$

as the *(optimal) value function*. For the purpose of this chapter, we tacitly assume that J and $\mathcal{U}$ are defined such that the minimum in (14.4) exists. In the general setting, the minimum needs to be replaced by the infimum. Sometimes, to be explicit about the dependence of the optimization problem (14.4) on the dynamics we write

$$V(x_0) = \min_{u(\cdot) \in \mathcal{U}} J(x_0, u(\cdot))$$
$$\text{subject to (14.1).} \tag{14.5}$$

It is important to note that the initial state, x_0, and the input signal $u(\cdot)$ are sufficient to describe the solution $x(\cdot)$. Alternatively, the optimization problem (14.4) can be written in terms of the optimal open-loop control law

$$u^\star(\cdot) = \arg \min_{u(\cdot) \in \mathcal{U}} J(x_0, u(\cdot)). \tag{14.6}$$

The definition of $u^\star(\cdot)$ indicates that we assume that the optimal open-loop control law is uniquely defined, which might not be the case in general. For simplicity we will use (14.6) instead of the more precise definition $u^\star(\cdot) \in \arg \min_{u(\cdot) \in \mathcal{U}} J(x_0, u(\cdot))$ throughout this chapter.

The best outcome of the problem (14.4) or (14.6), respectively, would be a state feedback, i.e., a feedback law $\mu : \mathbb{R}^n \to \mathbb{R}^m$ such that

$$\mu(x^\star(t)) = u^\star(t) \qquad \forall \, t \in \mathbb{R}_{\geq 0}. \tag{14.7}$$

Here, $x^\star(\cdot)$ denotes the solution uniquely defined through $u^\star(\cdot)$ and $x^\star(0) = x_0$, and thus $(x^\star(\cdot), u^\star(\cdot)) \in \mathcal{X} \times \mathcal{U}$ is an optimal solution pair. In some cases, the selection of μ is indeed possible. In order to explicitly obtain optimal state feedback solutions, we will make use of the concept of a *feedback invariant*.

Definition 14.1 (Feedback invariant). *Consider the system* (14.1). *A functional* $H : \mathcal{X} \times \mathcal{U} \to \mathbb{R}$ *is called feedback invariant with respect to the inputs and solutions* $\mathcal{X} \times \mathcal{U}$ *if for all solution pairs* $(x_1(\cdot), u_1(\cdot)), (x_2(\cdot), u_2(\cdot)) \in \mathcal{X} \times \mathcal{U}$ *with* $x_1(0) = x_2(0)$ *the equality*

$$H(x_1(\cdot), u_1(\cdot)) = H(x_2(\cdot), u_2(\cdot))$$

holds.

A feedback invariant is thus a functional $H(x(\cdot), u(\cdot))$ whose value, when computed along a solution to (14.1), depends only on the initial state x_0 and not the particular input $u(\cdot)$. We will see examples of such functions below.

Suppose we have computed a feedback invariant for (14.1). In this case, if there exists a function $\Lambda : \mathbb{R}^n \times \mathbb{R}^m \to \mathbb{R}$ such that we can decompose the cost function as

$$J(x_0, u(\cdot)) = H(x(\cdot), u(\cdot)) + \int_0^\infty \Lambda(x(\tau), u(\tau)) d\tau$$

where

$$\min_{u \in \mathbb{R}^m} \Lambda(x, u) = 0, \qquad \forall\, x \in \mathbb{R}^n,$$

then an optimal (state) feedback stabilizer is given by

$$\mu(x(t)) = \arg\min_{u \in \mathbb{R}^m} \Lambda(x(t), u). \tag{14.8}$$

To see that this is indeed the case, observe that since the feedback invariant H has the same value regardless of the specific input, the choice of u will not change the contribution of H to J. For the integral term, since the choice of u which results in the least cost is one which makes Λ zero, we can clearly do no better than (14.8). In particular, using the definitions introduced in this chapter, the chain of equalities

$$\begin{aligned} V(x_0) &= \min_{u(\cdot) \in \mathcal{U}} J(x_0, u(\cdot)) \\ &= \min_{u(\cdot) \in \mathcal{U}} \left(H(x(\cdot), u(\cdot)) + \int_0^\infty \Lambda(x(\tau), u(\tau))\, d\tau \right) \\ &= H(x(\cdot), u(\cdot)) + \int_0^\infty \min_{u(\cdot) \in \mathcal{U}} (\Lambda(x(\tau), u(\tau)))\, d\tau \\ &= H(x(\cdot), u(\cdot)) \end{aligned}$$

is satisfied. In the next two sections, we will look at specific examples of how this can be applied.

14.1.1 Linear Quadratic Regulator

For linear systems

$$\dot{x}(t) = Ax(t) + Bu(t), \quad x(0) = x_0 \in \mathbb{R}^n, \tag{14.9}$$

$A \in \mathbb{R}^{n \times n}$, $B \in \mathbb{R}^{n \times m}$, it is possible to derive explicit feedback invariants. However, to ensure that H is well defined, i.e., $H(x(\cdot), u(\cdot)) < \infty$, we need to restrict the sets (14.2) to solution pairs $(x(\cdot), u(\cdot)) \in \mathcal{X} \times \mathcal{U}$ with the additional property $\lim_{t \to \infty} x(t) = 0$. We thus define

$$\mathcal{X}_s = \left\{ x(\cdot) \in \mathcal{X} : \lim_{t \to \infty} x(t) = 0 \right\}. \tag{14.10}$$

Theorem 14.2 (Feedback invariant). *Consider the linear system (14.9) with solution pairs $(x(\cdot), u(\cdot)) \in \mathcal{X}_s \times \mathcal{U}$. Then, for any symmetric matrix $P \in \mathcal{S}^n$, the* *functional* $H : \mathcal{X}_s \times \mathcal{U}$ *defined as*

$$H(x(\cdot), u(\cdot))$$
$$= -\int_0^\infty (Ax(\tau) + Bu(\tau))^T Px(\tau) + x^T(\tau)P(Ax(\tau) + Bu(\tau)) \, d\tau \qquad (14.11)$$

is a feedback invariant.

Proof. Consider a solution pair $(x(\cdot), u(\cdot)) \in \mathcal{X}_s \times \mathcal{U}$. Then integral (14.11) can be rewritten as

$$-\int_0^\infty \tfrac{d}{d\tau}\left(x^T(\tau)Px(\tau)\right) d\tau = -x^T(\tau)Px(\tau)\big|_0^\infty$$
$$= x^T(0)Px(0) - \lim_{t\to\infty} x^T(t)Px(t). \qquad (14.12)$$

Since $x(\cdot) \in \mathcal{X}_s$ by assumption, the term $\lim_{t\to\infty} x^T(t)Px(t)$ vanishes and (14.12) depends only on $x(0)$. $\qquad\square$

We will use the feedback invariant (14.11) to derive an optimal feedback stabilizer for the linear system (14.9). In particular, we define a feedback stabilizer which is optimal with respect to the quadratic cost function

$$J(x_0, u(\cdot)) = \int_0^\infty \left(x^T(\tau)Qx(\tau) + u^T(\tau)Ru(\tau)\right) d\tau \qquad (14.13)$$

where $Q \in \mathcal{S}_{\geq 0}^n$ and $R \in \mathcal{S}_{>0}^m$. The linear dynamics (14.9) together with the quadratic cost function (14.13) used to derive a feedback which stabilizes (or regulates to) the origin, combine to give the nomenclature of the *linear quadratic regulator* (LQR).

In order to simplify the following expressions, we suppress the integration variable τ. We first add and subtract the feedback invariant (14.11) to the cost function (14.13):

$$J(x_0, u(\cdot)) = H(x(\cdot), u(\cdot))$$
$$+ \int_0^\infty x^T Qx + u^T Ru + (Ax + Bu)^T Px + x^T P(Ax + Bu) \, d\tau.$$

Since by assumption $R > 0$, i.e., R^{-1} is well defined, rearranging terms under the integral and completing the squares yields

$$\int_0^\infty x^T Q x + u^T R u + (Ax + Bu)^T P x + x^T P (Ax + Bu) \ d\tau$$

$$= \int_0^\infty x^T (Q + A^T P + PA)x + u^T R u + 2u^T B^T P x \ d\tau$$

$$= \int_0^\infty \left(x^T (Q + A^T P + PA)x + u^T R u + 2u^T B^T P x \right.$$
$$\left. + x^T P B^T R^{-1} B P x - x^T P B^T R^{-1} B P x \right) d\tau$$

$$= \int_0^\infty \left(x^T (Q + A^T P + PA - PBR^{-1}B^T P)x \right.$$
$$\left. + (u + R^{-1}B^T P x)^T R (u + R^{-1}B^T P x) \right) d\tau.$$

If $P \in \mathcal{S}_{\geq 0}^n$ can be chosen so that

$$A^T P + PA + Q - PBR^{-1}B^T P = 0$$

then we see that the above cost function reduces to

$$J(x_0, u(\cdot)) = H(x(\cdot), u(\cdot)) + \int_0^\infty (u + R^{-1}B^T P x)^T R (u + R^{-1}B^T P x) d\tau.$$

Having assumed that $R > 0$, we see that

$$\Lambda(x, u) \doteq (u + R^{-1}B^T P x)^T R (u + R^{-1}B^T P x) \qquad (14.14)$$

has a minimum at zero given by $\mu(x(t)) = u(t) = -R^{-1}B^T P x(t)$.

For $H(x(\cdot), u(\cdot))$ to be a feedback invariant, Theorem 14.2 requires that

$$\lim_{t \to \infty} x(t) = 0,$$

or, equivalently, that $(x(\cdot), u(\cdot)) \in \mathcal{X}_s \times \mathcal{U}$ is satisfied. This is guaranteed if the closed-loop system matrix $A - BR^{-1}B^T P$, obtained from the above feedback, is Hurwitz. We summarize the above derivation in the following theorem.

Theorem 14.3. *Consider the linear system* (14.9) *and the quadratic cost function* (14.13) *defined through* $Q \in \mathcal{S}_{\geq 0}^n$, $R \in \mathcal{S}_{>0}^m$. *If there exists a positive semidefinite matrix* $P \in \mathcal{S}_{\geq 0}^n$ *satisfying the continuous time algebraic Riccati equation*

$$A^T P + PA + Q - PBR^{-1}B^T P = 0 \qquad (14.15)$$

and if $A - BR^{-1}B^T P$ *is Hurwitz, then* $\mu(x) = -R^{-1}B^T P x$ *minimizes* (14.13) *and the optimal value function is given by*

$$V(x_0) = x_0^T P x_0.$$

While Theorem 14.3 relies on the existence of a solution to (14.15), the following result guarantees that such a solution exists, and hence guarantees the existence of an optimal feedback stabilizer. A derivation of the result can be found in [63, Chapter 20].

Theorem 14.4 (Linear quadratic regulator). *Consider the linear system* (14.9), *assume that* (A, B) *is stabilizable, let* $Q \in \mathcal{S}^n$, $R \in \mathcal{S}^m$, *and* $N \in \mathbb{R}^{n \times m}$ *be chosen such that* $R > 0$ *is positive definite and* $Q - NR^{-1}N^T \geq 0$ *is positive semidefinite. Moreover, consider the quadratic cost function*

$$J(x_0, u(\cdot)) = \int_0^\infty x(\tau)^T Q x(\tau) + u(\tau)^T R u(\tau) + 2x(\tau)^T N u(\tau) d\tau. \qquad (14.16)$$

Then the following properties are satisfied.

- *If the pair* $(A - BR^{-1}N^T, Q - NR^{-1}N^T)$ *is detectable, then there exists a symmetric matrix* $P \in \mathcal{S}^n$ *satisfying the algebraic Riccati equation*

$$A^T P + PA + Q - (PB + N)R^{-1}(B^T P + N^T) = 0 \qquad (14.17)$$

  so that the matrix

$$A - BR^{-1}(B^T P + N^T) \qquad (14.18)$$

  is Hurwitz, i.e., the feedback $\mu(x) = -R^{-1}(B^T P + N^T)x$ *asymptotically stabilizes the origin for* (14.9).
- *If the pair* $(A - BR^{-1}N^T, Q - NR^{-1}N^T)$ *is observable, then there exists a unique* $P \in \mathcal{S}^n_{>0}$ *satisfying the algebraic Riccati equation* (14.17). *Moreover, the state feedback* $\mu(x) = -R^{-1}(B^T P + N^T)x$ *ensures that the closed-loop matrix* (14.18) *is Hurwitz and the optimal value function minimizing* (14.16) *is given by* $V(x_0) = x_0^T P x_0$, *where* V *is a Lyapunov function of the closed-loop system.*

The condition $(A - BR^{-1}N^T, Q - NR^{-1}N^T)$ observable/detectable is for example satisfied for $N = 0$, $Q = C^T C$, $C \in \mathbb{R}^{p \times n}$, and (A, C) observable/detectable. In many applications Q and R are selected as diagonal matrices. However, N adds an additional degree of freedom in the cost function. In general, small values in R (compared to Q) lead to larger controller gains and faster convergence, while large values in R (compared to Q) lead to reduced control effort and slower convergence. In contrast to pole placement discussed in Section 3.5.3, where the closed-loop poles of $A + BK$ are assigned directly, the cost function (14.16) implicitly defines the closed-loop poles of $A + BK$ through the selection of Q, R, and N.

Example 14.5. As an example of the linear quadratic regulator, we consider again the dynamics of an inverted pendulum linearized in the upright position discussed in Example 6.1:

$$\dot{x} = \begin{bmatrix} 0 & 1 \\ 9.81 & -0.1 \end{bmatrix} x + \begin{bmatrix} 0 \\ 1 \end{bmatrix} u.$$

To apply Theorem 14.4 we select $Q = \text{diag}(1, 1)$, $R = 1$ (and $N = 0$). Through the solution of the Riccati equation

$$P = \begin{bmatrix} 61.66 & 19.67 \\ 19.67 & 6.45 \end{bmatrix},$$

the feedback gain $K = [-19.67 \;\; -6.45]$ is defined. The eigenvalues of the closed-loop matrix $A + BK$ are given by -2.70 and -3.65. The closed-loop solution corresponding to the initial condition $x = [1\ 1]^T$ is shown in Figure 14.1 on the left. In contrast the closed-loop solution of $\dot{x} = A + BKx$ with $K = [-42.92\ -33.05]$

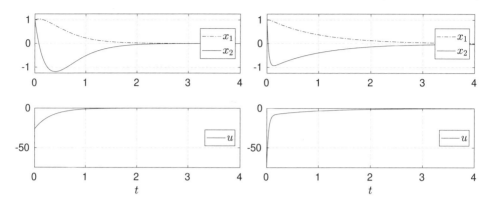

Figure 14.1: Closed-loop solutions corresponding to the linearized inverted pendulum using a linear quadratic regulator for the controller design. The control laws are defined through Theorem 14.4 for different selections of Q and R.

corresponding to the selection $Q = \mathrm{diag}(1, 1)$, $R = 10^{-3}$ (and $N = 0$) is illustrated in Figure 14.1 on the right. By penalizing the input less, a more aggressive feedback law and larger controller gains can be observed. We invite the reader to experiment with the selection of Q and R and its impact on the convergence rate of the state x, the magnitude of the input u, and the eigenvalues of the closed-loop matrix $A + BK$.

14.1.2   Control-Affine Nonlinear Systems

In this section, we extend the result for linear systems to control-affine nonlinear systems

$$\dot{x}(t) = f(x(t)) + g(x(t))u(t), \qquad x(0) = x_0 \in \mathbb{R}^n, \tag{14.19}$$

with an equilibrium pair at the origin. Following the arguments of Theorem 14.2, we can derive an expression for a feedback invariant for (14.19).

Theorem 14.6. *Consider the control affine nonlinear system* (14.19) *with solution pairs* $(x(\cdot), u(\cdot)) \in \mathcal{X}_s \times \mathcal{U}$. *Then for a continuously differentiable function* $V : \mathbb{R}^n \to \mathbb{R}$,

$$H(x(\cdot), u(\cdot)) = -\int_0^\infty \left(L_f V(x(\tau)) + L_g V(x(\tau))u(\tau) \right) d\tau$$

is a feedback invariant.

The proof is a straightforward modification of the proof of Theorem 14.2. We consider a cost function with a similar structure as (14.13). Consider $R : \mathbb{R}^n \to \mathbb{R}^{m \times m}$ and assume that $R(x)$ is positive definite and bounded away from zero for all $x \in \mathbb{R}^n$, i.e., there exists a $c > 0$ such that $R(x) - cI > 0$ for all $x \in \mathbb{R}^n$. Additionally, assume that $Q : \mathbb{R}^n \to \mathbb{R}_{\geq 0}$ is positive definite, and define the cost function

$$J(x_0, u(\cdot)) = \int_0^\infty \left(Q(x(\tau)) + u^T(\tau)R(x(\tau))u(\tau) \right) d\tau. \qquad (14.20)$$

As with the linear quadratic regulator, we can add and subtract the feedback invariant to the cost function and complete the squares to obtain

$$J(x(\cdot), u(\cdot)) = H(x(\cdot), u(\cdot))$$
$$+ \int_0^\infty Q(x) + L_f V(x) - \tfrac{1}{4} L_g V(x)(R(x))^{-1} L_g V(x)^T$$
$$+ \left(u + \tfrac{1}{2}(R(x))^{-1} L_g V(x) \right)^T R(x) \left(u + \tfrac{1}{2}(R(x))^{-1} L_g V(x) \right) d\tau.$$

This leads to a result similar to Theorem 14.3.

Theorem 14.7. *Consider the control-affine system* (14.19) *and the cost function* (14.20). *If there exists a continuously differentiable function* $V : \mathbb{R}^n \to \mathbb{R}$ *such that*

$$Q(x) + L_f V(x) - \tfrac{1}{4} L_g V(x)(R(x))^{-1} L_g V(x)^T = 0, \qquad \forall\, x \in \mathbb{R}^n, \qquad (14.21)$$

and if the feedback

$$\mu(x) = -\tfrac{1}{2}(R(x))^{-1} L_g V(x) \qquad (14.22)$$

asymptotically stabilizes the origin, then this feedback minimizes (14.20).

An important difference here is that the test for the feedback being asymptotically stabilizing is straightforward in the linear case, namely, by checking that the closed-loop system matrix $A - BR^{-1}B^T P$ is Hurwitz. In the nonlinear case considered in this section, it is clear that if the function V is a control Lyapunov function, then asymptotic stability of the origin follows.

14.1.3 Inverse Optimality

In the previous two sections we have followed the standard approach of optimal control whereby the designer specifies a cost function to be minimized and then looks to compute an optimal feedback stabilizer. However, it is possible to reverse this process as follows. Suppose we have a control Lyapunov function V and can write the stabilizing control in the form

$$\mu(x) = -\tfrac{1}{2}(R(x))^{-1} L_g V(x), \qquad (14.23)$$

where $R(x) - cI > 0$ for all $x \in \mathbb{R}^n$ and $c > 0$ selected appropriately. Then we can compute

$$Q(x) = -L_f V(x) + \tfrac{1}{4} L_g V(x)(R(x))^{-1} L_g V(x)^T,$$

where, since V is a control Lyapunov function, Q is positive definite. The calculations of the previous section then imply that the control (14.23) minimizes (14.20) with the computed functions Q and R.

The control (14.23) is referred to as *inverse optimal* since it is not the cost function that is specified by the designer, but rather the stabilizing feedback that is specified, from which the cost function is derived. In this context, recall the ISS redesign discussed in Section 9.2 and Sontag's universal formula discussed in Section 9.3. The concept of inverse optimality allows an analysis of the control laws obtained through the ISS redesign and Sontag's universal formula by calculating the performance criterion for which the controllers are optimal.

14.2   OPTIMAL CONTROL—DISCRETE TIME SETTING

As described in Chapter 5 on discrete time systems, many results for continuous time systems have equivalent or similar counterparts in the discrete time setting. In this section we introduce the basic optimal control concepts for general nonlinear discrete time systems and we discuss the linear quadratic regulator to derive the analogous results of Section 14.1.1 for linear discrete time systems.

14.2.1   Definitions and Notations

Consider

$$x(k+1) = f(x(k), u(k)), \qquad x(0) = x_0, \tag{14.24}$$

where $f : \mathbb{R}^n \times \mathbb{R}^m \to \mathbb{R}^n$, $x \in \mathbb{R}^n$, and $u \in \mathbb{R}^m$. Then the set of inputs and the set of solutions are defined in the same way as in (14.2), with the only difference being that the positive real numbers defining the time domain are replaced by the natural numbers

$$\mathcal{U} = \{u(\cdot) : \mathbb{N}_0 \to \mathbb{R}^m\}, \qquad \mathcal{X} = \{x(\cdot) : \mathbb{N}_0 \to \mathbb{R}^n\}. \tag{14.25}$$

Similarly the set of asymptotically stable solutions in (14.10) can be adapted to fit into the discrete time context

$$\mathcal{X}_s = \left\{x(\cdot) \in \mathcal{X} : \lim_{k \to \infty} x(k) = 0\right\}. \tag{14.26}$$

While the definition of the running cost $\ell : \mathbb{R}^n \times \mathbb{R}^m \to \mathbb{R}$ does not change in the discrete time setting, the integral in the cost functional (14.3) needs to be replaced by a sum

$$J(x_0, u(\cdot)) = \sum_{k=0}^{\infty} \ell(x(k), u(k)). \tag{14.27}$$

Finally, the optimal value function is defined as

$$V(x_0) = \min_{u(\cdot) \in \mathcal{U}} J(x_0, u(\cdot))$$
$$\text{subject to (14.24),} \tag{14.28}$$

where, compared to (14.4), the continuous time cost functional and the continuous time system are replaced by their discrete time counterparts.

Definition 14.1 of a feedback invariant equally applies to discrete time systems.

Therefore, we can again decompose the cost function

$$J(x_0, u(\cdot)) = H(x(\cdot), u(\cdot)) + \sum_{k=0}^{\infty} \Lambda(x(k), u(k))$$

with the property

$$\min_{u \in \mathbb{R}^m} \Lambda(x, u) = 0 \qquad \forall \, x \in \mathbb{R}^n,$$

to obtain an optimal feedback stabilizer

$$\mu(x(k)) = \arg \min_{u \in \mathbb{R}^m} \Lambda(x(k), u).$$

14.2.2   The Linear Quadratic Regulator

We conclude this section with a derivation of the linear quadratic regulator for discrete time linear systems

$$x(k+1) = Ax(k) + Bu(k), \quad x(0) = x_0 \in \mathbb{R}^n, \tag{14.29}$$

which are analogous to the results in Section 14.1.1.

Theorem 14.8. *Consider the discrete time linear system (14.29) with solution pairs $(x(\cdot), u(\cdot)) \in \mathcal{X}_s \times \mathcal{U}$. Then, for any symmetric matrix $P \in \mathcal{S}^n$, the functional $H : \mathcal{X}_s \times \mathcal{U} \to \mathbb{R}$ defined as*

$$H(x(\cdot), u(\cdot)) = -\sum_{k=0}^{\infty} (Ax(k) + Bu(k))^T P (Ax(k) + Bu(k)) - x(k)^T Px(k)$$
$$\tag{14.30}$$

is a feedback invariant.

The proof of Theorem 14.8 is left for the reader in Exercise 14.5. The form of the feedback invariant (14.30) for discrete time systems should not be a surprise when comparing the continuous time Lyapunov equation (3.4) with the discrete time Lyapunov equation (5.35).

Theorem 14.9. *Consider the discrete time linear system (14.29) and the quadratic cost function*

$$J(x_0, u(\cdot)) = \sum_{k=0}^{\infty} x(k)^T Qx(k) + u(k)^T Ru(k) \tag{14.31}$$

defined through $Q \in \mathcal{S}_{\geq 0}^n$, $R \in \mathcal{S}_{>0}^m$. If there exists a positive semidefinite matrix $P \in \mathcal{S}_{\geq 0}^n$ satisfying the discrete time algebraic Riccati equation

$$Q + A^T PA - P - A^T PB \left(R + B^T PB\right)^{-1} B^T PA = 0 \tag{14.32}$$

and if

$$A - B(R + B^T PB)^{-1} B^T PA$$

is a Schur matrix, then the state feedback

$$\mu(x) = -\left(R + B^T PB\right)^{-1} B^T PAx \qquad (14.33)$$

minimizes (14.31) and the optimal value function is given by $V(x_0) = x_0^T Px_0$.

Proof. The proof follows the same lines as that of Theorem 14.3 for the continuous time setting. We add and subtract the feedback invariant (14.30) to the cost function (14.31):

$$J(x_0, u(\cdot)) = H(x(\cdot), u(\cdot)) + \sum_{k=0}^{\infty} x(k)^T Qx(k) + u(k)^T Ru(k)$$

$$+ \sum_{k=0}^{\infty} (Ax(k) + Bu(k))^T P (Ax(k) + Bu(k)) - x(k)^T Px(k).$$

As a second step we rearrange terms in the sum. Additionally, we suppress the time index k and use the definition $\tilde{R} = R + B^T PB$ to shorten the expressions. Note that $\tilde{R}$ is invertible since $R > 0$ and $P \geq 0$ by assumption. This fact is implicitly used in (14.31). With these conventions, the cost function $J(x_0, u(\cdot))$ can be rewritten as

$$\sum_{k=0}^{\infty} x^T Qx + u^T Ru + (Ax + Bu)^T P (Ax + Bu) - x^T Px$$

$$= \sum_{k=0}^{\infty} x^T (Q + A^T PA - P)x + u^T \tilde{R}u + 2u^T B^T PAx$$

$$+ \sum_{k=0}^{\infty} x^T A^T PB\tilde{R}^{-1} B^T PAx - x^T A^T PB\tilde{R}^{-1} B^T PAx$$

$$= \sum_{k=0}^{\infty} x^T (Q + A^T PA - P - A^T PB\tilde{R}^{-1} B^T PA)x \qquad (14.34a)$$

$$+ \sum_{k=0}^{\infty} (\tilde{R}u + B^T PAx)^T \tilde{R}^{-1} (\tilde{R}u + B^T PAx). \qquad (14.34b)$$

The discrete time algebraic Riccati equation (14.32) implies (14.34a) is zero. By taking the state feedback $u = \mu(x)$ as in (14.33), we see that (14.34b) is also zero. Finally, the result follows from Theorem 14.8 since $A - B(R + B^T PB)^{-1} B^T PA$ is assumed to be a Schur matrix. $\square$

We conclude this section with a version of Theorem 14.4 for linear discrete time systems.

Theorem 14.10 (The discrete time linear quadratic regulator). *Consider the linear system (14.29), assume that* (A, B) *is stabilizable and let* $Q \in \mathcal{S}^n$, $R \in \mathcal{S}^m$, *and* $N \in \mathbb{R}^{n \times m}$ *be chosen such that* $R > 0$ *is positive definite and* $Q - NR^{-1}N^T \geq 0$ *is* *positive semidefinite. Moreover, consider the quadratic cost function*

$$J(x_0, u(\cdot)) = \sum_{k=0}^{\infty} x(k)^T Q x(k) + u(k)^T R u(k) + 2x(k)^T N u(k). \quad (14.35)$$

Then the following properties are satisfied.

- *If the pair $(A - BR^{-1}N^T, Q - NR^{-1}N^T)$ is detectable, then there exists a symmetric matrix $P \in \mathcal{S}^n$ satisfying the algebraic Riccati equation*

$$Q + A^T P A - P - (A^T P B + N) (R + B^T P B)^{-1} (B^T P A + N^T) = 0 \quad (14.36)$$

so that the matrix

$$A - (R + B^T P B)^{-1} (B^T P A + N^T) \quad (14.37)$$

is Hurwitz, i.e., the feedback

$$\mu(x) = - (R + B^T P B)^{-1} (B^T P A + N^T)x$$

asymptotically stabilizes the system (14.29).
- *If the pair $(A - BR^{-1}N^T, Q - NR^{-1}N^T)$ is observable, then there exists a unique $P \in \mathcal{S}^n_{>0}$ satisfying the algebraic Riccati equation (14.36). Moreover, the state feedback $\mu(x) = - (R + B^T P B)^{-1} (B^T P A + N^T)x$ ensures that the closed-loop matrix (14.37) is Schur and the optimal value function minimizing (14.35) is given by $V(x_0) = x_0^T P x_0$, where V is a Lyapunov function of the closed-loop system.*

As in the continuous time setting, condition $(A - BR^{-1}N^T, Q - NR^{-1}N^T)$ observable/detectable is for example satisfied for $N = 0$, $Q = C^T C$, $C \in \mathbb{R}^{p \times n}$, and (A, C) observable/detectable.

14.3   FROM INFINITE- TO FINITE-DIMENSIONAL OPTIMIZATION

As the name suggests, by designing an input based on the optimal solution of an optimization problem with a predefined cost function, the corresponding solution pair $(x^\star(\cdot), u^\star(\cdot)) \in \mathcal{X} \times \mathcal{U}$ is optimal with respect to the specific, defined measure.

However, to obtain the optimal solution pair an infinite-dimensional optimization problem (14.5) (or (14.28), respectively, in the discrete time setting) needs to be solved, which is in general only possible under specific assumptions on the system dynamics and the input and solution space $\mathcal{U}$ and $\mathcal{X}$. Even if it is possible to solve the optimal control problem (14.5) for a specific initial value $x_0 \in \mathbb{R}^n$, to obtain an optimal feedback law $\mu(x(t))$ (instead of an optimal open-loop input $u(t)$) it is necessary to solve (14.5) for all $x_0 \in \mathbb{R}^n$.

Under some conditions, the *principle of optimality* can be used to simplify the infinite-dimensional optimization problems (14.5) and (14.28). We illustrate this fact based on the example of an input constrained linear quadratic regulator and we briefly illustrate the application of *the principle of optimality* in *dynamic pro-*

gramming.

14.3.1   The Principle of Optimality

In words, the *principle of optimality* states that for any point on an optimal solution, the remaining control inputs are also optimal. More formally, we assume that solutions of the optimal control problem (14.5) are unique. Then, for an arbitrary initial condition $x(0) = x_0 \in \mathbb{R}^n$, suppose we find the optimal input $u^\star(\cdot; x_0) \in \mathcal{U}$ as a solution of the optimal control problem (14.5) and the corresponding solution $x^\star(\cdot; x_0) \in \mathcal{X}$. In particular, we assume that

$$u^\star(\cdot; x_0) = \arg \min_{u(\cdot) \in \mathcal{U}} J(x_0, u(\cdot))$$

subject to (14.1)

and the pair $(x^\star(\cdot; x_0), u^\star(\cdot; x_0))$ satisfies (14.1) for almost all $t \geq 0$. Additionally, for any $T \geq 0$ consider the optimal control problem

$$\bar{u}^\star(\cdot; x^\star(T; x_0)) = \arg \min_{u(\cdot) \in \mathcal{U}} J(x^\star(T; x_0), u(\cdot))$$

subject to (14.1),

whose solution pair is denoted by $(\bar{x}^\star(\cdot; x^\star(T; x_0)), \bar{u}^\star(\cdot; x^\star(T; x_0)))$. Then the principle of optimality states that

$$\begin{aligned} \bar{u}^\star(\cdot; x^\star(T; x_0)) &= u^\star(\cdot + T; x_0) \qquad \text{and} \\ \bar{x}^\star(\cdot; x^\star(T; x_0)) &= x^\star(\cdot + T; x_0). \end{aligned} \tag{14.38}$$

In words, suppose we solve the optimal control problem from an initial condition x_0 at time $t = 0$. This yields an optimal solution pair. If we solve a second optimal control problem from an initial condition that lies on the optimal solution at time $T \geq 0$, i.e., take the initial condition $x^\star(T; x_0)$, then the solution pairs to these two optimal control problems coincide. The definitions introduced here are illustrated in Figure 14.2.

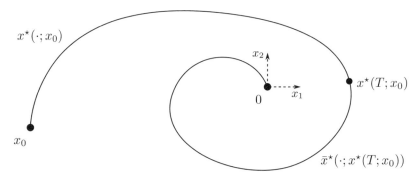

Figure 14.2: The principle of optimality states that tails of optimal trajectories are also optimal, i.e., $\bar{x}^\star(\cdot; x^\star(T; x_0)) = x^\star(\cdot + T; x_0)$ for all $T \geq 0$.

In the case where uniqueness of optimal solutions is not guaranteed, (14.38) does not necessarily need to be satisfied. However, even if (14.38) is not true, the optimal costs coincide, i.e.,

$$J(x^\star(T; x_0), \bar{u}^\star(\cdot; x^\star(T; x_0))) = J(x^\star(T; x_0), u^\star(\cdot + T; x_0))$$

holds.

In the discrete time setting, the same result can be formulated and we will indeed apply the principle of optimality to discrete time problems in the following.

Example 14.11. Assume you have to go from A to E and you are convinced that the best way to get to E is by passing through B and then C. Your friend, who is living in B, is sure that it is better to go from B to D and then to E. It is clear that only one of you can be correct. If the optimal way from A to E is to first go to B, then only one of the following options, namely through C or through D can be optimal (or both options are equal). The setting is illustrated in Figure 14.3.

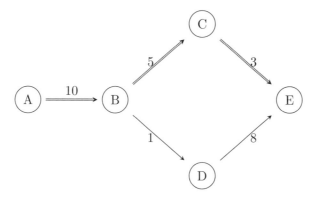

Figure 14.3: A graph connecting elements of the set $\{A, B, C, D, E\}$. The numbers on the arrows indicate the price to go from one element to another. The minimal costs from A to E are achieved through the path $A \to B \to C \to E$. Thus, the principle of optimality tells us that the optimal path from B to E is given by $B \to C \to E$.

14.3.2   Constrained Optimal Control for Linear Systems

In Section 14.2.2 we derived a stabilizing feedback law for linear systems (14.29), which is optimal with respect to a quadratic cost function (14.31) and whose derivation was based on the discrete time algebraic Riccati Equation (14.32). Due to the special structure of the dynamics, the cost function, and the input and solution spaces, an explicit solution of the infinite-dimensional optimization problem (14.28) was possible.

Here, we restrict the input space to

$$\mathcal{U}_{\mathbb{U}} = \{u(\cdot) : \mathbb{N}_0 \to \mathbb{R}^m | \, u(k) \in \mathbb{U} \; \forall k \in \mathbb{N}\} \tag{14.39}$$

for a closed and convex set $\mathbb{U} \subset \mathbb{R}^m$ containing the origin in its interior. By including this input constraint in the optimal control problem (14.28), i.e., by considering the optimization problem

$$V(x_0) = \min_{u(\cdot) \in \mathcal{U}_{\mathbb{U}}} J(x_0, u(\cdot))$$

$$\text{subject to (14.24)},$$

(14.40)

the approach followed in Section 14.2.2 is in general not possible. Moreover, standard (convex) optimization algorithms (such as those used in Section 3.4 and in Chapter 8) are not directly applicable since (14.40) defines an infinite-dimensional optimization problem in the unknowns $u(k)$ for $k \in \mathbb{N}$. However, we can combine the results from Section 14.2.2 with the principle of optimality, to rewrite the optimal control problem (14.40) as a finite-dimensional convex optimization problem which can be solved numerically (or shown to be infeasible) using standard software.

Consider the cost function

$$J(x_0, u(\cdot)) = \sum_{k=0}^{\infty} x(k)Qx(k) + u(k)Ru(k),$$

(14.41)

with $N = 0$ for simplicity, and as a first step, solve (14.36) to obtain $P_F \in \mathcal{S}_{>0}$ and hence the optimal value function $V(x) = x^T P_F x$ and the optimal state feedback law

$$\mu(x) = Kx = -\left(R + B^T P_F B\right)^{-1} B^T P_F A x.$$

Since V is a Lyapunov function for the closed-loop dynamics $x^+ = (A + BK)x$, the sublevel set

$$\mathbb{X}_F = \{x \in \mathbb{R}^n | V(x) \le c\}$$

is forward-invariant for $c > 0$. If the parameter c is selected such that $Kx \in \mathbb{U}$ for all $x \in \mathbb{X}_F$, then Theorem 14.10 holds for all $x_0 \in \mathbb{X}_F$.

As a second step we assume that for all $x_0 \in \mathbb{R}^n$ there exists an input $u(\cdot) \in \mathcal{U}_{\mathbb{U}}$ such that the origin of the linear system can be globally asymptotically stabilized. Then for all $x_0 \in \mathbb{R}^n$, there exists an $N \in \mathbb{N}$ and an optimal solution pair $(x^\star(\cdot), u^\star(\cdot)) \in \mathcal{X} \times \mathcal{U}_{\mathbb{U}}$ (with respect to (14.40)) such that $x^\star(k) \in \mathbb{X}_F$ for all $k \ge N$. Under the assumption that $x^\star(N) \in \mathbb{X}_F$ it holds that

$$\min_{u(\cdot) \in \mathcal{U}_{\mathbb{U}}} J(x_0, u(\cdot)) = \sum_{k=0}^{\infty} (x^\star(k))^T Q x^\star(k) + (u^\star(k))^T R u^\star(k)$$

$$= \sum_{k=0}^{N-1} (x^\star(k))^T Q x^\star(k) + (u^\star(k))^T R u^\star(k)$$

$$\quad + \sum_{k=N}^{\infty} (x^\star(k))^T Q x^\star(k) + (u^\star(k))^T R u^\star(k)$$

$$= \sum_{k=0}^{N-1} (x^\star(k))^T Q x^\star(k) + (u^\star(k))^T R u^\star(k) + (x^\star(N))^T P_F x^\star(N),$$

where the last equality follows from the principle of optimality. In particular, it holds that

$$\sum_{k=N}^{\infty} (x^\star(k))^T Q x^\star(k) + (u^\star(k))^T R u^\star(k) = (x^\star(N))^T P_F x^\star(N) = V(x^\star(N)).$$

If we adapt the definition of the cost function and the definition of the set of inputs to a finite horizon, i.e.,

$$\mathcal{U}^N = \{u_N(\cdot) = (u(0), \ldots, u(N-1)) | \ u(\cdot) \in \mathcal{U}\}$$

and

$$J_N(x_0, u_N(\cdot)) = \sum_{k=0}^{N-1} \ell(x(k), u(k)),$$

then we can further rewrite the optimal control problem (14.40) as

$$\min_{u(\cdot) \in \mathcal{U}_\mathbb{U}} J(x_0, u(\cdot)) = \min_{u_N(\cdot) \in \mathcal{U}_\mathbb{U}^N} \sum_{k=0}^{N-1} x(k)^T Q x(k) + u(k)^T R u(k) + x(N)^T P_F x(N)$$
$$= \min_{u_N(\cdot) \in \mathcal{U}_\mathbb{U}^N} J_N(x_0, u_N(\cdot)) + x(N)^T P_F x(N).$$

We have thus rewritten the infinite-dimensional problem (14.40) as a finite-dimensional optimization problem

$$V(x_0) = \min_{u_N(\cdot) \in \mathcal{U}_\mathbb{U}^N} J_N(x_0, u_N(\cdot)) + x(N)^T P_F x(N) \tag{14.42}$$
$$\text{subject to (14.24).}$$

The optimal open-loop input is given by

$$u^\star(\cdot) = (u_N^\star(0), \ldots, u_N^\star(N-1), K x^\star(N), K x^\star(N+1), \ldots). \tag{14.43}$$

Note that the optimal input and the optimal value function are only implicitly known as the solution of the optimization problem. They are not explicitly known as in the case of the linear quadratic regulator.

The equivalence of the optimal control problem (14.40) and (14.42) needs to be understood with caution since it relies on the nontrivial assumption that the optimal solution $x^\star(\cdot)$ reaches the set $\mathbb{X}_F$ in N discrete time steps. Alternatively this assumption can be included as a *terminal constraint* in the optimization problem

$$\min_{u_N(\cdot) \in \mathcal{U}_\mathbb{U}^N} J_N(x_0, u_N(\cdot)) + x(N)^T P_F x(N) \tag{14.44}$$
$$\text{subject to (14.24), } x(N) \in \mathbb{X}_F.$$

However, this approach has its disadvantages. Firstly, the optimal solution of (14.44) might not be optimal with respect to the cost function (14.41) since it might be cheaper to reach the set $\mathbb{X}_F$ in more than N steps. Thus, the optimal values of (14.42) and (14.44) do not need to coincide. The solution of (14.42) provides a lower bound for (14.44). Secondly, the optimization problem (14.44) will be infeasible if it is not possible to reach the set $\mathbb{X}_F$ in N steps. In contrast (14.42) is feasible by construction, but (14.43) might not satisfy $u^\star(\cdot) \in \mathcal{U}_\mathbb{U}$ since it is not guaranteed that $x^\star(N) \in \mathbb{X}_F$ is satisfied.

14.3.3   Dynamic Programming and the Backward Recursion

For $N \in \mathbb{N}$, we consider the finite horizon optimal control problem

$$. \quad V_N(x_0) = \min_{u_N(\cdot) \in \mathcal{U}_\mathbb{U}^N} J_N(x_0, u_N(\cdot))$$

$$\text{subject to (14.24), and } x(N) = 0, \tag{14.45}$$

which is again a special form of the optimal control problem (14.44) with $\mathbb{X}_F = \{0\}$. Due to the constraint $x(N) = 0$, (14.45) is not necessarily feasible for all $x_0 \in \mathbb{R}^n$. The set of initial states such that (14.45) is feasible, i.e.,

$$\mathbb{X}_{\{0\}}^N = \{x_0 \in \mathbb{R}^n | \; \exists u_N(\cdot) \in \mathcal{U}_\mathbb{U}^N \text{ such that } x(N) = 0 \text{ with respect to (14.24)}\},$$

depends on the inputs $\mathcal{U}_\mathbb{U}^N$, the dynamics, and on the horizon N.

If $(x^e, u^e) = (0, 0)$ is an equilibrium pair of (14.24) it is clear that $\mathbb{X}_{\{0\}}^N \subset \mathbb{X}_{\{0\}}^{N+1}$ for all $N \in \mathbb{N}$ since by definition of $\mathbb{X}_{\{0\}}^N$ it is possible to reach the origin in N steps and then stay there. Moreover, for $N = 0$, the set $\mathbb{X}_{\{0\}}^N$ can be initialized as $\mathbb{X}_{\{0\}}^0 = \{0\}$.

In this setting, the principle of optimality can be used to iteratively compute $\mathbb{X}_{\{0\}}^N$ as well as the optimal value function V_N for $N \in \mathbb{N}$. In particular, using the definition of the cost function, the optimization problem

$$V_N(x_0) = \min_{u_N(\cdot) \in \mathcal{U}_\mathbb{U}^N} J_N(x_0, u_N(\cdot))$$

$$\text{subject to (14.24), and } x(N) \in \{0\}$$

is equivalent to

$$V_N(x_0) = \min_{u_N(\cdot) \in \mathcal{U}_\mathbb{U}^N} \ell(x_0, u(0)) + J_{N-1}(f(x_0, u(0)), u_{N-1}(\cdot + 1))$$

$$\text{subject to (14.24), and } x(N) \in \{0\}.$$

Moreover, using the definition of the optimal value function, a second equivalent formulation of the optimization problem is given by

$$V_N(x_0) = \min_{u(0) \in \mathbb{U}} \ell(x_0, u(0)) + V_{N-1}(f(x_0, u(0)))$$

$$\text{subject to (14.24), and } f(x_0, u(0)) \in \mathbb{X}_{\{0\}}^{N-1}. \tag{14.46}$$

Thus, if V_{N-1} is known, V_N can be computed by minimizing (14.46) with respect to $u(0)$. Similarly, $\mathbb{X}_{\{0\}}^N$ can be constructed based on the knowledge of $\mathbb{X}_{\{0\}}^{N-1}$. Note that the condition $x(N) \in \{0\}$ can be replaced by alternative conditions on the final state. We illustrate the computations, also known as backward recursion in the literature, based on an example. (The example is a slight variation of [58, Chapter 3, Exercise 2].)

Example 14.12. Consider the discrete time system

$$x_1(k + 1) = x_1(k) + 2x_2(k)$$
$$x_2(k + 1) = -x_2(k) + 2u(k)$$

together with the running costs $\ell(x, u) = u^2$, i.e., we only penalize the control effort. We initialize the backward recursion with $V_0(x_0) = 0$ and $\mathbb{X}^0_{\{0\}} = \{0\}$.

For $N = 1$, optimization problem (14.46) is of the form

$$V_1(x_0) = \min_{u(0) \in \mathbb{R}} \quad u(0)^2$$
$$\text{subject to} \quad 0 = x_1(0) + 2x_2(0) \tag{14.47}$$
$$0 = -x_2(0) + 2u(0).$$

Rearranging the equality constraints shows that

$$u_1^\star(0) = \tfrac{1}{2}x_2(0) \quad \text{and thus} \quad V_1(x_0) = \tfrac{1}{4}x_2(0)^2.$$

Additionally, optimization problem (14.47) is feasible if the initial condition satisfies $x_1(0) = -2x_2(0)$. In particular, the set

$$\mathbb{X}^1_{\{0\}} = \{x \in \mathbb{R}^2 | \ x_1 = -2x_2\}$$

denotes the set of states which can be steered to the origin in one time step.

For $N = 2$, the optimal value function is defined through

$$V_2(x_0) = \min_{u(0), u(1) \in \mathbb{R}} \quad u(0)^2 + u(1)^2$$
$$\text{subject to} \quad 0 = -x_1(1) + x_1(0) + 2x_2(0)$$
$$0 = -x_2(1) - x_2(0) + 2u(0)$$
$$0 = x_1(1) + 2x_2(1)$$
$$0 = -x_2(1) + 2u(1).$$

Using the result from V_1, i.e., $u_1^\star(0) = u(1) = \tfrac{1}{2}x_2(1)$ and $x_1(1) = -2x_2(1)$, the optimization problem can be written as

$$V_2(x_0) = \min_{u(0) \in \mathbb{R}} \quad u(0)^2 + \tfrac{1}{4}x_2(1)^2$$
$$\text{subject to} \quad 0 = 2x_2(1) + x_1(0) + 2x_2(0)$$
$$0 = -x_2(1) - x_2(0) + 2u(0).$$

Finally, by eliminating $x_2(1)$, the optimization problem reduces to

$$V_2(x_0) = \min_{u(0) \in \mathbb{R}} \quad u(0)^2 + \tfrac{1}{4}(-x_2(0) + 2u(0))^2$$
$$\text{subject to} \quad 0 = x_1(0) + 4u(0). \tag{14.48}$$

Thus the optimal input $u_2^\star(\cdot)$ is given by

$$u_2^\star(0) = -\tfrac{1}{4}x_1(0),$$
$$u_2^\star(1) = \tfrac{1}{2}x_2(1) = \tfrac{1}{2}(-x_2(0) + 2u_2^\star(0)) = -\tfrac{1}{2}x_2(0) - \tfrac{1}{4}x_1(0)$$

and the optimal value function V_2 satisfies

$$V_2(x(0)) = \tfrac{1}{16}x_1(0)^2 + \left(\tfrac{1}{2}x_2(0) + \tfrac{1}{4}x_1(0)\right)^2$$
$$= \tfrac{1}{8}x_1(0)^2 + \tfrac{1}{4}x_1(0)x_2(0) + \tfrac{1}{4}x_2(0)^2.$$

The optimization problem (14.48) is feasible for all $x_0 \in \mathbb{X}_{\{0\}}^2 = \mathbb{R}^2$ and we can conclude that the origin can be reached from any point in two discrete time steps. This is consistent with the observation in Section 5.4 that in general n steps are necessary to reach the origin from an initial state.

We perform one final step to solve the optimization problem

$$V_3(x_0) = \min_{u_3(\cdot) \in \mathbb{R}^3} \quad u(0)^2 + u(1)^2 + u(2)^2$$

$$\text{subject to} \quad 0 = -x_1(1) + x_1(0) + 2x_2(0)$$
$$0 = -x_2(1) - x_2(0) + 2u(0)$$
$$0 = -x_1(2) + x_1(1) + 2x_2(1)$$
$$0 = -x_2(2) - x_2(1) + 2u(1)$$
$$0 = x_1(2) + 2x_2(2)$$
$$0 = -x_2(2) + 2u(2).$$

Using the results derived for V_2 reduces the optimization problem to

$$V_3(x_0) = \min_{u(0) \in \mathbb{R}} \quad u(0)^2 + V_2(x(1))$$

$$\text{subject to} \quad 0 = -x_1(1) + x_1(0) + 2x_2(0) \qquad (14.49)$$
$$0 = -x_2(1) - x_2(0) + 2u(0).$$

Thus, the cost function can be written as

$$u(0)^2 + V_2(x(1)) = u(0)^2 + \tfrac{1}{16}x_1(1)^2 + \left(\tfrac{1}{2}x_2(1) + \tfrac{1}{4}x_1(1)\right)^2$$
$$= u(0)^2 + \tfrac{1}{16}(x_1(0) + 2x_2(0))^2 + \left(u(0) + \tfrac{1}{4}x_1(0)\right)^2,$$

where we make use of the remaining equality constraints in (14.49). Taking the derivative with respect to $u(0)$ leads to the condition

$$0 = 2u_3^\star(0) + 2(u_3^\star(0) + \tfrac{1}{4}x_1(0))$$

and thus

$$u_3^\star(0) = -\tfrac{1}{8}x_1(0). \qquad (14.50)$$

We invite the reader to verify that the critical point (14.50) is indeed a minimizer of the optimization problem (and not a maximizer).

The results from $N = 2$ provide the missing components of the optimal input

$$u_3^\star(1) = -\tfrac{1}{4}x_1(1) = -\tfrac{1}{4}x_1(0) - \tfrac{1}{2}x_2(0),$$

$$u_3^\star(2) = \tfrac{1}{2}x_2(2) = \tfrac{1}{2}(-x_2(1) + 2u_3^\star(1)) = -\tfrac{1}{2}(x_2(1) + \tfrac{1}{2}x_1(1))$$

$$= -\tfrac{1}{2}(-x_2(0) - \tfrac{1}{4}x_1(0) + \tfrac{1}{2}x_1(0) + x_2(0)) = -\tfrac{1}{8}x_1(0),$$

from which the optimal value function $V_3(x_0) = u_3^\star(0)^2 + V_2(x(1))$ can be computed.

The example shows how the dynamic programming principle can be used to iteratively compute the optimal value function V_{N+1} based on the knowledge of V_N for all $N \in \mathbb{N}$. Similarly, the corresponding optimal input $u_{N+1}^\star(\cdot)$ can be computed based on $u_N^\star(\cdot)$. The approach is also applicable in the presence of additional input and state constraints. However, the calculations quickly become quite complicated since the set $\mathbb{X}_{\{0\}}^N$ also needs to be computed and considered.

Example 14.13. As a second example we consider a general linear system

$$x^+ = Ax + Bu,$$

$x \in \mathbb{R}^n$, $u \in \mathbb{R}^m$, together with quadratic running costs

$$\ell(x, u) = x^T Q x + u^T R u$$

for $Q \in \mathcal{S}_{\geq 0}^n$ and $R \in \mathcal{S}_{>0}^m$. The example can be found in [128, Example C.3], for example. Compared to Example 14.12 we drop the condition that the state needs to reach the origin in N time steps and instead assume that the system only needs to be controlled for $N \in \mathbb{N}$ discrete time steps and the behavior of the dynamics $x(k)$ for $k \geq N$ does not matter. Thus, for $N = 0$, the optimal value function is given by $V_0(x_0) = x_0^T P_0 x_0$ for $P_0 = 0 \in \mathcal{S}^n$ and the optimal input satisfies $u^\star(k) = 0$ for all $k \in \mathbb{N}$.

For $N = 1$, according to the dynamic programming principle we need to solve the optimization problem

$$V_1(x_0) = \min_{u_1(\cdot) \in \mathcal{U}^1} x_0^T Q x_0 + u(0)^T R u(0) + V_0(x^+)$$

$$\text{subject to } x^+ = Ax_0 + Bu(0),$$

which simplifies to

$$V_1(x_0) = \min_{u(0) \in \mathcal{U}^1} x_0^T Q x_0 + u(0)^T R u(0).$$

Since $R \in \mathcal{S}_{>0}^m$ the optimal input is given by $u_1^\star(\cdot; x_0) = 0$ and the optimal value function is given by

$$V_1(x_0) = x_0^T P_1 x_0,$$

where $P_1 = Q$.

The first interesting case occurs for $N = 2$. We need to solve the optimization problem

$$V_2(x_0) = \min_{u_2(\cdot) \in \mathcal{U}^2} x_0^T Q x_0 + u(0)^T R u(0) + V_1(x^+)$$

$$\text{subject to } x^+ = A x_0 + B u(0),$$

which can be rewritten as

$$V_2(x_0) = \min_{u \in \mathbb{R}^m} x_0^T Q x_0 + u^T R u + (A x_0 + B u)^T P_1 (A x_0 + B u)$$

$$= \min_{u \in \mathbb{R}^m} x_0^T Q x_0 + u^T R u + x_0^T A^T P_1 A x_0 + u^T B^T P_1 B u + 2 u^T B^T P_1 A x_0$$

$$= \min_{u \in \mathbb{R}^m} x_0^T (Q + A^T P_1 A) x_0 + u^T (R + B^T P_1 B) u + 2 u^T B^T P_1 A x_0$$

$$= \min_{u \in \mathbb{R}^m} x_0^T (Q + A^T P_1 A - A^T P_1 B (R + B^T P_1 B)^{-1} B^T P_1 A) x_0$$

$$+ ((R + B^T P_1 B) u - B^T P_1 A x_0)^T (R + B^T P_1 B)^{-1}$$

$$\cdot ((R + B^T P_1 B) u - B^T P_1 A x_0).$$

We can thus conclude that

$$u_2^\star(0; x_0) = -(R + B^T P_1 B)^{-1} B^T P_1 A x_0$$

and $V_2(x_0) = x_0^T P_2 x_0$ where

$$P_2 = Q + A^T P_1 A - A^T P_1 B (R + B^T P_1 B)^{-1} B^T P_1 A.$$

By inspecting the calculations corresponding to V_2 we observe that they are not specific to the case $N = 2$. In particular, for all $N \in \mathbb{N}$ it holds that

$$u_N^\star(0, x_0) = -(R + B^T P_{N-1} B)^{-1} B^T P_{N-1} A x_0$$

and $V_N(x_0) = x_0^T P_N x_0$ for

$$P_N = Q + A^T P_{N-1} A - A^T P_{N-1} B (R + B^T P_{N-1} B)^{-1} B^T P_{N-1} A. \qquad (14.51)$$

Compare the discrete time *dynamic Riccati equation* (14.51) with the derivation of the discrete time linear quadratic regulator in Section 14.2.2. If $P_N = P_{N-1}$ is satisfied, then the *algebraic Riccati equation* (14.32) is recovered.

14.4   EXERCISES

Exercise 14.1. Prove Theorem 14.6.

Exercise 14.2. Consider the linear system $\dot{x} = Ax + Bu$ describing the linearization of the pendulum on a cart in the upright position derived in Example 3.18

$$A = \begin{bmatrix} 0 & 0 & 1 & 0 \\ 0 & 0 & 0 & 1 \\ 0 & \frac{g}{MJ-1} & -\frac{\bar{J}\bar{c}}{MJ-1} & -\frac{\bar{\gamma}}{MJ-1} \\ 0 & \frac{Mg}{MJ-1} & -\frac{\bar{c}}{MJ-1} & -\frac{M\bar{\gamma}}{MJ-1} \end{bmatrix}, \qquad B = \begin{bmatrix} 0 \\ 0 \\ \frac{\bar{J}}{MJ-1} \\ \frac{1}{MJ-1} \end{bmatrix}.$$

For simplicity we assume that the constants introduced in Section 1.2.1 are defined as

$$M = J = m = \ell = 1, \quad \gamma = c = 0.1, \quad \text{and} \quad g = 9.81.$$

Use Theorem 14.4 and the matrices $Q = I \in \mathbb{R}^{4 \times 4}$ ($S = 0$) and $R = 1$ to define a stabilizing feedback law $\mu(x) = Kx$ and a Lyapunov function $V(x) = x^T P x$ of the closed-loop system $\dot{x} = (A + BK)x$.

Hint: In MATLAB, you can use the functions `care.m` (or `icare.m`) or `lqr.m` to solve the Riccati equation in Theorem 14.4.

Exercise 14.3. Use the Euler method with $\Delta = 0.01$ to discretize the continuous time system in Exercise 14.2.

Use Theorem 14.10 (and $Q = I \in \mathbb{R}^{4 \times 4}$ ($S = 0$) and $R = 1$) to compute a stabilizing feedback law $\mu(x) = Kx$ and a Lyapunov function $V(x) = x^T P x$ of the discrete time closed-loop system $\dot{x} = (A + BK)x$.

Hint: In MATLAB, you can use the functions `dare.m` (or `idare.m`) or `dlqr.m` to solve the Riccati equation in Theorem 14.10.

Exercise 14.4. Consider the system

$$\dot{x} = f_1(x, u) = 2x^2 + (x^4 + 1)u$$

together with the feedback law

$$\mu(x) = \frac{-2x^2 - x}{x^4 + 1} - x(x^4 + 1)$$

and the control Lyapunov function $V(x) = \frac{1}{2}x^2$ (see Exercise 9.1).

Apply the results discussed in Section 14.1.3 to calculate $R(x)$ and $Q(x)$ such that $\mu(x)$ is optimal with respect to the particular selection of $R(x)$ and $Q(x)$.

Exercise 14.5. Prove Theorem 14.8.

Hint: Construct a telescoping series.

Exercise 14.6. Perform an additional step in Example 14.12. In particular, compute $V_4(x_0)$ and $u_4^\star(\cdot)$.

14.5   BIBLIOGRAPHICAL NOTES AND FURTHER READING

Optimal control is a rich and interesting mathematical field, with many well-written books. The text [96] is an excellent starting point for further study. In contrast to the derivations here, which are based on feedback invariants, the monograph [96] uses the classical *calculus of variations* approach in the derivation of the linear quadratic regulator. For the mathematically advanced and adventurous student, there are deep connections between optimal control, nonsmooth analysis, and feedback stabilizers (as hinted at by the results on inverse optimality and the need to consider CLFs that are not continuously differentiable), and [40], [41] explore these issues in depth.

The use of feedback invariants to develop the linear quadratic regulators follows [63].

Dynamic programming suffers from the so-called *curse of dimensionality*, which refers to the fact that numerical computation quickly becomes intractable as the dimension of the problem grows. Nonetheless, for small problems, as a way of developing intuition, and as a proof technique, dynamic programming is frequently used. Dynamic programming was originally developed by Richard Bellman in the 1940s [17]. A modern treatment of dynamic programming can be found in [19] and [128].

Chapter Fifteen

Model Predictive Control

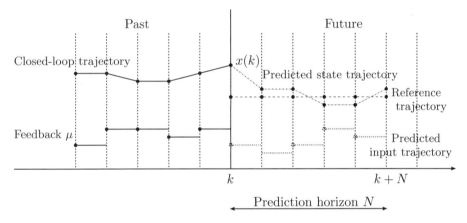

Figure 15.1: The *receding horizon principle* and the idea behind MPC.

The results derived in the previous chapter show how an infinite-dimensional optimal control problem can be approximated by a finite-dimensional optimization problem. Additionally, the principle of optimality allows us to make statements about the accuracy of the approximation. Moreover, the dynamic programming principle provides a method to iteratively solve an optimal control problem of increasing complexity. While these results play a crucial role in the derivation of theoretical results in optimal and *model predictive control* (MPC), they are in general not necessary to understand the main ideas of MPC and the related controller design.

MPC is one way to overcome the computational limitations of optimal control in designing controllers for nonlinear dynamical systems that explicitly incorporate state and input constraints. Since it is in general impossible to solve the infinite-dimensional optimal control problem (14.5) to obtain a closed expression for the optimal open-loop control law (14.6), MPC iteratively defines a feedback law $\mu : \mathbb{R}^n \to \mathbb{R}^m$ by solving finite-dimensional approximations of the optimal control problem (14.5) at every discrete time step $k \in \mathbb{N}$.

In particular, at a fixed time $k \in \mathbb{N}$, an optimal control problem depending on the current state $x(k)$ is solved over a given prediction horizon $N \in \mathbb{N}$, i.e., the system behavior from current time k up to time $k + N$ is optimized with respect to a given optimal reference trajectory. From the optimal open-loop control law, the first element is used to define a feedback law. Afterward, the feedback is applied to the system, the time is shifted from k to $k + 1$ and the process is repeated in a receding horizon fashion.

Due to this simple idea, MPC is also known as *receding horizon control* or *rolling horizon control* in the literature and the iterative process just described is known as the *receding horizon principle*. The receding horizon principle, which captures the idea behind MPC, is graphically shown in Figure 15.1. In the following we introduce the basic MPC algorithm before various aspects of MPC are discussed.

15.1   THE BASIC MPC FORMULATION

We consider discrete time systems

$$x(k+1) = f(x(k), u(k)), \qquad x(0) = x_0 \in \mathbb{R}^n \tag{15.1}$$

for $f : \mathbb{R}^n \times \mathbb{R}^m \to \mathbb{R}^n$ with $f(0,0) = 0$. The state $x \in \mathbb{R}^n$ and the input $u \in \mathbb{R}^m$ are possibly constrained; i.e., $x \in \mathbb{X} \subset \mathbb{R}^n$ and $u \in \mathbb{U}(x) \subset \mathbb{R}^m$, where $\mathbb{U}(x)$ may depend on x. If the set $\mathbb{U}(x)$ is independent of x we drop the argument and write $\mathbb{U} = \mathbb{U}(x)$ for simplicity. We combine the state and input constraints in the set $\mathbb{D} \subset \mathbb{R}^n \times \mathbb{R}^m$ defined by

$$\mathbb{D} = \mathbb{X} \times \mathbb{U}(x)$$

and we assume that $(0,0) \in \mathbb{D}$ is satisfied throughout this chapter. We give a simple example of polyhedral constraints in terms of the implicit definition $\mathbb{D}$ and an explicit description of the constraint set through inequalities. Polyhedral constraints are common in MPC applications and can be used to approximate more complicated constraint sets.

Example 15.1. For $r_1, r_2 \in \mathbb{N}$ consider

$$\Gamma_{x,1} \in \mathbb{R}^{n \times r_1}, \ \Gamma_{x,2} \in \mathbb{R}^{n \times r_2}, \ \Gamma_u \in \mathbb{R}^{m \times r_2}, \ \gamma_1 \in \mathbb{R}^{r_1}, \ \text{and} \ \gamma_2 \in \mathbb{R}^{r_2}.$$

Then, state constraints can be described by

$$\mathbb{X} = \{x \in \mathbb{R}^n : \ \Gamma_{x,1} x \le \gamma_1\}.$$

For a fixed $x \in \mathbb{X}$, we can define the set

$$\mathbb{U}(x) = \{u \in \mathbb{R}^m : \ \Gamma_u u \le \gamma_2 - \Gamma_{x,2} x\}.$$

Finally the state and input constraints can be combined in the definition

$$\mathbb{D} = \left\{ (x, u) \in \mathbb{R}^{n+m} \ \middle| \ \begin{bmatrix} \Gamma_{x,1} & 0 \\ \Gamma_{x,2} & \Gamma_u \end{bmatrix} \begin{bmatrix} x \\ u \end{bmatrix} \le \begin{bmatrix} \gamma_1 \\ \gamma_2 \end{bmatrix} \right\}. \tag{15.2}$$

Here the sets $\mathbb{X}$, $\mathbb{U}$, and $\mathbb{D}$ define *polyhedral sets* in $\mathbb{R}^n$, $\mathbb{R}^m$, and $\mathbb{R}^{n+m}$, respectively. As a simple example, consider the sets $\mathbb{X} = [-1,1]^2$ and $\mathbb{U} = [-\frac{1}{4}, \frac{1}{4}]$. Then we can define the matrices and the vectors

$$\Gamma_{x,1} = \begin{bmatrix} 1 & 0 \\ 0 & 1 \\ -1 & 0 \\ 0 & -1 \end{bmatrix} , \ \gamma_1 = \begin{bmatrix} 1 \\ 1 \\ 1 \\ 1 \end{bmatrix} , \ \Gamma_u = \begin{bmatrix} 1 \\ -1 \end{bmatrix} , \ \Gamma_{x,2} = \begin{bmatrix} 0 & 0 \\ 0 & 0 \end{bmatrix} , \ \gamma_2 = \begin{bmatrix} \frac{1}{4} \\ \frac{1}{4} \end{bmatrix} .$$

Combining these conditions into

$$\begin{bmatrix} 1 & 0 & 0 \\ 0 & 1 & 0 \\ -1 & 0 & 0 \\ 0 & -1 & 0 \\ 0 & 0 & 1 \\ 0 & 0 & -1 \end{bmatrix} \begin{bmatrix} x \\ u \end{bmatrix} \leq \begin{bmatrix} 1 \\ 1 \\ 1 \\ 1 \\ \frac{1}{4} \\ \frac{1}{4} \end{bmatrix}$$

provides a representation of $\mathbb{D}$ as in (15.2). Since $\mathbb{X}$ is independent of u, all entries of $\Gamma_{x,2}$ are zero.

For a given *prediction horizon* $N \in \mathbb{N} \cup \{\infty\}$ and an initial condition $x_0 \in \mathbb{R}^n$, we denote the set of feasible input trajectories of length N as

$$\mathcal{U}_{\mathbb{D}}^N = \left\{ u_N(\cdot) : \mathbb{N}_{[0,N-1]} \to \mathbb{R}^m \ \middle| \ \begin{array}{rcl} x(0) & = & x_0 \\ x(k+1) & = & f(x(k), u(k)) \\ (x(k), u(k)) & \in & \mathbb{D} \\ \forall \ k & \in & \mathbb{N}_{[0,N-1]} \end{array} \right\} . \tag{15.3}$$

To highlight that $u_N(\cdot)$ depends on the initial condition x_0, we will additionally use the notation $u_N(\cdot; x_0) = u_N(\cdot)$ if the initial condition is important or not clear from the context. We adapt the definition of the *cost function* introduced in the last chapter to be $J_N : \mathbb{R}^n \times \mathcal{U}_{\mathbb{D}}^N \to \mathbb{R} \cup \{\infty\}$,

$$J_N(x_0, u_N(\cdot)) = \sum_{i=0}^{N-1} \ell(x(i), u(i))$$

where $\ell : \mathbb{R}^n \times \mathbb{R}^m \to \mathbb{R}$ denotes again the *running cost*. In addition, we consider a *terminal cost* $F : \mathbb{R}^n \to \mathbb{R}$ and *terminal constraints* $\mathbb{X}_F \subset \mathbb{R}^n$ to define the optimal control problem

$$V_N(x_0) = \min_{u_N(\cdot) \in \mathcal{U}_{\mathbb{D}}^N} J_N(x_0, u_N(\cdot)) + F(x(N))$$
$$\text{subject to (15.1), and } x(N) \in \mathbb{X}_F, \tag{15.4}$$

which is a finite-dimensional optimization problem if N is finite.

Even if a closed expression of the the optimal value function $V_N : \mathbb{R}^n \to \mathbb{R} \cup \{\infty\}$ is not available, for a given initial value $x_0 \in \mathbb{R}^n$, the function $V_N(\cdot)$ can be evaluated at x_0 by solving the optimization problem (15.4). Here, the functions J_N and V_N are defined as *extended real-valued functions* which satisfy $J_N(x_0, u_N(\cdot)) = \infty$ and $V_N(x_0) = \infty$ whenever $\mathcal{U}_{\mathbb{D}}^N = \emptyset$. More precisely, $J_N(x_0, u_N(\cdot)) = \infty$ if and only if $u_N(\cdot) \notin \mathcal{U}_{\mathbb{D}}^N$ and $V_N(x_0) = \infty$ if and only if (15.4) is infeasible.

In (15.4) we have again tacitly assumed that the minimum is attained, which is not necessarily the case without further assumptions. However, we will use the minimum instead of the more general infimum here. If the minimum is attained, then there exists an *optimal open-loop input trajectory* $u_N^\star(\cdot\,; x_0) \in \mathcal{U}_{\mathbb{D}}^N$ such that

$$V_N(x_0) = J_N(x_0, u_N^\star(\cdot\,; x_0)) + F(x^\star(N))$$

and $x^\star(N) \in \mathbb{X}_F$ is defined through x_0 and $u_N^\star(\cdot\,; x_0)$. The optimal open-loop input $u_N^\star(\cdot\,; x_0)$ is used in MPC to iteratively define a *feedback law* $\mu_N : \mathbb{R}^n \to \mathbb{R}^m$. In particular, the *MPC feedback law* is defined as

$$\mu_N(x_0) = u_N^\star(0; x_0) \tag{15.5}$$

in general, leading to the *closed-loop solution*

$$x_{\mu_N}(k+1) = f(x_{\mu_N}(k), \mu_N(x(k))), \qquad x_{\mu_N}(0) = x_0. \tag{15.6}$$

The iterative computation of a feedback law using MPC is summarized in Algorithm 3. For a given *optimal open-loop input trajectory* $u_N^\star(\cdot\,; x_0)$, the corresponding

---

Algorithm 3: Model predictive control

Input: Measurement of the initial condition $x(0)$; prediction horizon $N \in \mathbb{N} \cup \{\infty\}$; running cost $\ell : \mathbb{R}^{n+m} \to \mathbb{R}$; constraints $\mathbb{D} \subset \mathbb{R}^{n+m}$; terminal cost $F : \mathbb{R}^n \to \mathbb{R}$ and terminal constraints $\mathbb{X}_F \subset \mathbb{R}^n$.

For $k = 0, 1, 2, \ldots$

1. Measure the current state of the system (15.1) and define $x_0 = x(k)$.
2. Solve the optimal control problem (15.4) to obtain the open-loop input $u_N^\star(\cdot\,; x_0)$.
3. Define the feedback law

$$\mu_N(x(k)) = u_N^\star(0; x_0).$$

4. Compute $x(k+1) = f(x(k), \mu_N(x(k)))$, increment k to $k+1$ and go to 1.

---

optimal open-loop solution is defined through

$$x_N^\star(k+1) = f(x_N^\star(k), u_N^\star(k; x_0)), \qquad x_N^\star(0) = x_0$$

for $k = 0, \ldots, N - 2$.

In many applications, the discrete time system (15.1) is only an approximation of a plant described by a continuous time system

$$\dot{x}_p = f_p(x_p, u), \qquad x_p(0) \in \mathbb{R}^n, \tag{15.7}$$

$f_p : \mathbb{R}^n \times \mathbb{R}^m \to \mathbb{R}^n$. In this setting the MPC feedback law is usually defined as a *sample-and-hold feedback*, illustrated in Figure 15.2. The continuous time application of the MPC Algorithm 3 is summarized in Algorithm 4.

Remark 15.2. Using Algorithm 4 does not guarantee that $x_p(\cdot)$ satisfies the state constraints $x_p(t) \in \mathbb{X}$ for all $t \in \mathbb{R}_{\geq 0}$ since the constraints are only enforced at discrete time steps in the optimal control problem (15.4).

Remark 15.3. Observe that in implementing Algorithm 3 or 4 on an actual, physical, plant, the compute and solve elements of Step 4 are not necessary as the (physical) plant itself will evolve from one time step to the next and then, returning to Step 1, its state will be measured. See the discussion around Figures 1.4 and 1.5.

We will illustrate a couple of properties of the MPC Algorithms 3 and 4 in the following example.

Example 15.4. Consider the discrete time system $x^+ = Ax + Bu$ where

$$A = \begin{bmatrix} \frac{6}{5} & \frac{6}{5} \\ -\frac{1}{2} & \frac{6}{5} \end{bmatrix} \quad \text{and} \quad B = \begin{bmatrix} 1 \\ \frac{1}{2} \end{bmatrix}. \tag{15.8}$$

The origin is unstable for the uncontrolled dynamics, which follows immediately from the fact that A is not a Schur matrix. In Figure 15.3 open-loop and closed-loop solutions of the dynamics defined through (15.8) and calculated via Algorithm 3 are visualized. Here, the prediction horizon is set to $N = 5$, the algorithm is simulated for 30 iterations, and initial condition $x(0) = [3,3]^T$ is used. The running cost is taken as $\ell(x, u) = x^T x + 5u^2$ and the input is constrained to satisfy $u \in \mathbb{U} = [-2.5, 2.5]$, while the states x are unconstrained. We can thus define

---

Algorithm 4: Model predictive control (continuous time application)

Input: Measurement of the initial condition $x_p(0)$; prediction horizon $N \in \mathbb{N} \cup \{\infty\}$; running cost $\ell : \mathbb{R}^{n+m} \to \mathbb{R}$; constraints $\mathbb{D} \subset \mathbb{R}^{n+m}$; terminal cost $F : \mathbb{R}^n \to \mathbb{R}$ and terminal constraints $\mathbb{X}_F \subset \mathbb{R}^n$; sampling time $\Delta > 0$.

For $k = 0, 1, 2, \ldots$

1. Measure the current state of the plant (15.1) and define $x_0 = x_p(k\Delta)$.
2. Solve the optimal control problem (15.4) to obtain the open-loop control law $u_N^\star(\cdot; x_0)$.
3. Define the feedback law

$$\mu_N(x_p(k\Delta)) = u_N^\star(0; x_0).$$

4. Apply the feedback law, i.e., for $t \in [k\Delta, (k+1)\Delta)$ solve

$$\dot{x}_p(t) = f_p(x_p(t), \mu_N(x_p(k\Delta))), \quad x_p(k\Delta) \in \mathbb{R}^n,$$

increment k to $k+1$ and go to 1.

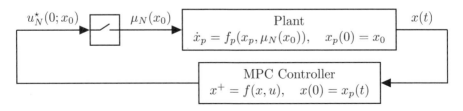

Figure 15.2: Block diagram of MPC Algorithm 4.

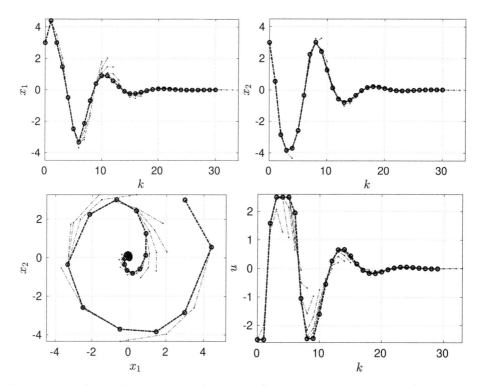

Figure 15.3: Closed-loop solution (thick lines) and open-loop solutions (thin lines) using the MPC Algorithm 3 for the dynamics defined in (15.8).

$\mathbb{D} = \mathbb{R}^2 \times \mathbb{U}$. Additionally, we define a terminal cost as $F(x) = x^T x$ and assume no terminal constraints, i.e., $\mathbb{X}_F = \mathbb{R}^2$. From Figure 15.3 we observe that the MPC controller drives the initial state to the origin even though the open-loop solutions do not seem to go to the origin.

Next, we additionally include the terminal constraint $\mathbb{X}_F = \{0\}$. Figure 15.4 shows open-loop and closed-loop solutions of Algorithm 3 that can be compared to Figure 15.3. Note that it is necessary to extend the prediction horizon to $N = 11$

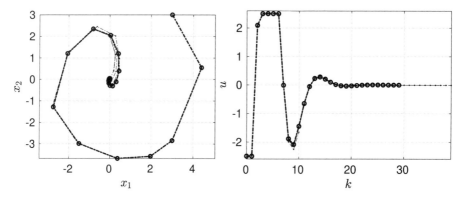

Figure 15.4: Closed-loop solution and open-loop solutions using the MPC Algorithm 3 for the dynamics defined through (15.8).

since (15.4) with respect to the initial condition $x_0 = [3,3]^T$ is infeasible for $N < 11$ (and the input constraints $u \in \mathbb{U}$). Note that the terminal constraints $\mathbb{X}_F = \{0\}$ make the terminal cost superfluous.

Example 15.5. The discrete time system (15.8) is an Euler approximation of the continuous time system

$$\dot{x}_p = A_p x + B_p x = \begin{bmatrix} \frac{1}{5} & \frac{12}{5} \\ -1 & \frac{1}{5} \end{bmatrix} x_p + \begin{bmatrix} 2 \\ 1 \end{bmatrix} u \qquad (15.9)$$

with a sampling rate $\Delta = 0.5$ (see Section 5.2.1). Figure 15.5 visualizes the closed-loop solution and the piecewise constant feedback law obtained through Algorithm 4. Here, the same setting as in Example 15.4, without terminal constraints, is used.

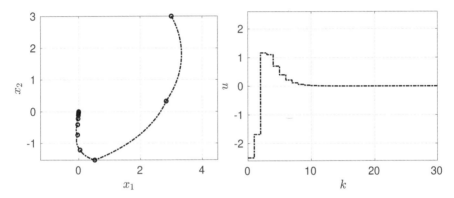

Figure 15.5: Closed-loop solution (left) and piecewise constant feedback law obtained through Algorithm 4.

Note that the apparently faster convergence seen in Figure 15.5 as compared to Figure 15.3 is not a general property and is merely an artifact of the particular system dynamics in this example.

Having computed an optimal control input over the prediction horizon $N = 5$, in Figure 15.6 an open-loop solution of the continuous time system is compared with the corresponding open-loop solution of the discrete time system. Note that for the continuous time system this input is implemented as a sample and hold, i.e., the input at time k is applied constantly to the system for $t \in [k\Delta, (k+1)\Delta)$. Since a rather large sampling rate Δ is used for the discretization of the continuous time system, the two solutions differ significantly. This highlights an important difference between a feedback law and an open-loop control law and provides one explanation why in MPC in general only the first piece of the optimal trajectory $u_N^\star(\cdot)$ is used to define a feedback law.

15.2   MPC CLOSED-LOOP ANALYSIS

Compared to many other control schemes, MPC has the advantage that it can be applied to general nonlinear systems and constraints can be taken into account directly in the controller design. However, since the control law is only implicitly known as the solution of a finite- (or infinite-)dimensional optimization problem,

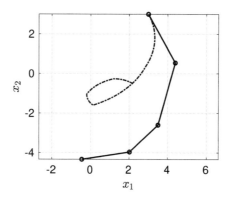

Figure 15.6: Open-loop solution of the discrete time system (15.8) (solid line) and the corresponding open-loop solution (dashed line) of the continuous time system (15.9).

the analysis of the closed-loop dynamics (15.6) is rather difficult. In this section we discuss several aspects in MPC design which need to be taken into consideration.

15.2.1 Performance Estimates

Often the optimal control problem (15.4) solved in every iteration of the MPC algorithm is a compromise between numerical complexity and optimality. In many applications, one is interested in solving the optimal control problem for $N = \infty$. However, as outlined in Chapter 14, the underlying infinite-dimensional optimization problem is usually intractable. Since it is thus necessary to solve (15.4) for finite $N \in \mathbb{N}$, it is reasonable to ask how the finite horizon cost $V_N(\cdot)$, $N \in \mathbb{N}$, relates to the infinite horizon cost $V_\infty(\cdot)$. In addition, when talking about MPC closed-loop performance, it is reasonable to compare $V_\infty(\cdot)$ with the MPC closed-loop cost

$$J_\infty(x_0, \mu_N(\cdot)) = \sum_{k=0}^{\infty} \ell(x_{\mu_N}(k), u_N^\star(0; x(k))), \qquad x(0) = x_0. \tag{15.10}$$

Here, we solely focus on positive semidefinite running costs $\ell : \mathbb{R}^n \times \mathbb{R}^m \to \mathbb{R}_{\geq 0}$. In the case that terminal costs and terminal constraints are not present in optimal control problem (15.4), i.e., $F(x) = 0$ and $\mathbb{X}_F = \mathbb{R}^n$, the inequalities

$$V_N(x_0) \leq V_\infty(x_0) \leq J_\infty(x_0, \mu_N(\cdot)) \qquad \forall\, x_0 \in \mathbb{R}^n \tag{15.11}$$

follow immediately from the definition of the functionals V_N and J_N, $N \in \mathbb{N} \cup \{\infty\}$. Indeed, the chain of inequalities

$$V_N(x_0) = J_N(x_0, u_N^\star(\cdot)) \leq J_N(x_0, (u_\infty^\star(0), \dots, u_\infty^\star(N-1)))$$
$$\leq J_\infty(x_0, u_\infty^\star(\cdot)) \leq J_\infty(x_0, \mu_N(\cdot))$$

is satisfied and $J_\infty(x_0, u_\infty^\star(\cdot))$ can be identified with $V_\infty(x_0)$.

While the above estimate is straightforward, it is in general more interesting   .

and useful to establish a relationship of the form

$$J_\infty(x_0, \mu_N(\cdot)) \leq \frac{1}{\alpha_N} V_\infty(x_0) \qquad \forall \, x \in \mathbb{R}^n \qquad (15.12)$$

for an $\alpha_N \in (0, 1]$. In particular, while the MPC closed-loop cost is necessarily at least as large as the open-loop infinite horizon cost according to (15.11), the bound (15.12) guarantees a level of suboptimality of the MPC control law.

For example, if (15.12) is true for $\alpha_N = \frac{1}{2}$, the MPC closed-loop cost is at most twice the infinite horizon optimal control cost. Moreover, while the infinite horizon optimal control problem might be intractable, the MPC feedback law relies only on the solution of finite-dimensional optimization problems. Under appropriate assumptions, one can expect that for $N \to \infty$ the parameter $\alpha_N \in (0, 1]$ satisfies $\alpha_N \to 1$. However, this comes at the price that the dimension of the optimal control problem (15.4) that needs to be solved at every time step of the MPC algorithm converges to infinity.

For a nonlinear system subject to nontrivial state and input constraints, it is in general very difficult to establish a bound of the form (15.12) for an $\alpha_N \in (0, 1]$, and it is out of the scope of this chapter. When the assumption on the definiteness of the running cost ℓ is dropped, performance bounds comparable to (15.12) can be derived using the framework of *economic MPC* (see Section 15.3.5). For further reading we refer to the monograph [58]. Here, we show how the MPC closed-loop cost (15.10) changes with respect to the prediction horizon $N \in \mathbb{N}$.

Example 15.6. Consider the discrete time system

$$x^+ = Ax + Bu = \begin{bmatrix} 1 & 4 & 0 & 3 & 2 \\ 2 & 4 & 2 & 4 & 2 \\ 3 & 3 & 3 & 0 & 4 \\ 3 & 1 & 3 & 0 & 3 \\ 2 & 3 & 1 & 4 & 4 \end{bmatrix} x + \begin{bmatrix} 2 \\ 3 \\ 1 \\ 2 \\ 3 \end{bmatrix} u \qquad (15.13)$$

together with the running cost $\ell(x, u) = x^T x + u^2$, the terminal cost $F(x) = x^T x$, the terminal constraints $\mathbb{X}_F = \{0\}$, and the input constraints $\mathbb{U} = [-40, 40]$.

Figure 15.7 shows the closed-loop MPC cost with respect to the prediction horizon N and for the initial value $x_0 = [1, 1, 1, 1, 1]^T$. Here, the setting with a terminal cost is compared to the setting without a terminal cost. Furthermore the infinite horizon optimal control cost $V_\infty(x_0)$ is shown, which provides a lower bound for the cost $J_\infty(x_0, \mu_N(\cdot))$.

To be precise, in the numerical simulations, $V_\infty(x_0)$ is approximated through $V_{1000}(x_0)$ and the MPC Algorithm 3 is simulated for 1,000 iterations. By inspecting $x(1000)$, which is in the magnitude of $|x(1000)| \leq 10^{-10}$, the error made through this approximation is negligible.

In the setting without terminal constraints, a prediction horizon $N \geq 5$ leads to finite closed-loop cost $J_\infty(x_0, \mu_N(\cdot)) < \infty$. In the setting with terminal constraints a prediction horizon of $N \geq 6$ is necessary to ensure that the optimal control problem (15.4) is feasible.

Note that Figure 15.7 shows the costs for a particular initial condition $x_0 = [1, 1, 1, 1, 1]^T$, and thus it does not provide an estimate of the form (15.12) which needs to be satisfied for all initial conditions. However, for the particular initial condition, Figure 15.7 shows that already a rather small prediction horizon $N = 10$

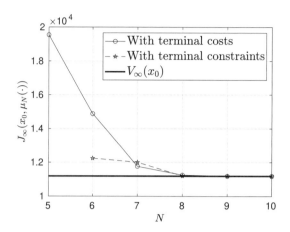

Figure 15.7: Closed-loop MPC cost with respect to the initial condition $x_0 = [1, 1, 1, 1, 1]^T$ depending on the prediction horizon N and compared to the infinite horizon optimal control cost.

leads to closed-loop performance which is indistinguishable from the infinite horizon optimal control costs.

Remark 15.7. MPC feedback laws are frequently referred to as "optimal." However, such a claim should be approached carefully as the feedback is optimal with respect to the specified cost function and the finite horizon. In the event that this is an approximation to a desirable infinite horizon optimal control problem, the MPC feedback solution can only, at best, be nearly optimal. How close the MPC approximation is to the desired optimal infinite horizon cost can be captured via the performance estimate (15.12). Of course, in all applications of optimal control methodologies one should always be conscious that the solutions are optimal with respect to the defined cost function, and that a poorly designed cost function will lead to undesirable performance even if that performance is "optimal."

15.2.2   Closed-Loop Stability Properties

A standard control application of MPC is the stabilization of an equilibrium pair $(x^e, u^e) \in \mathbb{X} \times \mathbb{U}$ by penalizing the deviation from (x^e, u^e) in the cost functional J_N. Following the presentation so far, the cost functional can then be defined through the quadratic cost

$$\ell(x, u) = (x - x^e)^T Q(x - x^e) + (u - u^e)^T R(u - u^e)$$

for positive semidefinite matrices $Q \geq 0$, $R \geq 0$, for example. To ensure asymptotic stability of $x^e \in \mathbb{X}$, the closed-loop dynamics (15.6) need to be investigated, with the difficulty that $\mu_N(\cdot)$ is not explicitly known in general.

A sufficient condition for x^e to be an asymptotically stable equilibrium of the closed-loop dynamics is that V_N is a Lyapunov function for (15.6). In particular, if

$$V_N(f(x, \mu_N(x))) < V_N(x) \tag{15.14}$$

holds for all $x \in \mathbb{X} \backslash \{x^e\}$, then asymptotic stability can be concluded using Theorem

5.5. Even though V_N and μ_N are only implicitly known as the solution of the optimization problem (15.4), it is possible to derive conditions on the system dynamics (15.1), the prediction horizon $N \in \mathbb{N} \cup \{\infty\}$, the running cost ℓ, the terminal cost F, and the terminal region $\mathbb{X}_F$ that ensure that V_N is indeed a Lyapunov function (see [58] or [128], for example). One way to derive these results relies on the *principle of optimality* and *dynamic programming*, briefly introduced in Section 14.3.

If terminal constraints of the form $\mathbb{X}_F = \{x^e\}$ are considered, then stability can be concluded quite easily. However, this comes with the price that it is usually difficult to ensure that the optimization problem (15.4) is feasible for all initial values $x_0 \in \mathbb{X}$. To illustrate that V_N is a Lyapunov function in the case of terminal constraint $\mathbb{X}_F = \{x^e\}$, we assume for simplicity that $(x^e, u^e) = (0, 0)$. Additionally, we assume that (15.4) is feasible for all initial values $x_0 \in \mathbb{X}$ and we assume that the running cost is defined such that $\ell(x, u) > \ell(0, 0) = 0$ for all $(x, u) \neq (x^e, u^e)$. (Note that due to the terminal constraint $\mathbb{X}_F = \{0\}$, terminal costs are not relevant.)

Consider an arbitrary initial value $x_0 \in \mathbb{X}$. Then, since $x(N) = 0$ is satisfied, it holds that

$$
\begin{aligned}
V_N(x_0) = J_N(x_0, u_N^\star(\cdot; x_0)) &= \sum_{i=0}^{N-1} \ell(x(i), u_N^\star(i; x_0)) \\
&= \ell(x(0), u_N^\star(0; x_0)) + \sum_{i=1}^{N-1} \ell(x(i), u_N^\star(i; x_0)) + \ell(x(N), 0) \\
&\geq \ell(x(0), u_N^\star(0; x_0)) + V_N(f(x_0, u_N^\star(0; x_0))).
\end{aligned}
$$

Thus, since $\ell(x_0, u) > 0$ for $x_0 \neq 0$ by assumption, inequality (15.14) is indeed satisfied and V_N is a Lyapunov function if V_N is additionally radially unbounded. This is for example the case if $\ell(x, 0) \to \infty$ is satisfied for $|x| \to \infty$.

Example 15.8. Consider again the discrete time system (15.13) together with the setting discussed in Example 15.6. In Figure 15.8 the open-loop costs $V_N(x(k))$ are visualized for different $N \in \mathbb{N}$. Figure 15.8, left, visualizes the setting without

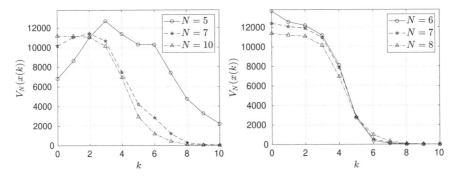

Figure 15.8: Open-loop costs $V_N(x(k))$ for different N. The setting without terminal constraints (left) is compared to the setting with terminal constraints (right).

terminal constraints while Figure 15.8, right, visualizes the setting with terminal constraints. As expected, in the setting with terminal constraints $V_N(x(k))$ is strictly decreasing independently of the prediction horizon $N \in \mathbb{N}$. (However, as mentioned in Example 15.6, N needs to satisfy $N \geq 6$ to ensure feasibility of the optimal control problem (15.4).) In the setting without terminal constraints this is not the case and for $N < 10$ and $x(0) = [1, 1, 1, 1, 1]^T$, the optimal value function $V_N(x(k))$ is not monotonically decreasing. However, for N large enough, i.e., $N \geq 10$ in this setting, $V_N(x(k))$ is strictly monotonically decreasing also in the setting without terminal constraints. While this suggests that $V_N(\cdot)$ is a (local) Lyapunov function for N sufficiently large, this cannot be definitively concluded from Figure 15.8, left, since the decrease (15.14) needs to be verified for all $x \in \mathbb{X}$.

15.2.3   Viability and Recursive Feasibility

So far we have ignored the fact that in the presence of state constraints $\mathbb{X} \neq \mathbb{R}^n$ the optimal control problem (15.4) may be infeasible. To ensure that Algorithms 3 and 4 define implementable feedback laws it is necessary that (15.4) be feasible for all discrete time steps $k \in \mathbb{N}$. We thus briefly discuss the terms *viability* and *recursive feasibility* in the context of MPC.

Definition 15.9 (Viability). *Consider the discrete time system (15.1) together with $\mathbb{X} \subset \mathbb{R}^n$ and $\mathbb{U}(x) \subset \mathbb{R}^m$ for all $x \in \mathbb{X}$. The set $\mathbb{X}$ is called viable if for all $x \in \mathbb{X}$ there exists $u \in \mathbb{U}(x)$ such that*

$$f(x, u) \in \mathbb{X}.$$

A viable set $\mathbb{X}$ is also sometimes called a *control-invariant set*. We illustrate viability via a simple one-dimensional example.

Example 15.10. For $a \in \mathbb{R}$, we consider the dynamics

$$x^+ = ax + u \tag{15.15}$$

together with the state and input constraints $\mathbb{X} = [-1, 1]$ and $\mathbb{U} = [-1, 1]$, respectively. For $|a| \leq 1$, the origin of the uncontrolled system (i.e., $u = 0$) satisfies $|x^+| \leq |x| \leq 1$. Thus, for all $x \in \mathbb{X}$, there exists $u \in \mathbb{U}$, namely $u = 0$, such that $x^+ \in \mathbb{X}$.

   For $|a| \in (1, 2]$ the input u can be defined as $u(x) = -\operatorname{sign}(a)x$, for example. Then, for all $x \in \mathbb{X}$, x^+ satisfies

$$|x^+| = |ax - \operatorname{sign}(a)x| = |a - \operatorname{sign}(a)| \cdot |x| = ||a| - 1| \cdot |x| \leq |x| \leq 1,$$

and thus $\mathbb{X}$ is viable.

   For $|a| > 2$, the set $\mathbb{X}$ is not viable. To see this, consider the point $x = \operatorname{sign}(a)$. Then, for $u = 0$, x^+ satisfies $x^+ = a \operatorname{sign}(a) = |a| > 2$. The best we can do with the input u is to select $u = -1$. This implies $x^+ = a \operatorname{sign}(a) - 1 = |a| - 1 > 1$, and thus, independently of the selection of u, the state x^+ is not contained in $\mathbb{X}$.

   If the set $\mathbb{X}$ is not viable, there exist initial conditions $x_0 \in \mathbb{X}$ such that every corresponding trajectory $x(\cdot; x_0)$ necessarily leaves the domain $\mathbb{X}$; i.e., there exists $K \in \mathbb{N}$ such that $x(k; x_0) \in \mathbb{X}$ for all $k \leq K$ and $x(K; x_0) \notin \mathbb{X}$. From a control perspective it is thus desirable to remove these states from the set $\mathbb{X}$ to be able to find a viable set of states $\mathbb{X}_v$ contained in $\mathbb{X}$.

Example 15.11. We continue with Example 15.10 and consider the dynamics $x^+ = 3x + u$, $\mathbb{X} = [-1, 1]$, and $\mathbb{U} = [-1, 1]$, where we know that $\mathbb{X}$ is not viable. However, since $x = 0 \in \mathbb{X}$ is an equilibrium of the system (and $u = 0 \in \mathbb{U}$), the set $\{0\} \subset \mathbb{X}$ is trivially a viable subset of $\mathbb{X}$. Is it possible to enlarge the viable set and what is its maximal size? Assume that $\mathbb{X}_v = [-c_1, c_2]$ for unknown constants $c_1, c_2 \in [0, 1]$. Then, in particular for $x = c_2$, there needs to exist a $u \in \mathbb{U}$ such that $x^+ \in [-c_1, c_2]$. Since $a > 0$, and $c_2 \geq 0$ by assumption, in the worst case it is only possible to guarantee that $x^+ = c_2$ but $x^+ \notin (-c_1, c_2)$. This leads to the condition

$$c_2 = 3c_2 - 1,$$

where $u = -1$ is the best input when x^+ is supposed to be minimized. Thus, $c_2 = \frac{1}{2}$ and the selection of $u = -1$ implies that

$$x^+ = 3x - 1 > c_2 \qquad \forall \, x > c_2 \qquad \text{and}$$
$$x^+ = 3x - 1 < c_2 \qquad \forall \, x < c_2.$$

Using the same arguments with $u = 1$ and $x \leq 0$ provides the lower bound $c_1 = c_2$. Thus the maximal viable set contained in $\mathbb{X}$ is given by $\mathbb{X}_v = [-\frac{1}{2}, \frac{1}{2}]$.

In Exercise 15.3 we invite the reader to compute $\mathbb{X}_v$ for the dynamics (15.15) depending on the parameter $a \in \mathbb{R}$.

For the one-dimensional linear system (15.15) it is relatively simple to check viability of $\mathbb{X}$, or to compute a viable set $\mathbb{X}_v$ for given system dynamics and input constraints $\mathbb{U}$. For higher dimensional linear systems subject to constraints $\mathbb{X}$ and $\mathbb{U}$ the verification of viability quickly becomes tedious, and for more general nonlinear systems subject to nonlinear constraints it might be impossible.

To illustrate how viable sets can be calculated in higher dimensions, we consider a two-dimensional example with polyhedral constraints. We apply the Fourier-Motzkin elimination (see [42], for example) to provide an analytic (but quickly tedious) way to compute viable sets $\mathbb{X}_v \subset \mathbb{X}$. In MATLAB the *MPT-toolbox* [62] can be used to perform these calculations.

Example 15.12. Consider the dynamics

$$\begin{bmatrix} x_1^+ \\ x_2^+ \end{bmatrix} = Ax + Bu = \begin{bmatrix} 1 & 1 \\ 0 & 1 \end{bmatrix} x + \begin{bmatrix} 0.5 \\ 1 \end{bmatrix} u \qquad (15.16)$$

and the constraint sets $\mathbb{X} = [-1, 1]^2$ and $\mathbb{U} = [-\frac{1}{4}, \frac{1}{4}]$. As discussed in Example 15.1 the constraints can alternatively be written as $\mathbb{X} = \mathbb{X}_0 = \{x \in \mathbb{R}^2 : \Gamma_0 x \leq \gamma_0\}$ and $\mathbb{U} = \{u \in \mathbb{R} : \Gamma_u u \leq \gamma_u\}$, where

$$\Gamma_0 = \begin{bmatrix} 1 & 0 \\ 0 & 1 \\ -1 & 0 \\ 0 & -1 \end{bmatrix}, \quad \gamma_0 = \begin{bmatrix} 1 \\ 1 \\ 1 \\ 1 \end{bmatrix}, \quad \Gamma_u = \begin{bmatrix} 1 \\ -1 \end{bmatrix}, \quad \gamma_u = \begin{bmatrix} \frac{1}{4} \\ \frac{1}{4} \end{bmatrix}. \qquad (15.17)$$

Using the notation $\mathbb{X}_0 = \mathbb{X}$, as a first approximation of the viable set, we could define the set

$$\mathbb{X}_1 = \{x \in \mathbb{X}_0 : \exists u \in \mathbb{U} \text{ such that } Ax + Bu \in \mathbb{X}_0\}.$$

However, this set may still contain points in $x \in \mathbb{X}_0 \backslash \mathbb{X}_1$ which are not viable, i.e., $Ax + Bu \in \mathbb{X}_0 \backslash \mathbb{X}_1$ holds for all $u \in \mathbb{U}$. Nevertheless, this idea can be used iteratively via

$$\mathbb{X}_{i+1} = \{x \in \mathbb{X}_i : \exists u \in \mathbb{U} \text{ such that } Ax + Bu \in \mathbb{X}_i\} \qquad (15.18)$$

for $i \in \mathbb{N}$, and a viable set $\mathbb{X}_v = \mathbb{X}_i$ is found if $\mathbb{X}_{i+1} = \mathbb{X}_i$ is satisfied.

To compute (15.18) based on the knowledge of $\mathbb{X}_i$, $i \in \mathbb{N}_0$, the Fourier-Motzkin elimination (see [42], for example) can be applied to the polyhedral representation (15.17) and $\mathbb{X}_i = \{x \in \mathbb{R}^2 : \Gamma_i x \le \gamma_i\}$. In particular, we define

$$\widetilde{\mathbb{X}}_i = \{[x^T, u]^T \in \mathbb{R}^3 : \Delta_i(x^T, u)^T \le \delta_i\} \qquad (15.19)$$

with

$$\Delta_i = \begin{bmatrix} \Gamma_i & 0 \\ \Gamma_i A & \Gamma_i B \\ 0 & \Gamma_u \end{bmatrix} \quad \text{and} \quad \delta_i = \begin{bmatrix} \gamma_i \\ \gamma_i \\ \gamma_u \end{bmatrix}.$$

Then, $\mathbb{X}_{i+1} = P_x(\widetilde{\mathbb{X}}_i)$ is obtained by projecting $\widetilde{\mathbb{X}}_i$ on the (x_1, x_2)-subspace. We illustrate the projection using the Fourier-Motzkin elimination for the dynamics (15.16) and the constraints $\mathbb{X} = [-1, 1]^2$ and $\mathbb{U} = [-\frac{1}{4}, \frac{1}{4}]$.

For the given parameters the set $\widetilde{\mathbb{X}}_0$ is defined through the inequalities $\Gamma_0 x \le \gamma_0$, which are independent of u, and the inequalities

$$\begin{array}{llll}
u & \le & 2 - 2x_1 - 2x_2, & \qquad u & \ge & -2 - 2x_1 - 2x_2, \\
u & \le & 1 - x_2, & \qquad u & \ge & -1 - x_2, \\
u & \le & \frac{1}{4}, & \qquad u & \ge & -\frac{1}{4},
\end{array} \qquad (15.20)$$

which are dependent on u. The inequalities (15.20) are equivalent to $\Gamma_0 Ax + \Gamma_0 Bu \le \gamma_0$ and $\Gamma_u u \le \gamma_u$, where we have isolated the input u. Following the Fourier-Motzkin elimination, the input u can be removed by projecting on the (x_1, x_2)-subspace by combining every $\le$-inequality with every $\ge$-inequality:

$$-2 - 2x_1 - 2x_2 \le 2 - 2x_1 - 2x_2, \qquad (15.21\text{a})$$

$$-1 - x_2 \le 2 - 2x_1 - 2x_2, \qquad (15.21\text{b})$$

$$-\tfrac{1}{4} \le 2 - 2x_1 - 2x_2, \qquad (15.21\text{c})$$

$$-2 - 2x_1 - 2x_2 \le 1 - x_2, \qquad (15.21\text{d})$$

$$-1 - x_2 \le 1 - x_2, \qquad (15.21\text{e})$$

$$-\tfrac{1}{4} \le 1 - x_2, \qquad (15.21\text{f})$$

$$-2 - 2x_1 - 2x_2 \le \tfrac{1}{4}, \qquad (15.21\text{g})$$

$$-1 - x_2 \le \tfrac{1}{4}, \qquad (15.21\text{h})$$

$$-\tfrac{1}{4} \le \tfrac{1}{4}. \qquad (15.21\text{i})$$

Inequalities (15.21a), (15.21e), and (15.21i) are trivially satisfied and can be neglected. The remaining conditions (15.21) are visualized in Figure 15.9 together with the constraints $\Gamma_0 x \le \gamma_0$. From Figure 15.9 we observe that the set $\mathbb{X}_1$ is uniquely defined through the conditions $\Gamma_0 x \le \gamma_0$ and (15.21c), (15.21g), i.e.,

$$\mathbb{X}_1 = \{x \in \mathbb{R}^2 : \Gamma_1 x \le \gamma_1\} \qquad (15.22)$$

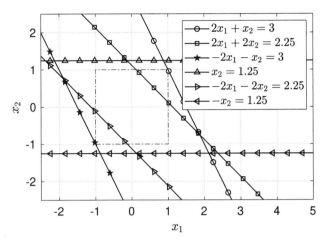

Figure 15.9: Visualization of the constraints (15.21) together with $\mathbb{X} = [-1,1]^2$.

where

$$\Gamma_1 = \begin{bmatrix} 1 & 0 \\ 0 & 1 \\ -1 & 0 \\ 0 & -1 \\ 2 & 2 \\ -2 & -2 \end{bmatrix}, \quad \gamma_1 = \begin{bmatrix} 1 \\ 1 \\ 1 \\ 1 \\ \frac{9}{4} \\ \frac{9}{4} \end{bmatrix}.$$

Note that this representation is not unique since rows of the matrix Γ_1 and corresponding entries in γ_1 can be multiplied by a positive constant without changing the condition $\Gamma_1 x \leq \gamma_1$. Based on the knowledge of $\mathbb{X}_1$, $\mathbb{X}_i$ for $i \in \mathbb{N}$ can be calculated iteratively by following the same steps. Since these calculations are quite tedious, we use the *MPT-toolbox* [62] in MATLAB instead. The *MPT-toolbox* can be used to calculate the viable set $\mathbb{X}_v$ by iteratively computing the sets $\mathbb{X}_i$ until $\mathbb{X}_{i+1} = \mathbb{X}_i$ is satisfied. Here the viable set $\mathbb{X}_v$ is defined through the inequalities

$$\begin{bmatrix} -0.24 & -0.97 \\ -0.32 & -0.95 \\ 0.71 & 0.71 \\ 0.32 & 0.95 \\ -0.45 & -0.89 \\ 0.24 & 0.97 \\ 0.45 & 0.89 \\ -0.71 & -0.71 \\ 1.00 & 0.00 \\ -1.00 & 0.00 \end{bmatrix} x \leq \begin{bmatrix} 0.73 \\ 0.67 \\ 0.80 \\ 0.67 \\ 0.67 \\ 0.73 \\ 0.67 \\ 0.80 \\ 1.00 \\ 1.00 \end{bmatrix}. \tag{15.23}$$

The sets $\mathbb{X}_1$ and $\mathbb{X}_v$ are visualized in Figure 15.10.

We now return to MPC and discuss viability in the context of Algorithm 3. Since the calculation of a viable set might be intractable, *recursive feasibility* ensures that the optimal control problem (15.4) involved in Algorithm 3 is feasible at every time

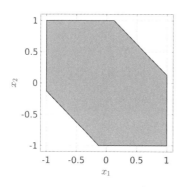 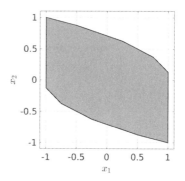

Figure 15.10: Visualization of the set $\mathbb{X}_1$ (left) and the viable set $\mathbb{X}_v$ (right).

step.

Definition 15.13 (Recursive feasibility). *Consider the MPC Algorithm 3 with input constraints $\mathbb{U}$ and a set of initial states $\mathbb{X}_{\mathrm{rf}}^N \subset \mathbb{X}$. The set $\mathbb{X}_{\mathrm{rf}}^N$ is called recursively feasible with respect to Algorithm 3 and the prediction horizon $N \in \mathbb{N}$ if feasibility of* (15.4) *for $x(0) = x_0 \in \mathbb{X}_{\mathrm{rf}}^N$ implies feasibility of* (15.4) *for all $k \in \mathbb{N}$.*

Definition 15.13 states that feasibility of the optimal control problem (15.4) at time $k = 0$ does not necessarily imply feasibility at time $k = 1$ with respect to the initial state $x(1)$. Additionally, the definition of the set $\mathbb{X}_{\mathrm{rf}}^N$ indicates that recursive feasibility depends on the prediction horizon N. We illustrate both properties by continuing Example 15.12 (cf. [58, Example 7.1]).

Example 15.14. Recall the dynamics (15.16) together with the constraint sets $\mathbb{X} = [-1, 1]^2$, $\mathbb{U} = [-\frac{1}{4}, \frac{1}{4}]$, discussed in Example 15.12. In addition we consider the running cost $\ell(x, u) = x^T x + 10u^2$ and consider the MPC Algorithm 3 without a terminal cost and without terminal constraints. The initial condition $x_0 = [-1, 1]^T \in \mathbb{X}_v$ is contained in the viable set (15.23).

Even though we start with a feasible point x_0, for $N = 3$, MPC Algorithm 3 arrives at an infeasible optimal control problem (15.4) after two iterations (see Figure 15.11, left). This is of course not the case if the constraints $\mathbb{X}$ in the MPC algorithm are replaced by the viable set $\mathbb{X}_v$. However, this implies that the viable set $\mathbb{X}_v$ needs to be known, which is not always the case. For $N = 4$ (and in fact also for $N \geq 4$), the MPC algorithm does not run into an infeasible optimization problem (see Figure 15.11, right).

Recursive feasibility shifts the problem of running into an infeasible optimization problem from viability to recursive feasibility. However, as with viability, recursive feasibility of a set $\mathbb{X}_{\mathrm{rf}}^N$ is in general nontrivial to establish. We conclude this section with two results using viability to guarantee recursive feasibility.

Lemma 15.15. *Consider the MPC Algorithm 3 defined through its input and assume that $\mathbb{U}(x) = \mathbb{U} \subset \mathbb{R}^m$ for all $x \in \mathbb{X}$. Additionally assume that terminal constraints are not present, i.e., $\mathbb{X}_F = \mathbb{R}^n$. If $\mathbb{X}$ is viable, then $\mathbb{X}_{\mathrm{rf}}^N = \mathbb{X}$ is recursively feasible for all $N \in \mathbb{N}$.*

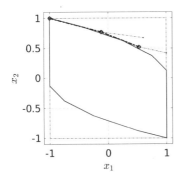
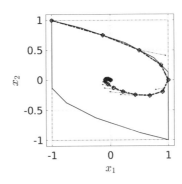

Figure 15.11: Closed-loop solution of Algorithm 3 for the dynamics (15.16) subject to the constraints $\mathbb{X} = [-1,1]^2$, $\mathbb{U} = [-\frac{1}{4}, \frac{1}{4}]$. For $N = 3$ and $x_0 = [-1,1]^T$ the optimal control problem (15.4) is infeasible after two iterations (left). For $N = 4$ the same setting does not run into an infeasible optimization problem (right).

Proof. The result follows immediately from viability of the set $\mathbb{X}$ and the definition of the optimal control problem (15.4). Since $\mathbb{X}$ is viable, for all $x(k) \in \mathbb{X}$ there exist $u(k) \in \mathbb{U}$ such that $x(k+1) \in \mathbb{X}$, $k = 0, \ldots, N-1$. Thus, if $x(0) = x_0 \in \mathbb{X}$ is satisfied, the optimal control problem (15.4) is feasible. At the next time step, the optimal control problem is initialized through $f(x_0, u^\star(0; x_0)) \in \mathbb{X}$ and the same argument can be applied iteratively. $\qquad\square$

Lemma 15.16. *Consider the MPC Algorithm 3 defined through its input and assume that $\mathbb{U}(x) = \mathbb{U} \subset \mathbb{R}^m$ for all $x \in \mathbb{X}$. Additionally assume that terminal constraints $\mathbb{X}_F \subset \mathbb{X}$ define a viable set. If the optimal control problem (15.4) is feasible for all $x_0 \in \mathbb{X}_{\mathrm{rf}}^N$, then $\mathbb{X}_{\mathrm{rf}}^N = \mathbb{X}$ is recursively feasible.*

Proof. Let the optimal control problem (15.4) be feasible for all $x_0 \in \mathbb{X}_{\mathrm{rf}}^N$. Then there exist $u(k) \in \mathbb{U}$ such that $x(k+1) \in \mathbb{X}$ for all $k = 0, \ldots, N-1$ and $x(N) \in \mathbb{X}_F$. Moreover, since $\mathbb{X}_F$ is viable, there exists $u(N) \in \mathbb{U}$ such that $x(N+1) \in \mathbb{X}_F$. In particular $u(1), \ldots, u(N)$ is feasible for (15.4) at time $k = 1$ initialized through $x_0 = x(1)$. This argument can be applied iteratively, showing recursive feasibility. $\qquad\square$

15.2.4 Hard and Soft Constraints

Infeasibility of the optimal control problem (15.4) solved at every time step of MPC Algorithm 3 can only occur in the presence of state constraints $\mathbb{X} \neq \mathbb{R}^n$. Since recursive feasibility is in general hard to establish, the possibility of an infeasible optimal control problem after some time steps can constrain the applicability of an MPC scheme. In some applications it is however possible and justifiable to circumvent infeasible optimization problems by rewriting *hard constraints* as *soft constraints*. In essence, hard constraints refer to constraints which cannot be violated under any circumstances, while soft constraints indicate that the constraints can be violated but will incur additional, possibly significant, costs in doing so.

Consider a closed set $\mathbb{D} \subset \mathbb{R}^{n+m}$ defining combined state and input constraints.

Then, the positive semidefinite function $d_{\mathbb{D}} : \mathbb{R}^{n+m} \to \mathbb{R}_{\geq 0}$,

$$d_{\mathbb{D}}(x, u) = \min_{(v,w) \in \mathbb{D}} \sqrt{|x - v|^2 + |u - w|^2}, \tag{15.24}$$

defines the distance to the set $\mathbb{D}$. Similarly, we can define the distance to the terminal set $\mathbb{X}_F$ through $d_F : \mathbb{R}^n \to \mathbb{R}_{\geq 0}$,

$$d_F(x) = \min_{v \in \mathbb{X}_F} |x - v|. \tag{15.25}$$

To incorporate the constraints $(x, u) \in \mathbb{D}$ in the objective function, we consider functions $\alpha, \alpha_F \in \mathcal{K}$ and define

$$\ell_{\mathrm{s}}(x, u) = \alpha(d_{\mathbb{D}}(x, u)) \qquad \text{and} \qquad F_{\mathrm{s}}(x) = \alpha_F(d_F(x)). \tag{15.26}$$

Here, the index $\cdot_{\mathrm{s}}$ refers to costs introduced by rewriting hard constraints into to soft constraints. If we now redefine the set of feasible input trajectories (15.3) by dropping the condition $(x, u) \in \mathbb{D}$, i.e.,

$$\mathcal{U}^N = \left\{ u_N(\cdot) : \mathbb{N}_{[0,N-1]} \to \mathbb{R}^m \;\middle|\; \begin{array}{rcl} x(0) & = & x_0, \\ x(k+1) & = & f(x(k), u(k)), \\ \forall & k \in \mathbb{N}_{[0,N-1]} \end{array} \right\},$$

we can consider the optimal control problem

$$V_N(x_0) = \min_{u_N(\cdot) \in \mathcal{U}^N} J_N(x_0, u_N(\cdot)) + F(x(N)) + \sum_{i=0}^{N-1} \ell_{\mathrm{s}}(x(i), u(i)) + F_{\mathrm{s}}(x(N))$$

subject to (15.1), $\tag{15.27}$

instead of the original optimal control problem (15.4).

Compared to (15.4), an optimal solution of (15.27) does not necessarily satisfy $(x_N^\star(k), u_N^\star(k)) \in \mathbb{D}$ for all $k \in \{0, \ldots, N - 1\}$. However, (15.27) is feasible for $x_0 \in \mathbb{X}$, which is not necessarily true for (15.4). The constraints $(x, u) \in \mathbb{D}$ and $x \in \mathbb{X}_F$ are hard constraints, which must be satisfied, while the formulation with $\ell_{\mathrm{s}}(x, u)$ and $F_{\mathrm{s}}(x)$ in the cost function define soft constraints. If $(x, u) \in \mathbb{D}$ and $x \in \mathbb{X}_F$ is satisfied then $\ell_{\mathrm{s}}(x, u) = 0$ and $F_{\mathrm{s}}(x) = 0$ holds. If $(x, u) \notin \mathbb{D}$ and $x \notin \mathbb{X}_F$ then $\ell_{\mathrm{s}}(x, u) > 0$ and $F_{\mathrm{s}}(x) > 0$ impose additional costs.

Note that the definitions (15.24) to (15.27) describe only one possible way to rewrite hard constraints as soft constraints. We illustrate a different approach based on polyhedral constraints.

Example 15.17. We continue with Example 15.14, where we have shown that for $x_0 = [-1, 1]^T$ and $N = 3$ recursive feasibility is not guaranteed with the constraints $\mathbb{X} = [-1, 1]^2$ and $\mathbb{U} = [-\frac{1}{4}, \frac{1}{4}]$. The state constraints $\mathbb{X} = [-1, 1]^2$ can be represented through the inequality constraints (15.17). Introducing a set of *slack variables* $s \in \mathbb{R}^4_{\geq 0}$, we can rewrite the hard state constraints as soft constraints

$$\begin{bmatrix} 1 & 0 \\ 0 & 1 \\ -1 & 0 \\ 0 & -1 \end{bmatrix} x - \begin{bmatrix} 1 & 0 & 0 & 0 \\ 0 & 1 & 0 & 0 \\ 0 & 0 & 1 & 0 \\ 0 & 0 & 0 & 1 \end{bmatrix} s \leq \begin{bmatrix} 1 \\ 1 \\ 1 \\ 1 \end{bmatrix}, \tag{15.28}$$

which can always be satisfied by selecting the entries of s sufficiently large. To incorporate the variables s in the cost function we penalize s through $10000s(i)^T s(i)$ for all time steps $i \in \mathbb{N}$ within the prediction horizon. In Figure 15.12 the closed-loop solution corresponding to this setting is visualized. Compared to Figure 15.11, left,

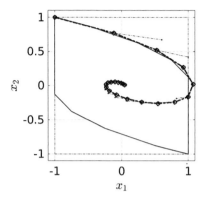

Figure 15.12: MPC closed-loop solution where, compared to Figure 15.11, left, the hard constraints are replaced by soft constraints.

the hard state constraints given by (15.17) have been replaced by soft constraints (15.28). Since the closed-loop solution leaves the viable set, the closed-loop solution necessarily leaves the set $\mathbb{X}$. Even though the solution leaves the set $\mathbb{X}$, the closed-loop solution converges to the origin.

Instead of the soft constraints (15.28), alternatively a single slack variable $s \in \mathbb{R}_{\geq 0}$ and

$$
\begin{bmatrix} 1 & 0 \\ 0 & 1 \\ -1 & 0 \\ 0 & -1 \end{bmatrix} x - \begin{bmatrix} 1 \\ 1 \\ 1 \\ 1 \end{bmatrix} s \leq \begin{bmatrix} 1 \\ 1 \\ 1 \\ 1 \end{bmatrix}
$$

could have been used.

15.3   MODEL PREDICTIVE CONTROL SCHEMES

Algorithm 3 presents a general MPC scheme with its core, the optimal control problem (15.4), which needs to be solved at every time step. In the literature many different MPC schemes have been established, which mainly differ in (15.4) and in the way (15.4) is solved. In this section we illustrate features of some commonly encountered MPC schemes without the goal of providing a comprehensive list or comparison of all schemes. In particular, *robust MPC schemes* are illustrated only on the example of *tube-based MPC*, and we do not discuss *stochastic MPC schemes* at all.

15.3.1   Time-Varying Systems and Reference Tracking

So far we have assumed a time-invariant setting. However, for Algorithm 3 there is no need to exclude time-varying system dynamics, time-varying constraints, or time-varying running costs. We can thus replace the time-invariant dynamics (15.1) with the time-varying dynamics

$$x(k+1) = f(k, x, u), \qquad x(k_0) = x_0 \in \mathbb{R}^n, \ k_0 \in \mathbb{N}_0, \tag{15.29}$$

for $f : \mathbb{N}_0 \times \mathbb{R}^n \times \mathbb{R}^m \to \mathbb{R}^n$. While we were primarily interested in stabilizing the origin of the time-invariant system (15.1), the origin $x = 0$ is not of special interest here. The sets $\mathbb{X} \subset \mathbb{R}^n$ and $\mathbb{U}(x) \subset \mathbb{R}^m$ can be replaced by time-varying sets $\mathbb{X}(k) \subset \mathbb{R}^n$ and $\mathbb{U}(k, x) \subset \mathbb{R}^m$ for all $k \in \mathbb{N}_0$. With

$$\mathbb{D}(k) = \mathbb{X}(k) \times \mathbb{U}(k, x) \subset \mathbb{R}^n \times \mathbb{R}^m, \qquad \forall \, k \in \mathbb{N}_0,$$

the set of feasible input trajectories becomes

$$\mathcal{U}_{\mathbb{D}}^N(k) = \left\{ u_N(\cdot; k) : \mathbb{N}_{[k, k+N-1]} \to \mathbb{R}^m \left| \begin{array}{rcl} x(k) & = & x_0, \\ x(i+1) & = & f(i, x(i), u(i)), \\ (x(i), u(i)) & \in & \mathbb{D}(i), \\ \forall \, i & \in & \mathbb{N}_{[k, k+N-1]} \end{array} \right. \right\}, \tag{15.30}$$

and $u_N(\cdot; k)$ may be denoted as $u_N(\cdot; k, x_0)$ to highlight its dependency on k and x_0. Moreover, the cost function $J_N : \mathbb{N}_0 \times \mathbb{R}^n \times \mathcal{U}_{\mathbb{D}}^N(k) \to \mathbb{R} \cup \{\infty\}$,

$$J_N(k, x_0, u_N(\cdot)) = \sum_{i=k}^{k+N-1} \ell(i, x(i), u(i)),$$

defined through the running cost $\ell : \mathbb{N}_0 \times \mathbb{R}^n \times \mathbb{R}^m \to \mathbb{R}$ as well as the terminal cost $F : \mathbb{N}_0 \times \mathbb{R}^n \to \mathbb{R}$ and terminal constraints $\mathbb{X}_F(k) \subset \mathbb{R}^n$ for all $k \in \mathbb{N}_0$, can be time-varying. The time-varying equivalent to (15.4) can thus be summarized through the optimal control problem

$$V_N(k, x_0) = \min_{u_N(\cdot; k) \in \mathcal{U}_{\mathbb{D}}^N(k)} J_N(k, x_0, u_N(\cdot)) + F(k, x(N)) \tag{15.31}$$

$$\text{subject to (15.29), and } x(N) = \mathbb{X}_F(k).$$

The algorithm extending the time-invariant setting in Algorithm 3 to the time-varying setting is given in Algorithm 5.

Typical running costs for reference tracking are quadratic costs of the form

$$\ell(k, x, u) = (x - x_{\text{ref}}(k))^T Q(x - x_{\text{ref}}(k)) + (u - u_{\text{ref}}(k))^T R(u - u_{\text{ref}}(k))$$

for positive semidefinite matrices $Q \in \mathcal{S}_{\geq 0}^n$, $R \in \mathcal{S}_{\geq 0}^m$ and *reference signals* $x_{\text{ref}}(\cdot) : \mathbb{N}_0 \to \mathbb{R}^n$, $u_{\text{ref}}(\cdot) : \mathbb{N}_0 \to \mathbb{R}^m$.

15.3.2   Linear MPC versus Nonlinear MPC

Linear MPC and *nonlinear MPC* are frequently discussed as separate topics, though the distinction is not always uniform.

Algorithm 5: Model predictive control, the time-varying setting

---

Input: Initial time $k_0 \in \mathbb{N}_0$ and measurement of the initial condition $x(k_0)$; prediction horizon $N \in \mathbb{N} \cup \{\infty\}$; running cost $\ell : \mathbb{N}_0 \times \mathbb{R}^{n+m} \to \mathbb{R}$; constraints $\mathbb{D}(k) \subset \mathbb{R}^{n+m}$, $k \in \mathbb{N}$; terminal cost $F : \mathbb{N}_0 \times \mathbb{R}^n \to \mathbb{R}$ and terminal constraints $\mathbb{X}_F(k) \subset \mathbb{R}^n$, $k \in \mathbb{N}$.

For $k = k_0, k_0 + 1, k_0 + 2, \ldots$

1. Measure the current state of the system (15.29) and define $x_0 = x(k)$.
2. Solve the optimal control problem (15.31) to obtain the open-loop control law $u_N^\star(\cdot; k, x_0)$.
3. Define the feedback law

$$\mu_N(k, x(k)) = u_N^\star(0; k, x_0).$$

4. Set $x(k+1) = f(k, x(k), \mu_N(k, x(k)))$, increment k to $k+1$ and go to 1.

For example, it is reasonable to refer to linear MPC in the case of linear system dynamics (15.1),

$$x^+ = Ax + Bu \tag{15.32}$$

for $A \in \mathbb{R}^{n \times n}$ and $B \in \mathbb{R}^{n \times m}$, and nonlinear MPC whenever the right-hand side of (15.1) is nonlinear. Alternatively, it is common to distinguish between linear and nonlinear MPC based on the optimal control problem (15.4) which needs to be solved at every time step $k \in \mathbb{N}$. If the optimal control problem (15.4), or equivalently the finite- or infinite-dimensional optimization problem, naturally leads to a *linear program* (LP) or a *quadratic program* (QP), the corresponding MPC scheme is referred to a linear scheme while a general *nonlinear program* leads to a nonlinear scheme. In this context the distinction between linear and nonlinear MPC is indirectly used to differentiate MPC schemes with relatively simple optimal control problems from schemes with optimal control problems which are hard to solve. It is thus also reasonable to identify linear MPC with schemes involving convex optimization problems and nonlinear MPC with schemes involving nonconvex optimization problems.

We will briefly discuss several examples leading to different optimal control problems (15.4).

Example 15.18. (Quadratic Programs) For linear dynamics (15.32), $Q, P \in \mathcal{S}_{\geq 0}^n$, $R \in \mathcal{S}_{\geq 0}^m$, and polyhedral constraints defined through $\Gamma_x \in \mathbb{R}^{r \times n}$, $\Gamma_u \in \mathbb{R}^{r \times m}$, $\gamma \in \mathbb{R}^r$, $\Gamma_N \in \mathbb{R}^{q \times n}$, $\gamma_N \in \mathbb{R}^q$, the optimal control problem (15.4) can be written as a QP of the form

$$
\min_{\substack{u(i)\in\mathbb{R}^m \\ i\in\mathbb{N}_{[0,N-1]}}} \sum_{i=0}^{N-1} x(i)^T Q x(i) + u(i)^T R u(i) + x(N)^T P x(N)
$$

$$
\begin{aligned}
\text{subject to} \quad 0 &= x(0) - x_0 & & (15.33)\\
0 &= x(i+1) - Ax(i) - Bu(i) & & \forall\, i = 0, \ldots, N-1\\
\gamma &\geq \Gamma_x x(i) + \Gamma_u u(i) & & \forall\, i = 0, \ldots, N-1\\
\gamma_N &\geq \Gamma_N x(N).
\end{aligned}
$$

Observe how the system dynamics appear as the second equality constraint in (15.33). Also note that replacing the quadratic cost by a cost related to the 1-norm or the ∞-norm allows the optimal control problem to be written as a linear program.

Example 15.19. (Convex programs) As discussed earlier, terminal constraints based on a quadratic Lyapunov function are commonly used. In particular, for $P \in \mathcal{S}_{>0}^n$ and $c \in \mathbb{R}_{>0}$, we can define the constraints

$$
\mathbb{X}_F = \{x \in \mathbb{R}^n : x^T P x \leq c\} \tag{15.34}
$$

representing a convex terminal set. Additionally, the quadratic cost function in (15.33) can be replaced with a non-quadratic running cost of the form $\ell(x, u) = (x^T x)^2 + (u^T u)^2$, for example, leading to the optimization problem

$$
\min_{\substack{u(i)\in\mathbb{R}^m \\ i\in\mathbb{N}_{[0,N-1]}}} \sum_{i=0}^{N-1} (x(i)^T x(i))^2 + (u(i)^T u(i))^2
$$

$$
\begin{aligned}
\text{subject to} \quad 0 &= x(0) - x_0 \\
0 &= x(i+1) - Ax(i) - Bu(i) & & \forall\, i = 0, \ldots, N-1\\
\gamma &\geq \Gamma_x x(i) + \Gamma_u u(i) & & \forall\, i = 0, \ldots, N-1\\
c &\geq x(N)^T P x(N).
\end{aligned}
$$

The optimization problem is convex, since the objective function is convex and the constraints define a convex set, but it is neither a quadratic nor a linear program. In particular, the objective function is neither a quadratic nor a linear function and the constraints are nonlinear due to the terminal constraints (15.34).

In the general nonlinear MPC setting the running costs as well as the terminal costs, the dynamics, and the constraints can be arbitrary. However, the optimization problem is in general more challenging.

Example 15.20. As a last example in this section we consider the optimization problem

$$\min_{\substack{u(i)\in\mathbb{R}^m \\ i\in\mathbb{N}_{[0,N-1]}}} \sum_{i=0}^{N-1} \ell(x(i),u(i))$$

$$
\begin{array}{rcll}
\text{subject to} \quad 0 &=& x(0) - x_0 & \\
0 &=& x(i+1) - x(i) - \Delta f(x(i),u(i)) & \forall\, i = 0,\dots,N-1 \\
c_u &\geq& u(i) & \forall\, i = 0,\dots,N-1 \\
c_u &\geq& -u(i) & \forall\, i = 0,\dots,N-1 \\
c_x &\geq& x_1(i) & \forall\, i = 0,\dots,N-1 \\
c_x &\geq& -x_1(i) & \forall\, i = 0,\dots,N-1.
\end{array}
$$

Here $\dot{x} = f(x,u)$,

$$
f(x,u) = \begin{bmatrix}
x_3 \\
x_4 \\
\dfrac{-\bar{J}\bar{c}x_3 - \bar{J}\sin(x_2)x_4^2 - \bar{\gamma}\cos(x_2)x_4 + g\cos(x_2)\sin(x_2) + \bar{J}u}{\bar{M}\bar{J} - \cos^2(x_2)} \\
\dfrac{-\bar{M}\bar{\gamma}x_4 + \bar{M}g\sin(x_2) - \bar{c}\cos(x_2)x_3 - \cos(x_2)\sin(x_2)x_4^2 + \cos(x_2)u}{\bar{M}\bar{J} - \cos^2(x_2)}
\end{bmatrix}, \qquad (15.35)
$$

with $f : \mathbb{R}^4 \times \mathbb{R} \to \mathbb{R}^4$, captures the dynamics of the inverted pendulum on a cart discussed in Section 1.2.1. In the optimization problem the dynamics are approximated through an Euler discretization with sampling rate $\Delta > 0$ (see Section 5.2). To stabilize the pendulum in the upright position we can define a running cost of the form

$$\ell(x,u) = c_1 x_1^2 + c_2(1 - \cos(x_2))^2 + c_3 x_3^2 + c_4 x^2 + c_5 u^2$$

with $c_1, c_2, c_3, c_4, c_5 \in \mathbb{R}_{>0}$, for example. To incorporate upper and lower bounds on the control input, the constraint $-c_u \leq u \leq c_u$ for $c_u \in \mathbb{R}_{>0}$ is included in the optimization problem. Similarly, for $c_x \in \mathbb{R}_{>0}$ the constraints $-c_x \leq x_1 \leq c_x$ ensure that the cart stays in a specified domain.

Based on these assumptions the optimization problem defines a general nonlinear and nonconvex optimization problem. In particular, through the cosine function in the running cost the objective function is nonconvex. Additionally, the dynamics are nonlinear and thus enforce nonconvex equality constraints.

In Appendix B.5 we present an implementation of Example 15.20 using CasADi [6] in MATLAB.

15.3.3 MPC without Terminal Costs and Constraints

In many MPC applications a finite horizon optimal control problem is used to approximate a corresponding infinite horizon optimal control problem. This compromise in numerical complexity versus optimality raises the question of how to define a "good" approximation of the infinite horizon optimal control problem. In particular, for a given running cost defining an infinite horizon optimal control problem, how should one select the prediction horizon, the terminal cost, and the terminal constraints of a finite horizon counterpart? Note that this question ignores the problem of how to define good running costs for an infinite horizon optimal control problem in the first place, which, depending on the application, might be challenging as well, but will not be addressed here.

For linear systems we have seen systematic ways to choose the terminal cost and the terminal constraints in terms of control Lyapunov functions and forward-invariant level sets of Lyapunov functions in Section 14.3.2. However, in Section 14.3.2 we have also discussed problems originating from these approximations. For nonlinear dynamics one might try to apply the same techniques by linearizing the dynamics around the target state, but the selection of a forward-invariant sublevel set defining terminal constraints with an appropriate size is again nontrivial. A natural question arising from these considerations, which is basically answered in works on *MPC without terminal costs and without terminal constraints* is: *Are terminal costs/constraints necessary?*

A simple answer to this question is that for sufficiently large prediction horizons $N \in \mathbb{N}$, the MPC closed loop calculated through Algorithm 3 without terminal costs/constraints approximates the corresponding infinite horizon solution arbitrarily well. We refer to the monograph [58] for details and technical assumptions making this statement more precise. While "sufficiently large $N \in \mathbb{N}$" does not provide a satisfying answer, and, if possible to derive, theoretical bounds on N are too large in general to be applicable in real-world applications, many applications of MPC schemes without terminal costs/constraints lead to almost indistinguishable closed-loop behavior compared with the corresponding infinite horizon optimal control solution for small or moderate selections of the prediction horizon N.

Some of the early works on MPC without terminal costs and terminal constraints are published under the misleading name *unconstrained MPC*. For detailed results on the topic discussed in this section, we refer to the monograph [58] and the references therein.

15.3.4   Explicit MPC

To apply the MPC Algorithms 3–5, at every time step an optimization problem needs to be solved *online*, and in the discussion so far we have tacitly assumed that an optimal solution (and thus the feedback law) is instantaneously available. While the feedback law is implicitly defined as the solution of an optimal control problem depending on the initial condition, an explicit formula, as in the case of the unconstrained linear quadratic regulator (see Theorems 14.4 and 14.10), for example, is usually not available.

In some cases it is possible to compute an *explicit* solution of the optimal control problem (15.4) as a function of the initial condition x_0. Corresponding MPC schemes are usually referred to as *explicit MPC* with the property that the feedback law is explicitly known, and in contrast to *implicit MPC*, where the feedback law needs to be computed at every time step.

Explicit MPC shifts the problem of solving an optimization problem online at every time step to a so-called *multiparametric program* which only needs to be solved once and can be solved offline. We illustrate the ideas of explicit MPC based on a simple example.

Example 15.21. Consider the one-dimensional dynamics

$$x^+ = x + 0.5u$$

with input constraints $u \in [-1, 1]$. The origin of the unforced ($u = 0$) discrete time system is stable but not asymptotically stable. It is straightforward to verify that every initial state $x_0 \in \mathbb{R}$ can be driven to the origin in finite time. The optimal control problem (14.5) with $\ell(x, u) = x^2 + u^2$ and prediction horizon $N = 2$ is given by

$$\min_{u(0), u(1)} x(0)^2 + x(1)^2 + u(0)^2 + u(1)^2$$

$$\begin{array}{rcll}
\text{subject to} \quad 0 & = & x(0) - x_0 \\
0 & = & x(1) - x(0) - 0.5u(0) \\
0 & = & x(2) - x(1) - 0.5u(1) \\
u(0), u(1) & \in & [-1, 1].
\end{array} \tag{15.36}$$

Equivalently, the optimization problem can be written as

$$\min_{u(0), u(1)} 2x_0^2 + x_0 u(0) + 1.25u(0)^2 + u(1)^2$$

$$\text{subject to} \quad u(0), u(1) \in [-1, 1].$$

The gradient of the objective function

$$J_2(u(\cdot); x_0) = 2x_0^2 + x_0 u(0) + 1.25u(0)^2 + u(1)^2$$

is given by

$$\nabla_u J_2(u(\cdot); x_0) = \begin{bmatrix} 2.5u(0) + x_0 \\ 2u(1) \end{bmatrix},$$

from which we observe that the minimum is attained for $u(0) = -0.4x_0$ and $u(1) = 0$ if we ignore the input constraints for a moment. The fact that $u(N - 1)$ is equal to zero is due to the selection of the objective function, where we have set the terminal cost to zero and thus $x(N)$ is not penalized. For $x_0 \in [-2.5, 2.5]$ the input constraints $u(0) \in [-1, 1]$ are not violated and thus the optimal feedback law with respect to the optimal control problem (15.36) is defined as

$$\mu(x_0) = -0.4x_0.$$

For $x_0 \geq 2.5$ the best one can do with respect to the objective function is to set $u(0)$ to -1, i.e., to define $\mu(x_0) = -1$. Similarly, for $x \leq -2.5$ we obtain $\mu(x_0) = 1$.
 For $i = 1, 2, 3$ the feedback law can thus be summarized as

$$\mu_2(x_0) = K_i x_0 + k_i \quad \text{whenever} \quad C_i x_0 \leq d_i. \tag{15.37}$$

Here, K_i, k_i, C_i, and d_i are defined as

$$\begin{array}{rcllrcl}
K_1 & = & -0.4, & \quad & k_1 & = & 0, \\
K_2 & = & 0, & \quad & k_2 & = & -1, \\
K_3 & = & 0, & \quad & k_3 & = & 1,
\end{array}$$

and

$$C_1 = \begin{bmatrix} 1 \\ -1 \end{bmatrix}, \qquad d_1 = \begin{bmatrix} 2.5 \\ 2.5 \end{bmatrix},$$
$$C_2 = \begin{bmatrix} -1 \end{bmatrix}, \qquad d_2 = \begin{bmatrix} -2.5 \end{bmatrix},$$
$$C_3 = \begin{bmatrix} 1 \end{bmatrix}, \qquad d_3 = \begin{bmatrix} -2.5 \end{bmatrix}.$$

Observe that the feedback law μ_2 is a continuous piecewise affine function. The sets $C_i x_0 \le d_i$, $i = 1, 2, 3$ define polyhedrons.

The optimal value function satisfies

$$V_2(x_0) = \begin{cases} 1.8x_0^2 & \text{if} \quad C_1 x_0 \le d_1 \\ 2x_0^2 - x_0 + 1.25 & \text{if} \quad C_2 x_0 \le d_2 \\ 2x_0^2 + x_0 + 1.25 & \text{if} \quad C_3 x_0 \le d_3, \end{cases}$$

which is a continuously differentiable piecewise quadratic function. We invite the reader to verify that the function is indeed continuously differentiable but not twice continuously differentiable.

Through these calculations, steps 2 and 3 in MPC Algorithms 3–5 reduce to the evaluation of the control law (15.37); i.e., for a given $x_0 \in \mathbb{R}$, one has to find the index $i \in \{1, 2, 3\}$ such that $C_i x_0 \le d_i$ is satisfied.

In Example 15.21 we started with a quadratic program (15.36) depending on the initial condition x_0. The optimal value function turned out to be a continuously differentiable piecewise quadratic function and the optimal feedback law turned out to be a continuous piecewise affine function defined over a polyhedral partition of the state space. This is not a coincidence and not a particular property of Example 15.21.

To illustrate that these properties hold for a general class of problems, consider again the quadratic program (15.33). The optimization problem can be rewritten in the form

$$\min_{U \in \mathbb{R}^{pN}} U^T \mathcal{H} U + U^T \left(\mathcal{G} x_0 + \mathcal{F} \right) + \widehat{c}(x_0)$$
$$\text{subject to } \widehat{\mathcal{A}} U \le \widehat{\mathcal{B}} + \widehat{\mathcal{C}} x_0, \tag{15.38}$$

where $U = \left[u(0)^T, \ldots, u(N-1)^T \right]^T$ denotes the input sequence to be optimized, $\widehat{c} : \mathbb{R}^n \to \mathbb{R}$ denotes a function only depending on the initial condition x_0, and $\mathcal{H}, \mathcal{G}, \mathcal{F}, \widehat{\mathcal{A}}, \widehat{\mathcal{B}}$, and $\widehat{\mathcal{C}}$ are matrices and vectors of appropriate dimension. In Exercise 15.5 we ask the reader to rewrite the optimization problem (15.33) in form of (15.38).

In the case that $\mathcal{H}$ is nonsingular, we consider the coordinate transformation

$$\tilde{U} = U + \tfrac{1}{2} \mathcal{H}^{-1} \left(\mathcal{G} x_0 + \mathcal{F} \right), \quad \text{i.e.,} \quad U = \tilde{U} - \tfrac{1}{2} \mathcal{H}^{-1} \mathcal{G} x_0 - \tfrac{1}{2} \mathcal{H}^{-1} \mathcal{F}. \tag{15.39}$$

Sufficient conditions for $\mathcal{H}$ to be positive definite, and thus in particular to be nonsingular, are that the pair (A, B) defining the linear system be controllable and that the weight matrices $Q, P \in \mathcal{S}_{\ge 0}^n$, $R \in \mathcal{S}_{> 0}^m$ be positive semidefinite and positive definite, respectively. Then the objective function of (15.38) can be rewritten as

$$U^T \mathcal{H} U + U^T \left(\mathcal{G} x_0 + \mathcal{F} \right) + \widehat{c}(x_0)$$

$$= \left(\tilde{U} - \tfrac{1}{2} \mathcal{H}^{-1} \mathcal{G} x_0 - \tfrac{1}{2} \mathcal{H}^{-1} \mathcal{F} \right)^T \mathcal{H} \left(\tilde{U} - \tfrac{1}{2} \mathcal{H}^{-1} \mathcal{G} x_0 - \tfrac{1}{2} \mathcal{H}^{-1} \mathcal{F} \right)$$

$$\quad + \left(\tilde{U} - \tfrac{1}{2} \mathcal{H}^{-1} \mathcal{G} x_0 - \mathcal{F} \right)^T \left(\mathcal{G} x_0 + \mathcal{F} \right) + \widehat{c}(x_0)$$

$$= \tilde{U}^T \mathcal{H} \tilde{U} + \tilde{U}^T \left(-\mathcal{G} x_0 - \mathcal{F} \right)$$

$$\quad + \left(\tfrac{1}{2} \mathcal{H}^{-1} \mathcal{G} x_0 + \tfrac{1}{2} \mathcal{H}^{-1} \mathcal{F} \right)^T \mathcal{H} \left(\tfrac{1}{2} \mathcal{H}^{-1} \mathcal{G} x_0 + \tfrac{1}{2} \mathcal{H}^{-1} \mathcal{F} \right)$$

$$\quad + \tilde{U}^T \left(\mathcal{G} x_0 + \mathcal{F} \right) - \left(\tfrac{1}{2} \mathcal{H}^{-1} \mathcal{G} x_0 + \tfrac{1}{2} \mathcal{H}^{-1} \mathcal{F} \right)^T \left(\mathcal{G} x_0 + \mathcal{F} \right) + \widehat{c}(x_0)$$

$$= \tilde{U}^T \mathcal{H} \tilde{U} + c(x_0)$$

and the last equality follows from the definition of the function $c : \mathbb{R}^n \to \mathbb{R}$,

$$c(x_0) = \widehat{c}(x_0) + \left(\tfrac{1}{2} \mathcal{H}^{-1} \mathcal{G} x_0 + \tfrac{1}{2} \mathcal{H}^{-1} \mathcal{F} \right)^T \mathcal{H} \left(\tfrac{1}{2} \mathcal{H}^{-1} \mathcal{G} x_0 + \tfrac{1}{2} \mathcal{H}^{-1} \mathcal{F} \right)$$

$$\quad - \left(\tfrac{1}{2} \mathcal{H}^{-1} \mathcal{G} x_0 + \tfrac{1}{2} \mathcal{H}^{-1} \mathcal{F} \right)^T \left(\mathcal{G} x_0 + \mathcal{F} \right),$$

collecting the terms which are independent of $\tilde{U}$. Since $c(x_0)$ is independent of U (and $\tilde{U}$) the term $c(x_0)$ in the objective function does not have an impact on the minimizer $U^\star$. With the coordinate transformation (15.39) we additionally transform the constraints in (15.38) by defining the matrices and vectors

$$\mathcal{A} = \widehat{\mathcal{A}}, \qquad \mathcal{B} = \widehat{\mathcal{B}} + \tfrac{1}{2} \mathcal{H}^{-1} \mathcal{F}, \qquad \mathcal{C} = \widehat{\mathcal{C}} + \tfrac{1}{2} \mathcal{H}^{-1} \mathcal{F}.$$

Combining these results, we can solve the optimization problem

$$\tilde{U}^\star(x_0) = \arg \min_{\tilde{U} \in \mathbb{R}^{pN}} \tilde{U}^T \mathcal{H} \tilde{U}$$

$$\text{subject to } \mathcal{A} \tilde{U} \leq \mathcal{B} + \mathcal{C} x_0 \tag{15.40}$$

to obtain a minimizer $U^\star$ of the optimization problem (15.38) through $U^\star(x_0) = \tilde{U}^\star(x_0) - \tfrac{1}{2} \mathcal{H}^{-1} \left(\mathcal{G} x_0 + \mathcal{F} \right)$. In the literature, problem (15.40) is called *multiparametric program*, depending on the unknown U and the parameter x_0.

Theorem 15.22 ([18, Theorem 4.2])**.** *Consider the multiparametric program defined in (15.40) and let $\mathcal{H} \in \mathcal{S}_{>0}^{pN}$ be positive definite. Then the set of feasible initial conditions*

$$\mathbb{X} = \left\{ x_0 \in \mathbb{R}^n \mid \exists \tilde{U} \in \mathbb{R}^{pN} \text{ s.t. } \mathcal{A} \tilde{U} \leq \mathcal{B} + \mathcal{C} x_0 \right\} \tag{15.41}$$

is convex. The minimizer $\tilde{U}^\star(\cdot) : \mathbb{X} \to \mathbb{R}^{pN}$ is continuous and piecewise affine and the optimal solution $V : \mathbb{X} \to \mathbb{R}$, $V(x_0) = (\tilde{U}^\star(x_0))^T \mathcal{H} \tilde{U}^\star(x_0)$ is continuous, convex, and piecewise quadratic.

Multiparametric programs (15.40) of moderate size can be solved efficiently through the *Multi-Parametric Toolbox 3*, [62], in MATLAB, for example. We illustrate the theorem and its connection to explicit MPC based on an example.

Example 15.23. Consider the linear system $x^+ = Ax + Bu$ defined by the matrices

$$A = \begin{bmatrix} 1 & 1 \\ -\frac{1}{4} & 1 \end{bmatrix} \quad \text{and} \quad B = \begin{bmatrix} -1 \\ 0 \end{bmatrix}$$

on the domain $x \in [-5,5]^2$. Additionally, assume that the input u is bounded, i.e., $u \in [-1,1]$. For the quadratic optimization problem (15.33), select the weight matrices

$$Q = P = \begin{bmatrix} 1 & 0 \\ 0 & 1 \end{bmatrix} \quad \text{and} \quad R = 1,$$

fix the prediction horizon $N = 5$, and note that the bounds on x and u define polyhedral constraints. The corresponding finite horizon optimal control problem can be written as a multiparametric program of the form (15.40).

The solution of the multiparametric program is shown in Figures 15.13–15.15.

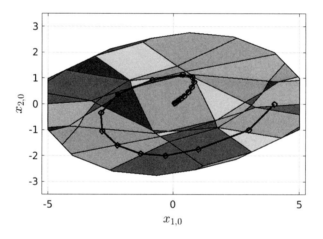

Figure 15.13: Visualization of the feasible region (15.41) as well as a partition into 53 polyhedral regions. In each region the optimal solution can be described through a quadratic function.

Figure 15.13 shows the feasible domain (15.41) of the multiparametric program.
In addition, in the different polyhedral regions

$$\mathbb{P}_i = \{x_0 \in \mathbb{R}^2 | \mathcal{A}_i x_0 \leq \mathcal{B}_i\}, \qquad i = 1, \dots, 53,$$

the first element of the optimal input $\mu(x_0) = u^\star(0)$ can be described through an affine function

$$\mu(x_0) = K_i x_0 + k_i, \qquad \forall\, x_0 \in \mathbb{P}_i, \tag{15.42}$$

$K_i \in \mathbb{R}^{1 \times 2}$, $k_i \in \mathbb{R}$, $i \in \{1, \dots, 53\}$. The MPC feedback law, i.e., the optimal input $\mu(x_0) = u^\star(0)$, is visualized in (15.14). As expected from Theorem 15.22 the feedback law is not only piecewise affine, but also continuous. The optimal cost

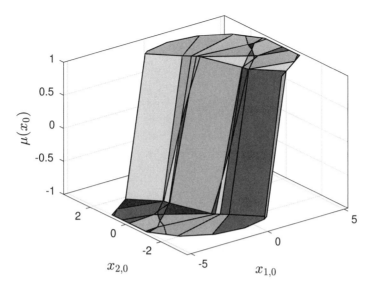

Figure 15.14: Visualization of the piecewise linear feedback law $\mu(x_0) = u^\star(0) = K_i x_0 + k_i$ on the partition of the feasible set into 53 polyhedral regions.

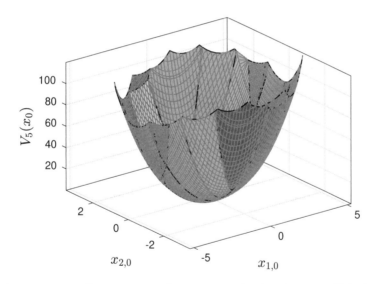

Figure 15.15: Piecewise quadratic optimal value function $V_5(x_0)$.

function $V_5(x_0)$, which is shown in Figure 15.15, is a piecewise quadratic continuous function

$$V_5(x_0) = x_0^T W_i x_0 + w_i x_0 + v_i, \qquad \forall\, x_0 \in \mathbb{P}_i,$$

$W_i \in \mathbb{R}^{2\times 2}$, $w_i \in \mathbb{R}^{1\times 2}$, $v_i \in \mathbb{R}$, $i \in \{1, \dots, 53\}$.

If the feedback law $\mu(x_0)$ is known and written as in (15.42), it is not necessary to solve an optimization problem at every time step of the MPC Algorithm 3. Instead, for a given initial state x_0, the index $i \in \{1, \dots, 53\}$ such that $x_0 \in \mathbb{P}_i$ is satisfied needs to be identified. Then the corresponding feedback law (15.42) can easily be applied. In Figure 15.13 a closed-loop MPC solution corresponding to the initial condition $x_0 = [4, 0]^T$ making use of this calculation of the feedback law is shown, in addition to the partition of the feasible domain.

Before we conclude this example, note that the feasible region $\mathbb{X}$ is not necessarily forward invariant under the feedback law μ. In particular, an initial value selected in the feasible domain does not imply that the corresponding closed-loop solution remains in the feasible region for all times $k \in \mathbb{N}$.

As indicated at the beginning of this section, in *explicit MPC*, the feedback law is explicitly known and does not need to be computed by solving an optimization problem at every time step. Thus, explicit MPC shifts the online computational burden to an offline optimization problem which only needs to be solved once. Some multiparametric problems, including quadratic problems of the form (15.40), for example, can be solved efficiently and an explicit control law can be obtained.

However, the number of regions necessary to represent the feedback law in general increases exponentially with the number of constraints, i.e., the number of rows in the matrix $\mathcal{A}$ in (15.40). Thus, the storage demand for the controller increases exponentially with the number of constraints and efficiently finding the correct control law for a given initial condition becomes a nontrivial task. While there are approaches to mitigate these problems in explicit MPC, one should be aware of them when one intends to derive a controller based on explicit MPC.

15.3.5   Economic MPC

So far in this chapter (at least in the applications) we have tacitly assumed that the running cost ℓ is a positive semidefinite function penalizing the distance to a reference point (x^e, u^e) or the distance to a reference trajectory $(x_{\mathrm{ref}}(t), u_{\mathrm{ref}}(t))$. This implies that the optimal behavior in terms of a reference point or reference trajectory needs to be known, which might not always be the case. For example, "maximizing the revenue of a plant" might be impossible to phrase using a running cost of this form. Moreover, if Algorithm 3 is used to maximize the revenue, does the closed-loop solution converge to an optimal reference point or optimal reference trajectory, even if the optimal point or the optimal trajectory is not known in advance? In *economic MPC*, the performance of MPC schemes with non-classical positive semidefinite running costs or simply more general running costs is investigated. While Algorithm 3 can be applied independently of the definition of the running cost ℓ, as long as an optimal solution of (15.4) can be obtained in every iteration, the methods to analyze control schemes with non-positive semidefinite running costs are significantly different.

We illustrate some of the differences of *economic MPC* in contrast to *classical MPC* based on a simple example. In particular, we consider linear dynamics

$$x(k + 1) = 2x(k) + u(k) \tag{15.43}$$

together with running cost

$$\ell(x, u) = u^2 - x. \tag{15.44}$$

Then, minimizing the running cost over a prediction horizon $N \in \mathbb{N}$ subject to the the dynamics (15.43) can be interpreted as maximizing the revenue x while simultaneously minimizing the operational cost u^2.

A first observation we can make is that without additional constraints, the optimal control problem

$$\min_{u(\cdot)} \sum_{i=0}^{N-1} \left(u(i)^2 - x(i) \right)$$

$$\text{subject to} \quad 0 = x(0) - x_0$$
$$0 = x(i+1) - 2x(i) - u(i) \qquad \forall i = 0, \dots, N-1$$

$$(15.45)$$

is only well defined for $N < \infty$. On an infinite horizon, $N = \infty$, the minimum is not attained.

Recall that one of our motivations for MPC was to approximate the feedback law corresponding to an infinite horizon optimal control problem by iteratively computing a feedback law through Algorithm 3. However, if the infinite horizon optimal control problem is not well defined, it is not clear what to compare the MPC closed-loop cost with to assess the performance of the MPC controller.

Thus, instead of the cost $V_\infty(x_0)$, the average cost

$$\bar{V}_\infty(x_0) = \limsup_{K \to \infty} \frac{1}{K} V_K(x_0)$$

$$(15.46)$$

is generally considered as a measure. In particular, even if $V_\infty(x_0)$ is not well defined, the right-hand side of (15.46) may still be finite and thus $\bar{V}_\infty(x_0) \in \mathbb{R}$.

This is however not the case in the optimization problem (15.45) since the running cost is not bounded from below. For example, in the case that $x_0 > 0$, it is easy to see that $\limsup_{K \to \infty} \frac{1}{K} V_K(x_0) \to -\infty$ and $\bar{V}_\infty(x_0)$ is not well defined since the running cost is not bounded from below.

We thus consider additional constraints

$$x \in \mathbb{X} = [-2, 2] \qquad \text{and} \qquad u \in \mathbb{U} = [-3, 3],$$

and note that the set $\mathbb{X}$ is viable with respect to the dynamics and the input $u \in \mathbb{U}$. With this definition the optimal average cost satisfies

$$\bar{V}_\infty(x_0) = \limsup_{K \to \infty} \frac{1}{K} V_K(x_0) \le \limsup_{K \to \infty} \frac{1}{K} \sum_{k=0}^{K-1} (3^2 + 2) = 11$$

and

$$\bar{V}_\infty(x_0) = \limsup_{K \to \infty} \frac{1}{K} V_K(x_0) \ge \limsup_{K \to \infty} \frac{1}{K} \sum_{k=0}^{K-1} (0^2 - 2) = -2.$$

Since with the additional constraints the average cost is well defined, we can investigate the solution $x^\star(k)$ corresponding to the running cost (15.46). To this end, we solve the optimization problem

$$\min_{u(\cdot)} \sum_{i=0}^{N-1} u(i)^2 - x(i)$$

$$
\begin{aligned}
\text{subject to} \quad 0 &= x(0) - x_0 \\
0 &= x(i+1) - 2x(i) - u(i) && \forall i = 0, \ldots, N-1 \\
x(i+1) &\in [-2, 2] && \forall i = 0, \ldots, N-1 \\
u(i) &\in [-3, 3] && \forall i = 0, \ldots, N-1
\end{aligned}
\tag{15.47}
$$

for different horizons $N \in \mathbb{N}$ and $x_0 = -2$. The corresponding open-loop solutions $x^\star(k)$, $k = 0, \ldots, N$, are visualized in Figure 15.16. Independently of the horizon N

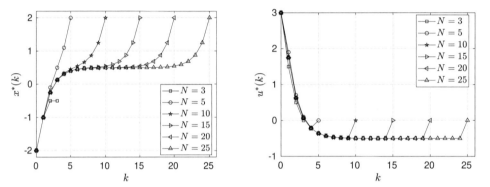

Figure 15.16: Open-loop solutions $x^\star(k)$ (left) and $u^\star(k)$ (right), $k = 0, \ldots, N$, corresponding to (15.47) for different $N \in \mathbb{N}$.

(except for $N = 3$, which is too short) we can subdivide the solution $x^\star$ into three different regions:

- an approaching arc, converging to a neighborhood around $x = 0.5$;
- a stable segment, staying in a neighborhood around $x = 0.5$;
- a leaving arc, diverging from a neighborhood around $x = 0.5$.

While the approaching and leaving arcs do not change in length, the stable segment increases with N. This so-called *turnpike property* in the economic MPC literature indicates that for the given running cost and the state and input constraints, the system is optimally operated at the steady state $x^e = 0.5$. The leaving arc is due to the finite horizon. For $x \in [-2, 2]$ and $u \in [-3, 3]$, the minimum of $\ell(x, u) = u^2 - x$ is attained for $x = 2$ and $u = 0$. However, this minimum is only optimal if the state at the next time step is not considered in the objective function, since the dynamics $x^+ = 2x + u \in [-2, 2]$ imply a higher cost at the next time step due to the selection of u to ensure that the trajectory stays feasible.

There is also an analytical approach which suggests that $x^e = 0.5$ is the optimal steady state with respect to the running cost (15.44). To illustrate this approach, assume that the dynamics reach a steady state, i.e.,

$$x = x^+ = 2x + u$$

or equivalently $x = -u$. Using this condition in the running cost leads to $\ell(x, u) =$

$u^2 - x = u^2 + u$. To compute the minimum of this function we take the derivative and set it to zero; i.e.,

$$\tfrac{d}{du}\left(u^2 + u\right) = 2u + 1 = 0,$$

and thus $u^e = -0.5$. We can thus identify the equilibrium pair $(x^e, u^e) = (0.5, -0.5)$. However, note that for general running costs the optimal solution does not need to converge to a steady state but can alternatively converge to a periodic orbit, for example.

As a next step we investigate the closed-loop solution of the MPC Algorithm 3 under the assumptions so far in this section. The MPC closed-loop solutions for different horizons using (15.47) in step 2 of Algorithm 3 are visualized in Figure 15.17.

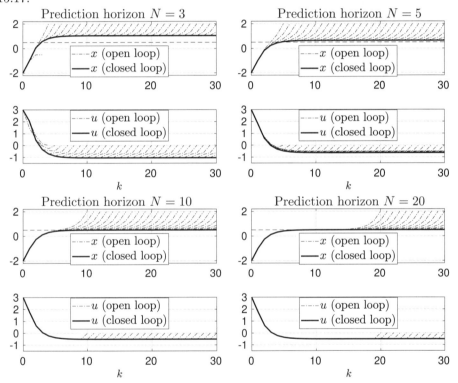

Figure 15.17: Closed-loop and open-loop solutions

We have seen in this example that running costs do not necessarily need to be positive semidefinite to obtain a well-defined optimal control problem. We have also pointed out that in contrast to *classical MPC*, where the running costs are designed to obtain a certain steady-state solution, economic MPC indirectly recovers the optimal steady-state solution for a given running cost and the optimal steady-state solution does not even need to be known.

However, if the steady-state solution is known, as is the case for our given example here, one might ask the question whether it is beneficial to replace the (positive semidefinite) running cost with a positive definite running cost. In particular, one could consider the quadratic cost

$$\tilde{\ell}(x, u) = c_1 |x - x^e|^2 + c_2 |u - u^e|^2$$

for positive constants $c_1, c_2 \in \mathbb{R}_{>0}$. With this definition, the same asymptotic behavior can be expected. Thus when evaluated with respect to

$$\bar{J}_N(x_0; u(\cdot)) = \limsup_{K \to \infty} \sum_{k=0}^{K-1} \ell(x(k), u(k)),$$

the two solutions $u^\star(\cdot)$ and $\tilde{u}^\star(\cdot)$ lead to the same value. Here $u^\star(\cdot)$ corresponds to the infinite horizon scaled optimization problem (15.47) and $\tilde{u}^\star(\cdot)$ corresponds to the same optimization problem where ℓ is replaced by $\tilde{\ell}$. However, the transient behavior and the transient costs (i.e., the behavior and the costs on a fixed time window) may vary significantly. In this case, only the trajectory obtained through the original running cost is optimal. This is relevant for plants that are only operated on a finite time window and where the cost to reach a neighborhood around the optimal steady state is not negligible.

We conclude the discussion on economic MPC with a slightly more complicated example.

Example 15.24. To illustrate some aspects of economic MPC we consider an application for mobile robots. Consider a simple extension of the model of a mobile robot introduced in Section 1.2.2 in the continuous time setting and its discrete time counterpart using a sampling time $\Delta > 0$:

$$\begin{bmatrix} \dot{p}_x \\ \dot{p}_y \\ \dot{\phi} \\ \dot{v} \\ \dot{w} \end{bmatrix} = \begin{bmatrix} v \cos(\phi) \\ v \sin(\phi) \\ w \\ a \\ q \end{bmatrix} \quad \text{and} \quad \begin{bmatrix} p_x^+ \\ p_y^+ \\ \phi^+ \\ v^+ \\ w^+ \end{bmatrix} = \begin{bmatrix} p_x + \Delta v \cos(\phi) \\ p_y + \Delta v \sin(\phi) \\ \phi + \Delta w \\ v + \Delta a \\ w + \Delta q \end{bmatrix}.$$

Here, a and q denote the input of the system. Our goal is to finish the "race track" in Figure 15.18, described through a sinusoid shape, as quickly as possible by driving from the left to the right. In particular, the track is defined by the set

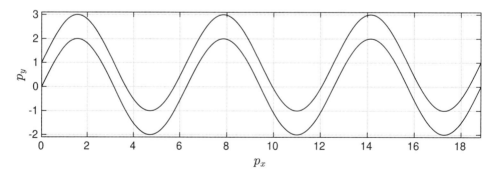

Figure 15.18: "Race track" with the shape of a sinusoid wave which needs to be completed from left to right.

$$\mathcal{P} = \{[p_x, p_y]^T \in \mathbb{R}^2 | p_y \geq 2\sin(p_x), \quad p_y \leq 2\sin(p_x) + 2\}$$

defining state constraints with respect to the position of the robot. Additionally, we only allow a movement in forward direction with respect to the p_x-component and thus enforce the constraints $p_x(t_1) \leq p_x(t_2)$ for all $t_1 \leq t_2$. The orientation $\phi \in \mathbb{R}$ is not constrained. The velocity and the angular velocity are bounded through $v \in [0, 4]$, $w \in [-2, 2]$, by assumption. For the inputs we consider the constraints $a \in [-4, 4]$ and $q \in [-6, 6]$.

Maximizing the distance traveled in the p_x-direction in a fixed time while staying on the track can be expressed through the objective

$$\max \int_0^{T_{end}} v(t)\cos(\phi(t))dt = -\min \int_0^{T_{end}} -v(t)\cos(\phi(t))dt. \tag{15.48a}$$

Here, $[0, T_{end}]$, $T_{end} > 0$, denotes a fixed time window. Similarly, through a simple discretization, we can consider the objective

$$\max \Delta \cdot \sum_{i=0}^{K_{end}} v(i)\cos(\phi(i)) = -\min \Delta \cdot \sum_{i=0}^{K_{end}} -v(i)\cos(\phi(i)) \tag{15.48b}$$

for $K_{end} \in \mathbb{N}$, in the discrete time setting. The running cost $\ell(x, u) = -v\cos(\phi)$ is not positive semidefinite, i.e., with this objective we are in the economic MPC setting. While we could approximate the indefinite running cost through a classical positive semidefinite time-varying cost, economic MPC provides the framework to optimize with respect to (15.48) directly. Additionally, while optimizing with respect to a racing line (i.e., penalizing the distance to a racing line representing positive semidefinite costs) is optimal when starting on the racing line, the transient behavior (i.e., reaching the racing line) is not optimal when compared to the original cost (15.48).

Figures 15.19 and 15.20 show the closed-loop MPC solution for different prediction horizons $N \in \mathbb{N}$. Additionally, the open-loop solution with $K_{end} = 300$ approximating the infinite horizon optimal control problem is included in the plots. Here, the discrete time system with $\Delta = 0.05$ is simulated to approximate the continuous time system and the continuous time problem. The initial condition is set to $x_0 = [0, 0.5, 0, 0, 0]^T$. The optimal control problems of this example are solved using *CasADi*, [6]. Compared to Figure 15.20, Figure 15.19 only shows the domain $p_x \in [0, 2\pi]$. While the closed-loop solution for small N differs significantly from the optimal open-loop solution, for $N = 35$ the closed-loop MPC solution is almost indistinguishable from the open-loop solution. This is the case even though the open-loop solutions for $N = 35$ differ significantly from the open-loop solution with $N = K_{end} = 300$ (see Figure 15.21). This is the case due to the fact that only the first piece of the open-loop solution is used to define the closed-loop.

In Figure 15.22, the MPC closed-loop solution for $N = 35$, together with the feedback law, is shown. Since the track described through a sine wave is periodic, the closed-loop solution (except for the state p_x), as well as the feedback, is periodic.

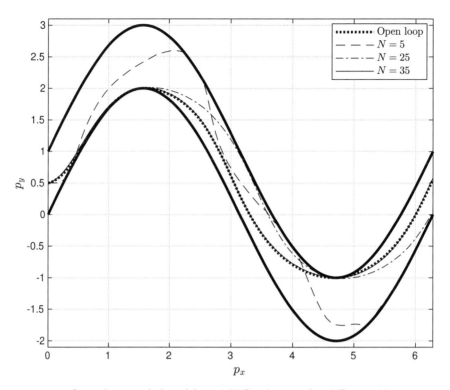

Figure 15.19: Open-loop and closed-loop MPC solutions for different N maximizing the velocity in p_x-direction. For the open-loop solution the horizon is selected as $K_{end} = 300$.

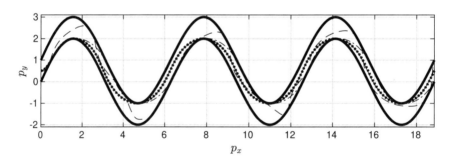

Figure 15.20: Visualization of the results in Figure 15.19 on a longer time window.

15.3.6   Tube-Based MPC

As a last example we briefly discuss a representative of *robust MPC*. Consider the discrete time nonlinear system

$$x^+ = f(x, u, w) \tag{15.49}$$

with state $x \in \mathbb{R}^n$, input $u \in \mathbb{R}^m$, unknown disturbances $w \in \mathbb{R}^q$, and constraints $(x, u, w) \in \mathbb{D} = \mathbb{X} \times \mathbb{U} \times \mathbb{W} \subset \mathbb{R}^n \times \mathbb{R}^m \times \mathbb{R}^q$.

For the perturbed dynamics we extend the definition of the feasible input tra-

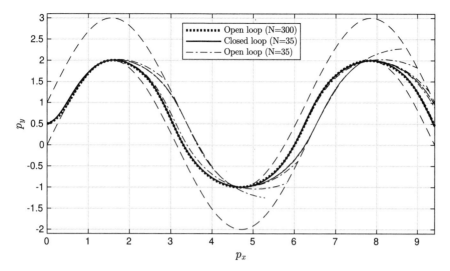

Figure 15.21: Comparison of the open-loop solution with $N = 300$ and the closed-loop MPC solution with $N = 35$. Additionally for some time instances the open-loop solutions for $N = 35$ are shown.

jectories (15.50) to incorporate the perturbations

$$
\mathcal{U}_{\mathbb{D}}^N = \left\{ u_N(\cdot) : \mathbb{N}_{[0,N-1]} \to \mathbb{R}^m \; \middle| \; \begin{array}{rcl} x(0) &=& x_0 \\ x(k+1) &=& f(x(k), u(k), w(k)) \\ (x(k), u(k), w(k)) &\in& \mathbb{D} \\ \forall \; k &\in& \mathbb{N}_{[0,N-1]} \end{array} \right\}.
$$
(15.50)

Similarly, the running cost $\ell : \mathbb{R}^n \times \mathbb{R}^m \times \mathbb{R}^q \to \mathbb{R}$ and the cost function $J_N : \mathbb{R}^n \times \mathcal{U}_{\mathbb{D}}^N \to \mathbb{R} \cup \{\infty\}$,

$$
J_N(x_0, u_N(\cdot), w_N(\cdot)) = \sum_{i=0}^{N-1} \ell(x(i), u(i), w(i)),
$$

are extended to the perturbed setting. Then, the optimal control problem (15.4) becomes

$$
V_N(x_0) = \min_{u_N(\cdot) \in \mathcal{U}_{\mathbb{D}}^N} \; \max_{w_N(\cdot) \in \mathbb{W}^N} \; J_N(x_0, u_N(\cdot)) + F(x(N))
$$
(15.51)
$$
\text{subject to (15.49), and } x(N) \in \mathbb{X}_F.
$$

In particular, the optimal input trajectory $u_N^\star(\cdot) \in \mathcal{U}_{\mathbb{D}}^N$ for all possible disturbance trajectories $w_N(\cdot) \in \mathbb{W}^N$ needs to be found to be able to define the optimal feedback law $\mu(x) = u_N^\star(0)$. Through the min-max structure the optimization problem (15.51) is in general significantly more challenging than the corresponding problem (15.4) with $w \equiv 0$ and many approaches to this problem have been proposed. Here, we limit our discussion to the idea behind *tube-based MPC*, one representative of robust MPC, on the example of linear systems.

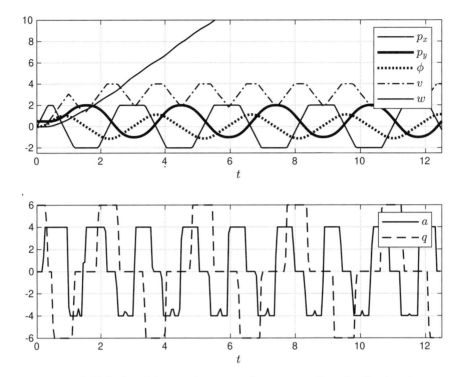

Figure 15.22: MPC closed-loop solution and corresponding feedback using a prediction horizon of $N = 35$.

Consider the linear dynamics

$$x^+ = Ax + Bu + \bar{B}w, \tag{15.52}$$

where $\bar{B} \in \mathbb{R}^{n \times q}$ and we assume that the state, the input, and the disturbances are bounded through polyhedral sets

$$\mathbb{X} = \{x \in \mathbb{R}^n : \Gamma_x x \leq \gamma_x\},$$
$$\mathbb{U}(x) = \{u \in \mathbb{R}^m : \Gamma_u(Kx + u) \leq \gamma_u\}, \tag{15.53}$$
$$\mathbb{W} = \{w \in \mathbb{R}^q : \Gamma_w w \leq \gamma_w\},$$

where $\Gamma_x \in \mathbb{R}^{n_x \times n}$, $\gamma_x \in \mathbb{R}^{n_x}$, $\Gamma_u \in \mathbb{R}^{m_u \times m}$, $\gamma_u \in \mathbb{R}^{m_u}$, $K \in \mathbb{R}^{m \times n}$, $\Gamma_w \in \mathbb{R}^{q_w \times q}$, $\gamma_w \in \mathbb{R}^{q_w}$, and $n_x, m_u, q_w \in \mathbb{N}$. Moreover, we assume that the matrix A is a Schur matrix. Note that, if the pair (A, B) is stabilizable, this can be assumed without loss of generality through a coordinate transformation $u = Kx + \bar{u}$, where $K \in \mathbb{R}^{m \times n}$ is defined such that $A + BK$ is a Schur matrix and $\bar{u} \in \mathbb{R}^m$ defines a new degree of freedom. Then the system dynamics can be written as $x^+ = (A + BK)x + B\bar{u} + \bar{B}w$ and we can consider the system in the new input $\bar{u}$.

We consider the known deterministic dynamics

$$x_d^+ = Ax_d + Bu, \qquad x_d(0) = x_{d,0} \tag{15.54}$$

and define the unknown error dynamics $e = x - x_d$ which satisfy

$$e^+ = Ae + \bar{B}w.$$

With this expression the error $e(k)$ for $k \in \mathbb{N}$ can be expressed through the initial error $e(0) = e_0$ and the unknown disturbance sequence $(w(k))_{k \in \mathbb{N}} \subset \mathbb{W}$ as

$$e(k) = A^k e_0 + \sum_{j=0}^{k-1} A^j B w(j). \tag{15.55}$$

While the exact value of $e(k)$ is not known due to the unknown disturbance sequence $(w(k))_{k \in \mathbb{N}}$, if the set $\mathbb{W}$ is bounded and the initial error e_0 is known to be contained in a bounded set $e_0 \in \Gamma \subset \mathbb{R}^n$, we can find bounded sets $\Phi_k \subset \mathbb{R}^n$ containing $e(k)$ for all $k \in \mathbb{N}$. For the calculation of $\mathbb{S}_k$ we introduce a definition from set-valued algebra.

Definition 15.25 (Set algebra, [128, Definition 3.10]). *Let $\mathbb{A}, \mathbb{B} \subset \mathbb{R}^p$ and let $K \in \mathbb{R}^{r \times p}$ for $p, r \in \mathbb{N}$. The Minkowski sum is defined as $\mathbb{A} \oplus \mathbb{B} = \{a + b \in \mathbb{R}^p : a \in \mathbb{A}, b \in \mathbb{B}\}$. The Minkowski difference is defined as $\mathbb{A} \ominus \mathbb{B} = \{a \in \mathbb{R}^p : \{a\} \oplus \mathbb{B} \subset \mathbb{A}\}$. Set multiplication is defined as $K\mathbb{A} = \{Ka \in \mathbb{R}^r : a \in \mathbb{A}\} \subset \mathbb{R}^r$.*

Using the Minkowski sum and set multiplication in (15.55) by replacing $w(j)$, $j \in \mathbb{N}$, with the set $\mathbb{W}$, we obtain

$$e(k) \in \{A^k e_0\} \oplus \sum_{j=0}^{k-1} \left(A^j \bar{B}\right) \Gamma \subset A^k \mathbb{E} \oplus \sum_{j=0}^{k-1} \left(A^j \bar{B}\right) \mathbb{W}.$$

Thus, the error $e(k)$, $k \in \mathbb{N}$, evolves in the sets

$$\Phi_k = A^k \Gamma \oplus \bar{\Phi}_k, \qquad \bar{\Phi}_k = \sum_{j=0}^{k-1} \left(A^j \bar{B}\right) \mathbb{W}. \tag{15.56}$$

If the matrix A is a Schur matrix, then $A^k \Gamma \to \{0\}$ for $k \to \infty$. Additionally, in the case that A is a Schur matrix, $\bar{\Phi}_\infty = \sum_{j=0}^\infty A^j \bar{B} \mathbb{W}$ is well defined and for all $w \in \mathbb{W}$ the condition $e \in \bar{\Phi}_\infty$ implies that $e^+ = Ae + \bar{B}w \in \bar{\Phi}_\infty$ (see [91]). In other words, if the initial error is in the bounded set $e_0 \in \bar{\Phi}_\infty$, the error will remain in this set, i.e., $e(k) \in \bar{\Phi}_\infty$ for all $k \in \mathbb{N}$. Finally note that from the definition of $\bar{\Phi}_k$ and the Minkowski sum it trivially follows that $\bar{\Phi}_k \subset \bar{\Phi}_{k+1}$, for $k \in \mathbb{N}$.

Example 15.26. Consider a Schur matrix A together with the sets Γ and $\mathbb{W}$ defined by

$$A = \begin{bmatrix} 0.7 & 0.5 \\ 0.1 & 0.6 \end{bmatrix}, \qquad \Gamma = [-1, 1]^2, \qquad \mathbb{W} = [-0.2, 0.2] \times [-0.1, 0.1].$$

The sets $A^k \Gamma$ and $\bar{\Phi}_k$, $k \in \{0, \ldots, 4\}$, calculated through the *MPT-toolbox* in MATLAB, are visualized in Figure 15.23.

With the observations on the error dynamics derived so far, we return to the original dynamics (15.52). Since $e(k) = x(k) - x_d(k)$ the state satisfies the dynamics

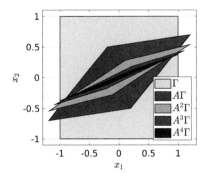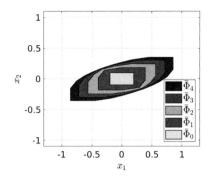

Figure 15.23: Visualization of the sets involved in (15.56) and calculated through the Minkowski sum.

$x(k) = x_d(k) + e(k)$ for all $k \in \mathbb{N}$ and in particular

$$x(k) \in \{x_d(k)\} \oplus \bar{\Phi}_\infty,$$

i.e., all possible solutions $x(k)$ evolve in the tube $\bar{\Phi}_\infty$ centered around the deterministic state $x_d(k)$, if the initial error satisfies $e_0 \in \bar{\Phi}_\infty$.

Thus, if the MPC Algorithm 3 is used to asymptotically stabilize the origin of the deterministic dynamics (15.54), then the actual state converges to a neighborhood around the origin. In particular, $x_d(k) \to 0$, implies that $x(k) \to \bar{\Phi}_\infty$ for $k \to \infty$.

The remaining task is to rewrite the constraints (15.53) in terms of the variables x_d to be able to apply Algorithm 3 to the dynamics (15.54). In this context, note that $x_d \in \mathbb{X}$ does not necessarily imply that $x \in \mathbb{X}$ and $u \in \mathbb{U}(x_d)$ does not imply that $u \in \mathbb{U}(x)$. Here, the set difference in Definition 15.25 can be used to obtain appropriate sets $\mathbb{X}_d$ and $\mathbb{U}_d(x_d)$ such that $x_d \in \mathbb{X}_d$ and $u \in \mathbb{U}_d(x_d)$ imply that $x_d + e \in \mathbb{X}$ and $u \in \mathbb{U}(x + e)$ for all $e \in \bar{\Phi}_\infty$, respectively. In particular the sets

$$\mathbb{X}_d = \mathbb{X} \ominus \bar{\Phi}_\infty \qquad \text{and} \qquad \mathbb{U}_d(x_d) = \mathbb{U}(x_d) \ominus K\bar{\Phi}_\infty \tag{15.57}$$

ensure feasibility of the state and the input.

Example 15.27. We continue with the setting from Example 15.26 and additionally define the set of feasible states $\mathbb{X} = [-2, 2]^2$. Figure 15.24 shows the set $\mathbb{X}$ together with $\bar{\Phi}_\infty$ on the left. The set $\mathbb{X}_d$ in (15.57) obtained through the Minkowski difference is visualized in Figure 15.24 on the right.

To conclude this section, tube-based MPC allows an application of Algorithm 3 to the deterministic dynamics (15.54). If the constraints are adapted appropriately, i.e., according to (15.57), for example, then the perturbed system (15.54) satisfies the original constraints. Moreover, if the deterministic states x_d converge to the origin, the state x converges to a neighborhood around the origin. This is achieved without increasing the online complexity of the MPC scheme, but at the cost of offline computations.

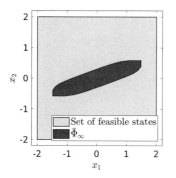
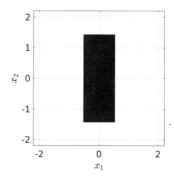

Figure 15.24: Visualization of $\mathbb{X}$ and $\bar{\Phi}_\infty$ on the left and $\mathbb{X}_d = \mathbb{X} \ominus \bar{\Phi}_\infty$ on the right.

15.4   IMPLEMENTATION ASPECTS OF MPC

In the discussion so far, we have presented an idealized version of MPC by assuming that every step in Algorithm 4 can be performed instantaneously and without any delay. This in particular implies that the amount of time to solve the optimization problem in Step 2 of Algorithm 4, the core of the MPC algorithm, does not take time.

To overcome this assumption, we revisit Figure 15.1 and introduce the parameter $\delta > 0$ in Figure 15.25. With the parameter $\delta > 0$ we denote the maximal time to

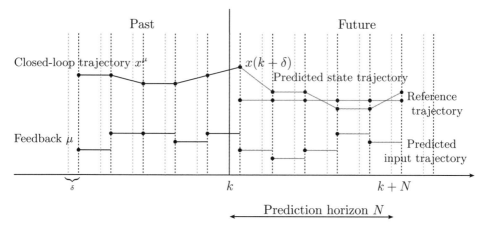

Figure 15.25: Revised Figure 15.1 to take computation delay into account. At time k, the optimal control problem (15.4) corresponding to time $k+\delta$ needs to be solved to ensure that the feedback law is available at time $k + \delta$.

perform Steps 2 and 3 in Algorithm 4, i.e., to solve the optimal control problem (15.4) and to compute a feedback action μ_N. At time $k\Delta \in \mathbb{R}_{\geq 0}$, instead of measuring the state $x_p(k\Delta)$ in Step 1 of Algorithm 4, we predict where the state $x_p(k\Delta + \delta)$ is going to be.

In the MPC algorithm, the time interval $[(k-1)\Delta + \delta, k\Delta)$ can be used to predict the state $x_p(k\Delta)$ and for preprocessing in the optimal control problem (15.4), while the optimal control problem is solved in the time interval $[k\Delta, k\Delta + \delta)$. If $\delta > \Delta$, i.e., the time to solve the optimal control problem is larger than the discretization time Δ, a common approach is to use multiple elements of the optimal open-loop trajectory to define the feedback law. In particular, the feedback law in Step 3 of Algorithm 4 is defined as

$$\mu_N(x_p(k\Delta + \delta)) = u_N^\star(j - 1; x_0)$$

for $t \in [(k + j - 1)\Delta + \delta, (k + j)\Delta + \delta)$, $j = 1, \ldots, m$, and $m \in \mathbb{N}_{\geq 1}$. For $m = 1$ the original feedback law is recovered. The case of $m > 1$ is usually called m-step MPC. Using m-step MPC, the maximal time to calculate an optimal solution of the optimal control problem is $m\Delta$, and if $\delta = m\Delta$ one has to assume that at time $k\Delta$ a prediction of the state $x_p((k + m)\Delta)$ is available.

In the case that explicit MPC is used to define the feedback law μ_N (see Section 15.3.4), an upper bound $\delta > 0$ to find μ_N can be defined based on the explicit knowledge of the optimal control law. In contrast, if an iterative optimization algorithm is used to solve (15.4) as is usually the case, an upper bound $\delta > 0$ which guarantees that within this time the optimal control problem (15.4) is solved with a prespecified accuracy is in general nontrivial or even intractable.

15.4.1   Warm-Start and Suboptimal MPC

MPC has several intrinsic properties which have motivated the development of many heuristic methods to reduce the numerical complexity of the optimization problem and to reduce the time δ to find an optimal solution.

As a first observation, note that the optimal control problems (15.4) at different time steps only differ in the initial state x_0 and possibly in the input constraints $\mathbb{U}(x)$ if the input constraints are state-dependent.

Let $u^\star(\cdot; k, x_0)$ and $x^\star(\cdot; k, x_0)$, with

$$u^\star(\cdot; k, x_0) = \begin{bmatrix} u_N^\star(0; k, x_0) \\ \vdots \\ u_N^\star(N - 1; k, x_0) \end{bmatrix}, \qquad x^\star(\cdot; k, x_0) = \begin{bmatrix} x_N^\star(0; k, x_0) \\ \vdots \\ x_N^\star(N - 1; k, x_0) \end{bmatrix},$$

denote the optimal open-loop input and state trajectories of (15.4) at time $k \in \mathbb{N}$ with respect to the initial state x_0. Then, in particular if the prediction horizon N is large, it is not unreasonable to assume that

$$u^\star(i + 1; k, x_0) \approx u^\star(i; k + 1, x^\star(1; k, x_0)) \qquad (15.58)$$

$$x^\star(i + 1; k, x_0) \approx x^\star(i; k + 1, x^\star(1; k, x_0)) \qquad (15.59)$$

is satisfied for $i = 0, \ldots, N - 2$.

Thus,

$$
u^0(\cdot; k+1, x^\star(1; k, x_0)) = \begin{bmatrix} u_N^\star(1; k, x_0) \\ \vdots \\ u_N^\star(N-1; k, x_0) \\ 0 \end{bmatrix} \quad \text{and}
$$

$$
x^0(\cdot; k+1, x^\star(1; k, x_0)) = \begin{bmatrix} x_N^\star(1; k, x_0) \\ \vdots \\ x_N^\star(N-1; k, x_0) \\ f(x_N^\star(N-1; k, x_0), 0) \end{bmatrix}
$$

(15.60)

can be used as an initial guess of the optimal solution of (15.4) at time $k+1$. Depending on (15.4) and the optimization algorithm used to solve the optimization problem, the initialization (15.60), in contrast to an arbitrary initialization, can reduce the necessary number of iterations of the optimization algorithm significantly. This so-called *warm-start* method in optimization in general reduces the numerical complexity of the optimization problem involved in the MPC Algorithms 3 and 4. The last elements of $u^0(\cdot; k+1, x^\star(1; k, x_0))$ and $x^0(\cdot; k+1, x^\star(1; k, x_0))$, for which no initial guess is available, might also be initialized differently.

By choosing the initial guess of the optimal open-loop solution at the next time in a smart way, it is not uncommon to fix the number of iterations performed by the optimization algorithm. This provides an upper bound δ on the computation time to derive the feedback μ_N. Even if in this case only a suboptimal feedback law is obtained, closed-loop stability of the corresponding MPC closed loop can in many cases be ensured or at least be verified numerically. In this context recall that in many applications, the optimal control problem (15.4) is only an approximation of the corresponding infinite horizon problem. In particular, the prediction horizon N is in general chosen heuristically as a compromise between optimality with respect to the infinite horizon optimal control problem and the numerical complexity of the optimization problem. Similarly, running costs ℓ are in general designed through heuristic methods. Thus, even if the feedback law is not optimal with respect to the particular selection of N and ℓ, it might be optimal with respect to a different N and ℓ, which could have been obtained through different heuristics.

Remark 15.28. If the optimal control problem (15.4) is nonconvex and has multiple local minima, warm-start may be counterproductive in finding a global minimum.

15.4.2   Formulation of the Optimization Problem

We conclude this section with a few words on the formulation of the optimal control problem. The optimal control problem (15.4) is only an abstract formulation which needs to be written in a particular form to be solved by a specific optimization algorithm or software.

A standard formulation of an optimization problem is given by

$$
y^\star = \arg \min_{y \in \mathbb{R}^{\alpha_1}} \; F(y) \tag{15.61a}
$$

$$
\text{s.t.} \quad G_i(y) \leq 0, \qquad i = 1, \dots, \alpha_2 \tag{15.61b}
$$

$$
H_j(y) = 0, \qquad j = 1, \dots, \alpha_3, \tag{15.61c}
$$

where $y \in \mathbb{R}^{\alpha_1}$ denotes the optimization variable, $F : \mathbb{R}^{\alpha_1} \to \mathbb{R}$ defines the objective function, $G : \mathbb{R}^{\alpha_1} \to \mathbb{R}^{\alpha_2}$ defines inequality constraints, $H : \mathbb{R}^{\alpha_1} \to \mathbb{R}^{\alpha_3}$ defines equality constraints, and $\alpha_1, \alpha_2, \alpha_3 \in \mathbb{N}_0$.

For the example of the QP (15.33) we give two equivalent representations of the optimization problem (15.61) to illustrate that the optimal control problem (15.4) can be represented differently. Depending on the software and the optimization algorithm one or the other representation may be beneficial. Here, we say two representations are equivalent if the optimal minimizer of the first representation can be easily obtained from the minimizer of the second representation and vice versa.

Example 15.29. Consider the vector of unknowns

$$\dot{y} = \begin{bmatrix} x(0)^T & u(0)^T & \cdots & x(N-1)^T & u(N-1)^T & x(N)^T \end{bmatrix}^T. \tag{15.62}$$

Then with the definition of the matrix

$$M = \begin{bmatrix} Q & 0 & \cdots & \cdots & \cdots & 0 \\ 0 & R & \ddots & & & \vdots \\ \vdots & \ddots & \ddots & \ddots & & \vdots \\ \vdots & & \ddots & Q & \ddots & \vdots \\ \vdots & & & \ddots & R & 0 \\ 0 & \cdots & \cdots & \cdots & 0 & P \end{bmatrix}$$

the objective function in (15.33) can be written as $F(y) = y^T M y$. Similarly, through definition of matrices and vectors

$$N = \begin{bmatrix} I & 0 & \cdots & \cdots & \cdots & \cdots & \cdots & \cdots & 0 \\ -A & -B & I & 0 & \cdots & & & & \vdots \\ 0 & 0 & -A & -B & I & 0 & \cdots & & \vdots \\ \vdots & & & \ddots & \ddots & \ddots & \ddots & \ddots & 0 \\ 0 & \cdots & \cdots & \cdots & \cdots & 0 & -A & -B & I \end{bmatrix}, \quad \eta = \begin{bmatrix} -x_0 \\ 0 \\ \vdots \\ \vdots \\ 0 \end{bmatrix}$$

and

$$O = \begin{bmatrix} \Gamma_x & \Gamma_u & 0 & \cdots & \cdots & \cdots & \cdots & \cdots & 0 \\ 0 & 0 & \Gamma_x & \Gamma_u & 0 & \cdots & \cdots & \cdots & 0 \\ \vdots & & & \ddots & \ddots & \ddots & \ddots & & \vdots \\ 0 & & & & & 0 & \Gamma_x & \Gamma_u & 0 \\ 0 & \cdots & \cdots & \cdots & \cdots & \cdots & \cdots & 0 & \Gamma_N \end{bmatrix}, \quad \theta = \begin{bmatrix} -\gamma \\ \vdots \\ \vdots \\ -\gamma \\ -\gamma_N \end{bmatrix},$$

the equality and inequality constraints $G(y) = Ny + \eta$ and $H(y) = Oy + \theta$ can be defined, respectively. Through the functions F, G, and H we recover the representation (15.61).

Example 15.30. Consider again the QP (15.33) and define the vector of unknowns

$$y = \begin{bmatrix} u(0)^T & \cdots & u(N-1)^T \end{bmatrix}^T,$$

instead of (15.62). For $i = 0, \ldots, N-1$, the dynamics can be written as

$$x(i+1) = Ax(i) + Bu(i) = A^{i+1}x_0 + \sum_{j=0}^{i} A^{i-j}Bu(j) \qquad (15.63)$$

and $x(i)$ can be expressed through the vector of unknowns and the initial condition x_0. With the definition

$$\mathcal{A} = \begin{bmatrix} I \\ A \\ A^2 \\ A^3 \\ \vdots \\ A^N \end{bmatrix}, \qquad \mathcal{B} = \begin{bmatrix} 0 & \cdots & \cdots & \cdots & \cdots & 0 \\ B & 0 & & & & \vdots \\ AB & B & 0 & & & \vdots \\ A^2B & AB & B & 0 & & \vdots \\ \vdots & \ddots & \ddots & \ddots & \ddots & \vdots \\ \vdots & & & \ddots & \ddots & 0 \\ A^{N-1}B & A^{N-2}B & \cdots & \cdots & AB & B \end{bmatrix}$$

the equality constraints can be summarized in terms of y:

$$\begin{bmatrix} x(0) \\ \vdots \\ x(N) \end{bmatrix} = \mathcal{A}x_0 + \mathcal{B}y. \quad \cdot$$

With this relation and with the definition of the matrices

$$M_1 = \begin{bmatrix} Q & 0 & \cdots & \cdots & 0 \\ 0 & \ddots & \ddots & & \vdots \\ \vdots & \ddots & \ddots & \ddots & \vdots \\ \vdots & & \ddots & Q & 0 \\ 0 & \cdots & \cdots & 0 & P \end{bmatrix}, \qquad M_2 = \begin{bmatrix} R & 0 & \cdots & \cdots & 0 \\ 0 & \ddots & \ddots & & \vdots \\ \vdots & \ddots & \ddots & \ddots & \vdots \\ \vdots & & \ddots & R & 0 \\ 0 & \cdots & \cdots & 0 & R \end{bmatrix},$$

the objective function can be written as

$$F(y) = y^T \left(\mathcal{B}^T M_1 \mathcal{B} + M_2\right) y + 2x_0^T \mathcal{A}^T M_1 \mathcal{B} y + x(0)^T \mathcal{A}^T M_1 \mathcal{A}x(0).$$

Finally, with

$$O_1 = \begin{bmatrix} \Gamma_x & 0 & \cdots & 0 \\ 0 & \ddots & \ddots & \vdots \\ \vdots & \ddots & \Gamma_x & 0 \\ 0 & \cdots & 0 & \Gamma_N \end{bmatrix}, \quad O_2 = \begin{bmatrix} \Gamma_u & 0 & \cdots & 0 \\ 0 & \ddots & \ddots & \vdots \\ \vdots & \ddots & \ddots & 0 \\ 0 & \cdots & 0 & \Gamma_u \end{bmatrix}, \quad \theta = \begin{bmatrix} -\gamma \\ \vdots \\ \vdots \\ -\gamma \\ -\gamma_N \end{bmatrix},$$

the inequality constraints can be represented through the function

$$G(y) = (O_1\mathcal{B} + O_2)\, y + (O_1 \mathcal{A} x_0 - \theta)\,.$$

Equality constraints are not present in this representation of the optimization problem (15.33).

The representation derived in Example 15.29 is equivalent to the representation in Example 15.30. The optimal solution in Example 15.29 contains the optimal solution of Example 15.30, and based on the optimal solution of Example 15.30 the missing optimal states of Example 15.29 can be easily obtained by iterating the dynamics $x^\star(i+1) = Ax^\star(i) + Bu^\star(i)$ for $i = 0, \ldots N - 1$ and $x^\star(0) = x_0$.

The optimization problem derived in Example 15.29 has more unknowns, more inequality constraints, and more equality constraints compared to the representation in Example 15.29. However, the matrices describing the objective function and the constraints are sparse, i.e., the number of elements equal to zero is large, which can be exploited in *interior point algorithms*, for example. On the other hand, the optimization problem derived in Example 15.30 has fewer unknowns, fewer inequality constraints, and no equality constraints. However, in this representation the matrices $\mathcal{B}^T M_1 \mathcal{B} + M_2$ and $O_1 \mathcal{B}_2 + O_2$ are dense, i.e., the number of elements equal to zero is small compared to the number of elements unequal to zero. Thus, sparsity cannot be exploited and *active set methods* often outperform interior point methods.

Example 15.30 describes a *recursive elimination* of the state variables x in the optimization problem. In Example 15.29, additionally the states are included in the vector of unknowns even though the optimal open-loop trajectory is not necessary to define the feedback law μ. This approach is usually referred to as *full discretization* and leads to additional equality constraints in the optimization problem. If the optimization problem is derived based on a continuous time system, it is not uncommon to include additional unknowns in the vector y due to the discretization of the continuous time dynamics. Corresponding approaches can be found under the name *multiple shooting* in the literature in this context.

15.5 EXERCISES

Exercise 15.1. Implement Algorithm 3 to recreate the results in Example 15.4.

Exercise 15.2. Implement Algorithm 4 to recreate the results in Example 15.5.

Exercise 15.3. Recall Example 15.11 and consider the discrete dynamics

$$x^+ = f(x, u) = ax + u, \qquad \mathbb{U} = [-1, 1].$$

Compute the maximal viable subset $\mathbb{X}_v \subset \mathbb{X} = [-1, 1]$ depending on the parameter $a \in \mathbb{R}$.

Exercise 15.4. Consider the optimal control problem

$$\min_{u(i)\in\mathbb{R}^m,\ i\in\mathbb{N}_{[0,N-1]}} \sum_{i=0}^{N-1} |x(i)|_\infty + |u(i)|_\infty + |x(N)|_\infty$$

$$
\begin{aligned}
\text{subject to}\quad 0 &= x(0) - x_0 \\
0 &= x(i+1) - Ax(i) - Bu(i) \quad &&\forall\, i = 0, \ldots, N-1 \\
\gamma &\geq \Gamma_x x(i) + \Gamma_u u(i) \quad &&\forall\, i = 0, \ldots, N-1 \\
\gamma_N &\geq \Gamma_N x(N),
\end{aligned}
$$

with a nondifferentiable objective function. Rewrite the optimal control problem in the form of a linear program by introducing a new set of variables and constraints.

Exercise 15.5. Consider the optimization problem (15.33) defined through linear dynamics (15.32), $Q, P \in \mathcal{S}_{\geq 0}^n$, $R \in \mathcal{S}_{\geq 0}^m$, polyhedral constraints $\Gamma_x \in \mathbb{R}^{r \times n}$, $\Gamma_u \in \mathbb{R}^{r \times m}$, $\gamma \in \mathbb{R}^r$, $\Gamma_N \in \mathbb{R}^{q \times n}$, $\gamma_N \in \mathbb{R}^q$, and a prediction horizon $N \in \mathbb{N}$. Rewrite the quadratic optimization problem in the form of the optimization problem (15.38). In particular, define the function $\hat{c} : \mathbb{R}^n \to \mathbb{R}$ as well as the matrices $\mathcal{H}$, $\mathcal{G}$, $\mathcal{F}$, $\widehat{\mathcal{A}}$, $\widehat{\mathcal{B}}$, and $\widehat{\mathcal{C}}$ in (15.38).

Exercise 15.6. Consider the optimal control problem

$$\min_{\substack{u(i)\in\mathbb{R}^2 \\ i\in\mathbb{N}_{[0,N]}}} \sum_{i=0}^{N-1} \ell(x(i), u(i)) \tag{15.64}$$

$$
\begin{aligned}
\text{subject to}\quad 0 &= x(0) - x_0 \\
0 &= x(i+1) - x(i) - \Delta f(x(i), u(i)) \quad &&\forall\, i = 0, \ldots, N-1 \\
x(i) &\in [-1,1]^2 \times \mathbb{R} \quad &&\forall\, i = 0, \ldots, N-1 \\
u(i) &\in [-1,1]^2 \quad &&\forall\, i = 0, \ldots, N
\end{aligned}
$$

where $\dot{x} = f(x, u)$,

$$f(x, u) = \begin{bmatrix} u_1 \cos(x_3) \\ u_1 \sin(x_3) \\ u_2 \end{bmatrix} \tag{15.65}$$

with $f : \mathbb{R}^3 \times \mathbb{R}^2 \to \mathbb{R}^3$, captures the dynamics of the mobile robot discussed in Section 1.2.2. In the optimization problem the dynamics are approximated through an Euler discretization with sampling rate $\Delta > 0$ (see Section 5.2). (Note that $x(N)$ and $u(N)$ are not relevant in the optimization problem since a terminal cost is not present.) To drive the mobile robot to the origin we define the running cost function

$$\ell(x, u) = x_1^4 + 10x_2^2 + (1 - \cos(x_3))^2 + 0.1u_1^4 + 0.1u_2^4.$$

Implement Algorithm 4 and visualize the closed-loop solution for different prediction horizons $N \in \mathbb{N}$, different $\Delta > 0$, and different initial conditions.

Hint: You can modify the example presented in Appendix B.5 to solve the exercise.

15.6 BIBLIOGRAPHICAL NOTES AND FURTHER READING

MPC has a long history, particularly in the realm of process control. Solving the required optimization problems online in real time takes a nontrivial amount of time and computation times available with early computers necessarily limited the application of MPC to systems with very slow time constants, for example, such as those found in various industrial chemical processing plants. However, as computation speeds have improved, the range of systems amenable to MPC techniques has increased to the point where, particularly using custom bespoke computational hardware (for example, via FPGAs, or Field Programmable Gate Arrays), MPC can be applied to even very fast systems.

Many texts on MPC are available, with classical monographs including [37] and [103]. Nonlinear MPC is covered in detail in [58]. The books [127] and [128] are quite comprehensive.

For a reference with a focus on robust and stochastic MPC, see [92]. A survey on the topic of economic MPC is [49]. Here, we limited our discussion of tube-based MPC to linear systems but the ideas can be applied to a broader class of problems; see [92], [128], and references therein.

The core of MPC is of course actually numerically solving an optimization problem. There is a vast literature on optimization. Particularly useful references in the context of MPC include [20], [29], [111], and again the reference [128].

Part III

Observer Design and Estimation

Chapter Sixteen

Observer Design for Linear Systems

The concepts discussed in the preceding chapters generally rely on the knowledge of the state $x \in \mathbb{R}^n$. Unfortunately, in real applications, the full state x is in general not known and only the output $y \in \mathbb{R}^p$ is available. Consequently, a controller design cannot, in general, rely on the full state x. Under certain observability or detectability conditions, an estimate of the state $\hat{x}(t)$ based on measurements $y(t)$ can be derived. Under appropriate assumptions, if the estimated state obtained through an *observer* satisfies $\hat{x}(t) \to x(t)$ for $t \to \infty$, then the estimated state can be used for the definition of a feedback controller.

The simplest case of this setting is visualized in Figure 16.1. Here, an observer driven by the input u and with state $\hat{x}$ that estimates the unknown plant state x is used to define a linear input $u = K\hat{x}$, which in turn is used to control the plant.

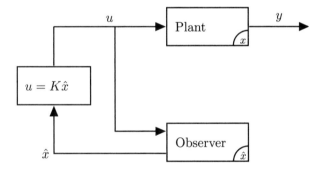

Figure 16.1: Controller based on estimated state $\hat{x}$ instead of true state x of the plant.

In this chapter, we introduce the fundamentals of observer design by focussing on linear systems

$$\dot{x} = Ax + Bu, \qquad x(0) \in \mathbb{R}^n, \tag{16.1a}$$

$$y = Cx + Du, \tag{16.1b}$$

where $A \in \mathbb{R}^{n \times n}$, $B \in \mathbb{R}^{n \times m}$, $C \in \mathbb{R}^{p \times n}$, and $D \in \mathbb{R}^{p \times m}$ for $n, m, p \in \mathbb{N}$. Differently from preceding chapters, we assume that only the output $y \in \mathbb{R}^p$ and the input $u \in \mathbb{R}^m$ are known, while the internal state $x \in \mathbb{R}^n$ and the initial condition $x(0)$ are unknown.

To derive an estimate of the state, we begin with a simple motivating observer for the linear system (16.1a), where we assume that the matrix A is Hurwitz. We introduce a copy of the system,

$$\dot{\hat{x}} = A\hat{x} + Bu, \qquad \hat{x}(0) \in \mathbb{R}^n, \qquad (16.2)$$

where $\hat{x} \in \mathbb{R}^n$ is an estimate of the true state $x \in \mathbb{R}^n$ and (16.2) denotes the observer dynamics. Additionally, we introduce the estimation error $e = x - \hat{x}$ as the difference between the estimated state and the true state. Even though, the estimation error e is unknown (since x is unknown), the error dynamics satisfy

$$\dot{e} = \dot{x} - \dot{\hat{x}} = Ax + Bu - A\hat{x} - Bu = A(x - \hat{x}) = Ae, \qquad (16.3)$$

which implies $\lim_{t\to\infty} e(t) = 0$ based on the assumption that A is Hurwitz and is noticeably independent of the input u. Therefore, if A is Hurwitz, the estimated state $\hat{x}$ converges to the true state x for all $x(0) \in \mathbb{R}^n$ and for all $\hat{x}(0) \in \mathbb{R}^n$. This also shows that the input of the system (16.1a) can be defined based on the estimated state; i.e., $u = u(\hat{x})$, when $\hat{x}$ is defined by the dynamics (16.2).

To generalize this simple result to dynamics where A is not Hurwitz, we make use of the definitions and results on observability and detectability introduced in Chapter 3 in the following sections.

In the discrete time setting, these ideas can be applied to the linear discrete time system

$$x^+ = Ax + Bu, \qquad x(0) \in \mathbb{R}^n, \qquad (16.4a)$$
$$y = Cx + Du, \qquad (16.4b)$$

where the matrices are defined in the same way as in (16.1). Again, through a copy of the dynamics

$$\hat{x}^+ = A\hat{x} + Bu, \qquad \hat{x}(0) \in \mathbb{R}^n, \qquad (16.5)$$

the discrete time error dynamics $e(k) = x(k) - \hat{x}(k)$, $k \in \mathbb{N}$, satisfy

$$e^+ = Ae$$

as in (16.3).

16.1   LUENBERGER OBSERVERS

Consider the linear system (16.1), or its discrete time counterpart (16.4), where A is not necessarily Hurwitz or Schur, respectively. Here, we focus on the continuous time setting. However, the discrete time results are immediately obtained by replacing the differential equations by difference equations and the continuous time $t \in \mathbb{R}_{\geq 0}$ with discrete time $k \in \mathbb{N}_0$.

As in (16.2) we use a copy of the system but with an additional *output injection term* $L \in \mathbb{R}^{n \times p}$ to describe the observer dynamics

$$\dot{\hat{x}} = A\hat{x} + Bu - L(y - \hat{y}), \qquad (16.6a)$$
$$\hat{y} = C\hat{x} + Du. \qquad (16.6b)$$

A state estimator or observer of this form is called a *Luenberger Observer*. A

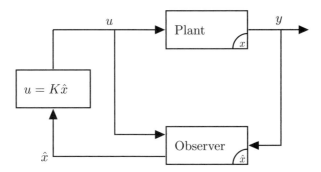

Figure 16.2: Observer design using the measured output y in the overall control loop.

diagram of this setup is shown in Figure 16.2, where we additionally assume that $u = K\hat{x}$, $K \in \mathbb{R}^{m \times n}$. In this case the estimation error $e = x - \hat{x}$ satisfies the dynamics

$$\begin{aligned}
\dot{e} = \dot{x} - \dot{\hat{x}} \\
= Ax + Bu - A\hat{x} - Bu + L(Cx + Du - C\hat{x} - Du) \\
= Ae + LCe = (A + LC)e.
\end{aligned} \tag{16.7}$$

If the pair (A, C) is observable, then the poles of $A + LC$ can be placed arbitrarily and, in particular, the matrix L can be defined such that the matrix $A + LC$ is Hurwitz. This follows from the fact that the matrix $A + LC$ has the same eigenvalues as the matrix $(A + LC)^T = A^T + C^T L^T$. Thus the results in Section 3.5.3 on pole placement can be applied to the pair (A^T, C^T) to compute the matrix L.

If the matrix L is defined such that the matrix $A + LC$ is Hurwitz, the error $e : \mathbb{R}_{\geq 0} \to \mathbb{R}^n$ satisfies $e(t) \to 0$ for $t \to \infty$ according to (16.7) for all $e(0) = e_0 \in \mathbb{R}^n$. The existence of an $L \in \mathbb{R}^{p \times n}$ with $A + LC$ Hurwitz is guaranteed if the pair (A, C) is detectable; i.e., the pair (A^T, C^T) is stabilizable.

These calculations show how the state x of a linear system can be approximated through $\hat{x}$. As a next step we turn to the simultaneous estimation of the state and the stabilization of the origin of (16.1) using a linear feedback law u based on the estimated state $u(\hat{x}) = K\hat{x}$, $K \in \mathbb{R}^{m \times n}$. With this definition, we rewrite the state dynamics (16.1a) in terms of the state x and the error e:

$$\begin{aligned}
\dot{x} = Ax + Bu(\hat{x}) = Ax + BK\hat{x} \\
= Ax + BK(x + e) \\
= (A + BK)x + BKe.
\end{aligned}$$

Combining the x-dynamics with the error dynamics (16.7) provides the overall closed-loop system

$$\begin{bmatrix} \dot{x} \\ \dot{e} \end{bmatrix} = \begin{bmatrix} A + BK & BK \\ 0 & A + LC \end{bmatrix} \begin{bmatrix} x \\ e \end{bmatrix}. \tag{16.8}$$

Therefore, if (A, B) is controllable and (A, C) is observable, and with the block triangular structure of the above matrix, we can place the poles of the above matrix arbitrarily by choosing K and L. The representation (16.8) in particular shows that the convergence $|x(t)| \to 0$ for $t \to \infty$ and the convergence $|e(t)| \to 0$ for $t \to \infty$ can be guaranteed by designing L and K individually. This property is referred to as the *separation principle* since the feedback controller and state observer can be designed separately. Note, however, that this only allows us to determine the asymptotic behavior of (16.8). When considering the transient behavior, the selection of L indeed impacts the convergence of the x-dynamics and the selection of K impacts the convergence of the e-dynamics.

Alternatively, (16.8) can be written in terms of x and $\hat{x}$:

$$\begin{bmatrix} \dot{x} \\ \dot{\hat{x}} \end{bmatrix} = \begin{bmatrix} A & BK \\ -LC & A + BK + LC \end{bmatrix} \begin{bmatrix} x \\ \hat{x} \end{bmatrix}. \tag{16.9}$$

While the separation is not visible in this representation, the dynamics (16.8) and (16.9) capture the same information.

Example 16.1. Recall the linearization of the pendulum on a cart in the upright position discussed in Example 3.38. In addition to the pair (A, B) defined as

$$A = \begin{bmatrix} 0 & 0 & 1.00 & 0 \\ 0 & 0 & 0 & 1.00 \\ 0 & 3.27 & -0.07 & -0.03 \\ 0 & 6.54 & -0.03 & -0.07 \end{bmatrix} \quad \text{and} \quad B = \begin{bmatrix} 0 \\ 0 \\ 0.67 \\ 0.33 \end{bmatrix},$$

we consider the output matrix

$$C = \begin{bmatrix} 1 & 0 & 0 & 0 \\ 0 & 1 & 0 & 0 \end{bmatrix},$$

i.e., only the position of the cart and the angle of the pendulum are available as measurements.

As discussed in Example 3.38, the feedback gain

$$K = \begin{bmatrix} 7.34 & -140.84 & 15.47 & -60.54 \end{bmatrix}$$

ensures that the closed-loop matrix

$$A + BK = \begin{bmatrix} 0 & 0 & 1.00 & 0 \\ 0 & 0 & 0 & 1.00 \\ 4.89 & -90.62 & 10.24 & -40.39 \\ 2.45 & -40.41 & 5.12 & -20.24 \end{bmatrix}$$

is Hurwitz and has the eigenvalues $\{-4, -3, -2, -1\}$.

Similarly, the observer gain

$$L = \begin{bmatrix} -2.90 & -1.07 \\ -3.75 & -6.49 \\ -2.58 & -6.96 \\ -8.53 & -16.64 \end{bmatrix}$$

is obtained through `place.m` in MATLAB and the matrix

$$A + LC = \begin{bmatrix} -2.90 & 0 & -0.07 & 0 \\ -3.75 & 0 & -6.49 & 1.00 \\ -2.58 & 3.27 & -7.03 & -0.03 \\ -8.53 & 6.54 & -16.67 & -0.07 \end{bmatrix}$$

has the same eigenvalues as $A + BK$.

Figure 16.3 shows the evolution of the states and the estimated states of the combined controller observer dynamics (16.9). Here, the plant is initialized at

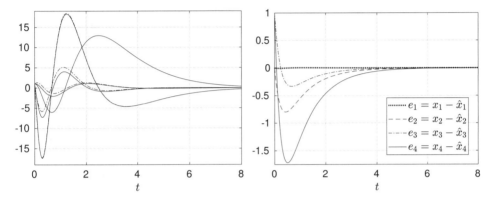

Figure 16.3: Combined observer and controller design. On the left, the state x (solid lines) together with the observed state $\hat{x}$ (dashed lines). On the right, evolution of the error dynamics $e = x - \hat{x}$.

the unknown state $x_0 = [1, 1, 1, 1]^T$. The estimated state is initialized at $\hat{x} = [1, 1, 0, 0]^T$. This captures the fact that x_1 and x_2 are contained in the output while x_3 and x_4 are not directly accessible.

As expected, the estimated state converges to the state of the plant, i.e., the error satisfies $e(t) \to 0$ for $t \to \infty$, and the origin of the plant is asymptotically stable. Note that this behavior is independent of the initial condition $x_0, \hat{x}_0$ since for linear systems local results are also global and the stability properties of the linear system solely depend on the properties of the matrix in (16.8) (or (16.9), respectively) defining the linear system.

16.2   MINIMUM ENERGY ESTIMATOR (CONTINUOUS TIME SETTING)

In Section 14.1.1 we derived an optimal feedback gain matrix based on the solution of an optimal control problem as an alternative to pole placement in Section 3.5.3. In the preceding section we have seen that the output injection term L of the Luenberger observer can be obtained through pole placement by considering the pair (A^T, C^T) in the controller design context. A similiar duality property can be used in the context of optimal control. In this section we build upon the ideas of Section 14.1.1 to derive an *optimal estimator* in terms of minimal energy.

Consider a perturbed version of the linear system (16.1) subject to *disturbance*

$v : \mathbb{R} \to \mathbb{R}^q$ and *measurement noise* $w : \mathbb{R} \to \mathbb{R}^p$,

$$\dot{x} = Ax + Bu + \bar{B}v \tag{16.10a}$$
$$y = Cx + Du + w, \tag{16.10b}$$

where $\bar{B} \in \mathbb{R}^{n \times q}$. Conceptually, (16.1) represents the *model of a plant*, while (16.10) are the actual dynamics of the *plant*. By choosing $v(\cdot)$ and $w(\cdot)$, an arbitrary state trajectory $x(\cdot)$ can be obtained from the dynamics (16.10) for a given input $u(\cdot)$ and a measured output $y(\cdot)$.

The *minimum energy estimation* problem is to find a particular solution of the perturbed dynamics (16.10) by using additional assumptions. In particular, we want to find a solution $\bar{x} : \mathbb{R}_{\leq t_0} \to \mathbb{R}^n$ for $t_0 \geq 0$, which satisfies the dynamics

$$\begin{aligned} \dot{\bar{x}} &= A\bar{x} + Bu + \bar{B}v \\ y &= C\bar{x} + Du + w \end{aligned} \tag{16.11}$$

for given $u(\cdot)$, $y(\cdot)$, and which minimizes the cost function

$$J_{\mathrm{MEE}}(\bar{x}(t_0), v(\cdot)) = \int_{-\infty}^{t_0} w(\tau)^T Q w(\tau) + v(\tau)^T R v(\tau) d\tau \tag{16.12}$$

subject to the dynamics (16.10). Here, the cost function (16.12) is defined through the positive definite matrices $Q \in \mathcal{S}_{>0}^p$, $R \in \mathcal{S}_{>0}^q$. Note that the cost function $J_{\mathrm{MEE}}(\bar{x}(t_0), v(\cdot))$ is only a function of $v(\cdot)$ and not a function of $w(\cdot)$, i.e.,

$$J_{\mathrm{MEE}}(\bar{x}(t_0), v(\cdot)) = \int_{-\infty}^{t_0} (C\bar{x}(\tau) + Du(\tau) - y(\tau))^T Q(C\bar{x}(\tau) + Du(\tau) - y(\tau))$$
$$+ v(\tau)^T R v(\tau) d\tau, \tag{16.13}$$

where the measurement noise is removed using (16.10b). In words, given past inputs $u(\cdot)$ and outputs $y(\cdot)$, the optimization problem (16.12) aims to find the disturbance $v(\cdot)$ with minimum energy and an estimated state $\bar{x}(t_0)$ that explains the observed inputs and outputs.

To ensure that the minimum energy estimation problem is well defined, which includes the fact that (16.13) needs to be finite, additional assumptions are necessary. In particular, we assume that

$$\lim_{t \to -\infty} \bar{x}(t) = 0, \qquad \lim_{t \to -\infty} u(t) = 0 \qquad \text{and} \qquad \lim_{t \to -\infty} w(t) = 0$$

holds. Note that this also implies that $\lim_{t \to -\infty} y(t) = 0$ is satisfied.

As in the derivation of the *linear quadratic regulator*, the matrices Q and R are design parameters that penalize the measurement noise and the disturbance, respectively. If Q and R are diagonal positive definite matrices, large entries in Q (compared to R) penalize the noise and neglect the disturbance. Thus $w(\cdot)$ is forced to be small when (16.12) is minimized. Vice versa, if the entries in R are large (compared to Q), $v(\cdot)$ is forced to be small. Looked at from a different perspective, the matrices Q and R can be selected based on assumptions which capture whether the model (16.10) is affected more by measurement noise or by disturbances, or, equivalently, whether the model (16.10) is less accurate with respect to measurement noise or disturbances.

In the following we derive a solution of the optimization problem

$$V_{\mathrm{MEE}}(\bar{x}_0) = \min_{v(\cdot):\mathbb{R}\to\mathbb{R}^q} J_{\mathrm{MEE}}(\bar{x}_0, v(\cdot))$$

$$\text{subject to (16.10a)} \tag{16.14}$$

to define the *minimum energy estimator gain*. Before we present the main result of this section, we introduce a *feedback invariant* similar to that in Definition 14.1 in the optimal control setting in Section 14.1. To this end we additionally use the definitions

$$\bar{\mathcal{X}} = \{x : \mathbb{R}_{\leq t_0} \to \mathbb{R}^n\} \qquad \text{and} \qquad \mathcal{V} = \{v : \mathbb{R}_{\leq t_0} \to \mathbb{R}^q\}.$$

Definition 16.2 (Feedback invariant). *Consider the system* (16.11), $t_0 \in \mathbb{R}_{\geq 0}$, *and fixed* $u(\cdot) : \mathbb{R}_{<t_0} \to \mathbb{R}^m$, $y(\cdot) : \mathbb{R}_{<t_0} \to \mathbb{R}^p$. *A functional* $H : \bar{\mathcal{X}} \times \mathcal{V} \to \mathbb{R}$ *is called feedback invariant if for all solution pairs* $(\bar{x}_1(\cdot), v_1(\cdot)), (\bar{x}_2(\cdot), v_2(\cdot)) \in \bar{\mathcal{X}} \times \mathcal{V}$ *with* $\bar{x}_1(t_0) = \bar{x}_2(t_0)$ *the equation*

$$H(\bar{x}_1(\cdot), v_1(\cdot)) = H(\bar{x}_2(\cdot), v_2(\cdot))$$

holds.

Theorem 16.3 (Feedback invariant, [63, Proposition 23.1]). *Consider the linear system* (16.10) *for a given output function* $y(\cdot) : \mathbb{R}_{\leq t_0} \to \mathbb{R}^p$ *and a given input function* $u(\cdot) : \mathbb{R}_{\leq t_0} \to \mathbb{R}^m$ *for* $t_0 > 0$. *Then, for every symmetric matrix* $P \in \mathcal{S}^n$, *differentiable signal* $\beta(\cdot) : \mathbb{R}_{\leq t_0} \to \mathbb{R}^n$, *and a scalar* $H_0 \in \mathbb{R}^n$ *(which does not depend on* $\bar{x}(\cdot)$ *and* $v(\cdot)$*), the functional*

$$\begin{aligned}
H(\bar{x}(\cdot), v(\cdot)) = {} & H_0 \\
& + \int_{-\infty}^{t_0} \left(A\bar{x}(\tau) + Bu(\tau) + \bar{B}v(\tau) - \dot{\beta}(\tau) \right)^T P\left(\bar{x}(\tau) - \beta(\tau)\right) \\
& + \left(\bar{x}(\tau) - \beta(\tau)\right)^T P\left(A\bar{x}(\tau) + Bu(\tau) + \bar{B}v(\tau) - \dot{\beta}(\tau) \right) d\tau \\
& - \left(\bar{x}(t_0) - \beta(t_0)\right)^T P\left(\bar{x}(t_0) - \beta(t_0)\right)
\end{aligned} \tag{16.15}$$

is a feedback invariant as long as $\lim_{t\to-\infty}(\bar{x}(t) - \beta(t)) = 0$.

Proof. Assume that the condition $\lim_{t\to-\infty}(\bar{x}(t) - \beta(t)) = 0$ is satisfied. Then the functional (16.15) can be rewritten as

$$\begin{aligned}
H(\bar{x}(\cdot), v(\cdot)) = {} & H_0 + \int_{-\infty}^{t_0} \frac{d}{d\tau}\left(\left(\bar{x}(\tau) - \beta(\tau)\right)^T P\left(\bar{x}(\tau) - \beta(\tau)\right) \right) d\tau \\
& - \left(\bar{x}(t_0) - \beta(t_0)\right)^T P\left(\bar{x}(t_0) - \beta(t_0)\right) \\
= {} & H_0 + \lim_{t\to-\infty}\left(\bar{x}(t) - \beta(t)\right)^T P\left(\bar{x}(t) - \beta(t)\right) \\
= {} & H_0,
\end{aligned}$$

which shows the assertion. $\square$

Theorem 16.4 (The minimum energy estimator). *Consider the linear system* (16.10) *and assume that the pair* $(A, \bar{B})$ *is controllable and the pair* (A, C) *is de-*

tectable. Additionally, consider the optimization problem (16.14) where the cost function is defined through positive definite matrices $Q \in \mathcal{S}_{>0}^p$ and $R \in \mathcal{S}_{>0}^q$. Then there exists a symmetric positive definite solution $S \in \mathcal{S}_{>0}^n$ to the dual algebraic Riccati equation

$$AS + SA^T + \bar{B}R^{-1}\bar{B}^T - SC^TQCS = 0 \tag{16.16}$$

such that $A - LC$ is Hurwitz, where $L = SC^TQ$. Furthermore, the minimum energy estimator for (16.10) for the criterion (16.14) is given by

$$\dot{\hat{x}} = A\hat{x} + Bu + L(y - C\hat{x} - Du) \tag{16.17}$$

and the initial condition $\hat{x}(t_0) = \bar{x}_0$, $t_0 \in \mathbb{R}_{\geq 0}$.

Proof. We start with the dual algebraic Riccati equation (16.16) and show that there exists a positive definite solution $S > 0$. If we define the matrices

$$\widehat{A} = A^T, \quad \widehat{C} = \bar{B}^T, \quad \widehat{B} = C^T, \quad \widehat{Q} = R^{-1} \quad \text{and} \quad \widehat{R} = Q^{-1}, \tag{16.18}$$

then (16.16) reads

$$\widehat{A}^T S + S\widehat{A} + \widehat{C}^T\widehat{Q}\widehat{C} - S\widehat{B}\widehat{R}^{-1}\widehat{B}^T S = 0, \tag{16.19}$$

which is a special form of the Riccati equation (14.17) in the unknown S instead of P. From $R > 0$ and $Q > 0$ it follows that $\widehat{Q}$ and $\widehat{R}$ are positive definite. Additionally, $(A, \bar{B})$ stabilizable and (A, C) detectable is equivalent to $(\widehat{A}, \widehat{C})$ detectable and $(\widehat{A}, \widehat{B})$ stabilizable. Thus, according to Theorem 14.4, the Riccati equation (16.19) has a unique positive definite solution $S > 0$.

For the next step, deriving a solution of the optimization problem (16.14), we use an approach similar to that in Section 14.1.1, where we provided a solution to the linear quadratic regulator problem. To this end, let the assumptions in Theorem 16.4 be satisfied and recall the definitions and notations in Theorem 16.3. To shorten the expressions we introduce the definitions

$$\phi(t) = A\bar{x}(t) + Bu(t) + \bar{B}v(t) - \dot{\beta}(t),$$
$$\eta(t) = \bar{x}(t) - \beta(t),$$
$$\bar{y}(t) = y(t) - Du(t).$$

Additionally we omit the time index, which allows us to compactly write the cost function (16.13) as

$$J_{\text{MEE}}(\bar{x}(t_0), v(\cdot)) = \int_{-\infty}^{t_0} (C\bar{x} - \bar{y})^T Q(C\bar{x} - \bar{y}) + v^T R v \, d\tau.$$

We add and subtract the feedback invariant (16.15) to and from the cost function

$$J_{\text{MEE}}(\bar{x}(t_0), v(\cdot)) = H(\hat{x}(\cdot), v(\cdot)) - H_0 + \eta^T P \eta$$
$$+ \int_{-\infty}^{t_0} (C\hat{x} - \bar{y})^T Q(C\hat{x} - \bar{y}) + v^T R v - (\phi^T P \eta + \eta^T P \phi) \, d\tau.$$

If we expand the term $\phi^T P \eta$ (and $\eta^T P \phi$), i.e.,

$$\phi^T P \eta = \left(A\bar{x} + Bu + \bar{B}v - \dot{\beta} \right)^T P \left(\bar{x} - \beta \right)$$
$$= \bar{x}^T A^T P \bar{x} + \bar{x}^T (-A^T P \beta + P B u - P \dot{\beta}) - \dot{\beta}^T P \left(B u - \dot{\beta} \right) + v^T B^T P(\bar{x} - \dot{\beta}),$$

then the cost function becomes

$$J_{\mathrm{MEE}}(\bar{x}(t_0), v(\cdot))$$
$$= H(\bar{x}(\cdot), v(\cdot)) - H_0 + \eta^T P \eta + \int_{-\infty}^{t_0} \bar{x}^T C^T Q C \bar{x} - 2x^T C^T P \bar{y} + \bar{y}^T P \bar{y}$$
$$+ v^T R v - \Big[\bar{x}(A^T P + PA)\bar{x} + 2\bar{x}^T (-A^T P \beta + P B u - P \dot{\beta})$$
$$- 2\dot{\beta}^T P \left(B u - \dot{\beta} \right) + 2v^T B^T P(\bar{x} - \beta) \Big] \, d\tau$$
$$= H(\bar{x}(\cdot), v(\cdot)) - H_0 + \eta^T P \eta + \int_{-\infty}^{t_0} \bar{x}^T (-A^T P - PA + C^T Q C)\bar{x}$$
$$- 2\bar{x}^T (C^T Q \bar{y} - A^T P \beta + P B u - P \dot{\beta}) + \bar{y}^T P \bar{y} + 2\dot{\beta}^T P \left(B u - \dot{\beta} \right)$$
$$+ v^T R v - 2v^T B^T P(\bar{x} - \beta) \, d\tau.$$

As a last step, we complete the square and rearrange the terms to be able to discuss every term in the integral individually:

$$J_{\mathrm{MEE}}(\bar{x}(t_0), v(\cdot))$$
$$= H(\bar{x}(\cdot), v(\cdot)) - H_0 + \eta^T P \eta + \int_{-\infty}^{t_0} \bar{x}^T (-A^T P - PA + C^T Q C)\bar{x}$$
$$- 2\bar{x}^T (C^T Q \bar{y} - A^T P \beta + P B u - P \dot{\beta}) + \bar{y}^T P \bar{y} + 2\dot{\beta}^T P \left(B u - \dot{\beta} \right)$$
$$+ (v - R^{-1} B^T P(\bar{x} - \beta))^T R(v - R^{-1} B^T P(\bar{x} - \beta))$$
$$- (\bar{x} - \beta)^T P B R^{-1} B^T P(\bar{x} - \beta) \, d\tau$$

$$= H(\bar{x}(\cdot), v(\cdot)) - H_0 + (\bar{x} - \beta)^T P(\bar{x} - \beta) \tag{16.20a}$$
$$+ \int_{-\infty}^{t_0} \bar{x}^T (-A^T P - PA + C^T Q C - P B R^{-1} B^T P)\bar{x} \tag{16.20b}$$
$$- 2x^T (C^T Q \bar{y} - (A^T P + P B R^{-1} B^T P)\beta + P B u - P \dot{\beta}) \tag{16.20c}$$
$$+ \bar{y}^T P \bar{y} + 2\dot{\beta}^T P \left(B u - \dot{\beta} \right) - \beta^T P B R^{-1} B^T P \beta \tag{16.20d}$$
$$+ (v - R^{-1} B^T P(\bar{x} - \beta))^T R(v - R^{-1} B^T P(\bar{x} - \beta)) \, d\tau. \tag{16.20e}$$

We first concentrate on the term (16.20b). We know that Equation (16.16) has a unique positive definite solution $S > 0$. Multiplying (16.16) by -1 and by S^{-1} from the left and the right leads to the equivalent equation

$$-S^{-1}A - A^T S^{-1} - S^{-1}\bar{B}R^{-1}\bar{B}^T S^{-1} + C^T Q C = 0.$$

Thus, for the selection $P = S^{-1}$ the term (16.20b) vanishes.

To get rid of the term (16.20c) we make use of the degree of freedom in the function β and define β implicitly through the differential equation

$$P\dot{\beta} = -(A^T P + PBR^{-1}B^T P)\beta + PBu + C^T Q\bar{y}.$$

Since P is positive definite (and thus invertible), the condition can be equivalently written as

$$\dot{\beta} = -P^{-1}(-PA + C^T QC)\beta + Bu + P^{-1}C^T Q\bar{y}$$
$$= (A - P^{-1}C^T QC)\beta + Bu + P^{-1}C^T Q\bar{y}. \tag{16.21}$$

For a given initial condition $\beta(t_0) = \beta_0 \in \mathbb{R}^n$ let $u(\cdot) : \mathbb{R}_{\leq t_0} \to \mathbb{R}^m$ and $y(\cdot) : \mathbb{R}_{\leq t_0} \to \mathbb{R}^p$ be defined such that $\lim_{t \to -\infty} u(t) = 0$ and $\lim_{t \to -\infty} y(t) = 0$ and the corresponding solution $\beta(\cdot) : \mathbb{R}_{\leq t_0} \to \mathbb{R}^n$ of (16.21) satisfies

$$\lim_{t \to -\infty} \beta(t) = 0.$$

To cancel the term (16.20d) we use the degree of freedom in H_0 and define

$$H_0 = \int_{-\infty}^{t_0} \bar{y}^T P\bar{y} + 2\dot{\beta}^T P\left(Bu - \dot{\beta}\right) - \beta^T PBR^{-1}B^T P\beta \, d\tau,$$

which is possible since (16.20d) does not depend on $\bar{x}(\cdot)$ and $v(\cdot)$.

With these choices, $J_{\text{MEE}}(\bar{x}(t_0), v(\cdot))$ simplifies to

$$J_{\text{MEE}}(\bar{x}(t_0), v(\cdot)) = H(\bar{x}(\cdot), v(\cdot)) + (\bar{x} - \beta)^T P(\bar{x} - \beta)$$
$$+ \int_{-\infty}^{t_0} (v - R^{-1}B^T P(\bar{x} - \beta))^T R(v - R^{-1}B^T P(\bar{x} - \beta)) \, d\tau. \tag{16.22}$$

Since $H(\bar{x}(\cdot), v(\cdot))$ is a feedback invariant, (16.22) is minimal if the integral and the quadratic term vanish, i.e., if

$$v(t) = R^{-1}B^T P(\bar{x}(t) - \beta(t))$$

is satisfied for all $t \in \mathbb{R}_{< t_0}$ and if the initial value of β is selected as $\beta(t_0) = \bar{x}(t_0)$.

As a final step of the proof, we need to obtain the observer dynamics (16.17). To this end we examine the dynamics (16.11) and the definition of the function β in (16.21). Using the definition of $L = SC^T Q$ and $S = P^{-1}$ the differential equation (16.21) can be rewritten as

$$\dot{\beta} = (A - LC)\beta + Bu + L\bar{y}$$
$$= (A - LC)\beta + Bu + L(y - Du)$$
$$= A\beta + Bu + L(y - C\beta - Du),$$

which recovers the dynamics of the minimum energy estimator (16.17)

$$\dot{\hat{x}} = A\hat{x} + Bu + L(y - C\hat{x} - Du). \tag{16.23}$$

This expression completes the proof. $\qquad\qquad\qquad\qquad\qquad\qquad\qquad\square$

Example 16.5. We consider again the inverted pendulum described by (1.45) and the parameters defined in Table 3.1 and where only the angle is measured. The linearization at the stable equilibrium $[x_1^e, x_2^e]^T = [\theta^e, \dot{\theta}^e]^T = [\pi, 0]^T$ is given by the linear system

$$\dot{x} = Ax + Bu + \bar{B}v$$
$$y = Cx + w,$$

where

$$A = \begin{bmatrix} 0 & 1 \\ -\frac{mg\ell}{J+m\ell^2} & -\frac{\gamma}{J+m\ell^2} \end{bmatrix}, \qquad B = \begin{bmatrix} 0 \\ \frac{\ell}{J+m\ell^2} \end{bmatrix}, \quad \text{and} \quad C = \begin{bmatrix} 1 & 0 \end{bmatrix}.$$

Additionally, v and w capture unknown model uncertainties and measurement noise with $\bar{B}$ defined as

$$\bar{B} = \begin{bmatrix} 1 & 0 \\ 0 & 1 \end{bmatrix}.$$

For simplicity we use the constants $m = \ell = 1$, $J = 0$, $g = 9.81$, and $\gamma = 0.1$.
Using the matrices

$$Q = 1 \qquad \text{and} \qquad R = \begin{bmatrix} 1 & 0 \\ 0 & 1 \end{bmatrix}$$

the observer gain

$$L = \begin{bmatrix} 0.9548 \\ -0.0441 \end{bmatrix}$$

is obtained. The corresponding matrix $A - LC$ has eigenvalues $\lambda_{1,2} = -0.53 \pm 3.10j$, i.e., as expected from Theorem 16.4 the matrix $A - LC$ is Hurwitz.

Figure 16.4 shows the evolution of the states of the plant x and the evolution of the observer states $\hat{x}$ corresponding to (16.17). Here, the plant is initialized at

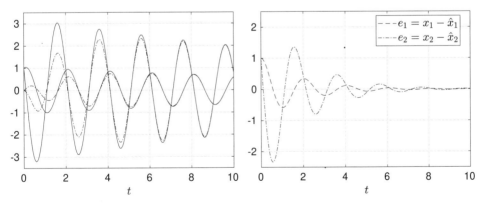

Figure 16.4: Visualization of the state dynamics x (solid lines) and the observer dynamics $\hat{x}$ (dashed lines). Since disturbances are set to zero, the error $e = x - \hat{x}$ converges to zero.

$x_0 = [1, 1]^T$ and $\hat{x}_0 = [0, 0]^T$. The input is defined as $u(t) = 0$ for all $t \in [0, 10]$. Similarly, $v(t)$ and $w(t)$ are set to zero for all t. As expected $\hat{x}$ converges to x, or equivalently, the error dynamics $e(t) = x(t) - \hat{x}(t)$ converge to the origin for $t \to \infty$ (see Figure 16.4, right).

Figure 16.5 shows the impact of $v(t)$ and $w(t)$ on the performance of the minimum energy estimator. Here, $v(t)$ and $w(t)$ are modelled as Gaussian white noise

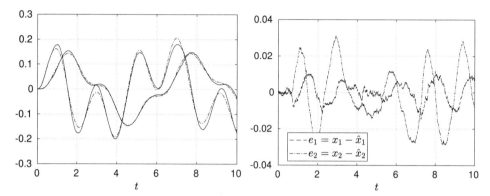

Figure 16.5: Visualization of the state dynamics x (solid line) and the observer dynamics $\hat{x}$ (dashed line) impacted by the disturbances in and the measurement noise in Figure 16.6. On the right, the error dynamics are shown.

with zero mean and standard deviation 0.1 generated through `randn.m` in MATLAB. The functions $v(t)$ and $w(t)$ are visualized in Figure 16.6. The input is defined

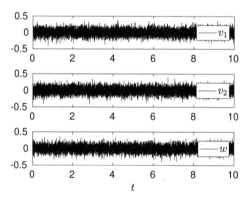

Figure 16.6: Gaussian white noise with zero mean and standard deviation 0.1 used to simulate the impact of disturbances in a minimum energy estimator.

as $u(t) = \sin(t)$. Differently from the simulation in Figure 16.4, for the setting in Figure 16.5, the plant and the observer are initialized at $x_0 = \hat{x}_0 = 0$.

Due to the disturbances and the measurement noise, the error $e = x - \hat{x}$ does not converge to the origin but only stays in a neighborhood around the origin. Nevertheless, a good performance of the observer can be observed since the mismatch between x and $\hat{x}$ is small (see Figure 16.6) compared to the values of v and w (see Figure 16.6).

Instead of the *deterministic setting* discussed above, the minimum energy estimator can alternatively be derived in the *stochastic setting*. In the stochastic setting the minimum energy estimator is better known as the (continuous time) *Kalman filter.* In particular, under certain assumptions on the disturbances $v(t)$ and $w(t)$ in the system dynamics (16.10), equivalences between the minimum energy estimator and the Kalman filter can be derived. To make this precise, assume that $v(\cdot)$ and $w(\cdot)$ represent functions of zero-mean Gaussian white noise with covariance matrices satisfying

$$\mathrm{E}[v(t)v(\tau)^T] = \delta(t - \tau)R^{-1}, \qquad \mathrm{E}[w(t)w(\tau)^T] = \delta(t - \tau)Q^{-1},$$

for all $t, \tau \in \mathbb{R}$ and for positive definite matrices $Q \in \mathcal{S}_{>0}^p$ and $R \in \mathcal{S}_{>0}^q$, and

$$\mathrm{E}[v(t)w(\tau)^T] = 0$$

for all $t, \tau \in \mathbb{R}$. A brief definition of the *expected value* $\mathrm{E}[\cdot]$ is given in Appendix A.6. Here, $\delta : \mathbb{R} \to \mathbb{R} \cup \{\infty\}$ denotes the *Dirac delta function*, which satisfies

$$\delta(t) = \left\{ \begin{array}{ll} \infty, & t = 0 \\ 0, & t \neq 0 \end{array} \right. \qquad \text{and} \qquad \int_{-\infty}^{\infty} \delta(t)dt = 1.$$

Under these conditions, the estimate $\hat{x}$ obtained through the minimum energy estimator also minimizes the expected value

$$\lim_{t \to \infty} \mathrm{E}\left[|x(t) - \hat{x}(t)|^2\right] \tag{16.24}$$

and is thus also optimal with respect to the measure (16.24). Instead of the derivations shown here, the Kalman filter is derived by minimizing (16.24). For the discrete time case discussed next, we follow this route in the derivation of an optimal observer.

16.3   THE DISCRETE TIME KALMAN FILTER

In this section we derive the *discrete time Kalman filter* for the linear system

$$\begin{aligned} x(k + 1) &= Ax(k) + Bu(k) + \bar{B}v(k), \\ y(k) &= Cx(k) + w(k). \end{aligned} \tag{16.25}$$

Here, $A \in \mathbb{R}^{n \times n}$, $B \in \mathbb{R}^{n \times m}$, $C \in \mathbb{R}^{p \times n}$, and $\bar{B} \in \mathbb{R}^{n \times q}$ are known matrices and $(v(k))_{k \in \mathbb{N}} \subset \mathbb{R}^q$ and $(w(k))_{k \in \mathbb{N}} \subset \mathbb{R}^p$ represent unknown disturbances and measurement noise, respectively.

We consider a finite set of measurements $y(0), \ldots, y(k)$, $k \in \mathbb{N}$, and a state observer satisfying the dynamics

$$\begin{aligned} \hat{x}(k + 1) &= A\hat{x}(k) + Bu(k) + \bar{B}\hat{v}(k), \quad \hat{x}(0) = \hat{x}_0 \\ y(k) &= C\hat{x}(k) + \hat{w}(k) \end{aligned} \tag{16.26}$$

for particular disturbance sequences $\hat{v}(\cdot)$, $\hat{w}(\cdot)$, to be determined. The sequences $\hat{v}(k)$, $\hat{w}(k)$ will be defined in such a way that the estimate $\hat{x}(k)$ is optimal with respect to certain assumptions on the unknown disturbances $v(\cdot)$ and $w(\cdot)$, and with respect to the measured output $y(0)$ to $y(k)$.

Assumption 16.6 ([39, Assumption 2.1]). *Let $v : \mathbb{N} \to \mathbb{R}^q$ and $w : \mathbb{N} \to \mathbb{R}^p$ be* *sequences of zero-mean Gaussian white noise such that* $\mathrm{Var}(v(k)) = Q^{-1} \in \mathcal{S}_{>0}^q$ *and* $\mathrm{Var}(w(k)) = R^{-1} \in \mathcal{S}_{>0}^p$ *are positive definite matrices and* $\mathrm{E}\left[v(k)w(j)^T\right] = 0$ *for all* $k, j \in \mathbb{N}_0$. *Additionally, the initial state is assumed to be independent of* $v(k)$ *and* $w(k)$ *in the sense that* $\mathrm{E}\left[x_0 v(k)^T\right] = 0$ *and* $\mathrm{E}\left[x_0 w(k)^T\right] = 0$ *for all* $k \in \mathbb{N}_0$.

Definitions of the *variance* $\mathrm{Var}(\cdot)$ and the *expectation* $\mathrm{E}[\cdot]$ are given in Appendix A.6. In addition to Assumption 16.6, we assume that the triple (A, B, C) is controllable and observable. For simplicity in the derivation, we also assume that A is nonsingular. Note that this can be assumed without loss of generality by considering the input $u = Kx + \tilde{u}$ with new degree of freedom $\tilde{u}$ and $A + BK$ Hurwitz (or nonsingular).

Since the dynamics (16.25) and the observer dynamics (16.26) are linear, the estimated state $\hat{x} = \hat{x}_d + \hat{x}_s$ can be split into a deterministic component and a stochastic component, satisfying the dynamics

$$\hat{x}_d(k+1) = A\hat{x}_d(k) + Bu(k), \qquad \hat{x}_{d,0} = \hat{x}_0$$
$$\hat{y}_d(k) = C\hat{x}_d(k) \tag{16.27}$$

and

$$\hat{x}_s(k+1) = A\hat{x}_s(k) + \bar{B}\hat{v}(k) \qquad \hat{x}_{s,0} = 0$$
$$\hat{y}_s(k) = C\hat{x}_s(k) + \hat{w}(k), \tag{16.28}$$

respectively.

For a given initial condition $\hat{x}_d(0) = \hat{x}_{d,0}$ and a given control sequence $(u(k))_{k\in\mathbb{N}}$ the deterministic state $\hat{x}_d(k)$ can be explicitly calculated as

$$\hat{x}_d(k) = A^k \hat{x}_{d,0} + \sum_{i=1}^{k} A^{k-i} Bu(i-1) \tag{16.29}$$

for all $k \in \mathbb{N}$.

The stochastic part of the state estimate $\hat{x}_s$, however, cannot be computed, since it depends on the selection of the disturbance sequences $\hat{v}(\cdot)$ and $\hat{w}(\cdot)$. To calculate $\hat{x}_s$, we look for sequences $\hat{v}(\cdot)$ and $\hat{w}(\cdot)$ that describe the mismatch,

$$\hat{y}_s(k) = y(k) - \hat{y}_d(k), \tag{16.30}$$

between the measured output $y(k)$ and the deterministic output $\hat{y}_d(k)$ in an optimal way.

16.3.1 Least Squares and Minimum Variance Solution

From (16.28) it holds that

$$\hat{x}_s(k) = A^{k-j} \hat{x}_s(j) + \sum_{i=j+1}^{k} A^{k-i} \bar{B}\hat{v}(i-1) \tag{16.31}$$

for all $0 \leq j \leq k \in \mathbb{N}$. Since we have assumed that A is invertible we can equivalently write equation (16.31) in the form

$$\hat{x}_s(j) = A^{j-k}\hat{x}_s(k) - \sum_{i=j+1}^{k} A^{j-i}\bar{B}\hat{v}(i-1).$$

In (16.28) this leads to the relation

$$\hat{y}_s(j) = C\hat{x}_s(j) + \hat{w}(j) = CA^{j-k}\hat{x}_s(k) + \hat{w}(j) - \sum_{i=j+1}^{k} CA^{j-i}\bar{B}\hat{v}(i-1)$$

for all $j \in \{0, \ldots, k\}$. In vector form, for $j \in \{0, \ldots, k\}$, this condition reads

$$\Lambda_k^j = \Phi_k^j \hat{x}_s^j(k) + \Gamma_k^j, \tag{16.32}$$

where

$$\Lambda_k^j = \begin{bmatrix} \hat{y}_s(0) \\ \hat{y}_s(1) \\ \vdots \\ \hat{y}_s(j) \end{bmatrix}, \qquad \Phi_k^j = \begin{bmatrix} CA^{-k} \\ CA^{1-k} \\ \vdots \\ CA^{j-k} \end{bmatrix}, \tag{16.33}$$

and

$$\Gamma_k^j = \begin{bmatrix} \hat{w}(0) - \sum_{i=1}^{k} CA^{1-i}\bar{B}\hat{v}(i-1) \\ \hat{w}(1) - \sum_{i=2}^{k} CA^{2-i}\bar{B}\hat{v}(i-1) \\ \vdots \\ \hat{w}(j) - \sum_{i=j+1}^{k} CA^{k-i}\bar{B}\hat{v}(i-1) \end{bmatrix}. \tag{16.34}$$

In (16.32) the index $j \in \{0, \ldots, k\}$ in $\hat{x}_s^j(k)$ indicates that the outputs $y(0)$ to $y(j)$ are taken into account to calculate the stochastic part of the state estimate. The matrices in (16.33) and (16.34) satisfy $\Lambda_k^j, \Gamma_k^j \in \mathbb{R}^{p(j+1)}$ and $\Phi_k^j \in \mathbb{R}^{p(j+1)\times n}$. For $j = k$, i.e., all the information from $y(0)$ to $y(k)$ is available at time k, we simply drop the index $j = k$, write $\Lambda_k = \Lambda_k^k$, $\Gamma_k = \Gamma_k^k$, and $\Phi_k = \Phi_k^k$, and consider the equation

$$\Lambda_k = \Phi_k \hat{x}_s(k) + \Gamma_k \tag{16.35}$$

with $\hat{x}_s(k) = \hat{x}_s^k(k)$ for simplicity of notation.

The vector Λ_k^j contains the mismatch between the actual output $y(\cdot)$ and the deterministic output $\hat{y}_d(\cdot)$ of the observer (16.28). Under Assumption 16.6 we are looking for $\hat{x}_s^j(k)$ which fits the data (16.32) in an optimal way, and thus provides an estimate of $x(k)$ through $\hat{x}(k) = \hat{x}_d(k) + \hat{x}_s^j(k)$. In (16.35) the vector Λ_k^j and the matrix Φ_k^j are known. Since Γ_k^j is not known, but $(v(k))_{k\in\mathbb{N}}$ and $(w(k))_{k\in\mathbb{N}}$ are assumed to be sequences of Gaussian white noise with zero mean, we are looking for $\hat{x}_s^j(k)$ that minimizes the expected value

$$
\begin{aligned}
F(\hat{x}_s^j(k), W_k^j) &= \mathrm{E}\left[|\Lambda_k^j - \Phi_k^j \hat{x}_s^j(k)|_{W_k^j}\right] \\
&= \mathrm{E}\left[(\Lambda_k^j - \Phi_k^j \hat{x}_s^j(k))^T W_k^j (\Lambda_k^j - \Phi_k^j \hat{x}_s^j(k))\right]
\end{aligned}
$$

for a positive definite matrix $W_k^j \in \mathcal{S}_{>0}^{p(j+1)}$. Under the assumption that the matrix $(\Phi_k^j)^T W_k^j \Phi_k^j$ is nonsingular, this expression can be rewritten as

$$
\begin{aligned}
F(\hat{x}_s^j(k), W_k^j) &= \mathrm{E}\left[(\Lambda_k^j - \Phi_k^j \hat{x}_s^j(k))^T W_k^j (\Lambda_k^j - \Phi_k^j \hat{x}_s^j(k))\right] \\
&= \mathrm{E}\left[[(\Phi_k^j)^T W_k \Phi_k^j)\hat{x}_s^j(k) - (\Phi_k^j)^T W_k^j \Lambda_k^j]^T ((\Phi_k^j)^T W_k^j \Phi_k^j)^{-1} \right. \\
&\quad \left. \cdot [(\Phi_k^j)^T W_k^j \Phi_k^j)\hat{x}_s^j(k) - (\Phi_k^j)^T W_k^j \Lambda_k^j]\right] \\
&\quad + \mathrm{E}\left[(\Lambda_k^j)^T (I - W_k^j \Phi_k^j ((\Phi_k^j)^T W_k^j \Phi_k^j)^{-1} (\Phi_k^j)^T) W_k^j \Lambda_k^j\right].
\end{aligned}
$$

Since the second term on the right-hand side is independent of $\hat{x}_s^j(k)$ and the first term is non-negative, the minimum is attained for

$$
\hat{x}_s^j(k) = \hat{x}_s^j(k; W_k^j) = ((\Phi_k^j)^T W_k^j \Phi_k^j)^{-1} (\Phi_k^j)^T W_k^j \Lambda_k^j. \tag{16.36}
$$

If the matrix $(\Phi_k^j)^T W_k^j \Phi_k^j$ is singular, $\hat{x}_s^j(k)$ can still be computed in a similar fashion by using the Moore-Penrose inverse instead of the inverse matrix, for example, but the optimal estimate $\hat{x}_s^j(k)$ will in general not be unique.

From (16.36), for a fixed positive definite matrix W_k^j the corresponding optimal estimate $\hat{x}_s^j(k)$ (where we omit the dependency on W_k^j for simplicity of notation) is defined. However, the question remains how to define W_k^j optimally with respect to Assumption 16.6. To find an optimal W_k^j we minimize the variance

$$
\mathrm{Var}(x(k) - \hat{x}^j(k)) = \mathrm{Var}(x(k) - \hat{x}_d(k) - \hat{x}_s^j(k)). \tag{16.37}
$$

It is clear that minimizing $F(\hat{x}_s^j(k), W_k^j)$ with respect to W_k^j leads to

$$
\inf_{W_k^j > 0} F(\hat{x}_s^j(k), W_k^j) = 0,
$$

but the minimum is not attained since $W_k^j = 0$ is not positive definite. This explains why a different measure, for example in the form of (16.37), is necessary.

With (16.36) and (16.35) it holds that

$$
\begin{aligned}
&x(k) - \hat{x}_d(k) - \hat{x}_s^j(k) \\
&= [(\Phi_k^j)^T W_k^j \Phi_k^j]^{-1} [(\Phi_k^j)^T W_k^j \Phi_k^j](x(k) - \hat{x}_d(k)) - [(\Phi_k^j)^T W_k^j \Phi_k^j]^{-1} (\Phi_k^j)^T W_k^j \Lambda_k^j \\
&= [(\Phi_k^j)^T W_k^j \Phi_k^j]^{-1} (\Phi_k^j)^T W_k^j [\Phi_k^j(x(k) - \hat{x}_d(k)) - \Lambda_k^j] \\
&= -[(\Phi_k^j)^T W_k^j \Phi_k^j]^{-1} (\Phi_k^j)^T W_k^j \Gamma_k^j.
\end{aligned}
$$

Moreover, the variance satisfies

$$
\begin{aligned}
\mathrm{Var}(x(k) - \hat{x}^j(k)) &= \mathrm{Var}(x(k) - \hat{x}_d(k) - \hat{x}_s^j(k)) \\
&= \mathrm{E}\left[[(\Phi_k^j)^T W_k^j \Phi_k^j]^{-1}(\Phi_k^j)^T W_k^j \Gamma_k^j (\Gamma_k^j)^T W_k^j \Phi_k^j [(\Phi_k^j)^T W_k^j \Phi_k^j]^{-1}\right] \\
&= [(\Phi_k^j)^T W_k^j \Phi_k^j]^{-1}(\Phi_k^j)^T W_k^j \, \mathrm{E}\left[\Gamma_k^j (\Gamma_k^j)^T\right] W_k^j \Phi_k^j [(\Phi_k^j)^T W_k^j \Phi_k^j]^{-1} \\
&= [(\Phi_k^j)^T W_k^j \Phi_k^j]^{-1}(\Phi_k^j)^T W_k^j \Xi_k^j W_k^j \Phi_k^j [(\Phi_k^j)^T W_k^j (\Phi_k^j)]^{-1},
\end{aligned}
$$
(16.38)

where the third equation follows from the definition of the expected value (see Appendix A.6) $\mathrm{E}[\Gamma_k^j (\Gamma_k^j)^T] = \Xi_k^j$ and $\Xi_k^j \in \mathcal{S}_{>0}^{p(j+1)}$ denotes a positive definite matrix which can be calculated using Assumption 16.6. According to Lemma 3.2 a positive definite matrix can be split $\Xi_k^j = H_k^j (H_k^j)^T$, similarly to the square root of a positive real number. Under the assumption that $(\Phi_k^j)^T (\Xi_k^j)^{-1}(\Phi_k^j)$ is nonsingular, we can define the matrices

$$
U = (H_k^j)^T W_k^j \Phi_k^j [(\Phi_k^j)^T W_k^j \Phi_k^j]^{-1}, \qquad V = (H_k^j)^{-1}\Phi_k^j,
$$

and with Lemma A.5 it holds that

$$
\begin{aligned}
\mathrm{Var}(x(k) - \hat{x}_d(k) - \hat{x}_s^j(k; W_k^j)) &= U^T U \geq (V^T U)^T (V^T V)^{-1}(V^T U) \\
&= \left[(\Phi_k^j)^T (H_k^j)^{-T}(H_k^j)^T W_k^j \Phi_k^j [(\Phi_k^j)^T W_k^j \Phi_k^j]^{-1}\right]^T \left[(\Phi_k^j)^T (H_k^j)^{-T}(H_k^j)^{-1}\Phi_k^j\right]^{-1} \\
&\quad \cdot \left[(\Phi_k^j)^T (H_k^j)^{-T}(H_k^j)^T W_k^j \Phi_k^j [(\Phi_k^j)^T W_k^j \Phi_k^j]^{-1}\right] \\
&= \left[[(\Phi_k^j)^T W_k^j \Phi_k^j]^{-1}(\Phi_k^j)^T W_k^j \Phi_k^j\right]\left[(\Phi_k^j)^T (\Xi_k^j)^{-T}\Phi_k^j\right]^{-1}\left[(\Phi_k^j)^T W_k^j \Phi_k^j [(\Phi_k^j)^T W_k^j \Phi_k^j]^{-1}\right] \\
&= \left((\Phi_k^j)^T (\Xi_k^j)^{-1}\Phi_k^j\right)^{-1}.
\end{aligned}
$$

Since additionally

$$
\begin{aligned}
\mathrm{Var}(x(k) - \hat{x}_d(k) - \hat{x}_s^j(k; (\Xi_k^j)^{-1})) &= [(\Phi_k^j)^T (\Xi_k^j)^{-1}\Phi_k^j]^{-1}(\Phi_k^j)^T (\Xi_k^j)^{-1}\Xi_k^j (\Xi_k^j)^{-1}\Phi_k^j [(\Phi_k^j)^T (\Xi_k^j)^{-1}\Phi_k^j]^{-1} \\
&= ((\Phi_k^j)^T (\Xi_k^j)^{-1}\Phi_k^j)^{-1},
\end{aligned}
$$

the minimum of (16.38) is attained for

$$
W_k^j = (\Xi_k^j)^{-1} = \left(\mathrm{E}\left[\Gamma_k^j (\Gamma_k^j)^T\right]\right)^{-1}.
$$
(16.39)

To sum up, the optimal least squares estimate of $x(k) - \hat{x}_d(k)$ taking the measurements $y(0)$ to $y(j)$ into account, which additionally minimizes the variance (16.37), is defined as

$$
\hat{x}_s^j(k) = [(\Phi_k^j)^T (\Xi_k^j)^{-1}\Phi_k^j]^{-1}(\Phi_k^j)^T (\Xi_k^j)^{-1}\Lambda_k^j,
$$

and $\Xi_k^j = \mathrm{E}\left[\Gamma_k^j (\Gamma_k^j)^T\right]$. Thus, with the observer dynamics (16.26), the decomposition (16.27) and (16.28), and the explicit calculation of $\hat{x}_d(k)$ in (16.29), the optimal estimate of the state $x(k)$, $k \in \mathbb{N}$, based on the measured output $y(0), \ldots, y(j)$,

$j \leq k$, is given by

$$\hat{x}^j(k) = \hat{x}_d(k) + \hat{x}_s^j(k) \tag{16.40}$$

$$= A^k \hat{x}_{d,0} + \sum_{i=1}^{k} A^{k-i} B u(i-1) + ((\Phi_k^j)^T (\Xi_k^j)^{-1} \Phi_k^j)^{-1} (\Phi_k^j)^T (\Xi_k^j)^{-1} \Lambda_k^j.$$

For $j = k$, the index is omitted: $\hat{x}(k) = \hat{x}^k(k)$.

16.3.2   A Prediction-Correction Formulation

The drawback of the estimate (16.40) derived in the previous subsection is that the dimension of the linear equation (16.35) grows linearly with $k \in \mathbb{N}$ if all measurements are taken into account, i.e., $j = k$, and thus the numerical complexity to calculate $\hat{x}(k)$ grows with k. However, it is possible to rewrite the problem in such a way that the complexity of the calculation of $\hat{x}(k)$ is independent of k. In particular, we derive a recursive formula to iteratively compute $\hat{x}(k)$ based on

$$\hat{\chi}(k) = A\hat{x}(k-1) + Bu(k-1), \tag{16.41a}$$
$$\hat{x}(k) = \hat{\chi}(k) + G_k(y(k) - C\hat{\chi}(k)), \tag{16.41b}$$

consisting of a *prediction step* (16.41a) with $\hat{\chi}(k) = \hat{x}^{k-1}(k)$ when the measurement $y(k)$ is not available yet and a *correction step* (16.41b), $\hat{x}(k) = \hat{x}^k(k)$ taking $y(k)$ into account. The matrices $G_k \in \mathbb{R}^{n \times p}$, $k \in \mathbb{N}$, are called *Kalman gain matrices* and will be defined in the following. The relation $\hat{\chi}(k) = \hat{x}^{k-1}(k)$ intuitively follows from Assumption 16.6. In fact, since $(v(k))_{k \in \mathbb{N}}$ has zero mean by assumption and $y(k)$ is not available, the optimal estimate of $v(k)$ is given by $\hat{v}(k) = 0$.

Instead of (16.41) we focus on the $\hat{x}_s$-dynamics in (16.28) in the derivation of G_k and in the derivation of the prediction and correction updates. The two equations in (16.41) can be combined as

$$\hat{x}(k) = A\hat{x}(k-1) + Bu(k-1) + G_k \left(y(k) - C \left(A\hat{x}(k-1) + Bu(k-1) \right) \right),$$

which, with the dynamics $\hat{x}_d(k)$ from (16.27), yields

$$\hat{x}(k) - \hat{x}_d(k) \tag{16.42}$$
$$= A\hat{x}(k-1) - A\hat{x}_d(k-1) + G_k \left(y(k) - C \left(A\hat{x}(k-1) + Bu(k-1) \right) \right).$$

With the definition of $\hat{y}_s(k) = y(k) - \hat{y}_d(k)$ in (16.30), it holds that

$$\hat{y}_s(k) = y(k) - \hat{y}_d(k) = y(k) - C(A\hat{x}_d(k-1) + Bu(k-1)).$$

Using this expression in (16.42), together with $\hat{x}_s = \hat{x} - \hat{x}_d$, leads to the equations

$$\hat{x}_s(k) = A\hat{x}(k-1) - A\hat{x}_d(k-1) + G_k \left(\hat{y}_s(k) + CA\hat{x}(k-1) - CA\hat{x}_d(k-1) \right)$$
$$= A\hat{x}_s(k-1) + G_k \left(\hat{y}_s(k) - CA\hat{x}_s(k-1) \right).$$

Finally, the optimal estimate of the state $\hat{x}(k)$ is described through (16.41) if we can show that the stochastic part of the estimate $\hat{x}_s(k)$ satisfies

$$\hat{x}_s^{k-1}(k) = A\hat{x}_s(k-1), \tag{16.43a}$$

$$\hat{x}_s(k) = \hat{x}_s^{k-1}(k) + G_k(\hat{y}_s(k) - C\hat{x}_s^{k-1}(k)). \tag{16.43b}$$

Let $\hat{x}_s(k)$ be the optimal least-squares estimate of $x(k) - \hat{x}_d(k)$ with minimum variance by choosing the weight matrix

$$W_k = (\Xi_k)^{-1} = (\text{Var}(\Gamma_k))^{-1} = \left(\text{E}\left[\Gamma_k \Gamma_k^T\right]\right)^{-1}$$

derived in (16.39). Due to the linearity of the expected value, the definition of Γ_k, and Assumption 16.6, it holds that

$$(W_k)^{-1} = \begin{bmatrix} (W_k^{k-1})^{-1} & 0 \\ 0 & R \end{bmatrix},$$

where

$$(W_k^{k-1})^{-1} = \begin{bmatrix} R^{-1} & 0 & \cdots & 0 \\ 0 & \ddots & \ddots & \vdots \\ \vdots & \ddots & \ddots & 0 \\ 0 & \cdots & 0 & R^{-1} \end{bmatrix} + \text{Var}\left(\begin{bmatrix} C\sum_{i=1}^{k} A^{1-i}\bar{B}v(i-1) \\ \vdots \\ C\sum_{i=k-1}^{k} A^{k-i}\bar{B}v(i-1) \\ CA^{-1}\bar{B}v(k-1) \end{bmatrix}\right) \tag{16.44}$$

and W_k^{k-1} and W_k are positive definite. Here W_k^{k-1} is positive definite since the first term on the right-hand side of (16.44) is positive definite according to Assumption 16.6 and the second term is positive semidefinite due to the definition of the variance.

To proceed, we assume that $((\Phi_k^{k-1})^T W_k^{k-1} \Phi_k^{k-1})$ and $(\Phi_k^T W_k \Phi_k)$ are nonsingular. This allows us to write

$$\hat{x}_s(k) = (\Phi_k^T W_k \Phi_k)^{-1} \Phi_k^T W_k \Lambda_k \tag{16.45a}$$

$$\hat{x}_s^{k-1}(k) = ((\Phi_k^{k-1})^T W_k^{k-1} \Phi_k^{k-1})^{-1}(\Phi_k^{k-1})^T W_k^{k-1} \Lambda_k^{k-1}, \tag{16.45b}$$

which are the optimal estimates $\hat{x}_s(k)$ and $\hat{x}_s^{k-1}(k)$ from the previous section obtained from (16.35) and (16.32) for $j = k$ and $j = k - 1$, respectively. To proceed note that

$$\Phi_k^T W_k \Phi_k = \begin{bmatrix} (\Phi_k^{k-1})^T & C^T \end{bmatrix} \begin{bmatrix} W_k^{k-1} & 0 \\ 0 & R \end{bmatrix} \begin{bmatrix} \Phi_k^{k-1} \\ C \end{bmatrix}$$

$$= (\Phi_k^{k-1})^T W_k^{k-1} \Phi_k^{k-1} + C^T RC \tag{16.46}$$

and

$$\Phi_k^T W_k \Lambda_k = (\Phi_k^{k-1})^T W_k^{k-1} \Lambda_k^{k-1} + C^T R\hat{y}_s(k) \tag{16.47}$$

hold. With (16.45) we obtain the equation

$$
\begin{aligned}
\left((\Phi_k^{k-1})^T W_k^{k-1}\Phi_k^{k-1} + C^T RC\right)\hat{x}_s^{k-1}(k) \\
= (\Phi_k^{k-1})^T W_k^{k-1}\Lambda_k^{k-1} + C^T RC\hat{x}_s^{k-1}(k)
\end{aligned}
\tag{16.48}
$$

and with (16.45), (16.46), and (16.47) the equations

$$
\begin{aligned}
\left((\Phi_k^{k-1})^T W_k^{k-1}\Phi_k^{k-1} + C^T RC\right)\hat{x}_s(k) \\
= \left(\Phi_k^T W_k \Phi_k\right)\hat{x}_s(k) = (\Phi_k^{k-1})^T W_k^{k-1}\Lambda_k^{k-1} + C^T R\hat{y}_s(k)
\end{aligned}
\tag{16.49}
$$

are satisfied. Subtracting (16.48) from (16.49) leads to

$$
\begin{aligned}
\left((\Phi_k^{k-1})^T W_k^{k-1}\Phi_k^{k-1} + C^T RC\right)(\hat{x}_s(k) - \hat{x}_s^{k-1}(k)) \\
= C^T R(\hat{y}_s(k) - C\hat{x}_s^{k-1}(k)),
\end{aligned}
$$

which relates the two estimates $\hat{x}_s(k)$ and $\hat{x}_s^{k-1}(k)$ in a single equation.

As a next step we focus on the Kalman gain matrix G_k and define

$$
\begin{aligned}
G_k &= \left((\Phi_k^{k-1})^T W_k^{k-1}\Phi_k^{k-1} + C^T RC\right)^{-1} C^T R \\
&= \left(\Phi_k^T W_k \Phi_k\right)^{-1} C^T R
\end{aligned}
\tag{16.50}
$$

and therefore

$$
\hat{x}_s(k) = \hat{x}_s^{k-1}(k) + G_k(\hat{y}_s(k) - C\hat{x}_s^{k-1}(k)),
$$

i.e., equation (16.43b) follows.

Having established the relation between $\hat{x}_s(k)$ and $\hat{x}_s^{k-1}(k)$, we continue with $\hat{x}_s^{k-1}(k)$ and $\hat{x}_s(k-1)$ and we show that (16.43a) is satisfied. To this end, note that with $\Phi_k^{k-1}A = \Phi_{k-1}$ the equations

$$
\begin{aligned}
(\Phi_k^{k-1})^T W_k^{k-1}\Phi_{k-1} &= [(\Phi_k^{k-1})^T W_k^{k-1}\Phi_k^{k-1}]A \\
&= [(\Phi_k^{k-1})^T W_k^{k-1}\Phi_k^{k-1}]A[\Phi_{k-1}^T W_{k-1}\Phi_{k-1}]^{-1}[\Phi_{k-1}^T W_{k-1}\Phi_{k-1}]
\end{aligned}
$$

are trivially satisfied. Thus, in particular,

$$
(\Phi_k^{k-1})^T W_k^{k-1} = (\Phi_k^{k-1})^T W_k^{k-1}\Phi_k^{k-1}A[\Phi_{k-1}^T W_{k-1}\Phi_{k-1}]^{-1}\Phi_{k-1}^T W_{k-1}
\tag{16.51}
$$

holds. This expression allows us to rewrite (16.45b) as

$$
\begin{aligned}
\hat{x}_s^{k-1}(k) &= [(\Phi_k^{k-1})^T W_k^{k-1}\Phi_k^{k-1}]^{-1}(\Phi_k^{k-1})^T W_k^{k-1}\Lambda_k^{k-1} \\
&= A[\Phi_{k-1}^T W_{k-1}\Phi_{k-1}]^{-1}\Phi_{k-1}^T W_{k-1}\Lambda_k^{k-1} \\
&= A\hat{x}_s(k-1),
\end{aligned}
$$

where the second and third equations follow from (16.51) and (16.45a), respectively, and (16.43a) is obtained.

To be able to iterate the dynamics (16.43), an iterative formula for G_k, $k \in \mathbb{N}$ is still missing. We rewrite the definition of G_k in (16.50) as

$$
G_k = P_k C^T R,
\tag{16.52}
$$

where

$$P_k = (\Phi_k^T W_k \Phi_k)^{-1}. \tag{16.53}$$

Additionally, consistent with the notation used in this chapter, we define

$$P_k^{k-1} = [(\Phi_k^{k-1})^T W_k^{k-1} \Phi_k^{k-1}]^{-1}$$

and with (16.50) it holds that

$$\begin{aligned}
[(P_k^{k-1})^{-1} + C^T RC]^{-1} &= [(\Phi_k^{k-1})^T W_k^{k-1} \Phi_k^{k-1} + C^T RC]^{-1} \\
&= (\Phi_k^T W_k \Phi_k)^{-1} = P_k,
\end{aligned} \tag{16.54}$$

which is equivalent to the expression $C^T RC = (P_k)^{-1} - (P_k^{k-1})^{-1}$. With these definitions, we can write $P_k = (P_k^{k-1} - P_k^{k-1}) + P_k$ and it holds that

$$\begin{aligned}
P_k &= P_k^{k-1} - P_k[(P_k)^{-1} - (P_k^{k-1})^{-1}]P_k^{k-1} = P_k^{k-1} - P_k[C^T RC]P_k^{k-1} \\
&= P_k^{k-1} - G_k C P_k^{k-1} = (I - G_k C)P_k^{k-1}.
\end{aligned} \tag{16.55}$$

However, the last expression cannot be used to calculate P_k yet, since G_k in (16.52) still explicitly depends on P_k. With Lemma A.6 the left-hand side of (16.54) can be rewritten and it holds that

$$P_k^{k-1} - P_k^{k-1} C^T [R^{-1} + C P_k^{k-1} C^T]^{-1} C P_k^{k-1} = P_k.$$

Rearranging terms we have

$$P_k^{k-1} - P_k = [I - P_k(P_k^{k-1})^{-1}]P_k^{k-1} = P_k^{k-1} C^T (R^{-1} + C P_k^{k-1} C^T)^{-1} C P_k^{k-1}$$

and with the definitions and calculations so far we can further manipulate the expressions to obtain

$$\begin{aligned}
P_k^{k-1} C^T &[R^{-1} + C P_k^{k-1} C^T]^{-1} C \\
&= I - P_k(P_k^{k-1})^{-1} = P_k[P_k^{-1} - (P_k^{k-1})^{-1}] = P_k[C^T RC] = G_k C.
\end{aligned}$$

In particular, G_k satisfies the equation

$$G_k = P_k^{k-1} C^T [R^{-1} + C P_k^{k-1} C^T]^{-1} C,$$

which allows a calculation of P_k based on P_k^{k-1} in (16.55).

It is left to show how P_k^{k-1} can be calculated from P_{k-1} and we focus again on W_k^{k-1} and W_k to obtain a corresponding expression. The equation

$$\Gamma_k^{k-1} = \Gamma_{k-1} - \Phi_k^{k-1} \bar{B} v(k-1)$$

follows from the definition of Γ_k^j and Φ_k^j in (16.33) and (16.34), respectively. With this relation it holds that

$$
\begin{aligned}
(W_k^{k-1})^{-1} &= \mathrm{Var}\left(\Gamma_k^{k-1}\right) \\
&= \mathrm{E}\left[[\Gamma_{k-1} - \Phi_k^{k-1}\bar{B}v(k-1)][\Gamma_{k-1} - \Phi_k^{k-1}\bar{B}v(k-1)]^T\right] \\
&= \mathrm{Var}\left(\Gamma_{k-1}\right) + \mathrm{E}\left[[\Phi_k^{k-1}\bar{B}v(k-1)][\Phi_k^{k-1}\bar{B}v(k-1)]^T\right] \\
&= (W_{k-1})^{-1} + \Phi_k^{k-1}\bar{B}\,\mathrm{E}\left[v(k-1)v(k-1)^T\right]\bar{B}^T(\Phi_k^{k-1})^T \\
&= (W_{k-1})^{-1} + \Phi_{k-1}A^{-1}\bar{B}Q^{-1}\bar{B}^T A^{-T}\Phi_{k-1}^T.
\end{aligned}
$$

Here, the equalities follow from the properties of the variance and the expected value as well as Assumption 16.6. Taking the inverse on both sides

$$
W_k^{k-1} = \left((W_{k-1})^{-1} - \Phi_{k-1}A^{-1}\bar{B}(-Q^{-1})\bar{B}^T A^{-T}\Phi_{k-1}^T\right)^{-1},
$$

and applying Lemma A.6 (and in particular equation (A.4)), it holds that

$$
\begin{aligned}
W_k^{k-1} = W_{k-1} - W_{k-1}\Phi_{k-1}A^{-1}\bar{B}\left(Q + \bar{B}^T A^{-T}\Phi_{k-1}^T W_{k-1}\Phi_{k-1}A^{-1}\bar{B}\right)^{-1} \\
\cdot \bar{B}^T A^{-T}\Phi_{k-1}^T W_{k-1}. \quad (16.56)
\end{aligned}
$$

With $\Phi_k^{k-1} = \Phi_{k-1}A^{-1}$ the last expression can be further rewritten as

$$
\begin{aligned}
(\Phi_k^{k-1})^T W_k^{k-1} = A^{-T}\Big(I - \Phi_{k-1}^T W_{k-1}\Phi_{k-1}A\bar{B} \\
\cdot \left(Q + \bar{B}^T A^{-T}\Phi_{k-1}^T W_{k-1}\Phi_{k-1}A^{-1}\bar{B}\right)^{-1}\bar{B}A^T\Big)\Phi_{k-1}^T W_{k-1},
\end{aligned}
$$

which provides a second representation of the term in (16.51). We multiply Equation (16.56) from the left by $(\Phi_k^{k-1})^T$ and from the right by Φ_k^{k-1}:

$$
\begin{aligned}
(\Phi_k^{k-1})^T W_k^{k-1}\Phi_k^{k-1} = (\Phi_k^{k-1})^T W_{k-1}\Phi_k^{k-1} - (\Phi_k^{k-1})^T W_{k-1}\Phi_{k-1}A^{-1}\bar{B} \\
\cdot [Q + \bar{B}^T A^{-T}\Phi_{k-1}^T W_{k-1}\Phi_{k-1}A^{-1}\bar{B}]^{-1}\cdot \bar{B}^T A^{-T}\Phi_{k-1}^T W_{k-1}\Phi_k^{k-1}.
\end{aligned}
$$

With the definition of P_k in (16.53) and $\Phi_k^{k-1} = \Phi_{k-1}A^{-1}$ this expression simplifies to

$$
\begin{aligned}
(P_k^{k-1})^{-1} = A^{-T}(P_{k-1})^{-1}A^{-1} \\
- A^{-T}(P_{k-1})^{-1}A^{-1}\bar{B}[Q + \bar{B}^T A^{-T}(P_{k-1})^{-1}A^{-1}\bar{B}]^{-1}\bar{B}^T A^{-T}(P_{k-1})^{-1}A^{-1}
\end{aligned}
$$

and according to Lemma A.6 applied to the right-hand side yields

$$
(P_k^{k-1})^{-1} = \left(AP_{k-1}A^T + \bar{B}Q^{-1}\bar{B}^T\right)^{-1}.
$$

Thus, P_k^{k-1} can be computed iteratively based on the knowledge of P_{k-1} through

$$
P_k^{k-1} = AP_{k-1}A^T + \bar{B}Q^{-1}\bar{B}^T \quad (16.57)
$$

for all $k \in \mathbb{N}$. Finally, under Assumption 16.6 and using the initial condition $\hat{x}_0 = \mathrm{E}[x_0]$,

$$
P_0 = \mathrm{E}\left[(x_0 - E[x_0])(x_0 - E[x_0])^T\right] = \mathrm{Var}\left(x_0\right),
$$

and it can be shown that the covariance matrices satisfy

$$P_k^{k-1} = \mathrm{E}\left[(x(k) - \hat{\chi}(k))(x(k) - \hat{\chi}(k))^T\right] = \mathrm{Var}\,(x(k) - \hat{\chi}(k))$$
$$P_k = \mathrm{E}\left[(x(k) - \hat{x}(k))(x(k) - \hat{x}(k))^T\right] = \mathrm{Var}\,(x(k) - \hat{x}(k))$$

(16.58)

for all $k \in \mathbb{N}$. For completeness, the corresponding calculations are given at the end of this section.

Remark 16.7. Note that $\mathrm{E}[x(k) - \hat{\chi}(k)] = 0$ and $\mathrm{E}[x(k) - \hat{x}(k)] = 0$ for all $k \in \mathbb{N}$, since $\hat{\chi}(k)$ and $\hat{x}(k)$ are optimal with respect to the expected value taking the past measurements $y(0), \ldots, y(k-1)$ and $y(0), \ldots, y(k)$, respectively, into account. Thus, $\mathrm{E}[(\cdot)(\cdot)^T]$ is justifiably used in (16.58) instead of $\mathrm{E}[(\cdot - \mathrm{E}[\cdot])(\cdot - \mathrm{E}[\cdot])^T]$.

The Kalman filtering process derived in this section is summarized in Algorithm 6. The estimates of the states $\hat{\chi}$ and $\hat{x}$ calculated in Steps 2 and 4 in the Kalman

---

Algorithm 6: Discrete time Kalman filter

Input: Linear system (16.25), deterministic model (16.27), a sequence of control inputs $(u(k))_{k \in \mathbb{N}}$, positive definite matrices $Q^{-1} = \mathrm{Var}(v(k))$, $R^{-1} = \mathrm{Var}(w(k))$, and initial estimates $\hat{x}(0) = \hat{x}_0$, $P_0 \in \mathcal{S}_{>0}^n$.

Output: Estimates $\hat{\chi}(k)$, $\hat{x}(k)$ of the state $x(k)$ for $k \in \mathbb{N}$.

Algorithm: For $k \in \mathbb{N}$:

1. Update the gain matrix G_k:

$$P_k^{k-1} = AP_{k-1}A^T + \bar{B}Q^{-1}\bar{B}^T,$$
$$G_k = P_k^{k-1}C^T[CP_k^{k-1}C^T + R^{-1}]^{-1},$$
$$P_k = [I - G_kC]P_k^{k-1}.$$

(16.59)

2. Update the optimal estimate (before $y(k)$ is available):

$$\hat{\chi}(k) = A\hat{x}(k-1) + Bu(k-1).$$

3. Measure the output:

$$y(k) = Cx(k) + w(k).$$

4. Update the optimal estimate (after $y(k)$ is available):

$$\hat{x}(k) = \hat{\chi}(k) + G_k(y(k) - C\hat{\chi}(k)),$$

set $k = k + 1$ and go to step 1.

---

filter, Algorithm 6, can be written as discrete time systems

$$\hat{\chi}(k+1) = A(\hat{\chi}(k) + G_k(y(k) - C\hat{\chi}(k))) + Bu(k)$$
$$= (A - AG_kC)\hat{\chi}(k) + AG_ky(k) + Bu(k)$$

(16.60)

and

$$\hat{x}(k+1) = A\hat{x}(k) + Bu(k) + G_{k+1}(y(k+1) - C(A\hat{x}(k) + Bu(k)))$$
$$= (I - G_{k+1}C)(A\hat{x}(k) + Bu(k)) + G_{k+1}y(k+1), \tag{16.61}$$

respectively.

Example 16.8. We continue with the dynamics from Example 16.5 but use the Euler discretization with sampling time $\Delta = 0.05$ discussed in Section 5.2 to obtain a discrete time system. In particular, the discrete time dynamics are described by the matrices

$$A = \begin{bmatrix} 1.000 & 0.050 \\ -0.491 & 0.995 \end{bmatrix}, \qquad B = \begin{bmatrix} 0 \\ 0.05 \end{bmatrix}, \qquad \bar{B} = \begin{bmatrix} 0.05 & 0 \\ 0 & 0.05 \end{bmatrix},$$

and $C = \begin{bmatrix} 1 & 0 \end{bmatrix}$. Note that the matrix A is not a Schur matrix, i.e., stability of the origin is lost through the discretization and the particular sampling time $\Delta = 0.05$. Additionally we consider the matrices

$$R^{-1} = \frac{1}{2} \qquad \text{and} \qquad Q^{-1} = \frac{1}{2} \begin{bmatrix} 1 & 0 \\ 0 & 1 \end{bmatrix}$$

defined based on Δ and based on the disturbances $v(k)$ and $w(k)$.

Figure 16.7 shows the results of the discrete time Kalman filter (Algorithm 6). Here, the difference between the state x and the estimates $\hat{\chi}$ and $\hat{x}$ are visualized.

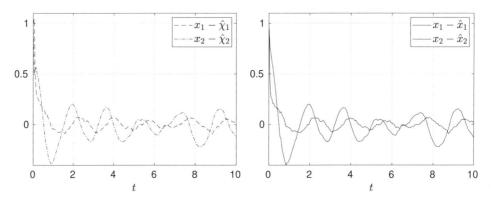

Figure 16.7: Error dynamics $x - \hat{\chi}$ (left) and $x - \hat{x}$ (right) of the discrete time Kalman filter.

The plant is initialized at $x_0 = [1, 1]^T$ and the Kalman filter is initialized at $\hat{\chi}_0 = \hat{x}_0 = [0, 0]^T$. The input is defined as $u(t) = 0$ for all $t \in [0, 10]$. The disturbances $v(k)$ and $w(k)$ are again taken as Gaussian white noise with zero mean and standard deviation 0.1 generated through `randn.m` in MATLAB. The disturbances $v(k)$ and $w(k)$ are visualized in Figure 16.8 on the right. The matrix P is initialized as the identity matrix. The evolution of the entries of P are shown in Figure 16.8 on the left.

The errors $x - \hat{\chi}$ and $x - \hat{x}$ converge to a neighborhood around the origin and then remain in this neighborhood. Due to the disturbances, convergence to the

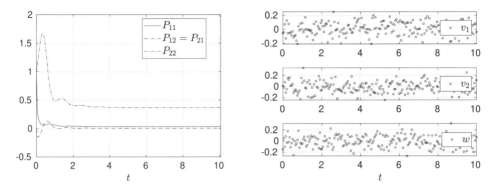

Figure 16.8: On the left, evolution of the entries of P defined through (16.59). On the right, disturbances used in the simulation of Algorithm 6 corresponding to the results in Figure 16.7.

origin is not achieved. The entries of the matrix P converge to a steady state. This is discussed in more detail in the next section.

We conclude this section by deriving several expressions for covariances, showing (16.58) and providing a different interpretation of (16.59). For the first expression in (16.58), it holds that

$$
\begin{aligned}
\text{Var} \, (x(k+1) &- \hat{\chi}(k+1)) \\
&= \text{E} \left[(A(x(k) - \hat{x}(k)) + \bar{B}v(k))(A(x(k) - \hat{x}(k)) + \bar{B}v(k))^T \right] \\
&= A \, \text{E} \left[(x(k) - \hat{x}(k))(x(k) - \hat{x}(k))^T \right] A^T + \bar{B} \, \text{E} \left[v(k)v(k)^T \right] \bar{B}^T \\
&= AP_k A^T + \bar{B}Q^{-1}\bar{B}^T,
\end{aligned}
$$

under the assumption that the second equation in (16.58) is satisfied. Here, the cross terms $\text{E} \left[(x(k) - \hat{x}(k))v(k)^T \right]$ cancel out due to Assumption 16.6.

The second expression in (16.58) expands to

$$
\text{Var} \, (x(k) - \hat{x}(k)) = \text{E} \left[(x - \hat{\chi} - G_k(Cx + w - C\hat{\chi}))(x - \hat{\chi} - G_k(Cx + w - C\hat{\chi}))^T \right],
$$

where we have omitted the time index k in the right-hand side for brevity. Under the assumption that the first expression in (16.58) is satisfied and by again using the fact that cross terms cancel out, the expression can be further rewritten as

$$
\begin{aligned}
\text{Var} \, (x(k) - \hat{x}(k)) &= \text{E} \left[((I - G_k C)(x - \hat{\chi}) - G_k w)((I - G_k C)(x - \hat{\chi}) - G_k w)^T \right] \\
&= (I - G_k C) \, \text{E} \left[(x - \hat{\chi})(x - \hat{\chi})^T \right] (I - G_k C)^T + G_k \, \text{E} \left[ww^T \right] G_k^T \\
&= (I - G_k C)P_k^{k-1}(I - G_k C)^T + G_k R^{-1} G_k^T.
\end{aligned}
$$

To complete the derivation, we expand the right-hand side, use the definition of G_k

in (16.59), and reorder the terms:

$$\text{Var}\,(x(k) - \hat{x}(k))$$
$$= (I - G_k C)P_k^{k-1} - P_k^{k-1}C^T G_k^T + G_k C P_k^{k-1}C^T G_k^T + G_k R^{-1}G_k^T$$
$$= (I - G_k C)P_k^{k-1} - P_k^{k-1}C^T G_k^T + G_k(C P_k^{k-1}C^T + R^{-1})G_k^T$$
$$= (I - G_k C)P_k^{k-1} - P_k^{k-1}C^T G_k^T$$
$$\qquad\qquad + P_k^{k-1}C^T(C P_k^{k-1}C^T + R^{-1})^{-1}(C P_k^{k-1}C^T + R^{-1})G_k^T$$
$$= (I - G_k C)P_k^{k-1}.$$

Additionally, we can derive an expression for the covariance of the output

$$\text{Var}(y(k) - \hat{y}(k)) = \text{E}[(y(k) - \hat{y}(k))(y(k) - \hat{y}(k))^T]$$
$$= \text{E}[(Cx(k) + w(k) - C\hat{\chi}(k))(Cx(k) + w(k) - C\hat{\chi}(k))^T]$$
$$= C\,\text{E}[(x(k) - \hat{\chi}(k))(x(k) - \hat{\chi}(k))^T]C^T + \text{E}[w(k)w(k)^T]$$
$$= C\,\text{Var}(x)C^T + R^{-1} = C P_k^{k-1}C^T + R^{-1}$$

and the following expected value

$$\text{E}[(x(k) - \hat{\chi}(k))(y(k) - \hat{y}(k))^T] \qquad\qquad\qquad (16.62)$$
$$= \text{E}[(x(k) - \hat{\chi}(k))(Cx(k) + w(k) - C\hat{\chi}(k))^T]$$
$$= \text{E}[(x(k) - \hat{\chi}(k))(x(k) - \hat{\chi}(k))^T]C^T + \text{E}[(x(k) - \hat{\chi}(k))w(k)^T]$$
$$= \text{Var}(x(k) - \hat{\chi}(k))C^T.$$

These calculations give an alternative representation of (16.59) in terms of

$$P_k^{k-1} = \text{Var}(x(k) - \hat{\chi}(k))$$
$$G_k = \text{E}[(x(k) - \hat{\chi}(k))(y(k) - \hat{y}(k))^T]\,(\text{Var}(y(k) - \hat{y}(k)))^{-1}$$
$$P_k = \text{Var}\,(x(k) - \hat{x}(k)) = P_k^{k-1} - G_k C P_k^{k-1} \qquad\qquad (16.63)$$
$$= P_k^{k-1} - G_k(C P_k^{k-1}C^T + R^{-1})(C P_k^{k-1}C^T + R^{-1})^{-1}(P_k^{k-1}C^T)^T$$
$$= \text{Var}(x(k) - \hat{\chi}(k)) - G_k(\text{Var}(y(k) - \hat{y}(k)))G_k^T.$$

These expressions will be used in Section 17.3.1 in the derivation of the *unscented Kalman filter*.

16.3.3   The Steady-State Kalman Filter

Under certain conditions the iteration (16.59) converges to a steady-state solution $G_k = G_\infty$, $P_k = P_\infty$ for $k \in \mathbb{N}$. To compute the steady-state Kalman gain matrix G_∞ and the corresponding matrix P_∞ we assume that $P_\infty = P_k = P_{k-1}$, use the notation $\Pi = P_k^{k-1}$, and combine the three equations (16.59) to remove G_k, P_{k-1}, and P_k:

$$\begin{aligned}
\Pi &= A(I - G_k C)P_k^{k-1}A^T + \bar{B}Q^{-1}\bar{B}^T \\
&= A(I - \Pi C^T(C\Pi C^T + R^{-1})^{-1}C)\Pi A^T + \bar{B}Q^{-1}\bar{B}^T \qquad (16.64)\\
&= A\Pi A^T - A\Pi C^T(C\Pi C^T + R^{-1})^{-1}C\Pi A^T + \bar{B}Q^{-1}\bar{B}^T.
\end{aligned}$$

Note that (16.64) is an algebraic Riccati equation in the unknown Π already discussed in (14.36) in the context of optimal control. In particular, we can identify the following relations between the parameters and unknowns in (14.36) and (16.64), respectively:

ARE (14.36)	P	A	B	C	S	R	Q
ARE (16.64)	Π	A^T	C^T	B^T	0	R^{-1}	Q^{-1}

$$(16.65)$$

The existence of a positive definite solution of the Riccati equation (16.64) can thus be guaranteed by adapting the assumptions of Theorem 14.10. With $\Pi \in \mathcal{S}_{>0}^n$ denoting a positive definite solution of the Riccati equation (16.64), based on the update equations (16.59) we can define the matrices

$$\begin{aligned}
G_\infty &= \Pi C^T(C\Pi C^T + R^{-1})^{-1}\\
\tilde{G}_\infty &= A\Pi C^T(C\Pi C^T + R^{-1})^{-1}
\end{aligned}$$

and

$$P_\infty = (I - G_\infty C)\Pi = (I - (\Pi C^T(C\Pi C^T + R^{-1})^{-1})C)\Pi. \qquad (16.66)$$

Existence of a positive definite solution $\Pi \in \mathcal{S}_{>0}^n$ of (16.64) can be ensured using Theorem 14.10 and the relations in (16.65).

Theorem 16.9. *Consider the linear system* (16.25) *and assume that the pair* $(A, \bar{B})$ *is stabilizable and the pair* (A, C) *is detectable. Additionally, let* $R \in \mathcal{S}_{>0}^p$ *and* $Q \in \mathcal{S}_{>0}^q$. *Then the Riccati equation* (16.64) *has a unique positive definite solution* $\Pi \in \mathcal{S}_{>0}^n$, *and the matrix*

$$A - \tilde{G}_\infty C = A - A\Pi C^T(C\Pi C^T + R^{-1})^{-1}C \qquad (16.67)$$

is a Schur matrix.

Proof. With the relations in (16.65) the proof follows immediately from Theorem 14.10. In particular, with (16.65) the assumptions on the stabilizability of $(A, \bar{B})$ and the detectability of (A, C) translate directly to the assumptions on stabilizability and detectability in Theorem 14.10. The existence of a unique positive definite solution of (14.36) is equivalent to the existence of a unique positive definite solution $\Pi \in \mathcal{S}_{>0}^n$ of (16.64).

The property that the matrix (16.67) is a Schur matrix follows again from (16.65) together with the second item in Theorem 14.10 and the fact that (16.67) and its transpose

$$A^T - C^T\tilde{G}_\infty^T = A^T - C^T(C\Pi C^T + R^{-1})^{-1}C\Pi A^T$$

have the same eigenvalues. $\qquad\qquad\square$

In the case that (16.59) converges, the steady-state Kalman filter estimation process reduces to the equations (16.60) and (16.61), respectively, which become

$$\hat{\chi}(k+1) = (A - \tilde{G}_\infty C)\hat{\chi}(k) + \tilde{G}_\infty y(k) + Bu(k) \qquad (16.68)$$

and

$$\hat{x}(k+1) = (I - G_\infty C)(A\hat{x}(k) + Bu(k)) + G_\infty y(k+1). \qquad (16.69)$$

In (16.68) we see that the steady-state Kalman filter recovers the structure of a Luenberger observer (16.9) as well as the structure of the minimum energy estimator (16.17).

16.3.4   A Hybrid Time Kalman Filter

In the derivation of the Kalman filter equations we have taken the time-invariant system (16.25) as the starting point for the calculations. However, similar results can be derived for time-varying systems

$$\begin{aligned} x(k+1) &= A(k)x(k) + B(k)u(k) + \bar{B}(k)v(k), \\ y(k) &= C(k)x(k) + w(k), \end{aligned} \qquad (16.70)$$

with $A(k) \in \mathbb{R}^{n\times n}$, $B(k) \in \mathbb{R}^{n\times m}$, $C(k) \in \mathbb{R}^{p\times n}$, and $\bar{B}(k) \in \mathbb{R}^{n\times q}$ for all $k \in \mathbb{N}_0$.

In this case the three steps in the update of the Kalman filter gain matrix (16.59) become

$$P_k^{k-1} = A(k-1)P_{k-1}A(k-1)^T + \bar{B}(k-1)Q^{-1}\bar{B}(k-1)^T, \qquad (16.71)$$

$$G_k = P_k^{k-1}C(k)^T(C(k)P_k^{k-1}C(k)^T + R^{-1})^{-1}, \qquad (16.72)$$

$$P_k = (I - G_kC(k))P_k^{k-1}, \qquad (16.73)$$

and the optimal state estimates are defined as

$$\hat{\chi}(k) = A(k-1)\hat{x}(k-1) + B(k-1)u(k-1) \qquad (16.74)$$

$$\hat{x}(k) = \hat{\chi}(k) + G_k(y(k) - C(k)\hat{\chi}(k)). \qquad (16.75)$$

These equations allow a nice interpretation of the discrete time Kalman filter applied to continuous time systems where measurements are only available at discrete, possibly non-equidistant, time instants.

To further investigate this setting, consider the continuous time system

$$\begin{aligned} \dot{x}_c(t) &= A_c x(t) + B_c u(t) + \bar{B}_c v_c(t), \\ y_c(t) &= C_c x(t) + w_c(t). \end{aligned} \qquad (16.76)$$

We assume that the output $y_c(t)$ is only available at discrete time steps τ_k, where $(\tau_k)_{k\in\mathbb{N}} \subset \mathbb{R}_{\geq 0}$ and $\tau_k < \tau_{k+1}$ for all $k \in \mathbb{N}_0$.

As a first step to establish a connection between the discrete time Kalman filter and the continuous time system (16.76) we assume for simplicity of presentation that $u(t) = u(\tau_k)$, $v(t) = v(\tau_k)$ for all $t \in [\tau_k, \tau_{k+1})$, for all $k \in \mathbb{N}_0$, i.e., we assume that $u(\cdot)$ and $v(\cdot)$ are constant between two measurements.

Using the discretization method described in Section 5.2.1, for all $k \in \mathbb{N}_0$, we can define the matrices $A(k) = e^{A_c(\tau_{k+1} - \tau_k)}$,

$$B(k) = \int_{\tau_k}^{\tau_{k+1}} e^{A_c(\tau_{k+1} - \tau)} d\tau B_c, \qquad \bar{B}(k) = \int_{\tau_k}^{\tau_{k+1}} e^{A_c(\tau_{k+1} - \tau)} d\tau \bar{B}_c, \qquad (16.77)$$

and $C(k) = C_c$. These matrices define a discrete time system of the form (16.25), and thus the equations (16.71)–(16.75) can be used to obtain estimates of the states $x(\tau_k)$.

If $u(\cdot)$ and $v(\cdot)$ are not constant on $[\tau_k, \tau_{k+1})$, $k \in \mathbb{N}_0$, an explicit calculation of $B(k)$ and $\bar{B}(k)$ as in (16.77) is in general not possible. Instead, based on the knowledge of $\hat{x}(\tau_{k-1})$, $k \in \mathbb{N}$, one can solve

$$\dot{\hat{\chi}}(t) = A_c \hat{\chi}(t) + B_c u(t), \qquad \hat{\chi}(\tau_{k-1}) = \hat{x}(\tau_{k-1}) \qquad (16.78)$$

for $t \in [\tau_{k-1}, \tau_k]$. Note that the differential equation (16.78) replaces the difference equation (16.74) in the calculation of $\hat{\chi}(\tau_k)$, and thus the discrete counterpart of B_c is not necessary.

The matrix $\bar{B}(k)$ is needed in (16.71), which is the discrete update of P_k. Observe that equation (16.71) is of the form of a *difference Lyapunov equation*

$$\Gamma(k+1) = \Lambda^T \Gamma(k) \Lambda + \Omega, \qquad \Gamma(0) = \Gamma_0 \in \mathbb{R}^{n \times n} \qquad (16.79)$$

for matrices $\Lambda, \Omega \in \mathbb{R}^{n \times n}$ and discrete state $\Gamma \in \mathbb{R}^{n \times n}$. The difference equation (16.79) is a linear discrete time system whose steady-state solutions satisfy the discrete time Lyapunov equation $\Gamma = \Lambda^T \Gamma \Lambda + \Omega$. The continuous time counterpart, the *differential Lyapunov equation*, is defined as

$$\dot{\Gamma}(t) = \Lambda^T \Gamma(t) + \Gamma(t) \Lambda + \Omega, \qquad (16.80)$$

whose steady-state solutions satisfy the continuous time Lyapunov equation $0 = \Lambda^T \Gamma + \Gamma \Lambda + \Omega$. The continuous time and the discrete time Lyapunov equations have been introduced in (3.4) and (5.35), respectively.

With these considerations, similarly to the replacement of (16.74) by (16.78) we can replace the discrete update (16.71) through the continuous update

$$\dot{P}(t) = A_c P(t) + P(t) A_c^T + \bar{B}_c Q^{-1} \bar{B}_c^T, \qquad P(\tau_{k-1}) = P_{k-1}.$$

These observations lead to the *hybrid time Kalman filter*, consisting of a continuous time filter and the discrete time updates summarized in Algorithm 7. Note that at the discrete times $\tau_k \in \mathbb{R}_{\geq 0}$, $k \in \mathbb{N}$, the state estimate $\hat{\chi}(\tau_k)$ is defined twice. In Algorithm 7, $\hat{x}(k)$, $k \in \mathbb{N}$, describe optimal estimates of the state $x_c(\tau_k)$, taking the most recent measurement $y_c(\tau_k)$ into account. Moreover, for $t \in [\tau_{k-1}, \tau_k)$, $\hat{\chi}(t)$ provides an optimal estimate of the state $x_c(t)$ based on the model (16.81) until the next measurement becomes available at time $t = \tau_k$.

16.4   EXERCISES

Exercise 16.1. Reproduce the results in Example 16.1 and numerically analyze the performance of the combined controller/observer for different selections of the eigenvalues of $A + BK$ and $A + LC$, respectively, and for different initial conditions

---

Algorithm 7: Hybrid time Kalman filter

Input: Linear system (16.76), control input $u : \mathbb{R}_{\geq 0} \to \mathbb{R}^m$, positive definite matrices Q, R, initial estimates $\hat{x}(0) = \hat{x}_0$, $P_0 \in \mathcal{S}^n_{>0}$, and a sequence of discrete time steps $(\tau_k)_{k \in \mathbb{N}} \subset \mathbb{R}_{\geq 0}$, $\tau_k < \tau_{k+1}$, for all $k \in \mathbb{N}_0$.

Output: Continuous time and discrete time estimates $\hat{\chi}(t)$ and $\hat{x}(\tau_k)$ of the state $x(t)$.

Algorithm: For $k \in \mathbb{N}$:

1. Continuous time update: For $t \in [\tau_{k-1}, \tau_k]$ solve

$$\dot{P}(t) = A_c P(t) + P(t) A_c^T + \bar{B}_c Q^{-1} \bar{B}_c^T, \qquad P(\tau_{k-1}) = P_{k-1},$$

and

$$\dot{\hat{\chi}}(t) = A_c \hat{\chi}(t) + B_c u(t), \qquad \hat{\chi}(\tau_{k-1}) = \hat{x}(k-1). \tag{16.81}$$

2. Measure the output:

$$y_c(\tau_k) = C x_c(\tau_k) + w(\tau_k).$$

3. Discrete time update:

$$G_k = P(\tau_k) C_c^T (C_c P(\tau_k) C_c^T + R^{-1})^{-1},$$
$$P_k = (I - G_k C_c) P(\tau_k),$$

and

$$\hat{x}(k) = \hat{\chi}(\tau_k) + G_k (y_c(\tau_k) - C_c \hat{\chi}(\tau_k)).$$

Set $k = k + 1$ and go to step 1.

---

x_0.

Exercise 16.2. Reproduce the results in Example 16.5 in MATLAB.

Exercise 16.3. Reproduce the results in Example 16.8 in MATLAB.

Exercise 16.4. Use a steady-state Kalman filter discussed in Section 16.3.3 for the setting in Example 16.8. In particular, replace Algorithm 6 with the steady-state Kalman filter.

16.5   BIBLIOGRAPHICAL NOTES AND FURTHER READING

The general structure of an observer as a copy of the system driven by an output injection term was proposed by Luenberger in [100] and subsequently came to be called the Luenberger Observer. The filter that came to be known as the Kalman filter was first formulated by Kalman in [77].

For the minimum energy estimator (MEE) we followed the development in [63]. The material on the discrete time Kalman filter follows [39] and [70]. See in particular [39, Chapters 1 and 2]. Note that [70] is an excellent reference for filtering in general.

Additional material on the discrete time Kalman filter is taken from [138]. In particular, the steady state solution of the discrete time Kalman filter can be found in [138, Chapter 7.3].

The discrete time Kalman filter equations can also be derived based on a deterministic approach (rather than the probabilistic development presented). See [128, Chapter 1.4.3], for example.

Chapter Seventeen

Extended and Unscented Kalman Filter and Moving Horizon Estimation

In this chapter, extensions of the Kalman filter for nonlinear systems are discussed. In particular, definitions and derivations in Chapter 16 discussing linear systems are extended to the nonlinear systems context.

The *extended Kalman filter* is examined in both the continuous and discrete time settings. The *unscented Kalman filter* for discrete time nonlinear systems is then introduced before the chapter concludes with a brief discussion on *moving horizon estimation* for discrete time systems. While moving horizon estimation is in general not a Kalman filter, as in the control context where we motivated MPC in Chapter 15 through limitations in optimal control, moving horizon estimation generalizes ideas from Kalman filtering by considering appropriately selected optimization problems over finite horizons.

17.1 EXTENDED KALMAN FILTER (CONTINUOUS TIME)

In this section we introduce the extended Kalman filter to estimate the state of the nonlinear plant

$$\dot{x}(t) = f(x(t), u(t)), \qquad x(0) \in \mathbb{R}^n \tag{17.1a}$$

from the (possibly nonlinear) output

$$y(t) = h(x(t)). \tag{17.1b}$$

As usual, $x \in \mathbb{R}^n$ denotes the state, $u \in \mathbb{R}^m$ denotes the input, and $y \in \mathbb{R}^p$ denotes the measured output. We assume that $f : \mathbb{R}^n \times \mathbb{R}^m \to \mathbb{R}^n$ and $h : \mathbb{R}^n \to \mathbb{R}^p$ are arbitrarily often continuously differentiable. Additionally, we assume that $f(x(t), u(t))$ satisfies (3.19) for continuous functions $u : \mathbb{R}_{\geq 0} \to \mathbb{R}^m$ so that Theorem 3.15 is applicable. While we start with the continuous time setting, the discrete time extended Kalman filter will be discussed in the next section.

Following the structure of the Luenberger Observer (c.f., equation (16.6a)), we start with observer dynamics that are a copy of the system model driven by an output injection term

$$\dot{\hat{x}}(t) = f(\hat{x}(t), u(t)) + L(t)(y(t) - h(\hat{x}(t))),$$

where $\hat{x} \in \mathbb{R}^n$ denotes the estimated state and $L : \mathbb{R}_{\geq 0} \to \mathbb{R}^{n \times p}$ represents a time-dependent output injection term to be designed.

With the definition of the error between the state and the estimated state

$$e = x - \hat{x}$$

the error dynamics satisfy

$$\dot{e} = f(x, u) - f(\hat{x}, u) - L(t)(h(x) - h(\hat{x})). \tag{17.2}$$

Since f and h are continuously differentiable by assumption, we can define

$$A(t) = \frac{\partial f}{\partial x}(\hat{x}(t), u(t)) \qquad \text{and} \qquad C(t) = \frac{\partial h}{\partial x}(\hat{x}(t))$$

as time-varying linearizations in $(\hat{x}, u)$.

By adding and subtracting $(A(t) - L(t)C(t))e$ on the right-hand side, the error dynamics (17.2) can be rewritten as

$$\dot{e} = (A(t) - L(t)C(t))e + \Delta(e, x, u), \tag{17.3}$$

where

$$\Delta(e, x, u) = f(x, u) - f(\hat{x}, u) - A(t)e - L(t)(h(x) - h(\hat{x}) - C(t)e). \tag{17.4}$$

Equation (17.3) can be interpreted as the Taylor approximation of (17.2) at $e = 0$ and with respect to $\hat{x}$. In particular, when $e = 0$, the term $f(e+\hat{x}, u) - f(\hat{x}, u)$ vanishes. Moreover, the derivative of the expressions on the right-hand side of (17.2) with respect to e and evaluated at $e = 0$ satisfy

$$\frac{\partial}{\partial e}\left(f(x, u) - f(\hat{x}, u)\right)\Big|_{e=0} = \frac{\partial}{\partial e}\left(f(e + \hat{x}, u) - f(\hat{x}, u)\right)\Big|_{e=0}$$
$$= \frac{\partial}{\partial e}f(e + \hat{x}, u)\Big|_{e=0} = A(t)$$

and

$$\frac{\partial}{\partial e}\left(h(x) - h(\hat{x})\right)\Big|_{e=0} = \frac{\partial}{\partial e}\left(h(e + \hat{x}) - h(\hat{x})\right)\Big|_{e=0} = C(t).$$

Hence, (17.3) represents the Taylor polynomial at $e = 0$, where the remainder is hidden in the function Δ.

To derive a time-dependent injection gain $L(t)$ which ensures that the origin of the error dynamics (17.3) is asymptotically stable we construct a Lyapunov function from which $L(t)$ can be deduced. Consider a continuously differentiable matrix function $P : \mathbb{R} \to \mathcal{S}^n$ where $P(t) > 0$ is positive definite for all $t \in \mathbb{R}$. In particular, we assume that there exists $\alpha_1, \alpha_2 \in \mathbb{R}_{>0}$ such that

$$\alpha_1 I \leq P(t) \leq \alpha_2 I, \qquad \forall\, t \in \mathbb{R}. \tag{17.5}$$

Then for $\alpha_3 = \frac{1}{\alpha_2} > 0$ and $\alpha_4 = \frac{1}{\alpha_1} > 0$, the inequalities

$$\alpha_3 I \leq P^{-1}(t) \leq \alpha_4 I, \qquad \forall\, t \in \mathbb{R}$$

are satisfied. With these definitions we consider the candidate Lyapunov function

$V : \mathbb{R} \times \mathbb{R}^p \to \mathbb{R}_{\geq 0}$ with $V(e(t)) = e(t)^T P^{-1}(t)e(t)$. To compute the time derivative of V, we first calculate an expression for the time derivative of $P^{-1}(t)$.

Lemma 17.1. *Consider* $P : \mathbb{R} \to \mathcal{S}_{>0}^n$ *continuously differentiable. Then the time derivative of* $P^{-1}(t)$ *satisfies*

$$\dot{P}^{-1}(t) = -P^{-1}(t)\dot{P}(t)P^{-1}(t). \tag{17.6}$$

Proof. Since $P(t) > 0$ for all $t \in \mathbb{R}$, the inverse $P^{-1}(t)$ is well defined. Additionally, from $I = P(t)P^{-1}(t) = P^{-1}(t)P(t)$ and the application of the chain rule to $\dot{P}^{-1}(t)$, we obtain

$$
\begin{aligned}
\dot{P}^{-1}(t) &= \tfrac{d}{dt}\left(P^{-1}(t)P(t)P^{-1}(t)\right) \\
&= \dot{P}^{-1}(t)P(t)P^{-1}(t) + P^{-1}(t)\dot{P}(t)P^{-1}(t) + P^{-1}(t)P(t)\dot{P}^{-1}(t) \\
&= 2\dot{P}^{-1}(t) + P^{-1}(t)\dot{P}(t)P^{-1}(t),
\end{aligned}
$$

which implies (17.6). $\qquad\qquad\square$

We proceed with the derivation of the time derivative of V and, for simplicity, omit the time dependence in the calculations. With (17.6) it holds that

$$
\begin{aligned}
\dot{V}(e) &= \dot{e}^T P^{-1}e + e^T \dot{P}^{-1}e + e^T P^{-1}\dot{e} \\
&= ((A - LC)e + \Delta)^T P^{-1}e + e^T P^{-1}((A - LC)e + \Delta) - e^T P^{-1}\dot{P}P^{-1}e \\
&= e^T(A - LC)^T P^{-1}e + e^T P^{-1}(A - LC)e + 2e^T P^{-1}\Delta - e^T P^{-1}\dot{P}P^{-1}e \\
&= e^T P^{-1}\left(P(A - LC)^T + (A - LC)P - \dot{P}\right)P^{-1}e + 2e^T P^{-1}\Delta. \tag{17.7}
\end{aligned}
$$

Under the assumption that $L(t)$ is of the form $L(t) = P(t)C(t)^T Q$ for a positive definite matrix $Q > 0$, similarly to the result in Theorem 16.4, we can further rewrite (17.7):

$$
\begin{aligned}
\dot{V}(e) &= e^T P^{-1}\left(P(A - PC^T QC)^T + (A - PC^T QC)P - \dot{P}\right)P^{-1}e + 2e^T P^{-1}\Delta \\
&= e^T P^{-1}\left(PA^T + AP - 2PC^T QCP - \dot{P}\right)P^{-1}e + 2e^T P^{-1}\Delta.
\end{aligned}
$$

If $P(t)$ satisfies the differential matrix equation

$$\dot{P}(t) = P(t)A(t)^T + A(t)P(t) - P(t)C(t)^T QC(t)P(t) + R^{-1} \tag{17.8}$$

for an arbitrary positive definite initial value $P(t_0) \in \mathcal{S}_{>0}^n$ and a fixed positive definite matrix $R \in \mathcal{S}_{>0}^n$ and for all $t \geq t_0$, then $\dot{V}(e)$ reduces to

$$\dot{V}(e) = -e^T P^{-1}\left(PC^T QCP + R^{-1}\right)P^{-1}e + 2e^T P^{-1}\Delta. \tag{17.9}$$

Equation (17.8) is called the *differential Riccati equation* (c.f., (16.16)). In Theorem 16.4 the matrices A, L, and C are constant and time invariant, i.e., also $P(t)$ is constant. In this case $\dot{P}(t) = 0$ and the differential Riccati equation reduces to the algebraic Riccati equation

$$0 = PA^T + AP - PC^T QCP + R^{-1} \tag{17.10}$$

whose solution P is a steady state (or equilibrium) of (17.8).

Since $P^{-1}R^{-1}P^{-1}$ is positive definite, $C^T Q C$ is positive semidefinite, and Δ represents the error term of the Taylor approximation in (17.9), we can expect that $\dot{V}(e) \leq 0$ for e sufficiently small. The decrease $\dot{V}$ is made more precise in the following lemma.

Lemma 17.2 ([87, Lemma 11.2]). *Consider the error dynamics* (17.2) *and let* $R \in \mathcal{S}_{>0}^n$ *and* $Q \in \mathcal{S}_{>0}^p$ *be given. Additionally, for* $P(t_0) = P_0 \in \mathcal{S}_{>0}^n$, *assume that the solution* $P : \mathbb{R} \to \mathbb{R}^{p \times p}$ *of* (17.8) *exists for all* $t \geq t_0$ *and satisfies* (17.5) *for* $\alpha_1, \alpha_2 \in \mathbb{R}_{>0}$. *Then, for* $L(t) = P(t)C(t)^T Q$ *the origin of* (17.2) *is locally exponentially stable, i.e., there exist* $\delta, \lambda, M > 0$ *such that if* $|e(t_0)| \leq \delta$ *then, for all* $t \geq t_0$,

$$|e(t)| \leq M|e(t_0)|\exp(\lambda(t - t_0)).$$

Proof. Before we begin the proof, recall that in this section the functions f and h are sufficiently often continuously differentiable by assumption. Here, f and h need to be three times continuously differentiable (with respect to x). This in particular implies that the derivatives $\frac{\partial f}{\partial x}$ and $\frac{\partial h}{\partial x}$ are locally Lipschitz continuous with Lipschitz constants denoted by L_1 and L_2, respectively.

With the Lipschitz continuity of $\frac{\partial f}{\partial x}$ and the mean value theorem for multivalued functions, the following estimate is satisfied:

$$
\begin{aligned}
|f(x, u) - f(\hat{x}, u) - A(t)e| &= \left| f(\hat{x} + e) - f(\hat{x}, u) - \tfrac{\partial f}{\partial x}(\hat{x}, u)e \right| \\
&= \left| \int_0^1 \tfrac{\partial f}{\partial x}(\sigma e + \hat{x}, u)d\sigma \cdot e - \int_0^1 \tfrac{\partial f}{\partial x}(\hat{x}, u)d\sigma \cdot e \right| \\
&\leq \left\| \int_0^1 \tfrac{\partial f}{\partial x}(\sigma e + \hat{x}, u) - \tfrac{\partial f}{\partial x}(\hat{x}, u)d\sigma \right\| |e| \\
&\leq \int_0^1 \sigma L_1 |e| d\sigma \, |e| = \frac{1}{2}L_1|e|^2.
\end{aligned}
$$

In the same way, the estimate

$$|h(x) - h(\hat{x}) - C(t)e| \leq \tfrac{1}{2}L_2|e|^2$$

is obtained. The function $C(t)$ can additionally be upper bounded by

$$
\begin{aligned}
\|C(t)\| &= \left\| \tfrac{\partial h}{\partial x}(x - \hat{x}) - \tfrac{\partial h}{\partial x}(0) + \tfrac{\partial h}{\partial x}(0) \right\| \\
&\leq L_2|x - \hat{x}| + \left\| \tfrac{\partial h}{\partial x}(0) \right\| \\
&\leq L_2(|x| + |\hat{x}|) + \left\| \tfrac{\partial h}{\partial x}(0) \right\|,
\end{aligned}
$$

which shows that

$$\|L(t)\| = \|P(t)C(t)^T Q\| \leq \alpha_2\|Q\| \left(L_2(|x| + |\hat{x}|) + \left\| \tfrac{\partial h}{\partial x}(0) \right\| \right).$$

Collecting these bounds and applying the triangle inequality to $|\Delta|$ defined in (17.4) guarantees the existence of $k_1, k_2 \in \mathbb{R}_{>0}$ such that

$$|\Delta(e, x, u)| \leq k_1|e|^2 + k_2|e|^3.$$

Since $Q > 0$, $R > 0$, and $P^{-1} \geq \frac{1}{\alpha_2}I$, there exists $c_1 > 0$ such that $C^T QC + P^{-1}R^{-1}P^{-1} > c_1 I$ for all $t \geq t_0$. In particular,

$$\dot{V}(e) \leq -c_1|e|^2 + 2\|P^{-1}\||e|(k_1|e|^2 + k_2|e|^3)$$
$$\leq -c_1|e|^2 + \frac{2}{\alpha_2}k_1|e|^3 + \frac{2}{\alpha_2}k_2|e|^4$$
$$= -\frac{1}{2}c_1|e|^2 + \left(-\frac{1}{2}c_1|e|^2 + c_2|e|^3 + c_3|e|^4\right),$$

with $c_2 = \frac{2}{\alpha_2}k_1$ and $c_3 = \frac{2}{\alpha_2}k_2$. Thus, local exponential stability of $e = 0$ follows from

$$\dot{V}(e) \leq -\frac{1}{2}|e|^2 \leq -\frac{1}{2\alpha_2}V(e), \qquad \forall\, |e|^2 \leq r^2 = \frac{c_2}{c_1} + \frac{\sqrt{c_2^2 + 2c_1c_3}}{c_1},$$

which shows the assertion. □

For nonlinear systems, only local exponential stability, i.e., only local convergence $\hat{x}(t) \to x(t)$, can be guaranteed in general. The equations of the extended Kalman filter are summarized by the ordinary differential equations

$$\dot{\hat{x}}(t) = f(\hat{x}(t), u(t)) + P(t)\left(\frac{\partial h}{\partial x}(\hat{x}(t))\right)^T Q(y(t) - h(\hat{x}(t))) \tag{17.11a}$$

$$\dot{P}(t) = P(t)\left(\frac{\partial f}{\partial x}(\hat{x}(t), u(t))\right)^T + \left(\frac{\partial f}{\partial x}(\hat{x}(t), u(t))\right)P(t)$$
$$- P(t)\left(\frac{\partial h}{\partial x}(\hat{x}(t))\right)^T Q\left(\frac{\partial h}{\partial x}(\hat{x}(t))\right)P(t) + R^{-1} \tag{17.11b}$$

with respect to initial conditions $\hat{x}(t_0) = \hat{x}_0 \in \mathbb{R}^n$ and $P(t_0) = P_0 \in \mathcal{S}_{>0}^n$.

Then, for given $u : \mathbb{R}_{\geq t_0} \to \mathbb{R}^m$, $y : \mathbb{R}_{\geq t_0} \to \mathbb{R}^p$, the solution of (17.11) provides an approximation of $x(t)$ with guaranteed convergence $\hat{x}(t) \to x(t)$ for $t \to \infty$ if the assumptions of Lemma 17.2 are satisfied. The system of ordinary differential equations (17.11) has to be solved in parallel. Since $P(t)$ is symmetric, (17.11b) can be written as an ordinary differential equation of dimension $n \cdot (n+1)/2$. We will illustrate the extended Kalman filter on an example.

Example 17.3. To analyze the performance of the extended Kalman filter consider the dynamics of the pendulum on a cart $\dot{x} = f(x, u)$ with output $y = h(x)$ defined by

$$f(x, u) = \begin{bmatrix} x_3 \\ x_4 \\ \frac{-\bar{J}\bar{c}x_3 - \bar{J}\sin(x_2)x_4^2 - \bar{\gamma}\cos(x_2)x_4 + g\cos(x_2)\sin(x_2) + \bar{J}u}{M\bar{J} - \cos^2(x_2)} \\ \frac{-\bar{M}\bar{\gamma}x_4 + \bar{M}g\sin(x_2) - \bar{c}\cos(x_2)x_3 - \cos(x_2)\sin(x_2)x_4^2 + \cos(x_2)u}{M\bar{J} - \cos^2(x_2)} \end{bmatrix} \tag{17.12}$$

and

$$h(x) = \begin{bmatrix} x_1 \\ x_2 \end{bmatrix} \tag{17.13}$$

(see Section 1.2.1). The partial derivatives of f satisfy $\frac{\partial f}{\partial x_1}(x, u) = 0 \in \mathbb{R}^{4 \times 1}$,

$$\frac{\partial f}{\partial x_3}(x, u) = \begin{bmatrix} 1 \\ 0 \\ \frac{\bar{c}\bar{J}}{\cos^2(x_2) - \bar{J}\bar{M}} \\ \frac{\bar{c}\cos(x_2)}{\cos^2(x_2) - \bar{J}\bar{M}} \end{bmatrix}, \qquad \frac{\partial f}{\partial x_4}(x, u) = \begin{bmatrix} 0 \\ 1 \\ \frac{\bar{\gamma}\cos(x_2) + 2\bar{J}x_4 \sin(x_2)}{\cos^2(x_2) - \bar{J}\bar{M}} \\ \frac{\bar{\gamma}\bar{M} + 2x_4 \cos(x_2)\sin(x_2)}{\cos^2(x_2) - \bar{J}\bar{M}} \end{bmatrix},$$

and

$$\frac{\partial f}{\partial x_2}(x, u) = \begin{bmatrix} 0 \\ 0 \\ \frac{2\cos(x_2)\sin(x_2)(\bar{J}\sin(x_2)x_4^2 + \bar{\gamma}\cos(x_2)x_4 - \bar{J}u + \bar{c}\bar{J}x_3 - g\cos(x_2)\sin(x_2))}{(\cos^2(x_2) - \bar{J}\bar{M})^2} \\ \frac{x_4^2 \sin(x_2)^2 - x_4^2 \cos(x_2)^2 - u\sin(x_2) + \bar{c}x_3 \sin(x_2) + g\bar{M}\cos(x_2)}{\cos^2(x_2) - \bar{J}\bar{M}} \end{bmatrix}$$

$$- \begin{bmatrix} 0 \\ 0 \\ \frac{-\bar{J}x_4^2 \cos(x_2) + \bar{\gamma}x_4 \sin(x_2) + g\cos(x_2)^2 - g\sin^2(x_2)}{\cos(x_2)^2 - \bar{J}\bar{M}} \\ \frac{x_4^2 \sin(x_2)^2 - x_4^2 \cos(x_2)^2 - u\sin(x_2) + \bar{c}x_3 \sin(x_2) + g\bar{M}\cos(x_2)}{\cos^2(x_2) - \bar{J}\bar{M}} \end{bmatrix}.$$

While in particular the partial derivative $\frac{\partial f}{\partial x_2}$ looks complicated, the expressions can be obtained simply through MATLAB, for example, by using symbolic variables defined through `syms` and symbolic differentiation `diff.m`.

With these calculations, we define the components

$$A(t) = \frac{\partial f}{\partial x}(\hat{x}(t), u(t))$$
$$= \begin{bmatrix} \frac{\partial f}{\partial x_1}(\hat{x}(t), u(t)) & \frac{\partial f}{\partial x_2}(\hat{x}(t), u(t)) & \frac{\partial f}{\partial x_3}(\hat{x}(t), u(t)) & \frac{\partial f}{\partial x_4}(\hat{x}(t), u(t)) \end{bmatrix}$$

and

$$C(t) = \frac{\partial h}{\partial x}(\hat{x}(t), u(t)) = \begin{bmatrix} 1 & 0 & 0 & 0 \\ 0 & 1 & 0 & 0 \end{bmatrix}$$

involved in the dynamical system (17.11).

For simplicity we use the constants $M = m = \ell = 1$, $J = c = \gamma = 0.1$, and $g = 9.81$ in the following simulations. Moreover, the matrices $R \in \mathbb{R}^{4 \times 4}$ and $Q \in \mathbb{R}^{2 \times 2}$ are defined as $R = I$ and $Q = I$, where I denotes the identity matrix of appropriate dimension.

In Figure 17.1 the evolution of the state of the plant x, the evolution of the observer state $\hat{x}$, as well as the error $x - \hat{x}$ with respect to the initial conditions $x(0) = [0, 0, 0, 0]^T$, $\hat{x}(0) = [2, 2, 2, 2]^T$, and $P(0) = 0.1 \cdot I \in \mathbb{R}^{4 \times 4}$ are visualized. For the simulations, the input is defined as

$$u(\hat{x}) = -\hat{x}_3(t) - \hat{x}_4(t).$$

Note that this choice of input u does not stabilize the origin and both $|x(t)| \to \infty$ and $|\hat{x}(t)| \to \infty$ for $t \to \infty$ even though this is not visible in Figure 17.1 where only the time interval from 0 to 10 is shown. Nevertheless, for the setting shown in Figure 17.1 the estimated state $\hat{x}$ converges to the state x.

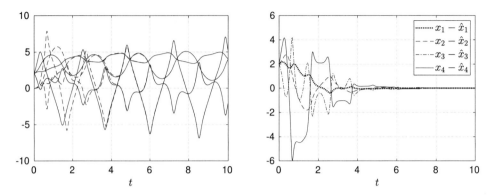

Figure 17.1: On the left, visualization of the states x (solid lines) and $\hat{x}$ (dashed lines); on the right the error dynamics $e = x - \hat{x}$. The states and the error dynamics correspond to the inverted pendulum on a cart and are obtained through the extended Kalman filter (17.11).

However, in contrast to the Kalman filter (for linear systems), the extended Kalman filter (for nonlinear systems) guarantees only local convergence of the error dynamics $x - \hat{x}$. Thus, for a different selection of the initial conditions $x(0)$, $\hat{x}(0)$, $P(0)$, as well as for a different selection of the matrices R and Q, the dynamics of the extended Kalman filter do not necessarily need to converge.

Figure 17.2 shows the results of the extended Kalman filter where the matrix Q used for the setting in Figure 17.1 is replaced by $Q = 0.1 \cdot I$. The other parameters

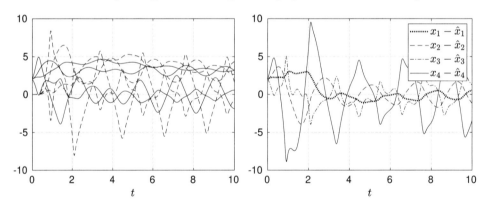

Figure 17.2: Illustration showing that the extended Kalman filter only guarantees local properties. Compared to 17.1 only the matrix Q in the simulations is changed.

are left unchanged. We observe that the error dynamics do not converge to the origin, which highlights that the extended Kalman filter is only a local tool, in general.

17.2   EXTENDED KALMAN FILTER (DISCRETE TIME)

In this section we extend the results of the discrete time Kalman filter discussed in Section 16.3 to nonlinear systems. In particular, we present an extended Kalman filter for systems of the form

$$x(k+1) = f(k, x(k)) + g(k, x(k))v(k), \tag{17.14a}$$
$$y(k) = h(k, x(k)) + w(k). \tag{17.14b}$$

Here, $x \in \mathbb{R}^n$ denotes the state, $y \in \mathbb{R}^p$ denotes the measured output, and $(v(k))_{k \in \mathbb{N}} \subset \mathbb{R}^q$ and $(w(k))_{k \in \mathbb{N}} \subset \mathbb{R}^p$ denote unknown disturbances. The time-dependent functions $f : \mathbb{N} \times \mathbb{R}^n \to \mathbb{R}^n$, $g = [g_1, \ldots, g_q]$, $g_1, \ldots, g_q : \mathbb{N} \times \mathbb{R}^n \to \mathbb{R}^n$, and $h : \mathbb{N} \times \mathbb{R}^n \to \mathbb{R}^p$ are continuously differentiable by assumption.

Note that an input $(u(k))_{k \in \mathbb{N}} \subset \mathbb{R}^m$ is not considered in the dynamics (17.14). Nevertheless, the input can be incorporated through the time index $k \in \mathbb{N}$. In particular, for a known input sequence $(u(k))_{k \in \mathbb{N}} \subset \mathbb{R}^m$ and functions $\tilde{f} : \mathbb{R}^n \times \mathbb{R}^m \to \mathbb{R}^n$, $\tilde{g} : \mathbb{R}^n \times \mathbb{R}^m \to \mathbb{R}^{n \times q}$, $\tilde{h} : \mathbb{R}^n \times \mathbb{R}^m \to \mathbb{R}^p$, we can consider the functions

$$\begin{aligned} f(k, x(k)) &= \tilde{f}(x(k), u(k)), \\ g(k, x(k)) &= \tilde{g}(x(k), u(k)), \\ h(k, x(k)) &= \tilde{h}(x(k), u(k)) \end{aligned} \tag{17.15}$$

in the dynamics (17.14) so as to include the input u.

As in Assumption 16.6, we suppose that $(v(k))_{k \in \mathbb{N}}$ and $(w(k))_{k \in \mathbb{N}}$ are sequences of zero mean Gaussian white noise, i.e., we assume that

$$\mathrm{E}\left[v(k)v(j)^T\right] = \begin{cases} Q^{-1}, & \text{if } k = j \\ 0, & \text{if } k \neq j \end{cases}, \quad \mathrm{E}\left[w(k)w(j)^T\right] = \begin{cases} R^{-1}, & \text{if } k = j \\ 0, & \text{if } k \neq j \end{cases}$$

for all $j, k \in \mathbb{N}$ and for positive definite matrices $Q \in \mathcal{S}^q_{>0}$, $R \in \mathcal{S}^p_{>0}$. Additionally, we assume that

$$\mathrm{E}\left[v(k)w(j)^T\right] = 0, \qquad \mathrm{E}\left[v(k)x_0^T\right] = 0, \qquad \text{and} \qquad \mathrm{E}\left[w(k)x_0^T\right] = 0$$

are satisfied for all $j, k \in \mathbb{N}$ and for all initial conditions $x_0 \in \mathbb{R}^n$.

For the linear system (16.43) we have derived the filter equation (16.41). If we project these updates to the nonlinear setting, the observer dynamics

$$\begin{aligned} \hat{\chi}(k) &= f(k-1, \hat{x}(k-1)) \\ \hat{x}(k) &= \hat{\chi}(k) + G_k\left(y(k) - h(k, \hat{\chi}(k))\right) \end{aligned} \tag{17.16}$$

for $k \in \mathbb{N}$ can be defined. What is missing is the definition of the Kalman gain matrix G_k for $k \in \mathbb{N}$, which indirectly defines disturbance sequences $(\hat{v}_k)_{k \in \mathbb{N}}$, $(\hat{w}_k)_{k \in \mathbb{N}}$ that explain the measured output $y(k)$, $k \in \mathbb{N}$, in an optimal way.

We follow the notation introduced for the discrete time Kalman filter for linear systems and decompose the estimate $\hat{x} = \hat{x}_d + \hat{x}_s$ and the output $y = \hat{y}_d + \hat{y}_s$ into a deterministic part and a stochastic part. We define the deterministic part through the known components

$$\hat{x}_d(k+1) = f(k, \hat{x}_d(k))$$
$$\hat{y}_d(k) = h(k, \hat{x}_d(k)).$$

(17.17)

Then the stochastic dynamics can be indirectly defined through the difference

$$\hat{x}_s(k+1) = \hat{x}(k+1) - \hat{x}_d(k+1)$$
$$= f(k, \hat{x}_d(k) + \hat{x}_s(k)) + g(k, \hat{x}(k))\hat{v}(k) - f(k, \hat{x}_d(k)).$$

Approximating $f(k, \cdot)$ through a first-order Taylor approximation around $\hat{x}_d(k)$ provides the estimate

$$\hat{x}_s(k+1) \approx f(k, \hat{x}_d(k)) + \frac{\partial f}{\partial x}(k, \hat{x}_d(k))\hat{x}_s(k) + g(k, \hat{x}(k))\hat{v}(k) - f(k, \hat{x}_d(k))$$
$$= \frac{\partial f}{\partial x}(k, \hat{x}_d(k))\hat{x}_s(k) + g(k, \hat{x}(k))\hat{v}(k).$$

(17.18)

For $\hat{y}_s(k) = y(k) - \hat{y}_d(k)$ we can use the Taylor approximation of $g(k, \cdot)$ around $\hat{x}_d(k)$ to define

$$\hat{y}_s(k) = y(k) - \hat{y}_d(k) = h(k, \hat{x}(k)) + \hat{w}(k) - h(k, x_d(k))$$
$$\approx h(k, \hat{x}_d(k)) + \frac{\partial h}{\partial x}(k, \hat{x}_d(k))\hat{x}_s(k) + \hat{w}(k) - h(k, \hat{x}_d(k))$$
$$= \frac{\partial h}{\partial x}(k, \hat{x}_d(k))\hat{x}_s(k) + \hat{w}(k)$$
$$= \frac{\partial h}{\partial x}(k, f(k-1, \hat{x}_d(k-1)))\hat{x}_s(k) + \hat{w}(k).$$

With these approximations, the dynamics of $\hat{x}_s(k)$ and $\hat{y}_s(k)$, $k \in \mathbb{N}$, are of the form (16.28) if we define the matrices A, $\bar{B}$, and C as the time-dependent matrices

$$A(k) = \frac{\partial f}{\partial x}(k, \hat{x}(k)), \quad \bar{B}(k) = g(k, \hat{x}(k)), \quad \text{and} \quad C(k) = \frac{\partial h}{\partial x}(k, \hat{\chi}(k)). \quad (17.19)$$

As in the linear setting in Section 16.3.2, equation (17.16) describes a one-step prediction-correction formulation of the Kalman filter. We thus can assume without loss of generality that (17.17) and (17.18) are reinitialized through $\hat{x}_d(k) = \hat{x}(k)$ and $\hat{x}_s(k) = 0$ after every iteration of the extended Kalman filter (17.16). This justifies the arguments used in the definition of the matrices $A(k)$ and $C(k)$.

With these observations we can adapt the calculation of the Kalman filter for linear systems defined in (16.59) for nonlinear systems (17.14) as

$$
\begin{aligned}
P_k^{k-1} &= \left[\frac{\partial f}{\partial x}(k-1, \hat{x}(k-1))\right] P_{k-1} \left[\frac{\partial f}{\partial x}(k-1, \hat{x}(k-1))\right]^T \\
&\quad + g(k-1, \hat{x}(k-1))Q^{-1}g(k-1, \hat{x}(k-1))^T, \\
G_k &= P_k^{k-1} \left[\frac{\partial h}{\partial x}(k, \hat{\chi}(k))\right]^T \\
&\quad \cdot \left(\left[\frac{\partial h}{\partial x}(k, \hat{\chi}(k))\right] P_k^{k-1} \left[\frac{\partial h}{\partial x}(k, \hat{\chi}(k))\right]^T + R^{-1}\right)^{-1}, \\
P_k &= \left(I - G_k \left[\frac{\partial h}{\partial x}(k, \hat{\chi}(k))\right]\right) P_k^{k-1}.
\end{aligned}
\tag{17.20}
$$

Hence, the extended Kalman filter to estimate the state $x(k)$, $k \in \mathbb{N}$, can be defined by the dynamics (17.16) and the Kalman gain is given by the iterates (16.59). The results of this section are summarized in Algorithm 8.

Algorithm 8: Discrete time extended Kalman filter

Input: System (17.14), positive definite weight matrices $Q^{-1} = \text{Var}(v(k))$, $R^{-1} = \text{Var}(w(k))$, and initial estimates $\hat{x}(0) = \hat{x}_0$, $P_0 \in \mathcal{S}_{>0}^n$.

Output: Estimates $\hat{x}(k)$ and $\hat{\chi}(k)$ of the state $x(k)$ for $k \in \mathbb{N}$.

Algorithm: For $k \in \mathbb{N}$:

1. Compute

$$
\hat{\chi}(k) = f(k-1, \hat{x}(k-1))
$$

   and

$$
\begin{aligned}
A(k-1) &= \frac{\partial f}{\partial x}(k-1, \hat{x}(k-1)), \\
\bar{B}(k-1) &= g(k-1, \hat{x}(k-1)), \\
C(k) &= \frac{\partial h}{\partial x}(k, \hat{\chi}(k)).
\end{aligned}
$$

2. Update the gain matrix

$$
\begin{aligned}
P_k^{k-1} &= A(k-1)P_{k-1}A(k-1)^T + \bar{B}(k-1)Q^{-1}\bar{B}(k-1)^T, \\
G_k &= P_k^{k-1}C(k)^T \left[C(k)P_k^{k-1}C(k)^T + R^{-1}\right]^{-1}, \\
P_k &= \left[I - G_k C(k)\right] P_k^{k-1}.
\end{aligned}
$$

3. Measure the output $y(k)$ and update the state estimate

$$
\hat{x}(k) = \hat{\chi}(k) + G_k \left(y(k) - h(k, \hat{\chi}(k))\right)
$$

   set $k = k + 1$ and go to step 1.

We conclude this section with an application of the extended Kalman filter by repeating Example 17.3 in the discrete time setting.

Example 17.4. We continue with the setting described in Example 17.3 and use the Euler discretization with $\Delta > 0$ to obtain a discrete time system.

In this case the discrete time counterparts of (17.12) and (17.13) are defined as

$$
f(x, u) = \begin{bmatrix} x_1 \\ x_2 \\ x_3 \\ x_4 \end{bmatrix} + \Delta \begin{bmatrix} x_3 \\ x_4 \\ \dfrac{-\bar{J}\bar{c}x_3 - \bar{J}\sin(x_2)x_4^2 - \bar{\gamma}\cos(x_2)x_4 + g\cos(x_2)\sin(x_2) + \bar{J}u}{M\bar{J} - \cos^2(x_2)} \\ \dfrac{-\bar{M}\bar{\gamma}x_4 + \bar{M}g\sin(x_2) - \bar{c}\cos(x_2)x_3 - \cos(x_2)\sin(x_2)x_4^2 + \cos(x_2)u}{M\bar{J} - \cos^2(x_2)} \end{bmatrix}
$$

and

$$
h(x) = \begin{bmatrix} x_1 \\ x_2 \end{bmatrix},
$$

respectively, which defines the discrete time system $x^+ = f(x, u)$, $y = h(x)$. Thus, the partial derivatives of f satisfy

$$
\frac{\partial f}{\partial x_1}(x, u) = \begin{bmatrix} 1 \\ 0 \\ 0 \\ 0 \end{bmatrix}, \qquad \frac{\partial f}{\partial x_3}(x, u) = \begin{bmatrix} \Delta \\ 0 \\ 1 + \Delta \dfrac{\bar{c}\bar{J}}{\cos^2(x_2) - \bar{J}\bar{M}} \\ \Delta \dfrac{\bar{c}\cos(x_2)}{\cos^2(x_2) - \bar{J}\bar{M}} \end{bmatrix},
$$

$$
\frac{\partial f}{\partial x_2}(x, u) = \begin{bmatrix} 0 \\ 1 \\ \Delta \cdot \dfrac{2\cos(x_2)\sin(x_2)(\bar{J}\sin(x_2)x_4^2 + \bar{\gamma}\cos(x_2)x_4 - \bar{J}u + \bar{c}\bar{J}x_3 - g\cos(x_2)\sin(x_2))}{(\cos^2(x_2) - \bar{J}\bar{M})^2} \\ \Delta \cdot \dfrac{x_4^2\sin(x_2)^2 - x_4^2\cos(x_2)^2 - u\sin(x_2) + \bar{c}x_3\sin(x_2) + g\bar{M}\cos(x_2)}{\cos^2(x_2) - \bar{J}\bar{M}} \end{bmatrix}
$$

$$
- \begin{bmatrix} 0 \\ 0 \\ \Delta \cdot \dfrac{\bar{J}x_4^2\cos(x_2) + \bar{\gamma}x_4\sin(x_2) + g\cos(x_2)^2 - g\sin^2(x_2)}{\cos(x_2)^2 - \bar{J}\bar{M}} \\ \Delta \cdot \dfrac{x_4^2\sin(x_2)^2 - x_4^2\cos(x_2)^2 - u\sin(x_2) + \bar{c}x_3\sin(x_2) + g\bar{M}\cos(x_2)}{\cos^2(x_2) - \bar{J}\bar{M}} \end{bmatrix},
$$

and

$$
\frac{\partial f}{\partial x_4}(x, u) = \begin{bmatrix} 0 \\ \Delta \\ \Delta \cdot \dfrac{\bar{\gamma}\cos(x_2) + 2\bar{J}x_4\sin(x_2)}{\cos^2(x_2) - \bar{J}\bar{M}} \\ 1 + \Delta \cdot \dfrac{\bar{\gamma}\bar{M} + 2x_4\cos(x_2)\sin(x_2)}{\cos^2(x_2) - \bar{J}\bar{M}} \end{bmatrix},
$$

which follows immediately from the continuous time setting and the Euler discretization. The time-dependent matrices $A(k)$ and $C(k)$ are then defined as

$$
A(k) = \begin{bmatrix} \frac{\partial f}{\partial x_1}(\hat{x}(k), u(k)) & \frac{\partial f}{\partial x_2}(\hat{x}(k), u(k)) & \frac{\partial f}{\partial x_3}(\hat{x}(k), u(k)) & \frac{\partial f}{\partial x_4}(\hat{x}(k), u(k)) \end{bmatrix}
$$

and

$$C(k) = \frac{\partial h}{\partial x}(\hat{\chi}(k), u(k)) = \begin{bmatrix} 1 & 0 & 0 & 0 \\ 0 & 1 & 0 & 0 \end{bmatrix}.$$

Additionally, we assume that the function g in the discrete time system (17.14a) is defined as $g(k, x) = \Delta \cdot I \in \mathbb{R}^{4\times 4}$, which implies that $\bar{B}(k) = g(k, x)$ for all $k \in \mathbb{N}$ and for all $x \in \mathbb{R}^n$. However, in the same way as in Example 17.3 we assume that the state x is unknown, but the dynamics are not affected by disturbances or measurement noise.

Figure 17.3 shows the results of the extended discrete time Kalman filter for $\Delta = 0.1$, and the matrices R and Q are selected as $R = I \in \mathbb{R}^{2\times 2}$ and $Q = I \in \mathbb{R}^{4\times 4}$. The initial conditions are set to $x_0 = [2, 2, 2, 2]^T$, $\hat{x}_0 = [0, 0, 0, 0]^T$, and $P_0 = I \in \mathbb{R}^{4\times 4}$. The input is defined as $u(t) = -\hat{x}_3(t) - \hat{x}_4(t)$. The remaining parameters are the same as in Example 17.3. For the selected set of parameters,

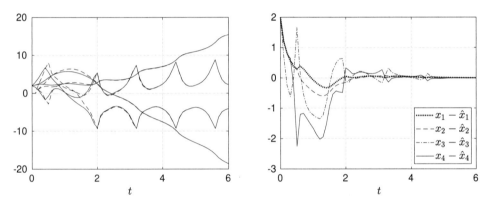

Figure 17.3: On the left, visualization of the states x (solid lines) and $\hat{x}$ (dashed lines); on the right the difference $e = x - \hat{x}$. The states correspond to the inverted pendulum on a cart and are obtained through the extended discrete time Kalman filter (17.11).

the state estimates $\hat{x}$ converge to x, which can be observed from Figure 17.3.

17.3 UNSCENTED KALMAN FILTER (DISCRETE TIME)

The extended Kalman filter relies on the propagation of the expected values $\hat{\chi}$ and $\hat{x}$ through the system dynamics (17.16), and the covariance matrices P_k and P_k^{k-1}, $k \in \mathbb{N}$, defining the Kalman gain G_k, are defined through a linearization of the dynamics (17.14) around $\hat{\chi}$ and $\hat{x}$, respectively. In this section we discuss the *unscented Kalman filter* for the discrete time dynamics (17.14) and indicate some of its advantages compared to the extended Kalman filter in Algorithm 8.

Instead of the propagation of the expected values $\hat{\chi}$ and $\hat{x}$, the unscented Kalman filter depends on the propagation of so-called *sigma points* introduced and discussed next in Section 17.3.1, describing the *unscented transformation*. Based on the unscented transformation the *unscented Kalman filter* is derived in Section 17.3.2. Here, we look at the same setting as in Section 17.2, i.e., we consider the dynamics (17.14), and Assumption 16.6 holds throughout Section 17.3.

17.3.1 Unscented Transformation

The unscented Kalman filter relies on the unscented transformation. Consider a *Gaussian* (or *normal*) *probability density function* $f : \mathbb{R} \to \mathbb{R}$,

$$f(x) = \frac{1}{\sqrt{2\pi}\sigma} e^{-\frac{1}{2\sigma^2}(x-\mu)^2} \tag{17.21}$$

with expected value $E[X] = \mu \in \mathbb{R}$ and standard deviation $\sigma \in \mathbb{R}_{>0}$, i.e., the variance is defined as $\mathrm{Var}(X) = E[(X - E[X])^2] = \sigma^2$, and X denotes a random variable. For a given invertible analytic function $\phi : \mathbb{R} \to \mathbb{R}$ we investigate how the normal distribution changes under the coordinate transformation $y = \phi(x)$ and $x = \phi^{-1}(y)$. In particular, we define the function $g : \mathbb{R} \to \mathbb{R}$,

$$g(y) = \frac{1}{\int_{\mathbb{R}} f(\phi^{-1}(y))\, dy} f(\phi^{-1}(y)), \tag{17.22}$$

under the additional assumption that $\int_{\mathbb{R}} f(\phi^{-1}(y))\, dy < \infty$ is satisfied. Here, we divide through the integral to ensure

$$\int_{\mathbb{R}} g(y)\, dy = \int_{\mathbb{R}} \frac{1}{\int_{\mathbb{R}} f(\phi^{-1}(\bar{y}))\, d\bar{y}} f(\phi^{-1}(y))\, dy = \frac{\int_{\mathbb{R}} f(\phi^{-1}(y))\, dy}{\int_{\mathbb{R}} f(\phi^{-1}(\bar{y}))\, d\bar{y}} = 1,$$

i.e., $g(\cdot)$ is a probability density function in the random variable Y. However, g is not necessarily a normal distribution. The expected value and the variance of the random variable Y are given by

$$E[Y] = \int_{\mathbb{R}} y g(y)\, dy \qquad \text{and}$$

$$\mathrm{Var}(Y) = E[(Y - E[Y])^2] = \int_{\mathbb{R}} y^2 g(y)\, dy - E[Y]^2, \tag{17.23}$$

respectively.

The calculation of $E[Y]$ and $\mathrm{Var}(Y)$ might be impossible or numerically expensive. Thus, the *unscented transformation* replaces the direct calculation of (17.23) and instead generates estimates $\mu_{\mathrm{ut}} \approx E[Y]$, $\sigma_{\mathrm{ut}}^2 \approx \mathrm{Var}(Y)$ and an estimate of g in terms of a normal probability density function (17.21) labeled f_{ut} in the following.

To define μ_{ut} and σ_{ut}, we consider three *sigma points*

$$\sigma_0 = \mu, \qquad \sigma_1 = \mu + \sigma\sqrt{1+c}, \qquad \text{and} \qquad \sigma_2 = \mu - \sigma\sqrt{1+c},$$

and corresponding weighting factors

$$\omega_0 = \frac{c}{1+c}, \qquad \omega_1 = \frac{1}{2(1+c)}, \qquad \text{and} \qquad \omega_2 = \frac{1}{2(1+c)}$$

for a fixed parameter $c \in \mathbb{R}\setminus\{-1\}$. Then, the estimated expected value and the estimated variance are defined through the transformed sigma points $\bar{y}_i = \phi(\sigma_i)$ for

$i \in \{0, 1, 2\}$ as

$$\mu_{\mathrm{ut}} = \sum_{i=0}^{2} \omega_i \bar{y}_i, \qquad \text{and} \qquad \sigma_{\mathrm{ut}}^2 = \sum_{i=0}^{2} \omega_i (\bar{y}_i - \mu_{\mathrm{ut}})^2. \tag{17.24}$$

The corresponding normal probability density function $f_{\mathrm{ut}} : \mathbb{R} \to \mathbb{R}$ is given by

$$f_{\mathrm{ut}}(y) = \frac{1}{\sqrt{2\pi}\sigma_{\mathrm{ut}}} e^{-\frac{1}{2\sigma_{\mathrm{ut}}^2}(y - \mu_{\mathrm{ut}})^2}. \tag{17.25}$$

The parameter $c \in \mathbb{R}\backslash\{-1\}$ is a degree of freedom which can be selected based on further knowledge of ϕ, μ, and σ.

In the case that ϕ is the identity $\phi(x) = x$, it holds that

$$\begin{aligned}
\mu_{\mathrm{ut}} &= \sum_{i=0}^{2} \omega_i \bar{y}_i = \frac{c}{1+c}\mu + \frac{1}{2(1+c)}\left(\mu + \sigma\sqrt{1+c}\right) + \frac{1}{2(1+c)}\left(\mu - \sigma\sqrt{1+c}\right) \\
&= \frac{c\mu}{1+c} + \frac{\mu}{2(1+c)} + \frac{\mu}{2(1+c)} = \mu
\end{aligned} \tag{17.26}$$

and

$$\begin{aligned}
\sigma_{\mathrm{ut}}^2 &= \sum_{i=0}^{2} \omega_i (\bar{y}_i - \mu_{\mathrm{ut}})^2 \\
&= \frac{c}{1+c}[\mu - \mu]^2 + \frac{1}{2(1+c)}\left([(\mu + \sigma\sqrt{1+c}) - \mu]^2 + [(\mu - \sigma\sqrt{1+c}) - \mu]^2\right) \\
&= \frac{(1+c)\sigma^2}{2(1+c)} + \frac{(1+c)\sigma^2}{2(1+c)} = \sigma^2.
\end{aligned} \tag{17.27}$$

Thus the expected value and the variance remain unchanged, showing that the selections of σ_i and ω_i, $i \in \{0, 1, 2\}$, are sensible.

Before we illustrate the transformation on a simple example, we discuss how the unscented transformation extends the propagation of the estimate of the mean value and the covariance matrix of the extended Kalman filter.

Remark 17.5. In the extended Kalman filter, the propagation of the estimate of the expected value solely depends on the mean value and thus satisfies the approximation

$$\mathrm{E}[Y] \approx \mu_{\mathrm{ekf}} = \phi(\mu). \tag{17.28}$$

The estimate of the variance in the extended Kalman filter can be explained through the linearization of the process or map ϕ around the expected value. This leads to the following approximation divided into two steps. As a first step, we use $\mu_{\mathrm{ekf}} = \phi(\mu)$ to get

$$\mathrm{Var}(Y) = \mathrm{E}[(Y - \mathrm{E}(Y))^2] \approx \mathrm{E}[(\phi(X) - \phi(\mu))^2].$$

As a second step, the linearization of ϕ around μ, i.e, $\phi(x) \approx \phi(\mu) + \frac{d}{dx}\phi(\mu)(x - \mu)$, leads to the approximation

$$\text{Var}(Y) \approx \text{E}[(\phi(\mu) + \tfrac{d}{dx}\phi(\mu)(X - \mu) - \phi(\mu))^2] = \text{E}[(\tfrac{d}{dx}\phi(\mu)(X - \mu))^2]$$
$$= (\tfrac{d}{dx}\phi(\mu))^2\, \text{E}[(X - \mu)^2] = (\tfrac{d}{dx}\phi(\mu))^2\, \text{Var}(X). \tag{17.29}$$

Even though we are not discussing the unscented transformation in the context of the Kalman filter yet, as indicated through the index, the estimated propagation of the expected value (17.28) and the variance (17.29) based on a single value and the linearization of ϕ correspond to updates performed in the extended Kalman filter. In contrast, the unscented transformation (17.24) for $\phi : \mathbb{R} \to \mathbb{R}$ relies on three sigma points centred around μ, which leads to superior performance as proven in [72] and [156], for example.

Example 17.6. Consider the normal probability density function (17.21) with $\sigma = 2$ and $\mu = 1$ together with the transformation

$$\phi(x) = \text{sign}(x)\sqrt{2|x|} \qquad \Longleftrightarrow \qquad \phi^{-1}(x) = \tfrac{1}{2}\text{sign}(x)x^2. \tag{17.30}$$

For $c = 1$, the sigma points and the weights are defined as

$$\sigma_0 = 1, \quad \sigma_1 = 3.83, \quad \sigma_2 = -1.83, \quad \text{and} \quad \omega_0 = 0.5, \quad \omega_1 = \omega_2 = 0.25,$$

respectively. The functions f, g, ϕ, and ϕ^{-1} are visualized in Figure 17.4. Ad-

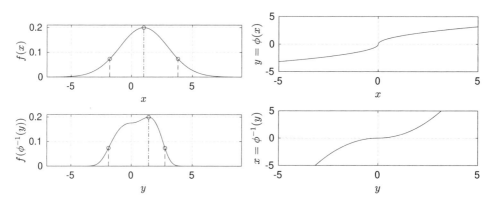

Figure 17.4: Normal distribution (17.21) with $\mu = 1$ and $\sigma = 2$ transformed through a nonlinear map $\phi(x)$ defined in (17.30). Additionally the σ-points and their transformation are shown.

ditionally, for $i \in \{0, 1, 2\}$, the sigma points σ_i and the transformed counterparts $\bar{y}_i = \phi(\sigma_i)$ are highlighted.

Using the formulas (17.24), the estimated expected value and the estimated variance satisfy

$$\mu_{\text{ut}} = 0.92 \qquad \text{and} \qquad \sigma^2_{\text{ut}} = 2.98$$

in this example. The true expected value and the true variance of the function g calculated from (17.23) are given by

$$\text{E}[Y] = \mu_g = 0.47 \qquad \text{and} \qquad \text{Var}(Y) = \sigma^2_g = 1.93.$$

The function g in (17.22) and its approximation f_{ut} in (17.25) are shown in Figure 17.5. As a comparison, additionally the normal probability density function

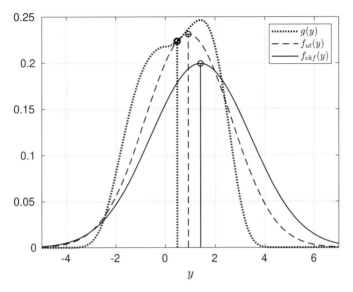

Figure 17.5: Probability density function g in (17.22) together with two normal approximations (17.25) and (17.31).

(17.21) with $\mu_{\mathrm{ekf}} = 1.41 = \phi(1)$ and $\sigma_{\mathrm{ekf}}^2 = 4$ (obtained through (17.28) and (17.29), respectively), i.e.,

$$f_{\mathrm{ekf}}(y) = \frac{1}{\sqrt{2\pi}\sigma_{\mathrm{ekf}}} e^{-\frac{1}{2\sigma_{\mathrm{ekf}}^2}(y-\mu_{\mathrm{ekf}})^2}, \tag{17.31}$$

is shown as a comparison.

Before we present the unscented Kalman filter, we extend the unscented transformation to the n-dimensional setting $\phi : \mathbb{R}^n \to \mathbb{R}^n$ and normal probability density function $f : \mathbb{R}^n \to \mathbb{R}$,

$$f(x) = \frac{1}{(2\pi)^{\frac{n}{2}}\sqrt{\det(P)}} e^{-\frac{1}{2}(x-\mu)^T P^{-1}(x-\mu)}, \tag{17.32}$$

with $\mathrm{E}[X] = \mu \in \mathbb{R}^n$, $\mathrm{Var}(X) = P \in \mathcal{S}_{>0}^n$.

Since P is symmetric and positive definite, the matrix is diagonalizable, i.e., $P = \Psi \Lambda \Psi^{-1}$. Here, $\Lambda \in R^{n \times n}$ is a diagonal matrix $\Lambda = \mathrm{diag}(\lambda_1^2, \dots, \lambda_n^2)$, $\lambda_i \in \mathbb{R}_{>0}$, $i \in \{1, \dots, n\}$, and $\Psi = [\psi_1, \dots, \psi_n] \in \mathbb{R}^{n \times n}$ contains the eigenvectors $\psi_i \in \mathbb{R}^n$ with $|\psi_i| = 1$ for all $i \in \{1, \dots, n\}$ without loss of generality.

In this case the estimates $\mu_{\mathrm{ut}} \approx \mathrm{E}[Y]$, $P_{\mathrm{ut}} \approx \mathrm{Var}(Y)$ of the *unscented transformation* are defined through $2n + 1$ *sigma points*

$$\sigma_0 = \mu, \qquad \sigma_i = \mu + \lambda_i \psi_i \sqrt{n + c}, \qquad \sigma_{n+i} = \mu - \lambda_i \psi_i \sqrt{n + c}, \tag{17.33}$$

for $i \in \{1, \ldots, n\}$, and the weighting factors

$$\omega_0 = \frac{c}{n+c}, \qquad \omega_i = \frac{1}{2(n+c)}, \qquad \forall\, i \in \{1, \ldots, 2n\},$$

for $c \in \mathbb{R}\backslash\{-n\}$. In particular, we define

$$\mu_{\mathrm{ut}} = \sum_{i=0}^{2n} \omega_i \bar{y}_i, \qquad \text{and} \qquad P_{\mathrm{ut}} = \sum_{i=0}^{2n} \omega_i (\bar{y}_i - \mu_{\mathrm{ut}})(\bar{y}_i - \mu_{\mathrm{ut}})^T, \qquad (17.34)$$

where $\bar{y}_i = \phi(\sigma_i)$ for $i \in \{0, 1, \ldots, 2n\}$.

17.3.2   Unscented Kalman Filter

As in the case of the extended Kalman filter we are looking for an observer with the structure

$$\begin{aligned} \hat{\chi}(k) &= F(k-1, \hat{x}(k-1)), \\ \hat{x}(k) &= \hat{\chi}(k) + G_k\left(y(k) - \hat{y}(k)\right) \end{aligned} \qquad (17.35)$$

consisting of a prediction step and a correction step, and the function F, the estimate of the output $\hat{y}(k)$, as well as the Kalman gain G_k, $k \in \mathbb{N}$, to be determined. To calculate $\hat{\chi}(k+1)$ through $\hat{x}(k)$ let

$$\hat{x}(k) = \mathrm{E}[x(k)], \qquad P_k = \mathrm{Var}(x(k) - \hat{x}(k)) = \mathrm{E}[(x(k) - \hat{x}(k))(x(k) - \hat{x}(k))^T].$$

For the unscented transformation, consider

$$\xi(k) = \begin{bmatrix} x(k) \\ v(k) \\ w(k) \end{bmatrix},$$

representing an augmented random variable $\xi \in \mathbb{R}^{n+p+q}$. With Assumption 16.6, it holds that

$$\hat{\xi}(k) = \mathrm{E}[\xi(k)] = \begin{bmatrix} \hat{x}(k) \\ 0 \\ 0 \end{bmatrix}$$

and

$$P_k^\xi = \mathrm{Var}(\xi(k) - \hat{\xi}(k)) = \mathrm{E}[(\xi(k) - \hat{\xi}(k))(\xi(k) - \hat{\xi}(k))^T] = \begin{bmatrix} P_k & 0 & 0 \\ 0 & Q^{-1} & 0 \\ 0 & 0 & R^{-1} \end{bmatrix}.$$

We apply the unscented transformation to the function

$$\phi(\xi) = \begin{bmatrix} \phi_x(\xi) \\ \phi_v(\xi) \\ \phi_w(\xi) \end{bmatrix} = \begin{bmatrix} f(k, x) + g(k, x)v \\ 0 \\ 0 \end{bmatrix}.$$

In particular, we define the sigma points σ_i, $i \in \{0, 1, \ldots, 2(n+p+q)\}$ according to (17.33). The sigma points can be decomposed into

$$\sigma_i = \begin{bmatrix} \sigma_i^x \\ \sigma_i^v \\ \sigma_i^w \end{bmatrix},$$

and due to the specific block structure of P_k^ξ, we can assume without loss of generality that

$$\begin{aligned}
\sigma_i^x &= \mathrm{E}[x(k)] = \hat{x}(k) & \forall\, i \in \{0, 2n{+}1, \ldots, 2(n{+}p{+}q)\}, \\
\sigma_i^v &= \mathrm{E}[v(k)] = 0 & \forall\, i \in \{0, \ldots, 2n, 2(n{+}p){+}1, \ldots, 2(n{+}p{+}q)\}, \quad (17.36) \\
\sigma_i^w &= \mathrm{E}[w(k)] = 0 & \forall\, i \in \{0, \ldots, 2(n{+}p)\}
\end{aligned}$$

after an appropriate permutation.

With these definitions, the estimate of the expected value satisfies

$$\mathrm{E}[\xi(k+1)] \approx \mu_{\xi_{\mathrm{ut}}} = \begin{bmatrix} \mu_{\xi_{\mathrm{ut}}}^x \\ \mu_{\xi_{\mathrm{ut}}}^v \\ \mu_{\xi_{\mathrm{ut}}}^w \end{bmatrix} = \sum_{i=0}^{2(n+p+q)} w_i \phi(\sigma_i)$$

$$= \sum_{i=0}^{2(n+p+q)} w_i \begin{bmatrix} f(k, \sigma_i^x) + g(k, \sigma_i^x)\sigma_i^v \\ 0 \\ 0 \end{bmatrix}.$$

Moreover, it holds that

$$\mu_{\xi_{\mathrm{ut}}}^x = \sum_{i=0}^{2(n+p+q)} w_i \left(f(k, \sigma_i^x) + g(k, \sigma_i^x)\sigma_i^v \right)$$

$$= \frac{p+q}{n+p+q+c} f(k, \hat{x}(k)) + \sum_{i=0}^{2n} w_i f(k, \sigma_i^x) + \sum_{i=2n+1}^{2(n+p)} w_i g(k, \hat{x}(k))\sigma_i^v$$

$$= \frac{\tilde{c}}{n+\tilde{c}} f(k, \hat{x}(k)) + \frac{1}{2(n+\tilde{c})} \sum_{i=1}^{2n} f(k, \sigma_i^x),$$

where the last equality follows from $\mathrm{E}[v(k)] = 0$, similar calculations as in (17.26), and the definition $\tilde{c} = p + q + c$. Finally, with

$$\tilde{w}_0 = \frac{\tilde{c}}{n+\tilde{c}}, \qquad w_i = \frac{1}{2(n+\tilde{c})}, \qquad \forall\, i \in \{1, \ldots, 2n\},$$

based on these calculations the prediction step of the unscented Kalman filter (17.35) is defined as

$$\hat{\chi}(k+1) = F(k, \hat{x}(k)) = \mu_{\xi_{\mathrm{ut}}}^x = \sum_{i=0}^{2n} \tilde{w}_i f(k, \sigma_i^x). \qquad (17.37)$$

To define $\hat{y}(k)$ in (17.35), we evaluate (17.14b) at the sigma points, i.e., we consider

$$\mathrm{E}[y(k)] \approx \mu_{y_{\mathrm{ut}}} = \sum_{i=0}^{2(n+p+q)} \omega_i \left[h(k, \phi_x(\sigma_i)) + \phi_w(\sigma_i) \right] = \sum_{i=0}^{2n} \tilde{\omega}_i h(k, \phi_x(\sigma_i)).$$

Here, the simplification follows similar steps as in the derivation of $\mu_{\xi_{\mathrm{ut}}}^x$. Note that in this definition, $\hat{y}(k) = \mu_{y_{\mathrm{ut}}}$ depends on the sigma points defined through $\hat{\xi}(k-1)$.

For the derivation of the Kalman gain G_k, we use approximations of the variance and relate the expressions to the linear Kalman filter in Algorithm 6 and in particular to the definitions in (16.63).

For sigma points σ_i, $i \in \{0, \ldots, 2(n+p+q)\}$ defined through $\hat{\xi}(k)$, the covariance matrix can be approximated through $\mathrm{Var}(x(k+1) - \hat{\chi}(k+1)) \approx P_\chi$,

$$P_\chi = \sum_{i=0}^{2(n+p+q)} \omega_i (\phi_x(\sigma_i) - \mu_{\xi_{\mathrm{ut}}}^x)(\phi_x(\sigma_i) - \mu_{\xi_{\mathrm{ut}}}^x)^T$$

$$= \sum_{i=0}^{2(n+p+q)} \omega_i (f(k, \sigma_i^x) + g(k, \sigma_i^x)\sigma_i^v - \mu_{\xi_{\mathrm{ut}}}^x)(f(k, \sigma_i^x) + g(k, \sigma_i^x)\sigma_i^v - \mu_{\xi_{\mathrm{ut}}}^x)^T$$

$$= \sum_{i=0}^{2(n+p+q)} \omega_i (f(k, \sigma_i^x) - \mu_{\xi_{\mathrm{ut}}}^x)(f(k, \sigma_i^x) - \mu_{\xi_{\mathrm{ut}}}^x)^T + \omega_i (g(k, \sigma_i^x)\sigma_i^v)(g(k, \sigma_i^x)\sigma_i^v)^T.$$

Here, the cross terms $(f(k, \sigma_i^x) - \mu_{\xi_{\mathrm{ut}}}^x)g(k, \sigma_i^x)\sigma_i^v$, $i \in \{0, 1, \ldots, 2(n+p+q)\}$, cancel out since either $\sigma_i^v = 0$ or $\sigma_i^x = \mathrm{E}[x(k)] = \hat{x}(k)$. In the case $\sigma_i^x = \hat{x}(k)$, the symmetry of the selection of σ_i^v leads to a cancellation. The last expression can be further rewritten as

$$P_\chi = \sum_{i=0}^{2n} \tilde{\omega}_i (f(k, \sigma_i^x) - \mu_{\xi_{\mathrm{ut}}}^x)(f(k, \sigma_i^x) - \mu_{\xi_{\mathrm{ut}}}^x)^T + g(k, \hat{x}(k))Q^{-1}g(k, \hat{x}(k))^T.$$

$$(17.38)$$

In Equation (17.38) the first term on the right-hand side follows again the same arguments as before. The second term on the right-hand side can be derived as follows:

$$\sum_{i=0}^{2(n+p+q)} \omega_i g(k, \sigma_i^x)\sigma_i^v (\sigma_i^v)^T g(k, \sigma_i^x)^T = \sum_{i=1}^{2q} \omega_i g(k, \sigma_i^x)\sigma_i^v (\sigma_i^v)^T g(k, \sigma_i^x)^T$$

$$= \sum_{i=1}^{2q} \frac{(\sqrt{n+p+q+c})^2}{2(n+p+q+c)} g(k, \hat{x}(k))(\lambda_i^v)^2 \psi_i^v (\psi_i^v)^T g(k, \hat{x}(k))^T$$

$$= g(k, \hat{x}(k)) \left(\sum_{i=1}^{2q} \frac{(\lambda_i^v)^2}{2} \psi_i^v \psi_i^v (\sigma_i^v)^T \right) g(k, \hat{x}(k))^T$$

$$= g(k, \hat{x}(k))Q^{-1}g(k, \hat{x}(k))^T.$$

As a next step, consider sigma points σ_i, $i \in \{0, \ldots, 2(n+p+q)\}$ defined through

$\hat{\xi}(k-1)$, which allow us to estimate the covariance matrix $\mathrm{Var}(y(k) - \hat{y}(k)) \approx P_y$,

$$P_y = \sum_{i=0}^{2(n+p+q)} \omega_i(h(k, \phi_x(\sigma_i)) + \sigma_i^w - \mu_{y_{\mathrm{ut}}})(h(k, \phi_x(\sigma_i)) + \sigma_i^w - \mu_{y_{\mathrm{ut}}})^T$$

$$= \sum_{i=0}^{2n} \tilde{\omega}_i(h(k, \phi_x(\sigma_i)) - \mu_{y_{\mathrm{ut}}})(h(k, \phi_x(\sigma_i)) - \mu_{y_{\mathrm{ut}}})^T + R^{-1}. \qquad (17.39)$$

Finally, inspired by (16.62), for sigma points σ_i, $i \in \{0, \ldots, 2(n+p+q)\}$, defined through $\hat{\xi}(k-1)$ we compute

$$P_{x,y} = \sum_{i=0}^{2(n+p+q)} \omega_i(\phi_x(\sigma_i) - \mu_{\xi_{\mathrm{ut}}}^x)(h(k, \phi_x(\sigma_i)) + \sigma_i^w - \mu_{y_{\mathrm{ut}}})^T$$

$$= \sum_{i=0}^{2n} \tilde{\omega}_i(\phi_x(\sigma_i) - \mu_{\xi_{\mathrm{ut}}}^x)(h(k, \phi_x(\sigma_i)) - \mu_{y_{\mathrm{ut}}})^T.$$

With these definitions, mirroring the update (16.63) we define

$$\begin{aligned} P_k^{k-1} &= P_\chi \\ G_k &= P_{x,y}(P_y)^{-1} \\ P_k &= P_k^{k-1} - G_k P_y G_k^T, \end{aligned} \qquad (17.40)$$

to obtain an iterative update of the Kalman gain G_k. The unscented Kalman filter is summarized in Algorithm 9.

Compared to the extended Kalman filter, the unscented Kalman filter does not rely on the derivative of the dynamics (17.14), which simplifies its implementation. The parameter $\tilde{c}$ is only necessary for the derivation of the unscented Kalman filter. Since the selection of c is arbitrary, we use c instead of $\tilde{c}$ in Algorithm 9. An optimal selection of c depending on the system dynamics and the unknown disturbances v and w is nontrivial. For the update of P_k^{k-1}, g is evaluated only at the mean value $\hat{x}$. Instead, g can also be evaluated at appropriately selected sigma points [161]. The interested reader is invited to compare the unscented Kalman filter with the extended Kalman filter on the system of Example 17.4 in Exercise 17.4.

17.4   MOVING HORIZON ESTIMATION

With the relation between optimal control and the Kalman filter (or minimum energy estimation in the continuous time setting) and extensions of optimal control in terms of MPC to handle state constraints, it is not surprising that there exists an equivalent extension in the context of state estimation. The basic concept of *Moving Horizon Estimation* (MHE), the dual problem of MPC, is briefly discussed here. Similarly to MPC, MHE relies on the solution of optimization problems at every discrete time step $k \in \mathbb{N}$, in general.

Algorithm 9: Unscented Kalman filter

Input: Parameter $c \in \mathbb{R}\backslash\{-n\}$, system (17.14), positive definite weight matrices $Q^{-1} = \text{Var}(v(k))$, $R^{-1} = \text{Var}(w(k))$, and initial estimates $\hat{x}(0) = \hat{x}_0$, $P_0 \in \mathcal{S}_{>0}^n$.

Output: Estimates $\hat{x}(k)$ and $\hat{\chi}(k)$ of the state $x(k)$ for $k \in \mathbb{N}$.

Algorithm: For $k \in \mathbb{N}$:

1. Based on $\hat{x}(k-1)$ and $P_{k-1} = \Psi\Lambda\Psi^{-1}$ define sigma points σ_i^x, $i \in \{0, \ldots, 2n\}$ (see (17.33)).

2. Compute

$$\phi_i^x = f(k-1, \sigma_i^x), \qquad i \in \{0, \ldots, 2n\},$$

and define (see (17.37) and (17.38), (17.39))

$$\hat{\chi}(k) = \sum_{i=0}^{2n} \omega_i \phi_i, \qquad \hat{y}(k) = \sum_{i=0}^{2n} \omega_i h(k, \phi_i^x),$$

$$P_\chi = \sum_{i=0}^{2n} \omega_i (\phi_i^x - \hat{\chi}(k))(\phi_i^x - \hat{\chi}(k))^T$$

$$+ g(k-1, \hat{x}(k-1))Q^{-1}g(k-1, \hat{x}(k-1))^T,$$

$$P_y = \sum_{i=0}^{2n} \omega_i (h(k, \phi_i^x) - \hat{y}(k))(h(k, \phi_i^x) - \hat{y}(k))^T + R^{-1}.$$

3. Measure the output $y(k)$, compute

$$P_{x,y} = \sum_{i=0}^{2n} \omega_i (\phi_i^x - \hat{\chi}(k))(h(k, \phi_i^x) - \hat{y}(k))^T.$$

$$G_k = P_{x,y}P_y^{-1}$$

$$\hat{x}(k) = \hat{\chi}(k) + G_k(y(k) - \hat{y}(k))$$

$$P_k = P_\chi - G_k P_y G_k^T,$$

and set $k = k+1$ to go to step 1.

Consider similar dynamics as in the preceding section defined as

$$x(k+1) = f(k, x(k), v(k)), \tag{17.41a}$$

$$y(k) = h(k, x(k)) + w(k). \tag{17.41b}$$

Again, $x \in \mathbb{X} \subset \mathbb{R}^n$ denotes the state, $y \in \mathbb{R}^p$ denotes the measured output, and $(v(k))_{k\in\mathbb{N}} \subset \mathbb{V} \subset \mathbb{R}^q$ and $(w(k))_{k\in\mathbb{N}} \subset \mathbb{W} \subset \mathbb{R}^p$ denote unknown disturbances. Inputs $u(k)$ can be included in $f: \mathbb{N} \times \mathbb{R}^n \times \mathbb{R}^q \to \mathbb{R}^n$ and $h: \mathbb{N} \times \mathbb{R}^n \to \mathbb{R}^p$ in the same way as in (17.15).

Based on measured output data $y(k)$, we are looking for "optimal" disturbances

$\hat{v}(k)$, $\hat{w}(k)$ such that

$$\hat{x}(k+1) = f(k, \hat{x}(k), \hat{v}(k)), \tag{17.42a}$$
$$y(k) = h(k, \hat{x}(k)) + \hat{w}(k) \tag{17.42b}$$

is satisfied and from which the optimal state estimates $\hat{x}(k)$ are obtained.

In Section 16.3.1 we introduced the discrete time Kalman filter through an optimization problem which grows with the available data $y(i)$, $0 \le i \le k \in \mathbb{N}$ at time $k \in \mathbb{N}$. However, in Section 16.3.2 we derived an iterative scheme to keep the numerical complexity constant at every time step. Nevertheless, this was only possible due to the unconstrained setting and due to the linearity of the dynamics. In contrast, to keep the numerical complexity of MHE constant at every time step k, only the most recent $\bar{N} \in \mathbb{N}$ data sets $y(i)$, $i \in \{k - \bar{N}, \ldots, k-1\}$ are taken into account for the state estimation of $x(k)$.

Before we introduce the MHE optimization problem and before we can specify what is meant by an optimal disturbance sequence, we require some definitions, for which we use notation similar to that in Section 15.1. We define the set

$$\mathbb{D} = \mathbb{X} \times \mathbb{V} \times \mathbb{W}$$

summarizing the state constraints and bounds of the disturbances. Similarly to (15.3), at time $k \in N$, for given $y(i)$ for $i \in \mathbb{Z}_{[k-\bar{N}, k-1]}$, we define the set of feasible disturbance trajectories

$$\mathcal{V}_{\mathbb{D}}^{\bar{N}} = \left\{ v_{\bar{N}}(\cdot) : \mathbb{Z}_{[k-\bar{N}, k-1]} \to \mathbb{R}^q \; \middle| \; \begin{array}{l} \hat{x}(i+1) = f(i, \hat{x}(i), v(i)), \\ y(i) = h(i, \hat{x}(i)) + w(i), \\ (\hat{x}(i+1), v(i), w(i)) \in \mathbb{D}, \\ \forall\, i \in \mathbb{Z}_{[k-\bar{N}, k-1]} \end{array} \right\}. \tag{17.43}$$

Note that $\mathcal{V}_{\mathbb{D}}^{\bar{N}} = \mathcal{V}_{\mathbb{D}}^{\bar{N}}(k, y_{\bar{N}})$ depends on the time index k and the sequence $y(\cdot) : \mathbb{Z}_{[k-\bar{N}, k-1]} \to \mathbb{R}^p$. For simplicity, the functional arguments are suppressed here. Additionally we define the *cost function* $\bar{J}_{\bar{N}} : \mathbb{R}^n \times \mathcal{U}_{\mathbb{D}}^N \to \mathbb{R} \cup \{\infty\}$,

$$\bar{J}_{\bar{N}}(\hat{x}(k-\bar{N}), v_{\bar{N}}(\cdot); y_{\bar{N}}(\cdot)) = F_{\bar{N}}(\hat{x}(k-\bar{N})) + \sum_{i=k-N}^{k-1} \ell(v(i), w(i)) \tag{17.44}$$

$$= F_{\bar{N}}(\bar{x}(k-\bar{N})) + \sum_{i=k-N}^{k-1} \ell(v(i), y(i) - h(i, \hat{x}(i))),$$

depending on the estimate of the state $\hat{x}(k-\bar{N})$, the disturbance sequence $v_{\bar{N}}(\cdot)$, and the measured output $y_{\bar{N}}(\cdot) : \mathbb{Z}_{[T-\bar{N}, T-1]} \to \mathbb{R}^p$. For given $\hat{x}(k-\bar{N})$, $v(\cdot)$, and $y(\cdot)$, the state trajectory $\hat{x}(\cdot)$ and the disturbances $w(\cdot)$ are implicitly defined through the dynamics (17.41).

Here, $\ell : \mathbb{R}^q \times \mathbb{R}^p \to \mathbb{R}$ denotes costs with respect to the disturbances and $F_{\bar{N}} : \mathbb{R}^n \to \mathbb{R}$ denotes costs with respect to the estimate of the state $x(k-N)$. The function $F_{\bar{N}}$ can be interpreted similarly to the terminal costs in MPC, capturing the costs which occur by neglecting the time steps $i \le k - \bar{N} - 1$ in the cost function.

From the optimization problem

$$\bar{V}_{\bar{N}}(k, y_{\bar{N}}(\cdot)) = \min_{\substack{v_{\bar{N}}(\cdot) \in \mathcal{V}_{\mathbb{D}}^{\bar{N}} \\ \hat{x}(k-\bar{N}) \in \mathbb{X}}} \bar{J}_{\bar{N}}(\hat{x}(k-\bar{N}), v_{\bar{N}}(\cdot); y_{\bar{N}(\cdot)})$$

(17.45)

$$\text{subject to (17.41a)},$$

the optimal sequence $\hat{v}_{\bar{N}}(\cdot)$ and the optimal state estimate $\hat{x}(k-\bar{N})$ can be obtained. Here, optimality is achieved with respect to the cost function (17.44). From $\hat{v}_{\bar{N}}(\cdot)$ and $\hat{x}(k-\bar{N})$, in particular the state estimate $\hat{x}(k)$ can be computed through the dynamics (17.41a).

The estimate $\hat{x}(k)$ can be used to design a state feedback $\mu(k) = u(\hat{x}(k))$ through MPC, for example. Similarly to MPC, after shifting the horizon by going from k to $k+1$, the shifted optimization problem (15.4) can be solved at the next time step to obtain $\hat{x}(k+1)$. In this context, for the first $k \in \{0, \ldots, \bar{N}-1\}$ time steps when only k measurements of y are available, $\bar{N}$ in the optimization problem (17.45) is replaced by k.

17.5   EXERCISES

Exercise 17.1. Reproduce the results in Example 17.3 in MATLAB.

Exercise 17.2. Embed the continuous time extended Kalman filter in the MPC implementation discussed in Appendix B.5. In particular, define the control law based on the estimated state $\hat{x}$.

Exercise 17.3. Reproduce the results in Example 17.4 in MATLAB.

Exercise 17.4. Apply the unscented Kalman filter to the setting discussed in Example 17.4 in MATLAB.

Exercise 17.5. Embed the discrete time extended Kalman filter in the MPC implementation discussed in Appendix B.5. In particular, define the control law based on the estimated state $\hat{\chi}$.

17.6   BIBLIOGRAPHICAL NOTES AND FURTHER READING

The story of the development of the extended Kalman filter is part of the broader story of the Apollo project. Specifically, the linearity assumed for the Kalman filter was too strict an assumption to solve a problem in mid-course navigation. The fortuitous timing of a visit by Kalman to NASA Ames in 1960, at precisely the time the mid-course navigation problem was being struggled with, and the subsequent development of the extended Kalman filter to account for nonlinearities in the problem, are described in [106].

The derivation of the extended Kalman filter equations in the continuous time setting follows the presentation in [87, Ch. 11.2].

For the discrete time extended Kalman filter we refer to [70] and [39]. For the unscented Kalman filter we refer to [156, 160, 161] and to the work by Julier and Uhlmann [72, 73, 74, 75], in which the unscented Kalman filter originates.

The unscented Kalman filter uses a particular selection of $2n + 1$ points to estimate the propagation of a normal probability density function in terms of the mean value and the covariance matrix. *Particle filters*, which are not discussed in this book but can be found in [130], for example, can be considered as an extension, where the propagation of a probability density function is estimated through an arbitrary number of particles instead of a fixed number of $2n + 1$ points.

Further extensions of the Kalman filter include *square-root information filter* [23] and *ensemble Kalman filter* [48], for example.

For further development of moving horizon estimation, see [128], for example.

Chapter Eighteen

Observer Design for Nonlinear Systems

In this chapter we discuss representatives of observers for nonlinear systems in terms of *high-gain observers* and *sliding mode observers*. In particular, in the first section of the chapter we give an introduction to *high-gain observers* while the second part discusses *sliding mode observers*. Sliding mode observers extend ideas from the controller design in Chapter 10.

18.1 HIGH-GAIN OBSERVERS

In this section we consider systems in (triangular) normal form

$$
\begin{aligned}
\dot{x}_1 &= x_2 + \psi_1(x_1, u) \\
\dot{x}_2 &= x_3 + \psi_2(x_1, x_2, u) \\
&\;\;\vdots \\
\dot{x}_{r-1} &= x_r + \psi_{r-1}(x_1, \ldots, x_{r-1}, u) \\
\dot{x}_r &= \phi(x, \eta, u) \\
\dot{\eta} &= q(x, \eta, u) \\
y &= x_1
\end{aligned}
\tag{18.1}
$$

with state $[x^T, \eta^T]^T \in \mathbb{R}^n$, $x \in \mathbb{R}^r$, input $u \in \mathbb{R}^m$, and output $y \in \mathbb{R}$. In Section 12.3 we have seen how to transform a system into this normal form in the case of single-input systems $u \in \mathbb{R}$. Here, $\psi_i : \mathbb{R}^i \times \mathbb{R}^m \to \mathbb{R}$ for $i = 1, \ldots, r-1$, $\phi : \mathbb{R}^r \times \mathbb{R}^{n-r} \to \mathbb{R}$, and $q : \mathbb{R}^r \times \mathbb{R}^{n-r} \to \mathbb{R}^{n-r}$. The η-dynamics represent the zero dynamics of the overall system.

We consider a *high-gain observer* of the form

$$
\begin{aligned}
\dot{\hat{x}}_1 &= \hat{x}_2 + \hat{\psi}_1(\hat{x}_1, u) + \frac{\alpha_1}{\varepsilon}(y - \hat{x}_1) \\
\dot{\hat{x}}_2 &= \hat{x}_3 + \hat{\psi}_2(\hat{x}_1, \hat{x}_2, u) + \frac{\alpha_2}{\varepsilon^2}(y - \hat{x}_1) \\
&\;\;\vdots \\
\dot{\hat{x}}_{r-1} &= \hat{x}_r + \hat{\psi}_{r-1}(\hat{x}_1, \ldots, \hat{x}_{r-1}, u) + \frac{\alpha_{r-1}}{\varepsilon^{r-1}}(y - \hat{x}_1) \\
\dot{\hat{x}}_r &= \hat{\phi}(\hat{x}, u) + \frac{\alpha_r}{\varepsilon^r}(y - \hat{x}_1),
\end{aligned}
\tag{18.2}
$$

where $\hat{\psi}_i : \mathbb{R}^i \times \mathbb{R}^m \to \mathbb{R}$ for $i = 1, \ldots, r-1$, $\hat{\phi} : \mathbb{R}^r \to \mathbb{R}$, $\alpha_1, \ldots, \alpha_r \in \mathbb{R}$, and $\varepsilon \in \mathbb{R}_{>0}$. In the observer dynamics, $\hat{\psi}_i$ and $\hat{\phi}$ are approximations of the possibly unknown functions ψ_i and ϕ, and α_i and ε are design parameters to ensure asymptotic stability of the origin for the error dynamics

$$e(t) = x(t) - \hat{x}(t) \to 0 \qquad \text{for } t \to \infty.$$

Note that the high-gain observer is only a partial state observer, i.e., only estimates of x, but not of η, are obtained. While ϕ might depend on η, $\hat{\phi}$ in the observer dynamics is independent of η. Nonetheless, the observer structure in (18.2) is similar to what we have seen previously in that it comprises a copy of the system dynamics (albeit without the zero dynamics) driven by an output injection term.

To be able to derive a convergence result for the error dynamics $e(t)$ for $t \to \infty$ we make the following assumptions on the functions involved in the plant dynamics (18.1) and the observer dynamics (18.2). We assume that the functions ψ_i, ϕ, and q are locally Lipschitz continuous in their arguments. Moreover, we assume that the input u as well as the states η are bounded and thus there exist compact sets $\mathcal{U} \subset \mathbb{R}^m$ and $\mathcal{Q} \subset \mathbb{R}^{n-r}$ such that $u(t) \in \mathcal{U}$ and $\eta(t) \in \mathcal{Q}$ for all $t \in \mathbb{R}_{\geq 0}$. Due to these assumptions, for $\mathcal{X} \subset \mathbb{R}^r$ compact, there exist $L_i \in \mathbb{R}_{>0}$ such that

$$|\psi_i(x_1, \ldots, x_i, u) - \psi_i(z_i, \ldots, z_i, u)| \leq L_i \sum_{k=1}^{i} |x_k - z_k| \qquad (18.3)$$

for all $x, z \in \mathcal{X}$, for all $u \in \mathcal{U}$, and for all $i \in \{1, \ldots, r\}$. Additionally, for a given compact set $\mathcal{X} \subset \mathbb{R}^r$ we define

$$M_i = \max_{x \in \mathcal{X}} |x_i|, \qquad \forall\, i \in \{1, \ldots, r\},$$

and define the functions $\hat{\psi}_i$ as

$$\hat{\psi}_i(x_1, \ldots, x_i, u) = \psi_i\left(M_1 \operatorname{sat}\left(\frac{x_1}{M_1}\right), \ldots, M_i \operatorname{sat}\left(\frac{x_i}{M_i}\right), u \right). \qquad (18.4)$$

Through this definition, $\hat{\psi}_i$ satisfies

$$\hat{\psi}_i(x_1, \ldots, x_i, u) = \psi_i(x_1, \ldots, x_i, u)$$

for all $x \in \mathcal{X}$ and

$$|\hat{\psi}_i(x_1, \ldots, x_i, u) - \hat{\psi}_i(z_i, \ldots, z_i, u)| \leq L_i \sum_{k=1}^{i} |x_k - z_k| \qquad (18.5)$$

for all $x, z \in \mathbb{R}^r$. In particular, while ψ_i might only be locally Lipschitz continuous, $\hat{\psi}_i$ is globally Lipschitz continuous with respect to x by design.

For the functions ϕ and $\hat{\phi}$, we assume that there exists an $L > 0$ and an $M_\phi > 0$ such that

$$|\phi(x, \eta, u) - \hat{\phi}(z, u)| \leq L|x - z| + M_\phi \qquad (18.6)$$

holds for all $x \in \mathcal{X}$, for all $z \in \mathbb{R}^r$, for all $\eta \in \mathcal{Q}$, and for all $u \in \mathcal{U}$.

There are several natural choices for $\hat{\phi}$. If the function ϕ is not known then

$\hat{\phi} \equiv 0$ is a valid selection. If ϕ is known, is globally Lipschitz continuous, and does not depend on the zero dynamics, then $\hat{\phi}$ can be defined as $\hat{\phi} = \phi$. In this case, (18.6) is satisfied for $M_\phi = 0$. If ϕ is only locally Lipschitz continuous the same result can be obtained through the definition

$$\hat{\phi}(x, u) = \phi \left(\left[M_1 \operatorname{sat} \left(\frac{x_1}{M_1} \right), \dots, M_r \operatorname{sat} \left(\frac{x_r}{M_r} \right) \right]^T, \eta, u \right),$$

similarly to (18.4).

18.1.1   Convergence Properties of High-Gain Observers

With the assumptions and notations introduced so far, the following results can be derived.

Theorem 18.1 ([88, Theorem 2.1]). *Consider the plant* (18.1) *together with the observer dynamics* (18.2) *and assume that the involved functions satisfy the assumptions* (18.3)–(18.6). *Additionally, let* $\mathcal{X}$, $\mathcal{U}$, *and* $\mathcal{Q}$ *be compact sets with* $x(t) \in \mathcal{X}$, $u(t) \in \mathcal{U}$, *and* $\eta(t) \in \mathcal{Q}$ *for all* $t \in \mathbb{R}_{\geq 0}$ *and assume that* α_i, $i \in \{1, \dots, r\}$, *are selected such that the polynomial*

$$\chi(s) = s^r + \alpha_1 s^{r-1} + \dots + \alpha_{r-1} s + \alpha_r \tag{18.7}$$

is Hurwitz. Then there is $\varepsilon^* \in (0, 1]$ *such that for* $0 < \varepsilon \leq \varepsilon^*$, *the estimation error* $e_i = x_i - \hat{x}_i$, $i = 1, \dots, r$, *of the high-gain observer satisfies*

$$|e_i| \leq \max \left\{ \frac{b}{\varepsilon^{i-1}} |e(0)| e^{-a\varepsilon/t}, \varepsilon^{r+1-i} c M_\phi \right\} \tag{18.8}$$

for some $a, b, c \in \mathbb{R}_{>0}$.

Before we prove the result, we highlight some properties of Theorem 18.1. Theorem 18.1 ensures that under technical assumptions, a high-gain observer ensures the following properties:

- For all $\varepsilon > 0$ sufficiently small, the error e and the observer states $\hat{x}$ remain bounded. This follows from the fact that the error e is bounded through (18.8) and the state x remains in the compact set $\mathcal{X}$.
- For any $\mu > 0$ there exists an $\varepsilon > 0$ so that after some time the error e is within μ of the origin, i.e., $|e(t)| \leq \mu$ for all t large enough. This follows from the fact that the first term on the right-hand side converges to zero for $t \to \infty$ and the second term can be made arbitrarily small by choosing ε arbitrarily small.

However, despite these beneficial properties the high-gain observer also needs to be applied with caution. For example, higher order systems can lead to numerical difficulties due to the term $\frac{1}{\varepsilon^r}$, which becomes very large if ε is small and the relative degree is large. For example $\varepsilon = 0.01$ and a relative degree of $r = 4$ generates values in the range of $\frac{1}{\varepsilon} = 100$ to $\frac{1}{\varepsilon^4} = 10^8$ in the observer dynamics. This additionally implies that measurement noise can be problematic since the observer will respond quickly and a peaking phenomenon might occur before the error converges to zero.

Proof of Theorem 18.1: We define the scaled estimation error

$$z_i = \frac{e_i}{\varepsilon^{r-i}}.$$

The scaled error satisfies

$$
\begin{aligned}
\varepsilon \dot{z}_i &= \frac{1}{\varepsilon^{r-i-1}} \left(\dot{x}_i - \dot{\hat{x}}_i \right) \\
&= \frac{1}{\varepsilon^{r-i-1}} \left(x_{i+1} + \psi_i(x_i, \ldots, x_i, u) - \hat{x}_{i+1} - \hat{\psi}_i(\hat{x}_1, \ldots, \hat{x}_i, u) - \frac{\alpha_i}{\varepsilon^i}(x_1 - \hat{x}_1) \right) \\
&= \frac{x_{i+1} - \hat{x}_{i+1}}{\varepsilon^{r-(i+1)}} - \frac{\alpha_i e_1}{\varepsilon^{r-1}} + \frac{1}{\varepsilon^{r-i-1}} \left(\psi_i(x_1, \ldots, x_i, u) - \hat{\psi}_i(\hat{x}_1, \ldots, \hat{x}_i, u) \right) \\
&= \dot{z}_{i+1} - \alpha_i \dot{z}_1 + \frac{1}{\varepsilon^{r-i-1}} \left(\psi_i(x_i, \ldots, x_i, u) - \hat{\psi}_i(\hat{x}_1, \ldots, \hat{x}_i, u) \right)
\end{aligned}
$$

for $i \in \{1, \ldots, r-1\}$ and

$$
\begin{aligned}
\varepsilon \dot{z}_r &= \frac{1}{\varepsilon^{-1}} \left(\phi(x, \eta, u) - \hat{\phi}(\hat{x}, u) - \frac{\alpha_r}{\varepsilon^r}(x_1 - \hat{x}_1) \right) \\
&= -\alpha_r z_1 + \frac{1}{\varepsilon^{-1}} \left(\phi(x, \eta, u) - \hat{\phi}(\hat{x}, u) \right).
\end{aligned}
$$

With the definitions

$$
F = \begin{bmatrix}
-\alpha_1 & 1 & 0 & \cdots & 0 \\
-\alpha_2 & 0 & 1 & \cdots & 0 \\
\vdots & \vdots & \ddots & \ddots & \vdots \\
-\alpha_{r-1} & \vdots & & 0 & 1 \\
-\alpha_r & 0 & \cdots & \cdots & 0
\end{bmatrix}
$$

and

$$
\delta(x, \hat{x}, \eta, u) = \begin{bmatrix}
\frac{1}{\varepsilon^{r-1}} \left(\psi_i(x_1, u) - \hat{\psi}_i(\hat{x}_1, u) \right) \\
\vdots \\
\frac{1}{\varepsilon^1} \left(\psi_{r-1}(x_1, \ldots, x_{r-1}, u) - \hat{\psi}_{r-1}(\hat{x}_1, \ldots, \hat{x}_{r-1}, u) \right) \\
\phi(x, \eta, u) - \hat{\phi}(\hat{x}, u)
\end{bmatrix}
$$

we can rewrite the z dynamics as $\varepsilon \dot{z} = Fz + \varepsilon \delta(x, \hat{x}, \eta, u)$. Since $\chi(s) = \det(sI - F)$ the matrix F is Hurwitz.

According to the definition of $\hat{\psi}_i$ and δ_i, it holds that

$$
\begin{aligned}
|\delta_i(x, \hat{x}, \eta, u)| &\leq \frac{1}{\varepsilon^{r-i}} \left| \hat{\psi}_i(x_1, \ldots, x_i, u) - \hat{\psi}_i(\hat{x}_1, \ldots, \hat{x}_i, u) \right| \\
&\leq \frac{L_i}{\varepsilon^{r-i}} \sum_{k=1}^{i} |x_k - \hat{x}_k| \leq \frac{L_i}{\varepsilon^{r-i}} \sum_{k=1}^{i} \varepsilon^{r-k} |z_k| \leq L_i \sum_{k=1}^{i} \varepsilon^{i-k} |z_k|
\end{aligned}
$$

for all $i = 1, \ldots, r-1$. Combining this estimate with (18.6) implies the existence of L_δ, which is independent of ε for all $\varepsilon \leq 1$, such that

$$|\delta| \leq \mathcal{L}_\delta |z| + M_\phi.$$

As a next step we consider the candidate Lyapunov function

$$V(z) = z^T P z,$$

where P is the solution of the Lyapunov equation $PF + F^T P = -I$. Note that P is positive definite since F is Hurwitz. The candidate Lyapunov function satisfies

$$\varepsilon \dot{V}(z) = -z^T z + 2\varepsilon z^T P \delta(x, \hat{x}, \eta, u),$$

which can be upper bounded by

$$\varepsilon \dot{V}(z) \leq -|z|^2 + 2\varepsilon |z| \|P\| (L_\delta |z| + M_\phi)$$
$$= -|z|^2 + 2\varepsilon \|P\| L_\delta |z|^2 + 2\varepsilon \|P\| M_\phi |z|.$$

Thus, for $\varepsilon \|P\| L_\delta \leq \frac{1}{4}$,

$$\varepsilon \dot{V}(z) \leq -\frac{1}{2}|z|^2 + 2\varepsilon \|P\| M_\phi |z|$$

and

$$\varepsilon \dot{z} \leq -\frac{1}{4}|z|^2 \qquad \forall \, |z| \geq 8\varepsilon M_\phi \|P\|.$$

This implies that $z(t)$ converges exponentially fast to the set $\{z \in \mathbb{R}^r : |z| \leq 8\varepsilon M \|P\|\}$ and

$$|z(t)| \leq \max\{b e^{-at/\varepsilon}|z_0|, \varepsilon c M_\phi\} \tag{18.9}$$

for some constants $a, b, c \in \mathbb{R}_{>0}$. With the scaling $|z_0| \leq \varepsilon^{1-r}|e_0|$ and $|e_i| = \varepsilon^{r-i}|z_i|$ this completes the proof. $\square$

Based on Theorem 18.1, only convergence to a neighborhood around the origin $e = 0$ can be guaranteed. The size of the neighborhood depends on $M_\phi \geq 0$ which was introduced to bound the mismatch between ϕ and $\hat{\phi}$ in (18.6). If the function ϕ is known and does not depend on η (in the case $r = n$, for example), then (18.6) is satisfied for $M_\phi = 0$ and $e(t) \to 0$ for $t \to \infty$ can be guaranteed. We summarize this fact in the following corollary.

Corollary 18.2. *Consider the plant (18.1) together with the observer dynamics (18.2) and assume that the involved functions satisfy the assumptions (18.3)–(18.6) with $M_\phi = 0$. Additionally, let $\mathcal{X}$, $\mathcal{U}$, and $\mathcal{Q}$ be compact sets with $x(t) \in \mathcal{X}$, $u(t) \in \mathcal{U}$, and $\eta(t) \in \mathcal{Q}$ for all $t \in \mathbb{R}$ and assume that α_i, $i \in \{1, \ldots, r\}$, are selected such that the polynomial (18.7) is Hurwitz. Then there is $\varepsilon^* \in (0, 1]$ such that, for $0 < \varepsilon \leq \varepsilon^*$, the estimation error $e = x - \hat{x}$ of the high-gain observer exponentially converges to zero.*

18.1.2   Examples

We now illustrate the performance of the high-gain observer via several examples.

Example 18.3. We begin with the simple linear dynamics

$$\dot{x}_1 = x_2$$
$$\dot{x}_2 = -x_1 - 2x_2 + u \tag{18.10}$$
$$y = x_1$$

in the form (18.1) with an asymptotically stable origin when the input is zero. Consequently, we can define the high-gain observer

$$\dot{\hat{x}}_1 = \hat{x}_2 + \frac{\alpha_1}{\varepsilon}(y - \hat{x}_1)$$
$$\dot{\hat{x}}_2 = \hat{\phi}(\hat{x}, u) + \frac{\alpha_2}{\varepsilon^2}(y - \hat{x}_1).$$

The polynomial

$$\chi(s) = s^2 + 2s + 1 = (s+1)^2$$

is Hurwitz and we can thus select $\alpha_1 = 2$ and $\alpha_2 = 1$ in accordance with (18.7). The function $\phi(x, u) = -x_1 - 2x_2 + u$ is globally Lipschitz continuous with respect to x and u and in particular satisfies

$$|\phi(x, u) - \phi(z, v)| \le \|A\| \cdot \left|[x^T, u]^T - [z^T, v]^T\right| \le \sqrt{6} \cdot \left|[x^T, u]^T - [z^T, v]^T\right|,$$

for all $x, z \in \mathbb{R}^2$ and $u, v \in \mathbb{R}$, where $A = \begin{bmatrix} -1 & -2 & 1 \end{bmatrix}$. Thus, for $\hat{\phi} = \phi$ the condition (18.6) is satisfied for $M_\phi = 0$ and for $\hat{\phi} \equiv 0$ the inequality

$$|\phi(x, u) - \hat{\phi}(z, u)| \le M_\phi$$

holds for an appropriately selected $M_\phi > 0$ if we restrict the domain of the states and the input to a compact set.

Figures 18.1 and 18.2 show the states of the plant as well as the states of the observer for $\hat{\phi} = \phi$ and for $\hat{\phi} \equiv 0$, respectively. The input u is defined as

$$u = \sin(2t),$$

for the parameter ε we use $\varepsilon = 0.1$, and the initial states of the plant and the observer are set to

$$x_0 = \begin{bmatrix} 1 \\ 1 \end{bmatrix} \quad \text{and} \quad \hat{x}_0 = \begin{bmatrix} 0 \\ 0 \end{bmatrix}.$$

In the case $\phi = \hat{\phi}$, in Figure 18.1, the error $x(t) - \hat{x}(t)$ converges to zero as expected from Corollary 18.2 since $M_\phi = 0$. In the second case, where $\hat{\phi} \equiv 0$, in Figure 18.2, only convergence of $x(t) - \hat{x}(t)$ to a neighborhood around the origin can be expected according to Theorem 18.1.

Figure 18.3 visualizes the performance of the high-gain observer with respect to the parameter ε. Here, $\hat{\phi}$ is defined as $\hat{\phi} = \phi$ and the parameter ε is varied. While small values of ε guarantee fast convergence of the error dynamics to the

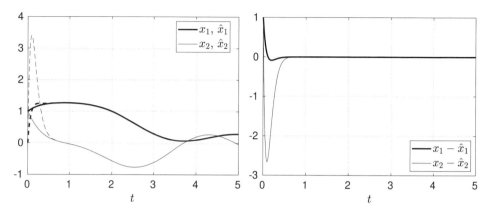

Figure 18.1: Performance of the high-gain observer in the case that the function ϕ is known, i.e., $\phi = \hat{\phi}$. On the left, the solid lines correspond to x while the dashed lines correspond $\hat{x}$.

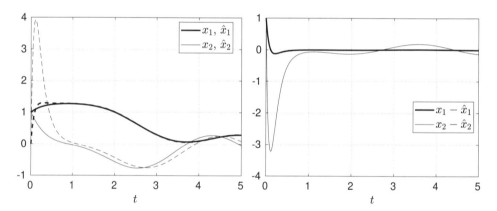

Figure 18.2: Performance of the high-gain observer in the case that the function ϕ is not known, i.e., $\hat{\phi}$ is selected as $\hat{\phi} \equiv 0$. On the left, the solid lines correspond to x while the dashed lines correspond $\hat{x}$.

origin, they also lead to a peaking phenomenon, i.e., a large short-term transient, for small t. In particular, for small parameters ε we observe in Figure 18.3 that the difference between the plant state and the observer state is large before exponential convergence to zero can be observed. We address this phenomenon again in the next example.

Example 18.4. Consider the nonlinear plant dynamics

$$
\begin{aligned}
\dot{x}_1 &= x_2, \\
\dot{x}_2 &= x_2^3 + u, \\
y &= x_1.
\end{aligned}
\tag{18.11}
$$

Since the input

$$
u = -x_2^3 - x_1 - x_2
\tag{18.12}
$$

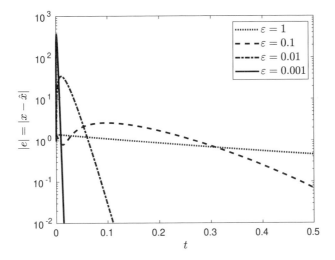

Figure 18.3: Performance of the high-gain observer with respect to ε.

leads to a linear system, it is straightforward to verify that (18.12) is a globally stabilizing state feedback controller for (18.11). A corresponding high-gain observer of (18.11) with $\alpha_1 = 2$ and $\alpha_2 = 1$ is given by the dynamics

$$\dot{\hat{x}}_1 = \hat{x}_2 + \frac{2}{\varepsilon}(y - \hat{x}_1)$$
$$\dot{\hat{x}}_2 = \frac{1}{\varepsilon^2}(y - \hat{x}_1),$$

where $\hat{\phi}$ is taken as $\hat{\phi} \equiv 0$.

The closed-loop solutions of the plant and the observer with respect to the initial conditions

$$x_0 = \begin{bmatrix} 0.5 \\ 0.5 \end{bmatrix} \qquad \text{and} \qquad \hat{x}_0 = \begin{bmatrix} 0 \\ 0 \end{bmatrix},$$

$\varepsilon = 0.01$, and input

$$u = -\hat{x}_2^3 - \hat{x}_1 - \hat{x}_2 \tag{18.13}$$

are visualized in Figure 18.4. We observe that the error $e = x - \hat{x}$ grows rapidly and eventually leads to finite escape of the overall plant-observer dynamics. This is due to the peaking phenomenon of the observer states in combination with the fact that the input (18.13) is defined based on the observer states $\hat{x}$ and not on the states of the plant.

For the setting discussed here, this problem can be avoided by considering a saturated input

$$u = \text{sat}(-\hat{x}_2^3 - \hat{x}_1 - \hat{x}_2) \tag{18.14}$$

instead. Figure 18.5 shows the plant states and the observer states for the case that the feedback law (18.13) is replaced by the saturated input (18.14). Here, the

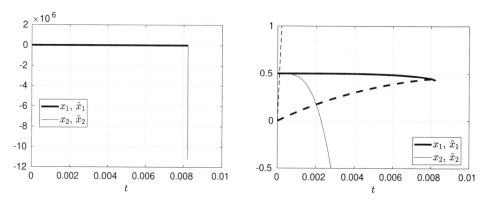

Figure 18.4: Closed-loop solutions of the plant states and the observer states. Due to the peaking phenomenon of the observer, the overall system has a finite escape time.

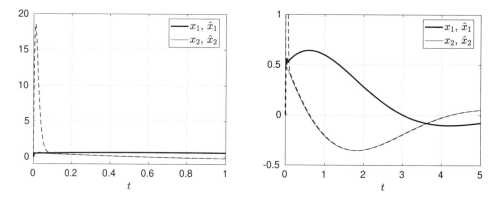

Figure 18.5: In contrast to Figure 18.4, the saturated input (18.14) achieves that the observer states track the plant states after the "peaking" seen in the estimate $\hat{x}_2$.

saturated input ensures that the states x are not pushed too far away from the origin even though $\hat{x}$ initially grows rapidly. Thus, for small t where $e(t)$ is large and the input (18.13) is meaningless to the plant, saturating the input ensures that the impact of the mismatch between x and $\hat{x}$ on the plant stays limited. The inputs for the two different simulations are shown in Figure 18.6.

Note that while saturating the input is a valid solution for the particular setting here, it in general requires that stabilization or tracking be possible with a saturated controller or, alternatively, that a global design not be necessary. However, this example contains ingredients that can be used to derive a *nonlinear separation principle* which, similarly to the (linear) separation principle derived in Section 16.1, allows the design of the controller and observer to be pursued separately. See Section 18.4 for further reading.

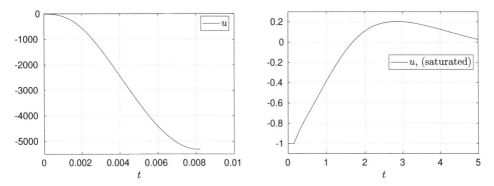

Figure 18.6: Input (18.13) and saturated input (18.14) corresponding to Figures 18.4 and 18.5, respectively.

18.1.3   Extension to Multi-Output Systems

The high-gain observer discussed for systems of the form (18.1) can be extended in a straightforward way to multi-output systems of the form

$$
\begin{aligned}
\dot{x}_{i,j} &= x_{i,j+1}, & i &= 1,\ldots,p, \quad j = 1,\ldots,r_i - 1, \\
\dot{x}_{i,r_i} &= \phi_i(x,\eta,u), & i &= 1,\ldots,p, \\
\dot{\eta} &= q(x,\eta,u), & & \\
y_i &= x_{i,1}, & i &= 1,\ldots,p,
\end{aligned} \tag{18.15}
$$

with $p \in \mathbb{N}$, $r_i \in \mathbb{N}_{\geq 2}$, $i \in \{1,\ldots,p\}$, and

$$
x = [x_{1,1} \quad \cdots \quad x_{1,r_1} \; x_{2,1} \quad \cdots \quad \cdots \quad x_{2,r_2} \quad \cdots \quad x_{p,1} \quad \cdots \quad x_{p,r_p}]^T \in \mathbb{R}^n.
$$

Here we define the corresponding high-gain observer through

$$
\begin{aligned}
\dot{\hat{x}}_{i,j} &= \hat{x}_{i,j+1} + \tfrac{\alpha_{i,j}}{\varepsilon^j}(y_i - \hat{x}_{i,1}), & i &= 1,\ldots,p, \; j = 1,\ldots,r_i - 1, \\
\dot{\hat{x}}_{i,r_i} &= \hat{\phi}_i(\hat{x},u) + \tfrac{\alpha_{i,j}}{\varepsilon^{r_j}}(y_i - \hat{x}_{i,1}), & i &= 1,\ldots,p,
\end{aligned} \tag{18.16}
$$

where we make the same assumptions on $\hat{\phi}_i$, $i \in \{1,\ldots,p\}$, as in (18.6). In this case, if

$$
\chi_i(s) = s^{r_i} + \alpha_{i,1}s^{r_i-1} + \ldots + \alpha_{i,r_i-1}s + \alpha_{i,r_i} \tag{18.17}
$$

are Hurwitz polynomials for $i = 1,\ldots,p$, then a bound similar to (18.9) can be established. In this case M_ϕ is defined as $M_\phi = \max_{i=1,\ldots,p} M_{\phi,i}$ and $M_{\phi,i}$ corresponds to (18.6) for ϕ_i and $\hat{\phi}_i$.

Note that the structure (18.15) is not uncommon and, for example, the inverted pendulum on a cart can be written in the form of (18.15).

Example 18.5. We consider again the dynamics of the inverted pendulum (17.12) and the parameters used in Example 17.3. After relabeling the states

$$
\begin{bmatrix} x_{1,1} & x_{1,2} & x_{2,1} & x_{2,2} \end{bmatrix}^T = \begin{bmatrix} x_1 & x_3 & x_2 & x_4 \end{bmatrix}^T,
$$

the dynamics (17.12) can be equivalently written as

$$
\begin{bmatrix} \dot{x}_{1,1} \\ \dot{x}_{1,2} \\ \dot{x}_{2,1} \\ \dot{x}_{2,2} \end{bmatrix} = \begin{bmatrix} x_{1,2} \\ \dfrac{-\bar{J}\bar{c}x_{1,2} - \bar{J}\sin(x_{2,1})x_{2,2}^2 - \bar{\gamma}\cos(x_{2,1})x_{2,2} + g\cos(x_{2,1})\sin(x_{2,1}) + \bar{J}u}{\bar{M}\bar{J} - \cos^2(x_{2,1})} \\ x_{2,2} \\ \dfrac{-\bar{M}\bar{\gamma}x_{2,2} + \bar{M}g\sin(x_{2,1}) - \bar{c}\cos(x_{2,1})x_{1,2} - \cos(x_{2,1})\sin(x_{2,1})x_{2,2}^2 + \cos(x_{2,1})u}{\bar{M}\bar{J} - \cos^2(x_{2,1})} \end{bmatrix},
$$

which is of the form (18.15). According to the definitions in this section, the corresponding high-gain observer (18.16) is given by

$$
\begin{bmatrix} \dot{\hat{x}}_{1,1} \\ \dot{\hat{x}}_{1,2} \\ \dot{\hat{x}}_{2,1} \\ \dot{\hat{x}}_{2,2} \end{bmatrix} = \begin{bmatrix} \hat{x}_{1,2} \\ \dfrac{-\bar{J}\bar{c}\hat{x}_{1,2} - \bar{J}\sin(\hat{x}_{2,1})\hat{x}_{2,2}^2 - \bar{\gamma}\cos(\hat{x}_{2,1})\hat{x}_{2,2} + g\cos(\hat{x}_{2,1})\sin(\hat{x}_{2,1}) + \bar{J}u}{\bar{M}\bar{J} - \cos^2(\hat{x}_{2,1})} \\ \hat{x}_{2,2} \\ \dfrac{-\bar{M}\bar{\gamma}\hat{x}_{2,2} + \bar{M}g\sin(\hat{x}_{2,1}) - \bar{c}\cos(\hat{x}_{2,1})\hat{x}_{1,2} - \cos(\hat{x}_{2,1})\sin(\hat{x}_{2,1})\hat{x}_{2,2}^2 + \cos(\hat{x}_{2,1})u}{\bar{M}\bar{J} - \cos^2(\hat{x}_{2,1})} \end{bmatrix}
$$
$$
+ \begin{bmatrix} \frac{\alpha_{1,1}}{\varepsilon}\left(x_{1,1} - \hat{x}_{1,1}\right) \\ \frac{\alpha_{1,2}}{\varepsilon^2}\left(x_{1,1} - \hat{x}_{1,1}\right) \\ \frac{\alpha_{2,1}}{\varepsilon}\left(x_{2,1} - \hat{x}_{2,1}\right) \\ \frac{\alpha_{2,2}}{\varepsilon^2}\left(x_{2,1} - \hat{x}_{2,1}\right) \end{bmatrix}.
$$

To illustrate the performance of the observer we use the parameters and the input defined in Example 17.3. Note that the input $u(\hat{x}) = -\hat{x}_{1,2} - \hat{x}_{2,2}$ does not stabilize the pendulum in the upright position and thus the states grow unboundedly for $t \to \infty$. Additionally we use the Hurwitz polynomial $\chi(s) = s^2 + 2s + 1$ to define the parameters $\alpha_{1,1} = \alpha_{2,1} = 2$ and $\alpha_{1,2} = \alpha_{2,2} = 1$ and select $\varepsilon = 0.1$. Figure 18.7 shows the closed-loop solution with respect to the initial conditions $x_0 = [2, 2, 2, 2]^T$, $\hat{x}_0 = [0, 0, 0, 0]^T$.

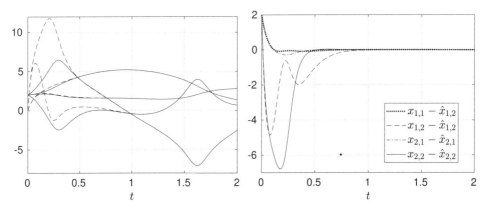

Figure 18.7: Closed-loop solutions of the inverted pendulum (solid lines) and the corresponding high-gain observer (dashed lines) together with the error dynamics on the right.

18.2   SLIDING MODE OBSERVERS

The technique of sliding mode control introduced in Chapter 10 can be modified to develop state observers. We first illustrate the ideas on linear systems before discussing the nonlinear setting.

18.2.1   Sliding Mode Observers for Linear Systems

Consider the linear system

$$\dot{x} = Ax + Bu$$
$$y = Cx$$

with state $x \in \mathbb{R}^n$, input $u \in \mathbb{R}^m$, and output $y \in \mathbb{R}^p$. Here, $A \in \mathbb{R}^{n \times n}$, $B \in \mathbb{R}^{n \times m}$, $C \in \mathbb{R}^{p \times n}$, and we assume that the pair (A, C) is observable and $\operatorname{rank}(C) = p$.

With these definitions, we can assume without loss of generality that A, B, and C are of the form

$$A = \begin{bmatrix} A_{11} & A_{12} \\ A_{21} & A_{22} \end{bmatrix}, \qquad B = \begin{bmatrix} B_1 \\ B_2 \end{bmatrix} \qquad \text{and} \qquad C = \begin{bmatrix} 0 & I \end{bmatrix}, \qquad (18.18)$$

where $I \in \mathbb{R}^{p \times p}$ denotes the identity matrix. In particular, through an appropriate coordinate transformation $x \mapsto Tx$, $T \in \mathbb{R}^{n \times n}$, the specific structure can always be obtained.

For the observer we consider the dynamics

$$\dot{\hat{x}} = A\hat{x} + Bu + G\nu$$
$$\hat{y} = C\hat{x}, \qquad\qquad (18.19)$$

where the output injection term $G \in \mathbb{R}^{n \times p}$ is of the form

$$G = \begin{bmatrix} -L \\ I \end{bmatrix}.$$

Here, the matrix $L \in \mathbb{R}^{(n-p) \times p}$ needs to be designed and $\nu \in \mathbb{R}^p$ is defined as

$$\nu = \rho \operatorname{sign}(Ce), \qquad (18.20)$$

where $e = x - \hat{x}$ denotes the error between the state of the plant and the state of the observer. In (18.20) the sign function is to be understood componentwise and $\rho > 0$ denotes a constant. Again note that the general structure of the observer (18.19) consists of a copy of the system dynamics driven by an output injection term.

Given the assumed structure of C we split the error

$$e = \begin{bmatrix} e_x \\ e_y \end{bmatrix},$$

where $e_x \in \mathbb{R}^{n-p}$ and $e_y \in \mathbb{R}^p$, which allows us to write (18.20) as

$$\nu = \rho \operatorname{sign}(e_y).$$

With these definitions, the error dynamics $\dot{e} = Ae - G\nu$ satisfy

$$
\begin{aligned}
\begin{bmatrix} \dot{e}_x \\ \dot{e}_y \end{bmatrix} &= \begin{bmatrix} A_{11} & A_{12} \\ A_{21} & A_{22} \end{bmatrix} \begin{bmatrix} e_x \\ e_y \end{bmatrix} - \begin{bmatrix} -L \\ I \end{bmatrix} \nu \\
&= \begin{bmatrix} A_{11}e_x + A_{12}e_y + L\nu \\ A_{21}e_x + A_{22}e_y - \rho\,\mathrm{sign}(e_y) \end{bmatrix}.
\end{aligned}
\tag{18.21}
$$

As a next step we investigate conditions to ensure that $e_y(t)$ converges to zero in finite time. Consider the candidate Lyapunov function $V(e_y) = \frac{1}{2}e_y^T e_y$ whose derivative satisfies

$$
\begin{aligned}
\dot{V}(e_y) &= e_y^T \dot{e}_y \\
&= e_y \left(A_{21}e_x + A_{22}e_y - \rho\,\mathrm{sign}(e_y) \right) \\
&= \sum_{i=1}^{p} e_{y,i} \left(A_{21,i}e_x + A_{22,i}e_y - \rho\,\mathrm{sign}(e_{y,i}) \right) \\
&\leq \sum_{i=1}^{p} |e_{y,i}| \left(|A_{21,i}e_x + A_{22,i}e_y| - \rho \right),
\end{aligned}
\tag{18.22}
$$

where $A_{21,i}$ and $A_{22,i}$ denote the ith row of A_{21} and A_{22}, respectively. If the constant $\rho > 0$ satisfies

$$
\rho \geq \eta + \max_{i=1,\dots,p} |A_{21,i}e_x + A_{22,i}e_y|
\tag{18.23}
$$

for $\eta > 0$, then (18.22) can be further rewritten as

$$
\dot{V}(e_y) = \sum_{i=1}^{p} -\eta|e_{y,i}| = -\eta|e_y|_1 \leq -\eta|e_y| = -\eta\sqrt{V(e_y)}.
\tag{18.24}
$$

The last inequality can easily be seen to be true by looking at the inequality $\sum_{i=1}^{p} e_{y,i}^2 \leq \left(\sum_{i=1}^{p} |e_{y,i}| \right)^2$. Additionally, recall that we use $|\cdot|_1$ to denote the 1-norm while we use $|\cdot|$ to denote the 2-norm (see Appendix A.1).

Thus, V is a finite-time Lyapunov function according to Theorem 10.4 and, consequently, the error e_y converges to zero in finite time. Using the terminology of Chapter 10, the *sliding surface* $\mathbb{R}^{n-p} \times \{0\} \subset \mathbb{R}^{n-p} \times \mathbb{R}^p$ is reached in finite time.

Note that condition (18.23) can only be satisfied locally, i.e., for every fixed ρ, η, there exists $e \in \mathbb{R}^n$ such that (18.23) is not satisfied. Thus, the observer developed here is only a local observer. However, for ρ arbitrarily large, the domain where (18.23) is satisfied is arbitrarily large. Finally, before we continue, note that (18.23) needs to be satisfied for all times $t \in \mathbb{R}_{\geq 0}$ and not only for the initial time $t = 0$ and the initial error $e(0)$ to ensure that the estimate (18.24) holds for all $t \in \mathbb{R}_{\geq 0}$.

Once the dynamics evolve along the sliding surface, i.e., $e_y(t) = 0$ and $\dot{e}_y(t) = 0$, the dynamics (18.21) collapse to

$$
\begin{bmatrix} \dot{e}_x \\ 0 \end{bmatrix} = \begin{bmatrix} A_{11}e_x + L\nu_{eq} \\ A_{21}e_x - \nu_{eq} \end{bmatrix}
\tag{18.25}
$$

and ν_{eq} is the *equivalent output injection* similar to the equivalent control (10.22).

As in (10.22), the equivalent output injection is only a tool to define the matrix

L. In numerical simulations $e_y(t) = 0$ and $\dot{e}_y(t) = 0$ will not be satisfied using the control law (18.20) implemented in MATLAB, for example. However, as in Section 10.4, ν_{eq} can be approximated through an observer. We will illustrate the approximation of ν_{eq} on an example at the end of this section.

Returning to (18.25), the error dynamics e_x satisfy

$$\dot{e}_x = A_{11}e_x + LA_{21}e_x = (A_{11} + LA_{21})e_x.$$

Thus $e_x(t) \to 0$ for $t \to \infty$ if the matrix L is selected such that $A_{11} + LA_{21}$ is Hurwitz. Note that observability of (A, C) implies observability of (A_{11}, A_{21}) and thus the selection of a matrix L stabilizing the error dynamics is always possible under the assumptions of this section. We ask the reader to verify this claim in Exercise 18.4.

Example 18.6. Consider the dynamics $\dot{x} = Ax + Bu$, $y = Cx$ defined by the matrices

$$A = \begin{bmatrix} 1 & 1 & -2 \\ 1 & 1 & 0 \\ -1 & -2 & 2 \end{bmatrix}, \quad B = \begin{bmatrix} 1 \\ 1 \\ 1 \end{bmatrix}, \quad C = \begin{bmatrix} -1 & -1 & 0 \\ -2 & 1 & -1 \end{bmatrix}. \quad (18.26)$$

We ask the reader to verify that the triple (A, B, C) is controllable and observable.

The matrices (18.26) are not in the form (18.18). The command $\texttt{null}(C)$ in MATLAB returns the (unit) vector

$$v = \frac{1}{\sqrt{11}} \begin{bmatrix} -1 \\ 1 \\ 3 \end{bmatrix}$$

defining the null space of C, and hence $v^T C = 0$ is satisfied. We can thus define the matrix

$$T = \begin{bmatrix} \sqrt{11}v^T \\ C \end{bmatrix} = \begin{bmatrix} -1 & 1 & 3 \\ -1 & -1 & 0 \\ -2 & 1 & -1 \end{bmatrix}.$$

whose inverse satisfies

$$T^{-1} = \frac{1}{11} \begin{bmatrix} -1 & -4 & -3 \\ 1 & -7 & 3 \\ 3 & 1 & -2 \end{bmatrix}.$$

Note that the constant $\sqrt{11}$ in the definition of T is not necessary but leads to a simpler representation of T. With T, we obtain the coordinate transformation $\tilde{x} = Tx$ and the matrices

$$\tilde{A} = TAT^{-1} = \frac{1}{11} \begin{bmatrix} 21 & 62 & -25 \\ 6 & 24 & -4 \\ 7 & -5 & -1 \end{bmatrix}, \quad \tilde{B} = TB = \begin{bmatrix} 3 \\ -2 \\ -2 \end{bmatrix},$$

$$\tilde{C} = CT^{-1} = \begin{bmatrix} 0 & 1 & 0 \\ 0 & 0 & 1 \end{bmatrix}.$$

Identify the error vector $e = [e_1\, e_2\, e_3]^T$ and then $e_x = e_1$ and $e_y = [e_2\, e_3]^T$.

For $\rho = 15$, $\eta = 1$, and $|e| \leq 6$ it holds that

$$\rho = 15 \geq \eta + 2.2782|e| = \eta + \frac{1}{11} \left\| \begin{bmatrix} 6 \\ 24 \\ -4 \end{bmatrix} \right\| \cdot |e| \geq \eta + \left(\begin{bmatrix} 6 & 24 & -4 \end{bmatrix} e \right) \quad \text{and}$$

$$\rho = 15 \geq \eta + 0.7873|e| = \eta + \frac{1}{11} \left\| \begin{bmatrix} 7 \\ -5 \\ -1 \end{bmatrix} \right\| \cdot |e| \geq \eta + \left(\begin{bmatrix} 7 & -5 & -1 \end{bmatrix} e \right).$$

Thus the input $\nu = \rho \, \mathrm{sign}(e_y)$ ensures finite-time convergence of $e(t)$ to the sliding surface for all e_0 in an appropriate neighborhood around the origin.

To ensure asymptotic convergence of $e(t)$ to the origin, we define the matrix

$$L = \begin{bmatrix} -3 & -2 \end{bmatrix},$$

which ensures that the matrix

$$\tilde{A}_{11} + L\tilde{A}_{21} = \frac{1}{11} \left(21 + \begin{bmatrix} -3 & -2 \end{bmatrix} \begin{bmatrix} 6 \\ 7 \end{bmatrix} \right) = -1$$

is Hurwitz.

In Figure 18.8 the solution of the error dynamics (18.21) for this setting are visualized. Here, the initial error is defined as $e_0 = [1, -1, 1]^T$. Visualizations of

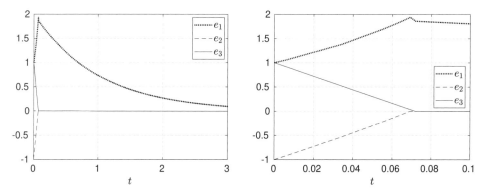

Figure 18.8: Convergence of the error dynamics using a sliding mode observer. After the sliding surface is reached in finite time, the remaining error variables converge to zero asymptotically.

the plant state and the observer state are omitted here, but can be easily obtained if an initial condition x_0 is fixed. As expected by construction, in Figure 18.8 we observe that the error e_y converges to zero in finite time before e_x asymptotically converges to zero. It can be verified that the condition $|e(t)| \leq 6$ is satisfied for all $t \in [0, 3]$.

The variables $\nu_1(t)$ and $\nu_2(t)$ corresponding to the visualization in Figure 18.8 are shown in Figure 18.9. Here, we observe that the sliding surface is reached approximately at $t = 0.07$ when ν starts chattering.

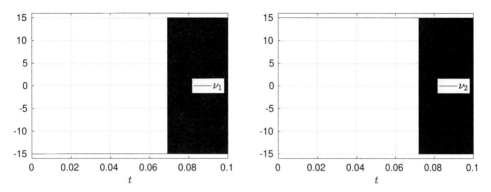

Figure 18.9: Visualization of $\nu(t)$ corresponding to Figure 18.8.

The equivalent output injection is approximated through a low pass filter

$$\dot{\nu}_{eq} = - \begin{bmatrix} \frac{1}{\tau} & 0 \\ 0 & \frac{1}{\tau} \end{bmatrix} \nu_{eq} + \begin{bmatrix} \frac{1}{\tau} & 0 \\ 0 & \frac{1}{\tau} \end{bmatrix} \nu.$$

The results are shown in Figure 18.10, where $\tau = 0.01$ is used and the initial value is set to $\nu_{eq}(0) = [-15, 15]^T$.

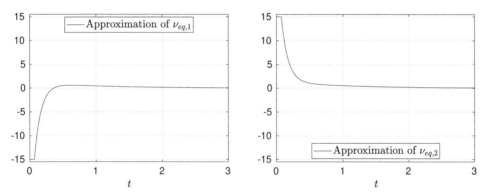

Figure 18.10: Approximation of ν_{eq} obtained through a low pass filter.

Compared to the Luenberger observer (16.6a) for linear systems, certain robustness properties of the sliding mode observer (18.19) can be established. To this end, consider the perturbed dynamics

$$\dot{x} = \begin{bmatrix} A_{11} & A_{12} \\ A_{21} & A_{22} \end{bmatrix} \begin{bmatrix} x_1 \\ x_2 \end{bmatrix} + \begin{bmatrix} B_1 \\ B_2 \end{bmatrix} u + \begin{bmatrix} \delta_1(t, x, u) \\ \delta_2(t, x, u) \end{bmatrix}$$

$$y = \begin{bmatrix} 0 & I \end{bmatrix} x$$

where $\delta_1 : \mathbb{R} \times \mathbb{R}^n \times \mathbb{R}^m \to \mathbb{R}^{n-p}$ and $\delta_2 : \mathbb{R} \times \mathbb{R}^n \times \mathbb{R}^m \to \mathbb{R}^p$ are unknown disturbances. In combination with the observer (18.19), the estimate (18.22) becomes

$$\dot{V}(e_y) \leq \sum_{i=1}^{p} |e_{y,i}| \left(|A_{21,i} e_x + A_{22,i} e_y + \delta_2(t, x, u)| - \rho \right).$$

Thus, if $\delta_2(t, x, u)$ is bounded, i.e., there exists $M > 0$ such $|\delta_2(t, x, u)| \le M$ for all $(t, x, u) \in \mathbb{R}_{\ge 0} \times \mathbb{R}^n \times \mathbb{R}^m$, then the selection of ρ with

$$\rho \ge \eta + M + \max_{i=1,\dots,p} |A_{21,i} e_x + A_{22,i} e_y|$$

ensures finite-time convergence of e_y similarly to the unperturbed case. With

$$\nu_{eq} = A_{21} e_x + \delta_2(t, x, u)$$

the e_x-dynamics satisfy

$$\dot{e}_x = (A_{11} + L A_{21}) e_x + \delta_1(t, x, u) + L \delta_2(t, x, u).$$

Hence, e_x does not necessarily converge to zero, but if $(A_{11} + L A_{21})$ is Hurwitz, e_x converges to a neighborhood around the origin whose size depends on the magnitude of $\delta_1(t, x, u) + L \delta_2(t, x, u)$.

To conclude this section we point out that chattering can be reduced as in the discussion in Section 10.2.2 by approximating the sign function with a continuous or smooth function. The parameter ρ should be selected based on a trade-off between guarantees on the size of the region of attraction (which means choosing ρ large) and minimizing the amplitude of the chattering on the sliding surface (which means choosing ρ small).

18.2.2   Nonlinear Systems

The results derived above for linear systems can be extended to nonlinear systems of the form

$$\begin{aligned}
\dot{x}_1 &= A_{11} x_1 + A_{12} x_2 + \phi_1(x, u), \\
\dot{x}_2 &= A_{21} x_1 + A_{22} x_2 + \phi_2(x, u) + D_2 \delta(t, u, y), \\
y &= C_2 x_2.
\end{aligned} \qquad (18.27)$$

In addition we use the shorthand notation

$$\begin{aligned}
\dot{x} &= A x + \phi(x, u) + D \delta(t, u, y), \\
y &= C x
\end{aligned} \qquad (18.28)$$

to represent the dynamics (18.27). Here, $\delta : \mathbb{R}_{\ge 0} \times \mathbb{R}^m \times \mathbb{R}^p \to \mathbb{R}^n$ is an unknown function, $\phi : \mathbb{R}^n \times \mathbb{R}^m \to \mathbb{R}^n$ is a known nonlinear function, and $A \in \mathbb{R}^{n \times n}$, $D \in \mathbb{R}^{n \times q}$, $C \in \mathbb{R}^{p \times n}$, and $q \le p < n$. Moreover, $C_2 \in \mathbb{R}^{p \times p}$ is invertible by assumption and the states x_1 and x_2 are of dimension $x_1 \in \mathbb{R}^{n-p}$ and $x_2 \in \mathbb{R}^p$, respectively. From these definitions the dimensions of the remaining matrices and functions can be deduced. Under mild technical observability conditions (see [137, Chapter 3, Assumption 1 and 2]) a generic system (18.28) can be transformed into a system of the form (18.27) using a coordinate transformation $x \mapsto T x$, $T \in \mathbb{R}^{n \times n}$.

While the function δ is unknown in this section, it is bounded by a known function $\rho : \mathbb{R}_{\ge 0} \times \mathbb{R}^p \times \mathbb{R}^m \to \mathbb{R}_{\ge 0}$ by assumption. In particular, we assume that

$$|\delta(t, u, y)| \le \rho(t, u, y) \qquad \forall\, (t, u, y) \in \mathbb{R}_{\ge 0} \times \mathbb{R}^m \times \mathcal{U}, \qquad (18.29)$$

where $\mathcal{U} \subset \mathbb{R}^m$ denotes the set of admissible inputs. The function ϕ is assumed to be Lipschitz continuous with respect to x, i.e., there exists $L_\phi \in \mathbb{R}_{\geq 0}$ such that

$$|\phi(x, u) - \phi(\tilde{x}, u)| \leq L_\phi |x - \tilde{x}|, \qquad \forall\, x, \tilde{x} \in \mathbb{R}^n, \quad \forall u \in \mathcal{U}. \tag{18.30}$$

To derive a sliding mode observer for the dynamics (18.27) we consider the coordinate transformation $z = Tx$ with

$$T = \begin{bmatrix} I & L \\ 0 & I \end{bmatrix} \qquad \text{and} \qquad T^{-1} = \begin{bmatrix} I & -L \\ 0 & I \end{bmatrix}, \tag{18.31}$$

where L has the specific form $L = \begin{bmatrix} L_x & 0 \end{bmatrix}$ and will be designed in the following. With this specific coordinate transformation, in the z-variables the dynamics (18.27) is of the form

$$\dot{z}_1 = (A_{11} + L A_{21})\, z_1 + (A_{12} + L A_{22} - (A_{11} + L A_{21})L)\, z_2 + \begin{bmatrix} I & L \end{bmatrix} \phi(T^{-1}z, u)$$
$$\dot{z}_2 = A_{21} z_1 + (A_{22} - A_{21}L) z_2 + \phi_2(T^{-1}z, u) + D_2 \delta(t, u, y)$$
$$y = C_2 z_2.$$

For the sliding mode observer we consider the corresponding dynamics

$$\dot{\hat{z}}_1 = (A_{11} + L A_{21})\, \hat{z}_1 + (A_{12} + L A_{22} - (A_{11} + L A_{21})L)\, C_2^{-1} y$$
$$\qquad + \begin{bmatrix} I & L \end{bmatrix} \phi \left(T^{-1} \begin{bmatrix} \hat{z}_1 \\ C_2^{-1} y \end{bmatrix}, u \right)$$
$$\dot{\hat{z}}_2 = A_{21} \hat{z}_1 + (A_{22} - A_{21}L) \hat{z}_2 - K(y - C_2 \hat{z}_2) + \phi_2(T^{-1}z, u) + \nu \tag{18.32}$$
$$\hat{y} = C_2 \hat{z}_2.$$

For the observer, the gain matrix $K \in \mathbb{R}^{p \times p}$, the function $\nu \in \mathbb{R}_{\geq 0} \to \mathbb{R}^p$, and the matrix L need to be designed.

As in the previous section, with the notation $e_x = z_1 - \hat{z}_1$ and $e_y = y - \hat{y} = C_2(z_2 - \hat{z}_2)$, we can define the error dynamics

$$\dot{e}_x = (A_{11} + L A_{21})\, e_x + \begin{bmatrix} I & L \end{bmatrix} \left(\phi(T^{-1}z, u) - \phi \left(T^{-1} \begin{bmatrix} \hat{z}_1 \\ C_2^{-1} y \end{bmatrix}, u \right) \right)$$
$$\dot{e}_y = C_2 A_{21} e_x + (C_2(A_{22} - A_{21}L)C_2^{-1} + C_2 K)e_y + C_2 D_2 \delta(t, y, u) \tag{18.33}$$
$$\qquad - C_2 \nu + C_2 \left(\phi_2(T^{-1}z, u) - \phi_2 \left(T^{-1} \begin{bmatrix} \hat{z}_1 \\ C_2^{-1} y \end{bmatrix}, u \right) \right).$$

Importantly, in this representation, the e_x-dynamics only implicitly depend on e_y, which we will use in the following steps.

Firstly, we define the matrix L, or more precisely the matrix L_x, such that

$$A_{11} + L A_{21}$$

is Hurwitz if possible. Secondly, for a symmetric positive definite matrix $A_s \in \mathbb{R}^{p \times p}$ we define the matrix K as

$$K = -(A_{22} - A_{21}L)C_2^{-1} - C_2^{-1} A_s. \tag{18.34}$$

With this L and K given, the linear term in (18.33) containing the matrix K

simplifies to

$$(A_{22} - A_{21}L)C_2^{-1} + C_2K$$
$$= (A_{22} - A_{21}L)C_2^{-1} + C_2(-(A_{22} - A_{21}L)C_2^{-1} - C_2^{-1}A_s) = -A_s,$$

which is Hurwitz due to the definition of A_s as a positive definite matrix. Thirdly, the output injection term is defined as

$$\nu = \begin{cases} k(t, y, u)C_2^{-1}\frac{e_y}{|e_y|} & \text{if} \quad e_y \neq 0, \\ 0 & \text{if} \quad e_y = 0. \end{cases} \tag{18.35}$$

Here, $\frac{e_y}{|e_y|}$ is a multi-dimensional extension of the sign function which satisfies $\frac{e_y}{|e_y|} = \text{sign}(e_y)$ if $e_y \in \mathbb{R}\backslash\{0\}$ is one-dimensional. It is however not the same as the componentwise definition in (18.20). The function $k : \mathbb{R}_{\geq 0} \times \mathbb{R}^p \times \mathbb{R}^m \to \mathbb{R}$ is still to be defined.

With the definitions of K and ν, the nonlinear system (18.33) can be rewritten in the form

$$\begin{bmatrix} \dot{e}_x \\ \dot{e}_y \end{bmatrix} = \begin{bmatrix} A_{11} + LA_{21} & 0 \\ C_2A_{21} & -A_s \end{bmatrix} \begin{bmatrix} e_x \\ e_y \end{bmatrix} \tag{18.36}$$

$$+ \begin{bmatrix} \begin{bmatrix} I & L \end{bmatrix} \left(\phi(T^{-1}z, u) - \phi \left(T^{-1} \begin{bmatrix} \hat{z}_1 \\ C_2^{-1}y \end{bmatrix}, u \right) \right) \\ C_2D_2\delta(t, u, y) - k(\cdot)\frac{e_y}{|e_y|} + C_2 \left(\phi_2(T^{-1}z, u) - \phi_2 \left(T^{-1} \begin{bmatrix} \hat{z}_1 \\ C_2^{-1}y \end{bmatrix}, u \right) \right) \end{bmatrix},$$

where the matrix defining the linear part is Hurwitz due to the construction of L and A_s.

As a next step in the observer design, we investigate the sliding motion of (18.33) on the sliding surface

$$\mathcal{S} = \left\{ \begin{bmatrix} e_x \\ e_y \end{bmatrix} \in \mathbb{R}^n \middle| e_y = 0 \right\}.$$

To this end, we assume that $P \in \mathcal{S}_{>0}^p$ is a positive definite matrix such that $\bar{P} = P\begin{bmatrix} I & L \end{bmatrix}$ and $\bar{P}$ is a solution of the matrix inequality

$$\bar{A}^T\bar{P}^T + \bar{P}\bar{A} + \frac{1}{\varepsilon}\bar{P}\bar{P}^T + \varepsilon L_\phi^2 I + \alpha P < 0 \tag{18.37}$$

for $\varepsilon, \alpha > 0$. Here $\bar{A}$ denotes the matrix

$$\bar{A} = \begin{bmatrix} A_{11} \\ A_{21} \end{bmatrix}.$$

We consider the candidate Lyapunov function $V(e_x) = e_x^T P e_x$ and assume that

$e_y = \dot{e}_y = 0$. Then V satisfies

$$\dot{V}(e_x) = e_x^T((A_{11} + LA_{21})^T P + P(A_{11} + LA_{21}))e_x$$
$$+ 2e_x^T P \begin{bmatrix} I & L \end{bmatrix} \left(\phi(T^{-1}z, u) - \phi\left(T^{-1}\begin{bmatrix} \hat{z}_1 \\ C_2^{-1}y \end{bmatrix}, u\right) \right)$$
$$= e_x^T(\bar{P}\bar{A} + \bar{A}^T\bar{P}^T)e_x + 2(\bar{P}^T e_x)\left(\phi(T^{-1}z, u) - \phi\left(T^{-1}\begin{bmatrix} \hat{z}_1 \\ C_2^{-1}y \end{bmatrix}, u\right) \right)$$
$$\leq e_x^T(\bar{P}\bar{A} + \bar{A}^T\bar{P}^T)e_x + \varepsilon e_x^T\bar{P}^T\bar{P}e_x + \frac{1}{\varepsilon}\left| \phi(T^{-1}z, u) - \phi\left(T^{-1}\begin{bmatrix} \hat{z}_1 \\ C_2^{-1}y \end{bmatrix}, u\right) \right|^2,$$

where the inequality follows from Young's inequality (Lemma A.4). Since the function ϕ is Lipschitz continuous according to (18.30), the last term on the right-hand side can be further estimated by

$$\left| \phi(T^{-1}z, u) - \phi\left(T^{-1}\begin{bmatrix} \hat{z}_1 \\ C_2^{-1}y \end{bmatrix}, u\right) \right| \leq L_\phi \left| T^{-1}\begin{bmatrix} z_1 \\ z_2 \end{bmatrix} - T^{-1}\begin{bmatrix} \hat{z}_1 \\ C_2^{-1}y \end{bmatrix} \right|$$
$$= L_\phi \left| \begin{bmatrix} I & -L \\ 0 & I \end{bmatrix}\begin{bmatrix} z_1 - \hat{z}_1 \\ 0 \end{bmatrix} \right| = L_\phi|e_x|. \tag{18.38}$$

Using this result, the derivative of the candidate Lyapunov function satisfies

$$\dot{V}(e_x) \leq e_x^T(\bar{P}\bar{A} + \bar{A}^T\bar{P}^T)e_x + \varepsilon e_x^T\bar{P}^T\bar{P}e_x + \frac{1}{\varepsilon}L_\phi^2|e_x|^2$$
$$= e_x^T\left(\bar{P}\bar{A} + \bar{A}^T\bar{P}^T + \bar{P}^T\bar{P} + \frac{1}{\varepsilon}L_\phi^2 I \right)e_x < -\alpha V(e_x),$$

where the last inequality follows from (18.37). Thus, on the sliding surface exponential stability of the motion e_x can be concluded under the assumptions made so far. In particular, from the proof of Theorem 3.3 it follows that

$$|e_x(t)| \leq M|e_x(0)|e^{-\alpha t/2}$$

and the constant M can be defined as

$$M = \sqrt{\frac{\lambda_{\max}(P)}{\lambda_{\min}(P)}}.$$

Moreover, for all initial values of the error $e_x(0) \in \mathcal{B}_w(0)$ for a positive constant $w > 0$ it holds that

$$|e_x(t)| \leq Mwe^{-\alpha t/2}.$$

To ensure that the sliding surface is reached in finite time, we concentrate on the definition of the gain $k(t, y, u)$ in (18.35). We consider the candidate Lyapunov function $V_y(e_y) = e_y^T e_y$. With (18.36) the time derivative of V_y satisfies

$$\dot{V}_y(e_y) = e_y^T(C_2 A_{21}e_x - A_s e_y + C_2 D_2 \delta(t, y, u))$$
$$+ e_y^T\left(-k(\cdot)\frac{e_y}{|e_y|} + C_2\left(\phi_2(T^{-1}z, u) - \phi_2\left(T^{-1}\begin{bmatrix} \hat{z}_1 \\ C_2^{-1}y \end{bmatrix}, u\right) \right) \right)$$
$$\leq -e_y^T A_s e_y + 2|e_y|(\|C_2 A_{21}\| \cdot |e_x| + \|C_2 D_2\|\rho(t, y, u) + \|C_2\|L_\phi|e_x|) - k(\cdot)|e_y|$$
$$\leq 2|e_y|((\|C_2 A_{21}\| + \|C_2\|L_\phi)|e_x| + \|C_2 D_2\|\rho(t, y, u)) - k(\cdot)|e_y|, \tag{18.39}$$

where we have used the fact that $-A_s$ is Hurwitz, Young's inequality, the estimate (18.38), and assumption (18.29). If the gain $k(t, y, u)$ is defined such that

$$k(t, y, u) \geq (\|C_2 A_{21}\| + \|C_2\| L_\phi) M w e^{-\alpha t/2} + \|C_2 D_2\| \rho(t, y, u) + \eta \qquad (18.40)$$

for $\eta > 0$, then the estimate (18.39) reduces to

$$\dot{V}_y(e_y) \leq -\eta |e_y| = -\eta \sqrt{V_y(e_y)},$$

for all $e_x(0) \in \mathcal{B}_w(0)$. In particular, if the initial error $e_x(0)$ is small enough, i.e., $e_x(0) \in \mathcal{B}_w(0)$, then the gain $k(t, y, u)$ guarantees finite-time convergence of $e_y(t)$ to the sliding surface. Together with the asymptotic convergence of $e_x(t) \to 0$ for $t \to \infty$ the sliding mode observer guarantees that $\hat{z}(t) \to z(t)$, from which the original state $x(t) = T^{-1} z(t)$ can be obtained.

Before we conclude this section we return to the (nonlinear) matrix inequality (18.37) and derive a linear matrix inequality whose feasibility can be verified through standard software. With the definitions of $\bar{P}$ and $\bar{A}$, (18.37) can be rewritten as

$$0 > \begin{bmatrix} A_{11}^T & A_{21}^T \end{bmatrix} \begin{bmatrix} I \\ L^T \end{bmatrix} P + P \begin{bmatrix} I & L \end{bmatrix} \begin{bmatrix} A_{11} \\ A_{21} \end{bmatrix} + \frac{1}{\varepsilon} P \begin{bmatrix} I & L \end{bmatrix} \begin{bmatrix} I \\ L^T \end{bmatrix} P + \varepsilon L_\phi^2 I + \alpha P$$

$$= A_{11}^T P + P A_{11} + A_{21}^T L^T P + P L A_{21} + \frac{1}{\varepsilon} P P + \frac{1}{\varepsilon} P L L^T P + \varepsilon L_\phi^2 I + \alpha P$$

$$= A_{11}^T P + P A_{11} + \varepsilon L_\phi^2 I + \alpha P + A_{21}^T L^T P + P L A_{21} + \frac{1}{\varepsilon} \begin{bmatrix} P & P L \end{bmatrix} \begin{bmatrix} P \\ L^T P \end{bmatrix},$$

which allows us to apply Lemma 8.3. In particular, using the Schur complement, the matrix inequality is equivalent to the condition

$$\left[\begin{array}{c|cc} A_{11}^T P + P A_{11} + A_{21}^T L^T P + P L A_{21} + \alpha P + \varepsilon L_\phi^2 I & P & P L \\ \hline P & -\varepsilon I & 0 \\ L^T P & 0 & -\varepsilon I \end{array} \right] < 0. \quad (18.41)$$

Thus, if feasible under the additional condition $P > 0$, the linear matrix inequality (18.41) can be solved to verify condition (18.37). If the Lipschitz constant L_ϕ is known, then (18.41) can be solved together with the condition $P > 0$ and for fixed constants $\alpha, \varepsilon \in \mathbb{R}_{>0}$. In the case L_ϕ is not known, (18.41) can be solved with respect to the unknowns $P > 0$ and $L_\phi > 0$ and with respect to the objective function max L_ϕ.

The sliding mode observer design for nonlinear systems derived in this section is summarized in Algorithm 10.

18.3 EXERCISES

Exercise 18.1. Reproduce the results in Example 18.3 in MATLAB.

Algorithm 10: Sliding mode observer design (for nonlinear systems)

Input: Nonlinear system of the form (18.27) as well as design parameters $\alpha, \varepsilon, w, \eta \in \mathbb{R}_{>0}$, and $A_s \in \mathbb{R}^{p \times p}$ Hurwitz.

Output: Matrices $T \in \mathbb{R}^{n \times n}$, $L \in \mathbb{R}^{(n-p) \times (n-q)}$, $K \in \mathbb{R}^{p \times p}$, $P \in \mathcal{S}_{>0}^{n-p}$, and the gain k defining the observer dynamics (18.32) and the error dynamics (18.33).

Algorithm:

1. If possible, define $L = \begin{bmatrix} L_x & 0 \end{bmatrix}$ such that $A_{11} + LA_{21}$ is Hurwitz.
2. Define T and K according to (18.31) and (18.34), respectively.
3. If (18.41) is feasible for $P > 0$ and

   - if L_ϕ is known, then solve the LMI subject to $P > 0$.
   - if L_ϕ is not known, then maximize L_ϕ subject to (18.41), $P > 0$ and $L_\phi > 0$.

4. Define $k(t, y, u)$ according to (18.40) and with respect to $w > 0$ and $\eta > 0$.
5. Define the output injection term ν according to (18.35).

Exercise 18.2. Consider the control system

$$
\begin{aligned}
\dot{x}_1 &= x_2 - \tfrac{1}{2}x_1 \\
\dot{x}_2 &= x_3 - x_2^3 \\
\dot{x}_3 &= \cos(x_3) + \sin^2(x_1) + u \\
y &= x_1
\end{aligned}
\tag{18.42}
$$

in normal form and the polynomial

$$
\chi(s) = s^3 + 1.5s^2 + 0.5s + 0.5.
\tag{18.43}
$$

1. Show that the polynomial (18.43) is Hurwitz and use the coefficients of the polynomial to set up a high-gain observer. In the observer design, assume that the dynamics (18.42) are known but the initial state is unknown.
2. Plot the states of the plant $(t, x(t))$, the states of the observer $(t, \hat{x}(t))$, and the input $(t, u(t))$ for $t \in [0, 10]$. Use the following control law, initial conditions, and parameters for the simulations:

(a) $x_0 = [1, 1, 1]^T$, $\hat{x}_0 = [0, 0, 0]^T$, $\varepsilon = 0.05$, and

$$
u = -\cos(\hat{x}_3) - \sin^2(\hat{x}_1) - \hat{x}_3.
$$

(b) $x_0 = [1, 1, 1]^T$, $\hat{x}_0 = [0, 0, 0]^T$, $\varepsilon = 0.05$, and

$$
u = 5 \operatorname{sat}(0.2(-\cos(\hat{x}_3) - \sin^2(\hat{x}_1) - \hat{x}_3)).
$$

Exercise 18.3. Embed the high-gain observer in the MPC implementation discussed in Appendix B.5. In particular, define the control law based on the estimated state $\hat{x}$.

Exercise 18.4. Consider the matrices

$$A = \begin{bmatrix} A_{11} & A_{12} \\ A_{21} & A_{22} \end{bmatrix} \qquad \text{and} \qquad C = \begin{bmatrix} 0 & I \end{bmatrix}, \tag{18.44}$$

with $A \in \mathbb{R}^{n \times n}$ $C \in \mathbb{R}^{p \times n}$. Show that observability of the pair (A, C) implies observability of the pair (A_{11}, A_{21}).

Exercise 18.5. Reproduce the results in Example 18.6 in MATLAB.

18.4   BIBLIOGRAPHICAL NOTES AND FURTHER READING

The material on high-gain observers is mainly taken from [88] and in particular from [88, Chapters 1–2]. A development of a nonlinear separation principle can be found in [88, Chapter 3]

As with the material on sliding mode control, our development of sliding mode observers closely follows [137] and in particular [137, Chapter 3].

Additional observers for nonlinear systems, which are not discussed here, include *nonlinear Luenberger observers* [80], [7], and *observers for systems with symmetry* [104], for example.

Appendix

Appendix A

Fundamental Definitions and Results

A.1 NORMS IN VECTOR AND FUNCTION SPACES

While the presentation of the results in this book mainly rely on the *Euclidean norm* or *2-norm* for finite-dimensional vector spaces, particular results occasionally necessitate the use of other norms. These differences are made precise here.

Definition A.1 (Norm). *Consider a space V over the real numbers $\mathbb{R}$ (or the complex numbers $\mathbb{C}$). A function $p : V \to \mathbb{R}_{\geq 0}$ is a norm on V if the following properties are satisfied.*

1. *Positive definiteness: $p(x) \geq 0$ for all $x \in V$ and $p(x) = 0$ if and only if $x = 0 \in V$;*
2. *Absolute homogeneity: $p(a \cdot x) = |a| \cdot p(x)$ for all $a \in \mathbb{R}$ (or $a \in \mathbb{C}$);*
3. *Triangle inequality: $p(x + y) \leq p(x) + p(y)$ for all $x, y \in V$.*

In the context of this book, V can either be a finite-dimensional vector space, $\mathbb{R}^n$ or $\mathbb{C}^n$, for example, or an infinite-dimensional function space. Here, three types of norms are especially relevant:

1. The *Euclidean norm* or *2-norm* (denoted by $|\cdot|$) for vectors $x \in \mathbb{R}^n$ (or $x \in \mathbb{C}^n$), $n \in \mathbb{N}$;
2. The *spectral norm* (denoted by $\|\cdot\|$) for matrices $A \in \mathbb{R}^{m \times n}$, $n, m \in \mathbb{N}$.
3. The $\mathcal{L}_2$-norm, the $\mathcal{L}_\infty$-norm, and the $\mathcal{H}_\infty$-norm (denoted by $\|\cdot\|_{\mathcal{L}_2}$, $\|\cdot\|_{\mathcal{L}_\infty}$, and $\|\cdot\|_\infty$, respectively) for functions $\psi : \mathbb{R}_{\geq 0} \to \mathbb{R}^n$ and $\hat{\psi} : \mathbb{C} \to \mathbb{C}^n$, $n \in \mathbb{N}$.

The Euclidean norm for $x \in \mathbb{C}^n$, $n \in \mathbb{N}$, is defined as

$$|x| = \sqrt{\sum_{i=1}^n \overline{x}_i x_i},$$

where $\overline{x}$ denotes the *complex conjugate*, i.e., for $a, b \in \mathbb{R}^n$, $\overline{a + jb} = a - jb$. Alternatively, the Euclidean norm, or the Euclidean norm squared, can be written as

$$|x|^2 = \overline{x}^T x = \langle x, x \rangle.$$

For vectors in $\mathbb{R}^n$, the complex conjugate is not necessary. In the one-dimensional case, i.e., $n = 1$, $|\cdot|$ reduces to the absolute value, which we have already used in the definition of the norm in Definition A.1. Alternative vector norms commonly used in $\mathbb{R}^n$ or $\mathbb{C}^n$ are the *1-norm*

$$|x|_1 = \sum_{i=1}^{n} |x_i|$$

and the ∞-*norm* or *maximum norm*

$$|x|_\infty = \max_{i=1,\dots,n} |x_i|.$$

For matrices $A \in \mathbb{R}^{m \times n}$, the spectral norm is defined using the Euclidean norm as

$$\|A\| = \max_{x \in \mathbb{R}^n \setminus \{0\}} \frac{|Ax|}{|x|} = \max_{|x|=1} |Ax|.$$

It can be shown that the equality

$$\|A\| = \sqrt{\lambda_{\max}(A^T A)}$$

is satisfied, where $\lambda_{\max}(A^T A)$ denotes the largest eigenvalue of $A^T A$.

The $\mathcal{L}_2$-norm, the $\mathcal{L}_\infty$-norm, and the $\mathcal{H}_\infty$-norm where the space V is an infinite-dimensional function space are defined and discussed in Section 4.1.3.

A.2   AUXILIARY RESULTS

Not every result we use is stated in detail in this book. However, some recurring results are given for completeness here. For example, a commonly used tool to upper bound the scalar product of two vectors is given by the *Cauchy-Schwarz inequality*.

Lemma A.2 (Cauchy-Schwarz inequality). *For all $x, y \in \mathbb{R}^n$ the inequality*

$$x^T y = \langle x, y \rangle \leq |x| \cdot |y| \tag{A.1}$$

is satisfied.

Proof. For $x = 0$ or $y = 0$ the inequality is trivially satisfied. Let $x \neq 0$, $y \neq 0$ and $\lambda \in \mathbb{R}$. Then it holds that

$$0 \leq \langle x - \lambda y, x - \lambda y \rangle = |x|^2 - 2\lambda \langle x, y \rangle + \lambda^2 |y|^2$$

and $2\lambda \langle x, y \rangle \leq |x|^2 + \lambda^2 |y|^2$. With $\lambda = |x|/|y|$ (which satisfies $\lambda \in \mathbb{R}$ since $y \neq 0$) the inequality becomes

$$2 \frac{|x|}{|y|} \langle x, y \rangle \leq |x|^2 + \frac{|x|^2}{|y|^2} |y|^2 = 2|x|^2,$$

from which the assertion follows. $\square$

The same estimate is satisfied for the $\mathcal{L}_2$-norm. This means two functions

$\psi_1, \psi_2 : \mathbb{R}_{\geq 0} \to \mathbb{R}^n$ satisfy the inequality

$$\int_0^t \psi_1(\tau)^T \psi_2(s) \, d\tau \leq \left(\int_0^t |\psi_1(\tau)|^2 \, d\tau \right)^{\frac{1}{2}} \cdot \left(\int_0^t |\psi_2(\tau)|^2 \, d\tau \right)^{\frac{1}{2}}$$
$$= \|\psi_1\|_{\mathcal{L}_2[0,t)} \|\psi_2\|_{\mathcal{L}_2[0,t)}$$

if all integrals are well defined.

With respect to the spectral norm, submultiplicativity combined with the Cauchy-Schwarz inequality can be used to find upper bounds for vector-matrix multiplications. In this context, submultiplicativity of the spectral norm means that

$$|Ax| \leq \|A\| \cdot |x|$$

is satisfied for all $A \in \mathbb{R}^{m \times n}$ and for all $x \in \mathbb{R}^n$. Combined with the Cauchy-Schwarz inequality this can be used to obtain the estimate

$$y^T A x \leq |y| \cdot |Ax| \leq |y| \cdot \|A\| \cdot |x|$$

for all $A \in \mathbb{R}^{m \times n}$, $x \in \mathbb{R}^n$, $y \in \mathbb{R}^m$.

The Cauchy-Schwarz inequality provides an upper bound in terms of the product of the norm of two vectors. A similar estimate in terms of the sum is given by *Young's inequality*.

Lemma A.3 (Generalized Young's inequality, [124],[81, Lemma 16]). *Let $\alpha \in \mathcal{K}_\infty$ be continuously differentiable such that $\alpha' \in \mathcal{K}_\infty$. Then for any $x, y \in \mathbb{R}^n$ the inequality*

$$x^T y \leq \alpha(|x|) + \int_0^{|y|} (\alpha')^{-1}(s) \, ds$$

is satisfied.

By selecting $\alpha(s) = \frac{1}{p} x^p$, $p \in \mathbb{R}_{>0}$, a less general form of Young's inequality is obtained. Even though it is less general, the inequality is sufficient for the examples in this book.

Lemma A.4 (Young's inequality). *Let $p, q \in \mathbb{R}_{>0}$ such that $\frac{1}{p} + \frac{1}{q} = 1$. Then for any $x, y \in \mathbb{R}^n$ the inequality*

$$x^T y \leq \frac{1}{p} |x|^p + \frac{1}{q} |y|^q$$

is satisfied.

Young's inequality is used repeatedly in Chapter 7 in the context of input-to-state stability.

For matrices, in terms of definiteness, the following inequality can be shown to be true.

Lemma A.5. *Let $P \in \mathbb{R}^{m \times n}$ and $Q \in \mathbb{R}^{m \times p}$ such that $P^T P$ is nonsingular. Then the matrix inequality*

$$Q^T Q \geq (P^T Q)^T (P^T P)^{-1} (P^T Q) \tag{A.2}$$

is satisfied.

Recall that for two matrices $Q_1, Q_2 \in \mathcal{S}^n$, by definition, the inequality $Q_1 \geq Q_2$ is satisfied if $Q_1 - Q_2$ is positive semidefinite. This is how inequality (A.2) is to be understood.

Proof of Lemma A.5: Consider the matrix

$$S = (P^T P)^{-1}(P^T Q). \tag{A.3}$$

Since $(Q - PS)^T(Q - PS) \geq 0$, it holds that

$$Q^T Q \geq S^T(P^T Q) + (P^T Q)^T S - S^T(P^T P)S.$$

With the definition of S the right-hand side satisfies

$$S^T(P^T Q) + (P^T Q)^T S - S^T(P^T P)S = (P^T Q)^T(P^T P)^{-1}(P^T Q),$$

which completes the proof. $\square$

In (1.33) we have given a simple formula to explicitly calculate the inverse of a 2×2 matrix. A more general result is stated in the following lemma. The result is used in the derivation of the discrete time Kalman filter equations in Chapter 16 and can in general be used to manipulate matrix equations.

Lemma A.6 (Matrix Inversion Lemma, [39, Lemma 1.2]). *Consider the matrix*

$$A = \begin{bmatrix} A_{11} & A_{12} \\ A_{21} & A_{22} \end{bmatrix},$$

where $A_{11} \in \mathbb{R}^{n \times n}$, $A_{22} \in \mathbb{R}^{m \times m}$, $n, m \in \mathbb{N}$, and the matrices

$$A_{11}, \qquad A_{22}, \qquad (A_{11} - A_{12}A_{22}^{-1}A_{21}), \quad \text{and} \quad (A_{22} - A_{21}A_{11}^{-1}A_{12})$$

are nonsingular by assumption. Then the matrix A is nonsingular and

$$A^{-1} = \begin{bmatrix} B_{11} & B_{12} \\ B_{21} & B_{22} \end{bmatrix},$$

where

$$
\begin{aligned}
B_{11} &= A_{11}^{-1} + A_{11}^{-1}A_{12}(A_{22} - A_{21}A_{11}^{-1}A_{12})^{-1}A_{21}A_{11}^{-1} \\
&= (A_{11} - A_{12}A_{22}^{-1}A_{21})^{-1}
\end{aligned} \tag{A.4}
$$

$$
\begin{aligned}
B_{12} &= -A_{11}^{-1}A_{12}(A_{22} - A_{21}A_{11}^{-1}A_{12})^{-1} \\
&= -(A_{11} - A_{12}A_{22}^{-1}A_{21})^{-1}A_{12}A_{22}^{-1}
\end{aligned} \tag{A.5}
$$

$$
\begin{aligned}
B_{21} &= -(A_{22} - A_{21}A_{11}^{-1}A_{12})^{-1}A_{21}A_{11}^{-1} \\
&= -A_{22}^{-1}A_{21}(A_{11} - A_{12}A_{22}^{-1}A_{21})^{-1}
\end{aligned} \tag{A.6}
$$

$$
\begin{aligned}
B_{22} &= A_{22}^{-1} + A_{22}^{-1}A_{21}(A_{11} - A_{12}A_{22}^{-1}A_{21})^{-1}A_{12}A_{22}^{-1} \\
&= (A_{22} - A_{21}A_{11}^{-1}A_{12})^{-1}.
\end{aligned} \tag{A.7}
$$

Additionally, the determinant of A satisfies

$$\det(A) = \det(A_{11}) \det(A_{22} - A_{21} A_{11}^{-1} A_{12})$$
$$= \det(A_{22}) \det(A_{11} - A_{12} A_{22}^{-1} A_{21}).$$

A.3 SELECTION OF COMPARISON FUNCTION RESULTS

Comparison functions introduced in Chapter 1 are a useful tool throughout the book to derive various results. Their simplicity in terms of being one-dimensional or two-dimensional functions allows a visualization and interpretation which is in general not possible in higher dimensions. Moreover, properties such as positivity and monotonicity allow for a wide range of structure-preserving manipulations. Making use of these properties, several results for comparison functions have been derived over the years and are scattered in various research papers. A collection of such results is given in [81]. Here, we present some of them which are used in this book.

Lemma A.7 ([81, Lemma 18]). *Let $\rho \in \mathcal{P}^n$. Then there exist functions $\alpha \in \mathcal{K}_\infty$ and $\sigma \in \mathcal{L}$ such that*

$$\rho(x) \geq \alpha(|x|)\sigma(|x|), \quad \forall x \in \mathbb{R}^n.$$

Lemma A.8 ([81, Lemma 24],[85, Lemma 5.4]). *For each $\rho \in \mathcal{P}$ there exists $\alpha \in \mathcal{K}_\infty$ such that $\alpha(\cdot)$ is locally Lipschitz on its domain, continuously differentiable on $(0, \infty)$, and satisfies $\alpha'(s) > 0$ for all $s > 0$ and*

$$\alpha(s) \leq \rho(s)\alpha'(s), \quad \forall s \in \mathbb{R}_{>0}.$$

Lemma A.9 (Weak triangle inequality, [81, Lemma 10],[71, Eq. (6)]). *Given $\alpha \in \mathcal{K}$ and any function $\varphi \in \mathcal{K}_\infty$ such that $\varphi - id \in \mathcal{K}_\infty$, for any $a, b \in \mathbb{R}_{\geq 0}$*

$$\alpha(a + b) \leq \alpha(\varphi(a)) + \alpha\left(\varphi \circ (\varphi - id)^{-1}(b)\right). \tag{A.8}$$

Here, $id : \mathbb{R}_{\geq 0} \to \mathbb{R}_{\geq 0}$ denotes the identity function $id(x) = x$. In this general form, the connection to the triangle inequality in Definition A.1 might not be obvious. For $\varphi(s) = 2s$ inequality (A.8) simplifies to

$$\alpha(a + b) \leq \alpha(2a) + \alpha(2b),$$

which resembles the structure of the triangle inequality but with the additional factor 2. The additional degree of freedom in the function φ allows us to shift the weight toward a or b. Through the selections $\varphi(s) = 3s$ and $\varphi(s) = \frac{3}{2}s$, the inequalities

$$\alpha(a + b) \leq \alpha(3a) + \alpha\left(3\tfrac{1}{2}b\right) = \alpha(3a) + \alpha\left(\tfrac{3}{2}b\right),$$
$$\alpha(a + b) \leq \alpha(\tfrac{3}{2}a) + \alpha\left(\tfrac{3}{2}2b\right) = \alpha(\tfrac{3}{2}a) + \alpha(3b)$$

are satisfied, for example.

The last result on comparison functions stated here is known as *Sontag's lemma on $\mathcal{KL}$-estimates* and was originally derived in [143, Proposition 7]. Here we present it in the way given in [81] and [85].

Lemma A.10 (Sontag's lemma on $\mathcal{KL}$-estimates, [81, Lem. 7],[85, Lem. 5.3]). *Let* $\beta \in \mathcal{KL}$ *and* $\lambda \in \mathbb{R}_{>0}$ *be given. Then there exist* $\alpha_1, \alpha_2 \in \mathcal{K}_\infty$ *such that* $\alpha_1(\cdot)$ *is Lipschitz on its domain, smooth on* $\mathbb{R}_{>0}$, $\alpha_1(s) \leq s\alpha_1'(s)$ *for all* $s \in \mathbb{R}_{>0}$, *and*

$$\alpha_1(\beta(s,t)) \leq \alpha_2(s)e^{-\lambda t}, \qquad \forall\, (s,t) \in \mathbb{R}_{\geq 0} \times \mathbb{R}_{\geq 0}.$$

Sontag's lemma on $\mathcal{KL}$-estimates provides an upper bound on $\mathcal{KL}$ functions in terms of a $\mathcal{K}_\infty$ function and an $\mathcal{L}$ function by separating the variables of the $\mathcal{KL}$ function. Moreover, it shows that through an appropriate selection of α_1, the decay with respect to t can be upper bounded by an exponential decay with arbitrary decay rate $\lambda > 0$.

A.4   BARBALAT'S LEMMA

The Lyapunov results discussed in Chapter 2 rely on some form of the decrease condition

$$\dot{V}(x(t)) \leq -\phi(x(t)),$$

with Lyapunov function $V(\cdot)$, solution $x(\cdot)$, and $\phi : \mathbb{R}^n \to \mathbb{R}_{\geq 0}$. Proofs rely on the manipulation of this inequality, i.e., by integrating both sides over t, for example, which leads to

$$V(x(t)) - V(x(0)) = \int_0^t \dot{V}(x(\tau))d\tau \leq \int_0^t \phi(x(\tau))d\tau. \qquad (A.9)$$

To investigate asymptotic properties of $x(t)$, the limit $t \to \infty$ needs to be considered which raises the question about whether all expressions in (A.9) are well defined for $t \to \infty$ and what implications this may have. The question is answered by *Barbalat's lemma* under appropriate continuity assumptions on the involved functions.

Definition A.11 (Uniform continuity). *Consider a function* $\phi : \mathcal{D} \to \mathbb{R}$, $\mathcal{D} \subset \mathbb{R}$. *The function* ϕ *is uniformly continuous if for all* $\varepsilon > 0$ *there exists* $\delta > 0$ *such that* $t_1, t_2 \in \mathcal{D}$, $|t_1 - t_2| \leq \delta$ *implies* $|\phi(t_1) - \phi(t_2)| \leq \varepsilon$.

Note that continuously differentiable functions ϕ are locally Lipschitz continuous and local Lipschitz continuity implies uniform continuity. The converse implications are in general not true.

Lemma A.12 (Barbalat's lemma, [93, Lemma A.6]). *Consider a function* $\phi :$ $\mathbb{R}_{\geq 0} \to \mathbb{R}$. *If* ϕ *is uniformly continuous, and* $\lim_{t \to \infty} \int_0^\infty \phi(\tau)d\tau$ *exists and is finite, then*

$$\lim_{t \to \infty} \phi(t) = 0. \qquad (A.10)$$

The statement can be shown by contradiction. Since the proof is quite illustrative, the proof given in [93] is reported for completeness here.

Proof. For the sake of a contradiction, assume that (A.10) is not satisfied. Hence, either the limit $\lim_{t \to \infty} \phi(t)$ does not exist or it is unequal to zero. Due to the uniform continuity of ϕ, there exist $\varepsilon > 0$ and $\delta(\varepsilon) > 0$ so that for every $T > 0$ there exists $t_1 \geq T$ with $|\phi(t_1)| > \varepsilon$ and $|\phi(t) - \phi(t_1)| < \frac{\varepsilon}{2}$ for all $|t - t_1| \leq \delta(\varepsilon)$. Thus, for all $t \in [t_1, t_1 + \delta(\varepsilon)]$, the estimate

$$|\phi(t)| = |\phi(t) - \phi(t_1) + \phi(t_1)| \geq |\phi(t_1)| - |\phi(t) - \phi(t_1)| > \varepsilon - \frac{\varepsilon}{2} = \frac{\varepsilon}{2}$$

is satisfied. Moreover, since $\phi(\cdot)$ does not change sign on $t \in [t_1, t_1 + \delta(\varepsilon)]$ it follows that

$$\left| \int_{t_1}^{t_1 + \delta(\varepsilon)} \phi(\tau) d\tau \right| = \int_{t_1}^{t_1 + \delta(\varepsilon)} |\phi(\tau)| d\tau > \frac{\varepsilon}{2} \delta(\varepsilon).$$

Since t_1 can be chosen arbitrarily large, the limit $\lim_{t \to \infty} \int_0^\infty \phi(\tau) d\tau$ cannot exist, which leads to a contradiction and which completes the proof. $\square$

A.5 CONVEXITY AND CONVEX OPTIMIZATION

While in many cases a problem or the verification of statements is in general intractable, convex problems, or settings admitting convexity properties, frequently add the structure necessary to derive a satisfying solution. Here, we give some basic definitions and insights to illustrate why convexity is a beneficial property. This particularly applies to controller designs based on optimization, i.e., Chapter 15 on MPC and derivations based on LMIs. We start with the general definitions of *convex sets* and *convex functions*.

Definition A.13 (Convex set). *Let $\mathcal{D} \subset \mathbb{R}^n$ for $n \in \mathbb{N}$. The set $\mathcal{D}$ is convex if $x, y \in \mathcal{D}$ implies*

$$\lambda x + (1 - \lambda)y \in \mathcal{D}$$

for all $\lambda \in [0, 1]$.

Note that this definition also covers sets $\mathcal{D} \subset \mathbb{R}^{n \times n}$, since $\mathbb{R}^{n \times n}$ can be interpreted as $\mathbb{R}^{n^2}$. For example the set of symmetric matrices $\mathcal{S}^n$ as a subset of $\mathbb{R}^{n \times n}$ ($\mathcal{S}^n \subset \mathbb{R}^{n \times n}$) is convex. Indeed, if $A_1, A_2 \in \mathcal{S}^n$ are symmetric, then $\lambda A_1 + (1 - \lambda)A_2 \in \mathcal{S}^n$ for all $\lambda \in [0, 1]$. Alternatively, $\mathcal{S}^n$ as a representation of the vector space $\mathbb{R}^{\frac{n(n+1)}{2}}$ is convex due to the properties of a vector space.

Definition A.14 (Convex function). *Let $\mathcal{D} \subset \mathbb{R}^n$ be convex for $n \in \mathbb{N}$. A function $f : \mathcal{D} \to \mathbb{R}$ is convex if*

$$f(\lambda x + (1 - \lambda)y) \leq \lambda f(x) + (1 - \lambda)f(y) \tag{A.11}$$

for all $x, y \in \mathcal{D}$ and for all $\lambda \in [0, 1]$. If (A.11) is satisfied with $<$ instead of $\leq$, then the function f is strictly convex.

It can be shown that convex functions are continuous on the interior of $\mathcal{D}$. Special classes of convex functions which are convex but not strictly convex are *affine functions* and *linear functions*.

Definition A.15 (Affine function). *Let $a \in \mathbb{R}^n$ and $b \in \mathbb{R}$. A function $f : \mathbb{R}^n \to \mathbb{R}$ defined through $f(x) = a^T x + b$ is called an affine function. If in addition $b = 0$, then the function f is linear.*

The advantage of the structure of affine functions over general nonlinear functions can be immediately seen by the presentation of the explicit results on linear systems presented in Chapter 3 in contrast to the significantly more complicated results discussed in various other chapters on nonlinear systems.

While inequality (A.11) might be hard to verify, in the case that f is continuously differentiable, convexity can be shown using the gradient or the Hessian of f.

Lemma A.16 (First-order convexity condition, [29, Section 3.1.3]). *Let $\mathcal{D} \subset \mathbb{R}$ be convex and let $f : \mathcal{D} \to \mathbb{R}$ be continuously differentiable. The function f is convex if and only if*

$$f(y) \geq f(x) + (\nabla f(x))^T (y - x)$$

for all $x, y \in \mathcal{D}$. Moreover, if

$$f(y) > f(x) + (\nabla f(x))^T (y - x)$$

for all $x, y \in \mathcal{D}$, $x \neq y$, then the function f is strictly convex.

Lemma A.17 (Second-order convexity condition, [29, Section 3.1.3]). *Let $\mathcal{D} \subset \mathbb{R}$ be convex and let $f : \mathcal{D} \to \mathbb{R}$ be twice continuously differentiable. The function f is convex if and only if the Hessian is positive semidefinite, i.e., $\nabla^2 f(x) \geq 0$ for all $x \in \mathcal{D}$. Moreover, if $\nabla^2 f(x) > 0$ for all $x \in \mathcal{D}$, then f is strictly convex.*

Note that quadratic Lyapunov functions $V(x) = x^T P x$, $P \in \mathcal{S}^n_{>0}$, are strictly convex functions, which follows immediately from Lemma A.17. Indeed, the gradient of V is given by $\nabla V(x) = 2Px$ and the Hessian $\nabla^2 V(x) = 2P$ is positive definite since P is positive definite.

Similarly, from P being symmetric and positive definite, it holds that

$$0 > -(y - x)^T P(y - x) = -y^T P y + 2x^T P y - x^T P x$$

for all $x \neq y$. Rearranging terms leads to the inequality

$$y^T P y > 2x^T P y - x^T P x + (2x^T P x - 2x^T P x)$$
$$= 2x^T P(y - x) + x^T P x$$

and thus strict convexity alternatively follows from Lemma A.16.

As a next step we discuss connections between convex functions defined on convex sets and the existence of local and global minima.

Definition A.18 (Local and global minima). *Let $\mathcal{D} \subset \mathbb{R}^n$ and $f : \mathcal{D} \to \mathbb{R}$. Then, $f(x^\star) \in \mathbb{R}$ is called local minimum of f if there exists an $\varepsilon > 0$ such that*

$$f(x^\star) \leq f(y) \tag{A.12}$$

holds for all $y \in \mathcal{B}_\varepsilon(x^\star) \cap \mathcal{D}$. If inequality (A.12) is satisfied for all $y \in \mathcal{D}$, then

$f(x^\star) \in \mathbb{R}$ *is called global minimum of f. For a local (global) minimum or* $f(x^\star)$,
$x^\star \in \mathcal{D}$ *is called a local (global) minimizer of the function f.*

Theorem A.19 ([20, Proposition B.10]). *Let* $\mathcal{D} \subset \mathbb{R}^n$ *be convex and let* $f : \mathcal{D} \to \mathbb{R}$.
If f is convex, then every local minimum of f is also a global minimum. Moreover,
if f is strictly convex, then f has at most one global minimum.

Note that Theorem A.19 does not guarantee the existence of a local, and thus
global, minimum of a function f. As an example consider the function $f : \mathbb{R} \to \mathbb{R}$,
$f(x) = e^x$. The set $\mathcal{D} = \mathbb{R}$ is convex and convexity of f follows immediately from
Lemma A.17. However, f does not have a local or global minimum on $\mathbb{R}$ since for
all $x \in \mathbb{R}$ the condition $f(\frac{1}{2}x) < f(x)$ is satisfied.

Nevertheless, since continuous functions attain their minimum on compact sets
the existence of local and global minima can be guaranteed through the following
result.

Corollary A.20. *Let* $\mathcal{D} \subset \mathbb{R}^n$ *be compact and convex and let* $f : \mathcal{D} \to \mathbb{R}$ *be*
continuous. If f is convex, then there exists $x^\star \in \mathcal{D}$ *such that* $f(x^\star) \leq f(y)$ *for*
all $y \in \mathcal{D}$. *Moreover, if f is strictly convex, then there exists* $x^\star \in \mathcal{D}$ *such that*
$f(x^\star) < f(y)$ *for all* $y \in \mathcal{D} \backslash \{x^\star\}$.

To put these results into context, we consider the following setting. For $f, g_i, h_j :$
$\mathbb{R}^n \to \mathbb{R}$, $i \in \{1, \dots, p\}$, $j \in \{1, \dots, q\}$ for $p, q \in \mathbb{N}$, we define the optimization
problem

$$\min_{x \in \mathbb{R}^n} f(x) \tag{A.13a}$$

$$\text{subject to } g_i(x) \leq 0, \qquad i = 1, \dots, p, \tag{A.13b}$$

$$h_j(x) = 0, \qquad j = 1, \dots, q. \tag{A.13c}$$

Here, f in (A.13a) denotes an *objective function*, (A.13b) denotes *inequality con-*
straints, and (A.13c) denotes *equality constraints*. The *feasible set* of the optimization problem (A.13) is defined through the constraints

$$\mathcal{D} = \left\{ x \in \mathbb{R}^n \ \middle| \ \begin{matrix} g_i(x) \leq 0 & \forall & i \in \{1, \dots, p\} \\ h_j(x) = 0 & \forall & j \in \{1, \dots, q\} \end{matrix} \right\}. \tag{A.14}$$

Thus the optimization problem (A.13) can be compactly expressed as

$$\min_{x \in \mathcal{D}} f(x). \tag{A.15}$$

If $g_i(\cdot)$ are convex functions for all $i \in \{1, \dots, p\}$ and $h_j(\cdot)$ are affine functions for
all $j \in \{1, \dots, q\}$ then the set $\mathcal{D}$ in (A.14) is convex. If additionally the function f
is convex (or strictly convex), then the optimization problem (A.13) or equivalently
(A.15) is called a *convex optimization problem*.

Compared to general nonconvex optimization problems, convex optimization
problem can be solved efficiently, in general, which is mainly due to the properties
of convex functions and convex sets highlighted in Theorem A.19 and Corollary
A.20. For specific algorithms and for further reading we refer to the monographs
[20], [29], and [111].

In Chapter 15 we have seen several optimization problems which fit into the
setting (A.13). We conclude this section by giving some intuition as to why opti-

mization problems corresponding to LMIs are convex problems.

Example A.21. Consider the function $g : \mathcal{S}^n \to \mathbb{R}$ as a mapping from the set of symmetric matrices $\mathcal{S}^n \subset \mathbb{R}^{n \times n}$ to $\mathbb{R}$ defined as

$$g(P) = \lambda_{\max}(P - B) \tag{A.16}$$

for a given symmetric matrix $B \in \mathcal{S}^n$. First note that the eigenvalues of a symmetric matrix are real and thus the mapping is well defined. Moreover, for every $P \in \mathcal{S}^n$ there exists an eigenvector $x_P \in \mathbb{R}^n$, $|x_P| = 1$, such that

$$x_P^T P x_P = x_P^T \lambda_{\max}(P) x_P = \lambda_{\max}(P).$$

With these definitions, for all $\mu \in [0, 1]$ and for all $P_1, P_2 \in \mathcal{S}^n$ it holds that

$$\begin{aligned}
g(\mu P_1 + (1 - \mu) P_2) &= \lambda_{\max}(\mu P_1 + (1 - \mu) P_2 - B) \\
&= x_{\mu P_1 + (1 - \mu) P_2 - B}^T (\mu P_1 + (1 - \mu) P_2 - B) x_{\mu P_1 + (1 - \mu) P_2 - B} \\
&= \mu \cdot x_{\mu P_1 + (1 - \mu) P_2 - B}^T (P_1 - B) x_{\mu P_1 + (1 - \mu) P_2 - B} \\
&\quad + (1 - \mu) \cdot x_{\mu P_1 + (1 - \mu) P_2 - B}^T (P_2 - B) x_{\mu P_1 + (1 - \mu) P_2 - B}.
\end{aligned}$$

Since x_P by definition denotes an eigenvector of P of length 1 corresponding to the largest eigenvalue, the last expression above can be estimated as

$$\begin{aligned}
g(\mu P_1 + (1 - \mu) P_2) & \\
&\leq \mu \cdot x_{P_1 - B}^T (P_1 - B) x_{P_1 - B} + (1 - \mu) \cdot x_{P_2 - B}^T (P_2 - B) x_{P_2 - B} \\
&= \mu g(P_1) + (1 - \mu) g(P_2),
\end{aligned}$$

which shows that the function (A.16) is convex.

Note that

$$g(P) = \lambda_{\max}(P - B) \leq 0 \tag{A.17}$$

is equivalent to $P - B$ being negative semidefinite. Thus a linear matrix inequality

$$P - B \leq 0$$

can be incorporated through convex inequality constraints (A.17) in the optimization problem (A.13). Similarly, the inequality constraint

$$g(-P) = \lambda_{\max}(-P - B) \leq 0$$

is equivalent to $P + B \geq 0$; i.e, if the constraints are satisfied, $P + B$ is positive semidefinite. In the same way it can be shown that the function $g : \mathcal{S}^n \to \mathbb{R}$,

$$g(P) = \lambda_{\max}(A^T P + PA - B), \tag{A.18}$$

is convex, for example.

A.6 PROBABILITY THEORY

For the derivation of the discrete time Kalman filter, basic results and definitions from *probability theory* are used and are given here for completeness.

Consider a random variable $X : \Omega \to \mathbb{R}$ from a sample space Ω (the set of all possible outcomes of an experiment) to the real numbers. Additionally, we denote the σ-algebra on the set Ω, i.e., the collection of all subsets of Ω, as Σ. The probability $P : \Sigma \to [0,1]$ of an event is the probability of a particular outcome. Here, for $A \subset \mathbb{R}$, an event is a set $\{s \in \Omega : X(s) \in A\}$, which is simplified to $\{X \in A\} = \{s \in \Omega : X(s) \in A\}$ by definition.

Definition A.22 (Probability distribution). *Consider a sample space Ω and its σ-algebra Σ. Then $P : \Sigma \to [0,1]$ is called a probability distribution of $X : \Omega \to \mathbb{R}$ if the following properties are satisfied.*

- *It holds that $P(X \in \mathbb{R}) = 1$ and $P(X \in A) \geq 0$ for any $A \subset \mathbb{R}$ measurable.*
- *For any countable sequence of pairwise disjoint measurable sets $A_i \subset \mathbb{R}$, $i \in \mathbb{N}$, it holds that*

$$P\left(X \in \cup_{i=1}^{\infty} A_i\right) = \sum_{i=1}^{\infty} P(X \in A_i).$$

With the definition of a probability distribution, a function $f : \mathbb{R} \to \mathbb{R}$ such that

$$P(X \in A) = \int_A f(x)dx \tag{A.19}$$

for all $A \in \mathbb{R}$ measurable denotes a *probability density function*. A *Gaussian probability density function*, also called a *normal probability density function*, is defined as

$$f(x) = \frac{1}{\sqrt{2\pi}\sigma} e^{-\frac{1}{2\sigma^2}(x-\mu)^2}$$

for given parameters $\sigma > 0$ and $\mu \in \mathbb{R}$.

The *expectation* E of a random variable X is defined as

$$E[X] = \int_{-\infty}^{\infty} x f(x)dx.$$

A normal distribution satisfies $E[X] = \mu$. The *variance* of a random variable X is defined as

$$\text{Var}(X) = E[(X - E[X])^2] = \int_{-\infty}^{\infty} (x - E[X])^2 f(x)dx,$$

and a normal distribution satisfies $\text{Var}(X) = \sigma^2$.

The *covariance* of two random variables X_1, X_2 is defined as

$$\text{Cov}(X_1, X_2) = E[(X_1 - E[X_1])(X_2 - E[X_2])]$$

and it holds that $\text{Cov}(X_1, X_2) = E[X_1 X_2] - E[X_1]E[X_2]$ as well as $\text{Cov}(X, X) =$

$\text{Var}(X)$ if $X = X_1 = X_2$.

The one-dimensional definitions of random variables can be extended to vectors of random variables $X : \Omega \to \mathbb{R}^n$, $n \in \mathbb{N}$. For a vector of random variables $X \in \mathbb{R}^n$, $n \in \mathbb{N}$, the probability density function in (A.19) extends to $f : \mathbb{R}^n \to \mathbb{R}$ such that

$$P(X \in A) = \int_{A_n} \cdots \int_{A_1} f(x)dx_1 \cdots dx_n$$

for all $A = A_1 \times \ldots \times A_n \in \mathbb{R}^n$ measurable. The normal probability density function in higher dimensions can be written as

$$f(x) = \frac{1}{(2\pi)^{\frac{n}{2}} \sqrt{\det(R)}} e^{-\frac{1}{2}(x-\mu)^T R^{-1} (x-\mu)}.$$

Here, $\mu \in \mathbb{R}^n$ is a vector, and the positive definite matrix $R \in \mathcal{S}^n_{>0}$ takes the role of the parameter σ. For a normal probability density function it holds that

$$\int_{\mathbb{R}^n} f(x)dx = 1,$$

the expectation is given by

$$\text{E}[X] = \int_{\mathbb{R}^n} x f(x)dx = \mu,$$

and the variance satisfies

$$\text{Var}(X) = \text{E}[(X - \mu)(X - \mu)^T] = R.$$

Similarly, for vectors of random variables X_1 and X_2, the covariance satisfies

$$\text{Cov}(X_1, X_2) = \text{E}[(X_1 - \text{E}[X_1])(X_2 - \text{E}[X_2])^T] = \text{E}[X_1 X_2^T] - \text{E}[X_1]\,\text{E}[X_2]^T$$

and $\text{Cov}(X, X) = \text{Var}(X)$ for $X = X_1 = X_2$.

Appendix B

MATLAB Implementations

In this section we provide basic code to illustrate how some concepts in this book can be implemented in Matlab. Note that some functions need specific toolboxes which are not specified here. The use of external toolboxes such as CVX [56], [57], SOSTOOLS [116], and CasADi [6], executed in Matlab, is however pointed out.

The code provided is intended to give examples of how some topics discussed in this book can be implemented and illustrated using Matlab, and to inspire further examples and experiments. The code is far from being optimized to take advantage of all the capabilities of Matlab specifically, and other programming languages in general. While we also believe that there are more efficient implementations possible (particularly in terms of the runtime), we believe the provided implementations are illustrative.

B.1 SOLVING (NONLINEAR) DYNAMICAL SYSTEMS IN MATLAB

In this section we show how ordinary differential equations can be solved numerically in Matlab. We use the dynamics of the two-dimensional inverted pendulum in Equation (1.45) as an example.

We start by defining the parameters involved in the dynamics in a *structure array* (**struct**). Additionally, we define a time-dependent input as an *anonymous function*.

```
1  params.J=1;
2  params.m=1;
3  params.ell=1;
4  params.g=9.81;
5  params.gamma=0.1;
6
7  u = @(t,x) sin(t);
```

The dynamics (1.45) are defined in a separate *m-file*.

```
1  function [dx]=ode_pendulum(t,x,u,params)
2
3  c1 = 1/(params.J+params.m*params.ell^2);
4  c2 = params.m*params.g*params.ell;
5  c3 = params.gamma;
6  c4 = params.ell;
7
8  dx1 = x(2);
9  dx2 = c1*(c2*sin(x(1))-c3*x(2)+c4*cos(x(1))*u(t,x));
10
11 dx = [dx1;
```

```
12        dx2 ] ;
13
14 end
```

A numerical solver is implemented through `ode45.m` in MATLAB, for example. The dynamics are solved with respect to an initial condition and over a given time span.

```
 8 x0 = [0;
 9        0];
10
11 tspan=[0 10];
12
13 [t,x]=ode45(@(t,x)ode_pendulum(t,x,u,params),tspan,x0);
14
15 h=figure(1);
16 plot(t,x,'-o')
17 legend('x_1','x_2','location','southwest')
18
19 saveas(h,'solution_ode','epsc')
```

The output of `ode45.m` contains the numerical solution of the ordinary differential equation which can be visualized through `plot.m` . The solution is shown in Figure B.1. Through `saveas` the figure is saved as an *eps-file* in the current folder.

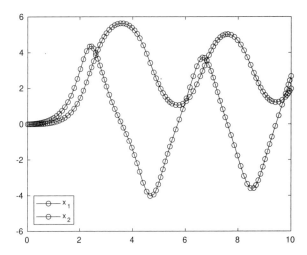

Figure B.1: Solution of the dynamical system (1.45) obtained through `ode45.m`.

A solution corresponding to a state-dependent input can be obtained by changing the definition of the function u. After solving the dynamical system, the feedback $u(x)$ can be recovered through a *for loop*, for example. The solution and the corresponding feedback law are shown in Figure B.2.

```
20 u = @(t,x) -x(1)-x(2);
21
22 x0 = [1;
23        1];
24
25 [t,x]=ode45(@(t,x)ode_pendulum(t,x,u,params),tspan,x0);
26
```

```
27 u_cl=zeros(length(t),1);
28 for i=1:length(t)
29     u_cl(i)=u(t(i),x(i,:));
30 end
31
32 h=figure(1);
33 subplot(2,1,1)
34 plot(t,x,'-o')
35 legend('x_1','x_2','location','southwest')
36 subplot(2,1,2)
37 plot(t,u_cl,'-o')
38 legend('u','location','northwest')
39
40 saveas(h,'solution_ode_feedback','epsc')
```

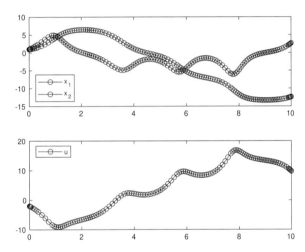

Figure B.2: Solution of the dynamical system (1.45) obtained through `ode45.m` and the corresponding feedback law.

Remark B.1. The function `ode45.m` can solve autonomous and non-autonomous, i.e., time-varying, ordinary differential equations. Nevertheless, the differential equation to be solved needs to be defined as a non-autonomous ordinary differential equation. In particular, as in the definition of the function `ode_pendulum.m`, the ordinary differential equation must have t and x as input variables.

B.2   LINEAR SYSTEMS

We again start with the nonlinear system (1.45) and define the right-hand side as a symbolic function. In particular, we define the state and the input x_1, x_2, and u, respectively, as real, i.e., not complex, symbolic variables. In the same way, the parameters J, m, ℓ, g, and γ are defined as positive symbolic variables, which allows us to define the right-hand side f.

```
1 syms x1 x2 u real
2 syms J m ell g gamma positive
3
4 f =[x2;
```

```
5       (1/(J+m* ell ^2))*((m*g* ell )* sin (x1)−gamma*x2+ell *cos (x1)*u)];
```

The derivative of f with respect to x and u can be obtained through symbolic differentiation `diff.m`.

```
6  dfdx = [diff(f,x1)  diff(f,x2)];
7  dfdu = diff(f,u);
```

One way to obtain the linearization of the nonlinear system at the origin, as described in (3.49), is achieved by substituting x and u with 0, which is done using the function `subs.m` here.

```
8  A = subs(dfdx ,[x1 ,x2 ,u] ,[0 ,0 ,0]);
9  B = subs(dfdu ,[x1 ,x2 ,u] ,[0 ,0 ,0]);
```

In the same way, for specific parameters the linear system $\dot{x} = Ax + Bu$ is obtained. In addition, we assume that $y = x_1$ and define $y = Cx + Du$ accordingly.

```
10  A = double(subs(A,[J,m, ell ,g,gamma] ,[1 ,1 ,1 ,9.81 ,0.1]));
11  B = double(subs(B,[J,m, ell ,g,gamma] ,[1 ,1 ,1 ,9.81 ,0.1]));
12
13  C = [1  0];
14  D = 0;
```

The function `double.m` is necessary to convert the symbolic expression to a numeric expression.

From the quadruple (A, B, C, D), the transfer function representing the system in the frequency domain is obtained through `ss2tf.m`.

```
>> [num,denom] = ss2tf(A,B,C,D)

num =

              0              0      0.5000

denom =

      1.0000       0.0500      −4.9050
```

In particular, the transfer function

$$G(s) = \frac{0.5}{s^2 + 0.05s - 4.905}$$

is defined by the numerator and the denominator polynomial.

Vice versa, a state space representation (A, B, C, D) based on a transfer function can be calculated through `tf2ss.m`.

```
>> [A,B,C,D] = tf2ss(num,denom)

A =

     −0.0500       4.9050
      1.0000            0

B =

          1
          0
```

```
C =
                0      0.5000

D =

        0
```

Note that the state space representation is not unique. Here, the quadruple (A, B, C, D) used to compute the transfer function does not coincide with the matrices returned by tf2ss.m.

The origin of the linear system is unstable, which can be verified by calculating the eigenvalues of A through eig.m.

```
>> Eval=eig(A)

Eval =

    -2.2399
     2.1899
```

Alternatively, the roots of the transfer function can be calculated through the command roots.m to check if the linear system is Hurwitz.

```
>> polesTF = roots(denom)

polesTF =

    -2.2399
     2.1899
```

Since the origin of the system is not asymptotically stable, we will construct a linear feedback law stabilizing the origin. However, we first verify that the linear system is controllable and thus ensure that the system can be stabilized through a linear feedback. The Kalman controllability matrix, defined in (3.57), can be calculated through ctrb.m.

```
>> Kal = ctrb(A,B)

Kal =

    1.0000   -0.0500
         0    1.0000
```

Since the obtained matrix is an upper triangular matrix with nonzero diagonal elements, the matrix has full rank, and thus the system is controllable.

If the rank cannot be read off immediately, the rank (or the numerical rank) can be checked through rank.m, qr.m (QR-decomposition), or svd.m (singular value decomposition), for example. Observability can be verified in the same way, by checking controllability of the pair (A^T, C^T).

```
>> Kal_obs = ctrb(A',C')

Kal_obs =

         0    0.5000
    0.5000         0
```

Alternatively, the observability matrix of a pair (A, C) can be directly calculated through obsv.m.

Since the linear system is controllable, we can use pole placement discussed in Section 3.5.3 to design a linear feedback law $u = Kx$ such that $A + BK$ is Hurwitz.

```
>> K = -place(A,B,[-1  -2])

K =

    -2.9500    -6.9050
```

For a controllable pair (A, B) the function place.m returns the feedback gain K such that $A + BK$ has the eigenvalues specified in the third argument. The minus sign is necessary here since MATLAB defines the closed-loop matrix as $A - BK$ instead of $A + BK$. To ensure we have not made a mistake, we check the eigenvalues of the closed-loop matrix through eig.m .

```
>> eig(A+B*K)

ans =

    -2.0000
    -1.0000
```

With the function place.m the selected eigenvalues need to have multiplicity less than or equal to the number of inputs. For single input systems, the function acker.m does not have this restriction. However, the function acker.m is numerically less reliable and is thus not recommended.

```
>> K2 = -acker(A,B,[-1  -1])

K2 =

    -1.9500    -5.9050

>> eig(A+B*K2)

ans =

    -1
    -1
```

Since $A + BK$ is Hurwitz, there exists a quadratic Lyapunov function $V(x) = x^T P x$, and $P \in \mathcal{S}_{>0}^2$ can be obtained by solving the Lyapunov equation (3.4) for a positive definite matrix Q. In MATLAB, (3.4) is solved through lyap.m .

```
>> P=lyap((A+B*K)',eye(2))

P =

    0.2500    0.2500
    0.2500    1.2500
```

Note the transpose in the input argument of the closed-loop matrix, which is necessary since in MATLAB the Lyapunov equation is defined slightly differently from the notation used in this book. The function eye.m defines the identity matrix in MATLAB.

It is straightforward to verify that the matrix P is indeed positive definite and satisfies the Lyapunov equation.

```
>> eig(P)
```

```
ans =

    0.1910
    1.3090

>> (A+B*K)'*P+P*(A+B*K)

ans =

   -1.0000         0
         0   -1.0000
```

To visualize the Lyapunov function we use the following code, which:

- (based on the matrix P) defines $V(x) = x^T P x$ as an anonymous function,
- defines a grid in the state space in the domain $[-2, 2]^2$,
- evaluates V at the grid points, and
- plots V.

The function visualized through surf.m is shown in Figure B.3.

```
1  V=@(x) x'*P*x;
2
3  x1=linspace(-2,2,41);
4  x2=linspace(-2,2,41);
5
6  [XX1,XX2]=meshgrid(x1,x2);
7
8  Vx=zeros(size(XX1));
9
10 for i=1:size(XX1,1)
11     for j=1:size(XX1,2)
12         Vx(i,j) = V([XX1(i,j); XX2(i,j)]);
13     end
14 end
15
16 h=figure(1);
17 surf(XX1,XX2,Vx)
18 xlabel('x_1')
19 ylabel('x_2')
20 zlabel('V(x)')
21
22 saveas(h,'visualize_Lyap_fcn','epsc')
```

A smoother plot is obtained by using a finer grid and by changing the line style (see Figure B.4). Additionally, we can restrict our attention to a certain sublevel set by setting values of $V(x)$ bigger than a certain threshold to infinity.

```
1  V=@(x) x'*P*x;
2
3  x1=linspace(-2,2,501);
4  x2=linspace(-2,2,501);
5
6  [XX1,XX2]=meshgrid(x1,x2);
7
8  Vx=zeros(size(XX1));
9
10 for i=1:size(XX1,1)
11     for j=1:size(XX1,2)
12         Vx(i,j) = V([XX1(i,j); XX2(i,j)]);
```

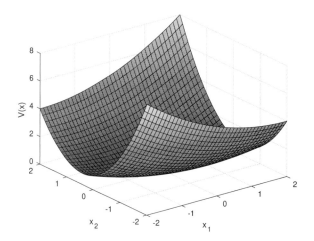

Figure B.3: Visualization of the quadratic Lyapunov function $V(x) = x^T P x$.

```
13        end
14 end
15 Vx(Vx>4)=inf;
16
17 h=figure(1);
18 surf(XX1,XX2,Vx,'linestyle','none')
19 xlabel('x_1')
20 ylabel('x_2')
21 zlabel('V(x)')
22
23 saveas(h,'visualize_Lyap_fcn2','epsc')
```

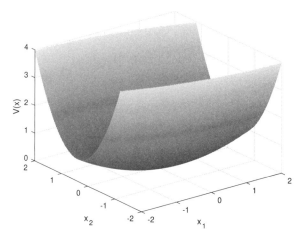

Figure B.4: Visualization of the quadratic Lyapunov function $V(x) = x^T P x$ restricted to $V(x) \leq 4$.

From the theory derived in Chapter 2 we know that $V(x(t))$ is strictly decreasing for all $x(t) \neq 0$. We verify this property for the solution corresponding to the initial value $x_0 = [1,1]^T$. To compute the solution of the differential equation $\dot{x} = (A + BK)x$ we could again use the function ode45.m. Alternatively, since the differential equation is linear, we know that the solution is given by

$$x(t) = e^{(A+Bk)t}x_0,$$

which we use here to compute the solution.

```
 1  V=@(x) x'*P*x;
 2  Acl=A+B*K;
 3  t=linspace(0,5,100);
 4  x0=[1;1];
 5
 6  Vx=zeros(size(t));
 7  norm_xt=zeros(size(t));
 8
 9  for i=1:length(t)
10      xt=expm(Acl*t(i))*x0;
11      Vx(i)=V(xt);
12      norm_xt(i)=norm(xt);
13  end
14  Vx(Vx>4)=inf;
15
16  h=figure(1);
17  plot(t,Vx,'-.')
18  hold on; grid on; box on;
19  plot(t,norm_xt)
20  xlabel('t')
21  legend('V(x(t))','|x(t)|')
22
23  saveas(h,'decrease_V','epsc')
```

Note that function `expm.m` needs to be used to compute the matrix exponential. Instead, the function `exp.m` calculates the exponential function for scalars and the scalar exponential function is calculated componentwise if a matrix is used as an input argument. The decrease of the function $V(x(t))$ is shown in Figure B.5, where additionally the norm $|x(t)|$ for $t \in [0,5]$ is visualized. The norm is not decreasing for all $t \in [0,5]$ and thus the norm does not define a Lyapunov function for these particular dynamics.

As a last MATLAB command in this section we briefly discuss `care.m` or `icare.m` (where `icare.m` is slated to replace `care.m` in future MATLAB releases).

In Chapter 14 we have discussed the *linear quadratic regulator* and in particular in Theorem 14.4 we have given a result which provides a stabilizing feedback law and a corresponding Lyapunov function of the closed-loop dynamical system as the solution of an algebraic Riccati equation. We can for example use `care.m` or `icare.m` to solve the equation (14.15) for $Q = C^T C$ and for $R = 1$.

```
>> [P,L,K]=care(A,B,C'*C,1)

P =

    4.3855      9.8354
    9.8354     22.1139


L =

   -2.2177 + 0.1099i
   -2.2177 - 0.1099i
```

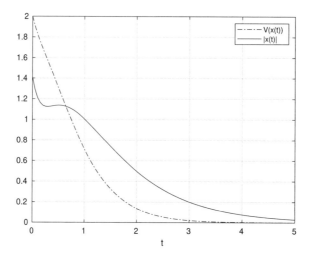

Figure B.5: Along a solution the Lyapunov function $V(x(t))$ is strictly decreasing. However, the solution is not strictly decreasing along arbitrary positive definite functions. In this particular example, the norm is not a Lyapunov function.

```
K =

    4.3855      9.8354
```

In addition to the solution of the algebraic Riccati equation P, the stabilizing feedback gain $K = R^{-1}B^T P$ is returned (see Theorem 14.4), i.e., $A - BK$ is Hurwitz, and L contains the eigenvalues of $A - BK$.

B.3 CVX

Based on two simple examples we illustrate how CVX can be used to solve linear matrix inequalities and convex optimization problems with a focus on model predictive control. We refer to [56], [57] for detailed documentation of CVX. Note that CVX needs to be installed separately before it can be used in MATLAB.

B.3.1 Linear Matrix Inequalities

Consider the linear system $\dot{x} = Ax$ defined through the matrix

$$A = \begin{bmatrix} 0 & 1 \\ -1 & -1 \end{bmatrix}. \tag{B.1}$$

We have seen in Section 3.4.1 that the matrix A is Hurwitz, and thus there exists a positive definite matrix $P \in \mathcal{S}_{>0}^2$ such that $A^T P + PA$ is negative definite (see Theorem 3.3), which means that P needs to satisfy the conditions

$$P > 0, \qquad A^T P + PA < 0.$$

Since CVX cannot handle strict inequalities we strengthen the inequality constraints to

$$P \geq \varepsilon I, \qquad A^T P + PA \leq -\varepsilon I \qquad (B.2)$$

for a small parameter $\varepsilon > 0$. Note that the selection of ε in (B.2) does not change the feasibility of (B.2) since by dividing by ε, the parameter can be incorporated in the unknown P.

The linear matrix inequality (B.2) can be solved in MATLAB through CVX using the following code, which returns the matrix

$$P = \begin{bmatrix} 6.3644 & 1.9088 \\ 1.9088 & 4.6670 \end{bmatrix}$$

as a solution. Since the solution is not unique, a different P may be returned as an output.

```
1  A=[0   1;
2     -1  -1];
3  epsi=0.01;
4
5  cvx_begin sdp %quiet
6
7      variable P(2,2) symmetric
8
9      P >= epsi*eye(2);
10     A'*P+P*A <= -epsi*eye(2);
11
12 cvx_end
```

CVX is initialized and ended in lines 5 and 12, respectively. LMIs are specified as semidefinite programs (sdp). Outputs in the workspace can be suppressed through the optional argument quiet. In line 7, the unknown P is defined as a 2×2 symmetric matrix, while the lines 9 and 10 implement (B.2).

The inclusion of additional variables and constraints is straightforward. For example,

```
13 cvx_begin sdp %quiet
14
15     variable P(2,2) symmetric
16     variable Q(2,2) symmetric
17
18     A'*P+P*A == Q;
19     Q <= -epsi*eye(2);
20     P >= epsi*eye(2);
21
22     for i=1:2
23         for j=1:2
24             -1 <= P(i,j) <= 1;
25         end
26     end
27     Q(1,1) == -1;
28
29 cvx_end
```

finds P and Q such that

$$P \geq \varepsilon I, \qquad A^T P + PA = Q, \qquad Q \leq -\varepsilon I,$$
$$|P_{ij}| \leq 1, \quad i, j \in \{1, 2\}, \qquad Q_{11} = -1, \tag{B.3}$$

is satisfied. Here, the matrices

$$P = \begin{bmatrix} 0.7922 & 0.5000 \\ 0.5000 & 0.9129 \end{bmatrix} \quad \text{and} \quad Q = \begin{bmatrix} -1.0000 & -0.6207 \\ -0.6207 & -0.8258 \end{bmatrix}$$

are returned.

Objective functions can also be included through a single command. The following lines find P and Q satisfying (B.3) while maximizing the entry Q_{22}.

```
30  cvx_begin sdp %quiet
31
32      variable P(2,2) symmetric
33      variable Q(2,2) symmetric
34
35      maximize(Q(2,2))
36 %    minimize(Q(2,2))
37
38      A'*P+P*A == Q;
39      Q <= -epsi*eye(2);
40      P >= epsi*eye(2);
41
42      for i=1:2
43          for j=1:2
44              -1 <= P(i,j) <= 1;
45          end
46      end
47      Q(1,1) == -1;
48
49  cvx_end
```

Through these commands, CVX returns the matrices

$$P = \begin{bmatrix} 1.0000 & 0.5000 \\ 0.5000 & 0.5050 \end{bmatrix} \quad \text{and} \quad Q = \begin{bmatrix} -1.0000 & -0.0050 \\ -0.0050 & -0.0100 \end{bmatrix}$$

as well as the optimal value -0.01.

B.3.2   Convex Optimization Problems

In addition to linear matrix inequalities, CVX allows a simple implementation of convex optimization problems using an intuitive syntax for users familiar with MATLAB. For example, the optimization problem

$$\min 2x^2 + y^2$$
$$\text{subject to } |x| \leq 5$$
$$|y| \leq 5$$
$$x + y \geq 5$$

can be encoded and solved through the following lines in MATLAB.

```
1  cvx_begin %quiet
```

```
 2      variable x
 3      variable y
 4
 5      val = 2*x^2+y^2;
 6
 7      minimize(val)
 8      subject to
 9
10      −5 <= x <= 5;
11      −5 <= y <= 5;
12       5 <= x+y;
13
14  cvx_end
```

In the context of this book, CVX is in particular helpful to quickly implement a model predictive control scheme as discussed in Chapter 15 for linear discrete time systems. As an example, we revisit Example 15.4 and indicate how a corresponding MPC application can be simulated.

Consider the discrete time system $x^+ = Ax + Bu$ defined through the matrices

$$A = \begin{bmatrix} \frac{6}{5} & \frac{6}{5} \\ -\frac{1}{2} & \frac{6}{5} \end{bmatrix} \qquad \text{and} \qquad B = \begin{bmatrix} 1 \\ \frac{1}{2} \end{bmatrix}. \tag{B.4}$$

Additionally, consider the constraints $u \in \mathbb{U} = [-2.5, 2.5]$ and select the running costs $\ell(x, u) = x^T x + 5u^2$. Then, a corresponding optimal control problem

$$\min_{\substack{u(i) \in \mathbb{R}^m \\ i \in \{0, \dots, N-1\}}} \sum_{i=0}^{N-1} x(i)^T x(i) + 5u(i)^2$$

$$\begin{aligned}
\text{subject to} \qquad x(0) &= x_0 \\
x(i+1) &= Ax(i) + Bu(i) & \forall \, i = 0, \dots, N-2 \\
|u(i)| &\leq 2.5 & \forall \, i = 0, \dots, N-1
\end{aligned}$$

can be implemented and solved through the following code, where N is defined as $N = 10$ and the initial condition is set to $x_0 = [3, 3]^T$.

```
 1  Q=eye(2); R=5;
 2  N=10;
 3
 4  A=[1.1   1.2;
 5     −0.5  1.1];
 6  B=[1;0.5];
 7
 8  lb_u=−2.5; ub_u=2.5;
 9
10  x0=[3;3];
11
12  cvx_begin %quiet
13      variable x1(N)
14      variable x2(N)
15      variable u(N)
16
17      x=[x1';
18         x2'];
19
20      val=0;
21      for i=1:N
22          val =val+ x(:,i)'*Q*x(:,i) + u(i)*R*u(i);
```

```
23    end
24
25    minimize( val )
26    subject to
27    % Initial condition
28    x1(1)==x0(1);
29    x2(1)==x0(2);
30    % Dynamics
31    for i=1:N-1
32        x(:,i+1)==A*x(:,i) + B*u(i);
33    end
34    % Constraints
35    for i=1:N
36        lb_u <= u(i) <= ub_u;
37    end
38 cvx_end
```

To implement Algorithm 3, we can simply embed the optimization problem in a for-loop and store the optimal open-loop and closed-loop solutions after every iteration. Moreover, the initial condition needs to be reinitialized in every iteration.

```
 1 N=10;
 2 iter=30;
 3
 4 Q=eye(2); R=5;
 5
 6 A=[1.1    1.2;
 7    -0.5   1.1];
 8 B=[1; 0.5];
 9
10 lb_u=-2.5; ub_u=2.5;
11
12 x0=[3;3];
13
14 Xcl=[x0];
15 Ucl=[];
16 Xol1=[];
17 Xol2=[];
18 Uol=[];
19
20 for k=1:1:iter
21
22 cvx_begin %quiet
23    variable x1(N)
24    variable x2(N)
25    variable u(N)
26
27    x=[x1';
28       x2'];
29
30    val=0;
31    for i=1:N
32        val =val+ x(:,i)'*Q*x(:,i) + u(i)*R*u(i);
33    end
34
35    minimize( val )
36    subject to
37
38    % Initial condition
39    x1(1)==x0(1);
40    x2(1)==x0(2);
41
```

```
42    % Dynamics
43      for i=1:N-1
44           x(:,i+1)=A*x(:,i) + B*u(i);
45      end
46    % Constraints
47      for i=1:N
48           lb_u <= u(i) <= ub_u;
49      end
50  cvx_end
51
52  x0=[x1(2);
53       x2(2)];
54
55  Xcl=[Xcl x0];
56  Ucl=[Ucl u(1)];
57  Xol1=[Xol1 x1];
58  Xol2=[Xol2 x2];
59  Uol=[Uol u];
60  end
```

Remark B.2. Note that the implementation of Algorithm 3 presented here is meant to be illustrative but is not designed to be computationally efficient.

B.4   SOSTOOLS

To illustrate SOSTOOLS, we solve the same problems as in Section B.3.1, where we used CVX instead. While we only discuss simple examples and commands here, we refer to [116] for a comprehensive introduction. Additionally, SeDuMi [149] needs to be installed as a solver in order to be able to use SOSTOOLS.

Consider the linear system $\dot{x} = Ax$ defined by the matrix

$$A = \begin{bmatrix} 0 & 1 \\ -1 & -1 \end{bmatrix} \tag{B.5}$$

and assume we wish to compute parameters $a, b, c \in \mathbb{R}$ of a Lyapunov function

$$V(x) = ax_1^2 + bx_1x_2 + cx_2^2 = x^T \begin{bmatrix} a & \frac{1}{2}b \\ \frac{1}{2} & c \end{bmatrix} x.$$

This means, for V to be a Lyapunov function,

$$P = \begin{bmatrix} a & \frac{1}{2}b \\ \frac{1}{2}b & c \end{bmatrix}$$

needs to be positive definite and $A^T P + PA$ needs to be negative definite.

Since A is Hurwitz (see Section 3.4.1), $P \in \mathcal{S}_{>0}^2$ satisfying

$$P > 0, \qquad A^T P + PA < 0$$

exists (see Theorem 3.3). Next, we strengthen the inequality constraints to

$$P - \varepsilon I \geq 0, \qquad -(A^T P + PA) - \varepsilon I \geq 0 \tag{B.6}$$

for a small parameter $\varepsilon > 0$ to transform the strict inequality constraints. Note that the selection of ε in (B.6) does not change the feasibility of (B.6) since by dividing through by ε, the parameter can be incorporated in the unknown P. Moreover, (B.6) is equivalent to

$$V(x) - \varepsilon(x_1^2 + x_2^2) \geq 0, \qquad -\langle \nabla V(x), Ax \rangle - \varepsilon(x_1^2 + x_2^2) \geq 0$$

for all $x \in \mathbb{R}^2$. These two constraints can be implemented and solved using SOSTOOLS.

```
1  syms x1 x2;
2  vars = [x1; x2];
3
4  % Define the right-hand side \dot{x}=Ax
5  Ax = [x2;
6        -x1-x2];
7
8  % Define a lower bound and an upper bound
9  epsil    = 0.01*(x1^2+x2^2);
10
11 % Initialize the sum of squares program
12 prog = sosprogram(vars);
13
14 % Define the Lyapunov function V(x):
15 [prog,V] = sospolyvar(prog,[x1^2; x2^2; x1*x2],'wscoeff');
16
17 % Define the constraints
18 % Constraint 1 : V(x) - alpha1 >= 0
19 expr1 = V-epsil;
20 prog  = sosineq(prog,expr1);
21
22 % Constraint 2: -dV/dx*f - epsil >= 0
23 expr2 = -[diff(V,x1) diff(V,x2)]*Ax - epsil;
24 prog  = sosineq(prog,expr2);
25
26 % Call solver
27 solver_opt.solver = 'sedumi';
28 prog = sossolve(prog,solver_opt);
29
30 % Get solution
31 SOL_V = sosgetsol(prog,V)
```

This returns the Lyapunov function .

$$V(x) = 0.9124x_1^2 + 0.5162x_1x_2 + 0.7244x_2^2.$$

Since the solution is not unique, different parameters a, b, and c may be returned as an output.

As a second example, we add additional constraints equivalent to the conditions in (B.3). In particular, we define the unknowns

$$V(x) = ax_1^2 + bx_1x_2 + cx_2^2, \qquad Q(x) = q_1x_1^2 + q_2x_1x_2 + q_3x_2^2$$

and consider the conditions

$$V(x) - \varepsilon(x_1^2 + x_2^2) \geq 0, \qquad -Q(x) - \varepsilon(x_1^2 + x_2^2) \geq 0$$
$$Q(x) - \langle \nabla V(x), Ax \rangle = 0, \qquad |a| \leq 1, \ |b| \leq 1, \ |c| \leq 1, \ q_1 = -1.$$

These inequalities and equations can be solved with the following code, and the functions

$$V(x) = 0.8252x_1^2 + 1.0x_1x_2 + 0.734x_2^2$$
$$Q(x) = -1.0x_1^2 - 0.8177x_1x_2 - 0.468x_2^2$$

are returned as a solution.

```
1  syms x1 x2;
2  syms a b c real;
3  syms q1 q2 q3 real;
4  vars = [x1; x2];
5
6  % Define the right-hand side \dot{x}=Ax
7  Ax = [x2;
8        -x1-x2];
9
10 % Define a lower bound and an upper bound
11 epsiI  = 0.01*(x1^2+x2^2);
12
13 % Initialize the sum of squares program
14 prog = sosprogram(vars,[q1 q2 q3 a b c]);
15
16 % Define the Lyapunov function V(x):
17 V = a*x1^2+b*x1*x2+c*x2^2;
18 % Define the function Q(x):
19 Q = q1*x1^2+q2*x1*x2+q3*x2^2;
20
21 % Define the constraints
22 % Constraint 1 : V(x) - alpha1 >= 0
23 expr1 = V-epsiI;
24 prog  = sosineq(prog,expr1);
25
26
27 % Constraint 2: -Q(x) - epsiI >= 0
28 expr2 = -Q - epsiI;
29 prog  = sosineq(prog,expr2);
30
31 % Constraint 3: Q - dV/dx*f == 0
32 expr3 = Q-[diff(V,x1) diff(V,x2)]*Ax;
33 prog  = soseq(prog,expr3);
34
35 % Additional constraints:
36 prog  = soseq(prog,q1+1);
37 prog  = sosineq(prog,a+1);
38 prog  = sosineq(prog,-a+1);
39 prog  = sosineq(prog,b+1);
40 prog  = sosineq(prog,-b+1);
41 prog  = sosineq(prog,c+1);
42 prog  = sosineq(prog,-c+1);
43
44 % Call solver
45 solver_opt.solver = 'sedumi';
46 prog = sossolve(prog,solver_opt);
47
48 % Get solution
49 SOL_V = sosgetsol(prog,V)
50 SOL_Q = sosgetsol(prog,Q)
```

As a last step, we include an objective function and maximize q_3. This is achieved through the following code, returning

$$V(x) = 1.0x_1^2 + 1.0x_1x_2 + 0.505x_2^2$$
$$Q(x) = -1.0x_1^2 - 0.01003x_1x_2 - 0.01003x_2^2,$$

and, in particular, $q_3 = -0.01003$. The objective function is defined in line 45.

```
1  syms x1 x2;
2  syms a b c real;
3  syms q1 q2 q3 real;
4  vars = [x1; x2];
5
6  % Define the right-hand side \dot{x}=Ax
7  Ax = [x2;
8        -x1-x2];
9
10 % Define a lower bound and an upper bound
11 epsiI   = 0.01*(x1^2+x2^2);
12
13 % Initialize the sum of squares program
14 prog = sosprogram(vars,[q1 q2 q3 a b c]);
15
16 % Define the Lyapunov function V(x):
17 V = a*x1^2+b*x1*x2+c*x2^2;
18 % Define the function Q(x):
19 Q = q1*x1^2+q2*x1*x2+q3*x2^2;
20
21 % Define the constraints
22 % Constraint 1 : V(x) - alpha1 >= 0
23 expr1 = V-epsiI;
24 prog  = sosineq(prog,expr1);
25
26
27 % Constraint 2: -Q(x) - epsiI >= 0
28 expr2 = -Q - epsiI;
29 prog  = sosineq(prog,expr2);
30
31 % Constraint 3: Q - dV/dx*f == 0
32 expr3 = Q-[diff(V,x1) diff(V,x2)]*Ax;
33 prog  = soseq(prog,expr3);
34
35 % Additional constraints:
36 prog  = soseq(prog,q1+1);
37 prog  = sosineq(prog,a+1);
38 prog  = sosineq(prog,-a+1);
39 prog  = sosineq(prog,b+1);
40 prog  = sosineq(prog,-b+1);
41 prog  = sosineq(prog,c+1);
42 prog  = sosineq(prog,-c+1);
43
44 % Define the objective function: max q3 <=> - min -q3
45 prog = sossetobj(prog,-q3);
46
47 % Call solver
48 solver_opt.solver = 'sedumi';
49 prog = sossolve(prog,solver_opt);
50
51 % Get solution
52 SOL_V = sosgetsol(prog,V)
53 SOL_Q = sosgetsol(prog,Q)
54 SOL_q3 = sosgetsol(prog,q3)
```

B.5 CASADI

In this section we present an implementation of Algorithm 4 from Chapter 15 on the example of the inverted pendulum on a cart (1.35) using CasADi [6] in MATLAB. Additionally, as a numerical solver, Ipopt [159] is used in CasADi.

Consider the optimal control problem

$$\min_{\substack{u(i)\in\mathbb{R}^m \\ i\in\mathbb{N}_{[0,N]}}} \sum_{i=0}^{N-1} \ell(x(i),u(i)) \tag{B.7}$$

$$
\begin{aligned}
\text{subject to} \quad 0 &= x(0) - x_0 \\
0 &= x(i+1) - x(i) - \Delta f(x(i),u(i)) && \forall\, i = 0,\dots,N-1 \\
c_u &\geq u(i) && \forall\, i = 0,\dots,N \\
c_u &\geq -u(i) && \forall\, i = 0,\dots,N \\
c_x &\geq x_1(i) && \forall\, i = 0,\dots,N-1 \\
c_x &\geq -x_1(i) && \forall\, i = 0,\dots,N-1,
\end{aligned}
$$

where $\dot{x} = f(x,u)$,

$$
f(x,u) = \begin{bmatrix}
x_3 \\
x_4 \\
\dfrac{-\bar{J}\bar{c}x_3 - \bar{J}\sin(x_2)x_4^2 - \bar{\gamma}\cos(x_2)x_4 + g\cos(x_2)\sin(x_2) + \bar{J}u}{\bar{M}\bar{J} - \cos^2(x_2)} \\
\dfrac{-\bar{M}\bar{\gamma}x_4 + \bar{M}g\sin(x_2) - \bar{c}\cos(x_2)x_3 - \cos(x_2)\sin(x_2)x_4^2 + \cos(x_2)u}{\bar{M}\bar{J} - \cos^2(x_2)}
\end{bmatrix}, \tag{B.8}
$$

with $f : \mathbb{R}^4 \times \mathbb{R} \to \mathbb{R}^4$, captures the dynamics of the inverted pendulum on a cart discussed in Section 1.2.1. In the optimization problem the dynamics are approximated through an Euler discretization with sampling rate $\Delta > 0$ (see Section 5.2). To stabilize the pendulum in the upright position we define the running cost

$$\ell(x,u) = c_1 x_1^2 + c_2(1 - \cos(x_2))^2 + c_3 x_3^2 + c_4 x^2 + c_5 u^2,$$

with $c_1, c_2, c_3, c_4, c_5 \in \mathbb{R}_{>0}$. To incorporate upper and lower bounds on the control input, the constraint $-c_u \leq u \leq c_u$ for $c_u \in \mathbb{R}_{>0}$ is included in the optimization problem. Similarly, for $c_x \in \mathbb{R}_{>0}$ the constraints $-c_x \leq x_1 \leq c_x$ ensure that the cart stays in a specified domain. Note that $x(N)$ and $u(N)$ are not relevant in the optimization problem (since a terminal cost is not present), but they are included to simplify the implementation.

To implement Algorithm 4, we define the continuous time dynamics and the discrete time dynamics of the pendulum as MATLAB functions.

```
1  function [dx]=dynamics_pendulum(t,x,u,params)
2
3  Mb   = params.Mb;
4  Jb   = params.Jb;
5  cb   = params.cb;
6  gamb = params.gamb;
7  g    = params.g;
8
9  denom=Mb*Jb-cos(x(2))^2;
10
11 dx1=x(3);
12 dx2=x(4);
13 dx3=-Jb*cb*x(3)-Jb*sin(x(2))*x(4)^2-gamb*cos(x(2))*x(4)...
```

```
14            +g*cos(x(2))*sin(x(2))+Jb*u;
15
16  dx4=-Mb*gamb*x(4)+Mb*g*sin(x(2))-cb*cos(x(2))*x(3) ...
17            -cos(x(2))*sin(x(2))*x(4)^2+cos(x(2))*u;
18
19  dx=[dx1;
20      dx2;
21      dx3/denom;
22      dx4/denom];
23
24  end
```

```
1  function [x_plus]=dynamics_pendulum_dt(t,x,u,params)
2
3  dx=dynamics_pendulum(t,x,u,params);
4
5  x_plus=x+params.Delta*dx;
6
7  end
```

As a next step we define the parameters of the pendulum and the parameter Δ in a main file. Additionally, options for CasADi, such as the maximum number of iterations, the level of information shown in the workspace, and the tolerance of the solver are defined.

```
1  % Set parameters
2  m=1;
3  M=2;
4  J=0.5;
5  l=1;
6  g=9.81;
7  c=0.1;
8  gam=0.1;
9
10  params.Jb    = (J+m*l^2)/(m*l);
11  params.Mb    = (M+m)/(m*l);
12  params.cb    = c/(m*l);
13  params.gamb  = gam/(m*l);
14  params.g     = g;
15
16  % Discretization stepsize
17  params.Delta=0.05;
18
19  % Parmeters of the running costs
20  params.cost.c1=10;
21  params.cost.c2=100;
22  params.cost.c3=1;
23  params.cost.c4=1;
24  params.cost.c5=1;
25
26  % Number of states and inputs
27  params.nx=4; % States
28  params.nu=1; % Inputs
29
30  % Parameters for CasADi
31  opts = struct;
32  opts.ipopt.max_iter     = 3000;
33  opts.ipopt.print_level = 5;%0,3,5     % Information displayed
34  opts.print_time         = 0;          % in the workspace
35  opts.ipopt.acceptable_tol =1e-15;
36  opts.ipopt.acceptable_obj_change_tol = 1e-15;
```

We continue with the definition of variables and parameters necessary for the MPC closed loop.

```
37  % Number of MPC iterations
38  mpcIter = 250;
39  % Prediction horizon
40  params.N=50;
41
42  % Define the initial condition
43  x0 = [1;
44        pi;
45        0;
46        0];
47
48  % Define upper and lower bound of the states/input
49  c_u =10;
50  c_x1 =5;
51  bounds_x=[-c_x1  c_x1;
52             -inf   inf;
53             -inf   inf;
54             -inf   inf];
55  bounds_u=[-c_u    c_u];
56
57  % Define bounds over the prediction horizon
58  Bounds.LB_X = [];
59  Bounds.UB_X = [];
60  Bounds.LB_U = [];
61  Bounds.UB_U = [];
62  for  i =1:params.N+1
63        Bounds.LB_X  =[Bounds.LB_X bounds_x(:,1)];
64        Bounds.UB_X  =[Bounds.UB_X bounds_x(:,2)];
65        Bounds.LB_U  =[Bounds.LB_U bounds_u(:,1)];
66        Bounds.UB_U  =[Bounds.UB_U bounds_u(:,2)];
67  end
68
69  % Initialize variables to store open-loop and closed-loop
70  %    solutions
71  Xcl=x0;
72  Ucl = [];
73  Tcl=0;
74  XX_ol=zeros(params.nx+params.nu,params.N+1);
75
76  % Generate an inital guess of the optimal solution
77  xu_guess=zeros(params.nx*(params.N+2)...
78        +params.nu*(params.N+1),1);
```

Before we continue with the main file and the loop of the MPC algorithm, we define a function solving the optimization problem (B.7) at a fixed time step.

```
1   function [xu_sol]=solve_OCP(params,x0_ini,xu_guess,Bounds,opts)
2
3   % Initialize CasADi
4   import casadi.*
5
6   %%%% Define OCP
7
8   %%% Define CASADI variables
9   % States (over the prediction horizon)
10  x     = SX.sym('x', params.nx, params.N+1);
11  % Inputs (over the prediction horizon)
12  u     = SX.sym('u', params.nu, params.N+1);
13  % Initial condition (is imlemented as an unknown)
14  %  (The initial condition is stored in the lower/upper bounds)
```

```
15  xini  = SX.sym('xini',params.nx, 1);
16  % Equality constraints
17  eq_con= SX.sym('eq_con', params.nx, params.N+1);
18
19
20  % Define equality constraints (generated through the dynamics)
21  for i=1:params.N
22      if  i==1 % Initial condition ...
23          eq_con(:,1) = x(:,1)-xini;
24      end
25              % ... and over the prediction horizon
26          eq_con(:,i+1) = x(:,i+1)-dynamics_pendulum_dt ...
27                          (0,x(:,i),u(:,i),params);
28  end
29
30  % Define the objective function
31  obj = 0;
32  for i=1:params.N
33      obj=obj+params.cost.c1*x(1,i)^2 ...
34          +params.cost.c2*(1-cos(x(2,i)))^2 ...
35          +params.cost.c3*x(3,i)^2 ...
36          +params.cost.c4*x(4,i)^2 ...
37          +params.cost.c5*u(i)^2;
38  end
39
40  % Create OCP
41  nlp = struct('x', [xini(:); x(:); u(:)],... % Unknowns
42                  'f', obj,...              % Objective function
43                  'g', [eq_con(:)]);        % Constraints
44
45  % Store the initial condition in upper and lower bounds
46  LBounds_ini=x0_ini;
47  UBounds_ini=x0_ini;
48  % Define overall upper and lower bounds
49  xu_LB=[LBounds_ini; Bounds.LB_X(:); Bounds.LB_U(:)];
50  xu_UB=[UBounds_ini; Bounds.UB_X(:); Bounds.UB_U(:)];
51
52
53  % Create IPOPT solver object and solve OCP
54  solver_ini=nlpsol('solver', 'ipopt', nlp, opts);
55  res=solver_ini('x0',xu_guess,...% guess of opt. solution
56                  'lbx',xu_LB,...  % lower .bound on x
57                  'ubx',xu_UB,...  % upper bound on x
58                  'lbg', 0, ...    % lower bound on g
59                  'ubg', 0);   .   % upper bound on g
60
61  % Obtain optimal solution
62  xu_sol=full(res.x);
63
64  end
```

With the function `solve_OCP.m` we can continue with the main file and the main loop of the MPC algorithm.

In line 82 the optimal control problem is solved and the open-loop solutions of x and u are returned. In line 100 the first element of the open-loop input trajectory is applied as a feedback to the continuous time dynamics of the pendulum and in line 103 the initial condition for the next iteration is updated. Based on the optimal open-loop solution, a simple initial guess of the optimal solution at the next time step is constructed starting in line 112.

The lines 120 to 142 visualize the open-loop and the closed-loop solutions of x

and u. The result of the last iteration is shown in Figure B.6.

```
79  % Main loop of the MPC algorithm
80  for kk=1:mpcIter
81    % Solve the OCP to obtain optimal open−loop solutions
82    [xu]=solve_OCP(params,x0,xu_guess,Bounds,opts);
83
84    %%% Extract parts of the solution
85    % Extract initial condition
86    x0_sol=xu(1:params.nx);
87    % Extract open−loop solution (states)
88    x_sol_ol=xu(params.nx+1:(params.N+2)*params.nx);
89    x_sol_ol=reshape(x_sol_ol,params.nx,params.N+1);
90    % Extract open−loop solution (input)
91    u_sol_ol=xu((params.N+2)*params.nx+1:end);
92    u_sol_ol=u_sol_ol';
93
94    % Store overall open−loop solution
95    XU_ol=[x_sol_ol;
96           u_sol_ol];
97
98    % Apply the input to the continuous time−dynamics ...
99    input_u=u_sol_ol(1);
100   [tt,x_cont]=ode45(@(t,x)dynamics_pendulum ...
101       (t,x,input_u,params),[0,params.Delta],x0);
102   % ... and update the initial condition
103   x0=x_cont(end,:)';
104
105   % Store the closed−loop dynamics
106   Xcl=[Xcl x0];
107   Ucl=[Ucl input_u];
108   Tcl=[Tcl Tcl(end)+params.Delta];
109
110   % Generate an initial guess for the next OCP
111   %    based on the open−loop solution
112   xu_guess(1:params.nx)=x0;
113   xu_guess(params.nx+1:(params.N+1)*params.nx) ...
114       =xu(2*params.nx+1:(params.N+2)*params.nx);
115   xu_guess((params.N+1)*params.nx+1:(params.N+2)*params.nx) ...
116       =zeros(params.nx,1);
117   xu_guess((params.N+2)*params.nx+1:end−params.nu) ...
118       =xu((params.N+2)*params.nx+1:end−params.nu);
119
120   % Visualize the closed loop solution
121   figure(1)
122   subplot(2,1,1)
123   plot(Tcl,Xcl')
124   xlim([0 mpcIter*params.Delta])
125   box on; grid on;
126   subplot(2,1,2)
127   stairs(Tcl,[Ucl Ucl(end)])
128   xlim([0 mpcIter*params.Delta])
129   xlabel('t')
130   box on; grid on;
131
132   %Visualize the open loop−solution
133   figure(2)
134   subplot(2,1,1)
135   plot((0:params.N)*params.Delta+Tcl(end−1),XU_ol(1:4,:)')
136   xlim([0 params.N]*params.Delta+Tcl(end−1))
137   box on; grid on;
138   subplot(2,1,2)
139   stairs((0:params.N)*params.Delta+Tcl(end−1),[XU_ol(5,:)])
```

```
140    xlim([0  params.N]*params.Delta+Tcl(end-1))
141    xlabel('t')
142    box on; grid on
143
144    drawnow
145 end
```

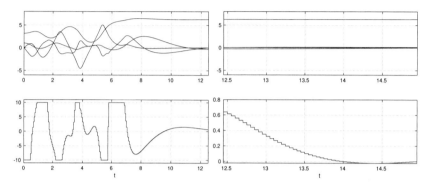

Figure B.6: On the top, closed-loop solution (left) and open-loop solution of the last iteration (right) of the MPC Algorithm 4 applied to the inverted pendulum on a cart. The lower graphs show the corresponding input signals.

To simulate and illustrate the closed loop of the pendulum the following code can be used after the main loop. The simple illustration generates figures as shown in Figure B.7. Here, only the initial and the final states of the pendulum are shown.

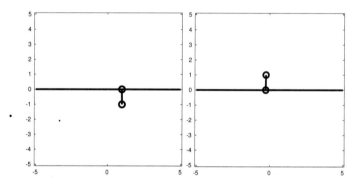

Figure B.7: Initial position and final position of the inverted pendulum using MPC as a controller.

```
146 % Simulate the closed-loop dynamics
147 P1=[Xcl(1,:);
148     zeros(1,length(Xcl(1,:)))];
149 P2=P1+[sin(-Xcl(2,:));
150     cos(-Xcl(2,:))];
151
152 for i=1:size(P1,2)
153     figure(3)
154     pause(params.Delta)
155     plot(P1(1,i),P1(2,i),'bo','MarkerSize',10,'LineWidth',3)
156     axis equal
```

```
157    xlim([-c_x1 -0.1  c_x1 +0.1])
158    ylim([-c_x1 -0.1  c_x1 +0.1])
159    hold on
160    plot(P2(1,i),P2(2,i),'bo','MarkerSize',10,'LineWidth',3)
161    plot([P1(1,i);P2(1,i)],[P1(2,i);P2(2,i)],'b','LineWidth',3)
162    plot([-c_x1  c_x1],[0;0],'b','LineWidth',3)
163    hold off
164    if i==1
165        disp('Press Return')
166        pause
167    end
168    drawnow
169 end
170 disp('Done')
```

Bibliography

[1] R. P. Agarwal. *Difference Equations and Inequalities: Theory, Methods, and Applications.* Marcel Dekker, 2nd edition, 2000.

[2] A. A. Ahmadi and B. E. Khadir. A globally asymptotically stable polynomial vector field with rational coefficients and no local polynomial Lyapunov function. *Systems & Control Letters,* 121:50–53, 2018.

[3] A. A. Ahmadi, M. Krstić, and P. A. Parrilo. A globally asymptotically stable polynomial vector field with no polynomial Lyapunov function. In *Proc. of the 50th IEEE Conference on Decision and Control and European Control Conference,* pages 7579–7580, 2011.

[4] M. Aicardi, G. Casalino, A. Bicchi, and A. Balestrino. Closed loop steering of unicycle like vehicles via Lyapunov techniques. *IEEE Robotics & Automation Magazine,* 2(1):27–35, 1995.

[5] J. Anderson and A. Papachristodoulou. Advances in computational Lyapunov analysis using sum-of-squares programming. *Discrete & Continuous Dynamical Systems-Series B,* 20(8):2361–2381, 2015.

[6] J. A. E. Andersson, J. Gillis, G. Horn, J. B. Rawlings, and M. Diehl. CasADi—A software framework for nonlinear optimization and optimal control. *Mathematical Programming Computation,* 11(1):1–36, 2019.

[7] V. Andrieu and L. Praly. On the existence of a Kazantzis–Kravaris/Luenberger observer. *SIAM Journal on Control and Optimization,* 45(2):432–456, 2006.

[8] D. Angeli, E. D. Sontag, and Y. Wang. A characterization of integral input-to-state stability. *IEEE Transactions on Automatic Control,* 45(6):1082–1097, 2000.

[9] P. J. Antsaklis and A. N. Michel. *Linear Systems.* Birkhäuser, 2006.

[10] P. J. Antsaklis and A. N. Michel. *A Linear Systems Primer.* Birkhäuser, 2007.

[11] Z. Artstein. Stabilization with relaxed controls. *Nonlinear Analysis,* 7(11):1163–1173, 1983.

[12] K. J. Åström and R. M. Murray. *Feedback Systems: An Introduction for Scientists and Engineers.* Princeton University Press, 2008.

[13] K. J. Åström and B. Wittenmark. *Adaptive Control.* Dover Publications, 2

edition, 2008.

[14] K. Atkinson, W. Han, and D. W. Stewart. *Numerical Solution of Ordinary Differential Equations*, volume 108. John Wiley & Sons, 2011.

[15] A. Bacciotti and L. Rosier. *Liapunov Functions and Stability in Control Theory*. Springer, 2nd edition, 2005.

[16] T. Basar, editor. *Control Theory: Twenty-Five Seminal Papers (1932–1981)*. IEEE Press, 2001.

[17] R. Bellman. *Dynamic Programming*. Dover, 1957.

[18] A. Bemporad, M. Morari, V. Dua, and E. N. Pistikopoulos. The explicit linear quadratic regulator for constrained systems. *Automatica*, 38(1):3–20, 2002.

[19] D. P. Bertsekas. *Dynamic Programming and Optimal Control*, volume 1. Athena Scientific Belmont, MA, 1995.

[20] D. P. Bertsekas. *Nonlinear Programming*. Athena Scientific, 3rd edition, 2016.

[21] S. Bhat and D. Bernstein. Finite-time stability of continuous autonomous systems. *SIAM Journal on Control and Optimization*, 38(3):751–766, 2000.

[22] N. P. Bhatia and G. P. Szegö. *Stability Theory of Dynamical Systems*. Springer, 1970.

[23] G. J. Bierman. *Factorization Methods for Discrete Sequential Estimation*. Academic Press, 1977.

[24] H. S. Black. Stabilized feedback amplifiers. *Bell System Technical Journal*, 13:1–18, 1934.

[25] F. Blanchini, G. Giordano, F. Riz, and L. Zaccarian. Solving nonlinear algebraic loops arising in input-saturated feedbacks. *IEEE Transactions on Automatic Control*, 2022. DOI:10.1109/TAC.2022.3170858.

[26] H. W. Bode. Relations between attenuation and phase in feedback amplifier design. *Bell System Technical Journal*, 19:421–454, 1940.

[27] A. Bogdanov. Optimal control of a double inverted pendulum on a cart. Technical report, Oregon Health and Science University, OGI School of Science and Engineering, 2004.

[28] S. Boyd, L. el Ghaoui, E. Feron, and V. Balakrishnan. *Linear Matrix Inequalities in System and Control Theory*. SIAM, 1994.

[29] S. Boyd and L. Vandenberghe. *Convex Optimization*. Cambridge University Press, 2004.

[30] V. O. Bragin, V. I. Vagaitsev, N. V. Kuznetsov, and G. A. Leonov. Algorithms for finding hidden oscillations in nonlinear systems. The Aizerman and Kalman conjectures and Chua's circuits. *Journal of Computer and Systems Sciences International*, 50(4):511–543, 2011.

[31] P. Braun, L. Grüne, and C. M. Kellett. *(In-)Stability of Differential Inclusions: Notions, Equivalences, and Lyapunov-like Characterizations.* Springer-Briefs in Mathematics. Springer, 2021.

[32] R. W. Brockett. On the stability of nonlinear feedback systems. *IEEE Transactions on Applications and Industry*, 83:443–448, 1964.

[33] R. W. Brockett. Asymptotic stability and feedback stabilization. In *Differential Geometric Control Theory*, volume 27, pages 181–191. Birkhauser, Boston, 1983.

[34] R. W. Brockett. *Finite Dimensional Linear Systems.* Classics in Applied Mathematics. Society for Industrial and Applied Mathematics, 2015.

[35] F. Bullo and A. D. Lewis. *Geometric Control of Mechanical Systems: Modeling, Analysis, and Design for Simple Mechanical Control Systems.* Springer, 2005.

[36] C. I. Byrnes, F. D. Priscoli, and A. Isidori. *Output Regulation of Uncertain Nonlinear Systems.* Birkhäuser, 1997.

[37] E. F. Camacho and C. B. Alba. *Model Predictive Control.* Springer Science & Business Media, 2nd edition, 2007.

[38] J. Carrasco, M. C. Turner, and W. P. Heath. Zames-Falb multipliers for absolute stability: From O'Shea's contribution to convex searches. *European Journal of Control*, 28:1–19, 2016.

[39] C. K. Chui and G. Chen. *Kalman Filtering with Real-Time Applications.* Springer, 1998.

[40] F. Clarke. *Functional Analysis, Calculus of Variations and Optimal Control.* Springer, 2013.

[41] F. H. Clarke, Y. S. Ledyaev, R. J. Stern, and P. R. Wolenski. *Nonsmooth Analysis and Control Theory.* Springer-Verlag, New York, 1998.

[42] G. B. Dantzig and M. N. Thapa. *Linear Programming 1: Introduction.* Springer Science & Business Media, 1997.

[43] S. Dashkovskiy, B. S. Rüffer, and F. R. Wirth. Small gain theorems for large scale systems and construction of ISS Lyapunov function. *SIAM Journal on Control and Optimization*, 48(6):4089–4118, 2010.

[44] C. A. Desoer and M. Vidyasagar. *Feedback Systems: Input-Output Properties.* Classics in Applied Mathematics. Society for Industrial and Applied Mathematics, 2009. Originally published 1975 by Academic Press.

[45] R. C. Dorf and R. H. Bishop. *Modern Control Systems.* Pearson, 13th edition, 2016.

[46] P. M. Dower and C. M. Kellett. Nonlinear $\mathcal{L}_2$-gain verification for bilinear systems. In *Proc. of the Australian Control Conference*, 2014.

[47] L. C. Evans and R. F. Gariepy. *Measure Theory and Fine Properties of*

Functions. CRC Press, 2015.

[48] G. Evensen. The ensemble Kalman filter: Theoretical formulation and practical implementation. *Ocean Dynamics*, 53(4):343–367, 2003.

[49] T. Faulwasser, L. Grüne, and M. A. Müller. Economic nonlinear model predictive control. *Foundations and Trends in Systems and Control*, 5(1):1–98, 2018.

[50] B. A. Francis and W. M. Wonham. The internal model principle of control theory. *Automatica*, 12:457–465, 1976.

[51] G. F. Franklin, J. D. Powell, and A. Emami-Naeini. *Feedback Control of Dynamic Systems*. Pearson, 7th edition, 2014.

[52] R. Freeman and P. Kokotović. *Robust Nonlinear Control Design: State-Space and Lyapunov Techniques*. Birkhäuser, 1996.

[53] R. Goebel, R. G. Sanfelice, and A. R. Teel. Hybrid dynamical systems. *IEEE Control Systems Magazine*, 29(2):28–93, 2009.

[54] R. Goebel, R. G. Sanfelice, and A. R. Teel. *Hybrid Dynamical Systems*. Princeton University Press, 2012.

[55] G. H. Golub and C. F. Van Loan. *Matrix Computations*. John Hopkins University Press, 4th edition, 2013.

[56] M. Grant and S. Boyd. Graph implementations for nonsmooth convex programs. In V. Blondel, S. Boyd, and H. Kimura, editors, *Recent Advances in Learning and Control*, Lecture Notes in Control and Information Sciences, pages 95–110. Springer-Verlag Limited, 2008.

[57] M. Grant and S. Boyd. CVX: Matlab software for disciplined convex programming, version 2.1. `http://cvxr.com/cvx`, 2014.

[58] L. Grüne and J. Pannek. *Nonlinear Model Predictive Control: Theory and Algorithms*. Springer, 2nd edition, 2017.

[59] W. Hahn. *Stability of Motion*. Springer-Verlag, 1967.

[60] E. Hairer, S. P. Noersett, and G. Wanner. *Solving Ordinary Differential Equations I*. Springer, 1987.

[61] S. Haykin and B. V. Veen. *Signals and Systems*. John Wiley & Sons Inc., 2nd edition, 2002.

[62] M. Herceg, M. Kvasnica, C. Jones, and M. Morari. Multi-Parametric Toolbox 3.0. In *Proc. of the European Control Conference*, pages 502–510, 2013. `http://control.ee.ethz.ch/~mpt`.

[63] J. P. Hespanha. *Linear Systems Theory*. Princeton University Press, 2009.

[64] R. A. Horn and C. R. Johnson. *Matrix Analysis*. Cambridge University Press, 1985.

[65] J. Huang. *Nonlinear Output Regulation: Theory and Applications*. Society

for Industrial and Applied Mathematics, 2004.

[66] S. Huang, M. R. James, D. Nešić, and P. M. Dower. Analysis of input-to-state stability for discrete time nonlinear systems via dynamic programming. *Automatica*, 41:2055–2065, 2005.

[67] P. A. Ioannou and J. Sun. *Robust Adaptive Control*. Prentice Hall, 1996.

[68] A. Isidori. *Nonlinear Control Systems*. Springer-Verlag, 3rd edition, 1995.

[69] A. Isidori, L. Marconi, and A. Serrani. *Robust Autonomous Guidance: An Internal Model Approach*. Springer, 2003.

[70] A. H. Jazwinski. *Stochastic Processes and Filtering Theory*. Academic Press, 1970.

[71] Z.-P. Jiang, A. R. Teel, and L. Praly. Small-gain theorem for ISS systems and applications. *Mathematics of Control, Signals, and Systems*, 7(2):95–120, 1994.

[72] S. Julier and J. K. Uhlmann. A general method for approximating nonlinear transformations of probability distributions. Technical report, University of Oxford, 1996.

[73] S. Julier and J. K. Uhlmann. Unscented filtering and nonlinear estimation. *Proceedings of the IEEE*, 92(3):401–422, 2004.

[74] S. J. Julier and J. K. Uhlmann. New extension of the Kalman filter to nonlinear systems. In *Signal Processing, Sensor Fusion, and Target Recognition VI*, volume 3068, pages 182–193. International Society for Optics and Photonics, 1997.

[75] S. J. Julier, J. K. Uhlmann, and H. F. Durrant-Whyte. A new approach for filtering nonlinear systems. In *Proceedings of 1995 American Control Conference*, volume 3, pages 1628–1632, 1995.

[76] V. Jurdjevic and J. P. Quinn. Controllability and stability. *Journal of Differential Equations*, 28:381–389, 1978.

[77] R. E. Kalman. A new approach to linear filtering and prediction problems. *Transactions of the ASME, Journal of Basic Engineering*, 82D:35–45, March 1960.

[78] R. E. Kalman. Canonical structure of linear dynamical systems. *Proceedings of the National Academy of Sciences of the United States of America*, 48:596–600, 1962.

[79] R. E. Kalman. Lyapunov function for the problem of Lur'e in automatic control. *Proceedings of the National Academy of Sciences of the United States of America*, 49(2):201–205, 1963.

[80] N. Kazantzis and C. Kravaris. Nonlinear observer design using Lyapunov's auxiliary theorem. *Systems & Control Letters*, 34(5):241–247, 1998.

[81] C. M. Kellett. A compendium of comparsion function results. *Mathematics*

of Controls, Signals and Systems, 26(3):339–374, 2014.

[82]   C. M. Kellett. Classical converse theorems in Lyapunov's second method. *Discrete and Continuous Dynamical Systems, Series B*, 20(8):2333–2360, 2015.

[83]   C. M. Kellett and P. M. Dower. Input-to-state stability, integral input-to-state stability, and $\mathcal{L}_2$-gain properties: Qualitative equivalences and interconnected systems. *IEEE Transactions on Automatic Control*, 61(1):3–17, January 2016.

[84]   C. M. Kellett, P. M. Dower, and H. Ito. Relationships between subclasses of integral input-to-state stability. *IEEE Transactions on Automatic Control*, 62(5):2476–2482, May 2017.

[85]   C. M. Kellett and A. R. Teel. Weak converse Lyapunov theorems and control Lyapunov functions. *SIAM Journal on Control and Optimization*, 42(6):1934–1959, 2004.

[86]   H. K. Khalil. *Nonlinear Systems*. Prentice Hall, 3rd edition, 2002.

[87]   H. K. Khalil. *Nonlinear Control*. Pearson, 2014.

[88]   H. K. Khalil. *High-Gain Observers in Nonlinear Feedback Control*. Society for Industrial and Applied Mathematics, 2017.

[89]   P. Kokotović and M. Arcak. Constructive nonlinear control: A historical perspective. *Automatica*, 37:637–662, 2001.

[90]   P. V. Kokotović. The joy of feedback: Nonlinear and adaptive. *IEEE Control Systems Magazine*, 12(3):7–17, June 1992.

[91]   I. Kolmanovsky and E. G. Gilbert. Theory and computation of disturbance invariant sets for discrete-time linear systems. *Mathematical Problems in Engineering*, 4(4):317–367, 1998.

[92]   B. Kouvaritakis and M. Cannon. *Model Predictive Control: Classical, Robust and Stochastic*. Springer, 2016.

[93]   M. Krstić, I. Kanellakopoulos, and P. Kokotović. *Nonlinear and Adaptive Control Design*. John Wiley and Sons, Inc., 1995.

[94]   S. Lefschetz. *Stability of Nonlinear Control Systems*. Academic Press, New York, 1965.

[95]   A. L. Letov. *Stability in Nonlinear Control Systems*. Princeton University Press, Princeton, New Jersey, 1961.

[96]   D. Liberzon. *Calculus of Variations and Optimal Control Theory: A Concise Introduction*. Princeton University Press, 2011.

[97]   Y. Lin, E. D. Sontag, and Y. Wang. A smooth converse Lyapunov theorem for robust stability. *SIAM Journal on Control and Optimization*, 34(1):124–160, 1996.

[98]   T. Liu, D. J. Hill, and Z.-P. Jiang. Lyapunov formulation of ISS cyclic-small-gain in continuous-time dynamical networks. *Automatica*, 47:2088–

2093, 2011.

[99]   J. Löfberg. Yalmip : A toolbox for modeling and optimization in MATLAB. In *In Proceedings of the CACSD Conference*, Taipei, Taiwan, 2004.

[100] D. G. Luenberger. Observing the state of a linear system. *IEEE Transactions on Military Electronics*, 8(2):74–80, April 1964.

[101] A. I. Lur'e and V. N. Postnikov. Stability theory of regulating systems. *Prikladnaya Matematika i Mekhanika*, 8:246–248, 1944. [Russian].

[102] A. M. Lyapunov. The general problem of the stability of motion. *Math. Soc. of Kharkov*, 1892. (Russian). (English Translation, *International J. of Control*, 55:531–773, 1992).

[103] J. M. Maciejowski. *Predictive Control: With Constraints*. Pearson Education, 2002.

[104] R. Mahony, T. Hamel, and J. Trumpf. Geometric observers on Lie groups. In *Proc. of the IEEE Conference on Decision and Control*, 2018.

[105] R. Marino and M. Spong. Nonlinear control techniques for flexible joint manipulators: A single link case study. In *Proc. of the IEEE International Conference on Robotics and Automation*, volume 3, pages 1030–1036, 1986.

[106] L. A. McGee and S. F. Schmidt. Discovery of the Kalman Filter as a practical tool for aerospace and industry. Technical Memorandum NASA-TM-86847, National Aeronautics and Space Administration, November 1985.

[107] A. Megretski and A. Rantzer. System analysis via integral quadratic constraints. *IEEE Transactions on Automatic Control*, 42(6):819–830, June 1997.

[108] A. N. Michel, L. Hou, and D. Liu. *Stability of Dynamical Systems: Continuous, Discontinuous, and Discrete Systems*. Birkhauser, 2008.

[109] P. J. Moylan. Dissipative systems and stability. Available at `http://www.pmoylan.org/pages/research/DissBook.html`.

[110] H. Nijmeijer and A. van der Schaft. *Nonlinear Dynamical Control Systems*. Springer, 1990.

[111] J. Nocedal and S. Wright. *Numerical Optimization*. Springer Science & Business Media, 2006.

[112] H. Nyquist. Regeneration theory. *Bell System Technical Journal*, 11:126–147, 1932.

[113] K. Ogata. *Discrete-Time Control Systems*. Pearson, 2nd edition, 1995.

[114] K. Ogata. *Modern Control Engineering*. Pearson, 5th edition, 2009.

[115] A. V. Oppenheim, A. V. Willsky, and S. H. Nawab. *Signals and Systems*. Prentice Hall, 2nd edition, 1997.

[116] A. Papachristodoulou, J. Anderson, G. Valmorbida, S. Prajna, P. Seiler, and P. A. Parrilo. SOSTOOLS: Sum of squares optimization toolbox for MAT-

LAB, 2013.

[117] P. A. Parrilo. *Structured Semidefinite Programs and Semialgebraic Geometry Methods in Robustness and Optimization*. PhD thesis, California Institute of Technology, 2000.

[118] A. Polyakov. Nonlinear feedback design for fixed-time stabilization of linear control systems. *IEEE Transactions on Automatic Control*, 57(8):2106–2110, 2012.

[119] V. M. Popov. Absolute stability of nonlinear systems of automatic control. *Automation and Remote Control*, 22:857–875, 1962.

[120] V. M. Popov. The solution of a new stability problem for controlled systems. *Automation and Remote Control*, 24:1–23, 1963.

[121] S. Prajna, A. A. Papachristodoulou, and F. Wu. Nonlinear control synthesis by sum of squares optimization: A Lyapunov-based approach. In *5th Asian Control Conference*, volume 1, pages 157–165, 2004.

[122] S. Prajna, A. Papachristodoulou, and P. A. Parrilo. Introducing SOSTOOLS: A general purpose sum of squares programming solver. In *Proc. of the 41st IEEE Conference on Decision and Control*, pages 741–746, 2002.

[123] S. Prajna, P. A. Parrilo, and A. Rantzer. Nonlinear control synthesis by convex optimization. *IEEE Transactions on Automatic Control*, 49(2):310–314, 2004.

[124] L. Praly and Z.-P. Jiang. Stabilization by output feedback for systems with ISS inverse dynamics. *Systems and Control Letters*, 21(1):19–34, July 1993.

[125] L. Praly, R. Ortega, and G. Kaliora. Stabilization of nonlinear systems via forwarding mod $L_g V$. *IEEE Transactions on Automatic Control*, 46(9):1461–1466, 2001.

[126] C. Prieur. A robust globally asymptotically stabilizing feedback: The example of the Artstein's circles. In *Nonlinear Control in the Year 2000*, volume 2, pages 279–300. Springer, 2001.

[127] S. V. Raković and W. S. Levine. *Handbook of Model Predictive Control*. Springer, 2018.

[128] J. B. Rawlings, D. Q. Mayne, and M. M. Diehl. *Model Predictive Control: Theory and Design*. Nob Hill Publishing, 2nd edition, 2017.

[129] L. Rifford. Existence of Lipschitz and semiconcave control-Lyapunov functions. *SIAM Journal on Control and Optimization*, 39(4):1043–1064, 2000.

[130] B. Ristic, S. Arulampalam, and N. Gordon. *Beyond the Kalman Filter: Particle Filters for Tracking Applications*. Artech House, 2003.

[131] N. Rouche, P. Habets, and M. Laloy. *Stability Theory by Liapunov's Direct Method*. Springer-Verlag, 1977.

[132] B. S. Rüffer. *Monotone Dynamical Systems, Graphs, and Stability of Large-*

Scale Interconnected Systems. PhD thesis, Universität Bremen, 2007.

[133] W. J. Rugh. *Linear System Theory*. Pearson, 2nd edition, 1995.

[134] I. W. Sandberg. On the $\mathcal{L}_2$-boundedness of solutions of nonlinear functional equations. *Bell System Technical Journal*, 43:1581–1599, 1964.

[135] R. Sepulchre, M. Janković, and P. Kokotović. *Constructive Nonlinear Control*. Springer-Verlag, 1997.

[136] L. F. Shampine, I. Gladwell, and S. Thompson. *Solving ODEs with MATLAB*. Cambridge University Press, 2003.

[137] Y. Shtessel, C. Edwards, L. Fridman, and A. Levant. *Sliding Mode Control and Observation*. Birkhäuser, 2014.

[138] D. Simon. *Optimal State Estimation: Kalman, H_∞, and Nonlinear Approaches*. John Wiley & Sons, 2006.

[139] J. Slotine and W. Li. *Applied Nonlinear Control*. Prentice-Hall, 1991.

[140] E. D. Sontag. A Lyapunov-like characterization of asymptotic controllability. *SIAM Journal on Control and Optimization*, 21:462–471, 1983.

[141] E. D. Sontag. Smooth stabilization implies coprime factorization. *IEEE Transactions on Automatic Control*, 34(4):435–443, April 1989.

[142] E. D. Sontag. A "universal" construction of Artstein's theorem on nonlinear stabilization. *Systems and Control Letters*, 13:117–123, 1989.

[143] E. D. Sontag. Comments on integral variants of ISS. *Systems and Control Letters*, 34(1–2):93–100, 1998.

[144] E. D. Sontag. *Mathematical Control Theory: Deterministic Finite Dimensional Systems*. Springer-Verlag, 2 edition, 1998.

[145] E. D. Sontag. Stability and stabilization: Discontinuities and the effect of disturbances. In *Proceedings NATO Advanced Study Institute "Nonlinear Analysis, Differential Equations, and Control"*, pages 551–598. Kluwer, 1999.

[146] E. D. Sontag and A. R. Teel. Changing supply functions in input/state stable systems. *IEEE Transactions on Automatic Control*, 40(8):1476–1478, 1995.

[147] E. D. Sontag and Y. Wang. On characterizations of the input-to-state stability property. *Systems and Control Letters*, 24:351–359, 1995.

[148] A. Stuart and A. Humphries. *Dynamical Systems and Numerical Analysis*. Cambridge University Press, 1996.

[149] J. F. Sturm. Using SeDuMi 1.02, a Matlab toolbox for optimization over symmetric cones. *Optimization Methods and Software*, 11(1-4):625–653, 1999.

[150] H. J. Sussman and V. Jurdjevic. Controllability of nonlinear systems. *Journal of Differential Equations*, 12:95–116, 1972.

[151] V. L. Syrmos, C. T. Abdallah, P. Dorato, and K. Grigoriadis. Static output

feedback—A survey. *Automatica*, 33(2):125–137, 1997.

[152] G. Tao and P. V. Kokotović. *Adaptive Control of Systems with Actuator and Sensor Nonlinearities*. John Wiley & Sons, Inc., 1996.

[153] A. R. Teel. *Feedback Stabilization: Nonlinear Solutions to Inherently Nonlinear Problems*. PhD thesis, University of California, Berkeley, June 1992.

[154] S. G. Tzafestas. *Introduction to Mobile Robot Control*. Elsevier, 2013.

[155] V. I. Utkin. Variable structure systems with sliding modes. *IEEE Transactions on Automatic Control*, AC-22(2):212–222, April 1977.

[156] R. Van der Merwe and E. A. Wan. The square-root unscented Kalman filter for state and parameter-estimation. In *IEEE International Conference on Acoustics, Speech, and Signal Processing*, volume 6, pages 3461–3464, 2001.

[157] A. J. van der Schaft. L_2-*Gain and Passivity Techniques in Nonlinear Control*. Springer, 3rd edition, 2017.

[158] M. Vidyasagar. *Nonlinear Systems Analysis: Second Edition*. Prentice-Hall, 1993.

[159] A. Wächter and L. Biegler. On the implementation of an interior-point filter line-search algorithm for large-scale nonlinear programming. *Mathematical Programming*, 106:25–57, 2006.

[160] E. A. Wan and R. Van Der Merwe. The unscented Kalman filter for nonlinear estimation. In *Proceedings of the IEEE Adaptive Systems for Signal Processing, Communications, and Control Symposium*, pages 153–158, 2000.

[161] E. A. Wan and R. Van Der Merwe. The unscented Kalman filter. In S. Haykin, editor, *Kalman Filtering and Neural Networks*, volume 5, pages 221–280. Wiley New York, 2001.

[162] W. Wang, C. Wen, and J. Zhou. *Adaptive Backstepping Control of Uncertain Systems with Actuator Failures, Subsystem Interactions, and Nonsmooth Nonlinearities*. CRC Press, 2017.

[163] G. Wanner and E. Hairer. *Solving Ordinary Differential Equations II*. Springer, 1996.

[164] J. C. Willems and J. W. Polderman. *Introduction to Mathematical Systems Theory: A Behavioral Approach*, volume 26. Springer Science & Business Media, 2013.

[165] W. M. Wonham. *Linear Multivariable Control: A Geometric Approach*. Springer-Verlag, 2nd edition, 1979.

[166] V. A. Yakubovich. The solution of certain matrix inequalities in automatic control theory. *Doklady Akademii Nauk SSSR*, 143:1304–1307, 1962.

[167] T. Yoshizawa. *Stability Theory by Liapunov's Second Method*. Mathematical Society of Japan, 1966.

[168] L. Zaccarian and A. R. Teel. *Modern Anti-windup Synthesis: Control Augmentation for Actuator Saturation.* Princeton University Press, 2011.

[169] G. Zames. On the input-output stability of time-varying nonlinear feedback systems part I: Conditions derived using concepts of loop gain, conicity, and passivity. *IEEE Transactions on Automatic Control*, 11:228–238, 1966.

[170] H. Zhang and P. M. Dower. Performance bounds for nonlinear systems with a nonlinear $\mathcal{L}_2$-gain property. *International Journal of Control*, 85:1293–1312, 2012.

[171] H. Zhang and P. M. Dower. Computation of tight integral input-to-state stability bounds for nonlinear systems. *Systems & Control Letters*, 62:355–365, 2013.

[172] J. Zhou and C. Wen. *Adaptive Backstepping Control of Uncertain Systems: Nonsmooth Nonlinearities, Interactions or Time-Variations.* Springer, 2008.

Index